HANDBOOK OF NUCLEOSIDE SYNTHESIS

HANDBOOK OF NUCLEOSIDE SYNTHESIS

HELMUT VORBRÜGGEN
CARMEN RUH-POHLENZ
Research Laboratories of Schering
Federal Republic of Germany

A WILEY-INTERSCIENCE PUBLICATION
John Wiley & Sons, Inc.
New York • Chichester • Weinheim • Brisbane • Singapore • Toronto

This text is printed on acid-free paper. 

Published by John Wiley & Sons, Inc.

Published simultaneously in Canada.

For ordering and customer service, call 1-800-CALL-WILEY.

Library of Congress Cataloging in Publication Data

Vorbrüggen, Helmut.
Handbook of nucleoside synthesis/Helmut Vorbrüggen and Carmen Ruh-Pohlenz.
p. cm.
Includes bibliographical references and index.
ISBN 0-471-09383-1 (paper)
1. Nucleosides—Synthesis—Handbooks, manuals, etc. I. Ruh-Pohlenz, Carmen, II. Title.

QP625.N88 V67 2001
572.8'545—dc21 2001024557

10 9 8 7 6 5 4 3 2 1

CONTENTS

CONTENTS

FOREWORD

Scientists employing the tools of organic synthesis in their research are often confronted with the problem of how best to carry out a well-known reaction in a new context. The aim of the *Organic Reactions* series since its formal inception in 1942 has been to assist organic chemists in accomplishing structural transformations by providing "critical discussions of the more important (synthetic) reactions." This publication is unique in several ways. Most apparent to our readers is the format of the series, which has changed little in nearly 60 years: authoritative discussion of the topic reaction accompanied by tables that organize essentially all published examples of the reaction being reviewed. This combination of critical discussion and thorough coverage is responsible for the leading position this series occupies for scientists interested in the reactions of organic chemistry. A second distinctive feature of this series is that it is assembled almost entirely through the volunteer efforts of our authors, editors and editorial assistants.

At the time of writing this foreword, 56 volumes of *Organic Reactions* have been published. A major change through the years has been the increasing size of chapters in *Organic Reactions,* a trend that reflects the dramatic growth in size and importance of the field of preparative organic chemistry. While the average length of the 12 chapters appearing in the first volume of *Organic Reactions* was 35 pages, the 12 most recent reactions surveyed in this series occupied six large volumes.

In recent years, the Editorial Board of *Organic Reactions* has been considering ways to make reviews of broadly utilitarian reactions more widely accessible to the organic chemistry community and practitioners of related disciplines. As the first step in 1988, the extensive review of the "Stille Reaction" by Vittorio Farina, Venkat Krishnamurthy, and William J. Scott was published in paperback. The current volume, a paperback edition of the important recent chapter surveying the "Synthesis of Nucleosides" by Helmut Vorbrüggen and Carmen Ruh-Pohlenz, continues this experiment. It is our hope that the availability of this lower-cost paperback edition will allow chemists engaged in nucleoside research, including those who do not regularly follow the *Organic Reactions* series, to obtain personal copies of this authoritative review co-authored by one of the pioneers in nucleoside synthesis.

Larry E. Overman
Editor-In-Chief
Irvine, California

PREFACE

Nucleosides and their analogues are of enormous importance. They are an established class of clinically useful medicinal agents possessing antiviral and anticancer activity. Recent interest in antiviral nucleosides has centered on the development of reverse transcriptase inhibitors as potential AIDS therapies. The nucleoside analogs AZT (3′-azido-3′-deoxythymidine), ddC (2′,3′-dideoxycytidine) and ddI (2′,3′-dideoxyinosine) are currently approved for the treatment of AIDS and are believed to work by termination of the replicating DNA chain since these nucleoside analogues lack the C-3′ hydroxy group. In addition to these three compounds, a number of nucleoside analogs have been synthesized and undergone preliminary clinical testing. Interest in nucleosides derived from L-ribose arises from the observation that the enantiomer of the 3-thianucleoside derivative 3TC, corresponding to the unnatural L-configuration, was a superior inhibitor of the HIV reverse transcriptase than the "natural" D-series. Importantly, the "unnatural" L-enantiomer was found to be less toxic than the "natural" enantiomer. It would appear that the L-enantiomer is phosphorylated by cellular kinases and is a substrate for reverse transcriptase, but with reduced affinity for other cellular enzymes. A number of other L-nucleoside derivatives have been prepared and shown to possess antiviral activity. These include L-(−)-2′,3′-dideoxythiacytidine, L-(−)-ddC, L-(−)-5-fluoro-2′,3′-dideoxycytidine and L-(−)-2′-fluoro-5-methylarabinouridine.

The biosynthesis of nucleotides is the limiting process in cell proliferation. Many of the enzymes involved in these pathways are highly active in cancer cells, but barely detectable in nonproliferating cells. Thus, nucleoside analogues that inhibit nucleotide biosynthesis are of great interest for cancer chemotherapy. The nucleoside analogue 5-fluoro-2′-deoxyuridine is an inhibitor of thymidylate synthase and has been used clinically for the treatment of leukemia and colorectal cancer for over three decades. More recently gemcitabine (2′,2′-difluoro-2′-deoxycytidine) has shown great promise for the treatment of pancreatic cancer. L-(−)-3′-Oxa-2′,3′-dideoxycytidine showed promising activity against hepatocellular and prostate tumors which are generally difficult to treat, and is thus a promising clinical candidate.

Modified nucleosides play a central role in the development of genetic therapies such as triplex (antigene) and antisense strategies. In these approaches, 2′-deoxyoligonucleotides can prevent the expression of specific genes, thus inhibiting the production of disease related proteins. For a protein to be expressed, the gene that codes for the protein must be transcribed from duplex DNA into single stranded messenger RNA, which in turn is translated into the protein. The triplex approach is aimed at preventing transcription of the gene by stabilizing DNA hybridization through triple helix formation. The antisense approach prevents translation of the gene by selective hybridization to a portion of messenger RNA, thus preventing its translation into a protein. Both forms of genetic therapy rely on an agent to bind to a specific nucleic acid sequence. The hybridization of the oligonucleotide to the target strand is an equilibrium process and the oligonucleotide is subject to degradation by cellular nucleases. Thus, designed nucleosides with enhanced hybridization properties and/or nuclease resistance could provide major advances in this area.

Modified nucleosides have also been developed as probes for biochemical processes. For instance, oligonucleotides labeled with fluorescent reporter groups have been utilized to probe the structure and dynamics of nucleic acids as well as the nature of DNA-protein interactions. Fluorescent nucleosides are indispensable reagents for modern DNA sequencing technologies. The preparation of modified nucleosides has also played a central role in the investigation of the molecular mechanism of chemical carcinogenesis and DNA repair. Such studies require the covalent modification of a specific DNA base within an oligonucleotide sequence with carcinogens. These carcinogens have included environmental toxins such as aflatoxin B_1, endogenous toxins such as lipid peroxidation products, and man-made and industrial toxins such as benzo[*a*]pyrene and vinyl chloride, among others. The synthesis of the nucleoside-carcinogen adducts are valuable for the verification of structure as well as for authentic analytical standards. In other studies, oligonucleotides containing modified sugars were designed and synthesized to generate a specific carbon-centered radical within the oligonucleotide. Such experiments served as models for oxidative free-radical damage to nucleic acids. So-called "caged" nucleotides have been widely employed to study the role of nucleotides in signaling biochemical events.

Clearly, the ability to synthesize modified nucleosides has favorably impacted a wide range of biomedical research including the clinical treatment of disease. They have advanced our current understanding of many vital biochemical processes involving nucleotides and nucleic acids. As such, methods for the synthesis of modified nucleosides are of general interest and widespread importance. Nucleoside chemistry has enjoyed a long and rich history. One can imagine two general strategies for the synthesis of nucleoside analogues. The first is to modify natural, readily available nucleosides. The second approach is the *de novo* synthesis of the desired nucleoside analogue by adding the nucleobase to a suitable sugar derivative. Both approaches have been enormously successful and are widely used. The current volume, "The Synthesis of Nucleosides," is a comprehensive review on the *de novo* synthesis of furanosyl, pyranosyl and acyclic nucleosides with heterocylic bases.

Within the *de novo* approach, two synthetic strategies are also possible. The first approach is to build the base onto a 1-amino sugar derivative. This approach is reminiscent of the biosynthesis of natural nucleotides. The second strategy is the glycosylation of a suitable sugar derivative with the desired heterocyclic base. Both strategies are covered in "The Synthesis of Nucleosides," although the main focus is the glycosylation of the sugar derivatives. Indeed, the nearly 500-page tabular survey is dedicated to the glycosylation of furanose and hexose derivatives or an acyclic carbohydrate surrogate.

The treatise begins with a brief overview of early synthetic approaches to nucleosides. Particular attention is focused on the Hilbert-Johnson and analogous reactions. An important modification of the Hilbert-Johnson reaction, which utilizes silylated nucleobases and strong Lewis acids, was developed by one of the authors (Vorbrüggen) and is the most widely used method for the preparation of ribonucleosides. This protocol is commonly referred to as the Vorbrüggen reaction and occupies an important place in modern nucleoside synthesis. The development of the Vorbrüggen reaction provided reproducibly high yields for the glycosylation with reliable and predictable stereochemical outcome for ribofuranose derivatives. Drawing together data from a variety of sources, a lucid discussion on the mechanism of the Vorbrüggen reaction, including the subtle aspects governing regioselectivity, is presented.

A chapter is devoted to the preparation of 2′-deoxyribonucleosides. Sections are dedicated to the glycosylation of 2-deoxyribose derivatives as well as the chemical conversion of ribonucleosides to 2′-deoxyribonucleosides via a deoxygenation reaction of the 2′-hydroxy group. The standard procedure for the 2′-deoxygenation of ribonucleosides was developed by Robins and involves a radical deoxygenation mechanism. The reagents developed for this modified Barton-McCombie reaction have subsequently found wide general use in organic synthesis. A broad overview of alternative chemical methods for nucleoside synthesis is discussed, including strategies for building a nucleobase onto a 1-amino sugar. Several interesting examples of enzymatic transglycosylation are also presented. This method, although limited by the ability of the enzymes to recognize substrates, has tremendous potential for the stereocontrolled synthesis of 2′-deoxy- and 2′,3′-dideoxynucleosides.

The text concludes with a survey of typical experimental protocols for nucleoside synthesis. Detailed instructions on preparing the various components for the reaction are included, such as the purification and drying of solvents, catalysts and the carbohydrate and nucleobase substrates. Further discussion on the silylation of the nucleobase, workup of the reaction and removal of protecting groups is also provided. Finally, a sampling of precise experimental procedures is given for the Vorbrüggen glycosylation reaction. Protocols for a variety of Lewis acid catalysts and carbohydrate precursors are given.

The treatise then concludes with an exhaustive tabular survey of glycosylation reactions with nucelobases. The tables cover approximately twenty years of literature through about 1994, a period of tremendous growth and innovation in this field of nucleoside chemistry. This compilation is nothing less than extraordinary. Eleven tables are organized according to the reagents used to effect the glycosylation reaction. Each table is then further organized according to the carbohydrate donor, then by heterocyclic base. The tables are well designed for quick visual inspection and will unquestionably be the primary literature resource for researchers working in this discipline.

A number of outstanding books and monographs exist on the chemistry of nucleosides and nucleotides. However, "The Synthesis of Nucleosides" is unique in that a broad perspective on nucleoside synthesis is given. This monograph is an essential resource to those working in the nucleoside and oligonucleotide area where the synthesis of labeled or modified nucleosides is necessary. With over 1400 references, knowledgeable researchers will find this book a thoughtful and well-organized accounting of the current state of nucleoside synthesis as a discipline. Those who are about to enter the area will also appreciate this volume as a single source for past accomplishments, and will find the detailed discussion on the experimental aspects of nucleoside synthesis indispensable. Nowhere else have such experimental details been gathered.

Many investigators have contributed greatly to the current state of nucleoside synthesis. For fear of omission, I will not attempt to list them. However, Helmut Vorbrüggen is widely recognized as one of the leading figures in this area over the past thirty years. Dr. Vorbrüggen has teamed with Carmen Ruh-Pohlenz to compile an exhaustive treatise on nucleoside synthesis. I believe it is fair to say that few people could have put together "The Synthesis of Nucleosides," and the nucleoside community is indebted to the authors for their heroic effort.

C. J. Rizzo
Nashville, TN USA

ACKNOWLEDGMENTS

We thank Schering AG and in particular Drs. U. Eder, B. Mueller, K. Specht, and G. Stock for their long-lasting and unfailing support of this review. We are furthermore obliged to Mr. W. Becker, Miss C. Heinz, Miss N. Schidniogrotski, and Mrs. M. Golde for patiently writing and rewriting the text and drawing all of the formulas for this review. H. Vorbrüggen also thanks his colleagues of the Department of Organic Chemistry, Free University of Berlin, for their kind hospitality since his retirement from Schering AG in 1995.

HANDBOOK OF NUCLEOSIDE SYNTHESIS

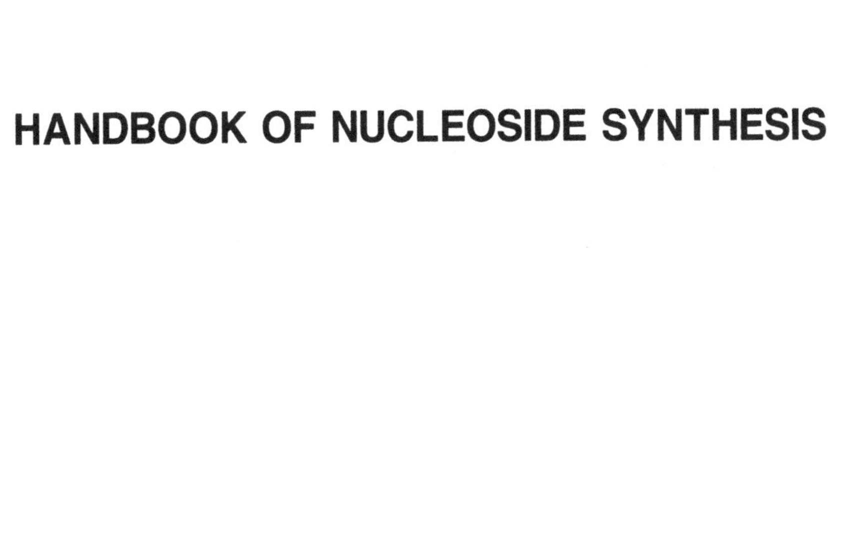

SYNTHESIS OF NUCLEOSIDES

INTRODUCTION

This chapter deals with the synthesis of nucleosides (e.g., the formation of *N*-glycosides of sugars such as D-ribose or 2-deoxy-D-ribose with heterocyclic nitrogen bases). The methods of nucleoside synthesis have been treated in a number of reviews and monographs.[1–7a]

It is now generally accepted that nucleosides were among the first organic compounds formed at the start of evolution in the early history of our planet earth. To support this point, guanine (**1**) and adenine (**2**) were heated with D-ribose (**3** or **4**) in sea water, which contains the Lewis acid magnesium chloride as catalyst.[8,9] One thus obtained the nucleosides guanosine (**5**, ca. 3%) and adenosine (**7**, 2.3%) together with comparable yields of the unnatural α-nucleosides **6** and **8**. The latter were gradually photoanomerized to the thermodynamically more stable **5** and **7** in overall yields of 5–6%. The furanose form of ribose (**3**) reacts faster than the pyranose form (**4**).

Corresponding syntheses of the pyrimidine nucleosides uridine (**11**) and cytidine (**12**) from uracil (**9**), cytosine (**10**) and ribose are more problematic and remain an enigma. The recent conversion of glycolaldehyde-*O*-phosphate and formaldehyde to ribose-2,4-di-*O*-phosphate[10] might give new insights into the prebiotic syntheses of **11** and **12**. The evidence and hypotheses for these prebiotic conversions and the evolution of RNA, as well as the implications of an "RNA World," have been reviewed.[11–20c]

These RNA nucleosides are reduced in vivo as 5′-*O*-diphosphates by ribonucleotide reductases[21–23] to the corresponding 2′-deoxynucleosides—the building blocks of DNA such as 2′-deoxyguanosine (see the atom numbering in **5**). The thermodynamically controlled synthesis of these four building blocks of RNA (**5**, **7**, **11**, and **12**) has implications for the design of efficient, high yielding, new methods for the synthesis of the naturally occurring nucleosides, nucleoside antibiotics,[24] and modified nucleosides that may serve as antimetabolites to fight viral and parasitic diseases and cancer.

The nucleoside rings in this chapter are depicted arbitrarily in the *anti* conformation, as occurs predominantly in the crystal[25–32] and solution (based primarily on NOE-^{1}H- and ^{13}C-NMR measurements)[33–38] forms of pyrimidine nucleosides. Only a few nucleosides, such as 6-methyluridine, occur with the heterocyclic ring predominantly in the *syn* conformation.[25,26]

The synthesis of *C*-nucleosides has been reviewed previously[39–44] and is not covered in this review.

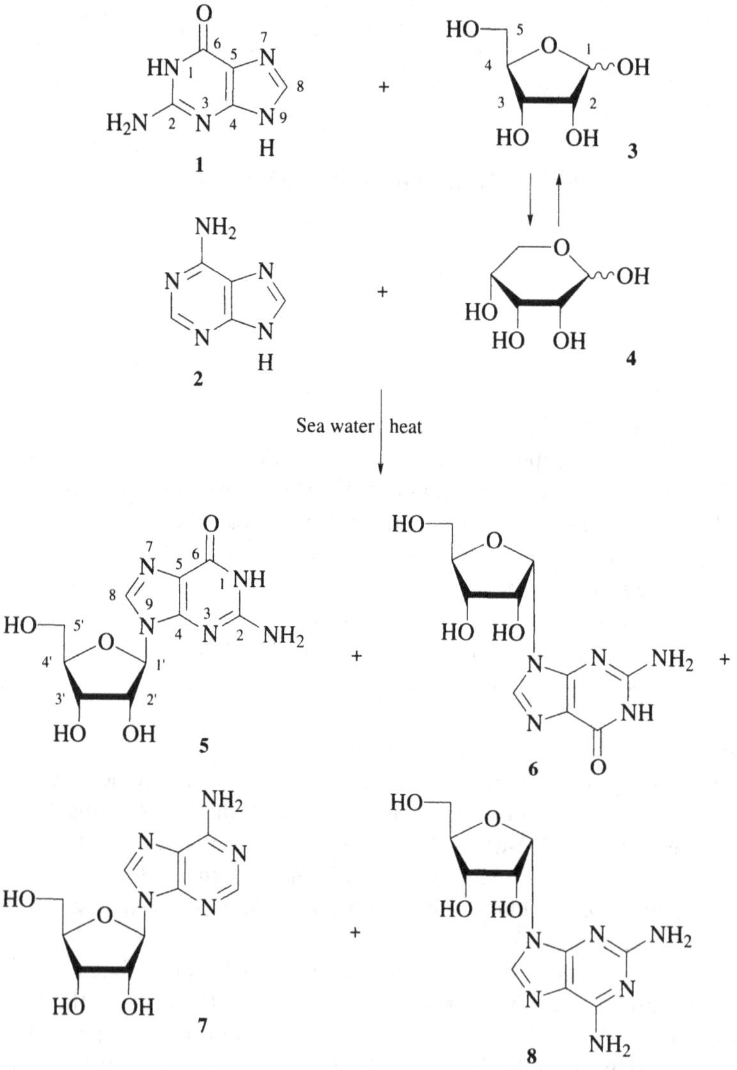

SCOPE AND LIMITATIONS

We describe the synthesis of nucleosides from compounds **13**, in which R^1 and R^2 are carbon or nitrogen moieties that usually form a heterocyclic ring, and sugar derivatives **14**, in which R^4 and R^5 normally form a ring that bears the leaving group Y, to give the nucleoside **15** and R^3Y (**16**).

$X = O, S, NCOR^{45a-c}$; $Y = Cl, Br, F, OAc, OBz, OC(NH)CCl_3, SOMe, OH$;
$R^3 = H, TMS, Na, Li, HgCl, Ag$

Sugar Moieties

Since we deal in this review with rather few protected sugar derivatives, these are introduced here.

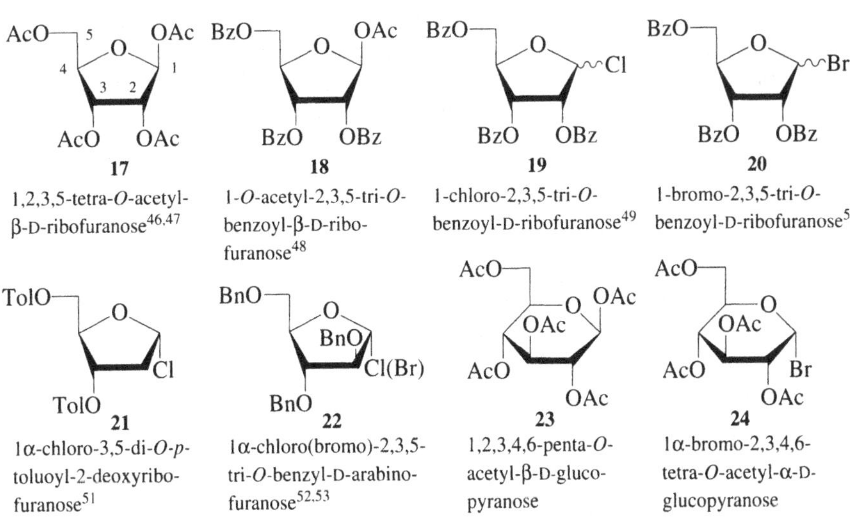

17 1,2,3,5-tetra-*O*-acetyl-β-D-ribofuranose[46,47]

18 1-*O*-acetyl-2,3,5-tri-*O*-benzoyl-β-D-ribofuranose[48]

19 1-chloro-2,3,5-tri-*O*-benzoyl-D-ribofuranose[49]

20 1-bromo-2,3,5-tri-*O*-benzoyl-D-ribofuranose[50]

21 1α-chloro-3,5-di-*O*-*p*-toluoyl-2-deoxyribofuranose[51]

22 1α-chloro(bromo)-2,3,5-tri-*O*-benzyl-D-arabinofuranose[52,53]

23 1,2,3,4,6-penta-*O*-acetyl-β-D-glucopyranose

24 1α-bromo-2,3,4,6-tetra-*O*-acetyl-α-D-glucopyranose

NUCLEOSIDE SYNTHESIS

There are three principal types of nucleoside-forming reactions: (a) The Fusion Reaction, (b) The Metal Salt Procedure, and (c) The Hilbert-Johnson Reaction.

The Fusion Reaction

In this method acidic heterocyclic systems such as 2,6-dichloropurine (**25**) react with peracylated sugars such as **17** at 150–155° in a melt to form the assumed intermediates **26** and **27**, which combine in 54% yield to give **28** and the volatile acetic acid.[54,55] This fusion reaction is usually performed in the presence of catalytic amounts of Lewis acids to promote the formation of the electrophilic sugar cation **27**; the reaction works with acidic systems such as substituted or annelated imidazoles, purines, triazoles, or pyrazoles. Yields, however, seldom exceed 60–70% (see Ref. 55 for a review).

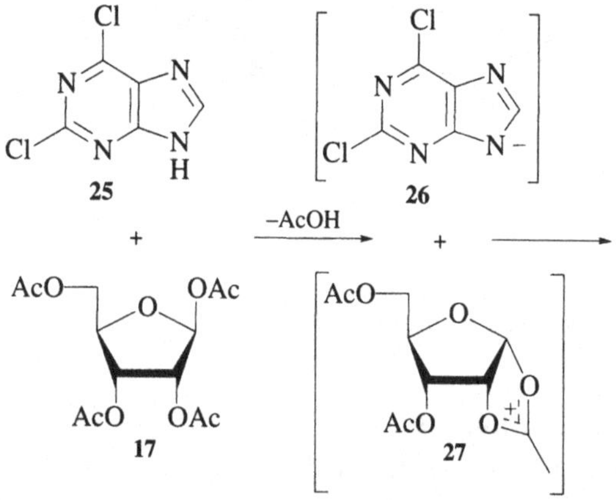

Cl N N Cl N N H **25** + AcO O OAc AcO OAc **17** —AcOH→ [Cl N N Cl N N⁻ **26**] + [AcO O AcO O O⁺ **27**] → Cl N N N Cl N AcO O AcO OAc **28**

The Metal Salt Procedure

In this procedure metal salts of heterocyclic systems are reacted with protected sugar halides. In the original procedure, the silver salt of 2,6,8-trichloropurine (**29**) was heated with acetobromoglucose (**24**)[56] in xylene to give glucopyranoside **30**.

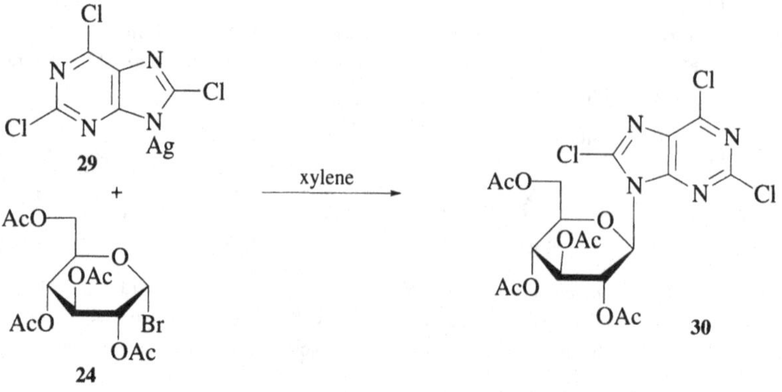

Cl N N Cl Cl N N Ag **29** + AcO O OAc AcO Br OAc **24** —xylene→ Cl N N Cl N Cl N AcO O OAc AcO OAc **30**

Because of the diminished polarity and thus better solubility (as well as reactivity) of mercuric salts compared with silver salts, investigators more recently have preferred the mercuric salts of heterocyclic bases, since the mercuric salt procedure often succeeds when the silver method fails.[1-5]

The initial products when using the mercuric salt procedure with uracils and cytosines are often *O*-glycosides such as **32**, which rearrange or react with another sugar halide to form the desired nucleoside.[57-59] Thus mercuric salt **31** reacts with **24** to give **32**, which is converted by excess **19** and added mercuric chloride in acetonitrile to nucleoside **33** and the N^1,N^3-bis(nucleoside) **34**.[60] Apparently, sugar cation **35** (derived from **19**) attacks the N^1 nitrogen in **32** to form intermediate **36**, which fragments to nucleoside **33**. The nucleophilic iminoether system in **33** can then react with additional cation **35** to give **34**.

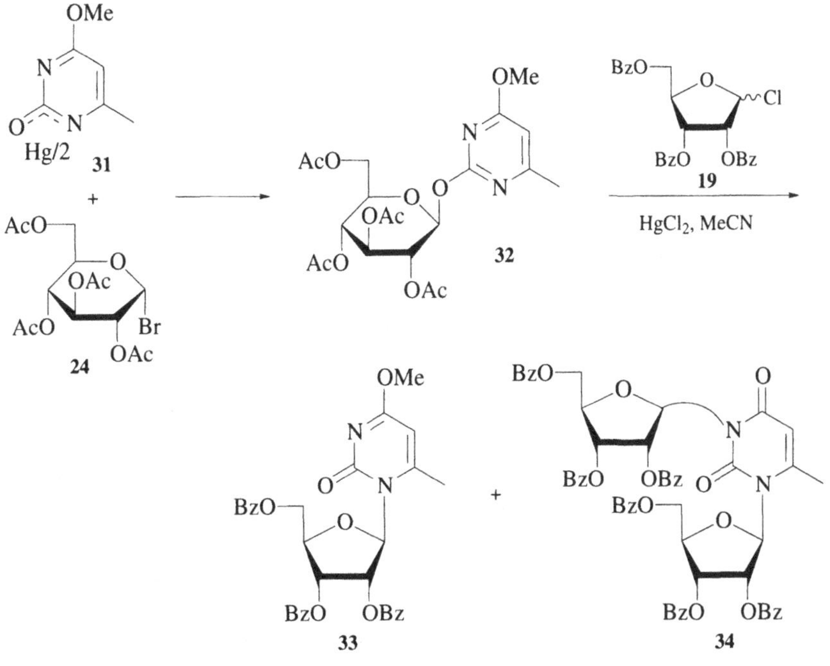

OMe N O N Hg/2 **31** + AcO O OAc AcO Br OAc **24** → AcO O OAc O OMe N N AcO OAc **32** BzO O Cl BzO OBz **19** $HgCl_2$, MeCN →

OMe N O N BzO O BzO OBz **33** + BzO O O N BzO OBz O N BzO O BzO OBz **34**

The mercuric salts of purines **37** can react with equivalent amounts of sugars such as **19** to give the kinetically controlled N^3-nucleosides **38**, which rearrange on heating with $HgBr_2$, $HgCl_2$, or $Hg(CN)_2$ to the thermodynamically controlled "natural" N^9-nucleosides **39** as well as some N^7-nucleosides.[61] As in the rearrangement of the *O*-glycoside **32** to nucleoside **33**, the presence of reactive sugar cations such as **35** promotes the rearrangement of **38** to **39**. All of these mechanisms have been summarized and discussed.[61a]

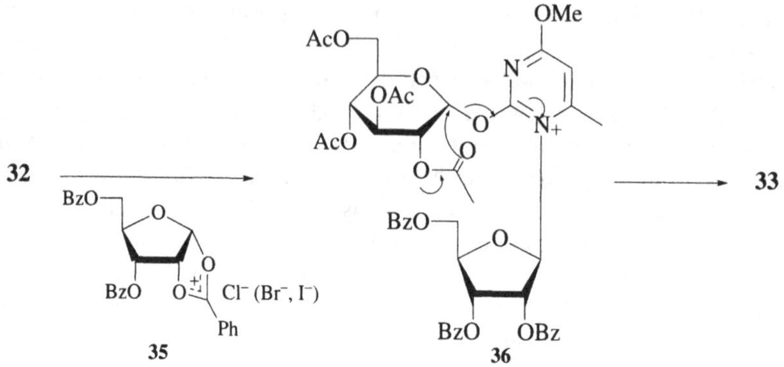

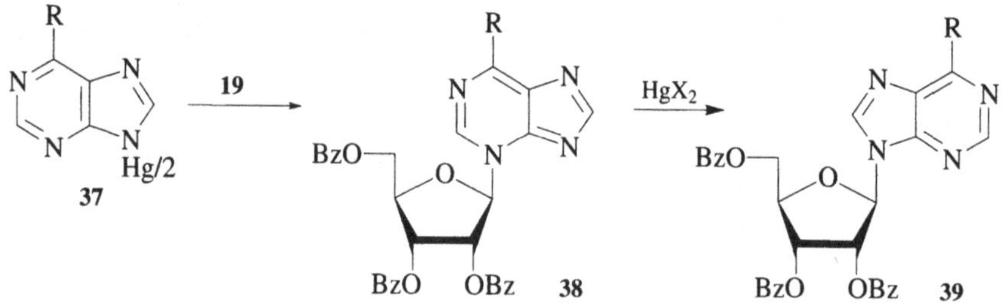

32 → 35 → 36 → 33

37 —19→ 38 —HgX_2→ 39

In addition to the often moderate yields and complicated mixtures obtained with the mercuric salt procedure, traces of mercuric impurities, which have strong biocidal properties, can falsify the biological data on the final nucleoside.[63,63a]

Recent important nucleoside syntheses employ the sodium salts of purines and related acidic heterocyclic systems, which are prepared in situ with NaH or analogous bases. These salts react with **21** in acetonitrile to give the corresponding β-nucleosides via an apparent S_N2 displacement of the 1α-chlorine by the heterocyclic base.[64] Thus, sodium salt **40** reacts with **21** in acetonitrile to afford 68% of 2′-deoxy-β-nucleoside **41**. The analogous reaction of the sodium salt of 6-chloropurine (**42**) with **21** gives 59% of the desired N^9-nucleoside **43** as well as 11% of the corresponding N^7-nucleoside **44**. Other authors report[65] that the isolated sodium salt of adenine (**45**) reacts best with **21** in acetone to afford directly

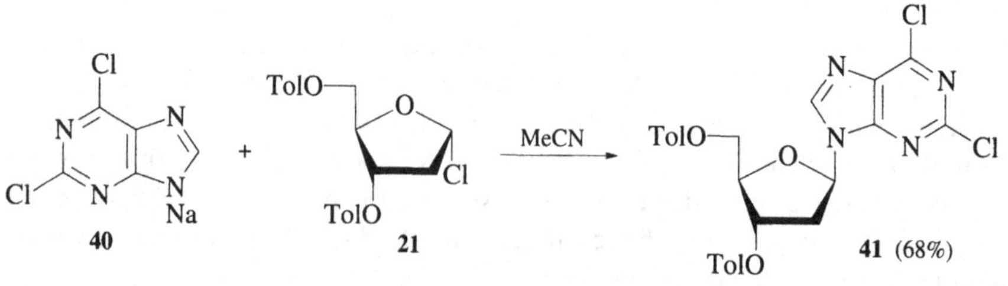

40 + **21** —MeCN→ **41** (68%)

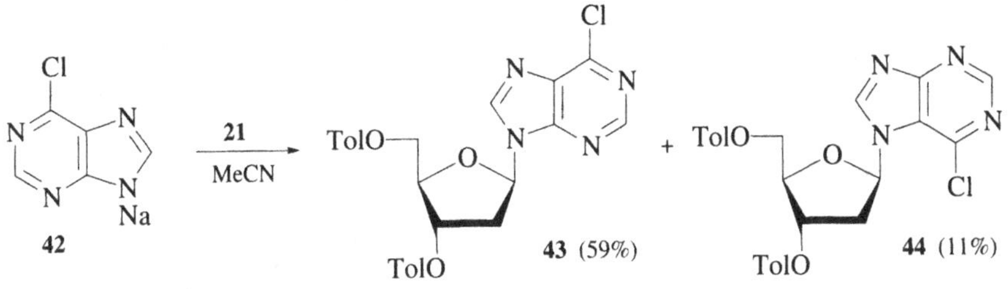

43% of the protected 2′-deoxyadenosine **46** plus 8% of the α-nucleoside **47**, thus obviating the need for aminating the 6-chloro compound **43** to the unprotected 2′-deoxyadenosine.

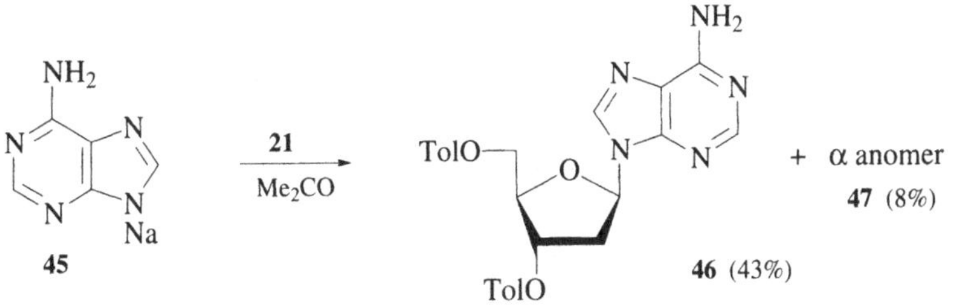

Equally effective are the phase-transfer glycosylations of heterocyclic bases such as 2-methylthio-4-methoxypyrrolo[2,3-*d*]pyrimidine (**48**), in which a standard N^7-glycosylation (e.g., by the Hilbert-Johnson method) is apparently not possible because the pyrimidine moiety is more nucleophilic than the pyrrole moiety. Thus, **48** is readily glycosylated by **22** in the presence of triethylbenzylammonium chloride (TEBA) and 50% aqueous sodium hydroxide to give 63% of β-nucleoside **49** as well as 21% of α-nucleoside **50**.[66] The acidic 4-nitro-1*H*-pyrrolo[2,3-*b*]pyridine **51** reacts analogously with **21** and tris[2-(2-methoxy)ethyl]amine (TDA-1) and KOH to afford **52** in 78% yield.[67] The sodium salt of 2-cyanopyrrole (**53**) reacts with **20** to give orthoamide **54**, whereas the reaction of **53** with ribofuranosyl chloride **55** affords 61% of nucleoside **56**.[68]

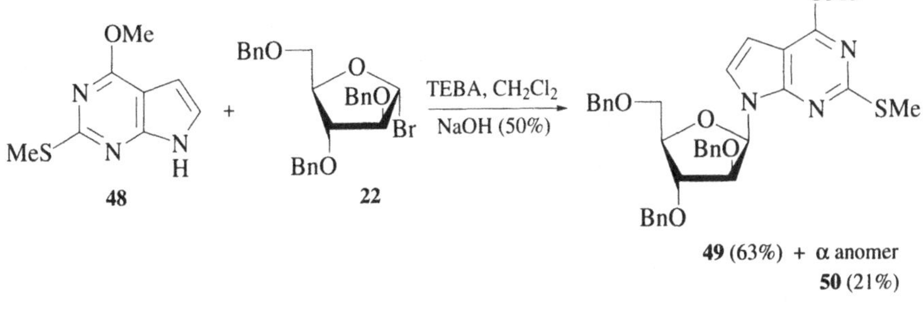

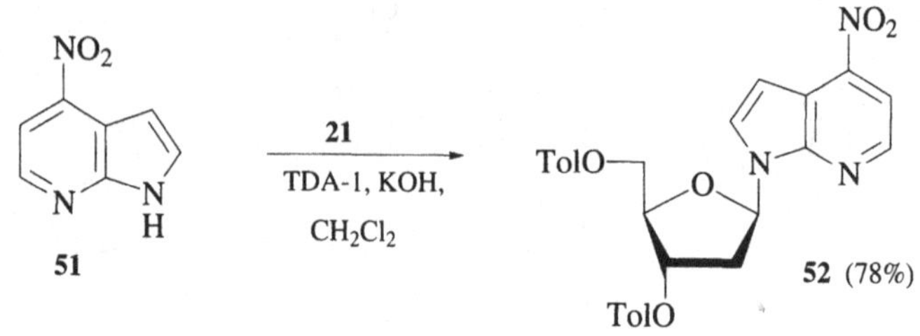

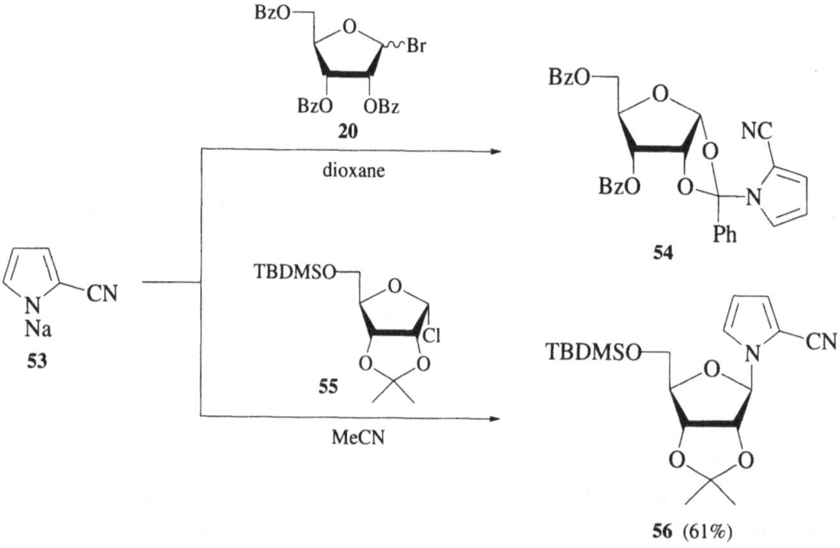

These methods employing sodium salts or phase-transfer reagents are quite efficient for the synthesis of 2′-deoxynucleosides (see the subsequent section on the synthesis of 2′-deoxynucleosides).

The Classical Hilbert-Johnson Procedure for the Preparation of Pyrimidine Nucleosides

In 1930 Hilbert and Johnson[69–71] reacted 2,4-diethoxypyrimidine (**57**) with **24** to give the protected intermediate **58** in ~30% yield. This intermediate can be saponified to the glycosylated uracil **59** or aminated by ammonia to the cytidine analog **60**. The liberated ethyl bromide can, however, convert **57** to the *N*-ethyl derivative **61**, thus diminishing the yields. It was later postulated that **57** and **24** react to give the N^1-alkylated intermediate **62**, which is then cleaved by bromide anion to afford **58** and ethyl bromide.[2,70]

Reaction of 2,4-dimethoxypyrimidine (**63**) with **21** affords a mixture of α-anomer **64** and β-anomer **65** in which the α-anomer predominates. Several explanations for this unexpected behavior have been advanced, since an S_N2 reaction of the nucleophilic pyrimidine should lead predominantly to the β-nucleoside **65**.[72]

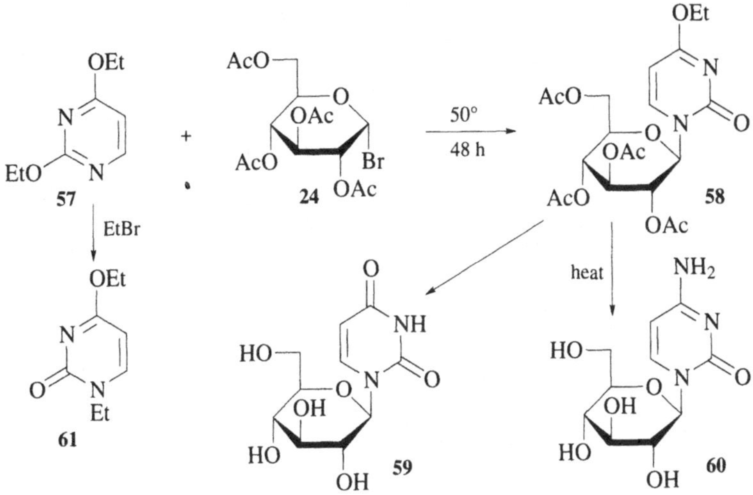

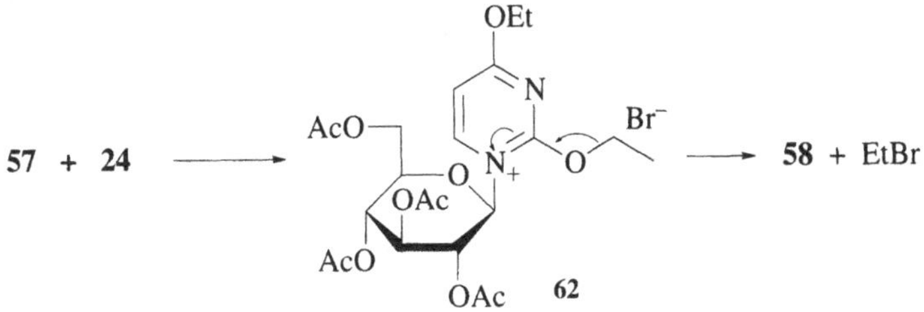

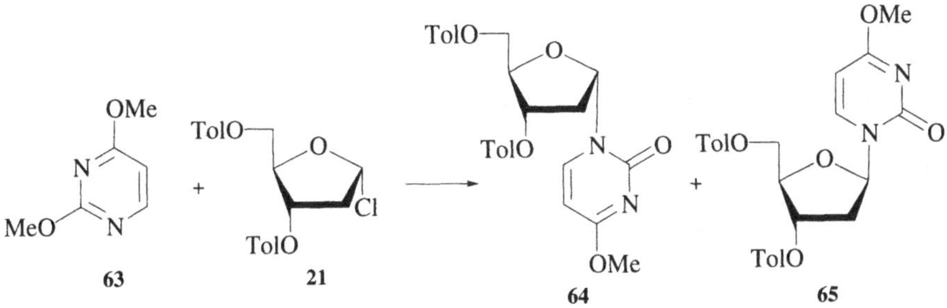

The classical Hilbert-Johnson reaction thus has a number of drawbacks:

(1) 2,4-dialkoxypyrimidines have to be prepared from the corresponding uracils via the corresponding 2,4-dichloropyrimidines; (2) yields are generally only moderate owing to the formation of byproducts such as *O*-glycosides and *N*-alkylated products such as **61**; and (3) dealkylation of the resulting 4-alkoxy compounds with acids can cause difficulties, whereas the transformation of **58** into the cytidine analog **60** with ammonia under pressure at elevated temperatures proceeds without complications.

The Silyl-Hilbert-Johnson Reaction

Introduction of the Silyl-Hilbert-Johnson reaction by Birkofer,[73,74] Nishimura,[75,76] and Wittenburg[77] was a major advance. Silylation (e.g., with HMDS) converts the polar, often rather insoluble pyrimidine bases into lipophilic silyl compounds, which can be distilled and which are readily soluble in organic solvents, permitting homogeneous reactions. Because of the electron-releasing property of silicon,[78,78a] the silylated heterocycles are better nucleophiles than the corresponding alkoxyheterocycles. The longer O-Si bond of 1.89 Å compared to the O-C bond of 1.53 Å makes the trimethylsilyl groups less bulky than a *tert*-butoxy group and results in the rapid solvolysis of remaining 4-*O*-trimethylsilyl groups. Because of the high mobility of the trimethylsilyl group one always obtains the thermodynamically most stable silylated heterocycle.[78a]

Silylated uracil **66** reacts with protected 1-halosugars such as **19** at room temperature in the presence of $HgCl_2$, $HgBr_2$, or $Hg(OAc)_2$ to afford the postulated intermediate **68**, which is cleaved by chloride or bromide ion to form 4-trimethylsilyloxy compound **71** and the volatile Lewis acid trimethylsilyl chloride (**69**) or trimethylsilyl bromide as leaving groups.[77,80] The 4-trimethylsilyloxy compound **71** can either be hydrolyzed in high yield to the 2,3,5-tri-*O*-benzoylated uridine **72** or reacted with primary or secondary amines to give the corresponding protected cytidines **73**.[79] Instead of adding mercuric salts, one can also heat **19** with **66** in absolute benzene or toluene or fuse **19** with **66** at temperatures up to 190° under reduced pressure in the absence of solvents.[76]

Finally, on reaction of **19** and **66** in benzene with silver perchlorate (or silver triflate) at room temperature, AgCl (or AgOTf) precipitates and the cyclic protected sugar perchlorate (or triflate) **67** is formed. Reaction of **67** with the silylated base **66** then affords intermediate **68**, which undergoes fragmentation by perchlorate (or triflate) anion to furnish **71** and the Lewis acid trimethylsilyl perchlorate (**70**) (or trimethylsilyl triflate).[74,80]

Because of the thermal instability of protected 1-halosugars such as **19**, **20**, or **21**, reactions of silylated bases with mercuric salts or silver perchlorate in benzene at room temperature usually afford the best results. However, as mentioned before, use of mercuric salts gives rise to toxic impurities.[63] Thus the $AgClO_4$ procedure is preferable, affording high yields of protected nucleosides and protected 2-thiouridines and cytidines.[81]

It should be emphasized that all perchlorates are potential explosives; neat trimethylsilyl perchlorate explodes above 50°.[82] Importantly, nonexplosive triflate salts analogous to **67** can be obtained efficiently by treatment of 1-*O*-acyl- or 1-*O*-alkylsugar benzoates with trimethylsilyl triflate (TMSOTf).

SILYL-HILBERT-JOHNSON REACTION IN THE PRESENCE OF FRIEDEL-CRAFTS CATALYSTS

Nucleoside Synthesis with $SnCl_4$ and Related Friedel-Crafts Catalysts

The Silyl-Hilbert-Johnson reaction of silylated 6-azauracil **74** with **19** in the presence of mercuric salts affords 6-azauridine-2′,3′,5′-tri-*O*-benzoate (**76**,

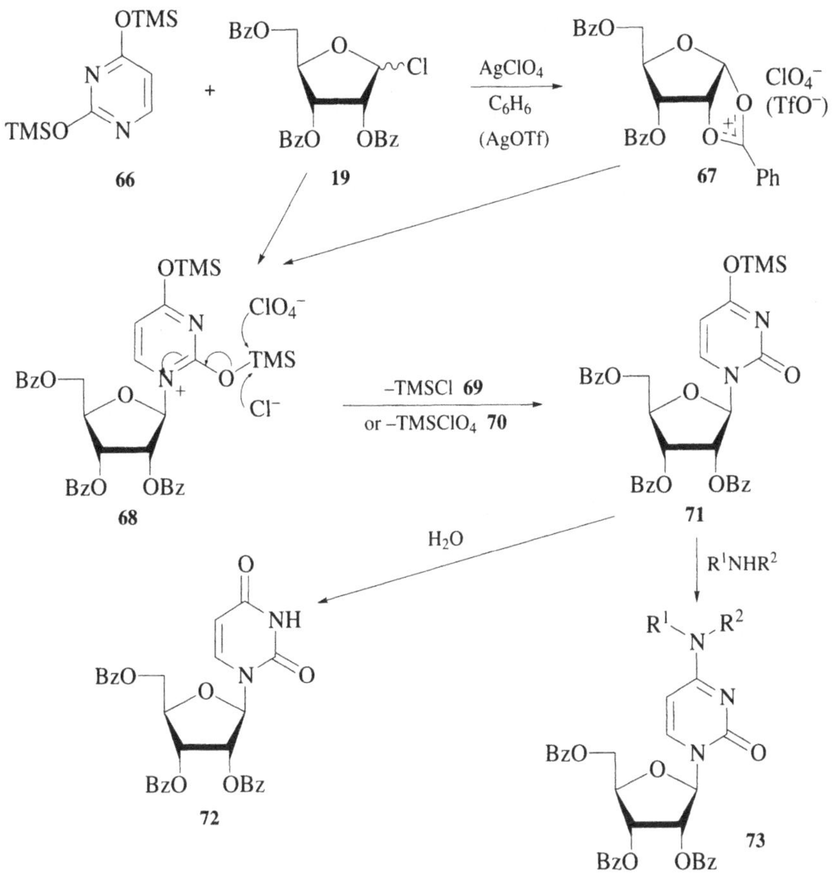

60%), as well as a series of colored impurities that apparently contain mercuric compounds. On the other hand, reaction of **74** with **18** in the presence of $SnCl_4$ or $TiCl_4$ in 1,2-dichloroethane at 20° affords crystalline **76** in 93% yield.[83,84] The reactive intermediate **75** can be converted by excess pyrrolidine into the cytidine analog **77** in 57% yield.[79]

Friedel-Crafts catalysts such as $SnCl_4$ or $TiCl_4$ had been previously employed for the synthesis of purine nucleosides.[85,86] For example, 1-*O*-acyl- or 1-*O*-alkyl-protected sugars were converted in situ into their corresponding reactive sugar cations such as **67** and then reacted with free purine bases, a technique which is still being applied.[86a,b] However, Friedel-Crafts catalysts had not previously been used in combination with silylated heterocycles or any other silylated compounds such as silyl enol ethers.[87]

The reaction of silylated heterocyclic bases such as 2-thiocytosine (**78**) with **18** and $SnCl_4$ affords 2-thiocytidine-2′,3′,5′-tri-*O*-benzoate (**79**) in 95% yield.[84]

Surprisingly, the weakly basic silylated 5-nitrouracil **80a** reacts with **18** in the presence of ~10 mol % of $SnCl_4$ to afford 5-nitrouridine-2′,3′,5′-tri-*O*-benzoate

(**81a**) in nearly quantitative yield,[84] whereas the corresponding 5-nitro-2,4-dimethoxypyrimidine does not react at all with **21** under classical Hilbert-Johnson conditions.[88]

Compared to **80a**, the more basic silylated 5-methoxyuracil **80b** and silylated 5-morpholinouracil **80c** react with **18** much more slowly and only in the presence of excess $SnCl_4$ in acetonitrile to give 90% of **81b** together with **81c**, **82b**, and **82c**, in 53, 3, and 32% yields, respectively, along with the corresponding N^1,N^3-bis(nucleosides) **83**. The same reaction in the less polar 1,2-dichloroethane af-

OTMS, R, TMSO, N — **18**, $SnCl_4$, MeCN, $Cl(CH_2)_2Cl$, 20° → **81** + **82** + N^1,N^3 bis(product) **83**

	R
80a	NO_2
80b	OMe
80c	*N*-morpholinyl

fords even less of the desired N^1-nucleosides **81b** and **81c**.[90] In contrast to $SnCl_4$, the weaker Lewis acid TMSOTf gives much higher yields of **81b** and **81c** (see the following section).

In the aforementioned examples, it is primarily electronic factors that determine the N^1/N^3 ratio. With silylated 6-methyluracil **84**, however, steric as well as electronic factors determine the formation of protected 6-methyluridine **85**[89] as well as the protected N^3 product **86** and the N^1,N^3-bis(nucleoside) **87**.[89a] Interestingly, nucleoside **85** in solution exists in the *syn* conformation[25,26] owing to the interaction of the 6-methyl group with the sugar moiety.

84 — **18**, $SnCl_4$, MeCN → **85** (41%) + **86** (52%) + N^1,N^3 bis(riboside) **87** (3%)

Under optimal conditions employing carefully redistilled $SnCl_4$ and purified acetonitrile, 41% of **85**, 52% of **86**, and 3% of **87** are obtained,[89a] whereas much higher yields of **85** are again obtained with TMSOTf as catalyst (see the following section). No regiochemical problems are encountered in the intramolecular cyclizations of **88** and **90** with $SnCl_4$ in acetonitrile to form the nucleosides **89** and **91** in good yields.[91-93] For analogous cyclizations of silylated bases with Lewis acids see refs 94–97a-f.

In the $SnCl_4$-catalyzed Hilbert-Johnson reaction, other sugar moieties such as **21** can be employed to give high yields of 1:1-mixtures of the β- and α-nucleosides. Analogous reactions with the arabinose derivative **22**, peracylated

β-D-ribopyranose **23**[84] and acylated di- and polysaccharides[98] are summarized in Tables III and IV.

Although reaction of 4-trimethylsilyloxypyridine (**92a**) with **18** in boiling 1,2-dichloroethane in the presence of $SnCl_4$ gives nucleoside **93a** in 63% yield, the corresponding silylated 4-aminopyridine **92b** does not react under these conditions. Analogously, silylated pyrimidin-4-one **94** furnishes a mixture of 26% **95** and 60% **96**.[99]

The reaction of silylated purine bases such as disilylated N^6-benzoyladenine **97** with peracylated sugars such as **98** affords the corresponding purine nucleoside **99** in up to 70% yield.[100,101]

In the reaction of silylated 6-azauracil **74** with **18**, other Friedel-Crafts catalysts such as $FeCl_3$, $BF_3 \cdot OEt_2$, $AlCl_3$, or $TiCl_4$ also give good yields of 6-azauridine-2′,3′,5′-tri-*O*-benzoate (**76**).[84] Some research groups have subsequently employed $BF_3 \cdot OEt_2$,[102–104] $SnCl_2$,[105,106,106a,b,c] $SbCl_5$,[107–109] $ZnCl_2$,[110–112] ZnI_2,[113] $EtAlCl_2$[114–117] or SiF_4. The latter boils at −86° and has been used in combination with protected 1-fluorosugars.[118] However, the majority of nucleoside chemists prefer either $SnCl_4$, which can be readily redistilled before use and gives homogeneous reaction mixtures in either 1,2-dichloroethane or acetonitrile, or the newer trimethylsilyl triflate (TMSOTf) or trimethylsilyl nonaflate. Newly developed Lewis acids for selective aldol reactions, such as $SnCl_4/Sn(OTf)_2/LiClO_4$,[119] $Sn(OTf)_2/Bu_2Sn(OAc)_2$,[120] $Cp_2ZrCl_2/AgClO_4$,[121] or Ph_3CClO_4[122,123] might also be applicable to nucleoside synthesis, as could the recently described combination of AgOTf and Ph_2SnS.[124] A direct comparison, however, shows that yields of nucleosides with other catalysts are essentially the same as those with trimethylsilyl triflate.[106c]

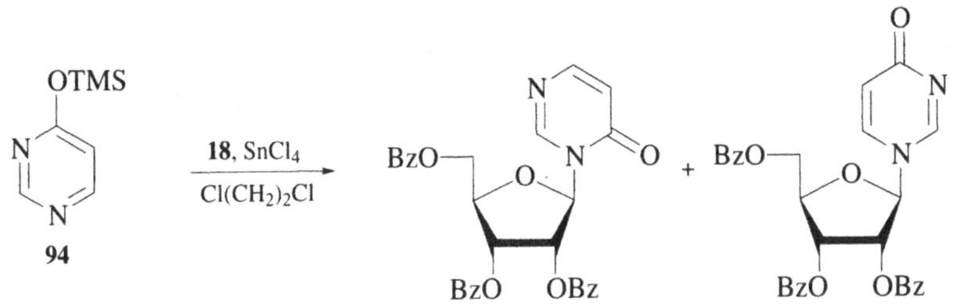

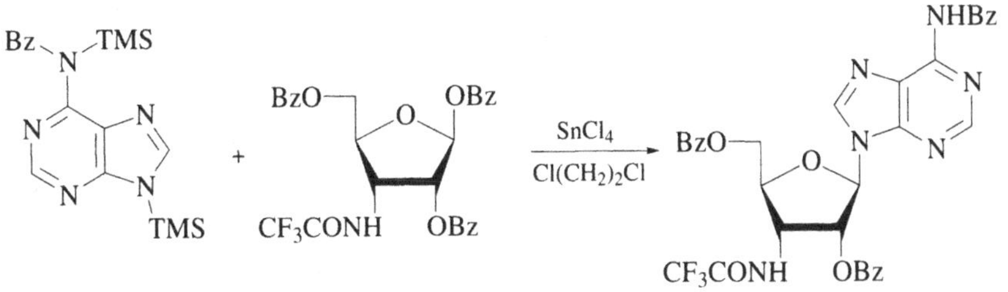

The Lewis acid should generally be just strong enough to convert protected 1-*O*-acyl-, 1-*O*-alkyl-, or 1-halosugars into their corresponding oxonium salts (such as **67**). Any additional acidic strength of the Lewis acid will only result in increased σ-complex formation with the silylated base, which in turn might lead to complications in regioselectivity or reaction rates.

Nucleoside Formation with TMSOTf

During the total synthesis of 5-methylaminomethyl-2-thiouridine,[125,126] a rare nucleoside from tRNA,[127] the silylated 2-thiouracil **100** was reacted with 1-chlorosugar **19** in the presence of $AgClO_4$ in benzene to give the substituted 2-thiouridine **101**, in which the *N-tert*-butoxycarbonyl group had been lost. The only strong Lewis acid that could have cleaved the BOC group was $TMSClO_4$

(**70**), whose formation as an intermediate during the Silyl-Hilbert-Johnson reaction in the presence of $AgClO_4$ had been postulated previously.[74,80] It was subsequently demonstrated that $TMSClO_4$, as well as TMSOTf, does indeed cleave *N*-BOC groups in amino acids and peptides.[128–130]

$$\mathbf{100} + \mathbf{19} \xrightarrow[\text{2. } H_2O]{\text{1. } AgClO_4,\ C_6H_6} \mathbf{101}\ (50\%)$$

$TMSClO_4$ and TMSOTf had previously been investigated using ^{29}Si NMR.[131] These studies showed that they are much stronger Lewis acids than, for example, $(TMS)_2SO_4$ or TMSCl. Whereas the explosive[82] $TMSClO_4$ is prepared from $AgClO_4$ and TMSCl in benzene, the chemically stable TMSOTf (**102**, bp 133–134°) is obtained by heating TfOH with TMSCl.[131] The higher boiling TMS nonaflate (**103**, bp 68–69°/11 torr) is formed on heating nonaflic acid with TMSCl or on reacting potassium nonaflate with TMSCl.

$$\text{TfOH} + \text{TMSCl} \xrightarrow[\text{heat}]{\text{neat}} \underset{\mathbf{102}}{\text{TMSOTf}} + \text{HCl}\uparrow$$

$$n\text{-}C_4F_9SO_3H + \text{TMSCl} \xrightarrow[\text{heat}]{\text{neat}} \underset{\mathbf{103}}{\text{TMSOSO}_2C_4F_9\text{-}n} + \text{H}$$

$$n\text{-}C_4F_9SO_3K + \text{TMSCl} \xrightarrow{\text{MeCN}} \text{TMSOSO}_2C_4F_9\text{-}n + \text{K}$$

$$AgClO_4 + \text{TMSCl} \xrightarrow{C_6H_6} \underset{\mathbf{70}}{TMSClO_4} + AgCl\downarrow$$

Since Friedel-Crafts catalysts such as $SnCl_4$ or $TiCl_4$ had been used successfully for the Silyl-Hilbert-Johnson nucleoside synthesis,[84] these new silylated Lewis acids $TMSClO_4$ and TMSOTf were reacted with silylated uracil **66** and **18**. It was found that catalytic amounts of $TMSClO_4$ or TMSOTf in 1,2-dichloroethane or acetonitrile were adequate for generating the reactive intermediate cation **67**, although the use of 1.1 equivalents was more efficient. Reaction of **67** with silylated uracil **66** leads to the silylated intermediate **71** and regenerated $TMSClO_4$ or TMSOTf.[132,133] Hydrolysis with aqueous $NaHCO_3$ in CH_2Cl_2 affords 2′,3′,5′-tri-*O*-benzoyluridine (**72**) in more than 80% yield. Most importantly, during hydrolysis no emulsions are formed[132,133] as are usually encountered on aqueous workup employing $SnCl_4$ as catalyst for nucleoside synthesis.

TMSO TMSO N N TMSO 66 + BzO OAc O BzO OBz 18 TMSOTf BzO O N N O BzO OBz 71 BzO O BzO O TfO⁻ Ph 67 NaHCO₃ CH₂Cl₂ O NH BzO O N O BzO OBz 72

In contrast to $TMSClO_4$ and TMSOTf, the weaker Lewis acids[131] $(TMS)_2SO_4$ and TMSCl do not promote nucleoside formation, since they apparently do not convert **18** into the sugar cation **67**. However, the even weaker Lewis acid TMSI, which can be prepared in situ from TMSCl and NaI in acetonitrile,[134] does catalyze the formation of nucleosides; TMSI is a combination of the hard trimethylsilyl cation and the soft iodide anion and effects nucleoside synthesis via an apparent push and pull mechanism.[135-139,139a,c,d] On reaction of 1-*O*-acetyl-5-*O*-pivaloyl-(3*S*)-2,3-dideoxyapiose with silylated thymine or N^6-benzoyladenine in acetonitrile at -5 to 0°, catalytic amounts of TMSI at $-5°$ seem to induce faster formation of the mixture of *syn* and *anti* nucleosides than equivalent amounts of TMSOTf at $-5°$.[139a] Pure TMSI is, however, quite unstable and thus must always be redistilled before use; it will also cleave ester or ether functionalities in the sugar or heterocyclic moieties.

The persilylated polymeric perfluorinated sulfonic acid Nafion®[139b] has not as yet been explored as catalyst,[133] although it could be recovered by filtration and regenerated by heating with excess TMSCl.

Importantly, use of TMSOTf (**102**) as catalyst dramatically increases the yields of the 5-methoxy- or 5-morpholino-2,3,5-uridine tri-*O*-benzoates (**81b** and **81c**) from the silylated uracils **80b,c** and **18**. Thus, even in 1,2-dichloroethane, 89% of **81b** (compared to 53% using $SnCl_4$) and 95% of **81c** (compared to 39% using $SnCl_4$) are obtained. Analogously, use of catalytic TMSOTf with the rather basic silylated 4-trimethylsilyloxypyridine (**92a**) and 4-trimethylsilylaminopyridine (**92b**) gives the corresponding nucleosides **93a** and **93b** in 87% and 80%

yields,[133] whereas **93a** is obtained only in 63% yield with $SnCl_4$.[99] For the explanation of these differences in chemical behavior see the following section.

Under carefully controlled conditions using TMSOTf in purified acetonitrile, silylated 6-methyluracil **84** affords the desired protected 6-methyluridine **85** in 71% yield compared to 41% with $SnCl_4$.[133] Furthermore, the undesired acylated N^3-nucleoside **86** rearranges on silylation to **104**, and heating with TMSOTf to yield 53% of the protected N^1-nucleoside **85** and 33% of the N^1,N^3-bis(riboside) **87**. The sterically hindered **87** reacts on heating with silylated 6-methyluracil **84** and TMSOTf to give the N^1-nucleoside **85**.[133]

On extended exposure of silylated 2′,3′,5′-tri-*O*-benzoyl-6-methyluridine **105** to TMSOTf in 1,2-dichlorethane, 24% of the benzoylated 2,2′-anhydronucleoside **106** is formed together with 23% of the N^1,N^3-bis(riboside) **87** and trimethylsilyl benzoate.[133] This cyclization to **106** might be favored by the *syn* conformation of **85** and **105**.

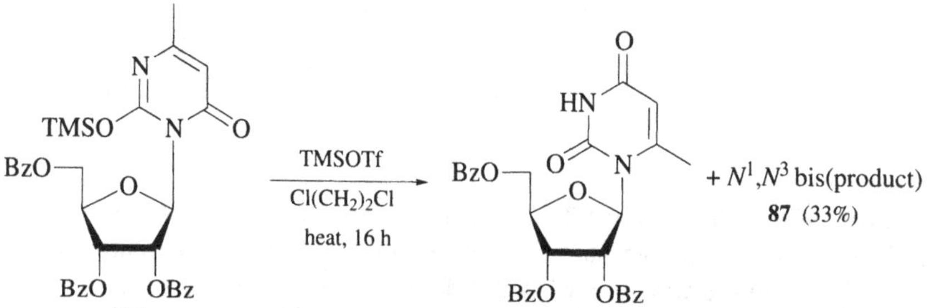

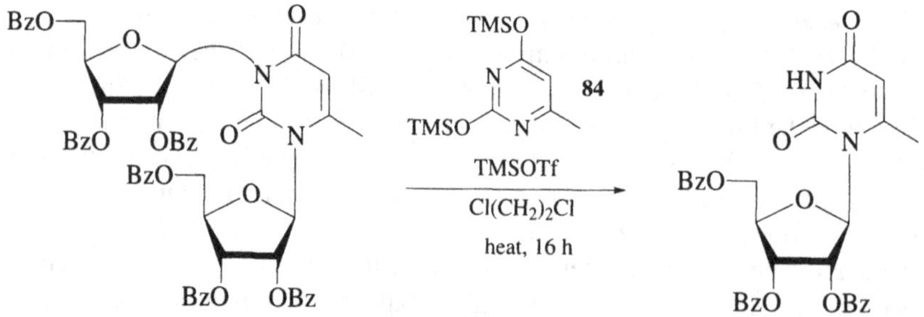

Silylated N^6-benzoyladenine **108a**, as well as silylated N^2-acetylguanine **108b** and xanthine **108c**, afford after saponification the corresponding crystalline purine nucleosides **109a–c** in 81, 66, and 49% yields, respectively.[133]

Following the TMSOTf catalyzed synthesis of N^6-benzoyladenosine-2′,3′,5′-tri-*O*-benzoate by TLC indicates that the reaction proceeds at least partially via the protected N^3-**38** or N^7-nucleoside **110a** to give adenosine after saponification.

TMSOTf, $Cl(CH_2)_2Cl$, 24°, 120 h: **105** → **106** (24%) + **87** (23%) + $PhCO_2TMS$ **107**

Thus the N^6-silylated N^3-adenosine **111** is smoothly rearranged in the presence of TMSOTf to 2′,3′,5′-tri-*O*-benzoyladenosine (**112**).[133]

The reaction of silylated N^2-acetylguanine (**108b**) with riboside **18** in the presence of TMSOTf at reflux in 1,2-dichlorethane affords, after workup and saponification of the protecting groups with methanolic ammonia, an overall yield of 66% of crystalline guanosine containing at most traces of the N^7-isomer **110b**[133] and not, as later claimed, a crystalline mixture of **109b** and **110b**.[140,141,141a] The guanosine synthesis apparently proceeds at least partially via the corresponding N^7-nucleoside, which can be isolated after saponification as the unprotected nucleoside **110b** along with the desired thermodynamically controlled **109b**. Heating of the reaction mixture with TMSOTf in boiling 1,2-dichloroethane affords a mixture of the protected N^7- and N^9-guanosines in which only 10–15% of the N^7-nucleoside can be detected, whereas use of $SnCl_4$ in acetonitrile results in predominant formation of the protected N^7-nucleosides.[142] These results seem to indicate that in 1,2-dichloroethane or acetonitrile, $SnCl_4$ forms a σ or chelate complex with the N^3- and N^9-nitrogen atoms of silylated N^2-acetylguanine, blocking the access of **67** to N^9 or the rearrangement of the silylated N^7-guanosine to the N^9-guanosine. However, reaction of **108b** with **17** in the presence of TMSOTf

108 + **18** → 1. TMSOTf; 2. MeOH, NH_3 → **109** + **110**

	R^1	R^2
108a	BzNTMS	H (= **97**)
108b	OTMS	AcNTMS
108c	OTMS	OTMS

109,110	R^1	R^2
a	NH_2	H
b	OH	NH_2
c	OH	OH

in 1,2-dichloroethane at reflux affords, after saponification, the N^9- and N^7-guanosines **109b** and **110b** in a ratio of 2:1. The sugar derivative **18**, which is converted into the more stable sugar cation **67** compared with **27** (p. 6) and will thus facilitate any rearrangement of the protected and silylated N^7-nucleoside into the corresponding N^9-nucleoside, furnishes **109b** and **110b** in a ratio of 6:1 under the same reaction conditions (cf p. 48).[142]

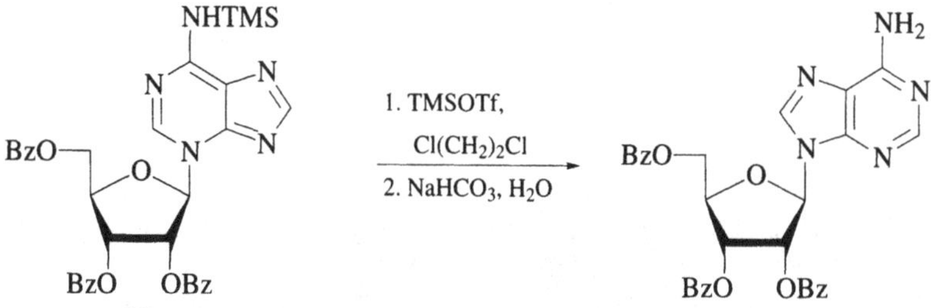
NHTMS
N
N
N
N
BzO
O
BzO
OBz
1. TMSOTf,
$Cl(CH_2)_2Cl$
2. $NaHCO_3$, H_2O
NH_2
N
N
N
N
BzO
O
BzO
OBz

111 **112**

Blocking the 6-oxygen in N^2-acetylguanine with the bulky diphenylcarbamoyl group followed by silylation to **113** and subsequent reaction with **17** in the presence of TMSOTf leads to the protected guanosine **114**, which can be saponified to the natural guanosine **109b** in 68% overall yield.[140] Analogously, in the TMSOTf catalyzed reaction with **17**, introduction of a 6-(4-nitrophenylethoxy) group into 2-bromoxanthine followed by silylation gives less than 5% of the undesired N^7-nucleoside.[143] Reaction of **113** with 1,2,3,4,6-penta-*O*-acetyl-*β*-D-galactopyranose in the presence of TMSOTf in toluene at 80° gives, in addition to the anticipated protected N^9-nucleoside, a rearranged protected nucleoside in which the O^6-diphenylcarbamoyl group has migrated to replace the N^2-acetyl group.[143a,143b]

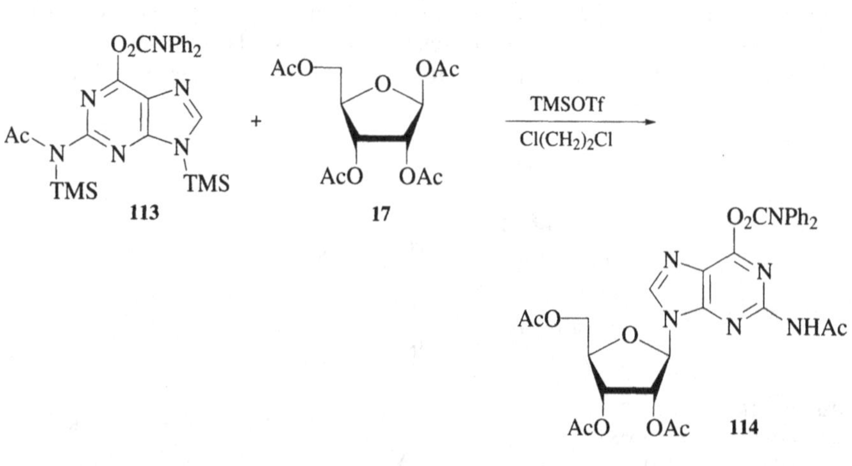
O_2CNPh_2
N
N
Ac
N
N
N
TMS
TMS
+
AcO
O
OAc
AcO
OAc
TMSOTf
$Cl(CH_2)_2Cl$
O_2CNPh_2
N
N
AcO
N
N
NHAc
O
AcO
OAc

113 **17** **114**

However, in view of the additional steps involved in preparing O^6-blocked guanine derivatives, the direct synthesis of guanosine (or analogs) starting with **108b** (or similar bases) and **18** to give crystalline guanosine in 66% yield[133] after saponification should always be considered. The synthesis of 9-substituted guanines has been reviewed.[144]

Peracylated pyranose **23** is transformed on heating with TMSOTf into the reactive pyranose cation **115**, which combines with silylated uracil **66** to give 1-[2′,3′,4′,6′-tetra-*O*-acetyl-*β*-D-glucopyranosyl]uracil (**116**) in 89% yield.[133] The lower reactivity of pyranose derivatives with TMSOTf permits differentiation between the furanose and pyranose forms. Brief treatment of 2-deoxyribose with absolute methanol-HCl gives primarily the two kinetically controlled 1-*O*-methyl-2-deoxyribofuranoses (**117**) and minimal amounts of the 1-*O*-methylpyranoses (**118**), which as the thermodynamically controlled products become the major products on longer exposure to methanolic HCl.[145]

Acylation of the mixture of **117** and **118** with *p*-toluolyl chloride and pyridine affords mainly a mixture of the two anomeric *O*-acylated furanoses **119** and a minor amount of the anomeric pyranoses **120**, which can be readily separated by chromatography on silica gel with hexane-diisopropyl ether.[146] This mixture of the acylated furanoses and pyranoses is commonly treated with anhydrous HCl in acetic acid to give the labile crystalline **21** on crystallization as well as some noncrystalline **121**.[51] The crystalline **21** has become the standard sugar for the preparation of 2′-deoxynucleosides since the 3′,5′-bis(*p*-toluoyl) *β*-nucleosides usually have higher melting points and lower solubility than the corresponding α-nucleosides, often permitting their separation by crystallization.

Because formation of the anomeric acylated *O*-methylfuranosides **119** is kinetically controlled, they are more readily converted to the corresponding reactive furanose cation intermediates by TMSOTf than the anomeric acylated 1-*O*-methylpyranosides **120**. Thus a mixture of **119** and **120** reacts with silylated 5-ethyluracil **122** at ambient temperature to give 58% of the desired β-nucleoside **123** as well as 31% of the undesired α-nucleoside **124**. The same mixture of **119** and **120** affords only 35% of the crystalline protected **21**[133] on treatment with HCl-AcOH. Because of the reversibility of nucleoside synthesis, an undesired 2′-deoxy-α-nucleoside such as **124**, when silylated and kept for 46 hours at 24° with TMSOTf, affords 27% of the desired β-nucleoside **123** as well as 67% of recovered α-nucleoside **124.**[133]

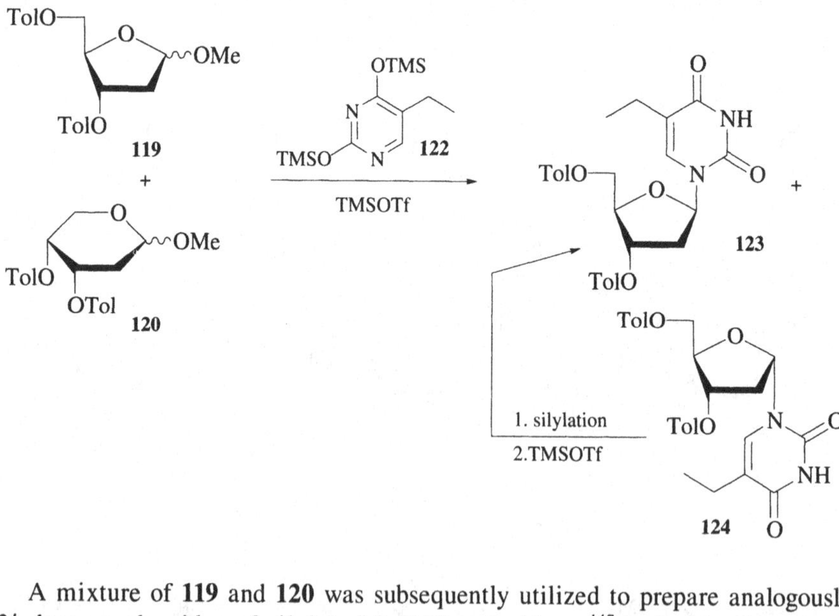

A mixture of **119** and **120** was subsequently utilized to prepare analogously 2′-deoxynucleosides of silylated 2-(1*H*)pyrimidone.[147] Related mixtures of 2-deoxyfuranosides and pyranosides were reacted with silylated uracil[148] and silylated 5-iodopyrimidin-2-one.[149]

Although longer reaction times increase the amount of β-anomer **123** obtained from **124**, they also lead to gradual decomposition of the sensitive 2-deoxyribose moiety in **123** and **124** to give 2-toluoyloxymethylfuran **259** (see section on 2′-deoxynucleosides).[150] Since there is always an equilibrium between the activated protected β- and α-anomers, silylation of β-anomers such as 3′,5′-di-*O*-acetyl-N^4-benzoyl-2′-deoxycytidine with *N,O*-bis(silylacetamide) (BSA), and subsequent heating with TMSOTf in acetonitrile for 3 hours at 80° affords 51% of the corresponding protected α-anomer.[151,151a] Treatment of 3′,5′-di-*O*-acetylthymidine with acetic anhydride and sulfuric acid leads to the predominant formation of the 3′,5′-di-*O*-acetyl-α-thymidine.[152,152a] On the other hand, free or

3′,5′-*O*-protected thymidines are cleaved to glycals on heating with HMDS and $(NH_4)_2SO_4$.[152b,152c]

On employing 1-*O*-methyl-3,5-di-*O*-toluoyl-2-deoxyribofuranoside **119** or 1-*O*-methyl-2,3,5-tri-*O*-benzoyl-D-arabinofuranoside **328** and insufficient amounts of TMSOTf or $SnCl_4$, seconucleosides such as **332** (which are probably derived via activated intermediates such as **333**) can be isolated (cf the section on arabinonucleosides). These seconucleosides undergo cyclization to the anticipated nucleosides upon introduction of additional amounts of catalyst.

Sugar moieties containing sensitive azide groups[153,279] as well as an assembly of complex functional groups have been employed. In the synthesis of octosyl acid A, silylated 5-carbomethoxyuracil **125** is reacted with the sugar moiety **126** to give the nucleoside intermediate **127**[154,155] in 90% yield (see a related approach to octosyl acid A[156]). Analogously, the reaction of silylated uracil **66** with the rather complex sugar moiety **128** yields intermediate **129** (91%) for the synthesis of ezomycin[157] (see a related approach to ezomycin A[158]). In the synthesis of hikizimycin, an even more complex sugar moiety was employed with TMSOTf in nitrobenzene at 127° (see **515** → **516**).[159,160] The synthesis of nucleoside antibiotics was reviewed recently.[160a]

125 + **126** —TMSOTf, $Cl(CH_2)_2Cl$→ **127** (90%)

66 + **128** —TMSOTf, MeCN→ **129** (91%)

On reacting complex sugar moieties containing many basic functional groups, one has to realize that all these groups form either weak σ complexes with TM-SOTf or stronger chelate-type complexes with $SnCl_4$, $TiCl_4$, or Et_2AlCl, which will slow down and perhaps even alter the course of the reaction. Thus, use of additional amounts of catalyst is necesssary.

Even very weakly basic systems can be silylated and converted into nucleoside-type structures. Thus 2,3-diaminomaleonitrile (DAMN) reacts in its disilylated form **130** with **18** and 2 equivalents of TMSOTf in CH_2Cl_2 to furnish 50% of nucleoside **131** and 24% of the seconucleoside **132**, whereas the same reaction in the presence of only 1 equivalent of TMSOTf gives only imine **132**.[161] Similarly, the silylated cyclic urea **133** or **134** affords only 16% of nucleoside **135** with **18** and TMSOTf as catalyst.[162] This low yield, however, might be due to the presence of monosilylated or free trifluoroacetamide derived from bis(trimethylsilyl)trifluoroacetamide (BSTFA) used for the silylation of the cyclic urea to **133** or **134**, since the trifluoroacetamide or its monosilyl derivative can compete with **133** → **134** for the reactive sugar intermediate **67**. This possibility was recently confirmed in the silylation of 1,2,4,6-thiatriazin-3-one-1,1-dioxides with BSA followed by reaction with peracylated sugars in the presence of TMSOTf in acetonitrile at reflux, where only moderate yields of protected nucleosides were obtained, along with up to 46% of 1-β-acetamides of the protected sugars.[162a]

130 + **18** → (TMSOTf, CH_2Cl_2) → **131** (50%) + **132** (24%)

133 ⇌ **134** → (**18**, TMSOTf, $Cl(CH_2)_2Cl$, 24°) → **135** (16%)

Transglycosylation with Lewis Acids

Since nucleoside formation in the presence of Lewis acids is a reversible process, a nucleoside base in a given nucleoside can be exchanged for another nucleoside base. This was first investigated by treating peracylated cytidine **136** with N^6-benzoyladenine (**137**) in the presence of $HgBr_2$ and DMA in xylene at reflux to afford after saponification 39% adenosine (**7**) and 6% α-adenosine (**8**).[163] An exchange of ribose in 2′,3′-isopropylideneinosine by acetobromoglucose gives the corresponding inosine analog in moderate yield.[164]

NHAc, AcO, AcO OAc, **136**; NHBz, **137**; $HgBr_2$, xylene, DMA, heat; MeOH, NH_3; NH2, HO, HO OH, **7** (39%) + α-adenosine **8** (6%)

Much more effective are transglycosylations of silylated nucleosides and bases in the presence of trimethylsilyl perchlorate or TMSOTf.[151,165–168,170–172,176–179] N^3-Benzoyl-2′,3′,5′-tri-*O*-acetyluridine (**138**) and excess persilylated N^6-benzoyladenine (**97**) furnish an 81% yield of N^6-benzoyl-2′,3′,5′-tri-*O*-acetyladenosine (**139**), whereas $SnCl_4$ as catalyst gives a much lower yield of a mixture of protected β- and α-adenosine.[166] N^4-2′,3′,5′-Tetraacetylcytidine (**136**) and persilylated N^2-acetylguanine **108b** afford peracetylated guanosine **140** in 66% yield as well as 2% of the corresponding α-anomer in the presence of TMSOTf.[166]

AcO, AcO OAc, **138**; TMS, Bz, **97** TMS; TMSOTf, MeCN, $Cl(CH_2)_2Cl$, heat; NHBz, AcO, AcO OAc, **139** (81%)

A derivative of octosyl acid, **141**, reacts with **97** in boiling 1,2-dichloroethane to give analog **142** in 60% yield.[165,166] Similarly, the transformation of the polyoxin derivative **143** affords **144**.[167] A similar transglycosylation of a complex uridine derivative leads to the corresponding adenosine derivative as an approach to a total synthesis of sinefugin[168] (see also the transglycosylation of protected griseolic acid[169]).

Reaction of persilylated 3′-azido-2′-deoxy-5′-*O*-acetylthymidine (**145**) with silylated N^6-octanoyladenine (**146**) affords after saponification 27% of the corresponding β-purine nucleoside **147** as well as 35% of the corresponding α-anomer **148**.[170] Reaction with persilylated N^2-palmitoylguanine gives 28% of the corresponding β-N^9-guanine nucleoside.[170]

The analogous reaction with silylated 2,6-diacetamidopurine affords the corresponding 2-aminoadenosine derivative.[171] Transformations of persilylated

3′-deoxy-3′-fluorothymidine and its 5′-acetyl derivative with persilylated N^2-acylguanine,[172,173] 2-fluoroadenine,[174] N^6-benzoyladenine,[175] and benzimidazole[176] as well as of uracil and of 5-substituted uracils,[177,178] furnish the corresponding protected β-2′-deoxynucleosides together with α-anomers in moderate yields.

TMS–N(COC$_7$H$_{15}$) **146** TMS; **145** → TMSOTf, MeCN → NH$_3$ → **147** (27%) + α anomer **148** (35%)

The transglycosylation of 2′-deoxy-2′-trifluoroacetylaminouridine (**149**) with N^2-palmitoylguanine **150** in the presence of BSA and TMSOTf and subsequent saponification with NH$_3$ affords guanine nucleoside **151** in 60% yield,[179] whereas **152** and **153** react in the presence of BSA and TMSOTf to give a 27% yield of **154** and 36% of the corresponding α-anomer **155**.[151]

149 + **150** (C$_{15}$H$_{31}$) → TMSOTf, BSA, MeCN → NH$_3$ → **151** (60%)

152 + **153** → TMSOTf, MeCN → **154** (27%) + α anomer **155** (36%)

Purine nucleosides can also be transformed into the corresponding pyrimidine nucleosides. Thus, protected purine nucleosides such as **156** and N^4-octanoylcytosine (**157**) on heating with BSA and $SnCl_4$ in 1,2-dichloroethane provide the corresponding cytidines **158** in 30–60% yield.[180]

Likewise N^2-2′,3′,5′-tetraacetylguanosine (**140**) reacts with 2-acetoxyethyl acetoxymethyl ether (**159**) in chlorobenzene at reflux in the presence of traces of *p*-toluenesulfonic acid to furnish the desired N^9-derivative **160** as well as the N^7-analog **161** in a 9:7 ratio.[181,181a,182]

157, $Cl(CH_2)_2Cl$, $SnCl_4$, 80°, 20 h: **156** → **158** (60%)

159, TsOH, PhCl: **140** → **160** + N^7-anomer **161**

On attempted transglycosylation of N^6-benzoyloxetanocin di-*O*-acetate (**162**) with persilylated uracil **66** in the presence of $SnCl_4$, the intermediate sugar cation rearranges to form furanose nucleosides **163** and **164.**[183] For further examples of transglycosylation, see Refs. 184–186. For a review of transglycosylation of purine nucleosides, see Ref. 186a.

66, $SnCl_4$, MeCN, $Cl(CH_2)_2Cl$, 24°: **162** → **163** + **164**

One Step–One Pot Nucleoside Syntheses

Since silylations are accelerated by Lewis acids,[187] and the silylation of heterocyclic bases is much more rapid in the presence of Friedel-Crafts catalysts, one can combine the different steps of nucleoside synthesis in a one step–one pot procedure in a polar solvent such as acetonitrile:[188,189] (a) silylation of the heterocyclic base, (b) silylation of the triflate or nonaflate salts to form TMSOTf or $(CH_3)_3SiOSO_2(CF_2)_3CF_3$ (if $SnCl_4$ is not used as a catalyst), and (c) nucleoside synthesis with acylated 1-*O*-acyl- or 1-*O*-alkylsugars in the presence of a Friedel-Crafts catalyst.

Such a one step–one pot preparation avoids the handling of the easily hydrolyzed silyl compounds and saves time since no prior silylation step is needed. Under these one step–one pot conditions the amounts of TMSCl and HMDS have to be chosen in such a way that all reactive heterocyclic hydroxy, thio, amino, or amido groups as well as the free triflate or nonaflate acids $C_nF_{2n+1}SO_3H$ or their alkali salts are silylated with formation of NH_4Cl and NaCl or KCl, since any free NH_3 would neutralize the Friedel-Crafts catalyst. Because all these salts are practically insoluble in acetonitrile, they precipitate and the equilibria are shifted toward the desired electrophilic silyl ester.[189]

$$3\ \text{TfOH} + \text{HMDS} + \text{TMSCl} \longrightarrow 3\ \text{TMSOTf} + NH_4Cl$$

$$n\text{-}C_4F_9SO_3K + \text{TMSCl} \longrightarrow \underset{\mathbf{103}}{\text{TMSOSO}_2C_4F_9\text{-}n} + \text{KCl}$$

Since potassium nonaflate ($C_4F_9SO_3K$) is only partially soluble in boiling acetonitrile, the KCl formed on its reaction with TMSCl to give **103** could occlude unreacted reagent. Therefore, an excess of finely powdered potassium nonaflate should be employed.

As described in Eq. 1, a mixture of 0.33–0.40 equivalent of TMSCl and HMDS has to be used for free TfOH, whereas for potassium nonaflate (Eq. 2) equimolar amounts of TMSCl have to be employed.[189]

For silylating uracil, cytosine, or a purine such as N^6-benzoyladenine containing two reactive oxygen or oxygen and nitrogen functionalities, a mixture of at least 0.7–0.8 equivalent each of TMSCl and HMDS is necessary to obtain the corresponding persilylated uracil, cytosine, or N^6-benzoyladenine with concomitant formation of 0.7–0.8 equivalent of NH_4Cl. For a heterocyclic base such as 4-pyridone with only one reactive oxygen group, only ca. 0.4 equivalents each of TMSCl and HMDS are needed to afford the silylated base **92a** (X = O). Uridine-2′,3′,5′-tri-*O*-benzoate (**166a** = **72**) is readily obtained in 80–84% yield starting from uracil (**165a**) and **18** employing free triflic acid or potassium nonaflate and TMSCl in boiling acetonitrile, whereas the stronger Friedel-Crafts catalyst $SnCl_4$ is effective at room temperature. Table A summarizes typical examples of one step–one pot syntheses of pyrimidine nucleosides.

1. Catalyst, TMSCl, HMDS, MeCN
2. $NaHCO_3$

165, 166	X	Y	R
a	O	O	H
b	S	O	H
c	O	O	OMe
d	O	NAc	H
e	O	NBz	H

2-Thiouracil (**165b**) with $SnCl_4$ as catalyst gives 2-thiouridine-2′,3′,5′-tri-*O*-benzoate (**166b**) in ~60% yield. The more basic 5-methoxyuracil (**165c**) reacts with **18** in the presence of potassium nonaflate/TMSCl/HMDS to afford crystalline 5-methoxyuridine-2′,3′,5′-tri-*O*-benzoate (**166c** = **81b**) in 71% yield. The analogous reaction of N^4-acetylcytosine (**165d**) with **18** followed by saponification with methanolic ammonia gives 56% of pure crystalline cytidine.[189]

Table A. One Step - One Pot Reactions with Sugar 18

Base	Acid or Salt	TCS/HMDS	Conditions	Acylated Nucleoside
165a	CF_3CO_3H	1.2/1.1	83°, 1 h	**166a** (81%)
165a	$C_4F_9SO_3K$	3.1/0.7	83°, 14 h	**166a** (84%)
165a	$SnCl_4$	0.8/0.8	24°, 2 h	**166a** (83%)
165b	$SnCl_4$	0.8/0.8	24°, 7 h	**166b** (59%)
165c	$C_4F_9SO_3K$	3.1/0.7	83°, 20 h	**166c** (71%)
165d	$C_4F_9SO_3K$	3.1/0.7	83°, 27 h	**166d** (56%)

6-Azauracil (**167**) reacts with **23** in the presence of $SnCl_4$/TMSCl/HMDS to furnish the crystalline protected nucleoside **168** in 42% yield. The rather basic 4-pyridone (**169**) and **18** are converted by potassium nonaflate/TMSCl/HMDS followed by saponification to the free nucleoside **170** in 50% yield.[189]

$SnCl_4$, TMSCl, HMDS, MeCN

168 (42%)

169 + 18 → (1. $C_4F_9SO_3K$, TMSCl, HMDS, MeCN; 2. MeOH, NH_3) → 170 (50%)

The oily mixture of anomers **119** and **120** furnishes with potassium nonaflate/TMSCl/HMDS primarily the kinetically controlled 1-cation of the furanose, which reacts in situ with 5-ethyluracil (**171**) to give 26% of the β-anomer **123** and 21% of the corresponding α-anomer **124**.[189]

The purine bases N^6-benzoyladenine (**137**) and N^2-acetylguanine (**138**) react with **18** in the presence of potassium nonaflate/TMSCl/HMDS, affording after saponification with methanolic ammonia, crystalline adenosine in 63% and crystalline guanosine in 44% yield, respectively.[189]

119 + 120 + 171 → ($C_4F_9SO_3K$, TMSCl, HMDS, MeCN) → 123 (26%) + α-anomer 124 (21%)

137 + 18 → ($C_4F_9SO_3K$, TMSCl, HMDS, MeCN) → (MeOH, NH_3) → 7 adenosine (63%)

138 + 18 → ($C_4F_9SO_3K$, TMSCl, HMDS, MeCN) → (MeOH, NH_3) → 5 guanosine (44%)

In 1,2-dichloroethane as solvent, increased salt concentrations cause increased formation of N^3-nucleosides as well as N^1,N^3-dinucleosides. Although these increased salt concentrations are usually irrelevant in acetonitrile, the reaction of 6-methyluracil with **18**, which is particularly sensitive to added salts or impurities, affords even in acetonitrile in the presence of potassium nonaflate/TMSCl/HMDS only 20–25% of 6-methyluridine-2′,3′,5′-tri-*O*-benzoate (**85**) besides the undesired N^3-nucleoside **86** and the N^1,N^3-bis(riboside) **87**.[189]

The one-pot reaction of 2-bromohypoxanthine with tetra-*O*-acetyl-β-D-ribofuranose (**17**) in acetonitrile can be performed by initially heating 2-bromohypoxanthine with BSA, then adding **17** and TMSOTf to give initially more N^7- than N^9-nucleoside (HPLC). This ratio changes rapidly within 15 minutes, especially in the presence of excess **17**, to afford 75% of the N^9-nucleoside and 10% of the N^7-nucleoside.[190]

The cytosine analog **172** gives the protected cytidine analog **173** in 41% yield.[191]

172 → (**18**; $C_4F_9SO_3K$, TMSCl, HMDS, MeCN) → **173** (41%)

Interestingly, the one-pot procedure employing $SnCl_4$ as catalyst applied to thymine or N^4-benzoylcytosine (**165e**) and methyl-4,6-di-*O*-acyl-2,3-dideoxy-α-D-glucopyranoside furnished 60–65% of the desired protected β-nucleosides and only 18–20% of the α-nucleosides. By comparison, the corresponding reaction of N^2-isobutyrylguanine with $SnCl_4$ or TMSOTf yields complicated α/β mixtures of the corresponding N^9- and N^7-nucleosides.[192]

Equally simple is an alternative one-pot synthesis of purine nucleosides[86,86a] and cytidines.[192e] In this procedure purines such as adenines,[86b,192a–j] N^6-benzoyl[192m,n,p] or N^6-octanoyladenine (**137**),[86,192g] 6-chloropurine,[192a] 2,6-dichloropurine (**25**),[86a,192a] N^2-palmitoylguanine,[86] or N^4-acylcytosines[180,192f,o] are reacted with 2α-acyloxysugars such as **18** in the presence of 1–2 equivalents of $SnCl_4$ or $AlCl_3$[86] in 1,2-dichloroethane or acetonitrile. The excess $SnCl_4$ apparently forms partially soluble σ complexes with the heterocyclic bases to afford, after the usual workup with aqueous $NaHCO_3$, the corresponding purine nucleosides[86,86a,b,192a–k] and cytidines[180,192e,f,o] in yields of up to 81%. In a recent synthesis of sinefugin, this one-step reaction failed with adenine and $SnCl_4$, whereas it succeeded with persilylated N^6-benzoyladenine in the presence of TMSOTf.[192l]

The one- or two-step methods with silylated heterocyclic bases virtually guarantee solubility of the silylated base in the reaction solvent, and thus a homogeneous, complete reaction. The free base, however, is apparently only partially soluble as its transient σ complex with $SnCl_4$, leading to incomplete reactions as well as to destruction of sugar moieties and thus lower yields.[1921]

Although the yields obtained in these one step–one pot reactions are usually somewhat lower compared to the conventional two-step procedure, the one step–one pot modification is so simple that it can also be used by investigators with limited practical experience in preparative organic chemistry.

MECHANISM OF NUCLEOSIDE FORMATION IN THE PRESENCE OF FRIEDEL-CRAFTS CATALYSTS

Experimental Results

The weakly basic silylated 5-nitrouracil **80a** reacts rapidly with **18** and small amounts of $SnCl_4$ in 1,2-dichloroethane to form 5-nitrouridine-2′,3′,5′-tri-*O*-benzoate (**81a**) in nearly quantitative yield,[84] whereas the much more basic silylated 5-methoxyuracil (**80b**) and 5-morpholinouracil (**80c**) do not react at all with **18** in the presence of less than 1 equivalent of $SnCl_4$[90,193,194] or TMSOTf.[133] Apparently, one equivalent of $SnCl_4$ (or TMSOTf) is inactivated or neutralized by σ-complex formation with the silylated bases, and only an excess of $SnCl_4$ (or TMSOTf) can lead to the formation of the electrophilic sugar cation and thus to nucleoside formation (albeit at a much lower rate than with **80a** since only small concentrations of free silylated bases **80b** and **80c** are available because of σ-complex formation).[90,193,194] Furthermore, besides the desired natural N^1-nucleosides **81b** and **81c**, large amounts of the undesired N^3-nucleosides **82b** and **82c** as well as the corresponding N^1,N^3-bisnucleosides **83b** and **83c** are obtained.

Replacing $SnCl_4$ by the weaker Friedel-Crafts catalysts TMSOTf or $(CH_3)_3SiOSO_2(CF_2)_3CF_3$ and switching from 1,2-dichloroethane to the more polar solvent acetonitrile, which competes with the basic silylated uracils for the Lewis acids, results in yields of up to 90% of the desired natural N^1-nucleosides **81b** and **81c**.[193,194]

These results can be reconciled if one assumes that three reversible processes occur during nucleoside formation: (a) reaction of the peracylated sugar **18** with the different Friedel-Crafts catalysts to give the rather stable electrophilic sugar cations **35** or **67**; (b) competing reversible formation of σ complexes between the silylated bases and the different Friedel-Crafts catalysts—with sugar moieties containing basic benzyl ether groups, TMSOTf will also form weak σ complexes whereas $SnCl_4$, $TiCl_4$, or Et_2AlCl will give stronger chelate-type complexes that consume additional amounts of catalyst; and (c) reaction of the electrophilic sugar cation with the uncomplexed free silylated bases to form the nucleoside bond.[193,194]

In the first process, the Friedel-Crafts catalysts $SnCl_4$, TMSOTf, $(CH_3)_3SiOSO_2(CF_2)_3CF_3$, or $TMSClO_4$ convert peracylated sugar **18** to the corresponding 1,2-acyloxonium salts **67** as the only electrophilic sugar moiety, with concomitant formation of either $SnCl_4OAc^-$ or $CF_3SO_3^-$, $C_4F_9SO_3^-$ or ClO_4^- and silylated acetic acid TMSOAc.[193,194]

These cyclic 1,2-acyloxonium salts should, however, always be generated in the presence of the nucleophilic silylated bases since all these salts derived from

1-acyloxy-, 1-alkoxy-, or 1-halofuranoses or pyranoses rearrange gradually in the presence of Lewis acids such as $SbCl_5$ or $BF_3 \cdot OEt_2$ in acetonitrile or nitromethane to isomeric cyclic 1,2-acyloxonium salts[195–198a] and might furthermore react with acetonitrile in a Ritter reaction[199–204] to give protected 1-acetylaminosugar derivatives.

Under these reversible and thus thermodynamically controlled conditions, the nucleophilic silylated bases can only attack the furanose (or pyranose) sugar cation **67** or **115** from the top (the β side) to afford the β-nucleosides with only minute amounts of the corresponding α-nucleosides as postulated by the Baker rule. The latter states that a base should approach the sugar ring from the side opposite the group at position 2, regardless of the relative configuration of C_1-C_2.[205] Peracylated 1-*O*-acyl- and in particular 1-*O*-acetyl-peracetylated di- and oligosaccharides[98] afford analogously the corresponding 1,2-cyclic acyloxonium salts with $SnCl_4$ or TMSOTf, which are converted by silylated bases to the corresponding acylated nucleosides.

Other nucleophiles such as alcohols or silylated alcohols react analogously under these reversible conditions with cyclic acyloxonium triflates **67** to give the corresponding β-glycosides **174** in often excellent yields,[206] which are thermodynamically favored compared to the orthoesters **175**.[210,211] Although this new methodology has found wide application,[207] only rarely is the origin of sugar cations such as **67** discussed.[208–209d]

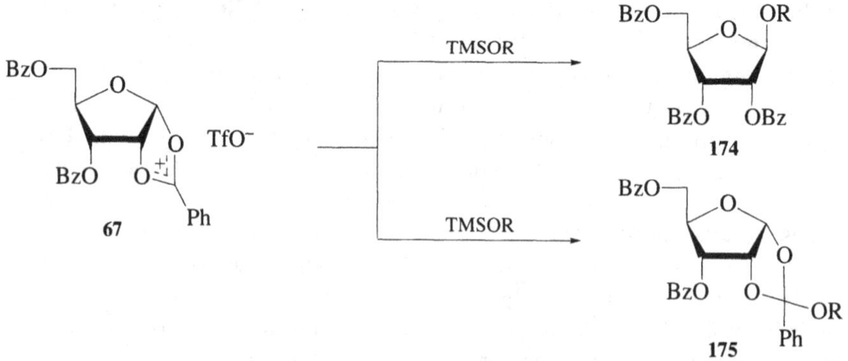

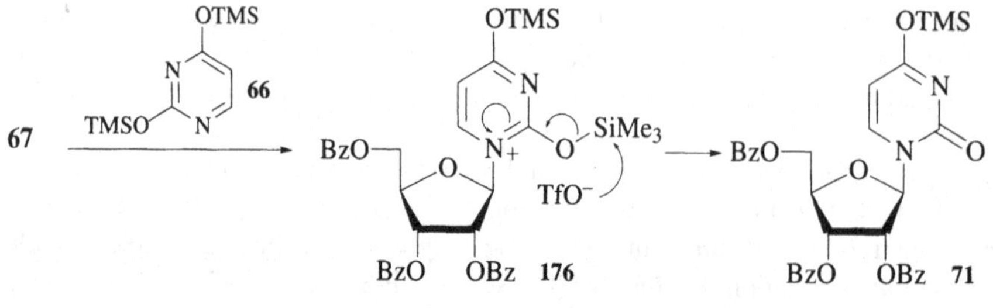

BzO— O, O TfO⁻, BzO O+, Ph, **67** — TMSOR → BzO— O OR, BzO OBz, **174**; TMSOR → BzO— O O, BzO O—OR, Ph, **175**

67 —(TMSO-pyrimidine-OTMS **66**)→ BzO— O N+ O SiMe₃ TfO⁻, OTMS, BzO OBz **176** → BzO— O N O, OTMS, BzO OBz **71**

After β attack of silylated uracil **66** on sugar cation **67**, the α-trimethylsilyloxy group in the intermediate salt **176** reacts with the triflate anion to regenerate

TMSOTf (or with $SnCl_4OAc^-$ to form trimethylsilyl acetate as well as regenerated $SnCl_4$), thus forming protected uridine **71**.

The classical 2,4-dialkoxypyrimidines such as 5-iodo-2,4-dimethoxypyrimidine (**177**) react analogously with sugar **18** in the presence of $SnCl_4$ with cleavage of the α-alkoxy group to give the corresponding 4-alkoxynucleosides **178**,[84] whereas with heterocyclic bases such as 4-trimethylsilyloxypyridine (**92a**) the 4-trimethylsilyloxy group reacts smoothly with triflate anion (or $SnCl_4OAc^-$) to give the protected nucleoside **93a**.[133]

On reaction of 2-methoxy-4-trimethylsilyloxypyridine (**179**) with **20** and silver triflate in nitromethane, the intermediate pyridinium triflate **180** is cleaved by attack of the hard triflate anion on the hard 4-trimethylsilyloxy group to furnish the protected 2-methoxy-4-pyridone nucleoside **181** in 51% yield. Reaction of **179** with **20** in the absence of silver triflate leads to the pyridinium bromide intermediate **180**, in which the soft 2-methoxy group is cleaved by the soft bromide ion to give, after hydrolysis of the 4-trimethylsilyloxy group, nucleoside **182** in 20% yield.[212,212a]

OMe, I, N, MeO, N, **177** + BzO, OAc, O, BzO, OBz, **18** $\xrightarrow[\text{(MeCN)}]{SnCl_4,\ Cl(CH_2)_2Cl}$ OMe, I, N, N, O, BzO, O, BzO, OBz **178**

OTMS, MeO, N, **179**; BzO, O, Br, BzO, OBz, **20** $\xrightarrow{\text{(TfOAg)},\ MeNO_2}$ $^-$OTf, O–$SiMe_3$, Br$^-$, Me, O, N_+, BzO, O, BzO, OBz **180** → O, BzO, N, OMe, O, BzO, OBz, **181** (51%); OH, BzO, N, O, O, BzO, OBz, **182** (20%)

[BzO, O, TfO$^-$, O, BzO, O$^+$, Ph, **67**]

As emphasized above, under reversible and thus thermodynamically controlled conditions, β-nucleosides are normally obtained nearly exclusively. The only exceptions to this near exclusive β attack on **67** (with TfO^-, $SnCl_4OAc^-$, or $C_4F_9SO_3^-$ as counterions) can apparently occur when:

(a) the heterocyclic base contains strongly polarized or negatively charged groups, which can associate with the positively charged α side of the sugar cation in **67**. This has been demonstrated in silylated 2-nitroimidazole[213,213a] or silylated 1,2,4-λ^3-diazaphosphole (**183a**), which is converted by TMSOTf to the α-anomer **185** in 16% yield via the orthoester intermediate **184**. The structure of **185** was confirmed by single crystal X-ray analysis.[214] Not unexpectedly, the analogous reaction of **183b**, which contains an electron-withdrawing ester group that impedes the formation of **184b**, furnishes the crystalline β-nucleoside **186** in 44% yield after saponification by methanolic ammonia;[214]

R: H **183a**; CO_2Et **183b**

TMSOTf / MeCN; MeOH NH_3

17 **184** **186** (44%) **185** (16%)

(b) the sugar cation is apparently not formed quantitatively as in the case of the 4-thiosugars;[215] or

(c) when the sugar 1,2-acyloxonium salt contains polar groups such as amide or nitro on the α side.[216,217] Peracylated glucosamides, however, in which the basic nitrogen moiety is protected or neutralized by a strongly electron-attracting *N*-trifluoroacetamido- or *N*-2,4-dinitrophenyl group give more than 80% of the β-nucleoside with persilylated uracil and $SnCl_4$;[218] or

(d) a stabilized cation can form above the plane of the sugar (e.g., as a chloronium cation in 1-*O*-acetyl-3,4-*O*-benzoyl-2-chloro-2-deoxy-α-D-arabinose), resulting exclusively in α-nucleoside formation.[219]

Two publications[220,221] report the isolation of ~5% α-nucleoside **190** in addition to the expected β-nucleoside **189** during the reaction of silylated 5-fluorouracil (**187**) with 5-deoxy-1,2,3-tri-*O*-acetyl-β-D-ribofuranose (**188**) in the presence of TMSOTf in CH_2Cl_2. The β-nucleoside **189** rearranges on standing for 1 week with TMSOTf in CH_2Cl_2 to form 2% of the α-nucleoside **190**. The assignment of the structure of the α anomer **190** is based on the identity of the free nucleoside **191** with the α-nucleoside obtained by reaction of **187** with 1-*O*-methyl-2,3-isopropylidene-5-deoxy-D-ribofuranose (**192**) in the presence of TMSOTf to give **193** and **191** on subsequent acid hydrolysis of the isopropylidene group.[221] The analogous reaction of silylated thymine with 2,3-*O*-isopropylidene-1,5-di-*O*-*p*-toluoyl-β-D-ribofuranose gives 65% of the protected α-nucleoside in the presence of TMSOTf.[222,223]

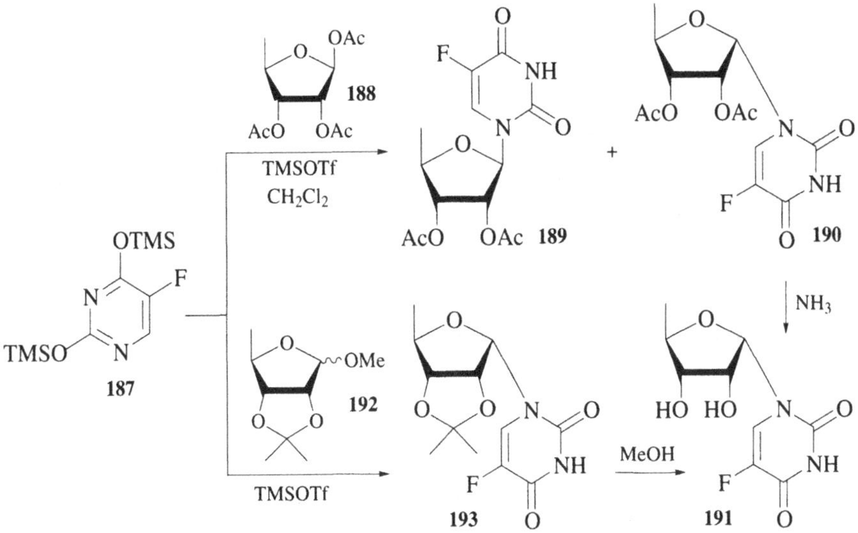

In the synthesis of nucleosides with 4-thiosugars, 2-α-benzoyloxy- or 2-α-(4-methoxy)benzoyloxy groups stabilize the cyclic cations (such as **67**) much better than a 2-α-acetoxy moiety to give more or exclusive formation of β-nucleosides.

More extensive application of HPLC to mother liquors of the desired β-nucleosides will reveal more cases where protected or free α-nucleosides can be detected in small amounts. During the synthesis of pteridine-ribofuranosides, the formation of small amounts of the corresponding α-nucleosides was observed.[223a]

The related reaction of silylated uracil **66** with 1-chloro-2,3-*O*-isopropylideneuronic acid methyl ester (**194**) in the presence of $SnCl_4$ furnishes the β-N^3-nucleoside **195** in 32% yield.[224] TMSOTf and $SnCl_4$ behave differently in the presence of polar groups such as ester and amide functionalities, since $SnCl_4$ binds more strongly to the carbonyl group and might conceivably bind more tightly than TMSOTf to the α-2,3-isopropylidene moiety in **194**, thus blocking the α side to result in formation of the β-N^3-nucleoside **195**.

Whereas the 2,3-*O*-isopropylidene group such as in **192** can only stabilize the 1-cation in the presence of chelate-forming $SnCl_4$, 2-α-methoxy,[225,226,226b,c] 2-α-silyloxy,[226,226a] 2-α-phenylsulfenyl,[227–229,229a] 2-α-phenylselenyl,[230,230a,231] 2-α-

tosyloxy,[192k] as well as 2,3-epimino groups[232] stabilize the 1-cation from the α side with $SnCl_4$ as well as with TMSOTf, resulting in the predominant formation of β-nucleosides. It should be realized, however, that reactions of **22** containing a 2-β-benzyloxy group give predominantly β-nucleosides with silylated bases in the presence of $SnCl_4$.[84]

In the synthesis of oxetanocin, the sugar derivative **196** with a stabilizing oxalyloxy group reacts with silylated *N*-benzoyladenine (**97**) in the presence of $SnCl_4$ to give via the bridged cation **197** (containing a less strained and therefore more stable bicyclo[5.2.0] system) after hydrolysis and *O*-benzoylation 16% of di-*O*-benzoyloxetanocin (**198**), 14% of the corresponding α-nucleoside **199**, as well as 9% of di-*O*-benzoylepioxetanocin (**200**).[233] The corresponding di-*O*-benzoylsugar **201** affords the cation **202** with **97** in the presence of $SnCl_4$ and thus exclusively the α-nucleoside **199**.[233]

Since the rather stable 1,2-acyloxonium ions **67** (from **18** and TMSOTf) are only weak electrophiles, they react only with electron-rich nucleophilic aromatic systems such as 1,3,5-trimethoxybenzene (**203**) to give 60% of the *C*-nucleoside **204** and 5% of the corresponding bis(nucleoside) **205**.[133] This reaction was later applied in a slightly modified form.[133a-c] Although *C*-nucleosides[39-44] are not the subject of this review, these results demonstrate the close relationship between the Friedel-Crafts catalyzed Silyl-Hilbert-Johnson reaction and the classical Friedel-Crafts reaction.

Reversible σ-Complex Formation between Silylated Bases and Friedel-Crafts Catalysts

The formation of σ complexes between Friedel-Crafts catalysts and silylated bases is dependent on the acidity of the Friedel-Crafts catalysts as well as on the basicity of the silylated heterocyclic bases. Although the pK_a values for silylated heterocyclic bases have not been determined, the pK_a values of the closely related methoxypyrimidines and pyridines[234] can be used for comparison purposes: the increase in basicity from 2-methoxypyridine (pK_a 3.2) to 4-methoxypyridine (pK_a 6.5) is striking and explains why 4-trimethylsilyloxypyridine (**92a**) forms such strong σ complexes and why **92a** is converted to the corresponding nucleoside **93a** only under forcing conditions.[99,133] The basicity of 2,4-dimethoxypyrimidine (pK_a 3.1) is increased to pK_a 3.63 on introduction of an electron-releasing

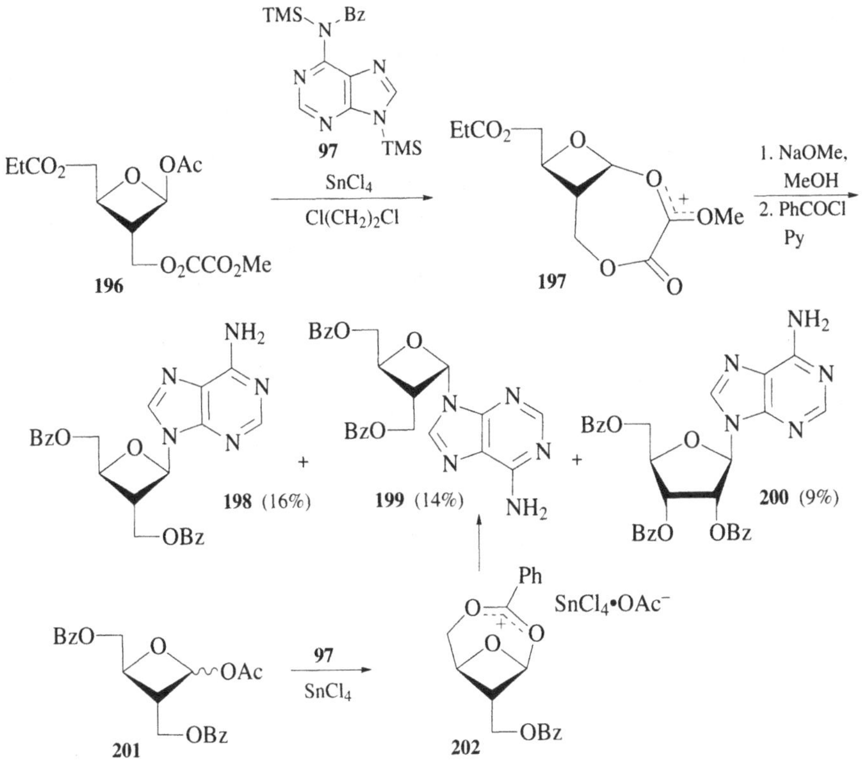

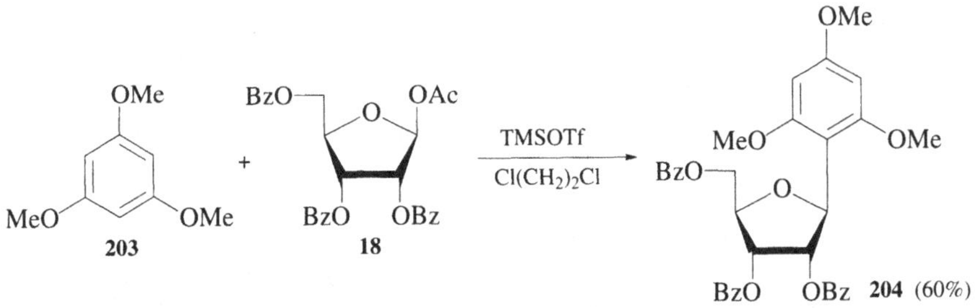

5-methyl group. Since the basicities of amino heterocycles such as cytosine and in particular adenine or guanine are decreased on acylation to the N^4-benzoyl or N^4-acetylcytosines, N^6-benzoyladenine or N^2-acetyl- or N^2-isobutyrylguanine, it is obvious that these *N*-acylated bases and their analogs are commonly employed in their silylated form for Friedel-Crafts catalyzed nucleoside synthesis in preference to the more basic silylated amino heterocycles, resulting in faster and cleaner nucleoside formation. With the more basic silylated cytosine **222**, at least two equivalents of $SnCl_4$ had to be employed to effect a smooth preparation of

224,[235] whereas other nucleoside forming reactions employing sensitive protected 2-deoxy sugars and silylated cytosine in the presence of *tert*-butyldimethylsilyl triflate failed.[236]

The structures of the σ complexes between silylated 2-pyridone and silylated 5-methoxyuracil and $SnCl_4$ as well as TMSOTf can be derived from their ^{13}C NMR spectra.[194] The resulting upfield shifts of the C^6-carbon atom adjacent to the quaternized nitrogen atoms in the pyridine and pyrimidine series give good information on the σ complexes between the silylated bases and the particular Lewis acids. Tin tetrachloride, as the stronger Lewis acid, binds more tightly to the N^1-nitrogen of silylated 5-methoxyuracil to give **206** than does TMSOTf to give **207**. Whereas one equivalent of $SnCl_4$ leads to practically quantitative formation of the σ complex **206**, nearly five equivalents of TMSOTf are needed to effect quantitative formation of the related σ complex **207**.[193,194]

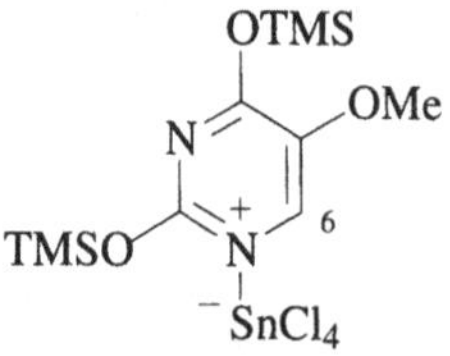

OTMS, OMe, N, TMSO, N+, 6, $^-SnCl_4$ — **206**

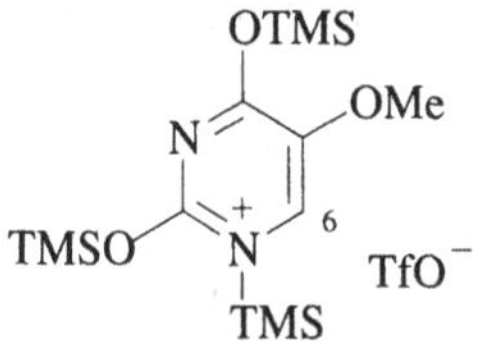

OTMS, OMe, N, TMSO, N+, 6, TMS, TfO^- — **207**

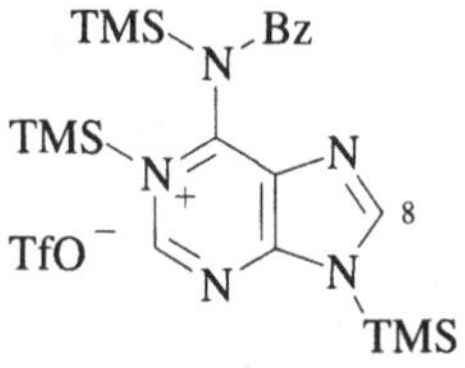

TMS, N, Bz, TMS, N+, TfO^-, N, N, N, 8, TMS — **208**

The pronounced upfield shift of the C^8-carbon atom in the ^{13}C NMR spectrum of silylated N^6-benzoyladenine (**97**) on addition of increasing amounts of TMSOTf can be interpreted by σ-complex formation at the N^1-nitrogen, the center of highest electron density as depicted in **208**.[194] These conclusions are in agreement with the ^{13}C NMR spectra of protonated adenine in solution.[237]

The interaction between the disilylated 5-methyl-5,6-dihydro-*sym*-triazine-2,4(1*H*,3*H*)dione (**209**), containing a basic tertiary nitrogen atom, and $SnCl_4$ leads apparently to a σ complex between this basic tertiary nitrogen atom and $SnCl_4$ to form **210**, in which the basicity of the N^3-nitrogen is decreased. Since the N^3-nitrogen in **210** is sterically more encumbered than the N^1-nitrogen, exclusive formation of the desired N^1-nucleoside is observed.[238,239] Thus the basic N^5-nitrogen group does not interfere with nucleoside synthesis. Likewise a free 5-methylaminomethyl group in silylated 2-thiouracil readily gives the anticipated protected nucleoside in the presence of excess catalyst.[126]

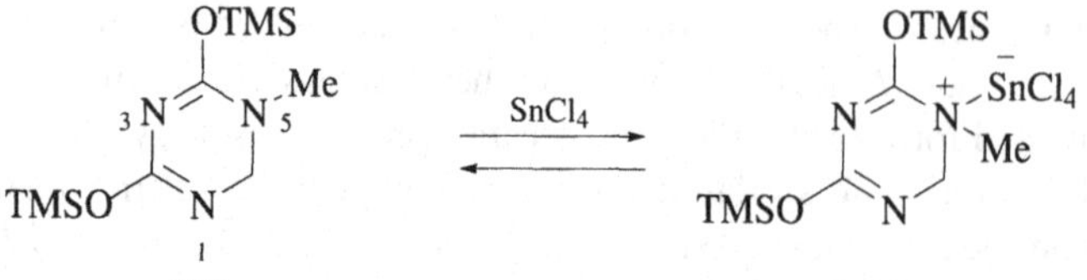

OTMS, Me, 3 N, N 5, TMSO, N, 1 — **209**

$SnCl_4$ ⇄

OTMS, $^-SnCl_4$, N, +N, Me, TMSO, N — **210**

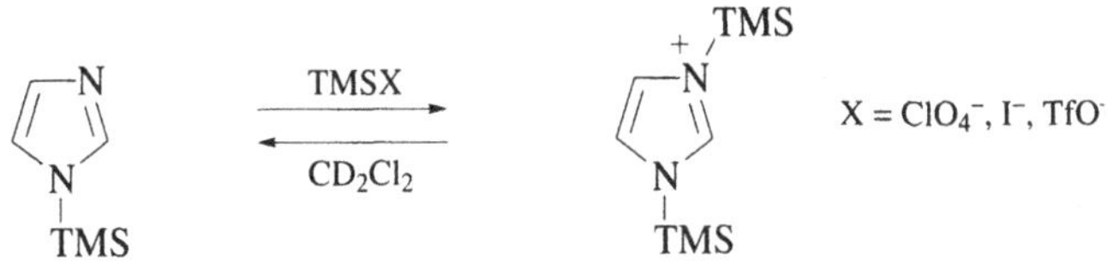

$$\mathbf{211} \underset{CD_2Cl_2}{\overset{TMSX}{\rightleftharpoons}} \mathbf{212} \quad X = ClO_4^-, I^-, TfO^-$$

Following the reaction of silylated imidazole **211** and TMSX ($X = ClO_4^-, I^-, TfO^-$) in CD_2Cl_2 by ^{29}Si-NMR measurements demonstrates that in the equilibrium between **211** and its σ complexes **212** the formation of **212** is most favored for $X = ClO_4^- > I^- > TfO^-$.[240] Thus $(CH_3)_3SiI$ apparently has a stronger tendency to form σ complexes with **211** than TMSOTf. No quantitative data, however, nor any discussion of earlier work[194] on such σ complexes are provided.[240]

During studies on the synthesis of bredinin,[241–243] the structure of persilylated imidazole **214** was established by MS and ^{13}C NMR and comparison with *N*-methylimidazole **213**. Imidazole **214** was then treated with increasing amounts of TMSOTf; the ^{13}C NMR spectra of the resulting σ complex **214** → **215** or **216** showed the expected upfield shift of the C-5 signal. The maximum shift, however, was not reached even on addition of 1.4 equivalents of TMSOTf.[242]

Addition of equivalent amounts of the stronger Lewis acid $SnCl_4$ to **214** leads to the fully complexed compound **217**, in which the C-5 signal is only slightly shifted downfield. In contrast, the carboxamide carbon undergoes a pronounced downfield shift, whereas the adjacent C-4 carbon is shifted upfield. These data support the structure of **217** as a σ complex between the silylated carboxamide group and $SnCl_4$.[242] Such σ complexes of carbonyl groups with $SnCl_4$ or $TiCl_4$ have recently been investigated and reviewed.[244–246] Complexes of Lewis acids such as $TiCl_4$ with peracylated sugars have also been described[247,248] and should be taken into account when reacting complex sugars such as peracylated disaccharides[98] or **515** (see p. 99).[160]

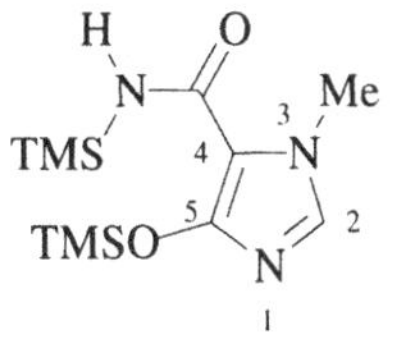

213

As a consequence of these different σ complexes between **214** and TMSOTf to give **215** → **216** and of **214** with $SnCl_4$ to yield **217**, the reaction of **214** with **17** in the presence of TMSOTf leads to only 17% of the desired protected bredinin **218** and 61% of the undesired N^3-nucleoside **219** as well as 9% of the N^1,N^3-bis-(riboside) **220**. On the other hand, use of 1 equivalent of $SnCl_4$ (to form **217**) followed by additional catalytic amounts of TMSOTf to generate the reactive sugar cation affords protected bredinin **218** in 83% yield.[242] It seems probable that the additional catalytic amount of TMSOTf can be replaced by a catalytic amount of $SnCl_4$.

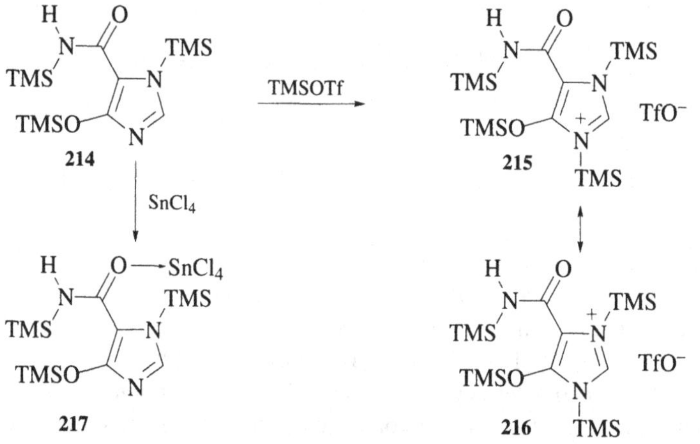

There are thus differences in catalytic behavior between $SnCl_4$ and TMSOTf if polar groups such as amides or esters are present in the silylated heterocyclic base or in the sugar. Note also the different reactions of persilylated N^2-acetylguanine (**108b**) with **18** in the presence of $SnCl_4$ and TMSOTf in dichloroethane.[142]

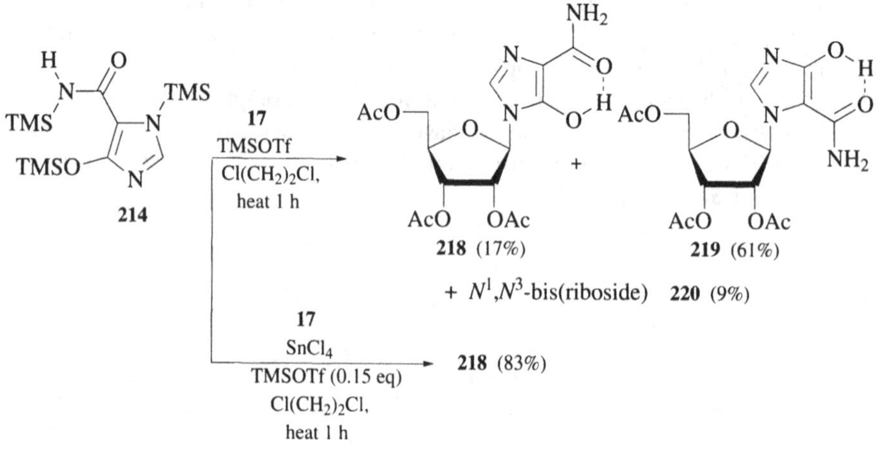

In a striking example of σ complexes of the sugar moiety with a Lewis acid, thiosugar **221** reacts with silylated cytosine (**222**) in the presence of 2 equivalents of $SnCl_4$ to form (via σ complex **223**) exclusively β anomer **224** in 80% yield, whereas with TMSOTf as catalyst a 1:1 β/α-mixture is produced.[235] In the case of the oxasugar **225**, dichlorotitanium diisopropoxide leads via **227** to exclusive formation of the β-nucleoside **228**, whereas TMSOTf gives again a 1:1 β/α-mixture.[235] Other studies, however, have demonstrated that during or prior to the formation of **223** or **227** a cation is also formed with $SnCl_4$ at the cyclic acetal or thioacetal carbon leading to completely racemized nucleosides .[249] In contrast to $SnCl_4$ or dichlorotitanium diisopropoxide as catalysts, TMSOTf converts the sugar derivative **225** into cation **229**, which reacts with silylated thymine **226** to give a 2:1 mixture of the optically active β-nucleoside **228** and the α anomer **230**. Thus reactive groups in the silylated base as well as in the sugar moiety have to be considered in choosing the optimal Lewis acid catalyst.

For additional publications on 3- or 2-oxa-, thia-, or selenanucleosides, see Refs. 249a–k.

TBDPSO, OAc, **221**; $SnCl_4$, CH_2Cl_2; TBDPSO, **223**; **222** (NHTMS, TMSO)

TBDPSO, NH_2, **224** (80%)

TBDPSO, OAc, **225** + OTMS, TMSO, **226**

$Cl_2Ti(OPr\text{-}i)_2$, CH_2Cl_2; **227**; **226**; TBDPSO, **228**

TMSOTf, $Cl(CH_2)_2Cl$; TBDPSO, TfO^-, **229**; **226**

228 + TBDPSO, **230**

Mechanism of Pyrimidine and Purine Nucleoside Synthesis

Taking all the above results into account, the mechanism of the Friedel-Crafts catalyzed Silyl-Hilbert-Johnson synthesis of pyrimidine nucleosides is straightforward as exemplified for the reaction of silylated 5-methoxyuracil **80b** in the presence of $SnCl_4$ and TMSOTf. The reaction of **80b**, **18**, and the stronger Lewis acid $SnCl_4$ sets up an equilibrium in which the σ complex **206** between the N^1 of **80b** (the center of highest electron density) predominates in nonpolar solvents such as 1,2-dichloroethane. Since only the free silylated base **80b** will react with the electrophilic sugar cation **67** (with $SnCl_4OAc^-$ as the counterion) to form the 4-*O*-trimethylsilylated *O*-benzoylated 5-methoxyuridine **231**, and the concentration of free silylated base **80b** is rather low in the equilibrium with $SnCl_4$, nucleoside formation will be rather slow.

Furthermore, in the equilibrium between σ complex **206**, silylated base **80b** and $SnCl_4$, the $SnCl_4$ will stay close to the electron-rich center at the N^1-nitrogen. It is this slightly dissociated form, in which the N^1-nitrogen is still blocked and the N^3-nitrogen is available, that reacts with **67** to form the undesired silylated *O*-protected N^3-nucleoside **232**. Both the silylated protected N^1-nucleoside **231** as well as the corresponding N^3-nucleoside **232** can react further with **67** (with $SnCl_4OAc^-$ as counterion) to give the protected N^1,N^3-bis(nucleoside) **83b**.

OTMS, OMe, N, TMSO, N
80b
67 ⇌
OTMS, MeO, N, BzO, N, O, O, BzO, OBz
231
67 ⇌ N^1,N^3-bis(riboside)
83b

$SnCl_4$ (TMSOTf) ⇅

BzO, O, O, BzO, O, (TfO⁻), Ph
67
$SnCl_4OAc^-$
← $SnCl_4$ (+TMSOTf) (−TMSOAc)
BzO, O, OAc, BzO, OBz
18

OTMS, OMe, N, TMSO, N, X
206 X = $SnCl_4^-$
207 X = TMS, TfO⁻
67 ⇌
N, OMe, TMSO, N, O, BzO, O, BzO, OBz
232
67 ⇌ N^1,N^3-bis(riboside)
83b

As already emphasized, the ratio of free **80b** and complexed form **206** is also dependent on the polarity of the solvent. The more nucleophilic solvent acetonitrile, which forms σ complexes with $TiCl_4$,[250] $SnCl_4$,[251] and BH_3,[252] competes with the silylated base for the electrophile and simultaneously favors the formation of the polar sugar cation **67**. Consequently, in acetonitrile there is more silylated base **80b** and more sugar cation **67** present, and thus more of the desired natural N^1-nucleoside **231** is formed.

The corresponding reaction with TMSOTf is analogous: since it is a weaker Lewis acid than $SnCl_4$, however, less σ complex **207** is formed in the equilibrium. Therefore, more free base **80b** is present and consequently more silylated N^1-nucleoside is obtained. Again, both the protected silylated N^1-nucleoside **231** as well as the protected silylated N^3-nucleoside **232** can react further with **67** to form the N^1, N^3-bis(nucleoside) **83b**. Experimental support for these equilibria is provided by the rearrangement of protected silylated N^3-nucleosides (**104** → **85**)[133] as well as by the different transglycosylations catalyzed by TMSOTf (see section on transglycosylations).

With respect to the reaction of silylated purines with peracylated sugars in the presence of TMSOTf (or $SnCl_4$) the aforediscussed σ complex between persilylated N^6-benzoyladenine **97** and TMSOTf yields the σ complex **208** as determined by ^{13}C NMR.[194] If **208** is assumed to be in equilibrium with the N^1-silyl compound **233**, then it should react readily with **67** to give the N^3-nucleoside **234** as a kinetically controlled intermediate, which can be isolated after a short reaction time[133] or rearranged in situ (**111** → **112**) to the thermodynamically most stable silylated natural N^9-nucleoside **235**.[133] Alternatively, **233** (or the isomeric **506**) could also react directly with **67** (compare the arrows in **233**) to the natural N^9-nucleoside **235** as well as to the N^7-nucleoside **236**, which would subsequently rearrange to the desired N^9-nucleoside **235**. If the reaction of **97** with **67** is followed by TLC, a number of intermediates can be discerned, which disappear during the course of the reaction.[133] Since the protected N^6-nitrogen in the N^7-nucleoside **236** interferes sterically with the sugar moiety, formation of **236** is disfavored, resulting in high yields of the protected natural adenosine **235**.

Because of the initial formation of the intermediates such as the N^3-nucleoside **234**, the formation of σ complexes between these intermediates with Lewis acids such as TMSOTf or $SnCl_4$ is necessary to induce the rearrangement of these intermediates to the thermodynamically favored protected adenosine **235** (or the corresponding protected N^9-guanosine). Consequently a less polar solvent such as 1,2-dichloroethane or toluene is advantageous for the synthesis of purine nucleosides. In contrast, the more polar acetonitrile competes with silylated purine bases for the Friedel-Crafts catalysts and impedes these rearrangements, leading to longer reaction times, some cleavage of the nucleosides, and destruction of the sugars.

In guanosine synthesis, however, the 6-trimethylsilyloxy group is less of a steric impediment so that formation of small amounts of N^7-guanosine (**110b**) is always observed in the equilibrium (cf. **108b** → **109b** + **110b** as well **113** → **114**). Consequently, introduction of a more bulky 6-triethylsilyloxy or 6-triisopropylsilyloxy group into N^2-acetylguanine might lead to formation of less N^7-guanosine.

For similar investigations compare the aforediscussed synthesis of *sym*-1,3,5-triazine nucleosides (**209** → **210**) as well as imidazole nucleosides (**214** → **220**).

Regioselectivity of Nucleoside Formation

When free or silylated organic bases contain more than one nitrogen atom, each of these nitrogen atoms can in principle become attached to a sugar derivative. The formation of the resulting different nucleosides is either kinetically or thermodynamically controlled.

Whereas the regioselectivity in the synthesis of pyrimidine nucleosides (e.g., formation of the undesired *O*-nucleosides and the N^3-nucleosides) has been discussed, consideration of some principles and of additional techniques such as blocking of nitrogen functions are in order. With heterocyclic bases containing a carbonyl or thiocarbonyl group, the nitrogen atom either α or γ to the carbonyl or thiocarbonyl group is preferred for nucleoside formation. Thus in 1,2,4-triazin-5-one (**237**) only the N^2-or N^4-nitrogen can serve as a nucleophile to give either **238** or **239**.[99] Thus silylated 1,2,4-triazin-5-one **240** affords **241** as well as **242.**

Reaction of silylated 4-cyano-2*H*-pyridazin-3-one **243** with **18** at 0° in the presence of $SnCl_4$ leads to the kinetically controlled rather stable mesoionic pyridazinium salt **244** (80%), which can be saponified to the free nucleoside.[253] Conducting the reaction of **243** with **18** and $SnCl_4$ at 80° or heating **243** with $SnCl_4$ in 1,2-dichloroethane affords the thermodynamically more stable nucleoside **245** in 76% yield.[253]

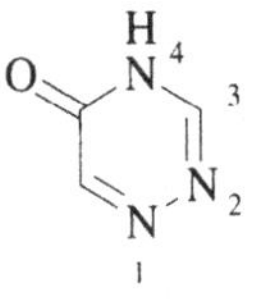

237 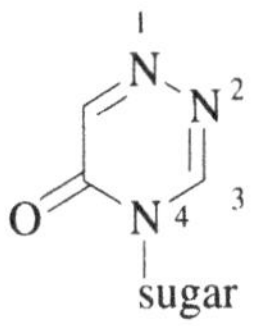 **238** 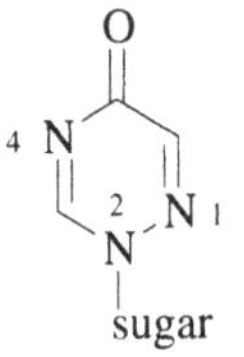 **239**

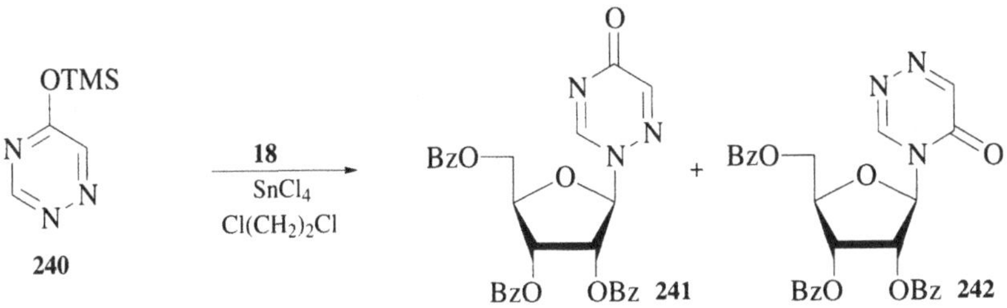

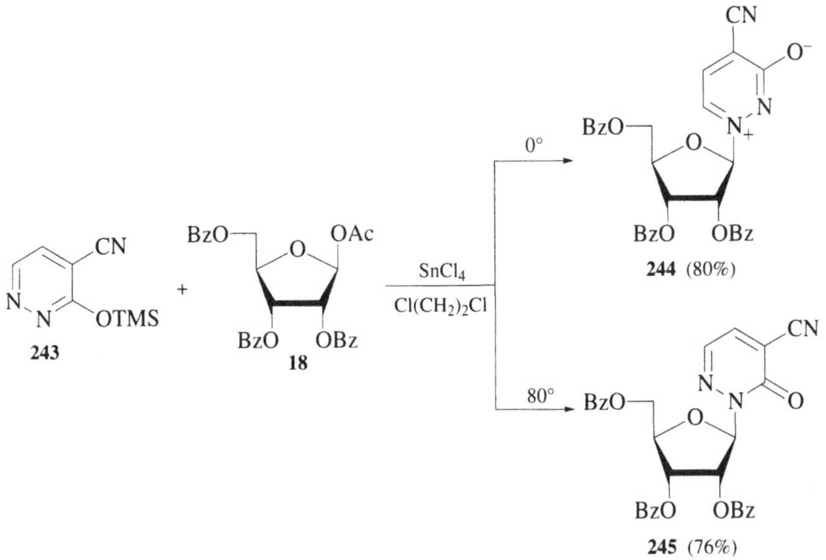

240 → (**18**, $SnCl_4$, $Cl(CH_2)_2Cl$) → **241** + **242**

243 + **18** → ($SnCl_4$, $Cl(CH_2)_2Cl$) → 0°: **244** (80%); 80°: **245** (76%)

To add further complications, the presence of carbonyl or thiocarbonyl groups in these nitrogen heterocycles can lead to the corresponding *O*- or *S*-glycosides via kinetic control . Since these *O*- or *S*-glycosides are normally not observed on using thermodynamically controlled conditions in the presence of Friedel-Crafts catalysts, they are not dealt with here.

On blocking the 1 position of *O*-alkylated or *O*-silylated uracils by a β-cyanoethyl substituent, as in **246** or by a benzyl group as in **248,** exclusive N^3-substitution can be achieved.[254,255] Alternatively, the silylated N^3-benzyl-6-methyluracil **250** reacts with the standard sugar **18** in the presence of TMSOTf in acetonitrile

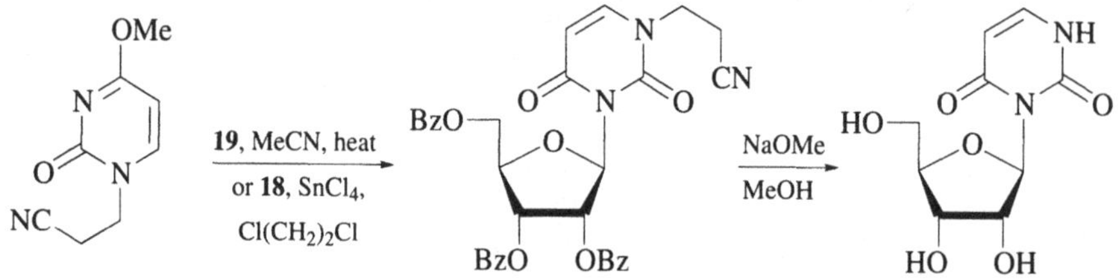

246 **247** (59%)

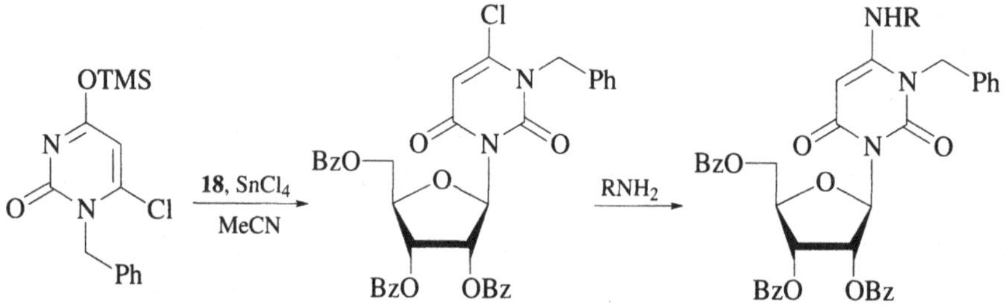

248 **249** (90%)

to give the protected 3-benzyl-6-methyluridine **251** (85%), from which the *N*-benzyl protecting group can be removed in 60% yield by treatment with BBr_3[256] to give the protected 6-methyluridine **85**. For the application of the N^1-*n*-octylthiocarbonyl-blocking group, see Ref. 256a. Analogously silylated 3,6-

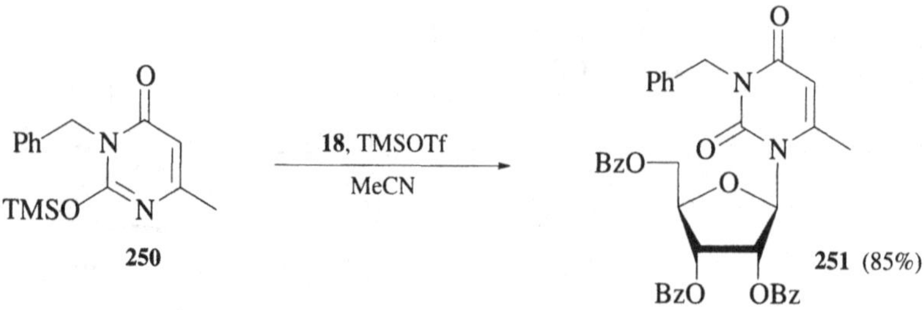

250 **251** (85%)

dimethyluracil or silylated 4-methylthio-6-methylpyrimidin-2-one afford the corresponding disubstituted uridines.[256]

The silylated heterocycle **252** reacts with 1-chlororibofuranose **253** in acetonitrile at room temperature to give a 3:2 mixture of **254** and **255** (38%), whereas its reaction with **17** in the presence of $SnCl_4$ affords an 84:16 mixture of **254** and **255** (55%).[257]

OTMS

252

253

MeCN, 22°

MeCN, $SnCl_4$

17

254

+

255

The sometimes complex regioselectivity observed with silylated imidazoles has already been discussed in the synthesis of protected bredinin (**218**). With 3-substituted 1,2,4-triazoles, the formation of N^1- or N^2-nucleosides can be only partially controlled.[133,258,259] Compare also the subsequently described reactions of benzotriazole **260** to give **261** and **262**.[271] Thus nucleoside formation of each new type of silylated heterocycle has to be investigated under a variety of reaction conditions to ascertain which conditions are optimal for the preparation of the kinetically or thermodynamically controlled nucleoside.

SPECIAL PREPARATIONS

Synthesis of 2′-Deoxynucleosides

Although the syntheses of 2′-deoxynucleosides from suitable derivatives of 2-deoxysugars and salts of acidic heterocyclic bases or silylated heterocyclic bases have already been discussed, the different methods for their preparation are summarized here. These syntheses are among the most difficult in nucleoside chemistry. There are three main problems:

(a) The derivatives of 2′-deoxysugars such as the commonly employed **21** are frequently rather unstable, although 1-alkoxysugars such as **119** are much more stable;

(b) the total yields of the mixtures of the desired β anomer and the undesired α anomer are frequently only moderate; and

(c) the ratios of the desired natural β anomer to the undesired α anomer are often difficult to control and to reproduce.

The standard sugar for the preparation of 2′-deoxyribonucleosides, the rather expensive crystalline **21** (prepared from **117** → **119** → **21**), is rather unstable. The 1-α-chloro moiety in **21** can isomerize,[260,261] especially in polar solvents in the presence of Lewis acids or excess phase-transfer catalysts/50% NaOH,[262] to the 1-β-chlorosugar, which results in the formation of predominantly α-nucleosides.

Furthermore, **21** can eliminate hydrochloric acid as well as *p*-toluic acid to form the crystalline 2-(*p*-toluoyloxy)methylfuran **259** as described in the classical Hilbert-Johnson reaction, whereupon only 3% of the desired β-anomer **257**, 20% of the α-anomer **258**, and up to 27% of **259** are isolated.[263] The analogous reaction of 5-iodouracilmercury (**256**) with **21** in DMF gave only low yields of the corresponding α- and β-nucleosides but a large amount of **259**.[263]

TolO 21 — MeCN (DMF), 24°, 10 h; 177 → 257 + 258 + 259

256

3′,5′-Di-*O*-*p*-toluoylated 2′-deoxy-β-nucleosides such as **257** or **123** crystallize better, melt at higher temperatures, are often less soluble than the corresponding α-nucleosides and can frequently be isolated by simple crystallization. In contrast, the corresponding 3′,5′-di-*O*-benzoylated, *p*-chloro-, or *p*-nitrobenzoylated derivatives often do not permit selective crystallization of the desired β anomer.[264–267]

The more acidic 3′,5′-*O*-acyl groups such as *p*-nitrobenzoyl, however, are much more readily removed by ammonia and primary or secondary amines than the *O*-*p*-toluoyl group. This can be of importance in the synthesis of nucleosides containing such base-sensitive groups as 5-trifluoromethyl. Thus only the 3′,5′-di-*O*-*p*-nitrobenzoyl-2′-deoxy-β-D-ribofuranosyl-5-trifluoromethyluridine could be saponified selectively to the desired free nucleoside, whereas other acyl groups are hydrolyzed only under more vigorous conditions, which cause concomitant saponification of the 5-trifluoromethyl group to yield 2′-deoxy-5-carboxyuridine.[268] Other investigators have used, however, the *p*-chlorobenzoate successfully to obtain, with sodium methylate and methanol, free 2-deoxy-5-trifluoromethyluridine.[269]

As described earlier, acidic heterocyclic systems such as 2,6-dichloropurine, 2,6-dichloro-4,6-imidazo[4,5-*c*]pyridine, 6-chloropurine, or even adenine form the sodium salts **40**, **42**, or **45** on treatment with NaH in acetonitrile or acetone. These salts give on reaction with **21** nearly exclusively the desired β-nucleosides **41**, **43**, and **46** (via S_N2 replacement of the 1α chlorine) as well as N^7-nucleoside **44** and, in the case of **45**, α-nucleoside **47**.

Analogously, pyrrolopyridine **51** is converted in situ by KOH and the phase-transfer catalyst TDA-1 into the corresponding potassium salt, which reacts with **21** to afford the desired β-nucleoside **52** in 78% yield.[67] Powdered K_2CO_3/TDA-1 in acetonitrile often leads to considerably higher yields than standard phase-transfer conditions such as 50% $NaOH/Bu_4NHSO_4/CH_2Cl_2$.[270] Benzotriazole (**260**) analogously affords 54% of the N^1-nucleoside **261** as well as 31% of the N^2-nucleoside **262**.[271]

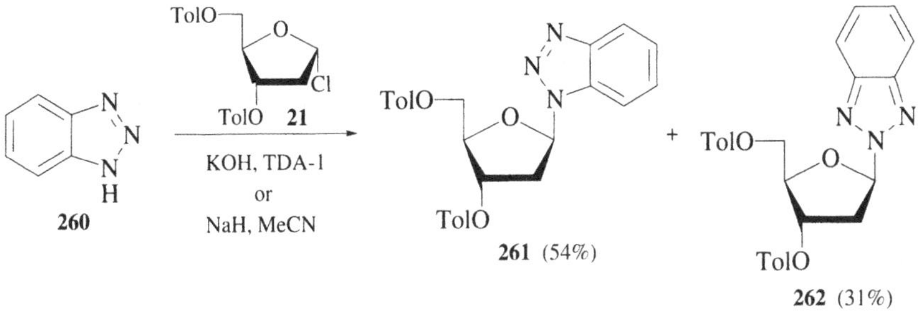

260 → (TolO–sugar–Cl **21**; KOH, TDA-1 or NaH, MeCN) → **261** (54%) + **262** (31%)

The Silyl-Hilbert-Johnson reaction of silylated substituted uracils and cytosines with the standard 1-halo-2-deoxysugar **21** in the presence of $HgBr_2$ or $AgClO_4$ furnishes the protected 2′-deoxynucleosides in moderate overall yields with α/β anomer ratios of 1:1-2.[80, 272, 272a]

The uncatalyzed as well as $ZnCl_2$-catalyzed reactions of silylated pyrimidines with **21** in anhydrous chloroform have been investigated.[261, 269] For example, the reaction of silylated thymine gave an α/β-ratio of 1:5 without catalyst. Importantly, the β-selectivity was increased in solvents with low dielectric constant, which promote S_{N2} displacement while minimizing anomerization of the 1α-chlorine. Catalysis by CuI in chloroform has been described[273] as promoting the formation of 90% of 2′-deoxypyrimidine nucleosides, favoring the β anomer over the α anomer by ratios of 9:1 – 9:3. These anomer ratios are seen even with the weakly nucleophilic silylated 5-nitrouracil, which reacts much slower than the more basic silylated uracils.[273, 273a] A possible explanation for the β-selectivity and high yield of desired natural 2′-deoxy-β-nucleoside might involve a push-pull process, in which nucleophilic attack of the silylated bases is assisted by CuI pulling at the α chlorine as depicted in **263**. This procedure has recently been applied successfully to reactions of silylated 6-azauracil and 5-methyl-6-azauracil with **21** to give the desired protected β-nucleosides in 70–75% yield, whereas silylated 5,6-dimethyluracil and **21** afforded the corresponding protected β-nucleoside in

50% yield.[274] For analogous reactions with CuCl in 1,2-dichloroethane, see Refs. 275, 275b. Addition of bases such as pyridine or triethylamine, as well as DBU, is reported to give an α/β ratio of 3:7 in reactions with 1-chloro-2,3-dideoxyribose derivatives.[276]

Pteridines have been reacted with 1-α-chloro-3,5-di-*O*-*p*-chlorobenzoyl-2-deoxyribofuranose in the presence of DBU to give the protected β-nucleosides in up to 54% yields.[276a]

The uncatalyzed Silyl-Hilbert-Johnson reaction has been applied to an anomeric mixture ($\alpha:\beta = 33:67$) of 1-chloro-2,3-dideoxy-3α-phenylthio-5-*O*-benzoyl-D-ribose (**264a**) to give on reaction with silylated uracil **66** in chloroform an anomeric mixture of the nucleosides **265** and **266** in a ratio of 32:68 as anticipated by a S_N2 mechanism.[229] The corresponding mixture of 1-*O*-acetates **264b** reacts with **66** in the presence of $SnCl_4$ in CH_2Cl_2 to give **265** and **266** in a 69:31 ratio in 73% yield owing to the coordination of $SnCl_4$ with the sulfur atom in **264**, whereas TMSOTf affords a 40:60 ratio of **265/266** in 87% yield.[229]

The sugar moieties related to **264** with phenylthio[227–229,229a] or phenylselenyl[230,230a,b] groups located at the 2α position give primarily the β-nucleosides with silylated bases **66** or **226** in the presence of $SnCl_4$.

264a X = Cl α:β = 33:67, $CHCl_3$
264b X = OAc α:β = 31:69, CH_2Cl_2, $SnCl_4$

An analogous complexation with a Lewis acid involving a 3α group is possible in sugar **267** to afford with **66** 90% of β-nucleoside **268** and the corresponding α anomer **269** in a ratio of 8:2.[277] As expected, a 3α-methoxythiocarbonyl methylene group complexes even better with $SnCl_4$ than with TMSOTf[278] to give the corresponding β anomers in more than 90% yield with silylated cytosine or thymine and $SnCl_4$ as catalyst. [Also see 3α-*O*-(*N*-benzoyl)carbamoyl derivatives,[278a] as well as 3α-*O*- or 5-*O*-thiocarbamates,[278b,c] which control the trans stereochemistry.]

BnO, O, OAc, O, S → O, Me **267**; **66** (OTMS, N, TMSO, N); TMSOTf, CH_2Cl_2; → BnO, O, N, NH, O, O, O, S → O, Me **268** + BnO, O, N, O, NH, O, O, O ← S, Me **269**

Since protected 1-halosugars such as crystalline **21** with an exclusively oriented 1α halogen (or mesylate or tosylate) are usually not available, whereas the chemically stable precursors for the preparation of these derivatives are readily available, protected 1-*O*-methyl or 1-*O*-acetyl derivatives of 2-deoxysugars such as **119** have been commonly reacted with silylated bases in the presence of TMSOTf or $SnCl_4$ to give 1:1 mixtures of the α/β anomers in high yields. The variations in the α/β ratio between 6:4 to 4:6 as observed by many groups in the Friedel-Crafts-catalyzed synthesis of nucleosides, which probably proceeds primarily via an S_N1 pathway, might be due to the different basicities of the applied silylated bases and might furthermore reflect the thermodynamic stabilities of the anomers. Reacting methyl-4,6-di-*O*-acetyl-2,3-dideoxy-α-D-glycopyranoside in the one step – one pot version of the Silyl-Hilbert-Johnson reaction (HMDS/TMSCl/$SnCl_4$ in CH_3CN) with silylated uracil, thymine, or N^4-benzoylcytosine affords the protected nucleosides in high yield with an α/β ratio of ~1:3.[192]

In mixtures of peracylated 1-*O*-alkylfuranosides and pyranosides, the furanosides react via kinetic control at ambient temperature in the presence of TMSOTf in preference to the pyranosides. In addition, the undesired *O*-acylated and persilylated α-nucleosides (see **124** → **123**) can be partially anomerized to the desired β-nucleosides in the presence of TMSOTf, thus raising the yields of the β-nucleosides.[133] The frequently obtained α/β ratio of 1:1 might also be due to the ready isomerization (anomerization) of the 4-*O*-trimethylsilylated intermediates in the presence of a Lewis acid such as TMSOTf, $SnCl_4$, and probably also $(NH_4)_2SO_4$.[151,152,152a] On heating 5′-*O*-(*tert*-butyldiphenylsilyl)thymidine with HMDS and 0.2 equivalent of $(NH_4)_2SO_4$ for 2 hours at 140°, the nucleoside bond is cleaved and 1,4-anhydro-2-deoxy-3,5-bis-*O*-(trimethylsilyl)-D-erythropent-1-enitol is isolated in 76% yield.[152a–c]

A series of 3α-substituted 2-deoxysugars **270** was reacted with silylated thymine **226** in the presence of TMSOTf to give protected AZT **271a**[153,279–286] or the related protected nucleosides **271b**,[288–291] **271c**,[114,281,287] **271d**,[276,292–295] **271e**,[296] and **271f**,[297] as well as the corresponding α anomers **272**. For reviews of the preparation of these drugs, see Refs. 293–295.

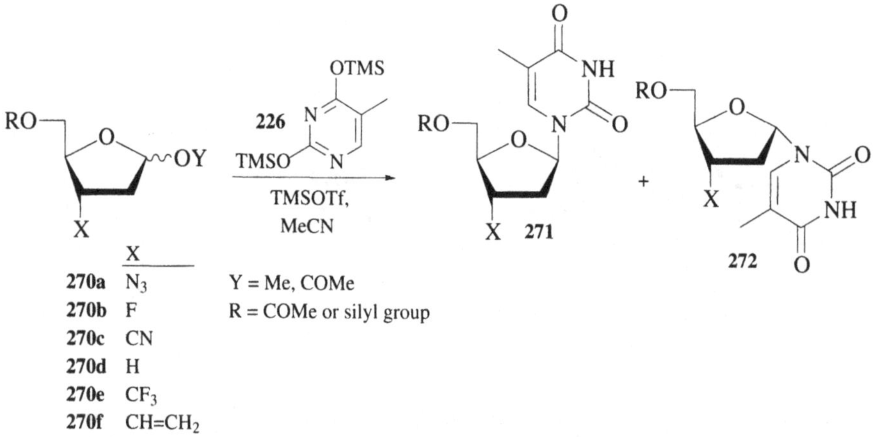
OTMS
RO
OY
226
TMSO
TMSOTf,
MeCN
RO
O
NH
O
X 271
+
RO
X
O
NH
272
X
270a N3
270b F
270c CN
270d H
270e CF3
270f CH=CH2
Y = Me, COMe
R = COMe or silyl group

Analogously, **270d** reacts with silylated 6-chloropurine in the presence of TMSOTf in 1,2-dichloroethane to give an $\alpha/\beta = 1:1$ mixture of the corresponding 2′,3′-dideoxypurine nucleosides.[298] Interestingly, using Et_2AlCl instead of TMSOTf affords the α/β mixture of the desired N^9-nucleosides as well as the N^7-nucleosides.[298]

Starting from inexpensive noncarbohydrate precursors, Sharpless oxidation and subsequent reactions afford the optically active dibenzylacetals **273**, which react with silylated thymine **226** in the presence of TMSOTf to give seconucleosides **274**, which were isolated in the case of **274a**. Preferential acid-catalyzed cyclization of **274** via transition state **275** (gauche effect) affords exclusively or predominantly the anti-AIDS drugs AZT **271a** and **271b**, as well as small amounts of the corresponding α-nucleosides **272a**[299] and **272b**.[300]

As described subsequently in the section on the synthesis of arabinofuranosyl nucleosides, seconucleoside **332** analogous to **274** and **275** was isolated.

The synthesis of 2′-deoxy-2′,2′-difluorocytidine ("gemcitabine") **278**[301,301a,302] poses analogous problems. Reaction of the 2,2-difluoro-1-mesylate **276** with **277** in the presence of TMSOTf in 1,2-dichloroethane at reflux for 18 hours gives a crystalline 1:1 mixture of the hydrochlorides **278** and **279** in 49% yield.[302,303] The α-nucleoside **279** can be anomerized to the β anomer **278** under basic conditions.[304] The two adjacent electron-attracting fluorine atoms, which deactivate any normal 1-acyloxy derivative, necessitate the 1-mesyloxy leaving group in **276** for a smooth reaction with TMSOTf to furnish the corresponding 1-cation.

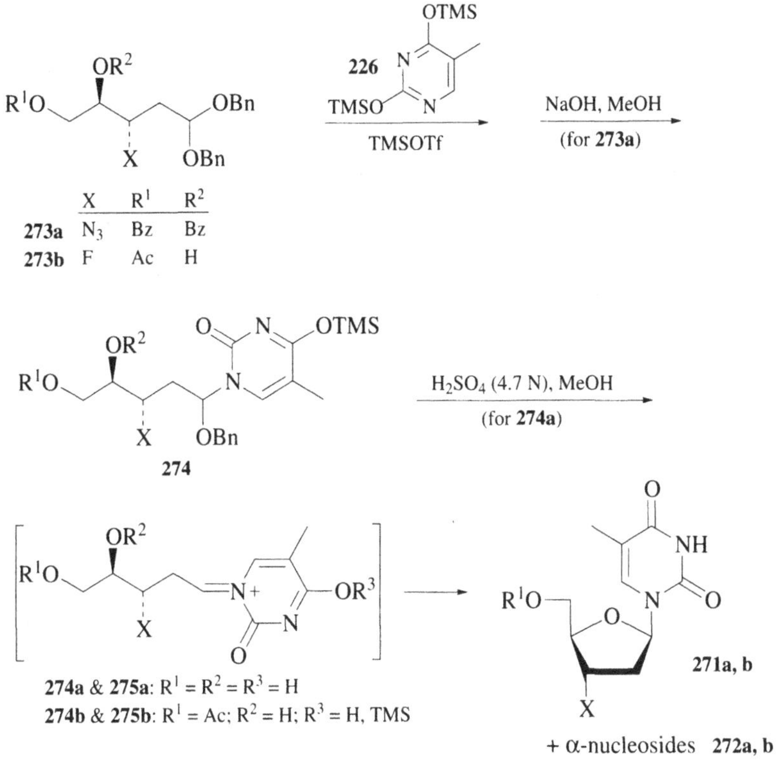

	X	R^1	R^2
273a	N_3	Bz	Bz
273b	F	Ac	H

274a & **275a**: $R^1 = R^2 = R^3 = H$
274b & **275b**: R^1 = Ac; R^2 = H; R^3 = H, TMS

+ α-nucleosides **272a, b**

In contrast to the 2,2-difluorosugar **276**, the corresponding 1-*O*-acetyl-2,2-diphenylthiosugar **280** reacts with silylated uracil **66** in the presence of TMSOTf to furnish a 4:1 mixture of **281** and **282** in 83% yield.[305]

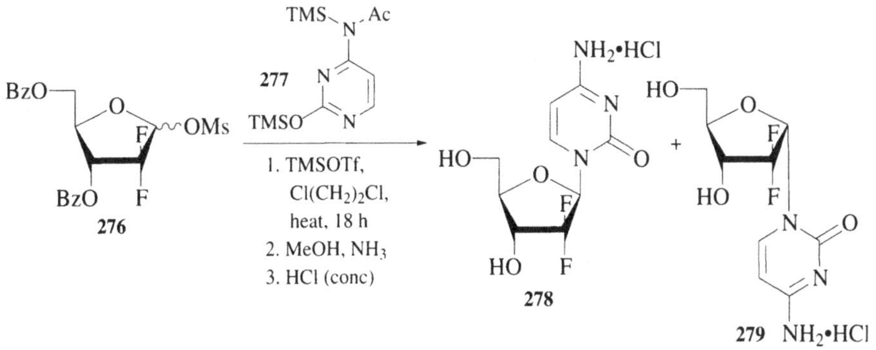

Acylated glucals such as 3,4,6-tri-*O*-acetyl-D-glucal **283** undergo a Ferrier reaction with Lewis acids[145] to yield conjugated cations such as **284** lacking an α- or β-directing group. The reactions of **283** with silylated uracils **66**, **80**, and **226**

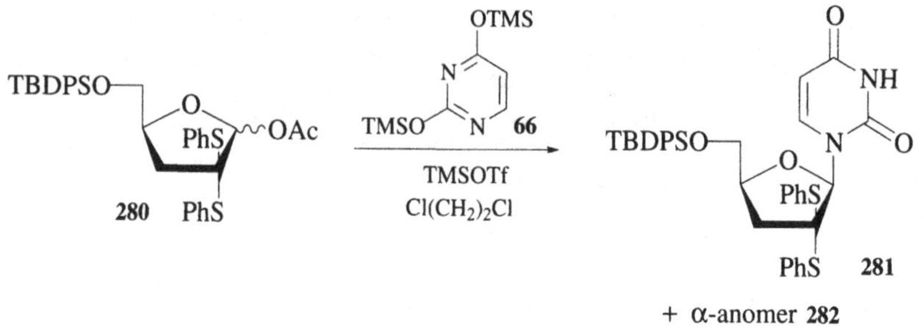
TBDPSO
O
PhS
OAc
280
PhS
OTMS
N
TMSO
N
66
TMSOTf
Cl(CH2)2Cl
O
NH
TBDPSO
N
O
O
PhS
PhS
281
+ α-anomer 282

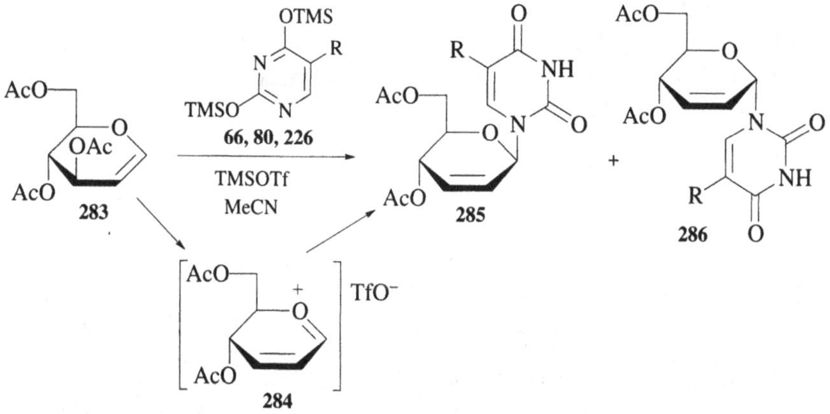
AcO
O
OAc
AcO
283
OTMS
N
R
TMSO
N
66, 80, 226
TMSOTf
MeCN
R
O
NH
AcO
O
N
O
AcO
285
+
AcO
O
AcO
N
O
NH
R
286
O
AcO
+
O
AcO
284
TfO⁻

in the presence of TMSOTf or $SnCl_4$ afford via **284** the 2′,3′-dideoxy-Δ^2 nucleosides **285** and **286** in ~1:1 ratio in moderate yields.[306] Compare also other papers on these Ferrier-type[145] reactions.[307–312]

The homologous Ferrier reaction of mesylate **287** with disilylated cytosine **222** in the presence of $EtAlCl_2$ gives an α,β mixture of the cyclopropyl nucleosides **288** in 81% yield.[115] An efficient Ferrier-type reaction was described employing *N,N*-dimethylformamide at reflux in the absence of a catalyst.[312a]

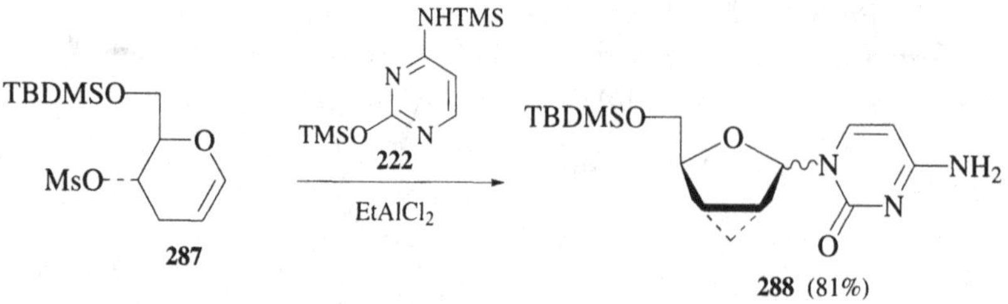
TBDMSO
O
MsO
287
NHTMS
N
TMSO
N
222
EtAlCl2
TBDMSO
O
N
NH2
N
O
288 (81%)

Chemical Conversion of Ribo- to 2′-Deoxyribonucleosides

As emphasized in the introduction, the building blocks of RNA—uridine, cytidine, adenosine, and guanosine—were apparently produced first during evolution and only subsequently reduced via their 5′-diphosphates by specific ribonucleotide reductases to the corresponding 2′-deoxynucleotides—the building blocks of DNA. Thus imitating nature, chemical methods have been developed to convert ribonucleosides into their corresponding 2′-deoxyribonucleosides.

Uridine and cytidine as well as their 5-substituted or 6-aza analogs can be readily transformed into their corresponding 2′-deoxy-2′-halonucleosides **290** with acetyl bromide[313–317] or propionyl bromide[318,319] in up to 94% yield via the protected 2,2′-anhydronucleosides **289**. The halogen atom X (Cl, Br) can subsequently be removed from **290** by hydrogenation[313,318,319,321] or by Bu_3SnH–AIBN[315,317] to give 2′-deoxynucleosides **291** in overall yields of up to 75%. Treatment with Zn or Zn/Cu[314,316,317] or electrochemical reduction[320,321] leads to the 2′,3′-didehydronucleosides. Uridine or cytidine can also be treated first with methyl orthoacetate followed by acetyl bromide to give an even higher yield of the 2′-bromo intermediate **290**.[316,317] Recently, 2′-*O*-(3-trifluoromethylbenzoyl-3′,5′-di-*O*-benzoates of nucleosides have been reduced photochemically in the presence of *N*-methylcarbazole to 2′-deoxynucleosides in up to 73% yields.[321a]

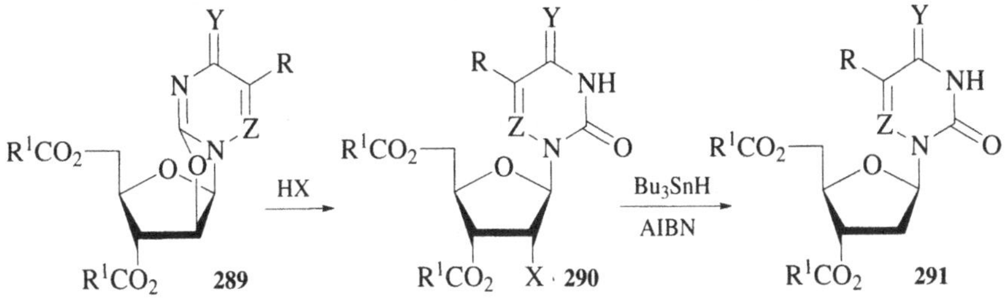

289 →(HX) **290** →(Bu_3SnH, AIBN) **291**

R = H, Me; Z = CH, N; Y = O, NH; R^1 = Me, Et; X = Cl, Br

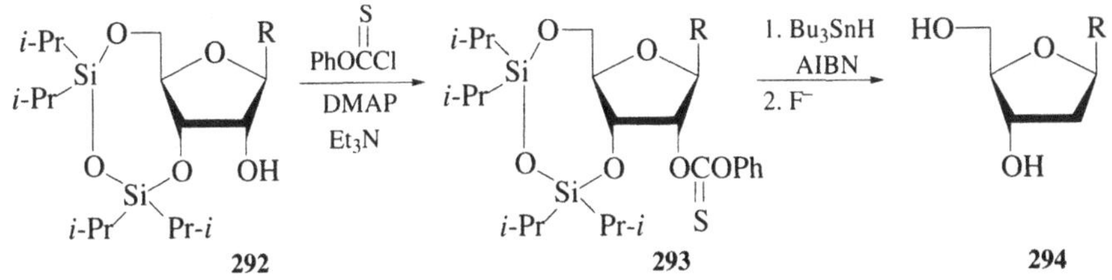

292 →(PhOC(S)Cl, DMAP, Et_3N) **293** →(1. Bu_3SnH, AIBN; 2. F^-) **294**

R = pyrimidine or purine base

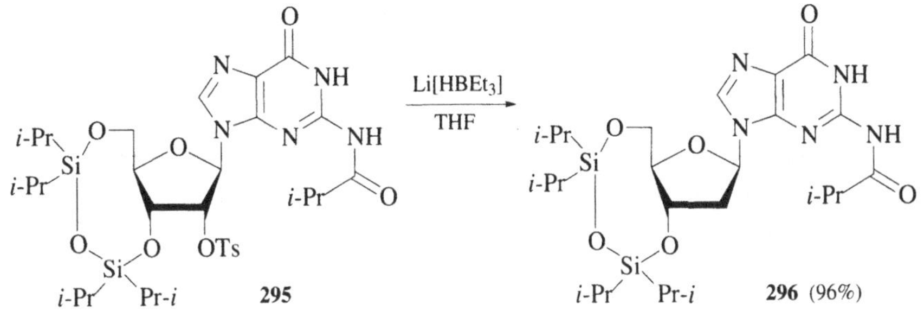

295 →(Li[$HBEt_3$], THF) **296** (96%)

Alternatively, the 3′,5′-hydroxy groups in ribo (or xylo) nucleosides as well as their analogs can be protected selectively by the bifunctional Markievicz reagent[322,323] TIPS-Cl_2 to **292**, which are readily transformed by a Barton reaction via **293** to **294** in high overall yields.[324,324a,325] The intermediate radical at C-2′ can be trapped by allyltributylstannane to introduce a carbon substituent at the 2′α position.[326] The analogous protection of N^2-isobutyrylguanosine and subsequent tosylation gives **295**, which can be reduced by Li($HBEt_3$) to **296** in 96% yield.[327] Furthermore, 2′-hydroxy-3′,5′-di-*O*-acetylribonucleosides can be obtained in up to 74% yield by selective saponification of the 2′-*O*-acetyl group in 2′,3′,5′-tri-*O*-acetates by hydrazine hydrate or hydroxylammonium acetate.[328,329]

Both the 2′- as well as the 3′-hydroxy groups of ribonucleosides can be removed by either applying DMF-dimethylacetal[330,330a,331] or triethyl orthoformate[330] followed by heating with acetic anhydride,[330,330a] reaction with methyl iodide,[331] or by treatment of the 2′,3′-di-*O*-xanthates with Ph_2SiH_2[332] to afford in high yields the 2′,3′-didehydronucleosides, which can be readily hydrogenated to the saturated nucleosides. Alternatively, reacting the ribonucleosides with 2-acetoxyisobutyryl chloride[333,334] followed by treatment of the resulting mixture of 2′,3′-chloroacyloxy isomers with Zn/Cu-acetic acid also yields the corresponding 2′,3′-didehydronucleosides.[335,335a,b] The different synthetic strategies for the preparation of the anti-AIDS drug 2′,3′-dideoxyinosine have been summarized.[336]

Starting with the readily available 8-bromopurine nucleosides, which are converted via the 2′,3′-dibutylstannylene derivatives to the corresponding 2′-*O*-tosyl derivatives such as **297**, heating with NaSH gives the 8,2′-*S*-cyclonucleoside **298** in 69% yield. Subsequent treatment with Ra-Ni affords 2′-deoxyadenosine (**299**).[337]

Synthesis of β-D-Arabinofuranosylpyrimidine and Purine Nucleosides

Pyrimidine nucleosides can be readily converted into protected β-D-arabinofuranosylpyrimidine nucleosides **300** via 2,2′-anhydronucleosides **289** and subsequent hydrolysis.[338] Furthermore, β-D-arabinofuranosylpurine nucleosides **302** are accessible by S_N2 displacements of 2′-*O*-triflates of purine nucleosides **301** by azide,[339] benzoate,[339a,b] or fluoride ions[339c] followed by deprotection. But the total synthesis of β-D-arabinofuranosylpurine and -pyrimidine nucleosides is more complicated and of particular importance because of their interesting biological properties.

289 $\xrightarrow{H_3O^+}$ **300**

R = H, F; Z = CH, N; Y = O, NH; R^1 = Me, Ph

301 $\xrightarrow[\text{heat}]{N_3^- (BzO^-, F^-)}$ $\xrightarrow[(OH^-)]{F^-}$ **302**

X = N_3, OH, F

In the first synthesis of such a nucleoside, N^6-benzoyladenine **303a** was reacted with the standard 2,3,5-*O*-benzyl-D-arabinofuranosyl chloride (**22**),[52,53] which consists of more than 90% of the 1α-chloro derivative,[340] in the presence of molecular sieves to give **304a** in 46% yield.[341] Hydrogenation of **304a** affords free 9-β-D-arabinofuranosyladenosine **305a** in more than 90% yield.[341]

303 + **22** $\xrightarrow{\text{MS, } CH_2Cl_2}$ **304** $\xrightarrow[(Na, NH_3)]{H_2, Pd (BCl_3)}$ **305**

	R	R^1	R^2	R^3	R^4
a	H	NHBz	H	NH_2	H
b	H	OH	NHAc	OH	NH_2
c	H	Cl	NHAc	Cl	NH_2
d	H	Cl	Cl	Cl	Cl
e	H	NHAc	NHAc	NH_2	NH_2
f	TMS	NHTMS	NHTMS	NH_2	NH_2
g	H	NH_2	F	NH_2	F
h	H	Cl	F	Cl	F
i	H	OH	Cl	OH	Cl

Subsequent studies established that the reactions of 2,6-disubstituted purines **303** with **22** proceeded to **304** in ~50% yield in nonpolar solvents in the presence of molecular sieves,[341–343] whereas with $Hg(CN)_2$ as catalyst the formation of the corresponding α-nucleoside ($\alpha/\beta = 2.5:1$) was favored.[343] In the reaction of 2,6-diacylaminopurine **303e** or the silylated base **303f** with **22**, the addition of molecular sieves,[342–345] distillative removal of TMSCl,[346] or addition of diisopropylethylamine[347] are recommended.

The benzyl ether groups in **304** can be readily removed by BCl_3,[344,345] by hydrogenation with H_2/Pd,[341,342,347] or by treatment with sodium in liquid ammonia[343] to give the corresponding free nucleosides **305**.

Alternatively, the acidic 7- and 3,7-deaza analogs[348,349] **306** and **22** are readily converted in the presence of phase-transfer catalysts predominantly into the corresponding β-nucleosides **307** and some α-nucleosides **308**. Interestingly, the concentrations of the phase-transfer catalyst[348] and NaOH[66,349] influence the ratio of **307** to **308**. A lower concentration of the phase-transfer catalyst tetrabutylammonium hydrogen sulfate (0.2 molar versus 1 molar) favors formation of the β anomer **307**.[348] The 1α-chlorine in **22**, which is displaced in a S_N2 reaction by the *N*-anion of **306**, apparently undergoes equilibration at higher concentrations of tetrabutylammonium hydroxide to give more of the 1β-chloride and consequently more of the corresponding α-nucleoside **308** (cf also the preparation of **49** and **50**[66,67]).

306 + BnO 22 → ($R_4N^+ HSO_4^-$, NaOH (50%), CH_2Cl_2) → **307** + **308**

	R^1	R^2	X
a	OMe	SMe	N
b	OH	NH_2	N
c	NH_2	H	CH
d	Cl	H	CH

Reaction of pyrazolopyrimidine **309** with **22** furnishes 43% of nucleoside **310**, 25% of the isomeric **311**, and 6% of the α anomer of **310**.[340] Imidazo[4,5-*d*]-isothiazoles give mixtures of the corresponding N^4- and N^6-regioisomers in good yields.[350]

The reaction of free or silylated imidazoles **312** with **22** proceeds with **312a** to give ~30% of **313a** in the presence of triethylamine in CH_3CN,[351–353] whereas

312b affords predominantly **314b** as well as nucleoside **313b** after pretreatment with NaH in DMF.[354]

After conversion to the sodium salt with NaH/CH_3CN, 2-carbomethoxy-3-cyanopyrrole reacts with **22** to give the corresponding β-D-aranucleoside **313** in 59% yield.[355]

After unsatisfactory reactions of 2,4-dimethoxypyrimidine **315a** (**63**) with **22**[356] to give **316a**, the 5-methyl derivative **315b** reacted with **22** in CH_2Cl_2 to give **316b** in 49% yield.[357] See also the reaction of 2,4-diethoxy-5-ethylpyrimidine with **22** to afford 80% of the corresponding 4-ethoxy-5-ethyl analog of **316**.[358]

Silylated pyrimidines **317b-c** or 1,2,4-triazine **317a** react with **22** in 1,2-dichloroethane and $SnCl_4$[84,358,359,363] to give β-nucleosides **318** in 40–60% yield. The analogous reaction of **317d** with **22** in CH_2Cl_2 and molecular sieves affords β-nucleoside **318d** in 58% yield.[360]

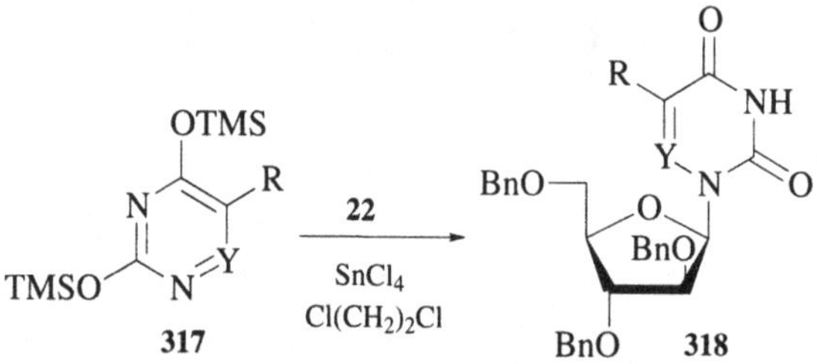

	R	Y
317a	H	N (**74**)
317b	Me	CH (**226**)
317c	Et	CH
317d	$(CH_2)_2OTMS$	CH

The silylated pyrazolouracils **319**[361] and **320**[362] react analogously with **22** in the presence of $SnCl_4$ as catalyst to give the corresponding β-nucleosides **321** and **323** as well as the isomeric β-nucleosides **322** and **324.**

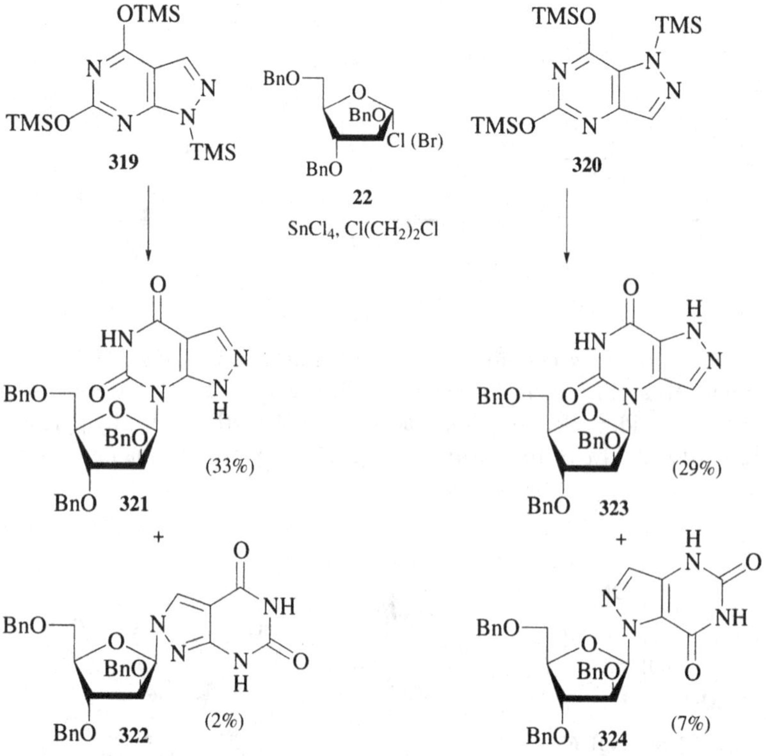

The formation of ara-β-nucleosides **318, 321-324** is surprising since a 2β-alkoxy group can coordinate with $SnCl_4$ and stabilize a cation at position 1 on the β side to result in the preponderant formation of the corresponding α-nucleosides. Thus it was demonstrated that reaction of 1-*O*-acetyl-3,5-di-*O*-benzoyl-2-*O*-methyl-β-D-ribofuranose with silylated uracil in the presence of $SnCl_4$ or TMSOTf[226b] afforded primarily the desired natural 2′-*O*-methyl-3,5-di-*O*-benzoyl-β-uridine.[225,226,226b] Reaction of a 2α-*tert*-butyldimethylsilyloxyribose derivative with silylated uracil and $SnCl_4$ furnished primarily the β-nucleoside.[226,226a] It is possible that $SnCl_4$ assists in the S_N2 reaction of the silylated base by coordination to the 1α-halogen from the α side as depicted in **325.** As a consequence, gradual addition of $SnCl_4$ (or of **22**) under diminished pressure in 1,2-dichloroethane (to remove the trimethylsilyl chloride formed) might give rise to increased amounts of the β-anomer **326**. The potential influence of CuI in an analogous reaction of silylated bases with **22** as indicated in **325** should also be considered.

OSiMe$_3$ ... N ... Cl$^-$ ... SiMe$_3$ ... BnO ... O ... BnO ... Cl ... BnO ... $SnCl_4$ (CuI)

325

1. $SnCl_4$, $Cl(CH_2)_2Cl$
2. $NaHCO_3$, H_2O

O ... NH ... N ... O ... BnO ... O ... BnO ... BnO

326

Silylated uracils **327** react with peracylated derivatives of D-arabinose **328** in the presence of $SnCl_4$ or TMSOTf via intermediate **329** in accordance with the Baker rule[205] to furnish exclusively the α-aranucleosides **330**[358,364–366] (see also a corresponding pyranose example[367] as well as a reaction with silylated theophyllin[368]). There is only one exception described: the reaction of **327e** with **328** in the presence of $SnCl_4$ in CH_2Cl_2 is claimed to give the β-nucleoside **331e** in 57% yield.[369]

Condensing **327d** with **328** in the presence of one equivalent of TMSOTf at 0° in acetonitrile provides (via **329**) 63% of the anticipated α-nucleoside **330d** as well as the crystalline seconucleoside **332d** in 3% yield, whereas **327f**, which contains a basic side chain, neutralizes the catalyst TMSOTf, giving a ~50% yield of the crystalline seconucleoside **332f**. Addition of excess catalyst TMSOTf to the reaction mixture leads to the anticipated ring closure of the persilylated seconucleosides **332** to **330**.[370] The formation of **332** is envisioned to take place by addition of TMSOTf to the furanoside oxygen atom in methyl 2,3,5-tri-*O*-benzoyl-α-D-arabinofuranoside **328** to give the intermediate secocation **333**, which reacts with the silylated bases **327** to give seconucleosides **332**.[370] Recently, on reaction of methyl 2,3-dideoxy-3-fluoro-5-(4-phenylbenzoyl)-β-D-erythropentafuranoside with silylated thymine **226** in the presence of TMSOTf in acetonitrile at −25° a seconucleoside corresponding to **332** was isolated in 21% yield. The authors postulated that the aforediscussed addition of TMSOTf to the furanoside oxygen is a general reaction resulting in secocations such as **333**.[370a] For further examples leading to analogous seconucleosides employing 1-*O*-methyl-3,5-di-*O*-*p*-toluoyl-2-deoxyribofuranoside **119** and other sugars in the presence of $SnCl_4$, see references 370b–f. On treatment of 3′,5′-diacetylthymidine with acetic anhydride/H_2SO_4 in acetonitrile, anomerization as well as formation of acetylated seconucleosides was observed.[152,152a]

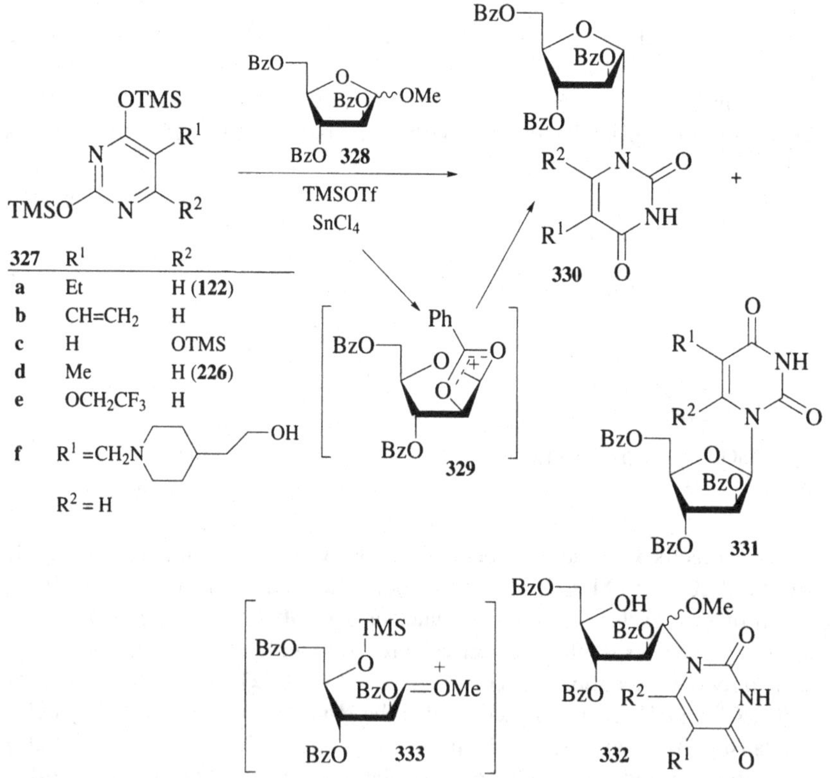

327	R^1	R^2
a	Et	H (**122**)
b	CH=CH_2	H
c	H	OTMS
d	Me	H (**226**)
e	OCH_2CF_3	H
f	R^1 = CH_2N(piperidine-4-yl)CH_2CH_2OH	R^2 = H

On fusion of 1,2,3,5-tetra-*O*-acetyl-D-arabinofuranose **334** with the acidic 2-fluoro-6-chloropurine **303h** at 150°, 85% of the anticipated α-nucleoside **335** is obtained.[343]

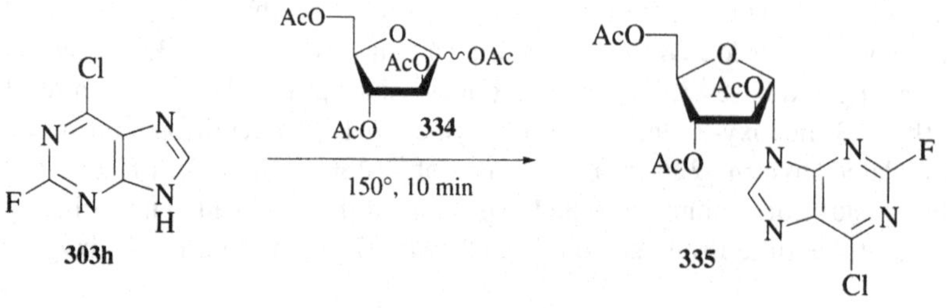

Apart from the hydrolysis of anhydronucleosides **289** to arapyrimidine nucleosides **300**, the Walden inversion of 3′,5′-*O*-protected 2′-*O*-triflates such as **301** with acetate or benzoate anions gives the corresponding 2′-*O*-acylated β-arabinofuranosylpurine nucleosides, and after hydrolysis araguanosine **302b** (X = OH). Because of the difficulties in preparing arapurine nucleosides with the expensive **22**, conversion of purine nucleosides such as guanosine (**5**) into arapurine nucleosides **303** should preferably be carried out via 3,5-protected 2′-*O*-triflates **301**.[339,339b]

Reaction of 8-bromo-2′-*O*-tosyladenosine (**297**) with sodium acetate in acetic acid/acetic anhydride followed by heating with sodium acetate in DMF affords the cyclonucleoside **336**, which is cleaved with hydrazine to lead to araadenosine (**305a**) after oxidative removal of the 8-hydrazino group with HgO. Araguanosine (**305b**) can be prepared analogously.[371]

A different approach to the synthesis of β-D-arabinopyrimidine nucleosides is the prebiotic type-synthesis of aracytidine (**340**). On heating D-arabinose (**337**) with cyanamide in aqueous methanol to 50°, the crystalline oxazoline **338** can be isolated in up to 70% yields.[372–374] Subsequent reaction of **338** with cyanoacetylene and hydrolysis with aqueous ammonia leads via **339** to β-D-arabinofuranosylcytosine (aracytosine) (**340**) in up to 70% yields.[372–374] A modified approach gives rise to arauridine[375,376] or arathymidine[377] as well as to 2′-deoxy-2′-chloro-6-methyluridine.[378,379]

In a similar fashion, condensation of D-fructose (**341**) with cyanamide affords the oxazoline **342**, which is converted via 2,2′-anhydronucleoside and mild hydrolysis to 1-(β-D-fructofuranosyl)uracil **343**.[380,381]

The synthesis of purine aranucleosides by enzymatic transglycosylation starting from uracil or cytosine arabinosides[382,382a] is discussed subsequently.

The synthesis of 2′-fluoro-2′-deoxy-β-D-arabinofuranosyl nucleosides is fraught with problems similar to those observed in the synthesis of β-D-arabinonucleosides. Whereas the conversion of 3′,5′-*O*-protected 2′-*O*-triflates such as **301** with fluoride anions to the corresponding 2′-fluoro-2′-deoxy-β-D-arabinofuranosyl nucleosides **302** (X = F) has already been mentioned, such nucleosides can also be synthesized by reaction of 2-fluorosugar **344** with silylated pyrimidine bases **66**, **226**, or **277**. These reactions proceed in CH_2Cl_2 for 4 days at 24°,[383] in acetonitrile in the presence of TMSOTf,[360] in CH_2Cl_2 in the presence of $Hg(CN)_2$[383a] at 24°, at 40° with molecular sieves,[384] in boiling chloroform,[385,386] or

337 $\xrightarrow[\text{MeOH, } H_2O]{H_2NCN}$ **338** $\xrightarrow{HC{\equiv}CCN}$ **339** $\longrightarrow$ **340** (70%)

341 $\xrightarrow{H_2NCN}$ **342** $\longrightarrow\longrightarrow$ **343**

in the presence of NaI in CH_2Cl_2/ CH_3CN[387] to give the desired β-nucleosides **345** and the α-nucleosides **346** in up to 60% yield with α/β ratios of up to 1:20. Analogous reactions of silylated 2-pyrimidone and its 5-fluoro analog with **344** in 1,2-dichloroethane at reflux are also described,[388] as are reactions of silylated pyrimidines,[389,390] cytosine, and free[391] or silylated 6-chloropurine[390–392] with the corresponding arabino- or 3-deoxysugar in the presence of TMSOTf (**102**).[392] For a mechanistic discussion of these reactions, see Ref. 385.

344 $\longrightarrow$ **345** + α anomer **346**

66, R = H; X = O
226, R = Me, X = O
277, R = H, X = NAc

Reaction of N^6-benzoyladenine (**303a** = **137**) with **344** for 3 days in CH_2Cl_2 at reflux in the presence of molecular sieves affords **347** (R^1 = NHBz; R^2 = H)

BzO AcO O F Br 344 → (303, R¹, R², N, H) BzO O F AcO N N N N R¹ R² 347 + α anomer 348

in 34% yield, as well as the α anomer **348**,[384] whereas silylated 2-acetamido-6-chloropurine (**303c**) in benzene at reflux for 4 hours in the presence of $Hg(CN)_2$ furnishes **347** (R^1 = Cl; R^2 = NHAc) in 32% yield.[384] Reaction of N^6-methyladenine with NaH in DMF followed by addition of **344** affords ~40% of **347** (R^1 = $NHCH_3$; R^2 = H).[298] Analogous reaction of 6-chloropurine with NaH in CH_3CN followed by **344** gives ~70% of **347** (R^1 = Cl; R^2 = H) as well as the corresponding α anomer **348** and minor amounts of the N^7 anomers.[393]

MISCELLANEOUS METHODS

Alternative Silyl-Hilbert-Johnson Procedures

Catalysis by Alkali Halides. A patent describes the smooth reaction of 2-acetoxytetrahydrofuran **349** with silylated 5-fluorouracil **187** in the presence of sodium iodide to give via **350** the anticancer drug Ftafur® **351** in 95% yield.[394] In the analogous reaction of the stable standard sugar **18** with silylated uracil **66** much more drastic reaction conditions are required to afford the *O*-benzoylated uridine **72** in ~60% yield via **35**.[395]

OAc O 349 → (OTMS, F, N, TMSO, N, 187; NaI, MeCN, 60°, 2 h) → H_2O → F O NH N O O 351 (95%)

349 NaI → 350 (+O, I^-)

It should be realized, however, that condensation of the reactive sugar intermediates **350** and **35** (iodide counterion) with the silylated bases **187** and **66** generates trimethylsilyl iodide, which is an effective Friedel-Crafts catalyst in nucleoside synthesis. Subsequently, potassium iodide/dibenzo-18-crown-6 as catalyst was used in the reaction of the 2-deoxysugar **352** with the silylated bases

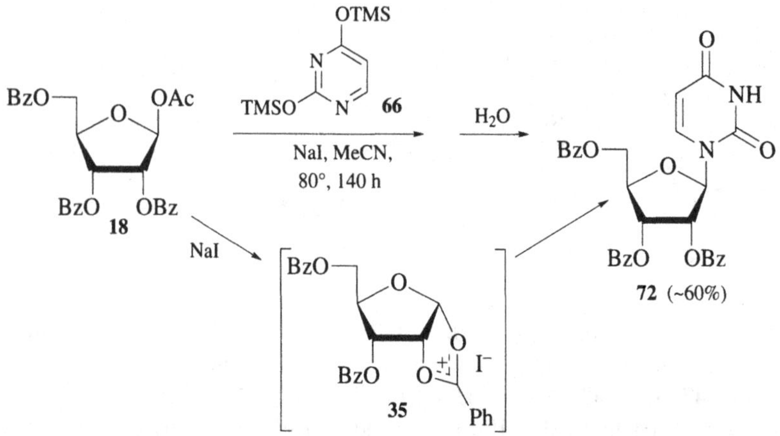

353 and 222 in acetonitrile : toluene (1 : 1) to give nucleosides 354 and 355 in 70–90% yield.[396] Nucleosides of secosugars were prepared analogously .[397–398]

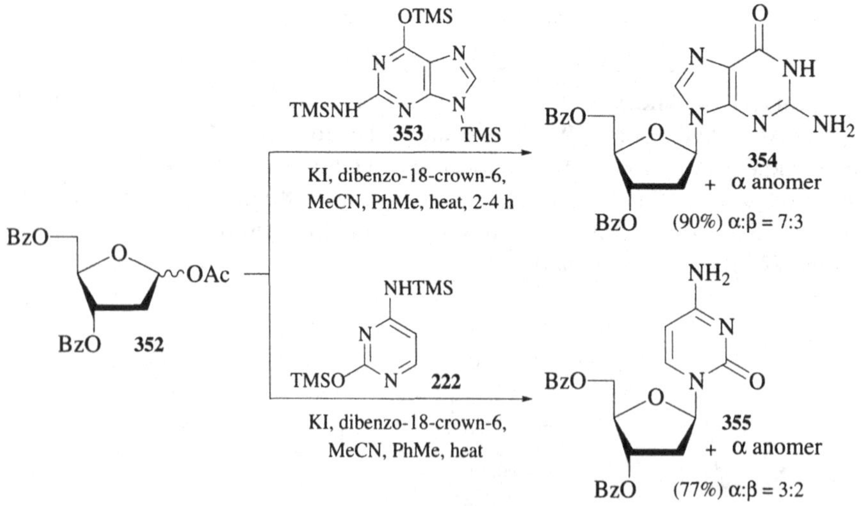

Silylated bases such as silylated thymine **226** or silylated guanine **353** react with the secosugar **159** in the presence of CsI in acetonitrile at reflux to afford **160** in 47% yield.[399,400] In addition to NaI, KI, and CsI, CsCl is also effective in the reaction of 2-acetoxytetrahydrofuran (**349**) with silylated 5-fluorouracil **187**, furnishing **351** in 87% yield.[401]

Activation of 1-Sulfur Groups. Parallel to publications on the activation of protected alkyl 1-thio sugars by *N*-iodosuccinimide (NIS)[402] in the presence of catalytic amounts of trifluoromethanesulfonic acid (TfOH) for glycoside synthesis, earlier studies showed that 1-phenylthiooxetane sugar derivative **356** reacts in CH_2Cl_2 with silylated N^6-benzoyladenine (**97**) in the presence of *N*-bromosuccinimide (NBS) and molecular sieves (4 Å) to give oxetanocin intermediate **357** in 58% yield.[403] The 1-phenylthio derivatives are often prepared via protected 1-*O*-acyl derivatives employing trimethylsilylated thiophenol in the presence of $BF_3 \cdot OEt_2$ or TMSOTf.[410] Thus the question arises, why not use the protected 1-*O*-acyl or 1-*O*-alkyl sugars for nucleoside synthesis instead of the corresponding 1-phenylthio sugars, which entail additional reaction steps and bad smelling thiophenols.

1-α-Phenylthiofuranoside **358a** reacts with silylated thymine **226** via **359** in 70–98% overall yield to form predominantly the α-nucleoside **272**, whereas with the 5-*tert*-butyldimethylsilyloxy derivative **358b** an α/β ratio of **272**/**271** = 3:1 to 2:1 is obtained.[404] The 2-deoxyxylofuranosyl derivatives **360** afford with **226** a ratio of **361**/**362** = 1:12 – 1:30 from benzylidene **360b** and isopropylidene derivative **360c** in the presence of molecular sieves. In contrast, the 5′-*O*-acetyl or *O*-benzoyl derivatives **360a** give α/β ratios of 1:1.5–1:2.[404a,405]

For further studies with 1-phenylthio-5-*O*-(*tert*-butyldiphenylsilyl)-2,3-dideoxyribofuranosyl derivatives, see Ref. 406. The D-arabinosyl derivative **363** affords predominantly the β-nucleoside **364** (91%) as well as some α-nucleoside **365** (**364**:**365** = 9:1) with **226** and NBS in $CHCl_3$.[407]

2,3,5-Tri-*O*-acetyl-1-phenylthio-D-ribofuranose (**366**) gives with NIS/CF_3SO_3H 56% of tri-*O*-acetyluridine (**368**) and, via **367**, 6% of the 2′-iodoacetyl derivative **369**, whereas silylated N^4-acetylcytosine **277** affords 81% of

	R
358a	COBu-*t*
358b	TBDMS

X = N_3, F

226: CH_2Cl_2, NBS, MS, 24°, 30 min → **271** + **272**

359

	R
360a	Ac, Bz
360b	PhCH
360c	Me_2C

360 + **226**: NBS, MS 4 Å, CH_2Cl_2, 24°, 20 min → **361** + **362**

363 + **226**: $CHCl_3$, NBS, MS 4 Å → **364** + **365**

tetraacetylcytidine.[408] For analogous reactions of 1-phenylthio-2,3,4,6-tetra-*O*-acetylglucopyranose with persilylated 2-*N*-acetyl-6-chloropurine, see Refs. 408a, 409.

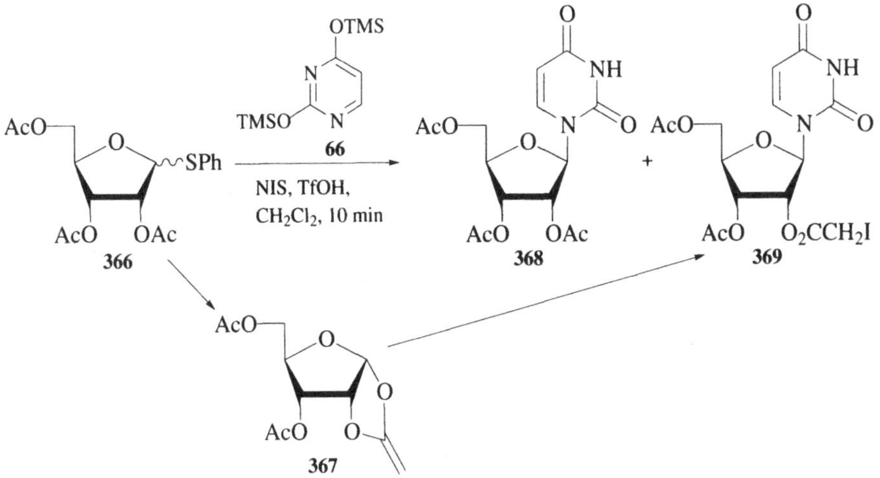

On oxidation of 1-phenylthio derivatives, which are readily available from the protected 1-*O*-methylsugars with thiophenol/$BF_3 \cdot OEt_2$,[411] with *m*-chloroperbenzoic acid one obtains the corresponding sulfoxides such as **370**. Treatment of these 1-sulfoxides with TMSOTf results in a Pummerer reaction that affords sugar cations such as **67**. Subsequent reaction with silylated bases provides the corresponding protected nucleosides in high yields.[412]

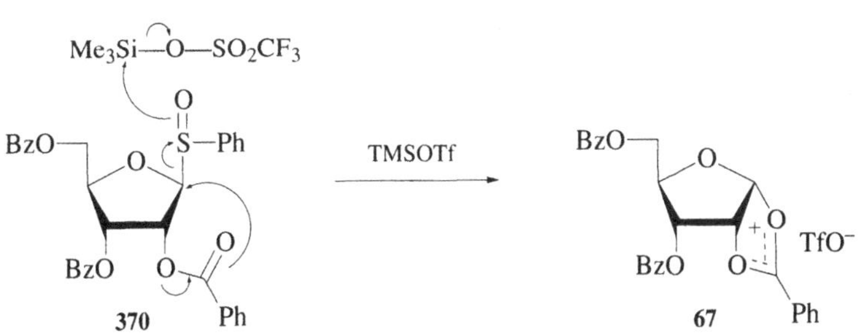

Displacement of the 1α-methylthio group in **371** by the 9-chloromercuric salt **372** of N^6-benzoyladenine affords via an S_N2 reaction the 5'-*O*-benzyl ether of 2'-deoxy-*N*-benzoyladenosine **373** in 20% yield, as well as 28% of the corresponding α anomer.[413]

The related Pummerer reaction of tetramethylene sulfoxide **374** with TMSOTf generates the electrophilic cation **375**, which reacts readily with silylated thymine **226** generated in situ from thymine to give the tetrahydrothiophene nucleoside

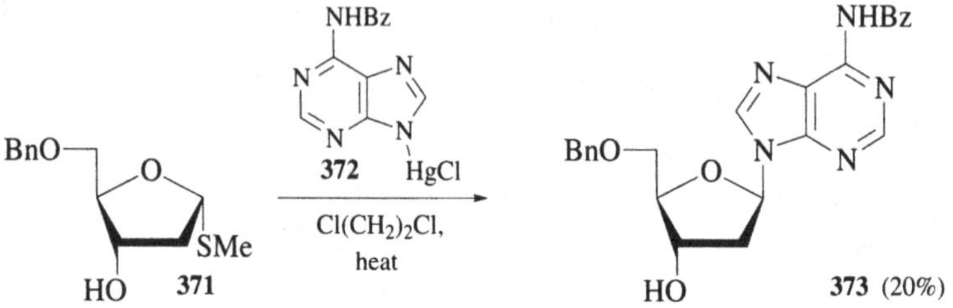

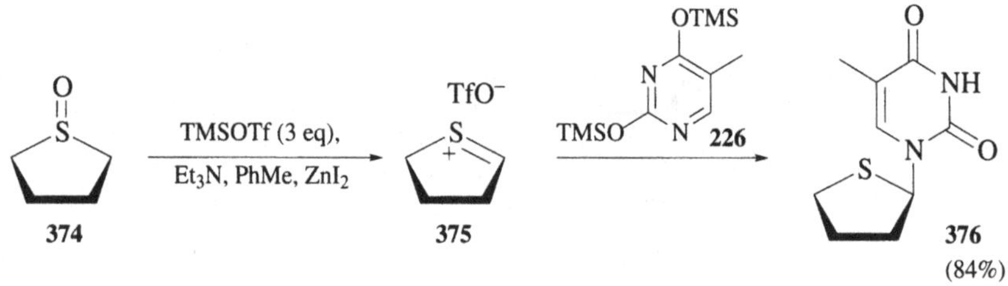

376 in 84% yield.[414] For other recent Pummerer reactions, see Ref. 414a–l, and for reviews on 4′-thionucleosides, see Ref. 414m, n.

Alternative Formation of 1-Sugar Cations. It was previously emphasized that any method that yields 1-sugar cations as intermediates is potentially useful for the synthesis of nucleosides or glycosides (and vice versa). Thus 4-pentenylglycosides **377** are converted by soft electrophiles such as NIS, IDCP, or NBS[415–419] in combination with a Lewis acid such as CF_3SO_3H or $Et_3SiOSO_2CF_3$ via **378** to the reactive sugar cations **379** and 2-iodomethyltetrahydrofuran **380**. In acetonitrile products of a Ritter reaction[199–204] between **379** and acetonitrile can be observed.[420]

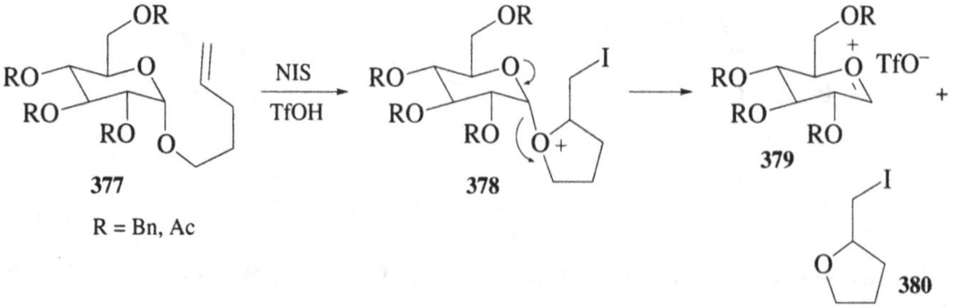

Reaction of glycoside **381** with 6-chloropurine (**42**) and iodonium reagents affords (apparently kinetically controlled) a 91% yield of the N^9- and N^7-nucleosides **43** and **44** as well as the corresponding α anomers, whereas **382** furnishes only **383** in 60% yield.[421]

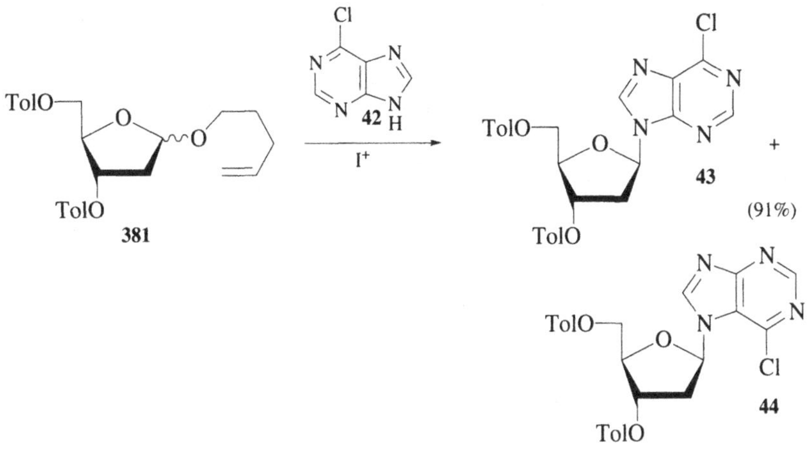

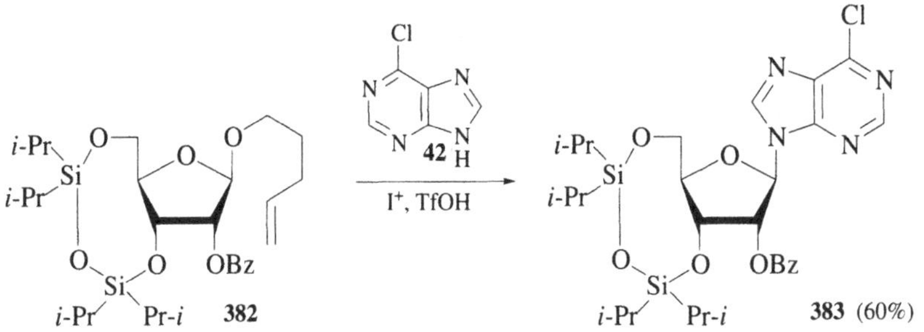

Transformation of the substituted dihydrofuran **384** with phenylsulfenyl chloride gives intermediate **385**, which affords with **277** in the presence of $SnCl_4$ or TMSOTf predominantly the β-nucleoside **386** in 60–65% yield. Since the soft sulfur in **385** complexes better with $SnCl_4$ than with the hard TMSOTf, $SnCl_4$ gives a β/α ratio of 18:1, whereas TMSOTf leads to a β/α ratio of only 6:1.[421a,b]

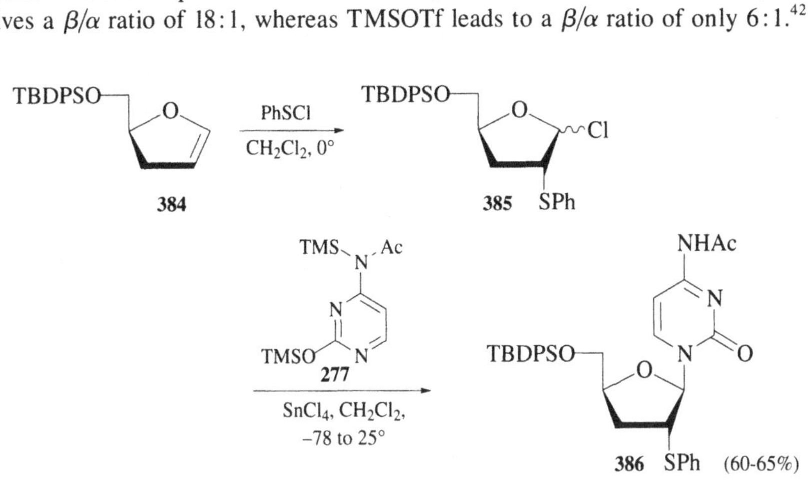

For analogous reactions with phenylselenyl chloride or *N*-iodosuccinimide, see Refs. 421c and 422.

Reaction of dihydrofuran **387a** or dihydropyran **387b** with *N*-fluoropyridinium triflate (**388**) in CH_2Cl_2 produces 2-fluoro-1-pyridinium triflates **389**, which react with silylated 5-fluorouracil **187** in DMF at 120° to give a 3:2 mixture of **390** and **391** in 50–60% yield.[422]

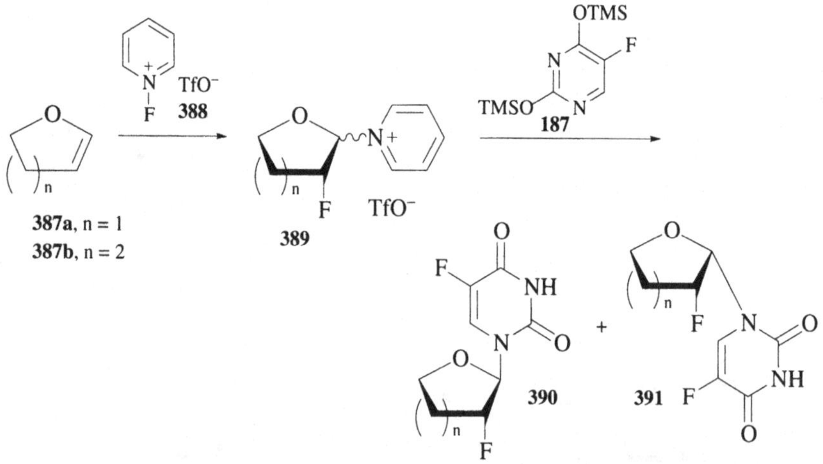

387a, n = 1
387b, n = 2

Iodination of dihydrofuran **392** with NIS in acetic acid produces a 14:1 mixture of iodoacetate **393** and its 1,2-epimer, which are reacted without purification with silylated N^6-benzoyladenine (**97**) in the presence of $SnCl_4$ to give the adenosine analog **394** in 45% overall yield.[423] Analogous iodination of **392** in the presence of silylated thymine **226** in CH_2Cl_2 furnishes the thymidine analog **395**.

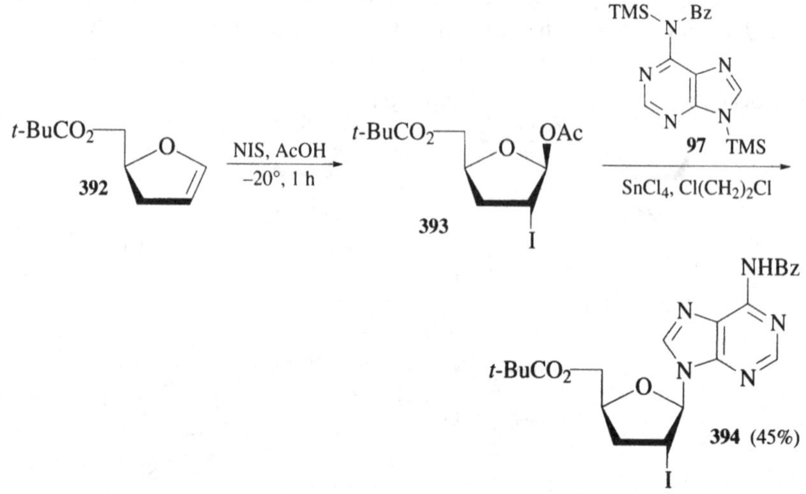

394 (45%)

Treatment of **395** with DBU affords the corresponding 2,2′-anhydronucleoside **396** (52%), which is transformed by potassium *tert*-butoxide in 82% yield into 2′,3′-didehydronucleoside **397**.[423] For analogous reactions of **392** with PhSeCl, see Ref. 423a, b.

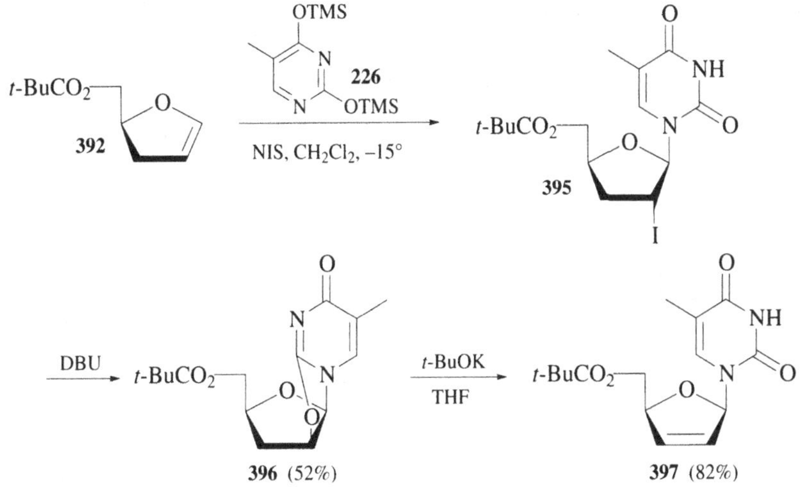

Variations of the Hilbert-Johnson Reaction. A patent claims the superiority of *N,O*-acylated cytosines compared to the corresponding silylated cytosines, since the *N,O*-acylated cytosines are not moisture labile.[424]

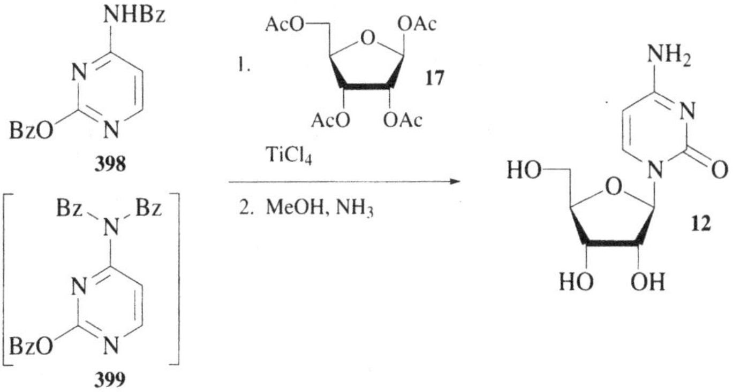

On acylation of cytosine with benzoyl chloride/triethylamine in 1,2-dichloroethane, cooling to 10° and filtering the precipitated triethylamine hydrochloride, the filtrate contains 2,4-di-*O,N*-benzoylcytosine (**398**) or 2,4,4-tri-*O,N*-benzoylcytosine (**399**), both of which can be isolated in crystalline form or reacted as a mixture in situ with **17** in the presence of $TiCl_4$, TMSOTf, or $SnCl_4$ to give on workup and saponification free cytidine in up to 90% yield.[424] Other publications describe, however, that during benzoylation of pyrimidine bases with benzoyl chloride-pyridine, the N^1,N^3-dibenzoylpyrimidines are obtained predominantly.[425,425a–c]

One should realize, however, that in nucleoside synthesis the cost of the silylation of the heterocycles is minute compared to the preparation of the sugar synthons or special heterocyclic bases.

Heating of **400a** and **400b** with bis(tributyltin)oxide in benzene with removal of water gives 2,4-bis(tributylstannyloxy)-5-bromopyrimidine (**401a**) or 2-tributylstannyloxy-5-bromopyrimidine (**401b**), which react readily with the standard sugar **18** in the presence of $SnCl_4$ or with reactive α-haloethers such as benzyloxymethyl chloride under mild conditions to afford 85–97% of the corresponding nucleoside **402a**[426] or the nucleoside analog **402b**.[427]

(Scheme: **400a**/**400b** → $(Bu_3Sn)_2O$, C_6H_6 → **401a**/**401b** → **18** or $BnOCH_2Cl$, CH_2Cl_2, heat, 3 h → **402a**/**402b**)

400a, R^1 = OH
400b, R^1 = H

401a, R^1 = $OSnBu_3$
401b, R^1 = H

402a, R^1 = OH,
R^2 = 1-β-D-2′,3′,5′-tri-*O*-benzoyl-D-ribofuranose
402b, R^1 = H,
R^2 = benzyloxymethyl

Although these alternatives to the Silyl-Hilbert-Johnson reaction seem to be effective, the enviromental problems connected with the large-scale use of expensive and toxic tin compounds may restrict their use to special cases.

ALTERNATIVE NUCLEOSIDE-FORMING REACTIONS

Nucleoside Synthesis with Protected 1-Hydroxysugars

Protected 1-hydroxysugars such as **403** can be activated by *N*-methyl-2-fluoropyridinium tosylate (**404**) to **405**, which reacts with silylated heterocycles such as silylated thymine **226** or 5,6-benzimidazole (**407**) to give primarily the α-nucleosides (α/β = 98:2-8:2) **406** and **408** in 71 and 82% yield, respectively.[428] Activation of theobromine (**409**) with **410** followed by reaction with triethyloxonium tetrafluoroborate gives **411**, which on addition of **412** affords N^7-nucleoside **413** in 93% yield.[429]

Activation of protected 1-hydroxysugars under Mitsunobu conditions permits the reaction of glucopyranose **414** with **42** to give nucleoside **415** in 66% yield,[430] whereas Mitsunobu coupling of difluororibose **416** with **25** affords 50% of a 1:1 anomeric mixture of **417** and **418**.[431] The 1-*O*-mesylate **276**, however, reacts smoothly with silylated cytosine in the presence of TMSOTf to give 2′-deoxy-2′,2′-difluorocytidines **278** and **279** in high yield (Eq. 97, p. 57). Likewise, the

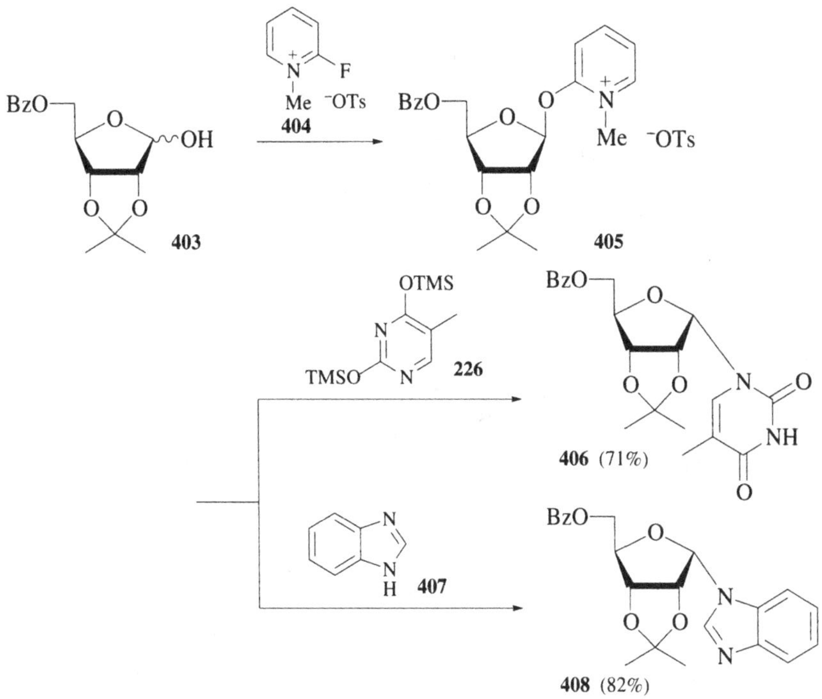

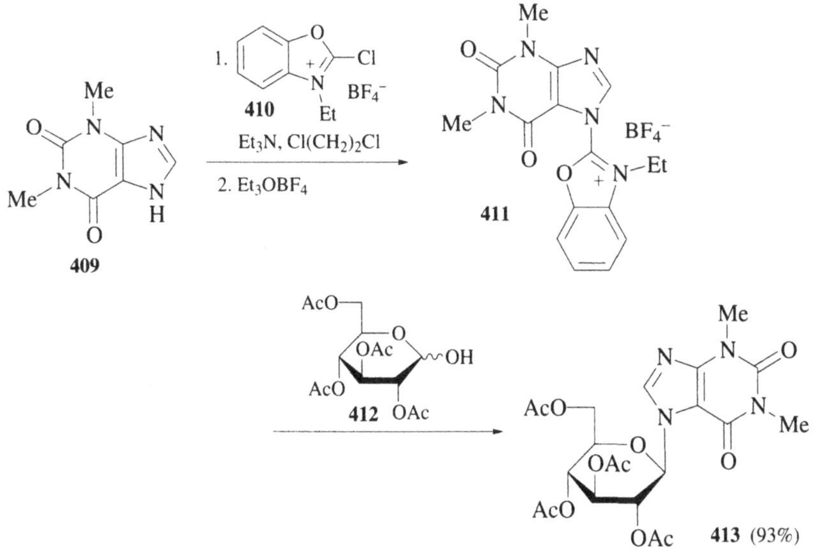

fructose derivative **419** analogously furnishes with **42** an α/β anomeric mixture of nucleosides **420** and **421** in 32% yield.[432] For recent applications of the Mitsunobu reaction, see Refs. 432a–h.

Related to the Mitsunobu reaction, which gives rise to activated 1-*O*-triphenylphosphonium intermediates, reaction of 2,3-*O*-isopropylidene-D-ribofuranose **422** with 2-fluorophenyl phosphorodichloridate **423** affords the activated cyclic 1,5-arylphosphate intermediate **424**. Reaction of **424** with Sn(II) derivatives of heterocyclic bases such as 5,6-dimethylbenzimidazole **425** furnishes the α-nucleoside **426** in 81% yield, whereas the analogous derivative of 6-(1-piperidino)purine affords only 30% of the corresponding α-purine nucleoside. Tin derivatives such as **425** are prepared by treatment of the heterocyclic bases with butyllithium followed by $SnBr_2$.[433]

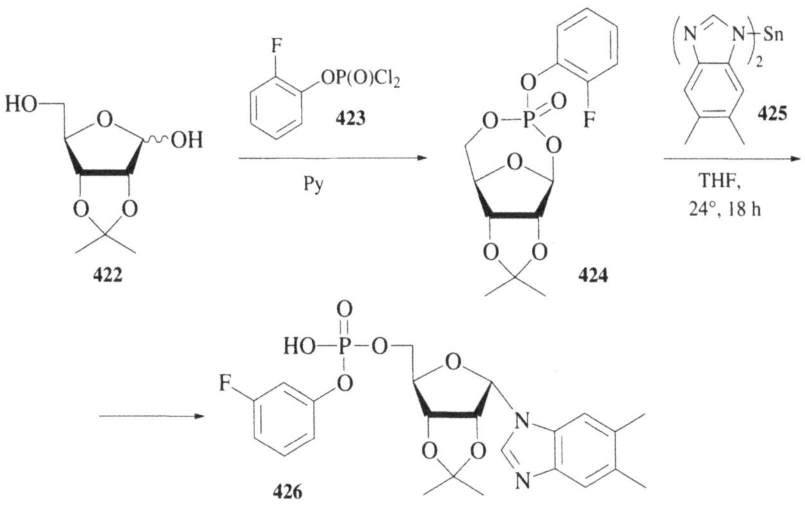

HO 422 O OH O O F OP(O)Cl2 423 Py O O P O O F O O O O 424 N N Sn 2 425 THF, 24°, 18 h

O HO—P—O O F O 426 O O N N

Radical rearrangement of 3,5-di-*O*-benzoyl-2-*O*-(diphenylphosphoryl)-D-ribofuranosyl bromide (**427**) with tributyltin hydride affords the sensitive activated α-D-ribofuranose **428**, which reacts with N^6-benzoyladenine to give 42% of a 1:1 mixture of the anomeric nucleosides **429** and **430**.[434]

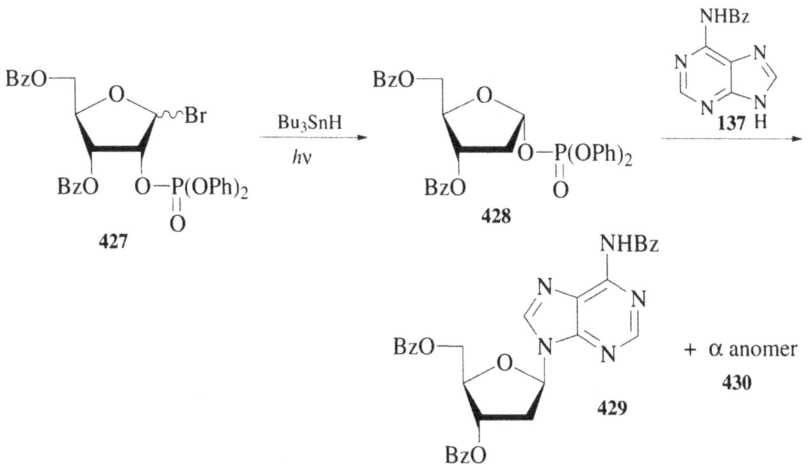

BzO O Br BzO O—P(OPh)2 O 427 Bu3SnH hν BzO O O—P(OPh)2 O BzO 428 NHBz N N N N 137 H

NHBz N N BzO N N O BzO 429 + α anomer 430

Following the pioneering studies by E. Fischer[435] in 1909 on the activation of 1-hydroxysugars with P_4O_{10}, G. Schramm[436] and other investigators[437–439] studied the reactions of protected 1-hydroxysugars with P_4O_{10} in DMF or $CHCl_3$. Thus D-glucopyranose **412** condenses with N^6-benzoyladenine (**137**) with P_4O_{10} in DMF to give after ion-exchange chromatography 9-β-D-glucopyranosyladenine (**431**) in 10% yield.[437] Condensations of purines with 2-deoxy-D-ribose in the presence of P_4O_{10}/Bu_3N give rise to 2,3-dideoxy-3-purine-substituted *C*-nucleosides.[438,439]

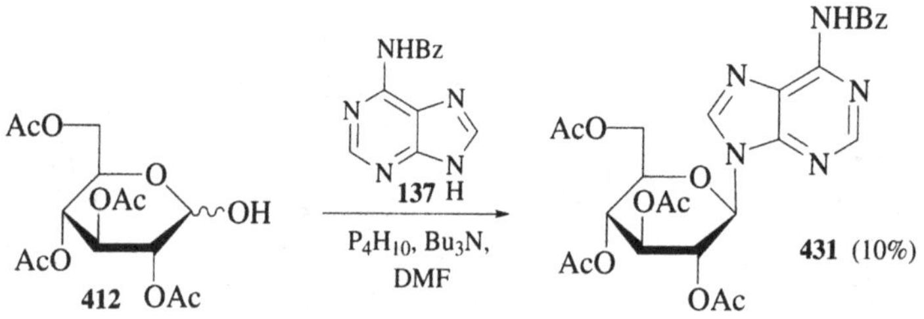

412 → 431 (10%): 137, P_4H_{10}, Bu_3N, DMF

Reaction of Ferrier intermediate **432** with adenine or uracil in the presence of $Pd(dba)_3$ and bis-1,4-(diphenylphosphino)butane (dppb) in THF furnishes α-nucleosides **433** and **434** in 80% and 36% yield, respectively.[440] For analogous reactions with furanoses, cf Ref. 440a.

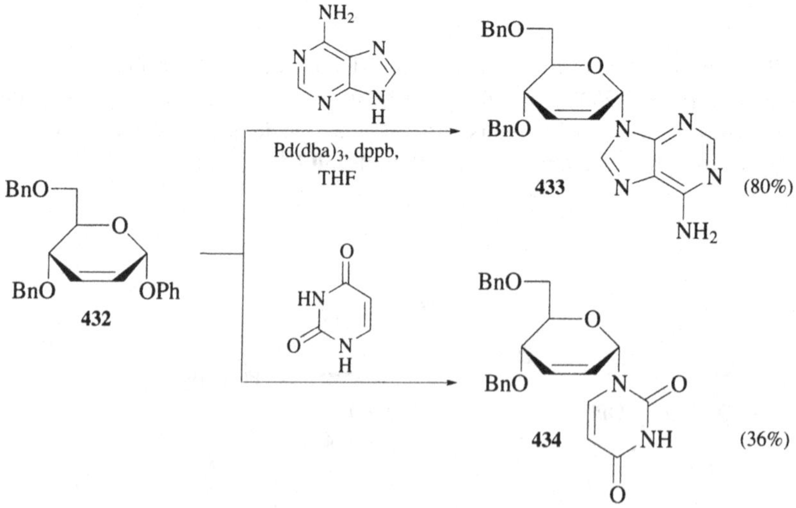

432 → 433 (80%), 434 (36%): adenine or uracil, $Pd(dba)_3$, dppb, THF

Construction of the Heterocyclic Base to Unsaturated Systems

Treatment of D-ribose with methanolic ammonia gives 90% of D-ribopyranosylamine **435**, which is converted by 2,2-dimethoxypropane, acetone, and *p*-toluenesulfonic acid into the crystalline D-ribofuranosylamine tosylate **436** in 80% yield.[441] Reaction of **436** with ethyl *N*-(α-cyano-β-ethoxyacryloyl)carbamate

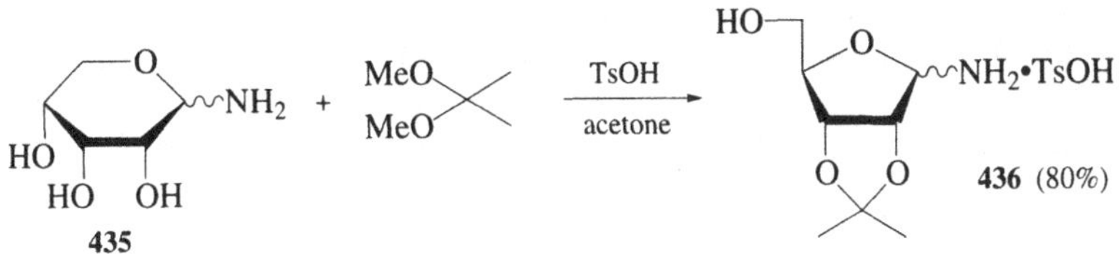

437 furnishes an anomeric mixture of 8% of the β anomer **438** and 15% of the α anomer **439**. Analogous reaction of **436** with ethyl formimidate hydrochloride followed by heating with ethyl α-amino-α-cyanoacetate affords 37% of the β anomer **440** as well as the α anomer **441**.[441] For an analogous reaction of 1-amino-2-deoxy-2-α-fluoro-3,5-di-*O*-benzoylribose with a derivative of ethyl 2-aminocyanoacetate, see Ref. 442.

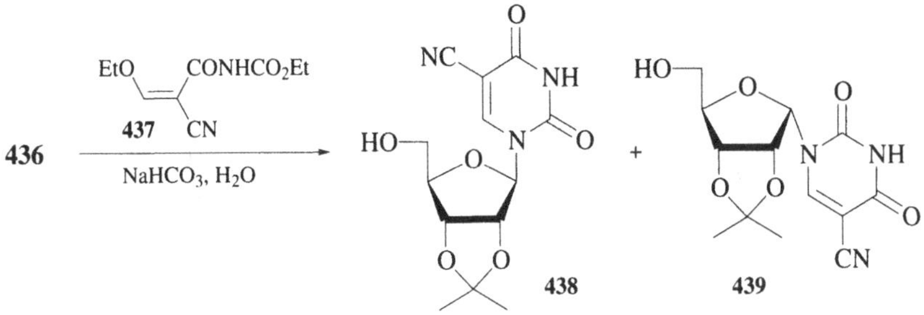

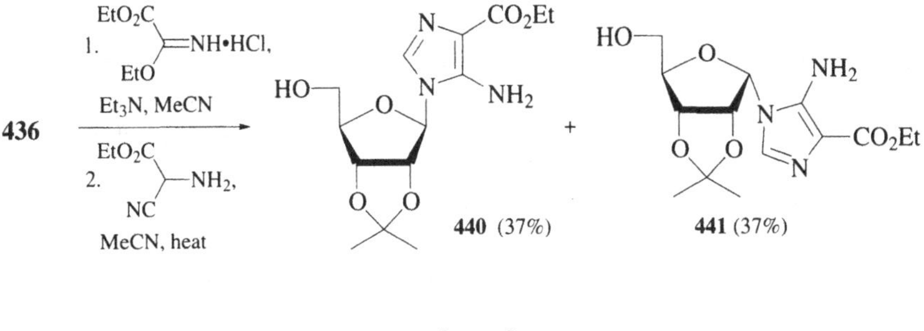

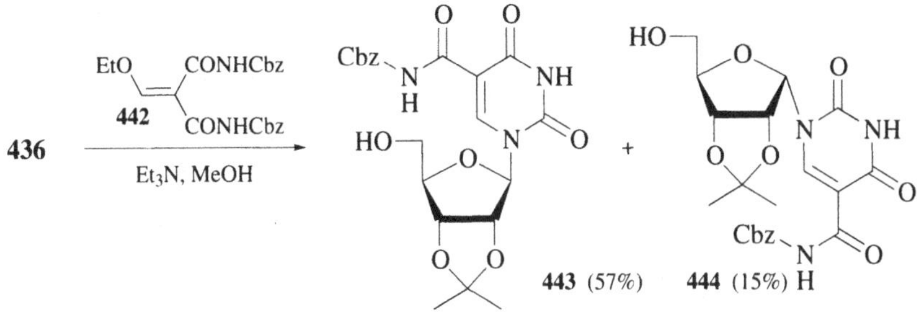

Condensation of **436** with *N,N'*-(benzyloxycarbonyl)ethoxymethylenemalonamide (**442**) furnishes 57% of the β anomer **443** and 15% of the α anomer **444**.[443] On using 3-methoxy-2-methacryloyl isocyanate (**446**), **445** is converted into the thymidine analog **447** in 88% yield.[444] See also the reaction of protected 1-isocyanatosugars.[445]

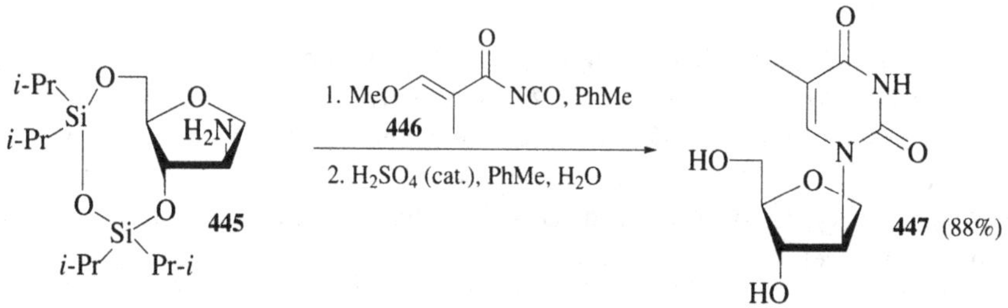

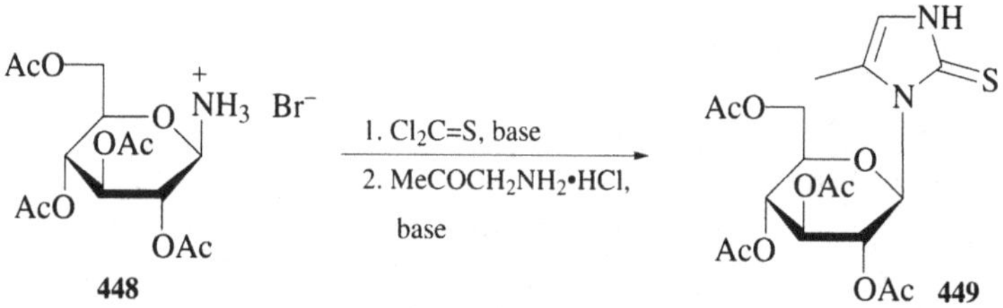

Treatment of β-D-glucopyranosylamine hydrobromide **448** with thiophosgene affords an intermediate 1-isothiocyanate, which condenses with aminoacetone hydrochloride to the nucleoside **449**.[446] Reaction of D-ribofuranosylhydrazine **450** with ketenethioacetal **451** furnishes the pyrazole nucleoside **452** in 62% yield,[447,448] whereas the protected hydrazine derivative **453** gives with thiouronium salt **454** 23% of mesoionic β anomer **455** and 7% of α anomer **456**.[449] For analogous cycloadditions of protected 1-azidosugars, see Ref. 450.

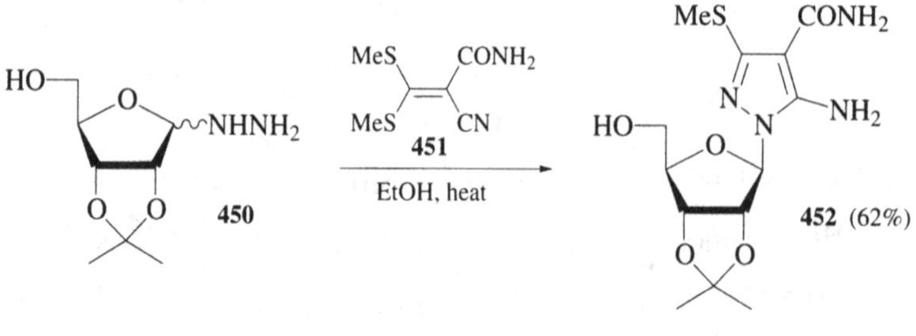

The easily accessible protected *E*/*Z* mixture of 1-oximes **457** reacts with formaldehyde and methyl methacrylate (**459**) to give via **458** a mixture of the stereoisomers **460** in good yield.[451,452]

The condensation of the 6-aminopyrimidin-4-one **461** with glucose to yield **462** followed by nitrosation and reduction affords intermediate **463**,[453,454] which is converted by formamidine acetate[454] to the corresponding purine nucleoside or with nitrous acid to the corresponding 8-azapurine nucleoside. It should be emphasized, however, that silylated amino heterocycles such as **464** react with ribose derivative **18** in the presence of TMSOTf in acetonitrile to give adenosine precursor **465** in 72% yield.[455]

Furthermore, D-fructose (**341**) condenses with KSCN in 10% HCl at +4° to give after subsequent treatment with TBDMSCl in DMF besides the pyranoside **466** also ca 40% of the desired crystalline furanoside **467**. The latter compound affords after RaNi treatment to the corresponding intermediate oxazoline,

Me N O MeO N NH_2 **461** —glucose, AcOH, EtOH→ HO HN N Me O N OMe O OH HO OH **462** —1. HNO_2 2. $Na_2S_2O_4$→

H_2N O N Me HO HN N OMe O OH HO OH **463**

NHTMS N NO_2 N NHTMS **464** —BzO O OAc BzO OBz **18**, TMSOTf, MeCN→ NH_2 O_2N N BzO HN N O BzO OBz **465** (72%)

subsequent condensation with α-amino-α-cyanoacetamide and finally acid treatment the unprotected fructofuranosylimidazole **468** in 43% yield.[455a]

HO O OH HO OH HO **341** —1. KSCN, 10% HCl, 4°, 8 weeks 2. TBDMSCl, DMF→ O H N S O TBDMSO TBDMSO $CH_2O(TBDMS)$ **466** +

TBDMSO O H N S O TBDMSO $CH_2O(TBDMS)$ **467** —1. RaNi 2. H_2N CN $CONH_2$→ HO O N N O NH_2 NH_2 HO OH HO **468** (43%)

Conjugate Additions of Heterocyclic Bases to Unsaturated Systems

Two recent examples of such conjugate additions may suffice. Purines **469** react with methyl acrylate in the presence of K_2CO_3 in DMF to give the corresponding N^9-Michael adduct **471** and small amounts of the N^7-adduct **472** in 85% yield.[456] Ethyl 2-bromoethylenemalonate (**473**) affords the substituted cyclopropanes **474** and **475** in an 8:1 ratio in 87% yield.[456] For additional examples, see Refs. 457–460, 460a.

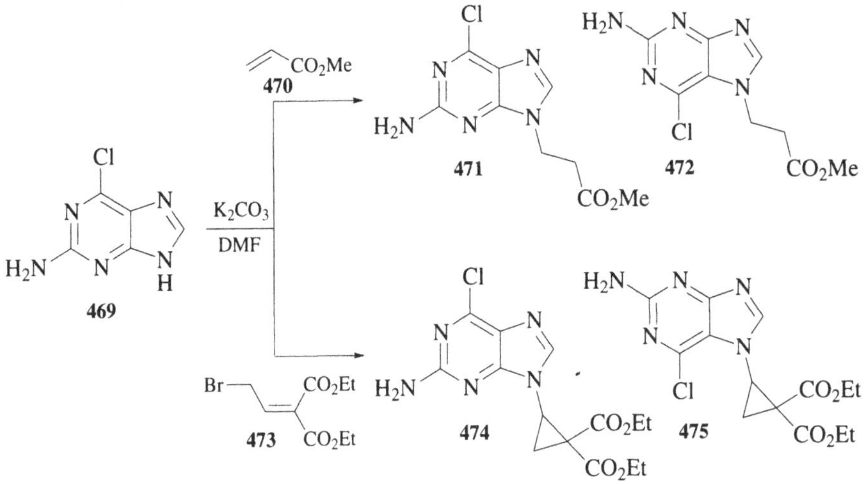

Enzymatic Transglycosylations

Uridine phosphorylase (EC 2.4.2.3) or thymidine phosphorylase (EC 2.4.2.4) degrades uridine, 1-β-D-arabinofuranosyluracil (ara-U) (**476**) as well as thymidine (**477**) in the presence of phosphate to the corresponding pentose-1-phosphates **478**, **479** and **480**, which are transformed in situ by added purines **303** and purine nucleoside phosphorylase (EC 2.4.2.1) to the corresponding purine nucleosides **109**, **305**, and **481**. The application of all these methods in the synthesis of antiviral agents has been reviewed recently.[461]

Whereas normal purine nucleosides can be readily synthesized with $SnCl_4$ or TMSOTf as catalyst, enzymatic methodology also permits the synthesis of imidazo[4,5-*c*]pyridine nucleosides (i.e., 3-deazapurine nucleosides).[462] Of particular interest is the transglycosylation of ara-U **476**,[463–467] which is readily available from uridine via 2,2′-anhydrouridine, to give arapurine nucleosides **305**, since these nucleosides are accessible in only moderate yields by chemical synthesis. Equally important are the transglycosylations of thymidine **477** with bases such as 6-dimethylaminopurine to give the corresponding 2′-deoxypurine nucleoside **481** (R^1 = $N(CH_3)_2$; R^2 = H) in 81% yield.[468]

The combination of thymidine phosphorylase and purine nucleoside phosphorylase from *E. coli*[468,469] can also be used to transform 2′-deoxy-2′-fluorouridine[470–472] (**482**) or 2′,3′-deoxy-3′-fluorothymidine (**271b**)[473] with purines **303** to the corresponding purine nucleosides **483** and **484**. Enzyme preparations from *Erwinia herbicola* permit the similar transformation of 2′-deoxy-2′-aminouridine (**485**) with 2-chlorohypoxanthine to **486** in 32% yield.[474,475]

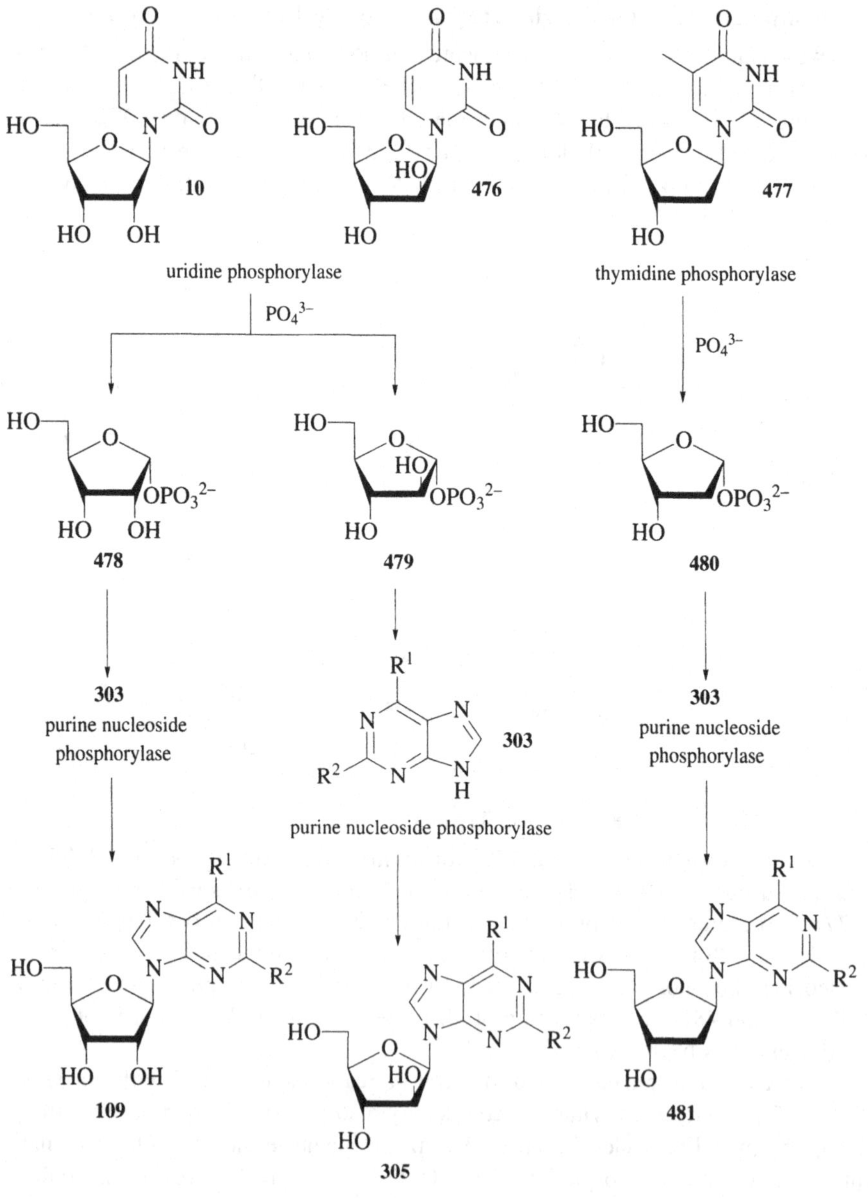

Interestingly, purine nucleosides can also be used as sources of pentose-1-phosphates. Thus inosine (**487**) can be transformed by a purine nucleoside phosphorylase from *Enterobacter aerogenes*[476] in the presence of 1,2,4-triazole-3-carboxamide (**488**) to virazole (**489**).[477]

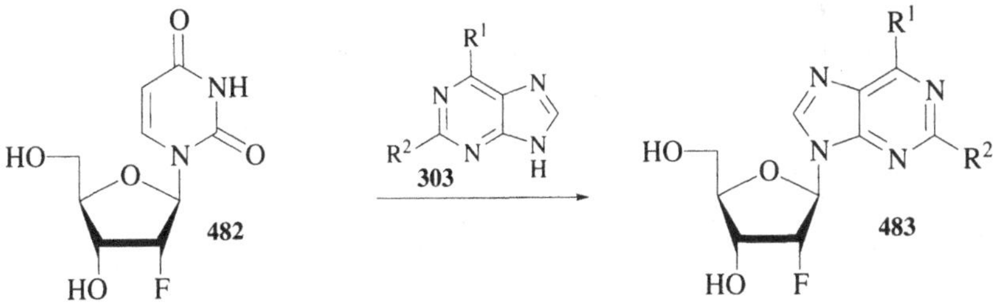

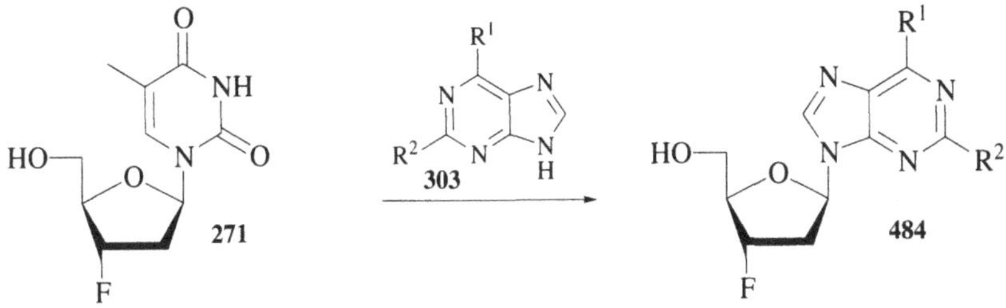

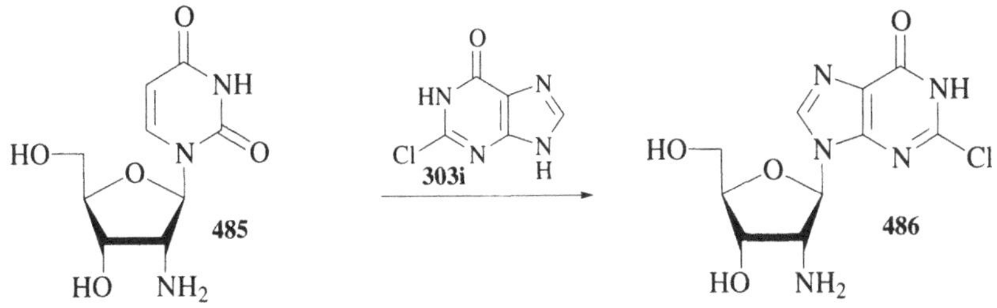

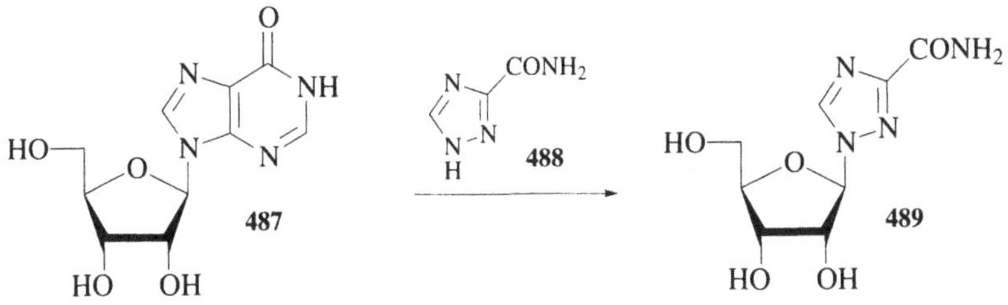

The enzymatic phosphorolysis of purine nucleosides such as inosine (**487**) or guanosine can be made irreversible and thus much more efficient by methylation of the N^7-nitrogen in **487** or **5** to afford in the presence of 1,2,4-triazole-3-carboxamide (**488**) 44% of virazole (**489**) or with 3-deazaadenine 53% of 3-deazaadenosine.[478]

In earlier studies, the enzyme nucleoside 2′-deoxyribosyltransferase (EC 2.4.2.6) from *Lactobacillus leichmannii* was used to catalyze large-scale transglycosylation reactions of excess thymidine (**477**) with 4-aminopyrazolo[3,4-*d*]-

pyrimidine (**490**) to **491** in 83% yield.[479] See also an analogous reaction with 6-dimethylaminopurine.[480] Excess 2′-deoxycytidine (**492**) and 6-thioguanine (**493**) or 6-azathymine (**495**) afford **494** and **496** in 67% and 21% yield, respectively.[479] The same enzyme system permits the preparation of a series of substituted 2′-deoxypurine nucleosides as well as of 1-deazapurine nucleosides.[481,482] Other bases used for enzymatic transglycosylations are benzimidazole and 5-aminoimidazole-4-carboxamide.[481,483]

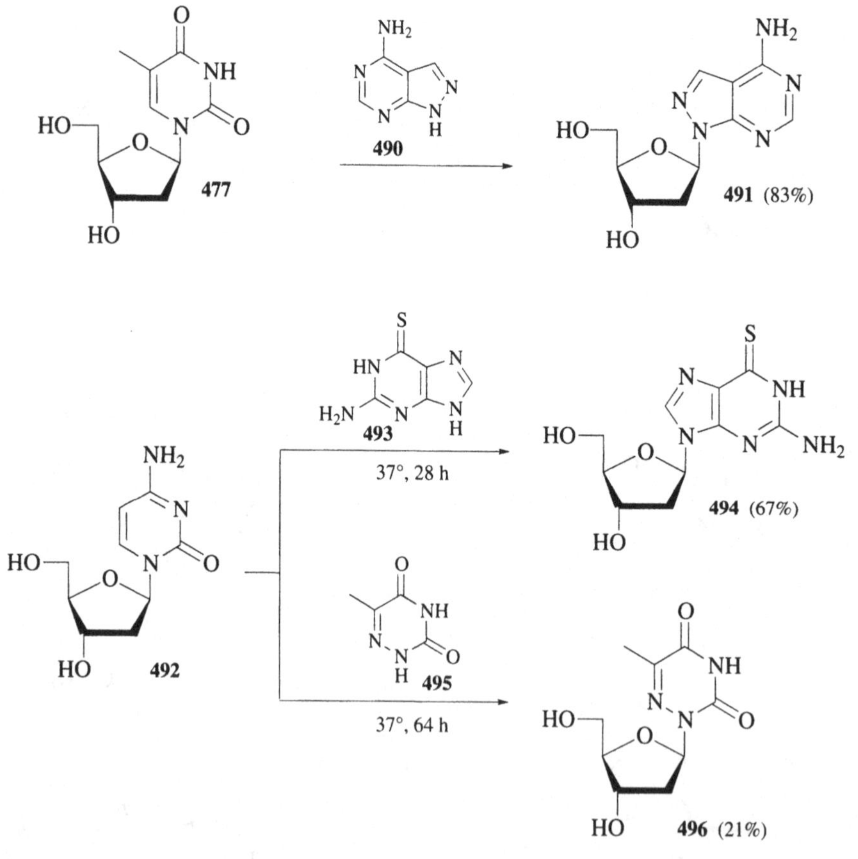

The nucleoside deoxyribosyltransferase II from *Lactobacillus leichmannii* also effects the transfer of the 2′,3′-dideoxyribose moiety from 2′,3′-dideoxycytidine (**497a**) to purine bases such as 6-dimethylaminopurine (**498**) to give **499** in 78% yield[484,485] and other bases such as 4-aminopyrazolo[3,4-*d*]pyrimidine (**490**)[484] to give **500**, whereas 6-alkoxypurines react with 3′-deoxythymidines in the presence of thymidine and purine phosphorylases to give the corresponding 6-alkoxypurine nucleosides.[486] Further, purine bases such as 1,7-dimethylguanine can also be used.[487] The analogous transglycosylation of 2′,3′-dideoxycytidine **497a**, 2′,3′-dideoxyuridine **497b**, or 2′,3′-dideoxythymidine **497c** with adenine (**2**) to 2′,3′-

dideoxyadenosine **501** in 71% yield can also be catalyzed by transferase enzymes from *E. coli*[488–491] or *Lactobacillus helveticus*.[492] Starting from the easily accessible uracil or cytosine arabinosides **300**, analogous enzymatic transglycosylations with a variety of purines such as **498** or 2-fluoro-6-aminopurine proceed in up to 60% yield.[382,382a]

	X	R
497a	NH	H
497b	O	H
497c	O	Me

498 → **499** (78%)

490 → **500**

2 → **501** (71%)

Although the preparation of these transferase enzymes is quite elaborate, once an enzyme preparation is isolated, it becomes possible to transfer the 2′-deoxyribose moiety of thymidine (**477**) or the 2′,3′-dideoxyribose moiety in 2′,3′-dideoxyuridine (**497b**) to a series of other bases without the formation of unnatural (and usually biologically inactive) α-nucleosides. Furthermore, modern techniques of attaching enzymes to polymers permit the extensive reuse of these polymeric enzymes.[492a] (For a review on enzyme catalysis, see Ref. 493.)

EXPERIMENTAL CONDITIONS

Sugar Moieties

The crystalline and commercially available sugar moieties such as **18** (mp 133–134°), **17** (mp 81–83°), **24** (mp 110–112°), and **23** (mp 130–132°) as well as 2,3,5-tri-*O*-benzyl-1-*O*-4-nitrobenzoyl-D-arabinofuranose (mp 89–91°) should be checked (mp, TLC) and carefully powdered and dried for 4–18 hours at 40–50°/0.1 mm to remove the last traces of solvent or acetic acid, which can interfere with nucleoside synthesis. If necessary, the crystalline sugar moieties should be recrystallized from methanol and dried subsequently. Sugar **18** should usually be preferred to **17**, since cyclic salts such as **67** with a phenyl substituent are much more stable than those with a methyl substituent as in **27**. Consequently, the thermodynamically controlled formation of *O*-acylated β-nucleosides is more favored with **18**. Furthermore *O*-benzoylated nucleosides usually crystallize much better than the corresponding *O*-acetylated nucleosides. The precious and sensitive **21** gives on recrystallization from anhydrous CCl_4 (100 mL/g) fine colorless needles, which can apparently be stored in a desiccator for months.[494] A new preparation of 1,3,5-tri-*O*-acetyl-2-deoxy-D-ribofuranose was recently described.[495]

On reacting 1-*O*-acyl-2-*O*-benzylated D-ribose **502** with $SnCl_4$ in the absence of nucleophilic silylated bases, the resulting 1-cation can undergo a Friedel-Crafts cyclization with the ortho position of the 2-*O*-benzyl group to form the corresponding tricyclic sugar derivative **503**[496,496a] (for other side reactions of sugar moieties in the presence of Lewis acids see Refs. 152b, 152c, and 256).

BnO— O O_2CR BnO O— **502** ⟶ BnO— O BnO O— **503**

Heterocycles

Whereas *N*-heterocycles such as uracil, thymine, or hypoxanthine are silylated as such and then transformed into the corresponding nucleosides, amino-substituted heterocycles such as cytosine, adenine, or guanine are much more basic so that the silylated amino-substituted bases form stronger σ complexes with Lewis acid catalysts such as TMSOTf or $SnCl_4$ and thus react more slowly or not at all with sugar moieties than do silylated uracil or thymine.

Consequently, the less basic *N*-acylated heterocycles such as N^4-acetyl (or benzoyl)cytosine,[497] N^6-benzoyladenine (mp 243°),[498] or the commercially available (Pharma Waldhof) N^2-acetyl (or N^2-isobutyryl)guanine should usually be preferred since they are also more lipophilic and thus more easily converted to their corresponding silylated derivatives **97**, **108b**, and **277**, which then form weaker σ complexes with Lewis acids with resulting faster reactions to the desired nucleosides. To enhance the solubility even further, silylated N^2-palmitoylguanine has been employed for transglycosylations[180] instead of silylated N^2-acetyl (or N^2-isobutyryl)guanine. Recently it was observed that 1-*O*-methyl-2-deoxypyranosides condense readily with silylated uracil (**66**) or thymine (**226**) but fail to react with silylated cytosine **222** in the presence of *tert*-butyldimethylsilyl triflate in CH_2Cl_2/acetonitrile.[236] One can assume that, owing to strong σ-complex formation of the basic silylated cytosine **222** with *tert*-butyldimethylsilyl triflate, the reaction of **222** with the intermediate sugar cation was so slow that the sugar cation was converted into the corresponding glycal.[152b,c,236] In the synthesis of hikizimycin, silylated cytosine (**222**) reacts only sluggishly in the presence of TMSOTf.[160] On working with such sensitive 2-deoxysugar moieties, the less basic *N*-acylated aminoheterocycles should always be employed using not only *N*-acetyl or benzoyl substituents but also *N*-*p*-nitrobenzoyl or *N*-trifluoroacetyl groups. The purification of crude heterocycles by silylation and subsequent distillation is briefly discussed in the following section on silylation.

Silylation

The polar, high melting and rather insoluble heterocyclic bases such as uracil, thymine, N^4-acetylcytosine, N^6-benzoyladenine, N^2-acetylguanine, or N^2-isobutyrylguanine are transformed on silylation into thermodynamically more stable[78a] lipophilic, basic, and nonpolar volatile silyl derivatives, which are thermally stable but very sensitive to moisture. Of the different silylating agents, HMDS, bp 126° is the most practical and commonly used since only ammonia is evolved on silylation, and excess reagent can be readily removed by evaporation and repeated codistillation with xylene. Even more effective is subsequent Kugelrohr (short path) distillation to give the pure silylated base. With 5-fluorocytosine, the silylated heterocycle was distilled and then recrystallized from heptane.[499] On silylation, distillation and subsequent desilylation with excess water or methanol, crude heterocyclic bases such as adenine can be readily purified.[500] The relatively low rate of silylation with HMDS is accelerated by adding catalytic amounts of acidic catalysts such as $(NH_4)_2SO_4$, TMSCl, or TMSOTf, whereupon ammonium salts such as NH_4Cl will show up in the reflux condenser as well as in the distilled silylated base as a turbid impurity. On silylation of N^4-acetylcytosine or N^6-benzoyladenine with HMDS in pyridine or in the presence of catalytic amounts of TMSCl, part of the *N*-acyl groups can be lost.[146] Thus samples of silylated *N*-acylated bases should be treated with methanol and subsequently checked by TLC for the potential loss of part of the *N*-acyl groups.

Very insoluble heterocyclic bases such as certain purines can often be silylated only by adding pyridine. Thus if a heterocyclic base does not dissolve after extended heating with HMDS containing catalytic amounts of TMSCl, a polar solvent such as acetonitrile or pyridine should be added, which can then be removed on codistillation with xylene. Other polar solvents such as DMF or *N*-methylpyrrolidone should also be considered as solvents for the silylation of very insoluble heterocyclic bases and the subsequent reaction with protected sugar derivatives such as **18** in the presence of Friedel-Crafts catalysts (see Solvents).[84] It should be realized, however, that DMF and probably also *N*-methylpyrrolidone can react with TMSCl at elevated temperatures (see Solvents).

The Langer method of silylation, employing equimolar amounts of HMDS and TMSCl,[501] is also very efficient and quite fast, particularly in acetonitrile at room temperature. The equivalent amounts of NH_4Cl formed are precipitated indicating the progress of silylation.[146] Subsequent filtration of the NH_4Cl with exclusion of moisture and washing with acetonitrile will remove practically all of the NH_4Cl. On Langer silylations in acetonitrile at reflux, the NH_4Cl sublimes nearly quantitatively into the reflux condenser and is thus removed from the reaction.[146] The Langer method transforms *N*-acylated purines such as N^6-benzoyladenine to the corresponding silylated bases such as **97** with minimal cleavage of the corresponding *N*-acyl groups.[146] Thus silylation of N^2-acetylguanine (**303b**) with HMDS and equivalent amounts of TMSCl or TMSOTf gives the silylated N^2-acetylguanine **108b** and NH_4Cl (or ammonium triflate). Silylations with TMSOTf/triethylamine[502] or TMSOTf/DBU[503] have also been described.

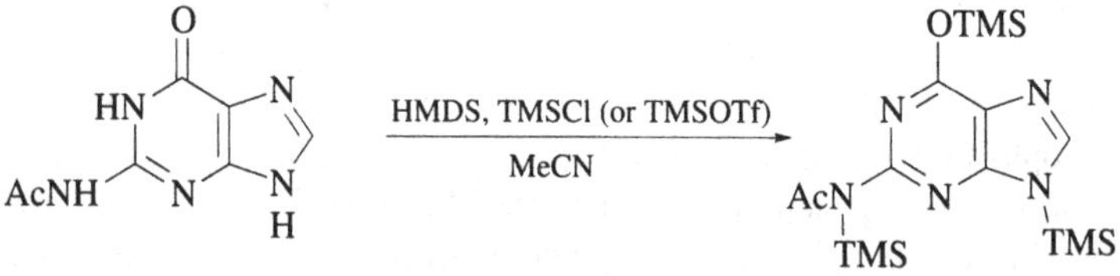

Other silylating agents such as the more expensive BSA or BSTFA silylate heterocyclic bases much faster than HMDS and have thus been used rather frequently.[151,162,329,504–507] During silylation with these reagents, however, *N*-monosilylated or free acetamide or trifluoroacetamide is formed, which can interfere with nucleoside synthesis by competing with persilylated bases for the sugar cation to form protected *N*-acetamides. This problem is particularly important in silylation and nucleoside synthesis in weakly basic systems, such as the preparation of silylated 2,3-diaminomaleodinitrile (DAMN, **130**) or silylated 4,6-diamino-5-nitropyrimidine **464**. Such a side reaction of BSA or *N*-trimethylsilylacetamide was recently observed on silylation of 1,2,4,6-thiatriazin-3-one 1,1-dioxides with BSA followed by reaction with peracylated sugars in the presence of TMSOTf in boiling acetonitrile, which gave only moderate yields of the protected nucleosides and up to 46% of the 1-*N*-β-acetamides of the protected sugars.[162a] Likewise, reaction of 1,3-bis(trimethylsilyl)-1,3,4,7-tetrahydro-2*H*-1,3-diazepin-2-one, prepared in situ with BSTFA, with 1-*O*-methyl-2,3-dideoxy-5-*O*-*p*-toluoyl-D-glyceropentofuranose in the presence of TMSOTf afforded the seconucleoside containing a 1-trifluoroacetamido group in 45% yield.[370f]

Whereas the structures of silylated uracil **66** or thymine **226** are unambiguous, the structures of the silylated N^4-acyl groups in cytosine or the N^2- and N^6-acyl moieties in purines can be formulated as either *N*-silyl **504** or *O*-silyl **505**.[78a,142,508,509] The structure of persilylated N^6-benzoyladenine was recently determined by NMR measurements as having one TMS group at N^9 and the other at the oxygen of the N^6-benzoyl group as in **505**.[509a]

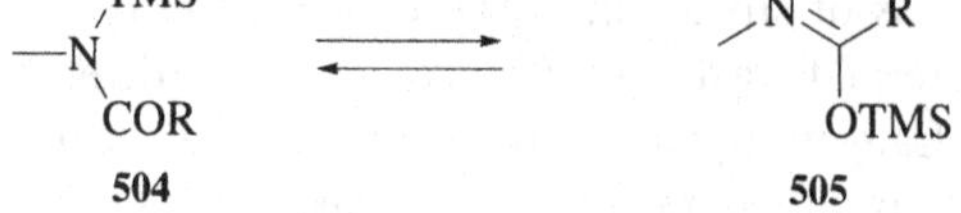

Although silylated purines are usually formulated as the N^9-silylated isomer **108**, it might well be that besides the N^9-silylated purines **108** there are also small amounts of N^7-silylated purines **506** present at equilibrium,[78a,144] which might be the intermediates in the reaction with sugar cations **67** to result in the formation of the N^9-nucleoside.

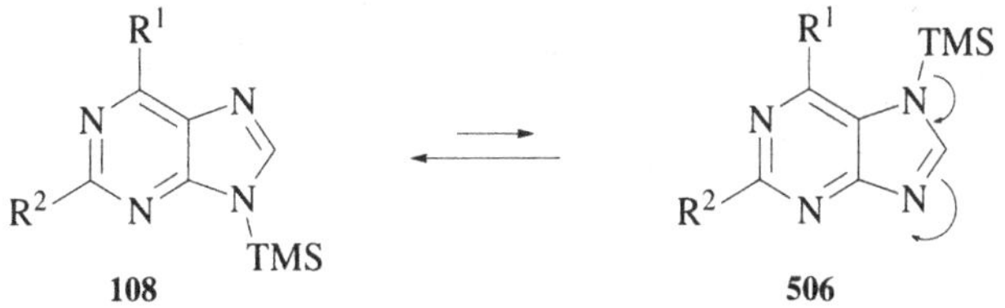

108 ⇌ **506**

For a detailed UV, ^{1}H NMR and ^{13}C NMR investigation of the structure of silylated allopurinol, see Ref. 510. Although the structure of the silylated diazepine was determined as **507** by ^{13}C NMR,[511] the heterocycle probably reacts via **508** with sugar cations **67**. For determination of the structure of bis(silylated) 5-methyl-5,6-dihydro-*sym*-triazine-2,4(1*H*,3*H*)dione (**209**), see Ref. 238. (For an alternative formulation of the bis(silylated) heterocycle see Ref. 512.)

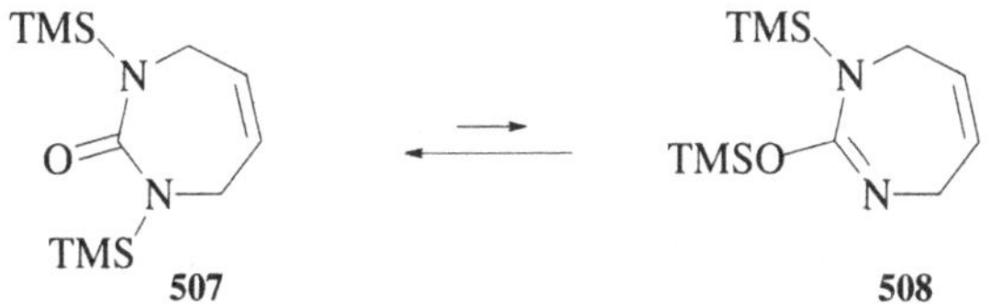

507 ⇌ **508**

Friedel-Crafts Catalysts

Commercial $SnCl_4$ should be redistilled (bp 114°) with careful exclusion of moisture if the reagent has been stored, to avoid slow nucleoside synthesis or failures. TMSOTf, which can be readily prepared from trifluoromethanesulfonic acid by heating with trimethylsilyl chloride (TMSCl)[131] until HCl evolution ceases, or with tetramethylsilane[513] until methane evolution ceases, should also be redistilled (bp 133–134° or 77°/80 mm) with careful exclusion of moisture, if the reagent has been stored. Trimethylsilyl nonaflate (bp 70°/15 mm) is prepared in similar fashion from nonaflic acid or by in situ reaction of potassium nonaflate with TMSCl in acetonitrile. Other reagents, such as *tert*-butyldimethylsilyl triflate (bp 65–67°/12 mm),[514] are apparently equivalent to TMSOTf in nucleoside synthesis but much more expensive.

Other Friedel-Crafts catalysts such as $TiCl_4$, $BF_3 \cdot OEt_2$, or TMSI should be redistilled before use. Humid samples of $ZnCl_2$ or $SnCl_2$ or their hydrates can be dehydrated by refluxing with TMSCl, whereupon HCl and hexamethyldisiloxane (bp 101°) are formed.[514a] For comparisons of Friedel-Crafts catalysts in nucleoside synthesis, see Refs. 84, 106b, 106c, 249c, and 514b.

The amount of Friedel Crafts catalyst used depends on the basicity of the silylated base and the sugar (cf **510**) and usually does not exceed 1.2–1.4 equivalents of TMSOTf or $SnCl_4$. Recently it was claimed, however, that 10 equivalents of $SnCl_4$ were necessary for the reaction of 1,3,5-tri-*O*-acetyl-2-deoxy-D-ribofuranose with silylated 2-*N*-acetyl-6-*O*-diphenylcarbamoylguanine **113** for optimal yields of 2′-deoxysugars.[515] But in this reaction BSA was used for silylation and it was not stated whether the $SnCl_4$ had been freshly distilled. Since 3,4- or 5-dialkoxyphosphonomethylribofuranosides are partially dealkylated at phosphorus on extended heating with persilylated bases in the presence of TMSOTf in acetonitrile, it is claimed that $SnCl_4$ gives higher yields of di-*O*-alkylated phosphonomethylnucleosides.[515a–d]

Solvents

Acetonitrile is the most commonly used low-boiling solvent (bp 82°) and is readily purified by heating at reflux over P_2O_5, subsequent distillation, and then distillation over CaH_2. As emphasized in the preparation of protected 6-methyluridine **85**, any impurity will favor the formation of the undesired N^3-nucleoside **86.** There are, however, some good commercial brands of absolute acetonitrile available such as from Merck AG Darmstadt, No. 100004 containing less than 0.005% H_2O, which can be used as such for nucleoside synthesis.

Owing to its polarity, acetonitrile permits homogeneous reactions of polar sugar moieties or silylated pyrimidine bases as well as of their corresponding salts. Acetonitrile also competes with silylated pyrimidine bases for the Lewis acids TMSOTf or $SnCl_4$ to form the corresponding σ complexes. Consequently, the most electron-rich N^1-nitrogen in silylated pyrimidines is only partially blocked by complex formation with TMSOTf or $SnCl_4$, so that the nucleophilic N^1-nitrogen can react with the sugar cation **67** to the desired protected natural N^1-pyrimidine nucleosides. If, however, TMSOTf or $SnCl_4$ blocks most of the N^1 nitrogen of silylated pyrimidines by σ-complex formation, only the less basic and reactive N^3 nitrogen is available to condense with **67** to give the undesired protected N^3-pyrimidine nucleosides.

The much less polar 1,2-dichloroethane, which is readily purified by distillation from P_2O_5 and permits reactions at the boiling point (83°), favors complex formation between the silylated purine moieties and TMSOTf so that the undesired protected N^3- and N^7-purine nucleosides are readily rearranged at 83° to the thermodynamically controlled desired protected natural N^9-purine nucleosides. Since these rearrangements proceed via dissociation of the protected N^3- and N^7-purine nucleosides to the persilylated purine bases and the sugar cations, the more stable sugar cation **67** derived from sugar **18** (compared to the less stable sugar cation **27** derived from sugar **17**) favors these dissociations and resulting rearrangements in 1,2-dichloroethane to the thermodynamically controlled N^9-purine nucleosides. The even less polar toluene (compared to 1,2-dichloroethane) was employed for the rearrangement of the N^7-purine nucleoside **509** to the N^9-purine nucleoside **114.**[141] It can furthermore be assumed that seconucleosides such as **132** or **332** will cyclize on extended heating with TMSOTf in boiling 1,2-

dichloroethane to the ring closed products **131** and **331.**[370] In contrast to these reactions and rearrangements in 1,2-dichloroethane and toluene at higher temperatures, silylated purine bases condense with sugar moieties such as **17** or **18** in acetonitrile in the presence of $SnCl_4$ at ambient temperature to give predominantly the undesired N^7-purine nucleosides.[142] See section on "Mechanism of Pyrimidine and Purine Nucleoside Synthesis."

Thus acetonitrile is the favored solvent for the TMSOTf or $SnCl_4$ catalyzed reaction of silylated pyrimidine bases with protected sugar moieties such as **18** to give the corresponding protected natural N^1-pyrimidine nucleosides, whereas 1,2-dichloroethane is the preferred solvent for the TMSOTf catalyzed reaction of silylated purine bases with **18** to the protected N^9-purine nucleosides.

Since the reaction temperature in methylene chloride can only be raised to 40°, this solvent is less suitable for the synthesis of purine nucleosides. Chloroform, which is the preferred solvent for the uncatalyzed,[261] CuI,[273] or $ZnCl_2$[261] catalyzed Silyl-Hilbert-Johnson synthesis of protected 2′-deoxynucleosides, should be freshly filtered over a small column of activated SiO_2 to remove the alcohol additives and water. Nitromethane, which should be dried with $CaCl_2$ and distilled (bp 101°), has been used as solvent for reactions of protected 1-halosugars with silylated bases in the presence of $AgClO_4$,[74,80] with 2,4-dimethoxypyrimidine[150] without catalysts as well as with free bases in the presence of $Hg(CN)_2$.[516–518] Nitromethane is comparable to acetonitrile in the reaction of silylated 5-fluorouracil and **18** in the presence of $SnCl_4$,[518] as well as for the synthesis of hikizimycin (**516**).[160] For a recent reaction of 4-amino-3-iodopyrazolo[3,4-*d*]pyrimidine with **18** and $BF_3 \cdot OEt_2$ in nitromethane, see Ref. 519. Applications of the very polar solvents DMF or *N*-methylpyrrolidone are discussed at the end of this section.

Whereas the rate of the reaction of sugar cations such as **67** with silylated heterocyclic bases is apparently only slightly affected by the solvent,[519a] there are additional factors that can influence nucleoside synthesis. Consequently, reactions of silylated pyrimidine bases in the presence of TMSOTf or $SnCl_4$ are often attempted in acetonitrile as well as in 1,2-dichloroethane or toluene, since some S_N2 reactions of protected D-arabinose **22** give much higher yields of the desired β-nucleosides in the nonpolar solvents 1,2-dichloroethane[359] and toluene. Likewise, the basic sugar moiety **510** reacts with silylated thymine **84** in 1,2-dichloroethane in the presence of excess TMSOTf to give 31% of the β-nucleoside **511** and 38% of the α-anomer **512**, whereas the same reaction in acetonitrile affords less than 5% of **511** and **512**.[520]

TMSOTf, $Cl(CH_2)_2Cl$, $(SnCl_4)$

+ α derivative **512**

In the reaction of the basic triazine **209** with the standard 2-deoxy sugar **21**, acetonitrile is deemed to be the optimal solvent to obtain the desired protected β-nucleoside **513**: 1,2-dichloroethane and nitromethane give lower yields of **513** and more α-nucleoside **514**.[239]

+ α derivative **514**

Protected gemcytabine (**278**) is prepared by heating the sugar moiety **276** with silylated N^4-acetylcytosine (**277**) in the presence of TMSOTf in 1,1,2-trichloroethane for 18 hours at 113°, whereas in xylene the reaction is complete after 3 hours at 125°.[303] In the synthesis of hikizimycin, the sugar moiety **515** reacts with excess silylated cytosine **222** and TMSOTf in nitrobenzene at 127° to furnish 76% of the protected nucleoside **516**.[160] This reaction, however, might succeed in boiling acetonitrile on employing the less basic silylated N^4-acetylcytosine **277** instead of **222**.

Silylated 6-chloro-2-acetylaminopurine (**517**) reacts with secosugar **518** at −30° in 1,2-dichloroethane to give after 2 hours the N^7-nucleoside **520** in 58% yield as well as ca. 6% of the N^9-nucleoside **519**, whereas 6-chloro-N^2,N^9-diacetylpurine affords with secosugar **518** in N-methylpyrrolidone or DMF in the presence of $BF_3 \cdot OEt_2$ or $SnCl_4$ after 4 hours at 100° 58% of the desired N^9-nucleoside **519**.[521] Longer exposure or heating with TMSOTf in 1,2-dichloroethane

NHTMS, N, N, OTMS, 222, TMSOTf, $PhNO_2$, 127°, 3.5 h

AcO AcO BnO H O OAc AcO AcO BnO N_3 OAc OAc **515**

NH_2 N N O AcO AcO BnO H O AcO AcO BnO N_3 OAc OAc **516** (76%)

might presumably result in the preferential formation of **519**. For recent reviews on the synthesis of seconucleosides such as **160** or **519**, see Refs. 521b–d.

Cl N N TMS N N N Ac TMS **517**

i-PrO OPr-*i* *i*-PrO O O OPr-*i* **518** Lewis acid $Cl(CH_2)_2Cl$, N Me O, DMF

Cl N N *i*-PrO N N NHAc *i*-PrO O **519** + N N NHAc *i*-PrO N N *i*-PrO O Cl **520**

Reaction of neat 1-α-bromo-2-deoxy-2-fluoro-3,5-di-*O*-benzoyl-4-thio-D-arabinofuranoside with neat silylated N^4-acetylcytosine (**277**) for 8 hours at 80° affords the corresponding protected β-cytidine analog in 33% yield besides 8% of the α-anomer, whereas the same reaction for 48 hours at 80° in 1,2-dichloroethane did not give any nucleoside.[521e]

Possible side reactions of solvents, such as Ritter reactions[199–204] of sugar cations with acetonitrile, should always be considered. Likewise, the very polar solvent DMF[84,86,90,312b,521a] reacts on extended heating with TMSCl to give vinylogous amidinium salts and hexamethyldisiloxane.[522] Finally, one should be aware that 1,2-dichloroethane is carcinogenic to rats and mice at high doses[523] and should therefore only be used in a well ventilated hood. Acetonitrile is likewise considered a toxic hazard[523] and should therefore also be handled with care.

For a general review on the influence of solvents on reactivity, see Ref. 523a.

Workup of Friedel-Crafts-Catalyzed Silyl-Hilbert-Johnson Reactions

TMSOTf or trimethylsilyl nonaflate are converted on aqueous workup with ice-cold saturated sodium bicarbonate solution into the corresponding sodium salts, which are water soluble and do not interfere with the subsequent extraction of the aqueous phase with CH_2Cl_2. On employing potassium bicarbonate (or carbonate), crystalline potassium nonaflate can be recovered on concentration of the aqueous solution for reuse in the one step-one pot nucleoside synthesis.

In contrast to the use of TMSOTf or nonaflate, the workup of reactions employing $SnCl_4$ (or $TiCl_4$) with ice-cold aqueous sodium bicarbonate usually gives rise to emulsions, which are often difficult to extract with CH_2Cl_2. In these cases, the crude reaction mixture should be filtered through a layer of Celite (or Kieselguhr) to remove the insoluble tin salts, which should be washed thoroughly with CH_2Cl_2. The combined filtrates can then be separated and the aqueous phase extracted with additional volumes of CH_2Cl_2. In certain cases the precipitated tin salts obtained on workup with ethanol/aqueous $NaHCO_3$ solution were subsequently extracted with CH_2Cl_2 in a Sohxlet extractor to avoid any loss of precious substance.[524]

If the $SnCl_4$-catalyzed reactions are run in 1,2-dichloroethane, addition of equivalent amounts of pyridine leads to a colorless precipitate of a pyridine-$SnCl_4$ σ complex, which can be readily filtered through a layer of Celite and washed with 1,2-dichloroethane or CH_2Cl_2. The subsequent workup with ice-cold $NaHCO_3$-CH_2Cl_2 proceeds without complications.[525,526] Other authors have added an ethanolic solution of triethylamine to the reaction mixture of $SnCl_4$ in 1,2-dichloroethane and then evaporated the volatile fraction. The resulting syrup was stirred in chloroform and evaporated with silica gel, which was then placed on top of a silica gel column for subsequent chromatography.[527,528] For the synthesis of very acid-sensitive 2-oxo-6-chloropurine nucleosides employing TMSOTf in CH_2Cl_2, pyridine was added to quench the Lewis acid before chromatography on silica.[523b,c]

Removal of *O*-Acyl, *N*-Acyl, *O*-Benzyl, or *O*-Silyl Protecting Groups

After nucleoside bond formation, workup and crystallization or chromatography followed by crystallization (if they crystallize), the *O*-acyl, *N*-acyl, *O*-benzyl, or *O*-silyl groups are normally removed. *O*-Benzyl groups can be removed by BCl_3[66,344,349] or BBr_3[256] at −78°, by sodium in liquid ammonia,[343] or by hydrogenation.[68,341] *O*-Silyl groups are usually cleared by treatment with TBAF in tetrahydrofuran,[339,529] by triethylamine hydrofluoride, by pyridine hydrofluoride, or by treatment with trifluoroacetic acid.[68] For a review of the removal of silyl groups, see Ref. 529a.

The most common procedure used for the removal of *O*- and *N*-acyl groups is transesterification-saponification with saturated methanolic ammonia. On saponification of a protected nucleoside with methanolic ammonia, the progress of the saponification can be followed by TLC or HPLC, whereupon the heterocyclic chromophore can be detected by UV light and the sugar moiety by spraying with 10% H_2SO_4 in ethanol and subsequent heating to 140° to induce darkening of spots containing the carbohydrate moiety. After 72 hours at 24° normally all the *O*- and *N*-acyl groups are removed to give the free nucleoside, which can be checked for purity on SiO_2-TLC plates using the upper phase of *n*-butanol/acetic acid/H_2O (4:1:5)[81] or by RP-HPLC.

In addition to the commonly used methanolic (or ethanolic) ammonia, methylamine, diisopropylamine, triethylamine, and hydrazine have also been employed.[221,298,409,475,530–532] Saponification of 6-(2′,3′,5′-tri-*O*-acetyl-β-D-ribofuranosyl)-6-aza-5,6-dihydro-5,5-dimethyluracil with methanolic ammonia leads to rearrangement of the sugar moiety and gives 6-(β-D-ribopyranosyl)-6-aza-5,6-dihydro-5,5-dimethyluracil.[533]

However, saponification-transesterification with methanolic ammonia can sometimes be quite slow so that even after 72 hours/24° *O*-acylated nucleosides can still be detected by TLC. In the case of 2′,3′,5′-tri-*O*-benzoyl-5-fluorocytidine, the ion exchanger Amberlyst A-26 (OH^- form) in methanol was used to effect a more rapid removal of the *O*-benzoyl groups.[534] Compare also the analogous saponification-transesterification of *O*-acetyl groups with IRA-400 (OH^- form) in methanol.[535] The saponification-transesterification of *O*-acetyl or *O*-benzoyl groups in nucleosides with $NaHCO_3$, Na_2CO_3, or K_2CO_3 in absolute methanol at −50° was recently suggested.[536,536a] Alternatively, $ZnBr_2$ in chloroform-methanol removes *N*-acyl groups selectively,[536b] whereas lipases selectively saponify *O*-acyl groups.[541]

After completion of the saponification-transesterification, the methanolic (ethanolic) ammonia (methylamine, diisopropylamine) is evaporated in vacuo. In the case of *O*-benzoyl, 4-nitro, 4-chloro, or 4-methylbenzoyl groups the rates of the saponification-transesterification differ, and the crude product contains the corresponding methyl (or ethyl) benzoates, as well as the corresponding amides, which can be readily extracted with diethyl ether or methyl *tert*-butyl ether.

On using hydrazine benzoate or hydroxylamine acetate in pyridine the more acidic 2′-*O*-acetates or benzoates can be selectively cleaved to give, with adenosine-2′,3′,5′-tri-*O*-benzoate, a 74% yield of adenosine-3′,5′-di-*O*-benzoate.[328,329,532,537] Heating with hydrazine to 100° leads to rapid removal of all *O*-acyl groups.[475]

The standard procedure with methanolic ammonia fails, however, with 5-nitrouridine-tri-*O*-benzoate,[538] which apparently decomposes via addition of NH_3 to the 6-position and subsequent ring opening. Sodium methylate in methanol, however, gives a high yield of 5-nitrouridine.[538,538a] Ammonia and primary or secondary amines in methanol (or ethanol) can aminate (or methoxylate) a 2-chloro or 2-fluoro group,[475,539] and under forcing conditions, a 6-chloro-,[298] 6-bromo-, or 6-fluoro group in purine nucleosides as well as a 6-fluoro group[540] in pyrimidine nucleosides, whereas NaOH in methanol-H_2O will introduce a methoxy group.[539] Likewise, a 5-trifluoromethyl group in pyrimidine nucleosides is transformed into the corresponding ester moiety,[268] whereas a 5-iodo substituent in a uracil moiety can be displaced by ethanolic methylamine after 20 hours at 20°.[530] Heating 2-chlorohypoxanthine nucleosides for only 5 minutes at 100° with hydrazine hydrate replaces the chloro group by a hydrazino group.[475] Ester groups react with methanolic ammonia to give the corresponding amide groups as in the synthesis of virazole (**489**)[133,259] and **186**. 5-Ethoxycarbonyluridine-2′,3′,5′-tri-*O*-acetate is also converted into 5-aminocarbonyluridine by methanolic ammonia.[541] It is of interest that cytidine catalyzes the aminolysis of esters by *n*-butylamine by stabilizing the tetrahedral transition state of aminolysis.[542]

When sodium methylate in methanol is used, treatment of the crude reaction mixture with CO_2 or filtration over a column of Dowex 50 (H^+ form) will neutralize the reaction mixture. The ion exchanger will also remove the sodium ions. On using the sodium methylate/methanol or the sodium benzyl oxide/benzyl alcohol procedure, reactive halogens as in 5,6-dichloropyridazin-3-one[543] or analogous nucleosides[544] are replaced by methoxy or benzyloxy groups.

Since the 2-fluorine moiety in 3′,5′-di-*O*-acyl-2,2′-difluoroadenosine **347** (R^1 = NH_2; R^2 = F) is readily aminated by methanolic ammonia, the more selective saponification of the *O*-acyl groups with lithium hydroxide in acetonitrile-H_2O was applied to afford the free nucleoside in 59% yield.[539] *O*-Benzoylated ribosides of barbituric acid can only be saponified without cleavage of the nucleoside bond by using a very dilute solution of sodium methylate in methanol.[544]

In contrast to methanolic ammonia, saponification with 0.2 N NaOH in THF/MeOH/H_2O (5:4:1) selectively removes the *O*-acetyl moieties in protected cytidines and adenosines, while not affecting the N^4- or N^6-benzoyl groups in the cytosine or adenine moieties.[192]

In 2′-*O*-acetyl-3′-*O*-mesyl-5′-*O*-methoxycarbonyl-1-β-D-xylofuranosylthymine the 2′-*O*-acetoxy group can be selectively removed in 75% yield by treatment with methanolic HCl at 24° for 72 hours.[545] Methanolic HCl also selectively removes the 3′- and 5′-*O*-acetyl groups in 2′-deoxy-5-(2,2-difluorovinyl)uridine, whereas methanolic K_2CO_3 leads to addition of methanol to the 5-(2,2-difluorovinyl) group.[546,547] Since methanolic ammonia cleaves the *N*-substituted succinimide ring in 1β-(2′,3′,5′-tri-*O*-acetyl-D-ribofuranosyl)-1*H*-pyrrole-2,5-dione, methanolic HCl had to be used for the selective removal of the *O*-acetyl protecting groups.[548] The reaction of ammonia with *O*-acyl groups[549] as well as the selective removal of the protecting groups[550] in carbohydrate chemistry has been reviewed.

Melting Points of Free Nucleosides

Because of the many possible hydrogen bonds between nucleoside molecules in the crystalline state, repeated recrystallization from different solvents or solvent mixtures can give rise to different crystal forms (polymorphism), which have different melting points. Thus one should not be surprised if one obtains crystals with different and usually higher melting points on repeating certain nucleoside syntheses (see the preparation of 6-methyluridine, lumazine riboside, or ribavirin in Ref. 133).

EXPERIMENTAL PROCEDURES

Unless indicated otherwise, the following experiments were carried out in a three-neck round-bottom flask equipped with a magnetic stirring bar, dropping funnel, thermometer, reflux condenser, and nitrogen inlet.

A standard TLC system for analysis of crude protected nucleosides is ethyl acetate : methanol (5 : 1). Optimal separations of crude protected nucleosides are often observed with two-phase partition systems such as toluene : acetic acid : H_2O (5 : 5 : 1) or *n*-butyl acetate : methyl glycol : H_2O (4 : 1 : 2) for the protected nucleosides, whereas the free nucleosides often show good separations with the two-phase system *n*-butanol : acetic acid : H_2O (5 : 1 : 4). In order to obtain reproducible results, the two-phase systems should be placed into the chromatography vessel. To avoid any contact of the silica layer of the TLC plates with the stationary aqueous lower phase, the lower part of the silica plates should be scraped off so that the silica layer will only come in contact with the upper, mobile phase of these partition systems.[81,84]

The best analytical, as well as preparative, separations of mixtures of free, unprotected nucleosides (especially mixtures of α/β anomers) can be achieved by HPLC on reversed phase (RP) columns (e.g., Nucleosil 7 C_{18} columns) using linear gradients of 100 to 50% water with 0 to 50% methanol or acetonitrile.

For supplies of special carbohydrate building blocks or larger amounts of particular carbohydrate derivatives such as **18**, **17**, or N^2-acetylguanine (**303b**) contact:

Ajinimoto Co., Inc. 1–15–1 Kyobashi, Chuo-ku, Tokyo 104, Japan
Phone +81–3–5250–8111; Fax +81–3–5250–8293

Berry & Assoc., Inc. P.O. Box 1071, Ann Arbor, MI 48106, USA
Phone +1–734–426–3787, Fax +1–734–426–9077

Boehringer Ingelheim KG, Chem. Division, Bingerstr. 137, 55216 Ingelheim, Germany
Phone +49–6132–772192; Fax +49–6132–773755

Cambridge Research Biochemicals, Gadbrook Park Northwich, Cheshire CW9 7RA, U.K.
Phone +44–606–41100; Fax +44–606–49366

Omnichem S. A., Indust. Res. Park-Fleming, 1348 Louvain-La Neuve, Belgium
Phone +32–10–483111; Fax +32–10–450693

Pharma-Waldhof GmbH, Postfach 110732, 40507 Düsseldorf, Germany
Phone +49–211–526020; Fax +49–211–5260211

Pro.Bio.Sint S.r.I., Via Valverde 20/22, 21100 Varese, Italy
Phone +39–332–218136; Fax +39–332–212575

Pfanstiehl Laboratories, Inc., 1219 Glen Rock Avenue, Waukegan, IL 60085–0439, USA
Phone +1–800–383–0126, Fax +1–708–623–9173

Reliable Biopharmaceutical Corp., P.O. Box 2517, St. Louis, MO 63114, USA
Phone +1–314–429–7700, Fax +1–314–429–0937

Yamasa Corporation, 1–23–8 Kakigarocho, Nihonbashi, Chuo-Ku, Tokyo 103, Japan
Phone +81–3-3668–3366; Fax +81–3-3668–3177

OTMS, NO_2, N, TMSO, N, **80a** → **18**, $Cl(CH_2)_2Cl$, $SnCl_4$, 20°, 1 h → O, O_2N, NH, BzO, N, O, O, BzO, OBz, **81a** (96%)

2′,3′,5′-Tri-*O*-benzoyl-5-nitrouridine (81a).[84] A mixture of 7.85 g (50 mmol) of 5-nitrouracil, 42 mL (200 mmol) of HMDS, and 1 mL of TMSCl was heated at reflux for 1 hour, whereupon the 5-nitrouracil had passed into solution. The excess HMDS (bp 126°) was removed by distillation and the residue was distilled at 110–111° (0.2 Torr) in a Büchi-Kugelrohr short-path distillation apparatus with careful exclusion of moisture to afford 14.3 g (95%) of silylated 5-nitrouracil[80a] as a viscous liquid. This product was dissolved in 100 mL of 1,2-dichloroethane (distilled from P_4O_{10}) to give a 475 mM standard solution.

To a solution of 5.04 g (10 mmol) of **18** in 75 mL of 1,2-dichloroethane, 23 mL (11 mmol) of a standard solution of silylated 5-nitrouracil in 1,2-dichloroethane and 0.35 mL (3 mmol) of redistilled (bp 114°) $SnCl_4$ in 40 mL of 1,2-dichloroethane were added with stirring. After 1 hour the reaction mixture was diluted with 150 mL of CH_2Cl_2 and shaken with 200 mL of saturated $NaHCO_3$. After filtration through a layer of Celite to remove the tin salts and repeated washings of the Celite layer with CH_2Cl_2, the layers were separated and the aqueous phase was extracted with 3 × 100 mL of CH_2Cl_2. The combined organic phase was dried (Na_2SO_4) and evaporated to give 6.9 g of crude product, which was recrystallized from 200 mL of ethanol to furnish in four crops 5.79 g (96%) of pure 2′,3′,5′-tri-*O*-benzoyl-5-nitrouridine (**81a**), mp 184–185°.

N, OTMS → **18**, $Cl(CH_2)_2Cl$, $SnCl_4$, 20°, 4 h → HO, N, O, O, HO, OH (85%)

1-β-D-Ribofuranosyl-2-pyridin-2-one.[99] A mixture of 9.51 g (100 mmol) of purified pyridin-2-one and 80 mL (400 mmol) of HMDS was heated at reflux for 1 hour in an oil bath with magnetic stirring and evolution of NH_3, whereupon the heterocyclic base passed into solution. After distillation of the excess HMDS, the residue was distilled in a Büchi short-path Kugelrohr apparatus at 65°/12 mm to afford 15.9 g (95%) of 2-trimethylsilyloxypyridine, which was dissolved in 200 mL of 1,2-dichloroethane. After addition of 50.4 g (100 mmol) of **18**, a solution of 14.0 mL (120 mmol) of redistilled (bp 114°) $SnCl_4$ in 300 mL of 1,2-dichloroethane was added rapidly with stirring and exclusion of moisture. After 4 hours the reaction was practically complete according to TLC [upper phase of toluene:acetic acid:H_2O (5:5:1)]. The reaction mixture was poured onto a mixture of 600 mL of ice-cold CH_2Cl_2 and 750 mL of saturated $NaHCO_3$ solution. The crude reaction mixture was filtered through a layer of Celite, and the Celite layer was carefully washed with CH_2Cl_2. The filtrate layers were separated and the aqueous phase was extracted with 4 × 250 mL of CH_2Cl_2. The combined organic phase was dried (Na_2SO_4) and evaporated to give 62 g of a crude oil, which was recrystallized from 900 mL of CCl_4 to give in several crops 45.6 g (85%) of pure 1-(2′,3′,5′-tri-*O*-benzoyl-β-D-ribofuranosyl)pyridin-2-one, mp 140–142°.

The 2′,3′,5′-tri-*O*-benzoate (2.75 g, 5 mmol) was dissolved in 150 mL of saturated methanolic ammonia, stirred for 24 hours at 24°, and evaporated. The residue was dissolved in 100 mL of H_2O and the aqueous phase was extracted with 3 × 75 mL of methyl *tert*-butyl ether to remove the methyl benzoate and benzamide. The aqueous solution was evaporated and the residue was recrystallized from 20 mL of ethanol: 2-propanol (3:1) to give in two crops 0.9 g (80%) of 1-β-D-ribofuranosyl-2-pyridin-2-one, mp 147–150°.

OTMS, OMe, TMSO — **18**, $Cl(CH_2)_2Cl$, TMSOTf, 24°, 4 h → MeO, NH, O, BzO, N, O, BzO, OBz **81b** (89%)

5-Methoxyuridine-2′,3′,5′-tri-*O*-benzoate (81b).[133,189] (a) To a stirred solution of 5.04 g (10 mmol) of **18** in 75 mL of 1,2-dichloroethane, 34 mL of a 0.356 N standard solution (11 mmol) of silylated 5-methoxyuracil (**80b**) in 1,2-dichloroethane and 22.8 mL of a 0.522 N standard solution of TMSOTf in 1,2-dichloroethane were added under nitrogen with exclusion of moisture. After 4 hours at 24° and dilution with CH_2Cl_2, the solution was shaken with excess ice-cold saturated $NaHCO_3$ solution and the aqueous phase was extracted with 3 × 50 mL of CH_2Cl_2. The combined organic phase was dried (Na_2SO_4) and evaporated to afford the crude protected nucleoside, which crystallized from ethyl acetate-hexane to give 5.24 g (89%) of pure crystalline **81b**, mp 205–207°, homogeneous on TLC (ethyl acetate:methanol 5:1). (b) To a mixture of 0.53 g (5 mmol) of 5-methoxyuracil, 2.52 g of **18**, and 3.84 g (12 mmol) of potassium nonaflate ($C_4F_9SO_3K$) in 50 mL of acetonitrile were added 0.74 mL (3.5 mmol) of HMDS and 1.89 mL (15 mmol) of trimethylchlorosilane. The mixture was heated at reflux for 20 hours. After workup with ice-cold $KHCO_3/CH_2Cl_2$, the biphasic mixture was filtered to afford ~3 g (80%) of recovered potassium nonaflate. The combined organic phase was dried (Na_2SO_4) and evaporated, and the crude product was recrystallized from ethyl acetate-hexane to give 2.09 g (71%) of pure crystalline 5-methoxyuridine-2′,3′,5′-tri-*O*-benzoate (**81b**), mp 205–207°.

108b → 1. **18**, $Cl(CH_2)_2Cl$, TMSOTf, reflux; 2. NH_3, MeOH → **5** (66%)

Guanosine (5).[133] To a stirred mixture of 13.5 mL (4.09 mmol) of a 0.303 N standard solution of silylated N^2-acetylguanine (**108b**) in 1,2-dichloroethane and 1.86 g (3.7 mmol) of **18** in 35 mL of 1,2-dichloroethane was added 6.32 mL (4.46 mmol) of a 0.705 N standard solution of TMSOTf in 1,2-dichloroethane. The reaction mixture was heated at reflux for 1.5–4 hours, and then diluted with CH_2Cl. On workup with ice-cold $NaHCO_3$ solution, there was obtained 2.32 g of crude product, which was kept for 42 hours in 125 mL of methanolic ammonia at 24°. After workup, recrystallization from H_2O gave, in two crops, 0.69 g (66%) of pure guanosine 5 which was homogeneous (R_f 0.3) in the partition system *n*-butanol: acetic acid: H_2O (5:1:4) and whose ^1H NMR spectrum at 400 MHz in D_2O showed only traces of the undesired N^7-anomer of guanosine.

HMDS, TMSCl, $SnCl_4$, MeCN, 45°, 20 h → (65%) + α-isomer (18%)

1-(4′,6′-Di-*O*-acetyl-2′,3′-dideoxy-β-and-α-D-glucopyranosyl)thymine.[192] To a stirred suspension of 2.15 g (8.75 mmol) of pure oily methyl 4,6-di-*O*-acetyl-2,3-dideoxy-α-D-glucopyranoside, 1.21 g (9.5 mmol) of thymine, and 2.0 mL (9.59 mmol) of HMDS in 70 mL of acetonitrile, 1.35 mL (10.5 mmol) of trimethylchlorosilane, and 1.3 mL (11 mmol) of redistilled $SnCl_4$ were added carefully. The reaction mixture was heated at 44° for 22 hours, then evaporated in vacuo. The crude product was taken up with 150 mL of ice-cold ethyl acetate and 100 mL of saturated $NaHCO_3$ solution. After filtration through a layer of Celite, the aqueous phase was extracted with 2 × 150 mL of ethyl acetate, and the combined ethyl acetate solution was washed with 150 mL of saturated NaCl solution, dried (Na_2SO_4), and evaporated to give 2.5 g of crude product, which contained four compounds (TLC, EtOAc). Chromatography on a 5 × 36-cm column of SiO_2 and elution with hexane:ethyl acetate (3:1) afforded 1.93 g (65%) of 1-(4,6-di-*O*-acetyl-2′,3′-dideoxy-β-D-glucopyranosyl)thymine and 0.53 g (18%) of 1-(4,6-di-*O*-acetyl-2′,3′-dideoxy-α-D-glucopyranosyl)thymine as well as 4% of the corresponding N^3-β and 4% of the N^3-α-nucleosides.

Cl N N N Na Cl + TolO O Cl TolO **21** —MeCN, 50°, 2 h→ TolO O TolO N N N Cl Cl **41** (66%)

2,6-Dichloro-1-(2′-deoxy-3′,5′-di-*O*-*p*-toluoyl-β-D-*erythro*-pentofuranosyl)imidazo-[4,5-*c*]pyridine (41).[64] To a stirred suspension of 0.61 g (3.2 mmol) of 2,6-dichlorimidazo-[4,5-*c*]pyridine in 25 mL of acetonitrile was added 0.17 g (3.5 mmol) of 50% oily immersion of NaH under nitrogen. After 30 minutes at room temperature, 1.38 g (3.5 mmol) of **21** was added and the reaction mixture was stirred for 2 hours at 50°. Filtration and evaporation of the filtrate afforded the crude oily product, which was chromatographed in toluene:acetone (95:5) on a 4 × 40-cm column of SiO_2 to give 1.15 g (66%) of pure **41**, mp 165–167°.

OMe N MeS N N H **48** + BnO O BnO Br BnO **22** —TEBA, CH_2Cl_2, NaOH (50%)→ BnO O BnO BnO N N N OMe SMe **49** (63%) + α anomer (21%)

4-Methoxy-2-(methylthio)-7-(2′,3′,5′-tri-*O*-benzyl-β-D-arabinofuranosyl)pyrrolo-[2,3-*d*]pyrimidine (49).[66] Dry hydrogen bromide was bubbled into a solution of 3.5 g (6.15 mmol) of commercial 1-*O*-*p*-nitrobenzoyl-2,3,5-tri-*O*-benzyl-D-arabinofuranose (Pfanstiehl Laboratories, Inc.) in 15 mL of dry CH_2Cl_2 until no further *p*-nitrobenzoic acid precipitated. The acid was filtered and washed with a small volume of dry CH_2Cl_2 and the combined filtrate was evaporated to give **22**. The 1-bromosugar **22** was dissolved in 10 mL of CH_2Cl_2 and added to a suspension of 1.0 g (5.1 mmol) of 4-methoxy-2-(methylthio)-7*H*-pyrrolo[2,3-*d*]pyrimidine (**48**) in 10 mL of CH_2Cl_2. After addition of 0.3 g (1.1 mmol) of benzyltriethylammonium chloride and 30 mL of 50% NaOH, the mixture was stirred vigorously for 15 minutes with a vibromixer. The organic layer was separated, washed with water, dried (Na_2SO_4), and evaporated. The viscous residue was chromatographed using $CHCl_3$ on a 5 × 70-cm column of SiO_2. After elution of 0.63 g (21%) of the pure α anomer **50**, mp 62–63° (methanol), 1.94 g (63%) of the desired viscous β anomer **49** was obtained.

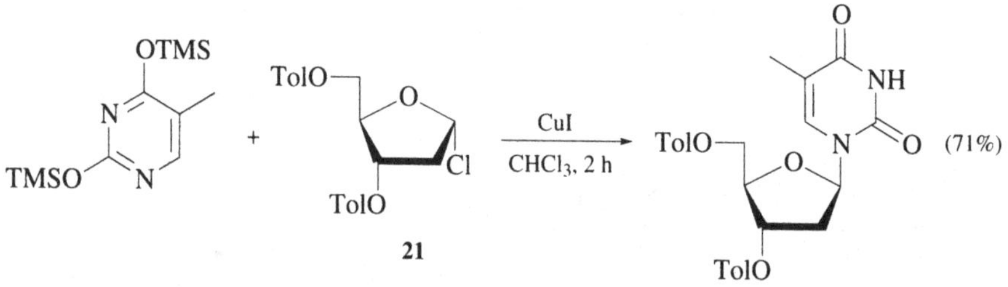

3′,5′-Di-*O*-toluoylthymidine.[273] To a stirred solution of 1.2 g (3.1 mmol) of **21** and 1.0 g (3.4 mmol) of bis(trimethylsilyloxy)thymine in 80 mL of dry $CHCl_3$, 0.60 g (3.1 mmol) of anhydrous CuI was added. After 2 hours the reaction mixture was treated with 60 mL of saturated $NaHCO_3$ solution and filtered through a layer of Celite. After reextraction of the aqueous phase with 50 mL of CH_2Cl_2, the combined organic phase was dried (Na_2SO_4) and concentrated to give 1.4 g (92%) of a white solid with a $\beta:\alpha$ ratio of 93:7 (^{1}H NMR). Stirring of the solid with 40 mL of ethanol, filtration, and washing with 2 × 15-mL portions of ethanol gave 1.1 g (71%) of the pure β anomer, mp 195–196°.

Tabular Survey

The following tables include examples of nucleoside synthesis collected during the last 20 years up to the middle of 1994. Subsequently only a few additional papers and patents are cited. Searches were conducted in the *Chemical Abstracts* databases and the *Science Citation Index*. Because of the enormous extent of literature dealing with the synthesis of nucleosides, however, this comprehensive collection of data cannot be guaranteed complete.

The tables include reactions of free sugars or sugar derivatives with heterocyclic bases and their derivatives. Transglycosidation reactions as well as reactions of sugars to build up the pyrimidine or purine part of the nucleosides, such as the "Sanchez-Orgel"-type reaction, are **not** part of the tables but are covered in the text.

The tabular survey is divided in the following headings:

Table I: One Step-One Pot Silylation/Coupling of Heterocyclic Bases with Sugars–Friedel-Crafts Catalysts

Table II: One Step-One Pot Coupling of Heterocyclic Bases with Sugars–Friedel-Crafts Catalysts

Table III: Reactions of Silylated Heterocyclic Bases with Protected Sugars–$SnCl_4$ Catalyst

Table IV: Analogous Reactions with Trimethylsilyl and Silver Triflates and Perchlorates

Table V:	Reactions with Titanium Tetrachloride as Catalyst
Table VI:	Reactions with Boron Trifluoride Etherate as Catalyst
Table VII:	Reactions with Miscellaneous Friedel-Crafts Catalysts
Table VIII:	Reactions of Silylated Bases with Protected Sugars with or without Catalysts
Table IX:	Fusion Reactions
Table X:	Miscellaneous Reactions of Heterocyclic Bases with Protected Sugars
Table XI:	Reactions of Acidic Heterocycles with 1-Halosugars in the Presence of Bases

Within each table, sugar derivatives reacting with heterocyclic bases are listed according to increasing carbon number, and increasing hydrogen number within a given carbon number.

For sugar derivatives having the same carbon and hydrogen number, the sugars are listed according to increasing ring size. The order of the reactants (heterocycles) is as follows:

Number of Rings

⇓

Ring size (5-ring > 6 -ring ..)

⇓

Number of N atoms in the Ring (pyridine > pyrimidine...)

⇓

N-pattern in the heterocycles

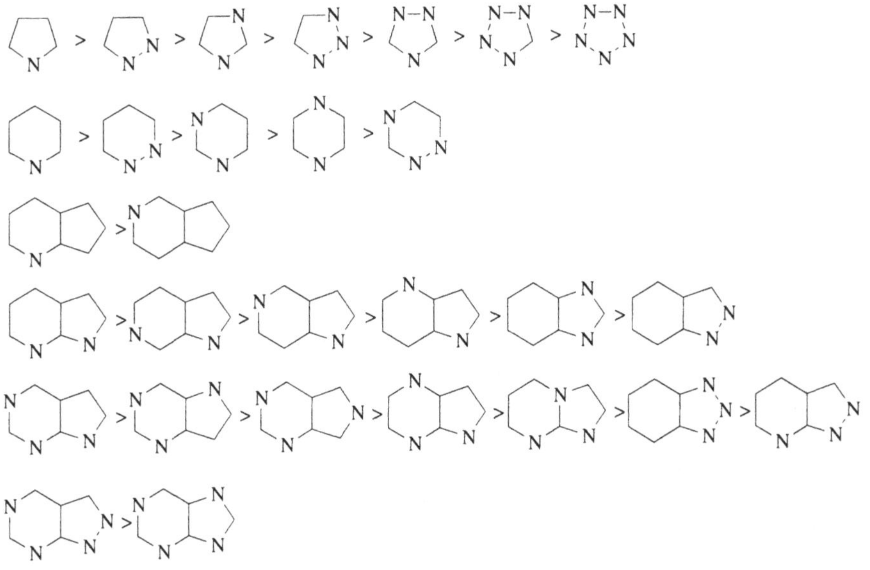

⇓

Number of substituents

⇓

Kind of substituents:

F > Cl > Br > I > CO > NHR > NR_2 > OH > NO_2 > CN > Alkyl > Aryl > O-Alkyl > CONR > CO_2R > S-Alkyl > Acetyl > Benzoyl

Yields are given in parentheses in the product column of the tables.

Abbreviations:

Aliquat 336	methyltrioctylammonium chloride
An	anisoyl
BETF	2-(1-benzimidazoyl)-3-ethylbenzoxazolium tetrafluoroborate
Bn	benzyl
BOC	*tert*-butoxycarbonyl
BSA	*N,O*-bis(trimethylsilyl)acetamide
Bz	benzoyl
t Bz	*p-tert*-butylbenzoyl
Cbz	carbobenzyloxy
CETF	2-chloro-3-ethylbenzoxazolium tetrafluoroborate
DEAD	diethyl azodicarboxylate
DIAD	diisopropyl azodicarboxylate
DMAP	4-dimethylaminopyridine
DME	1,2-dimethoxyethane
DMTST	dimethyl(methylthio)sulfonium triflate
FMPT	2-fluoro-1-methylpyridinium tosylate
HMDS	hexamethyldisilazane
HMPA	hexamethylphosphortriamide
IDCP	iodonium dicollidine perchlorate
MMTrO	*p*-monomethoxytrityl
Ms	mesyl
MS	molecular sieves
NBS	*N*-bromosuccinimide
NIS	*N*-iodosuccinimide
NMP	*N*-methylpyrrolidone
NPht	phthalimide
Piv	pivaloyl
Pht	phthaloyl
Py	pyridine
TBDMSO	*tert*-butyldimethylsilyloxy
TBDPSO	*tert*-butyldiphenylsilyloxy
TEBA	benzyltriethylammonium chloride
TFA	trifluoroacetic acid
Tf	trifluoromethylsulfonyl (triflyl)
Tf_2O	trifluoromethylsulfonic acid anhydride
$TIPDSCl_2$	1,3-dichloro-1,1,3,3-tetraisopropyldisiloxane

TMS	trimethylsilyl
TMSCl	trimethylsilyl chloride
TMSI	trimethylsilyl iodide
TMSOTf	trimethylsilyl triflate
Tol	toluoyl
Tr	triphenylmethyl
TsOH	*p*-toluenesulfonic acid
Ts	tosyl

TABLE I. ONE STEP-ONE POT SILYLATION/COUPLING OF HETEROCYCLIC BASES WITH SUGARS - FRIEDEL-CRAFTS CATALYSTS

Sugar	Base	Conditions	Product(s) and Yield(s) (%)	Refs.
C_4 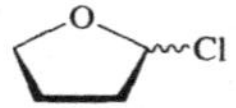	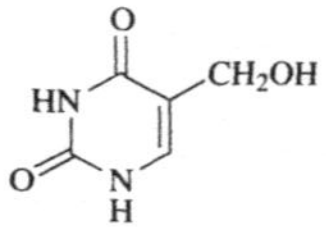	TCS, HMDS 1. 160° 2. rt, 24 h	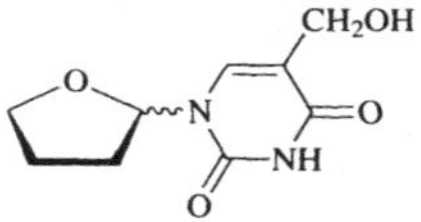 (40)	551
C_5 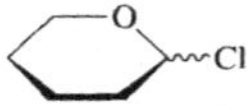	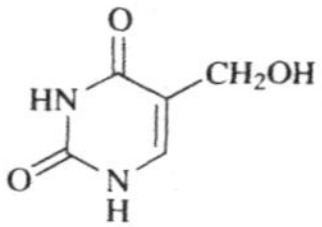	TCS, HMDS 1. 160° 2. rt, 24 h	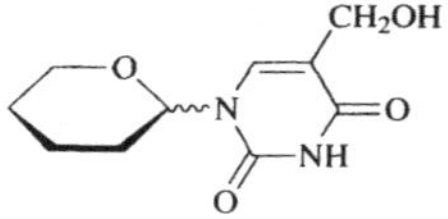 (61)	551
C_6 	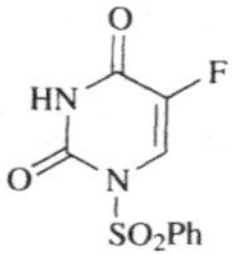	BSA, $SnCl_4$, CH_2Cl_2, rt, 4 h	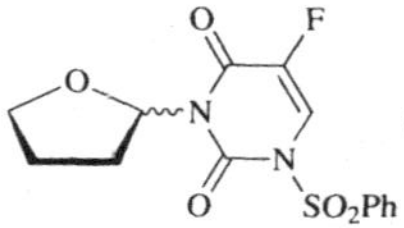 (88)	552
	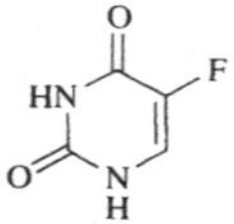	TCS, HMDS, $SnCl_4$, MeCN, 0°, 30 min	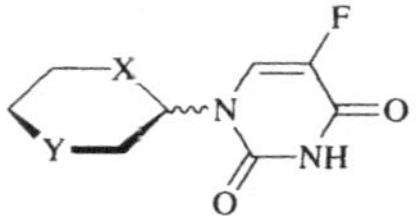 X Y O S (65) S O (82)	553
C_7 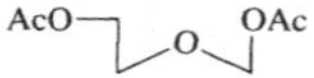	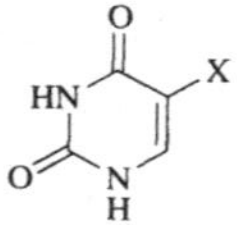	BSA, $SnCl_4$, CH_2Cl_2, rt, 12 h	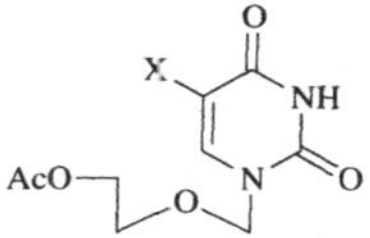 X F (53) Br (57) I (75)	554
	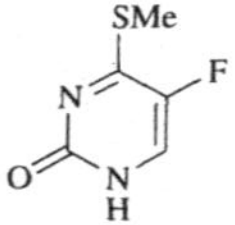	BSA, $SnCl_4$, CH_2Cl_2, rt, 12 h	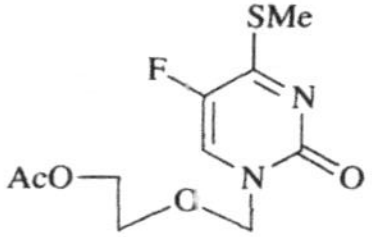 (84)	554

		Conditions		Yield (%)	Refs.
C_9 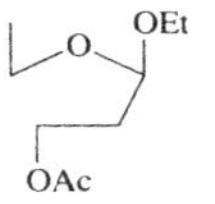	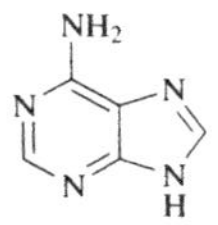	1. HMDS, $(NH_4)_2SO_4$, TMSOTf, MeCN, 20°, 16 h 2. NH_3, MeOH	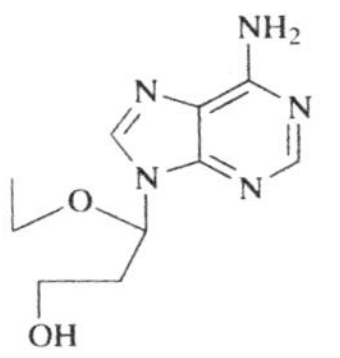	(50)	555
	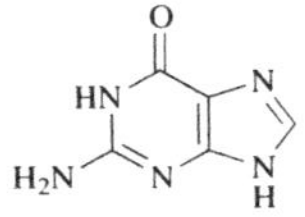	1. HMDS, $(NH_4)_2SO_4$, TMSOTf, MeCN, 20°, 16 h 2. NH_3, MeOH	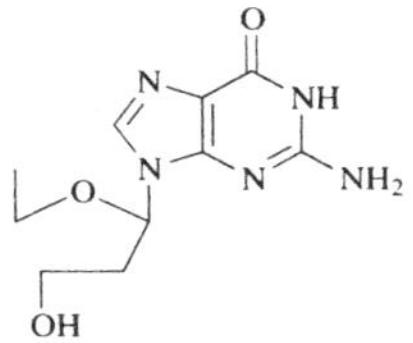	(50)	555
C_{10} 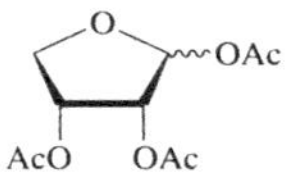	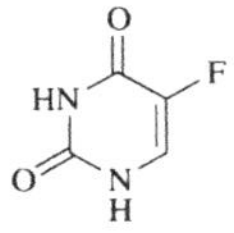	1. BSA, MeCN, 50°, 3 h 2. $SnCl_4$, rt, 12 h	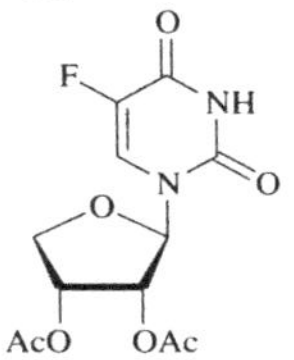	(28)	556
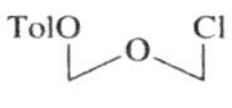	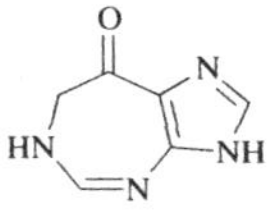	BSA, Py, MeCN, $SnCl_4$, $Cl(CH_2)_2Cl$	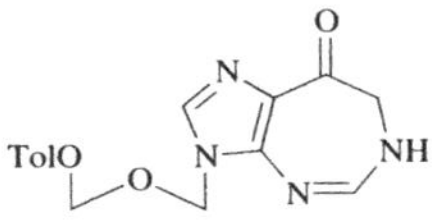	(27)	557
C_{11} 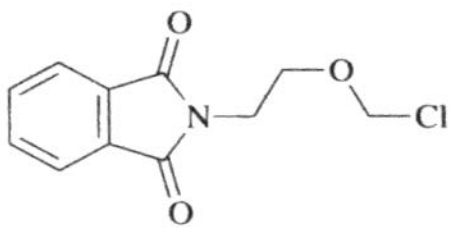	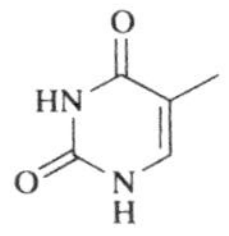	TCS, HMDS, TsOH, MeCN, reflux 19 h	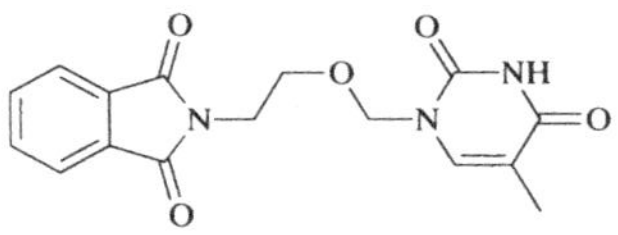	(88)	558

TABLE I. ONE STEP-ONE POT SILYLATION/COUPLING OF HETEROCYCLIC BASES WITH SUGARS - FRIEDEL-CRAFTS CATALYSTS (*Continued*)

Sugar	Base	Conditions	Product(s) and Yield(s) (%)	Refs.
OAc, OAc, O, H, O, O	HN, O, O, N, H	TCS, HMDS, $SnCl_4$, MeCN, 120°, 30 min	O, NH, N, O, O, H, O, O, OAc (60)	559
O, OBz, O	HN, O, O, F, N, H	TCS, HMDS, $SnCl_4$, MeCN, 0°, 30 min	F, O, NH, N, O, O, O (70)	553
S, OBz, S	HN, O, O, R, N, H	TCS, HMDS, $SnCl_4$, MeCN, 0°, 30 min	S, S, N, R, O, NH, O R: H (59); Me (67)	553
AcO, O, AcO, OMe	HN, O, O, N, H	TCS, HMDS, $SnCl_4$, MeCN, 45°, 20 h	O, NH, N, O, AcO, O, AcO (60) + α anomer (18)	192
	HN, O, O, N, H	TCS, HMDS, $SnCl_4$, MeCN, 45°, 20 h	O, NH, N, O, AcO, O, AcO (65) + α anomer (18) + N^3-β isomer (4) + N^3-α isomer (4)	192

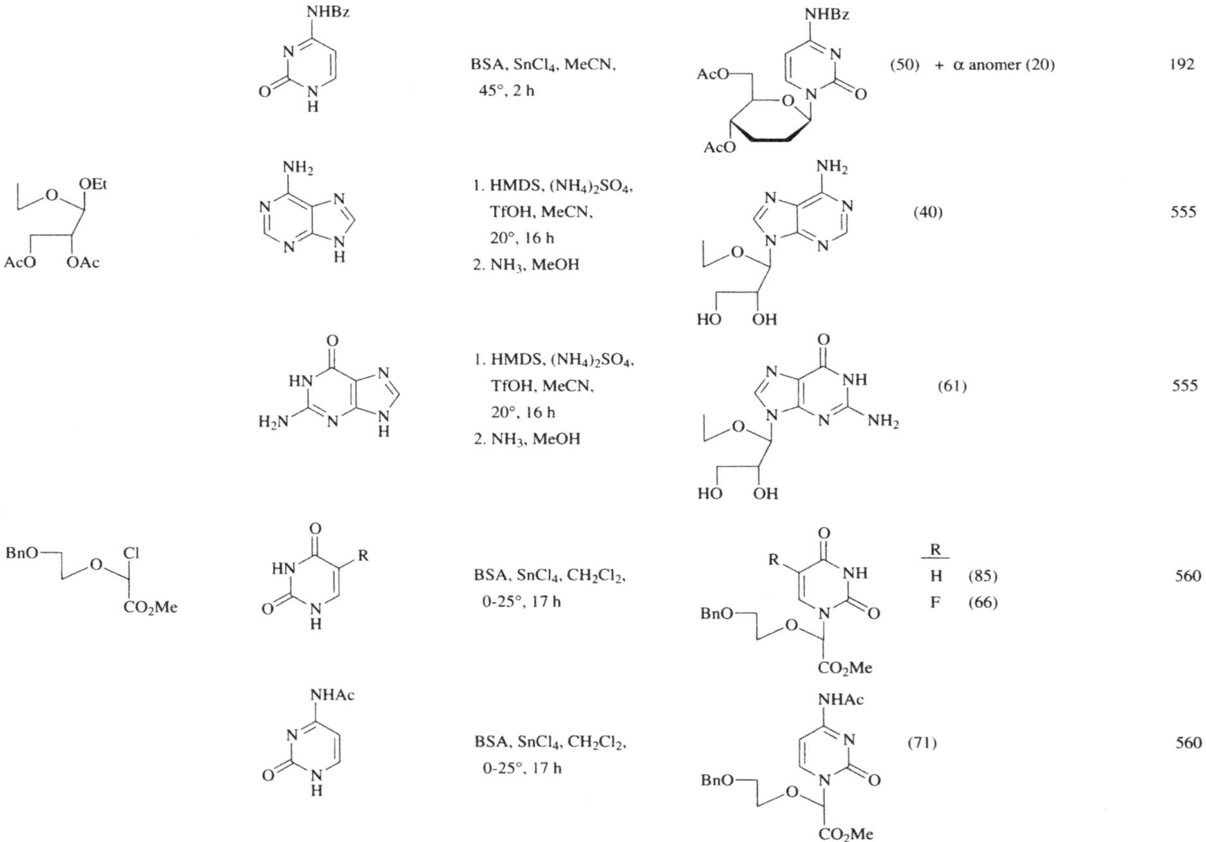

Reactant	Conditions	Product(s) and Yield(s) (%)	Refs.
	BSA, $SnCl_4$, MeCN, 45°, 2 h	(50) + α anomer (20)	192
	1. HMDS, $(NH_4)_2SO_4$, TfOH, MeCN, 20°, 16 h 2. NH_3, MeOH	(40)	555
	1. HMDS, $(NH_4)_2SO_4$, TfOH, MeCN, 20°, 16 h 2. NH_3, MeOH	(61)	555
C_{12}	BSA, $SnCl_4$, CH_2Cl_2, 0-25°, 17 h	R: H (85); F (66)	560
	BSA, $SnCl_4$, CH_2Cl_2, 0-25°, 17 h	(71)	560

TABLE I. ONE STEP-ONE POT SILYLATION/COUPLING OF HETEROCYCLIC BASES WITH SUGARS - FRIEDEL-CRAFTS CATALYSTS (*Continued*)

Sugar	Base	Conditions	Product(s) and Yield(s) (%)	Refs.
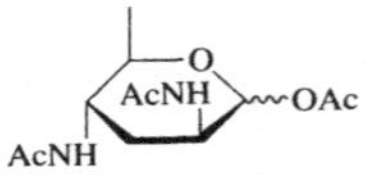 AcNH, AcNH, OAc	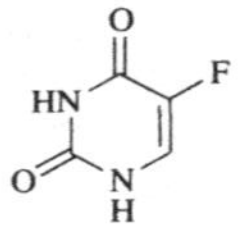 HN, O, F, O, N H	BSA, $SnCl_4$, MeCN, rt, 1 h	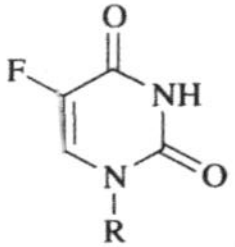 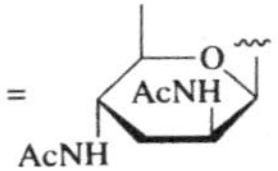 F, O, NH, O, N, R (71) R = AcNH, AcNH, O	561
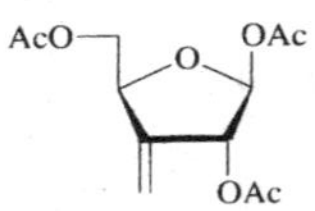 AcO, O, OAc, OAc	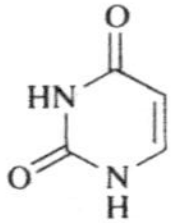 HN, O, O, N H	TCS, HMDS, TMSOTf, $Cl(CH_2)_2Cl$, 60°, 2 h	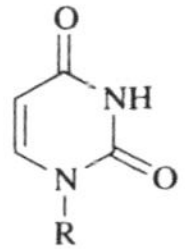 O, NH, O, N, R (64-70) R 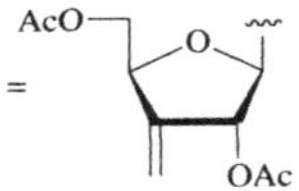 = AcO, O, OAc	562
C_{13}				
AcO, O, OAc, F_3CCONH, OAc	HN, O, N, AcHN, N, N H	TCS, HMDS, $C_4F_9SO_3K$, MeCN, reflux, 24 h	AcO, O, N, N, F_3CCONH, OAc, N, O, NH, AcHN (70) α:β = 1:2.5	563
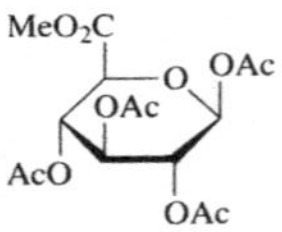 MeO_2C, O, OAc, OAc, AcO, OAc	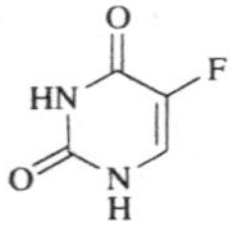 HN, O, F, O, N H	TCS, HMDS, $SnCl_4$, MeCN, rt, 10 h	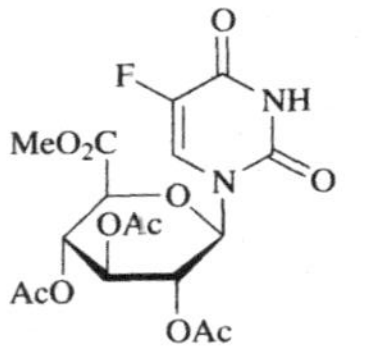 F, O, NH, MeO_2C, O, N, O, OAc, AcO, OAc (80)	767
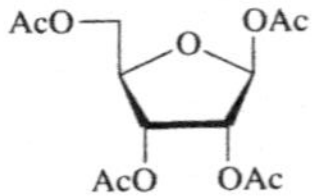 AcO, O, OAc, AcO, OAc	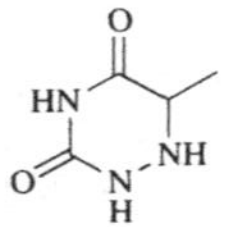 HN, O, O, NH, N H	TCS, HMDS, TMSOTf, $Cl(CH_2)_2Cl$, 0°, 24 h (Ar)	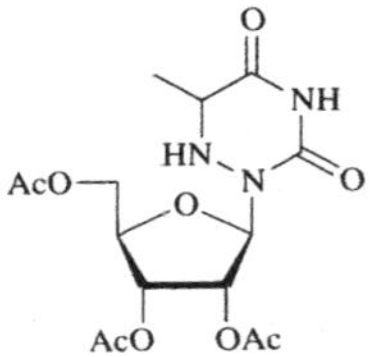 O, NH, HN, N, O, AcO, O, AcO, OAc (10)	565

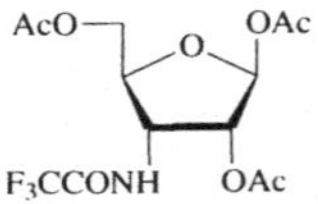

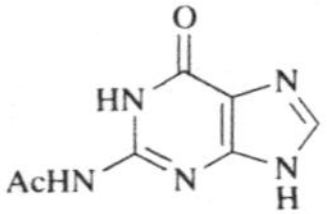

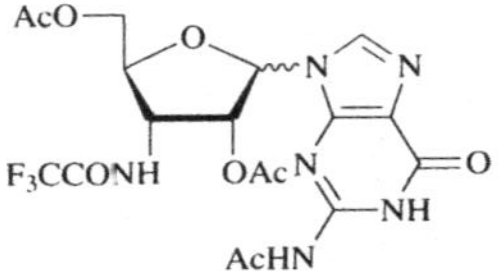

Reactant	Conditions	Product(s) and Yield(s) (%)	Refs.
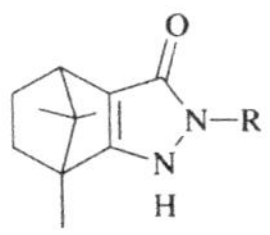 O; N—R; N; H	TCS, HMDS, $SnCl_4$, MeCN, rt, time	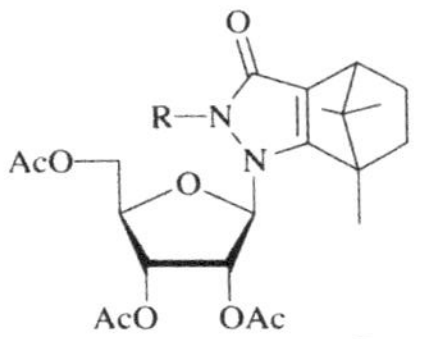 O; R—N; N; AcO; O; AcO; OAc	
		R = Ph, time 19 h (30)	566
		R = p-ClC_6H_4, time 13 h (33)	
O; HN; N; R; N; N; H	BSA, TMSOTf, MeCN	O; N; NH; AcO; N; N; R; O; AcO; OAc	
R = AcNH	heat, 8 h	(70) + 5% N^7-β isomer	190
R = Br	reflux, 15 min	(75)	190
R = p-n-BuC_6H_4NH	heat, 8 h	(69) + 11% N^7-β anomer + 11% N^7-β isomer	190
R = p-n-BuC_6H_4NH	reflux, 7 h	(69) + 11% N^7 isomer	505
HN; N; N; N; N; H	TMSOTf, dioxane (N_2), rt, 1 h	NH; N; N; AcO; N; N; O; AcO; OAc (41)	567
O; NH; X; N; H	TCS, HMDS, $SnCl_4$, MeCN, rt, time	O; HN; X; AcO; X; N; O; AcO; OAc	
		X = O, time 18 h (31)	566
		X = S, time 20 h (28)	566

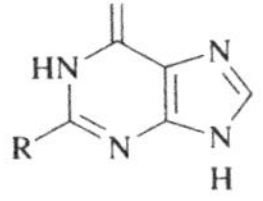

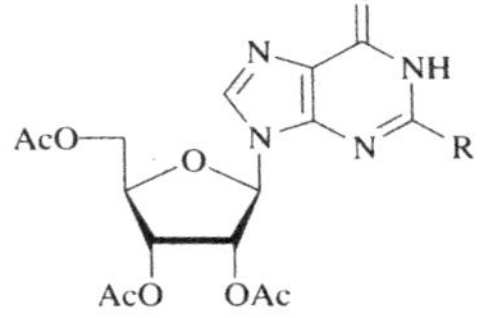

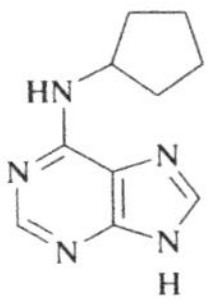

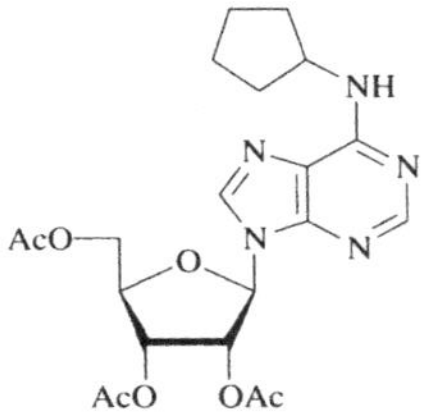

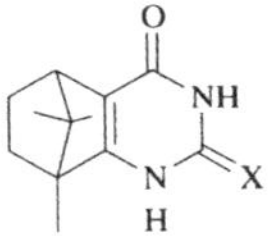

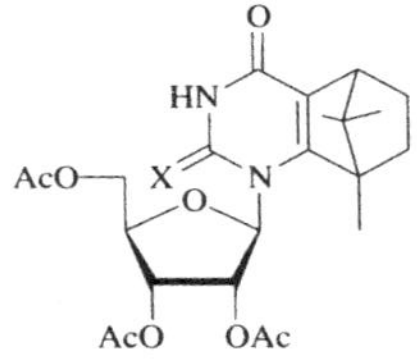

TABLE I. ONE STEP-ONE POT SILYLATION/COUPLING OF HETEROCYCLIC BASES WITH SUGARS - FRIEDEL-CRAFTS CATALYSTS (*Continued*)

Sugar	Base	Conditions	Product(s) and Yield(s) (%)	Refs.
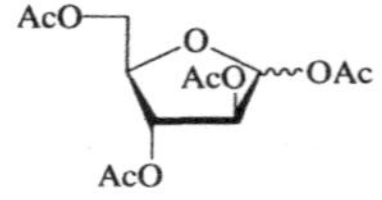 AcO, AcO, AcO, OAc	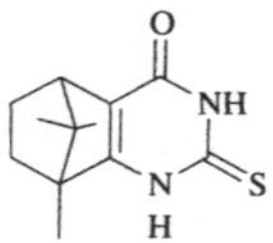 O, NH, N, H, S	TCS, HMDS, $SnCl_4$, MeCN, rt, 17 h	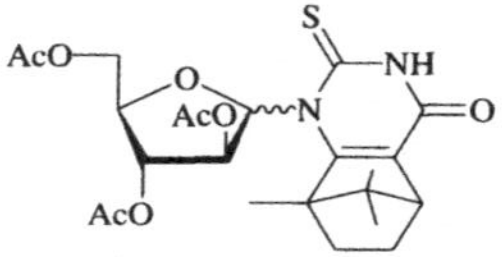 AcO, AcO, AcO, S, NH, N, O (26)	566
AcO, S, OAc, AcO, OAc	O, HN, R, O, N, H	TCS, HMDS, CF_3SO_3H, MeCN	R, O, NH, N, O, AcO, S, AcO, OAc; R = H: 80°, 3 h (41)	568
			R = F: 25°, 24 h (48)	568
			R = Me: 80°, 3 h (38)	568
			R = NO_2: 25°, 24 h (55)	568
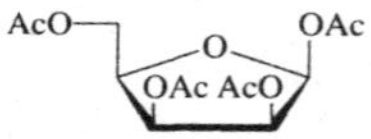 AcO, O, OAc, OAc AcO	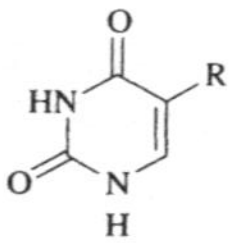 O, HN, R, O, N, H	TCS, HMDS, $SnCl_4$	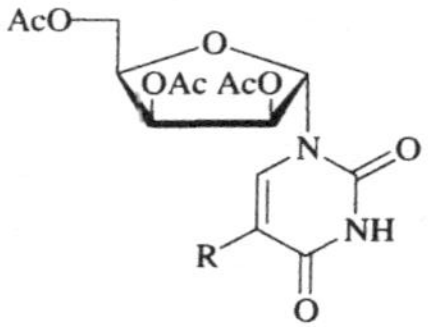 AcO, O, OAc AcO, N, O, NH, R, O **I**	
	R = H	MeCN, rt, 22 h	(49) + 14% N^3-isomer + 11% N^1,N^3-bisisomer	537
	R = Me	MeCN, heat, 25 min	(—)	537
 AcO, O, OAc AcO, OAc	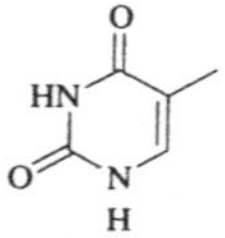 O, HN, O, N, H	TCS, HMDS, $SnCl_4$, MeCN, reflux 1 h	**I**, R = Me, (—) + N^3-isomer	537

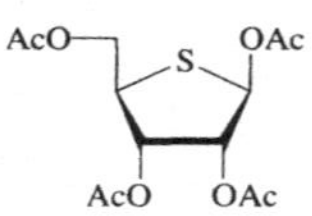

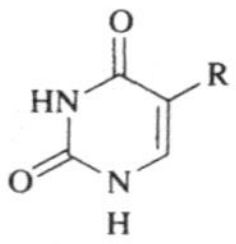

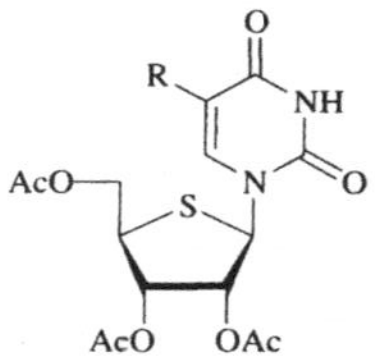

Sugar	Base	Conditions	Product (Yield %)	Refs.
		TCS, HMDS, $SnCl_4$, MeCN, rt, 24 h	(31)	566
		TCS, HMDS, $SnCl_4$, MeCN, rt, 21 h	X: O (29); S (31)	566
		TCS, HMDS, $SnCl_4$, MeCN, rt, 17 h	(37)	566
		TCS, HMDS, $SnCl_4$, MeCN, rt, 16 h	(26)	566
		BSA, $SnCl_4$, MeCN, rt, 0.8 h	(76) α:β = 2:3	569

TABLE I. ONE STEP-ONE POT SILYLATION/COUPLING OF HETEROCYCLIC BASES WITH SUGARS - FRIEDEL-CRAFTS CATALYSTS (*Continued*)

Sugar	Base	Conditions	Product(s) and Yield(s) (%)	Refs.
	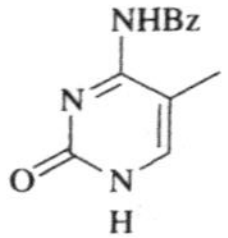 NHBz	TCS, HMDS, $SnCl_4$, MeCN, 0-55°, 70 min	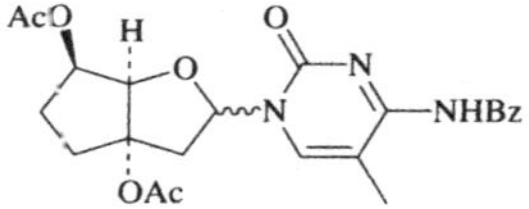 AcO, H, OAc, NHBz (76) α:β = 2:3	569
	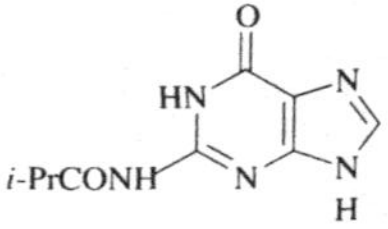 i-PrCONH	BSA, TMSOTf, MeCN 1. rt, 60 h 2. 40°, 6 h	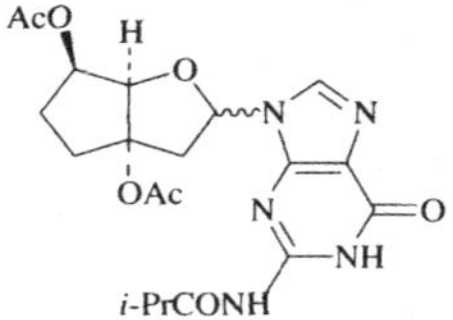 AcO, H, OAc, i-PrCONH (45) α:β = 8:5 + N^7-isomer (37) α:β = 3:2	569
	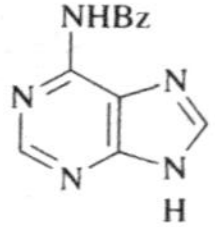 NHBz	BSA, TMSOTf, MeCN 1. 80°, 2 h 2. rt, 3 h	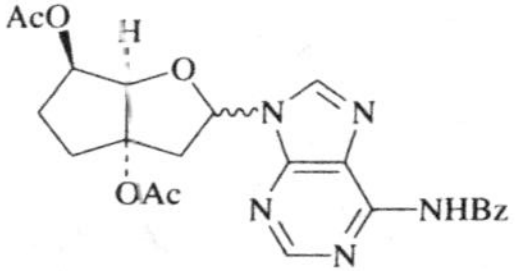 AcO, H, OAc, NHBz (78) α:β = 3:2	569
C_{14} 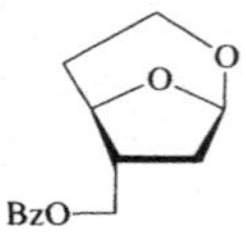 BzO	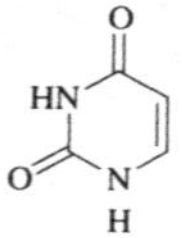	BSA, TMSOTf, MeCN, rt, 15 min	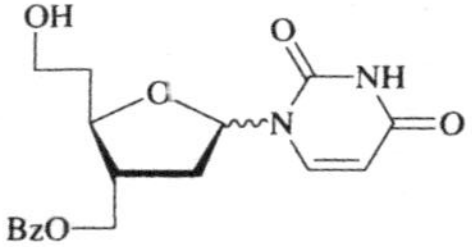 OH, BzO (75) α:β = 1:1	570
	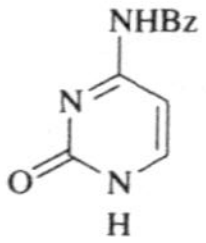 NHBz	BSA, TMSOTf, MeCN, 40°, 1 h	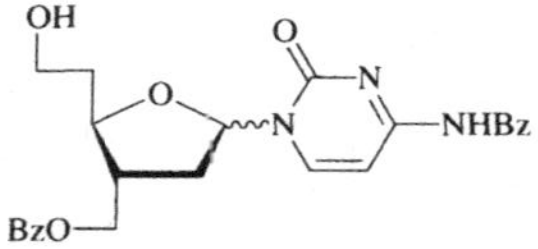 OH, BzO, NHBz (64) α:β = 1:1	570

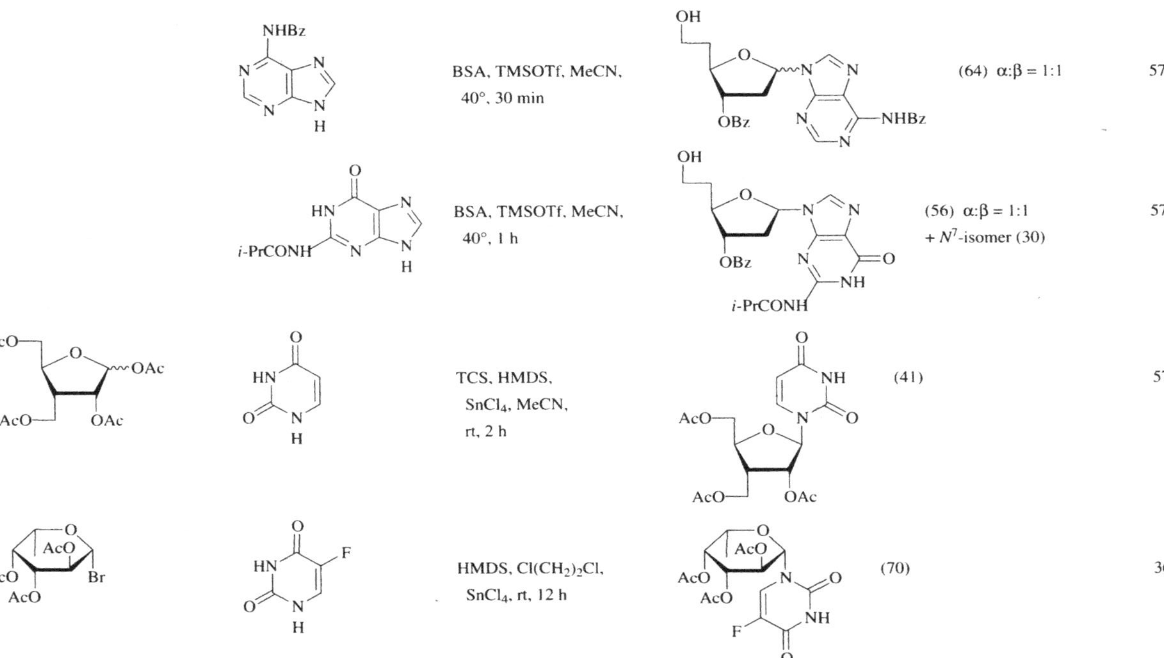
NHBz
N
N
N
N
H
BSA, TMSOTf, MeCN,
40°, 30 min
OH
O
OBz
N
N
N
N
NHBz
(64) α:β = 1:1
570
O
HN
N
N
N
H
i-PrCONH
BSA, TMSOTf, MeCN,
40°, 1 h
OH
O
OBz
N
N
N
O
NH
i-PrCONH
(56) α:β = 1:1
+ N^7-isomer (30)
570
AcO
O
OAc
AcO
OAc
O
HN
O
N
H
TCS, HMDS,
SnCl$_4$, MeCN,
rt, 2 h
O
NH
O
N
AcO
O
AcO
OAc
(41)
571
O
AcO
AcO
AcO
Br
O
HN
F
O
N
H
HMDS, Cl(CH$_2$)$_2$Cl,
SnCl$_4$, rt, 12 h
O
AcO
AcO
AcO
N
O
NH
F
O
(70)
367

TABLE I. ONE STEP-ONE POT SILYLATION/COUPLING OF HETEROCYCLIC BASES WITH SUGARS - FRIEDEL-CRAFTS CATALYSTS (*Continued*)

Sugar	Base	Conditions	Product(s) and Yield(s) (%)	Refs.
C_{15} 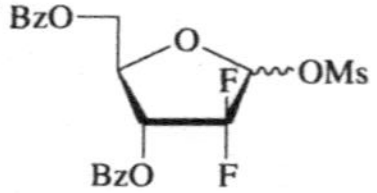	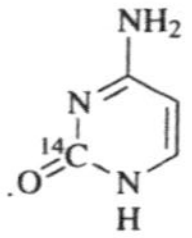	TCS, HMDS, TMSOTf, $Cl(CH_2)_2Cl$, reflux, 48 h	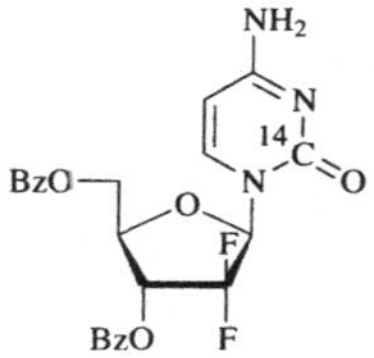 (16)	572
	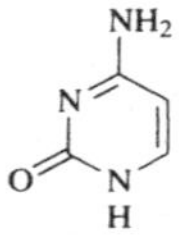	TCS, HMDS, TMSOTf, $Cl(CH_2)_2Cl$, reflux, 48 h	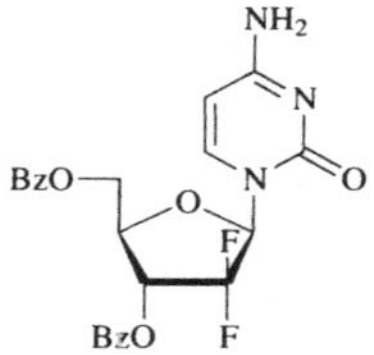 (16)	572
C_{16} 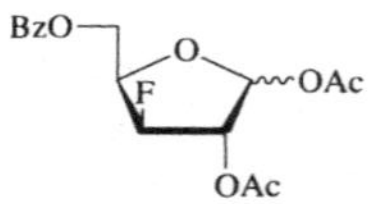	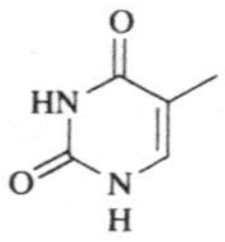	BSA, TMSOTf, $Cl(CH_2)_2Cl$, MeCN, reflux 2 h	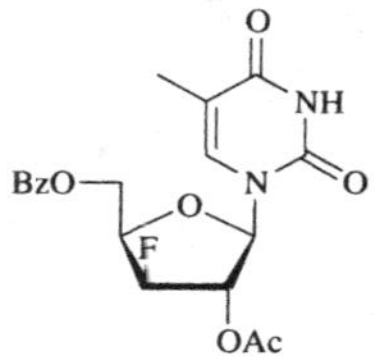 (63)	564
	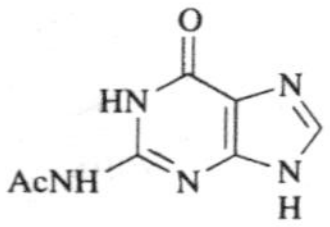	1. BSA, TMSOTf, MeCN, reflux 1 h 2. $NH_2NH_2 \cdot H_2O$, Py/AcOH (1:4) rt, 5 h	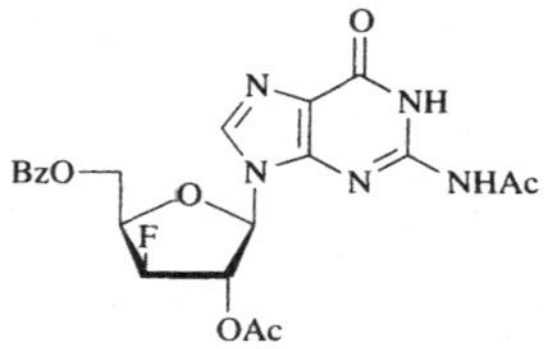 (58) + N^7-isomer (19)	573

Sugar	Base	Conditions	Product (Yield, %)	Refs.
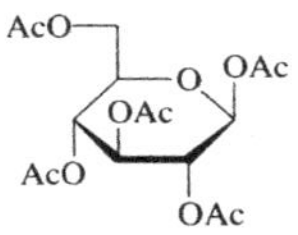	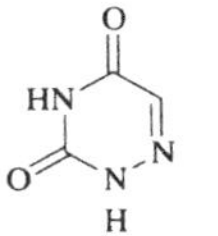	TCS, HMDS, $SnCl_4$, MeCN, 24°, 7 h	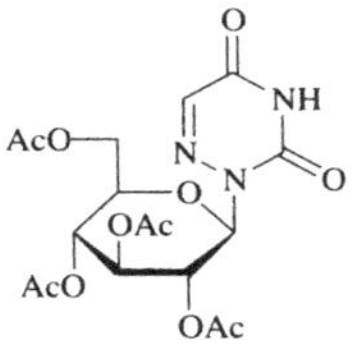 (42)	189
C_{17} 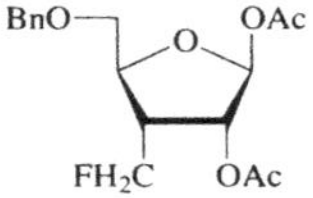	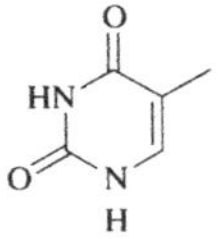	TCS, HMDS, $C_4F_9SO_3K$, MeCN, reflux, 24 h	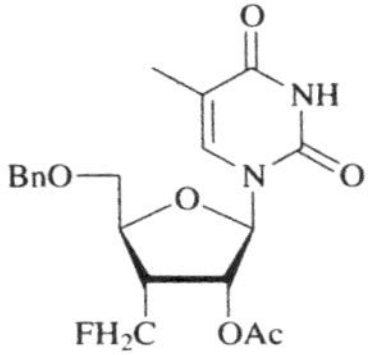 (84)	574
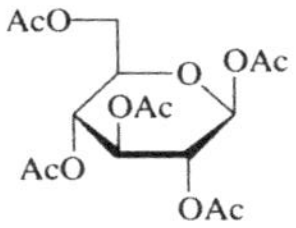	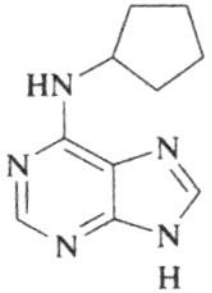	TCS, HMDS, TMSOTf, dioxane, rt, 12 h	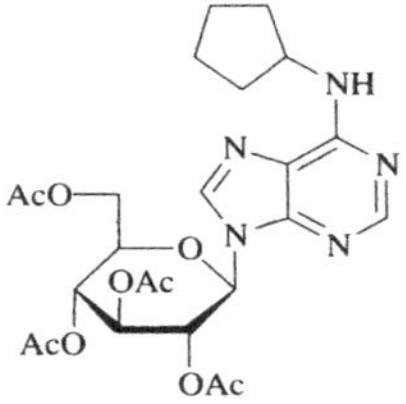 (35)	567
C_{18} 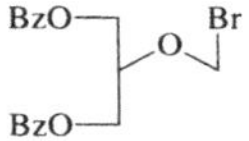	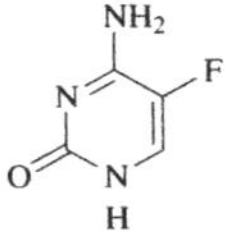	TCS, HMDS, $C_4F_9SO_3K$, MeCN, heat, 6 h	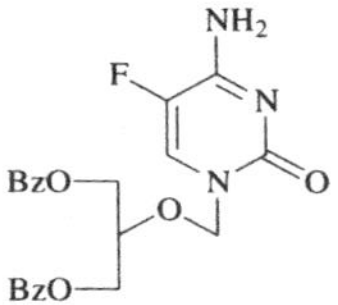 (56)	575

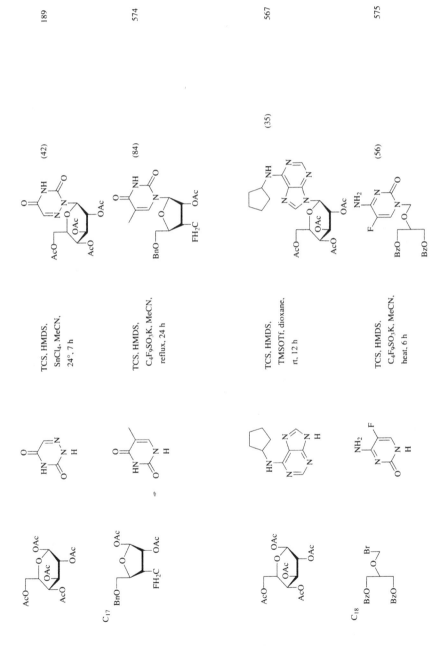

TABLE I. ONE STEP-ONE POT SILYLATION/COUPLING OF HETEROCYCLIC BASES WITH SUGARS - FRIEDEL-CRAFTS CATALYSTS (*Continued*)

Sugar	Base	Conditions	Product(s) and Yield(s) (%)	Refs.
p-$MeOC_6H_4CO_2$ ... OAc, CH_2, OAc	HN, O, R, N, H	TCS, HMDS, TMSOTf, $Cl(CH_2)_2Cl$, 60°, 2 h	R: H (70); Me (47)	576, 577
OBn, OAc, OAc	HN, O, N, H	BSA, $SnCl_4$, $Cl(CH_2)_2Cl$ 1. 5°, 30 min 2. rt 30 min 3. reflux, 30 min	(87)	578
AcO, OAc, AcO, OAc	NHBz, N, N, N, N, H	1. HMDS, TMSOTf, MeCN, 60°, 1 h 2. MeOH, NH_3, rt, 48 h	(65)	579
BzO, OAc, AcO, OAc	Cl, CN, N, N, N, H	TCS, HMDS, TMSOTf, MeCN, heat, 20-24 h	(31)	580

Reactant	Conditions	Product(s) and Yield(s) (%)	Refs.
	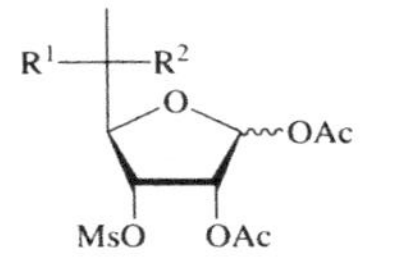 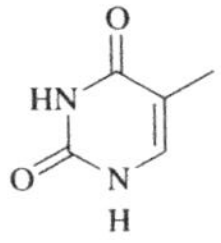 TCS, HMDS, CF_3SO_3H, MeCN, rt, time	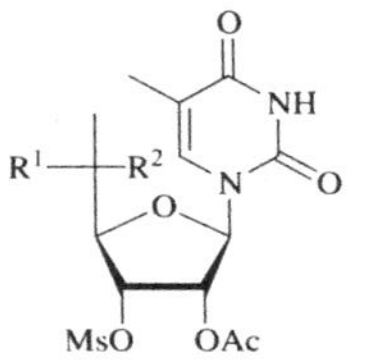 R^1 / R^2 / time: BzO / H / 10 h (68); H / BzO / 7 h (53)	581
	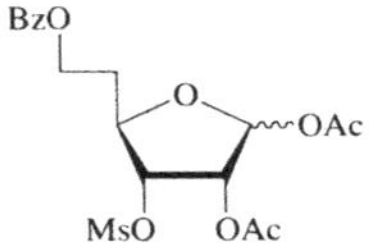 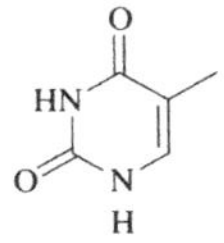 TCS, HMDS, CF_3SO_3H, MeCN, rt, 4 h, or 80°, 5 h	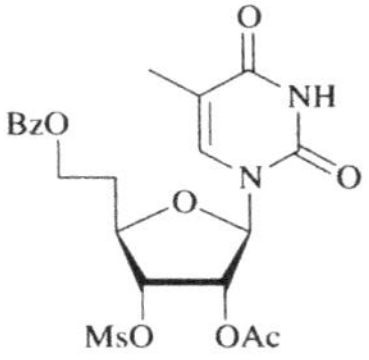 (46)	581-583
C_{19} 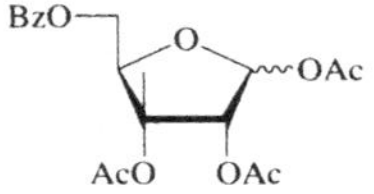	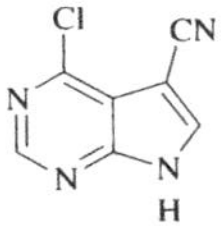 TCS, HMDS, TMSOTf, MeCN, heat, 20-24 h	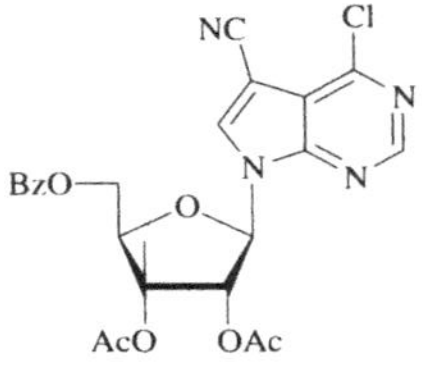 (28)	580
	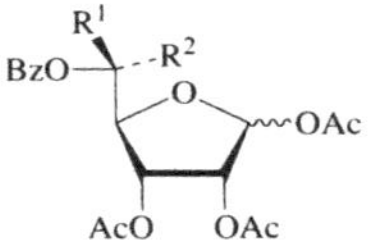 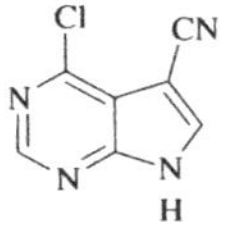 TCS, HMDS, TMSOTf, MeCN, heat, 20-24 h	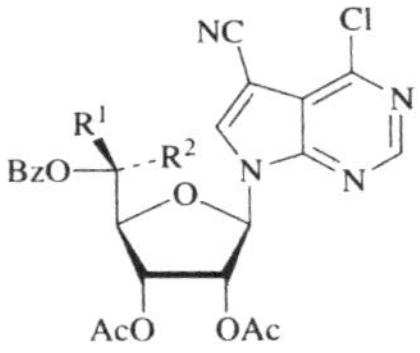 R^1 / R^2: H / Me (47); Me / H (32)	580

TABLE I. ONE STEP-ONE POT SILYLATION/COUPLING OF HETEROCYCLIC BASES WITH SUGARS - FRIEDEL-CRAFTS CATALYSTS (*Continued*)

Sugar	Base	Conditions	Product(s) and Yield(s) (%)	Refs.
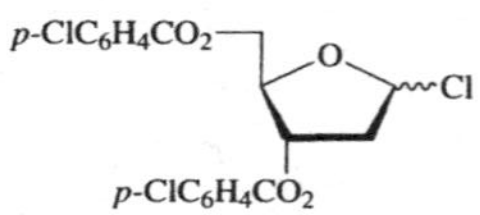 p-$ClC_6H_4CO_2$, Cl, p-$ClC_6H_4CO_2$	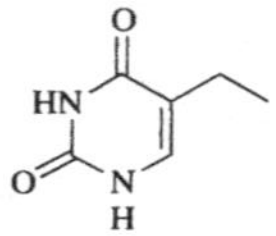 O, HN, O, N, H	TCS, HMDS, $SnCl_4$, MeCN	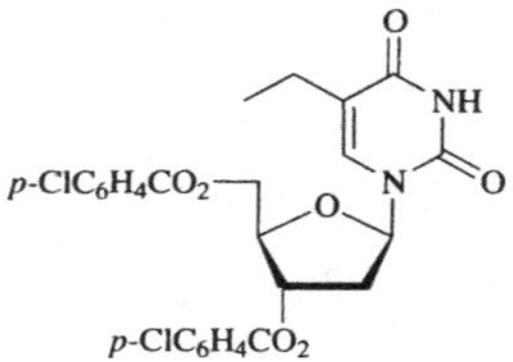 O, NH, p-$ClC_6H_4CO_2$, N, O, p-$ClC_6H_4CO_2$ (46) + α-anomer	584
C_{20}				
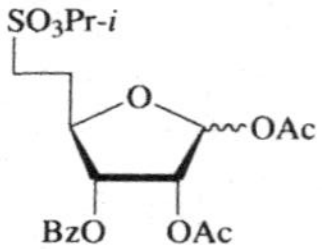 SO_3Pr-i, OAc, BzO, OAc	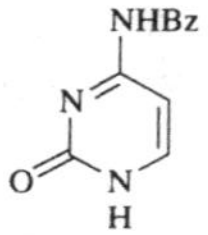 NHBz, N, O, N, H	HMDS, Py, $SnCl_4$, $Cl(CH_2)_2Cl$, 0°, 20 h	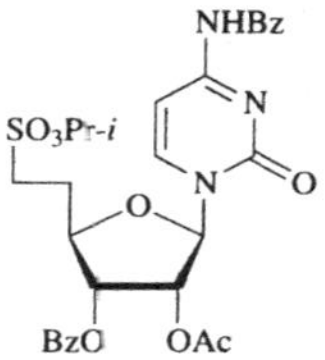 NHBz, SO_3Pr-i, N, N, O, BzO, OAc (76)	508
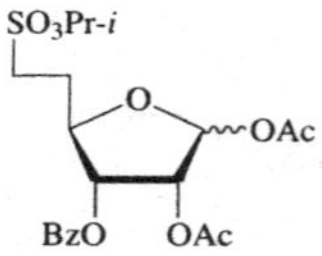 SO_3Pr-i, OAc, BzO, OAc	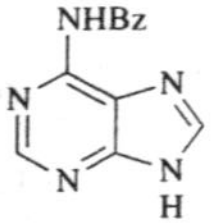 NHBz, N, N, N, N, H	HMDS, Py, TMSOTf, $Cl(CH_2)_2Cl$, 40°, 20 h	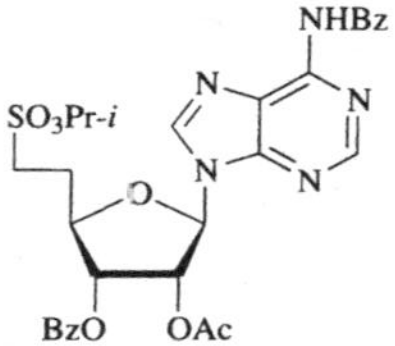 NHBz, SO_3Pr-i, N, N, N, N, BzO, OAc (51)	508
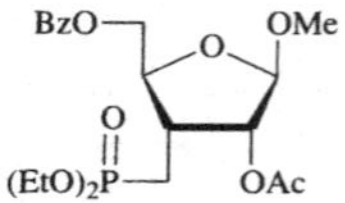 BzO, OMe, O, $(EtO)_2P$, OAc	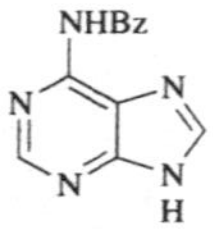 NHBz, N, N, N, N, H	TCS, HMDS, CF_3SO_3K, MeCN	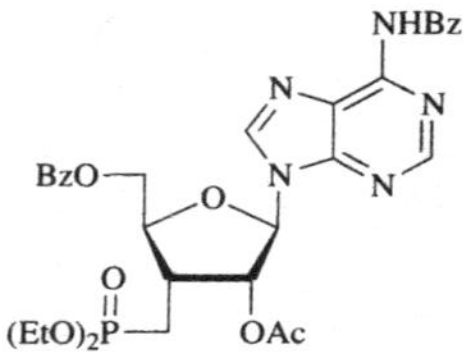 NHBz, N, N, BzO, N, N, O, $(EtO)_2P$, OAc (60)	585

C_{21} 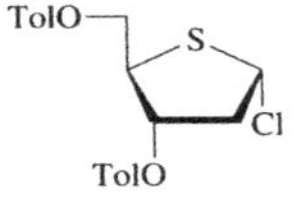	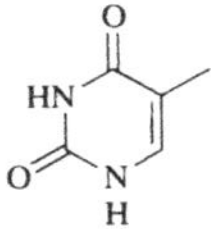	TCS, HMDS, TMSOTf, $Cl(CH_2)_2Cl$, –78°, 1.5 h	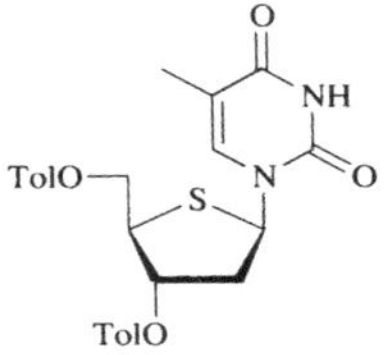	(—)	586
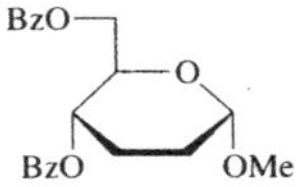	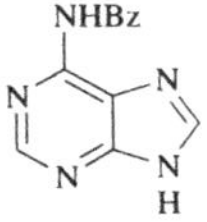	TCS, HMDS, $SnCl_4$, MeCN, rt, 20 h	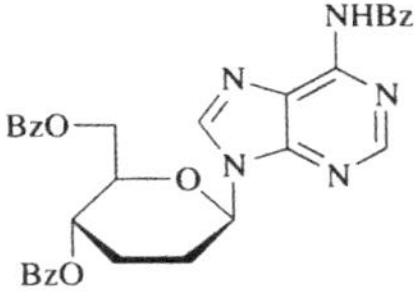	(35)	192
	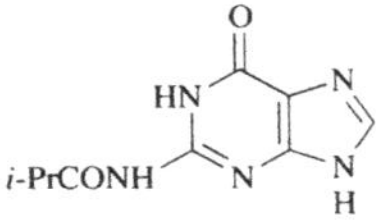	BSA, TMSOTf, MeCN, 80°, 16 h	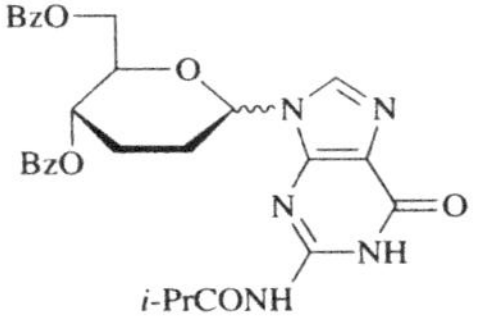	(40) N^7:N^9 = 1:4	192
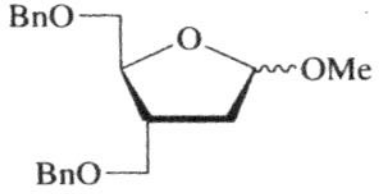	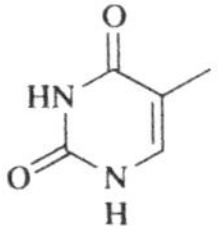	"Vorbrüggen conditions"	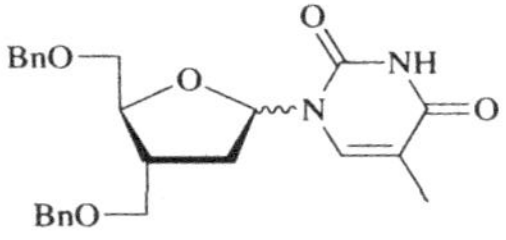	(—)	587
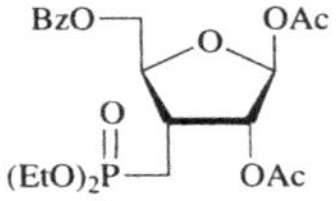	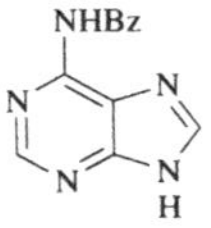	TCS, HMDS, $C_4F_9SO_3K$, MeCN, reflux, 10 h	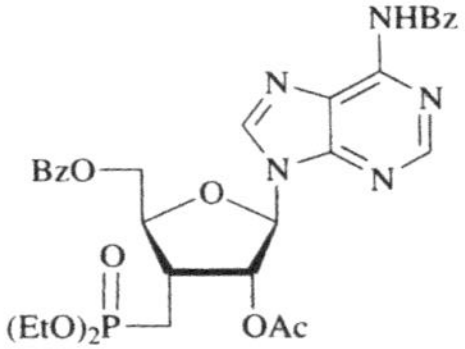	(61)	588

TABLE I. ONE STEP-ONE POT SILYLATION/COUPLING OF HETEROCYCLIC BASES WITH SUGARS - FRIEDEL-CRAFTS CATALYSTS (*Continued*)

Sugar	Base	Conditions	Product(s) and Yield(s) (%)	Refs.
C_{22} BzO, OAc, BzO, OMe	NH_2, N, O, N, H, NH_2	BSA, TMSOTf, MeCN, 70°	NH_2, N, H_2N, N, O, BzO, O, BzO, OMe (67) + α-anomer (8)	589
BzO, O, OAc, BzO	NHBz, N, O, N, H	BSA, TMSOTf, $Cl(CH_2)_2Cl$, 70°	NHBz, N, BzO, O, N, O, BzO (56) + α-anomer (8)	590
	O, HN, O, N, H	BSA, TMSOTf, $Cl(CH_2)_2Cl$, Cond.	O, NH, BzO, O, N, O, BzO Cond. / / α-anomer rt, 16 h (71) (9) 0°, 1 h, rt, 5 h (55) (22)	590
	NHBz, N, N, N, N, H	BSA, TMSOTf, MeCN, 50°, 64 h	NHBz, N, N, BzO, O, N, N, BzO (30)	590

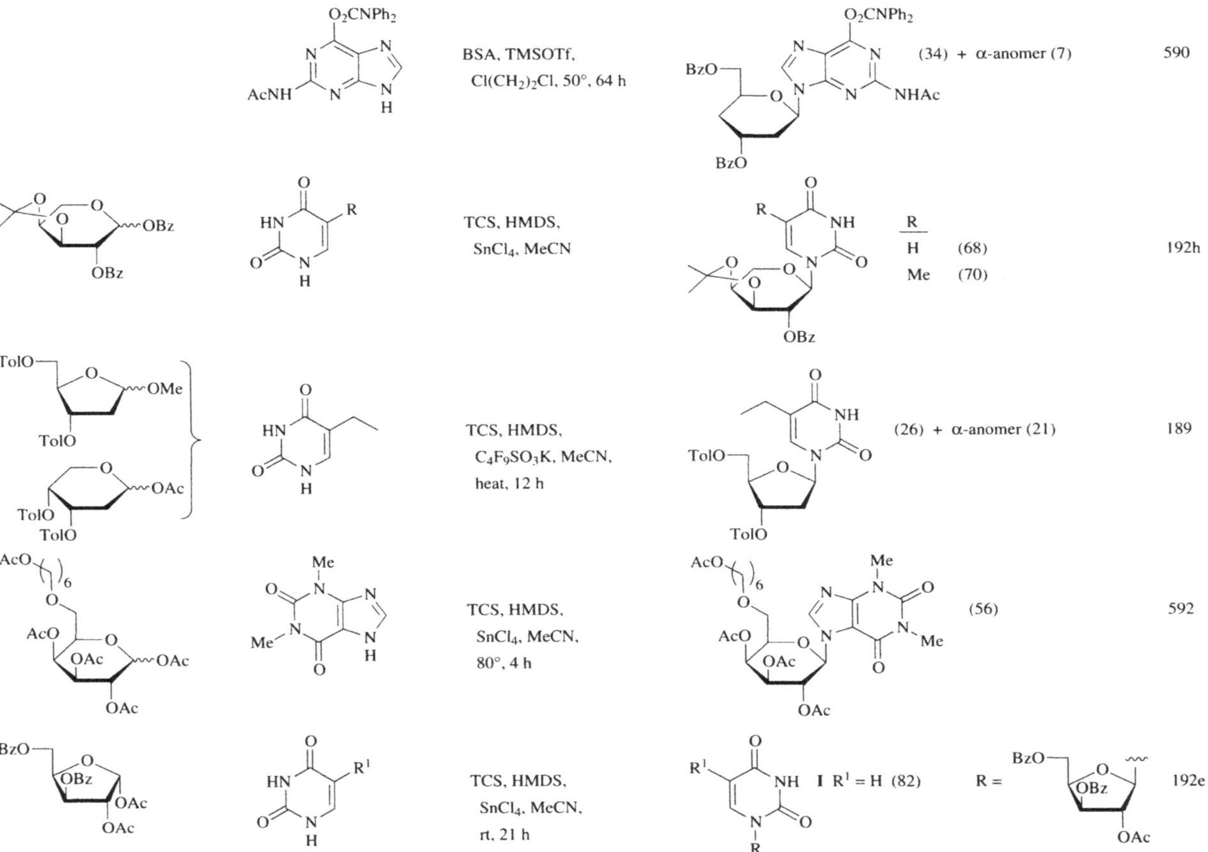

	Conditions	Product(s) and Yield(s) (%)	Refs.
	BSA, TMSOTf, $Cl(CH_2)_2Cl$, 50°, 64 h	(34) + α-anomer (7)	590
	TCS, HMDS, $SnCl_4$, MeCN	R: H (68); Me (70)	192h
	TCS, HMDS, $C_4F_9SO_3K$, MeCN, heat, 12 h	(26) + α-anomer (21)	189
	TCS, HMDS, $SnCl_4$, MeCN, 80°, 4 h	(56)	592
C_{23}	TCS, HMDS, $SnCl_4$, MeCN, rt, 21 h	I R^1 = H (82)	192e

TABLE I. ONE STEP-ONE POT SILYLATION/COUPLING OF HETEROCYCLIC BASES WITH SUGARS - FRIEDEL-CRAFTS CATALYSTS (*Continued*)

Sugar	Base	Conditions	Product(s) and Yield(s) (%)	Refs.
		TCS, HMDS, $SnCl_4$, MeCN	**I**, R^1 = H (82)	192f
		TCS, BSA, $SnCl_4$, MeCN, reflux, 5 min	**I**, R^1 = Me (46)	593
BzO, OBz, OAc, OAc	Cl, N, N, N, N, H	TCS, HMDS, $SnCl_4$, MeCN, rt, 1 h	Cl, N, N, N, N, R (54) + N^7 isomer (20)	594
	O, HN, N, AcNH, N, N, H	BSA, TMSOTf, MeCN, reflux, 6 h	N, N, NHAc, N, NH, R, O (9) + N^9 isomer (59)	537
BzO, OBz, OAc, OAc	O, HN, R, O, N, H	TCS, HMDS, $SnCl_4$ Cond.	R, O, NH, N, O, O, OBz, BzO, OAc	
			R = H, Cond. = rt, 60° (69); N^7-isomer (11)	595
			R = Me, Cond. = MeCN (—); N^7-isomer (—)	596
			R = Me, Cond. = reflux, 1.5 h (64); N^7-isomer (—)	595
	O, HN, N, HN, N, N, H, O, $C_{15}H_{31}$-*n*	1. BSA, TMSOTf, MeCN, reflux, 6 h 2. $NH_2NH_2{\cdot}H_2O$, rt, 7 h	O, N, NH, N, N, NH, O, OBz, BzO, OAc, O, $C_{15}H_{31}$-*n* (54) + N^7-isomer (16)	595

Glycosyl donor	Base	Conditions	Product(s) and Yield(s) (%)	Refs.
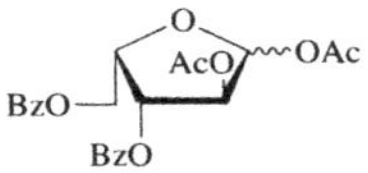 BzO, BzO, AcO, OAc (furanose)	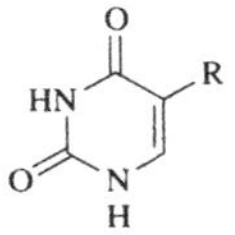 5-R-uracil (HN, O, N, H, R)	TCS, HMDS, $SnCl_4$, MeCN, rt, 24 h	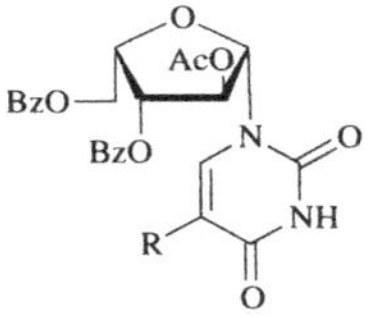 R: H (66); Me (64)	596
	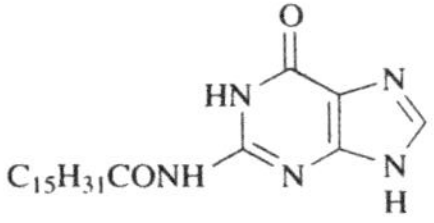 $C_{15}H_{31}CONH$-guanine	1. BSA, TMSOTf, MeCN, reflux, 6 h 2. H_2NNH_2•H_2O, Py, AcOH, rt, 3 h	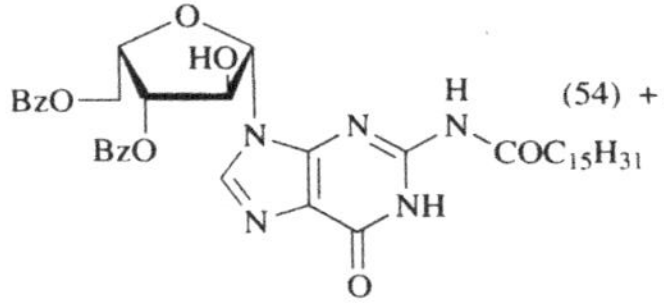 BzO, BzO, HO, N-$COC_{15}H_{31}$ (54) + N^7-isomer (16)	595
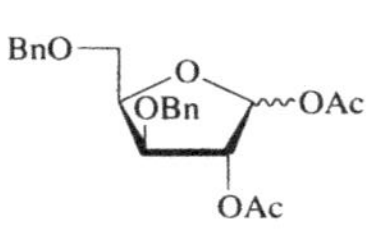 BnO, OBn, OAc, OAc (furanose)	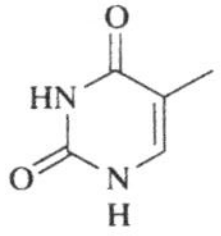 thymine	TCS, HMDS, $SnCl_4$, MeCN, reflux, 0.5 h	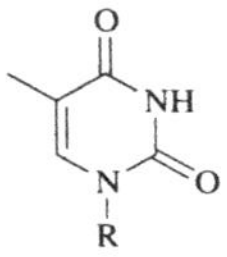 (70) R 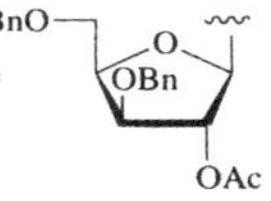 = BnO, OBn, OAc (furanosyl)	598, 599
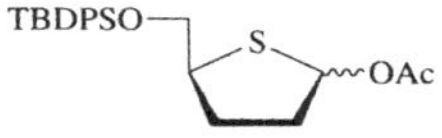 TBDPSO, S, OAc	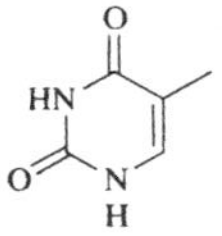 thymine	TCS, HMDS, $C_4F_9SO_3K$, MeCN, rt, 12 h	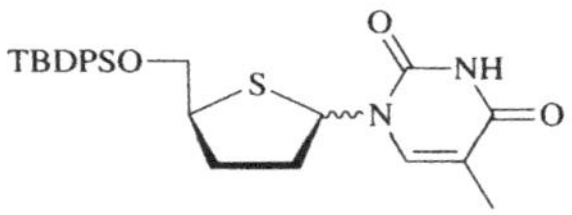 TBDPSO (thiolane)-thymine (90) α:β = 4:3	600
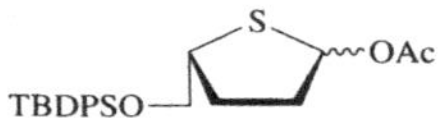 TBDPSO, S, OAc	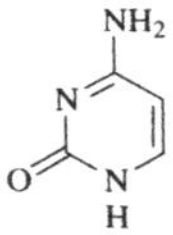 cytosine (NH_2)	TCS, HMDS, $C_4F_9SO_3K$, MeCN, 25°, 12 h	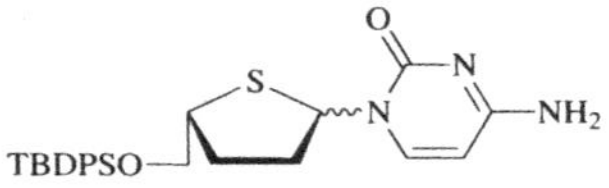 TBDPSO (thiolane)-cytosine (NH_2) (68) 1:1	597

TABLE I. ONE STEP-ONE POT SILYLATION/COUPLING OF HETEROCYCLIC BASES WITH SUGARS - FRIEDEL-CRAFTS CATALYSTS (*Continued*)

Sugar	Base	Conditions	Product(s) and Yield(s) (%)	Refs.
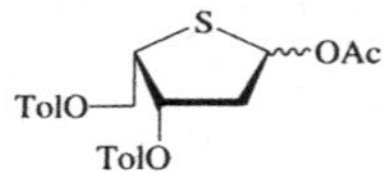 TolO, TolO, S, OAc	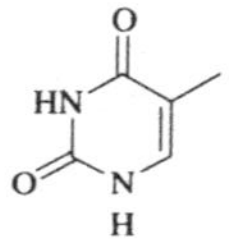 HN, O, O, N, H	TCS, HMDS, TMSOTf, MeCN, –78°, 1.5 h	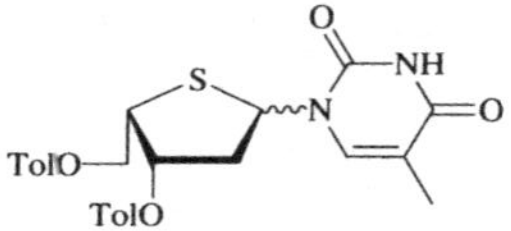 TolO, TolO, S, N, NH, O, O (—)	601
C_{24} 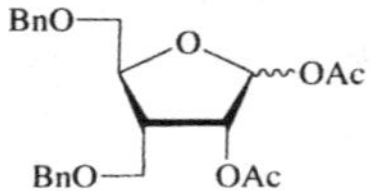 BnO, BnO, O, OAc, OAc	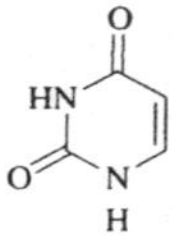 HN, O, O, N, H	TCS, HMDS, $C_4F_9SO_3K$, MeCN	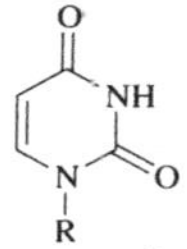 NH, O, O, N, R (—); R = BnO, BnO, O, OAc 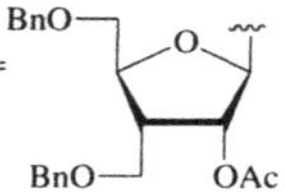	587
	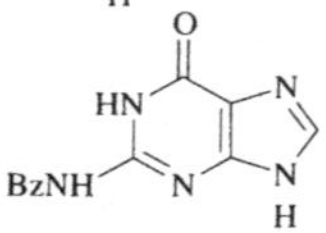 HN, O, N, N, BzNH, N, H	TCS, HMDS, $C_4F_9SO_3K$, MeCN	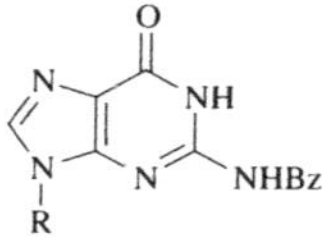 N, O, NH, N, N, NHBz, R (—)	587
	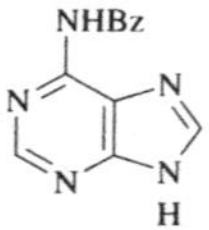 NHBz, N, N, N, N, H	TCS, HMDS, $C_4F_9SO_3K$, MeCN	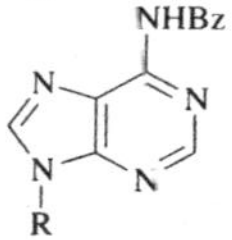 NHBz, N, N, N, N, R (—)	587
	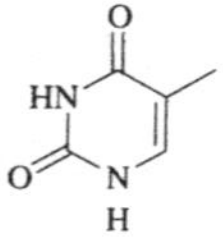 HN, O, O, N, H	TCS, HMDS, $SnCl_4$, MeCN, rt, 4 h	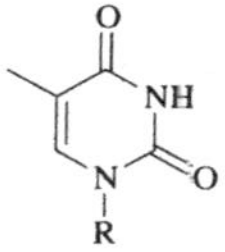 O, NH, N, O, R (91)	571
	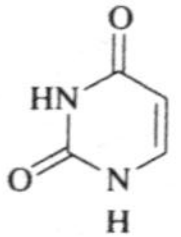 HN, O, O, N, H	TCS, HMDS, CF_3SO_3H, MeCN 1. rt, 3.5 h 2. 60°, 40 min	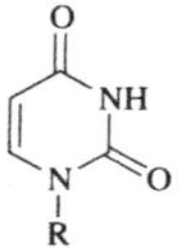 O, NH, N, O, R **I** (78)	571

Substrate	Base	Conditions	Product(s) and Yield(s) (%)		Refs.
		TCS, HMDS, $SnCl_4$, MeCN, rt, 4 h	**I** (79)		571
		TCS, HMDS, CF_3SO_3K, MeCN, reflux	**I** (87) + α-anomer (6)		574
C_{26}		TCS, HMDS, $SnCl_4$, MeCN		R =	
	R^1				
	Et	—	(—)		602
	n-Pr	—	(—)		602
	n-Bu	—	(—)		602
	CH=CH_2	—	(65)		602
	CH=CH_2	or $Cl(CH_2)_2Cl$	(65)		363
	i-Pr	rt, 24 h	(61)		359
		TCS, HMDS, TMSOTf, $Cl(CH_2)_2Cl$, rt, 10 h	(33) + N^7-isomer (2)		361
C_{27}		HMDS, $(NH_4)_2SO_4$, $SnCl_4$, $Cl(CH_2)_2Cl$, reflux, 1 h	(74)	R =	226
		BSA, $SnCl_4$, MeCN, 80°, 11 h	(73) α:β = 1:1	R =	226a

TABLE I. ONE STEP-ONE POT SILYLATION/COUPLING OF HETEROCYCLIC BASES WITH SUGARS - FRIEDEL-CRAFTS CATALYSTS (*Continued*)

Sugar	Base	Conditions	Product(s) and Yield(s) (%)	Refs.
	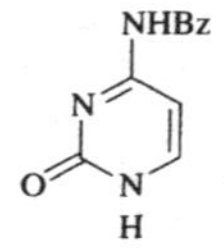 NHBz-cytosine (N-H)	HMDS, $(NH_4)_2SO_4$, $SnCl_4$, $Cl(CH_2)_2Cl$, reflux, 1 h	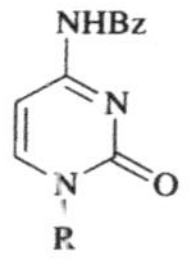 N^4-benzoylcytosine, N1-R (70)	226
C_{28} 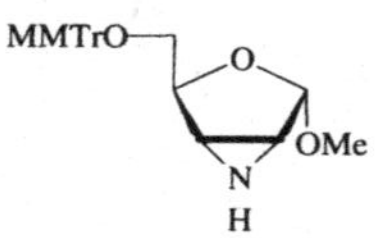 MMTrO-aziridine-fused furanoside, OMe	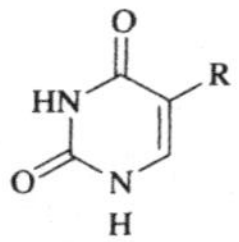 HN, O, R (5-R uracil)	TCS, HMDS, TBDMSOTf, MeCN, 124°	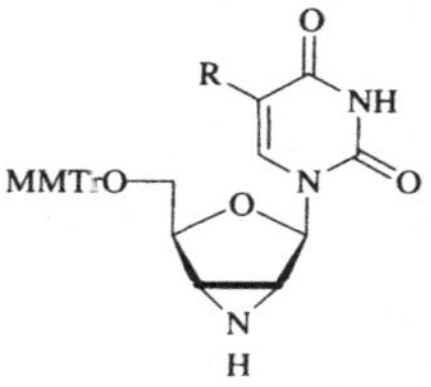 MMTrO- aziridine nucleoside; R: H (20), Me (20)	232
	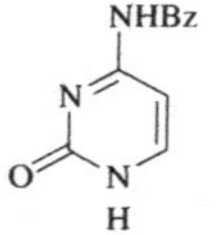 NHBz-cytosine	TCS, HMDS, TMSOTf, MeCN, 124°	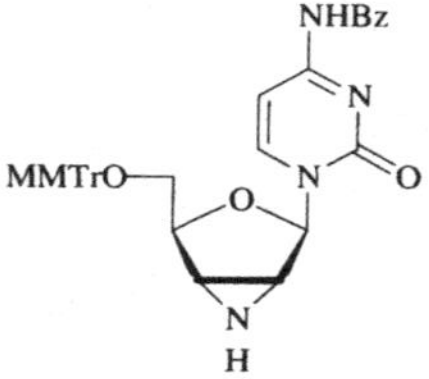 MMTrO- aziridine N^4-benzoylcytidine analog (20)	232
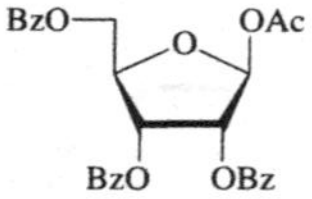 BzO, OAc, BzO, OBz ribofuranose	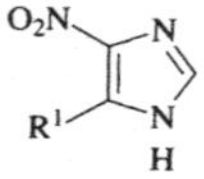 O_2N, R^1 imidazole (N-H)	TCS, HMDS, $SnCl_4$, MeCN, rt	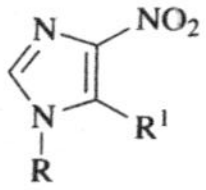 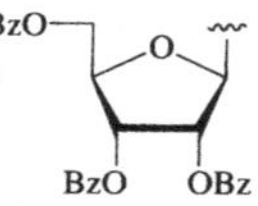 1-R-4-NO_2-5-R^1-imidazole; R = 2,3,5-tri-O-benzoyl-ribofuranosyl (BzO, BzO, OBz)	
	R^1 = Cl	2 h	(77)	603
	CO_2H	0.5 h	(>90)	604
	Br	2 h	(85)	603
	I	2 h	(61)	603

Substrate	Conditions	Product(s) and Yield(s) (%)	Refs.
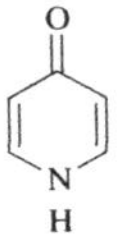 4-pyridone (structure)	1. TCS, HMDS, $C_4F_9SO_3K$, MeCN, heat, 24 h 2. NH_3, MeOH	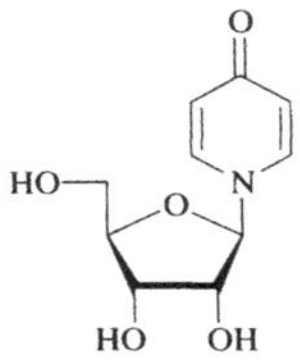 HO, HO, OH (structure) (50)	189
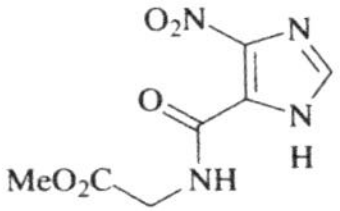 O_2N, MeO_2C, NH, H (structure)	TCS, HMDS, CF_3SO_3H, MeCN, rt, 0.5 h	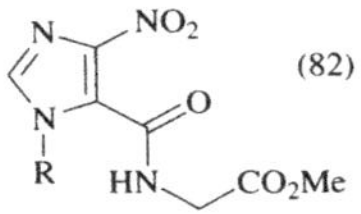 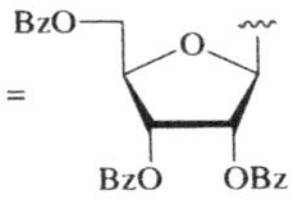 NO_2, CO_2Me, HN, R (structure) (82) R = BzO, BzO, OBz (structure)	604
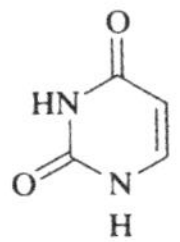 uracil (structure)	TCS, HMDS, MeCN	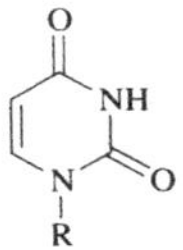 NH, R (structure)	
	$SnCl_4$, rt, 2 h	(83)	189
	NaI	(66)	137
	$NaBF_4$, 83°, 2 h	(42)	188
	NH_4ClO_4, 83°, 19 h	(40)	188
	$NH_4ClO_4 \cdot H_2O$, 83°, 19 h	(58)	188
	CF_3SO_3H, rt, 4 h, reflux 0.5 h	(81)	189
	$C_4F_9SO_3K$, heat	(84)	189
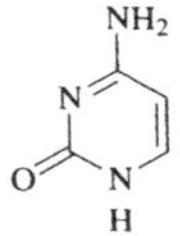 cytosine, NH_2 (structure)	TCS, HMDS, $C_4F_9SO_3K$, MeCN, heat, 26 h	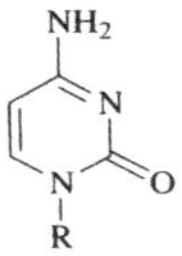 NH_2, R (structure) (56)	189

TABLE I. ONE STEP-ONE POT SILYLATION/COUPLING OF HETEROCYCLIC BASES WITH SUGARS - FRIEDEL-CRAFTS CATALYSTS (*Continued*)

Sugar	Base	Conditions	Product(s) and Yield(s) (%)	Refs.
	HN, O, N, H, O (thymine)	TCS, HMDS, CF_3SO_3H, MeCN, rt, 4 h	O, NH, O, N, R (88); R = BzO, O, BzO, OBz	605, 606
	HN, O, R[1], N, H, O	TCS, HMDS, MeCN	R[1], O, NH, O, N, R	
	R[1]			
	Me	$SnCl_4$	(—)	586
	Me	CF_3SO_3H, heat, 1 h	(88)	607
	Me	TMSOTf, reflux, 0.5 h	(82)	606
	OMe	$C_4F_9SO_3K$, reflux, 20 h	(70)	189, 608, 609
	SMe	$C_4F_9SO_3K$, reflux, 20 h	(48)	610
	S (propenyl)	$C_4F_9SO_3K$, reflux, 20 h	(70)	610
	S (allyl)	$C_4F_9SO_3K$, reflux, 20 h	(72)	610
	S (propargyl)	$C_4F_9SO_3K$, reflux, 20 h	(48)	610
	SBn	$C_4F_9SO_3K$, reflux, 20 h	(51)	610
	NH_2, N, F, O, N, H	1. TCS, HMDS, MeCN, $C_4F_9SO_3K$, reflux, 24 h 2. Amberlyst A-26, MeOH	NH_2, F, N, O, N, R (55); R = BzO, O, HO, OH	534

Substrate	Conditions	Product(s) and Yield(s) (%)		Refs.
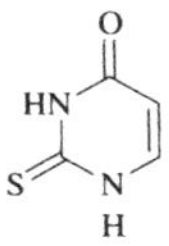 2-thiouracil	TCS, HMDS, $SnCl_4$, MeCN, rt, 6.5 h	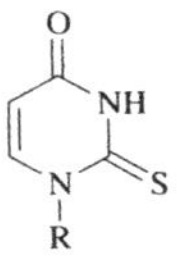 N-R product (59)	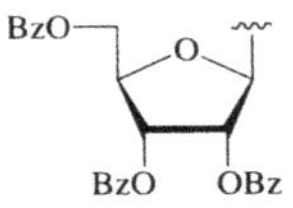 R = 2,3,5-tri-O-benzoyl-ribofuranosyl (BzO, BzO, OBz)	189
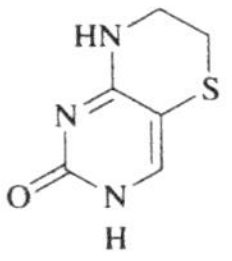 pyrimido-thiazine (HN, S, N, O, NH)	HMDS, TMSOTf, MeCN	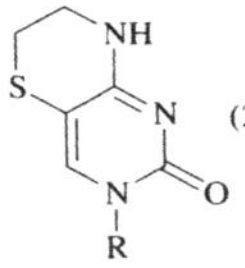 N-R product (28)		610
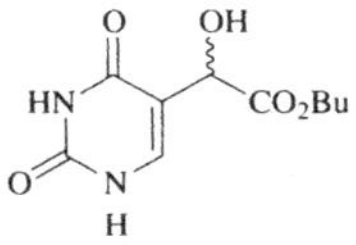 5-[CH(OH)CO_2Bu]uracil	TCS, HMDS, CF_3SO_3H	N-R product (76%) 1:1 mixture		611
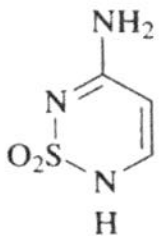 4-amino-thiadiazine dioxide (NH_2, O_2S)	TCS, HMDS, $C_4F_9SO_3K$, MeCN, heat, 24 h	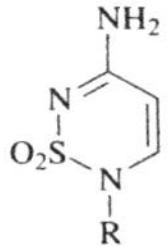 N-R product (41)		191
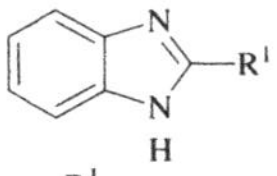 2-R^1-benzimidazole	TCS, HMDS, CF_3SO_3H, MeCN	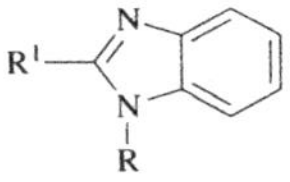 1-R-2-R^1-benzimidazole		612

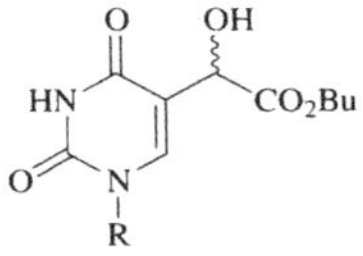

R^1	
H	(—)
CF_3	(—)
CH_2CN	(—)
Bn	(—)
SCF_3	(—)
CH_2SCF_3	(—)

TABLE I. ONE STEP-ONE POT SILYLATION/COUPLING OF HETEROCYCLIC BASES WITH SUGARS - FRIEDEL-CRAFTS CATALYSTS (*Continued*)

Sugar	Base	Conditions	Product(s) and Yield(s) (%)	Refs.
	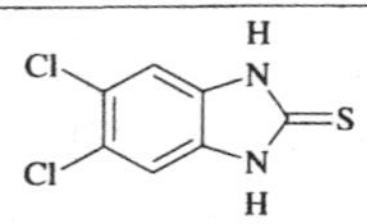	BSA, TMSOTf, MeCN, reflux, 18 h	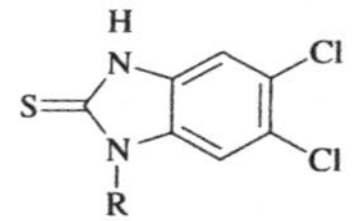 (74) R = ; + *N*,*N*-bisriboside (3) 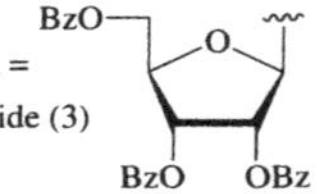	613
	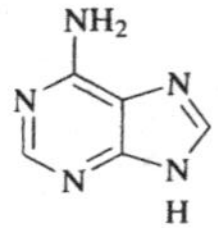	TCS, HMDS, $C_4F_9SO_3K$, MeCN	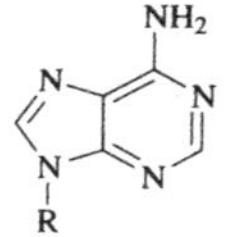 (62)	614
	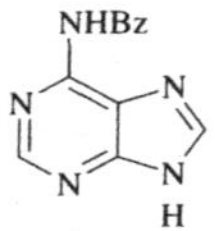	1. TCS, HMDS, $C_4F_9SO_3K$, MeCN, heat, 21 h 2. NH_3, MeOH	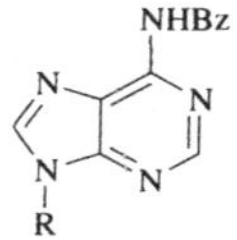 (63) R 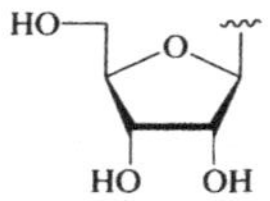 =	189
	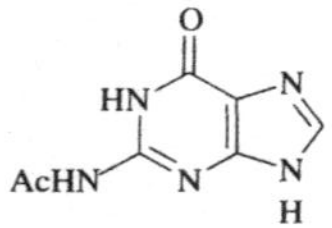	1. TCS, HMDS, $C_4F_9SO_3K$, MeCN, heat, 21 h 2. NH_3, MeOH	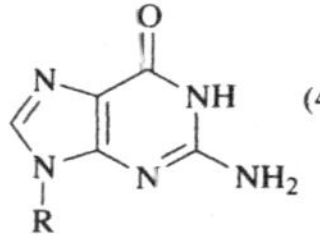 (44)	189
	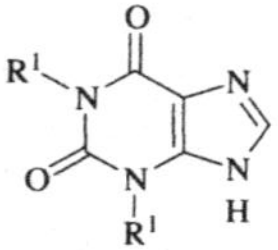	1. TCS, HMDS, $C_4F_9SO_3K$, MeCN, reflux, 30 min 2. NH_3, MeOH, rt, 2 days	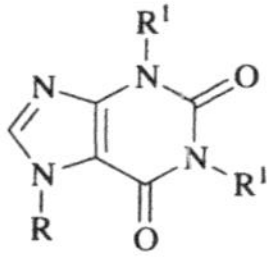 R^1: Me (—); *n*-Pr (—); *n*-Bu (—)	615
	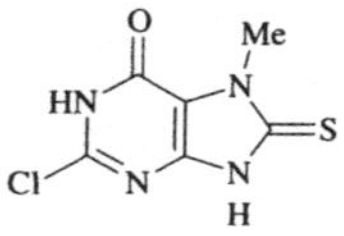	TCS, HMDS, TMSOTf, MeCN, reflux, 3.5 h	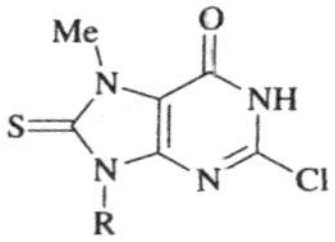 (76) R 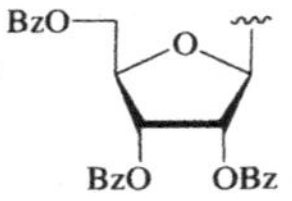 =	616

Sugar	Base	Conditions	Product (Yield %)	Ref.
	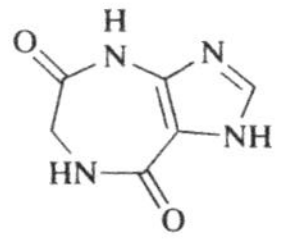	TCS, HMDS, CF_3SO_3H, MeCN rt, 17 h	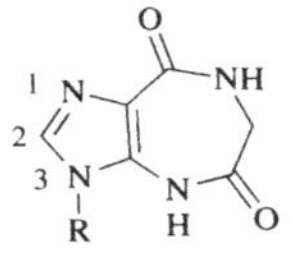 (14) + 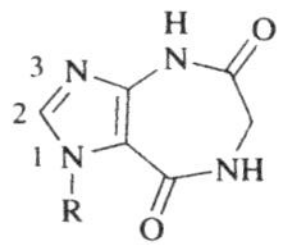 (51)	617
	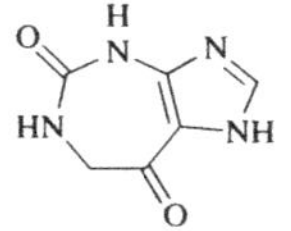	TCS, HMDS, CF_3SO_3H, MeCN rt, 1 h	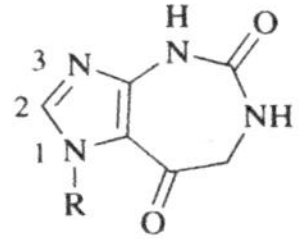 (35) + isomer (15)	618
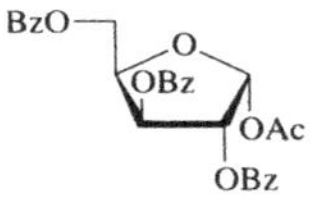 BzO, OBz, OAc, OBz	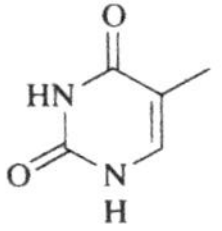	TCS, HMDS, $SnCl_4$, MeCN, reflux, 20 min	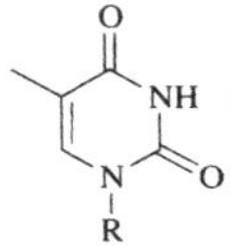 (72) R 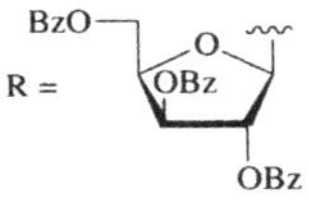 = (BzO, OBz, OBz)	192e
	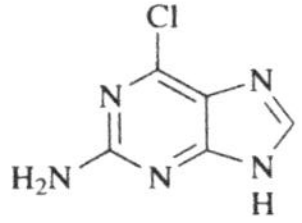 Cl, H_2N	TCS, HMDS, $SnCl_4$, MeCN, reflux, 20 min	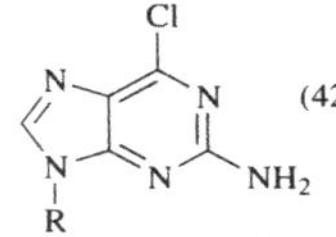 (42) + N^7-isomer (34)	192e
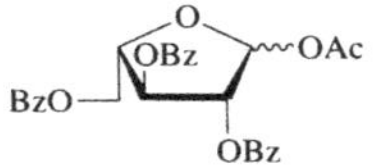 OBz, OAc, BzO, OBz	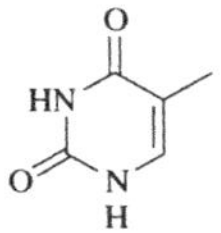	TCS, HMDS, $SnCl_4$, MeCN, reflux, 1 h	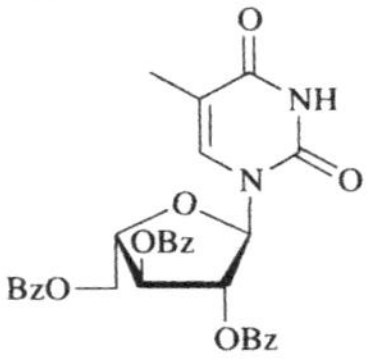 (60)	619
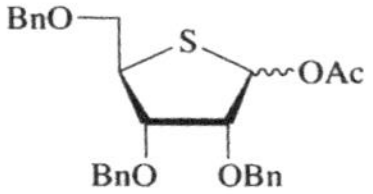 BnO, S, OAc, BnO, OBn	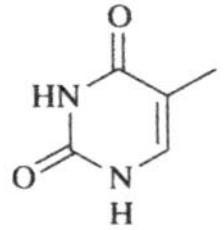	TCS, HMDS, $SnCl_4$, MeCN, reflux, 20 min	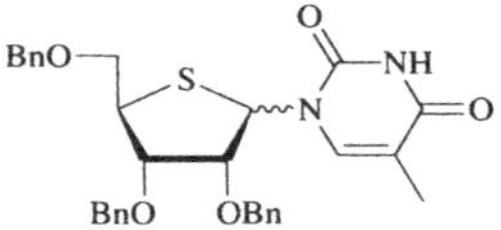 (39)	620

TABLE I. ONE STEP-ONE POT SILYLATION/COUPLING OF HETEROCYCLIC BASES WITH SUGARS - FRIEDEL-CRAFTS CATALYSTS (*Continued*)

Sugar	Base	Conditions	Product(s) and Yield(s) (%)	Refs.
	HN, O, N, H, R	BSA, TMSOTf, MeCN, rt, 2.5 h	R, O, NH, N, O, BnO, S, BnO, OBn; R: H (42), α-anomer (31); Me (41), α-anomer (37)	620
C_{30} BzO, BzO, O, OAc, MsO, OBz	HN, O, N, H, O	TCS, HMDS, TMSOTf, MeCN, rt, 12 h	O, NH, N, O, BzO, BzO, O, MsO, OBz (69)	581
BzO, OBz, O, OAc, MsO, OBz	HN, O, N, H, O	TCS, HMDS, TMSOTf, MeCN, rt, 8 h	O, NH, N, O, BzO, OBz, O, MsO, OBz (69)	581
C_{32} BzO, O, OBz, BzO, OTBDMS	HN, O, N, H, O	HMDS, $(NH_4)_2SO_4$, $SnCl_4$, $Cl(CH_2)_2Cl$, reflux, 1 h	O, NH, N, O, R (95); R = BzO, O, BzO, OTBDMS	226

C_{34}	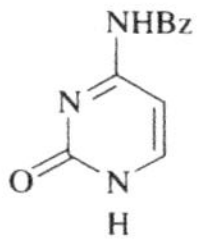	HMDS, $(NH_4)_2SO_4$, $SnCl_4$, $Cl(CH_2)_2Cl$, reflux, 1 h	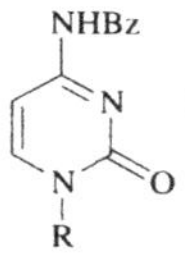 (86)	226
	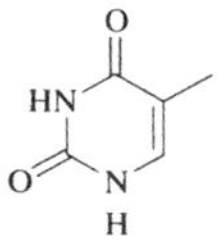	TCS, HMDS, $SnCl_4$, MeCN, reflux, 20 min	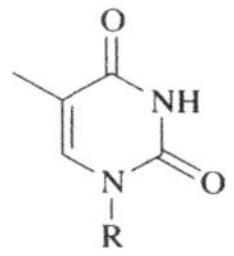 (87)	621, 226
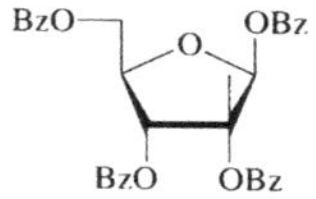	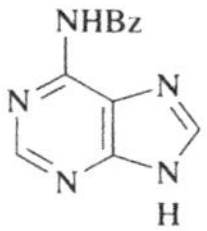	BSA, TMSOTf, MeCN, 80°, 3 h	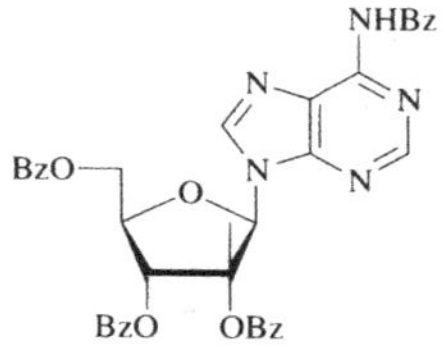 (86)	622, 623
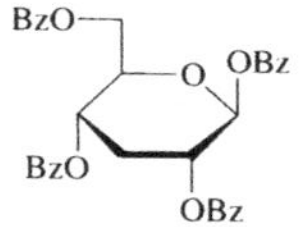	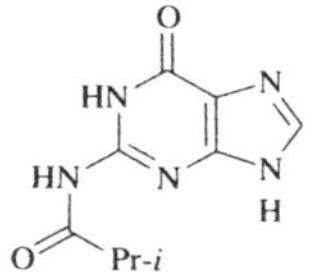	BSA, TMSOTf, MeCN, 80°, 16 h	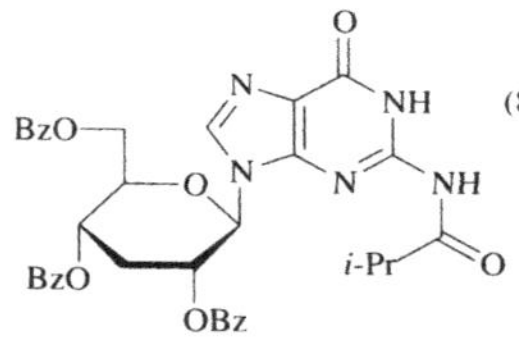 (89) β-N^9:β-N^7 = 3:2	192

TABLE II. ONE STEP-ONE POT COUPLING OF HETEROCYCLIC BASES WITH SUGARS - FRIEDEL-CRAFTS CATALYSTS

Sugar	Base	Conditions	Product(s) and Yield(s) (%)	Refs.
C_{11} 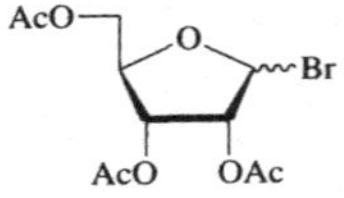 AcO, Br, AcO, OAc	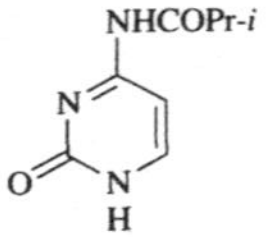 NHCOPr-*i*	$SnCl_4$, $Cl(CH_2)_2Cl$, 50°, 3 h	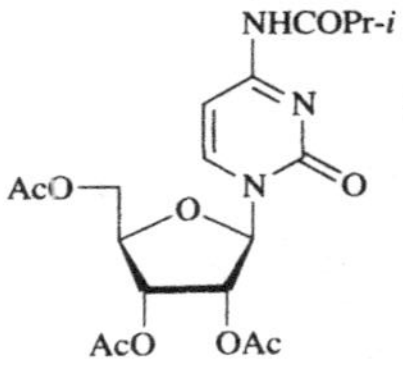 NHCOPr-*i*, AcO, AcO, OAc (50)	424a
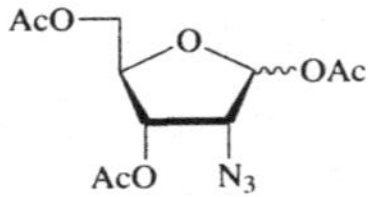 AcO, OAc, AcO, N_3	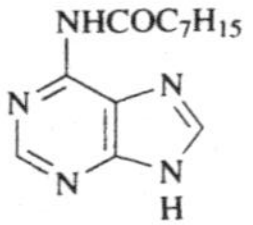 $NHCOC_7H_{15}$	1. $SnCl_4$, $Cl(CH_2)_2Cl$, reflux, 15 h 2. NaOMe	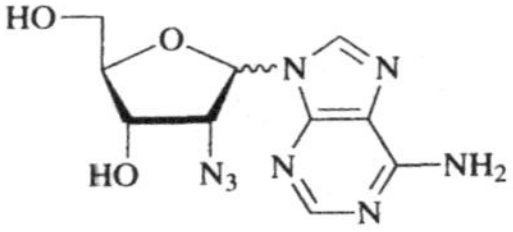 HO, HO, N_3, NH_2 (60-65) α:β = 1:2	624
	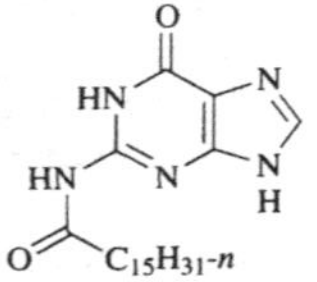 O, HN, HN, $C_{15}H_{31}$-*n*	1. $SnCl_4$, $Cl(CH_2)_2Cl$, reflux, 90 min 2. NaOMe, MeOH, 37°, 12 h	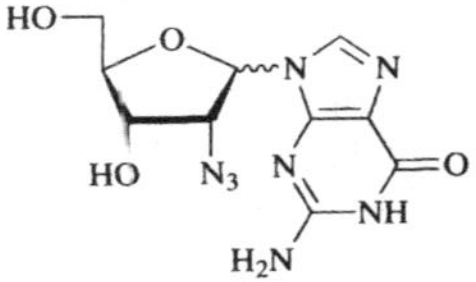 HO, HO, N_3, H_2N (42) α:β = 1.2:3 + N^9-isomer (58), α:β = 0.7:5.1	624
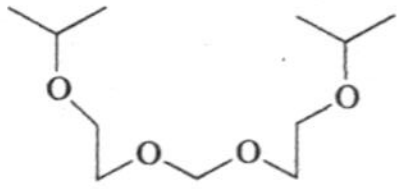	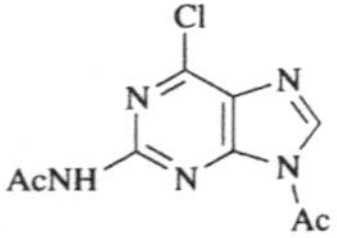 Cl, AcNH, Ac	TMSOTf, HMDS, $Cl(CH_2)_2Cl$, −30°	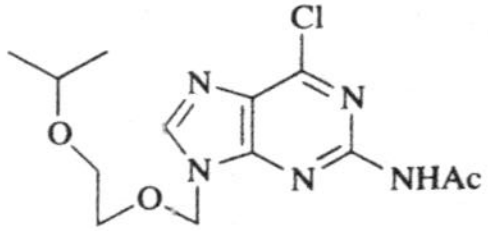 Cl, NHAc (58)	521
C_{13} 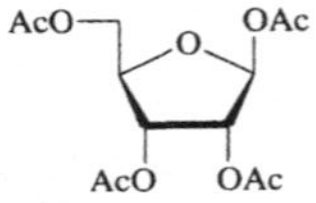 AcO, OAc, AcO, OAc	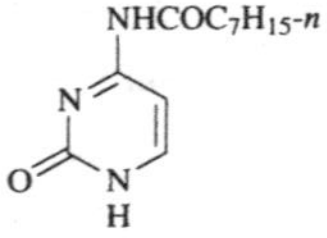 $NHCOC_7H_{15}$-*n*	$SnCl_4$, $Cl(CH_2)_2Cl$, rt, 24 h	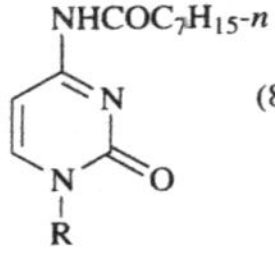 $NHCOC_7H_{15}$-*n*, R (80) R = AcO, AcO, OAc 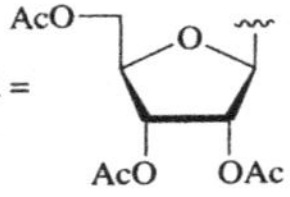	180

Substrate	Conditions	Product(s) and Yield(s) (%)	Refs.
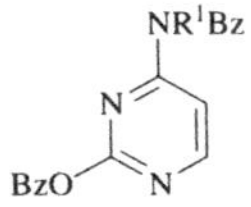 2-BzO-4-(NR^1Bz)pyrimidine	$SnCl_4$, $Cl(CH_2)_2Cl$	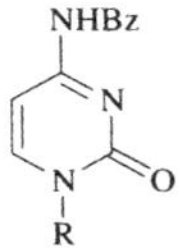 1-R-4-NHBz-pyrimidin-2-one; R^1 = H: 0°, 5 min, rt, 12 h (83); R^1 = Bz: 0°, 20 min, rt, 2 h (59)	424
6-R^1-purine (X)	$SnCl_4$, MeCN, rt	9-R-6-R^1-purine; R^1 = H, X = CH, 5 h (57)	625
		R^1 = NO_2, X = CH, 7 h (82)	626
		R^1 = NH_2, X = N, (—) (100)	625
		R^1 = NH_2, X = N, 12 h (77)	192a
		R^1 = SMe, X = N, 12 h (—)	192a
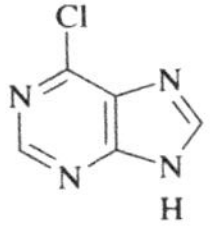 6-Cl-purine	1. $SnCl_4$, MeCN, rt, 12 h 2. thiourea, EtOH, reflux 3. rt, 12 h	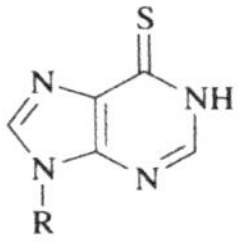 9-R-6-thioxopurine (78)	192a
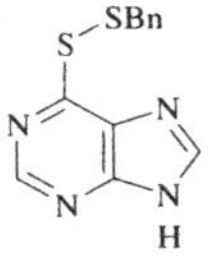 6-(BnS–S)-purine	1. $SnCl_4$, MeCN, rt, 12 h 2. $HS(CH_2)_2OH$, EtOH, rt, 3 h	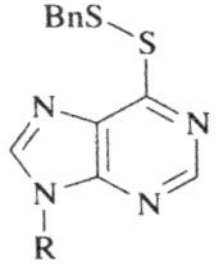 9-R-6-(BnS–S)-purine (85)	192a
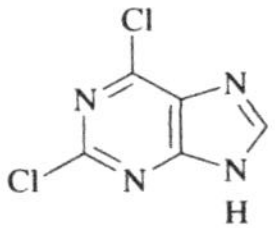 2,6-diCl-purine	$SnCl_4$, MeCN, rt, 12 h	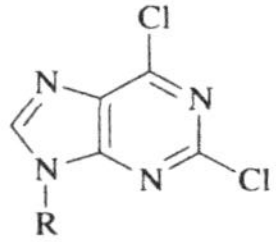 9-R-2,6-diCl-purine (81)	192a
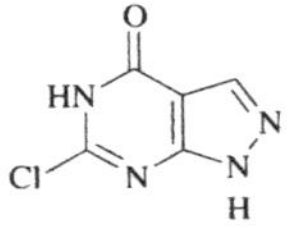 6-Cl-pyrazolo[3,4-d]pyrimidin-4-one	$SnCl_4$, MeCN, rt, 16 h	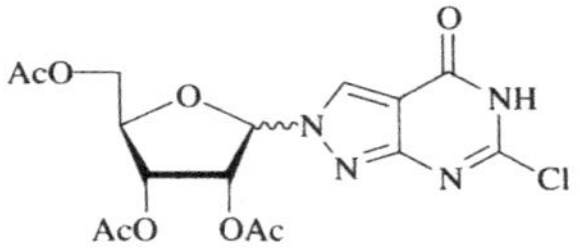 2-(2,3,5-tri-O-Ac-ribofuranosyl) product (AcO, AcO, OAc) (50)	627

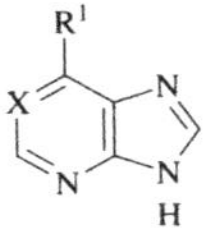

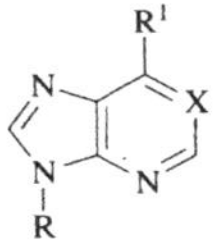

TABLE II. ONE STEP-ONE POT COUPLING OF HETEROCYCLIC BASES WITH SUGARS - FRIEDEL-CRAFTS CATALYSTS (*Continued*)

Sugar	Base	Conditions	Product(s) and Yield(s) (%)	Refs.
 AcO, O, OAc AcO, OAc	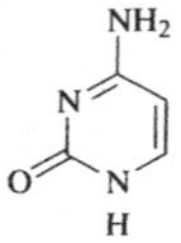 NH₂, N, O, N, H	$SnCl_4$, MeCN, rt, 23 h	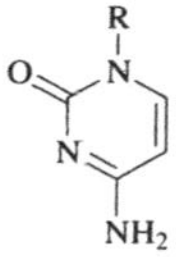 R, 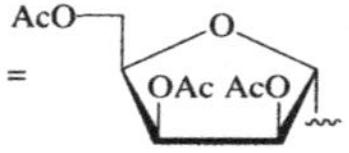 N, O, N, NH₂ (—) R = AcO, O, OAc AcO	537
	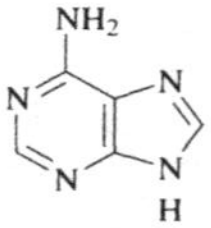 NH₂, N, N, N, N, H	$SnCl_4$, MeCN, rt, 23 h	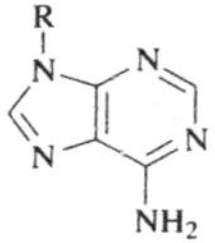 R, N, N, N, N, NH₂ (—)	537
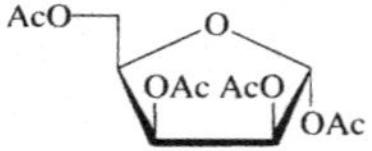 AcO, O, OAc AcO, OAc	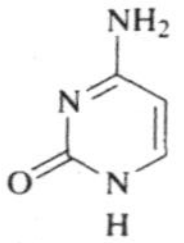 NH₂, N, O, N, H	$SnCl_4$, MeCN	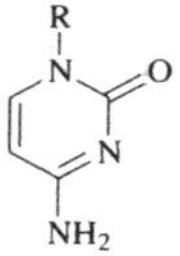 R, N, O, N, NH₂ (31)	192f
	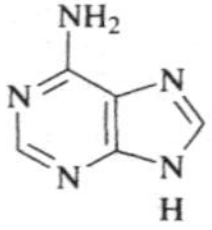 NH₂, N, N, N, N, H	$SnCl_4$, MeCN	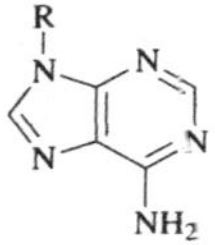 R, N, N, N, N, NH₂ (100)	192f
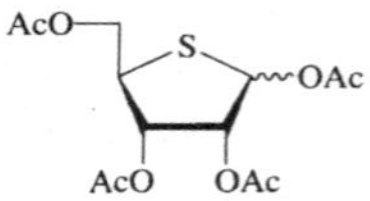 AcO, S, OAc, AcO, OAc	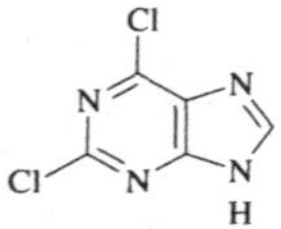 Cl, N, N, Cl, N, N, H	$SnCl_4$, MeCN or CH_2Cl_2, rt, 12 h	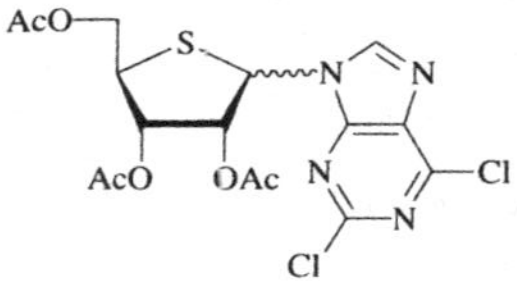 AcO, S, N, N, AcO, OAc, N, Cl, N, Cl (70) α:β = 1:3	86a

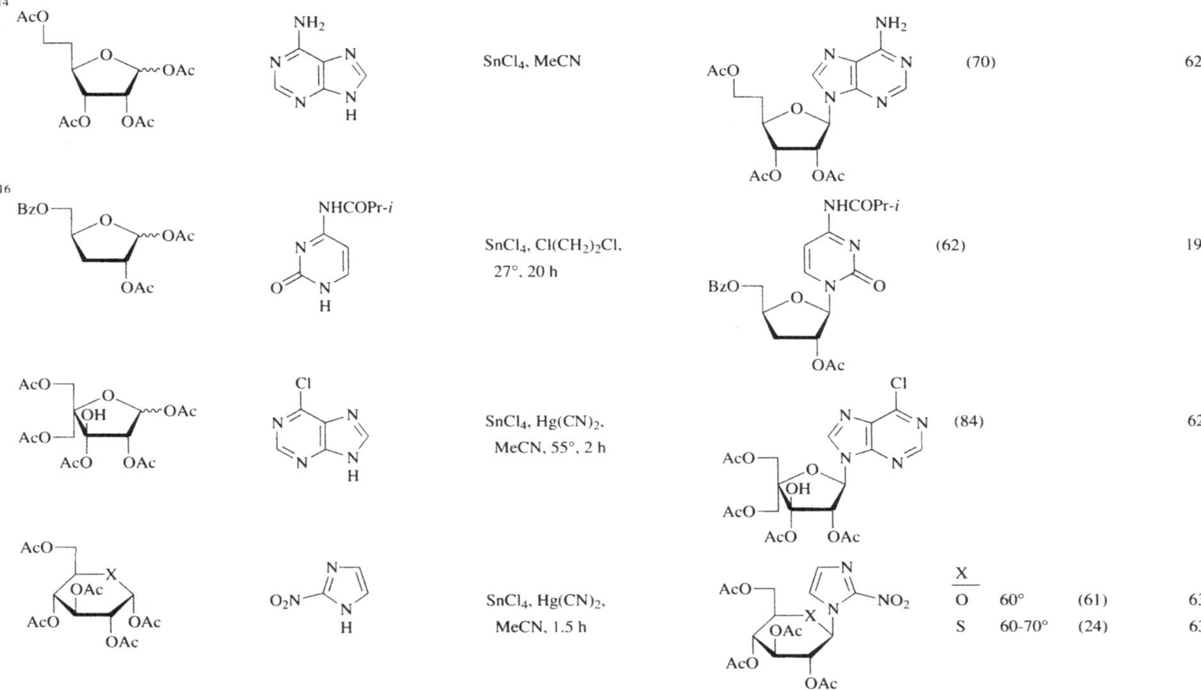

	Conditions	Product(s) and Yield(s) (%)	Refs.
C_{14}	$SnCl_4$, MeCN	(70)	628
C_{16}	$SnCl_4$, $Cl(CH_2)_2Cl$, 27°, 20 h	(62)	192o
	$SnCl_4$, $Hg(CN)_2$, MeCN, 55°, 2 h	(84)	629
	$SnCl_4$, $Hg(CN)_2$, MeCN, 1.5 h	X = O, 60° (61)	630
		X = S, 60-70° (24)	630

TABLE II. ONE STEP-ONE POT COUPLING OF HETEROCYCLIC BASES WITH SUGARS - FRIEDEL-CRAFTS CATALYSTS (*Continued*)

Sugar	Base	Conditions	Product(s) and Yield(s) (%)	Refs.
C_{18} 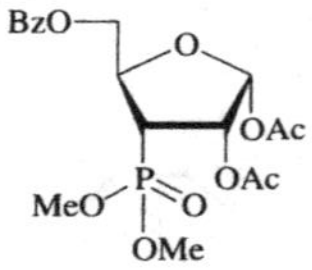 BzO, OAc, OAc, MeO–P(=O)–OMe	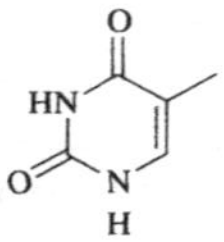 HN, O, O, N, H	$SnCl_4$, MeCN, rt, 5 d	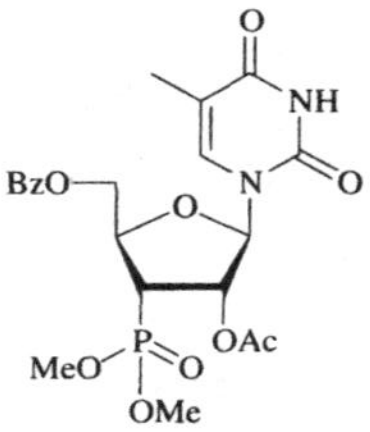 O, NH, O, N, BzO, O, OAc, MeO–P(=O)–OMe (29)	192j
	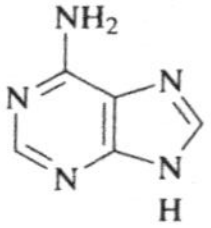 NH_2, N, N, N, N, H	$SnCl_4$, MeCN, rt, 5 d	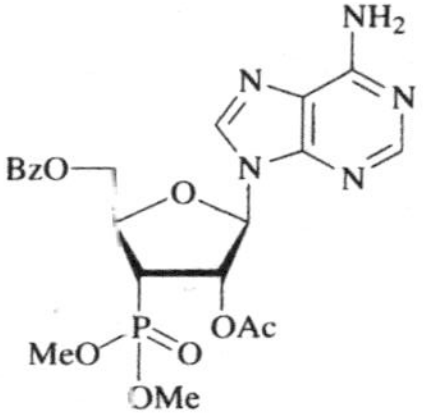 NH_2, N, N, N, N, BzO, O, OAc, MeO–P(=O)–OMe (28)	192j
C_{19} 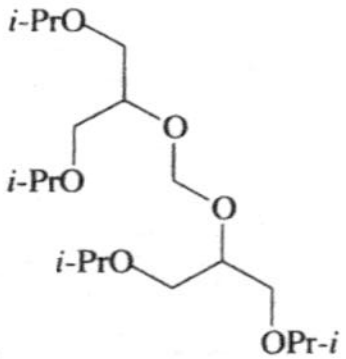 *i*-PrO, *i*-PrO, O, O, *i*-PrO, OPr-*i*	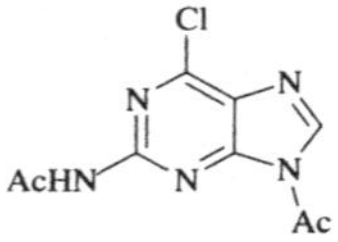 Cl, N, N, N, N, AcHN, Ac	$SnCl_4$, 100°, 8 h	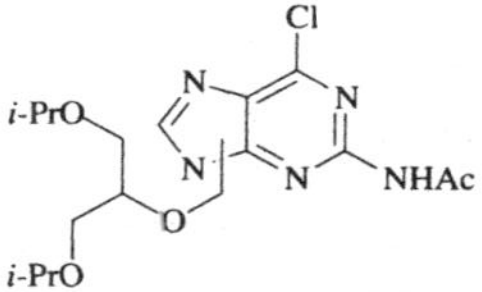 Cl, N, N, N, N, NHAc, *i*-PrO, O, *i*-PrO DMF (64) N^9:N^7 = 93.2:2.5 NMP (75) —	521
C_{22} 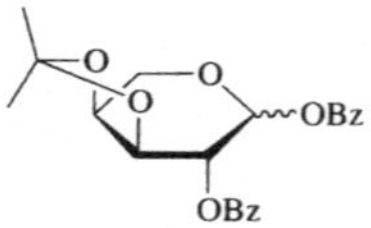 O, O, O, OBz, OBz	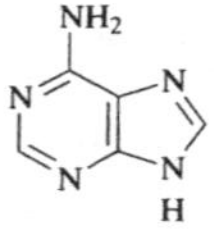 NH_2, N, N, N, N, H	$SnCl_4$, MeCN	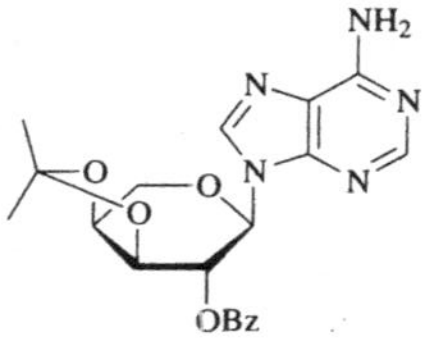 NH_2, N, N, N, N, O, O, O, OBz (74)	192h

Entry	Sugar	Base	Conditions	Product (Yield, %)	Refs.
C_{23} 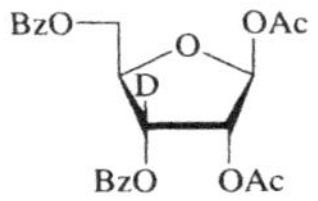			$SnCl_4$, MeCN, rt	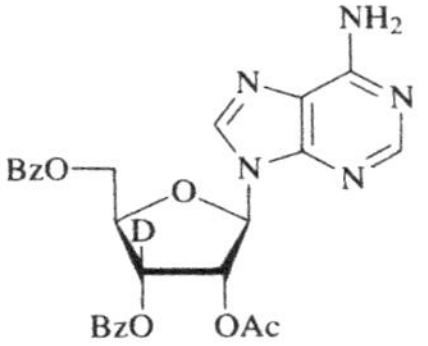 (75)	192d
			$SnCl_4$, MeCN, rt, 15 h	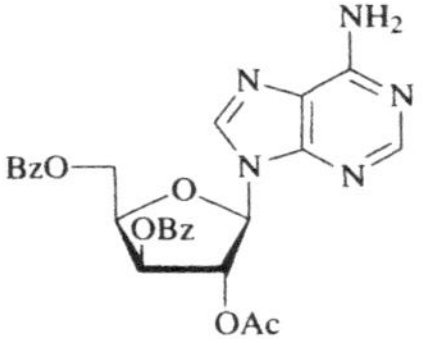 **I** (70)	192f
			$SnCl_4$, MeCN, rt, 15 h	**I** (70)	192e
			$SnCl_4$, MeCN	**I** (70)	
			$SnCl_4$, MeCN, rt, 18 h	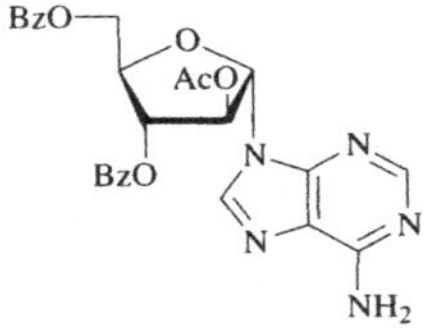 (66)	595
			$SnCl_4$, MeCN, rt, 20 h	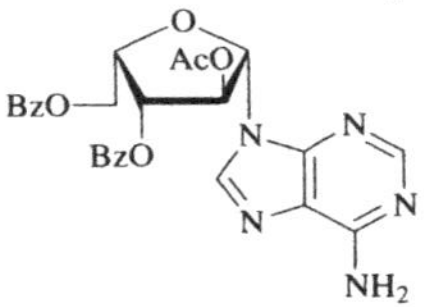 (61)	192i

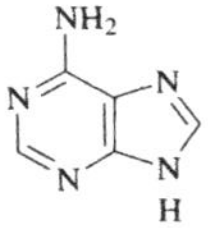

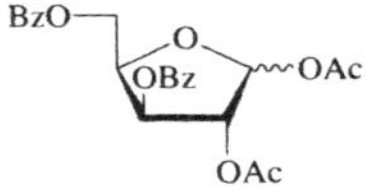

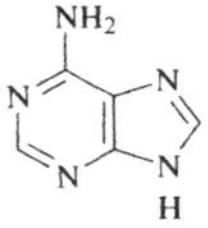

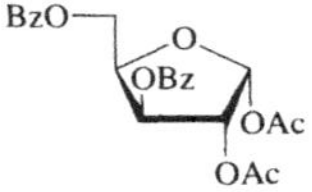

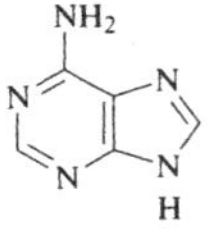

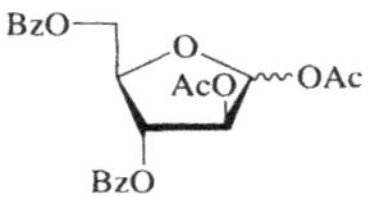

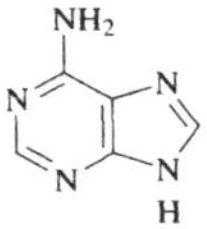

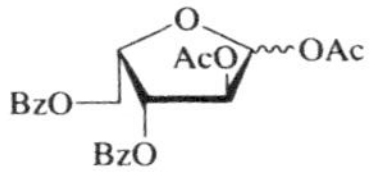

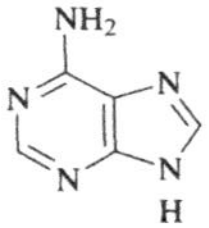

TABLE II. ONE STEP-ONE POT COUPLING OF HETEROCYCLIC BASES WITH SUGARS - FRIEDEL-CRAFTS CATALYSTS (*Continued*)

Sugar	Base	Conditions	Product(s) and Yield(s) (%)	Refs.
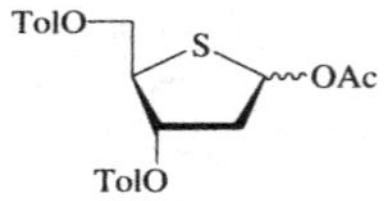 TolO, S, OAc, TolO	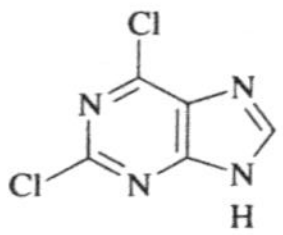 Cl, N, N, N, Cl, N, H	$SnCl_4$, MeCN, 0°, 1.5 h	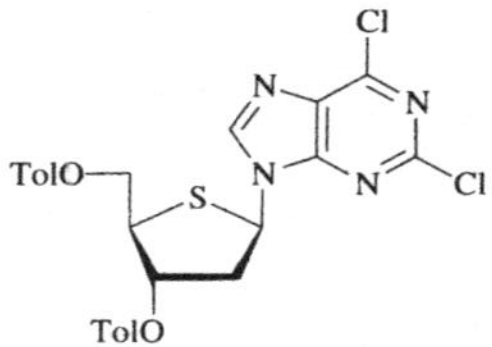 Cl, N, N, TolO, S, N, N, Cl, TolO (71)	631
C_{26} 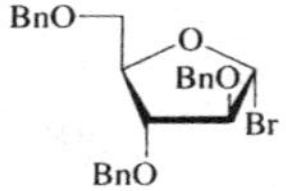 BnO, O, BnO, Br, BnO	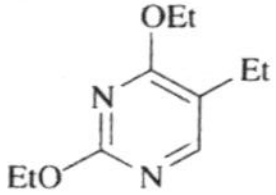 OEt, Et, N, EtO, N	$SnCl_4$, MeCN, rt, 6 d	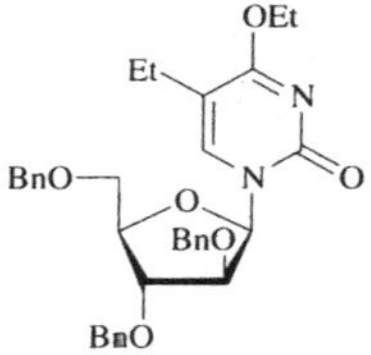 OEt, Et, N, BnO, N, O, O, BnO, BnO (49)	358
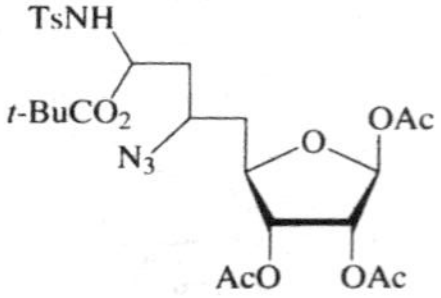 TsNH, *t*-$BuCO_2$, N_3, O, OAc, AcO, OAc	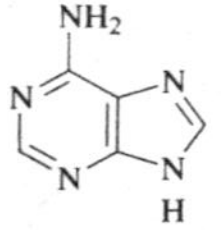 NH_2, N, N, N, N, H	$SnCl_4$, $Cl(CH_2)_2Cl$, MeCN, rt, 16 h	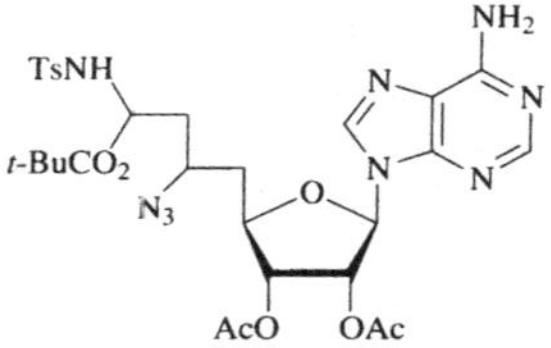 NH_2, TsNH, N, N, *t*-$BuCO_2$, N_3, O, N, N, AcO, OAc (59)	192g, 86b
C_{27} 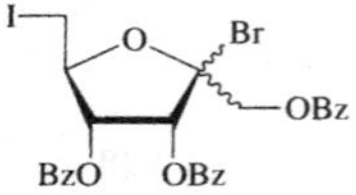 I, O, Br, OBz, BzO, OBz	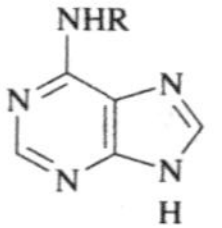 NHR, N, N, N, N, H	$SnCl_4$, $Hg(CN)_2$, MeCN, 60°, 2 h	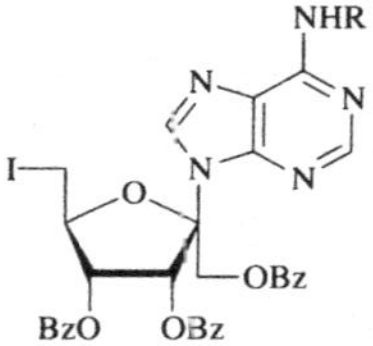 NHR, N, N, I, O, N, N, OBz, BzO, OBz R: H (21), α-anomer (7); COC_5H_{11}-*n* (46), α-anomer (12)	632

Substrate	Reagent	Conditions	Product(s) and Yield(s) (%)	Refs.
C_{28} 2,3,5-tri-O-benzoyl-1-O-acetylribofuranose (BzO, OAc, BzO, OBz)	2-nitroimidazole (O_2N, N, H)	$SnCl_4$, $Hg(CN)_2$, MeCN, 60°, 75 min	BzO, BzO BzO, NO_2 (61)	213
	NHR^1, N, O, N, H	$SnCl_4$, $Cl(CH_2)_2Cl$, rt, 20 h	NHR^1, N, O, N–R; R = (BzO, OBz, BzO furanosyl)	180
	R^1 = COPr-*i*		(64)	
	COC_7H_{15}		(80)	
	COCH(Et)Bu		(72)	
	$COC_{15}H_{31}$		(81)	
	$OSnBu_3$, Bu_3SnO, N, N	$SnCl_4$, CH_2Cl_2, 10°, 6 d	NH, O, O, N–R (60) + N^1-isomer (40)	426
	$OSnBu_3$, Bu_3SnS, N, N	$SnCl_4$, $Cl(CH_2)_2Cl$, 20°, 19 h	O, NH, S, N–R (90)	426
	$OSnBu_3$, R^1, Bu_3SnO, N, N	$SnCl_4$, $Cl(CH_2)_2Cl$	R^1, O, NH, O, N–R; R = Br, rt, 12 h (97)	426
			Me, 20°, 2 d (63)	426

TABLE II. ONE STEP-ONE POT COUPLING OF HETEROCYCLIC BASES WITH SUGARS - FRIEDEL-CRAFTS CATALYSTS (*Continued*)

Sugar	Base	Conditions	Product(s) and Yield(s) (%)	Refs.
	[structure: 2,4-dimethoxy-5-R¹-pyrimidine; OMe, R¹, N, MeO, N]	$SnCl_4$, MeCN	[structure: OMe, R¹, N, OMe, N-R] R = [structure: BzO, O, BzO, OBz]	
	R¹			
	F	50-60°, 10 h	(87)	514b
	Me	rt, 12 h	(99)	633
	Et	rt, 12 h	(92)	633
	Pr-*n*	rt, 12 h	(90)	633
	Bu-*n*	rt, 12 h	(94)	633
	C_5H_{11}-*n*	rt, 12 h	(95)	633
	C_6H_{13}-*n*	rt, 12 h	(99)	633
	[structure: OBu-*t*, F, N, *t*-BuO, N]	$SnCl_4$, MeCN, rt, 30 min	[structure: O, F, NH, N-R, O] (97)	634
	[structure: OMe, N, O, N, CN]	$SnCl_4$, $Cl(CH_2)_2Cl$, rt, 12 h	[structure: NC, N, O, N-R, O] (21)	254
	[structure: H_2N, H, N, N, N, N, N]	$SnCl_4$, MeCN, rt, 24 h	[structure: H_2N, N, N, N, N, N-R] (7) + N^1-α-isomer (—) + N^2-β-isomer (30)	635

Sugar	Base	Conditions	Product (Yield)	Ref.
		$SnCl_4$, $MeNO_2$, rt, 18 h	2-(ethylthio)adenine, N-9 R (63)	591
		1. $SnCl_4$, MeCN, rt, 24 h 2. NH_3, MeOH	(78)	192a
		1. $SnCl_4$, $Cl(CH_2)_2Cl$, rt 2. NaI	(35)	636
		$SnCl_4$, MeCN, rt, 15 h	(33)	192e, 192i
		$SnCl_4$, MeCN, rt, 30 min	(68)	192b

TABLE II. ONE STEP-ONE POT COUPLING OF HETEROCYCLIC BASES WITH SUGARS - FRIEDEL-CRAFTS CATALYSTS (*Continued*)

Sugar	Base	Conditions	Product(s) and Yield(s) (%)	Refs.
C_{29} 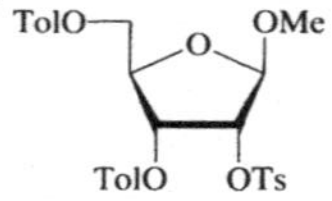 TolO, OMe, TolO, OTs	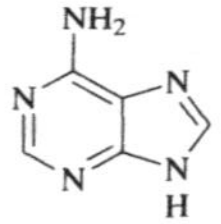 NH_2, N, N, N, N, H	$SnCl_4$, MeCN	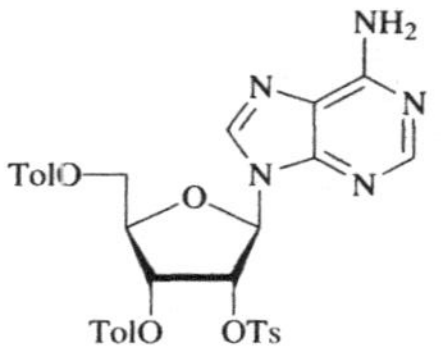 NH_2, N, N, N, N, TolO, TolO, OTs (—)	192k
C_{30} 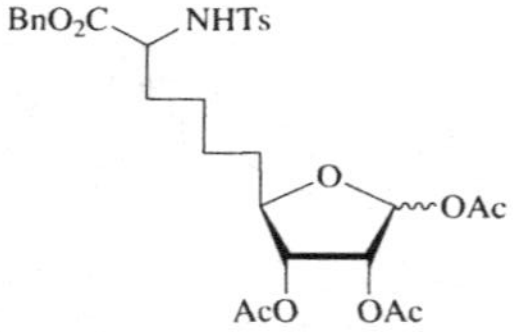 BnO_2C, NHTs, OAc, AcO, OAc	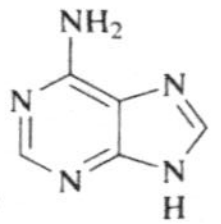 NH_2, N, N, N, N, H	$SnCl_4$, MeCN, 22°, 4.5 h	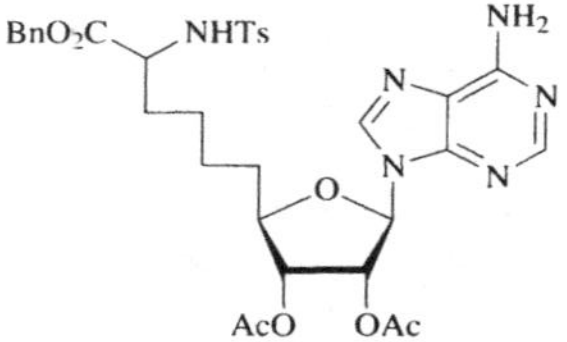 BnO_2C, NHTs, NH_2, N, N, N, N, AcO, OAc (61)	86b
C_{31} 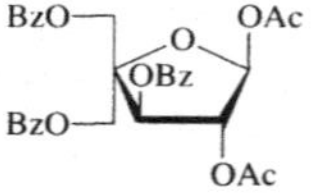 BzO, OAc, OBz, BzO, OAc	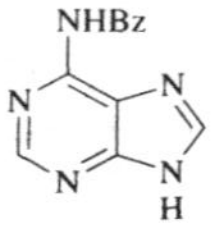 NHBz, N, N, N, N, H	$SnCl_4$, MeCN, 24°, 18 h	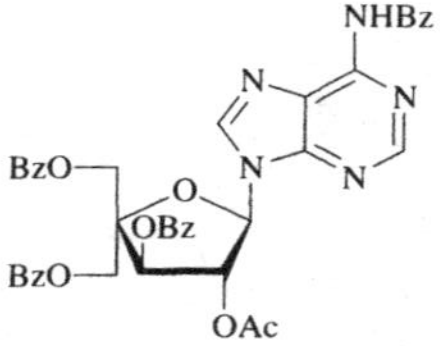 NHBz, N, N, N, N, BzO, OBz, BzO, OAc (73)	192n
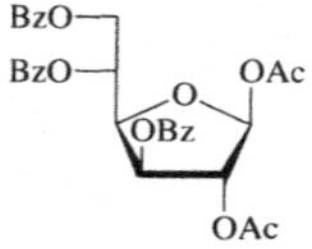 BzO, BzO, OAc, OBz, OAc	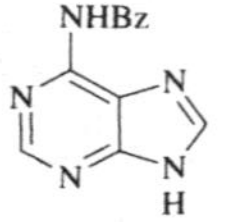 NHBz, N, N, N, N, H	$SnCl_4$, MeCN, 24°, 18 h	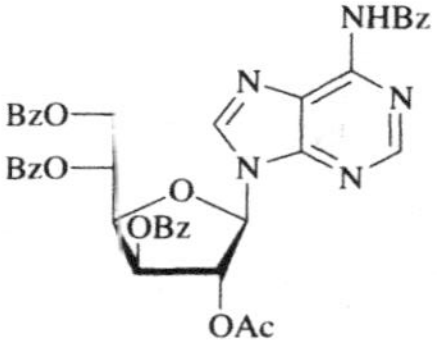 NHBz, N, N, N, N, BzO, BzO, OBz, OAc (79)	192n

Substrate	Nucleophile	Conditions	Product (Yield, %)		Ref.
		$SnCl_4$, MeCN, rt, 18 h		(64)	192c
		$SnCl_4$, MeCN, rt, 17 h		(67)	86b

TABLE III. REACTIONS OF SILYLATED HETEROCYCLIC BASES WITH PROTECTED SUGARS - $SnCl_4$ CATALYST

Sugar	Base	Conditions	Product(s) and Yield(s) (%)	Refs.
C_3	OTMS, F, N, N, TMSO	1. $SnCl_4$, CH_2Cl_2, rt, 1 h 2. $NaHCO_3$, MeOH, rt, 1 h	HO, O, R (23) R = F, O, NH, N, O	637
C_4 OMe, OMe		$SnCl_4$, $CHCl_3$, rt, 20 min	R, OMe (62)	638
O		1. H_2O 2. $SnCl_4$, dioxane, rt, 3 h	O, R **I** (26)	639
		1. H_2O 2. $SnCl_4$, PhMe, rt, 3 h	**I** (86)	639
O, X, X		$SnCl_4$, CH_2Cl_2, rt, 12 h	O, R, X X: Cl (51) + *cis* isomer (23) Br (51) + *cis* isomer (22)	640 640
O, O	OTMS, R, N, N, TMSO	$SnCl_4$, CH_2Cl_2, rt, 1 h	R, O, NH, N, O, HO, O **I** R: H (65) F (33) Me (54)	641
		1. $SnCl_4$, $CHCl_3$, rt, 0.5 h 2. $NaHCO_3$, MeOH	**I** R = H (—)	637
C_5 H, O, N, OMe	OTMS, R^1, N, N, TMSO	$SnCl_4$, CH_2Cl_2, MeCN, 0-5°, 6 h; rt, 6 h	H, O, N, R^2 R^1 = F, (41) R^2 = R^1, O, NH, N, O	642

Substrate	Nucleophile	Conditions	Product(s) and Yield(s) (%)	Refs.
		$SnCl_4$, CH_2Cl_2, rt, 48 h	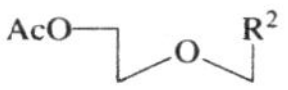 AcO–O–R^2 $R^1 = CF_3$, (55)	643
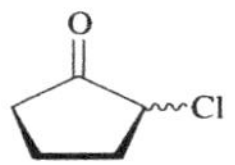 O, Cl	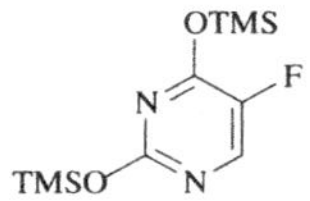 TMSO, N, N, OTMS, F	$SnCl_4$, CH_2Cl_2 reflux, 12 h	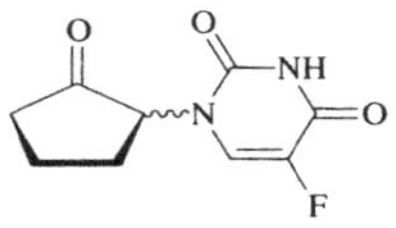 O, N, NH, O, O, F (10)	644
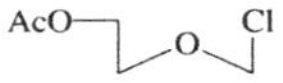 AcO–O–Cl	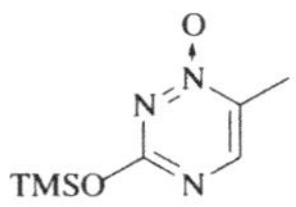 TMSO, N, N, N, O	$SnCl_4$, CH_2Cl_2, rt, 16 h	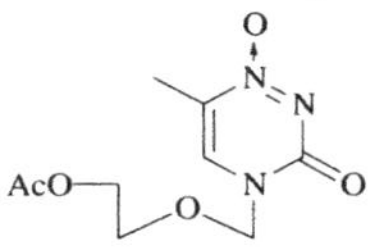 O, N, N, N, O, AcO–O (44)	645
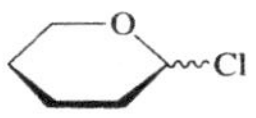 O, Cl	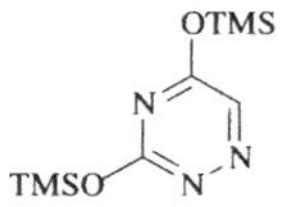 TMSO, N, N, N, OTMS	$SnCl_4$, CH_2Cl_2, MeCN, $Cl(CH_2)_2Cl$, 4°, 2 h	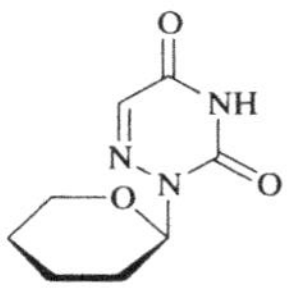 O, NH, N, N, O, O (92)	84
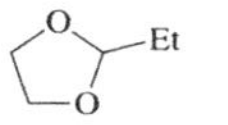 O, O, Et	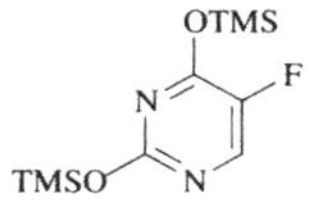 TMSO, N, N, OTMS, F	$SnCl_4$, CH_2Cl_2, rt, 1 h	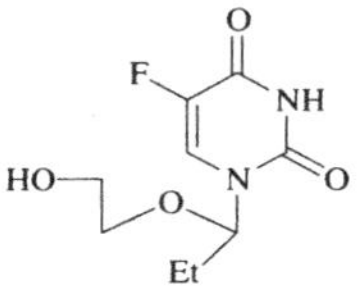 F, O, NH, HO–O, N, O, Et **I** (75)	641
		1. $SnCl_4$, $CHCl_3$, rt, 3 h 2. $NaHCO_3$, MeOH	**I** (—)	637
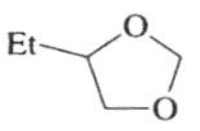 Et, O, O		1. $SnCl_4$, $CHCl_3$, rt, 3 h 2. $NaHCO_3$, MeOH	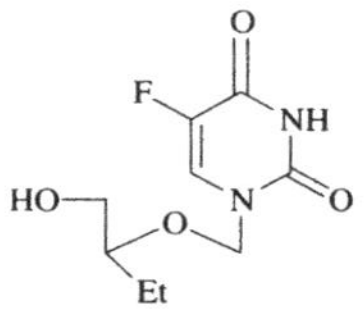 F, O, NH, HO, O, N, O, Et (—)	637

TABLE III. REACTIONS OF SILYLATED HETEROCYCLIC BASES WITH PROTECTED SUGARS - $SnCl_4$ CATALYST (*Continued*)

	Sugar	Base	Conditions	Product(s) and Yield(s) (%)	Refs.
	EtO⁀OEt		$SnCl_4$, $CHCl_3$, rt, 20 min	(14) + N^3-isomer (15) + N^1,N^3-bis(isomer) (9)	638
C_6			$SnCl_4$, $Cl(CH_2)_2Cl$, rt, 48 h	(77)	646
			$SnCl_4$, CH_2Cl_2	**I** (—)	647
			$SnCl_4$ (0.01 eq), CH_2Cl_2, rt, 3 h	**I** (96)	528
			$SnCl_4$	**I** (82)	648
			$SnCl_4$ (0.1 eq), CH_2Cl_2, rt, 8 h	**I** (61)	528
			$SnCl_4$ (0.1 eq), CH_2Cl_2, rt, 1.5 h	**I** (81)	528
			$SnCl_4$ (0.01 eq), CH_2Cl_2, rt, 3 h	**I** (93)	528

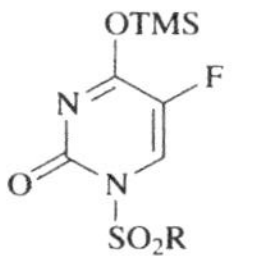

Substrate	Reagent	Conditions	Product(s) and Yield(s) (%)	Refs.
		$SnCl_4$, CH_2Cl_2	R: Me (72)	649
			R: Ph (62)	649
		$SnCl_4$	(80)	648
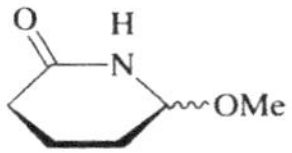	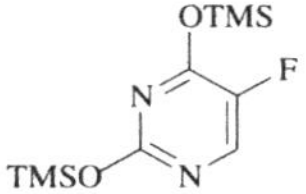	$SnCl_4$, CH_2Cl_2, MeCN, −20 to −25°, 6 h; rt, 6 h	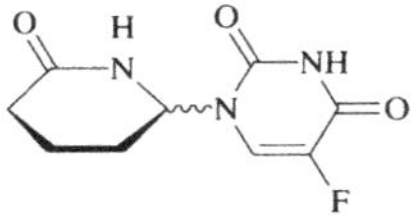 (40)	642
		$SnCl_4$, $Cl(CH_2)_2Cl$, −10°	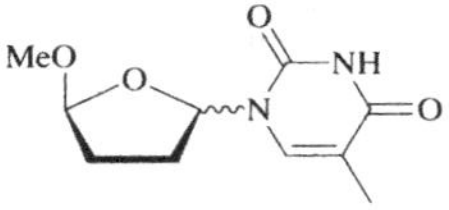 (75) *cis:trans* = 1:1	650
		$SnCl_4$, MeCN, $Cl(CH_2)_2Cl$, 80°, 24 h	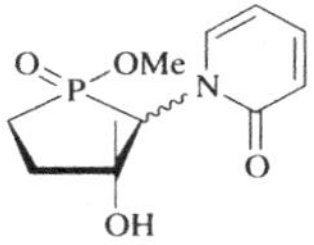 (28)	651
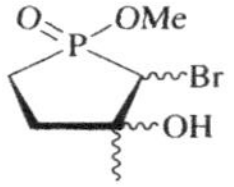	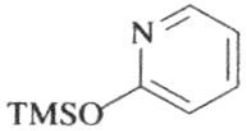	$SnCl_4$, MeCN, reflux, 24 h	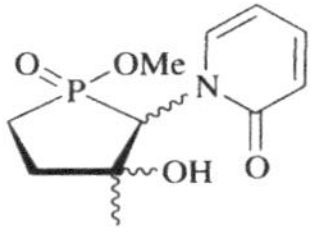 **I** (28)	652
		$SnCl_4$, MeCN, $Cl(CH_2)_2Cl$, reflux, 24 h	**I** (4)	652
		$SnCl_4$, MeCN, DMF, reflux, 24 h	**I** (11)	652

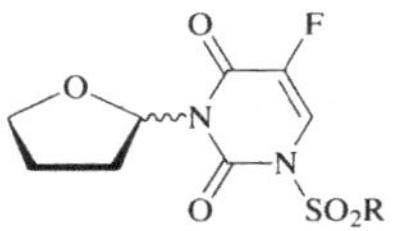

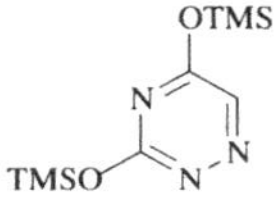

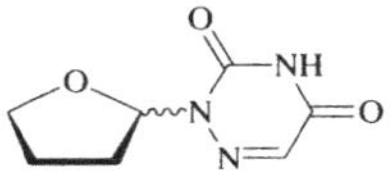

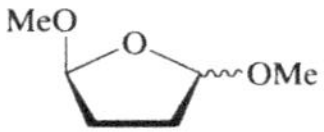

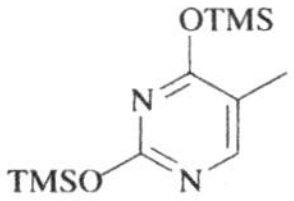

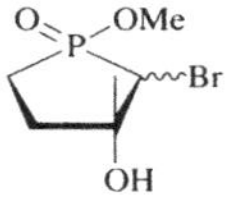

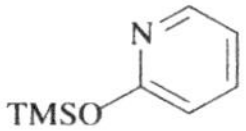

TABLE III. REACTIONS OF SILYLATED HETEROCYCLIC BASES WITH PROTECTED SUGARS - $SnCl_4$ CATALYST (*Continued*)

Sugar	Base	Conditions	Product(s) and Yield(s) (%)	Refs.
C_7				
		$SnCl_4$, CH_2Cl_2, rt, 18 h	(46)	643
		1. $SnCl_4$ 2. NH_3, MeOH	(—)	653
		1. $SnCl_4$ 2. NH_3, MeOH	(—)	653
		$SnCl_4$, MeCN, rt, 48 h	(41)	654
		$SnCl_4$, $Cl(CH_2)_2Cl$, rt, 18 h	(19) + N^3-isomer (9) + N^1,N^3-bis(isomer) (13)	655
		$SnCl_4$, $Cl(CH_2)_2Cl$, rt, 18 h	n: 1 (40); 2 (74)	656

Substrate	Silylated base	Conditions	Product(s) and Yield(s) (%)	Refs.
	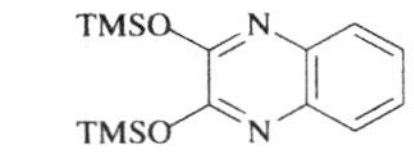 TMSO, TMSO-quinoxaline	$SnCl_4$, MeCN, rt, 48 h	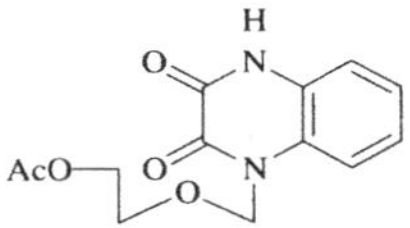 (18) + N^1,N^4-bis(isomer) (23)	657
	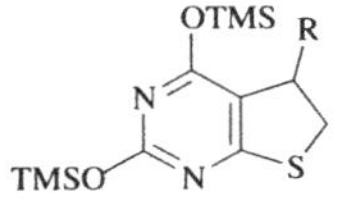 OTMS, TMSO, R	1. $SnCl_4$, $Cl(CH_2)_2Cl$, rt, 18 h 2. NH_3, MeOH	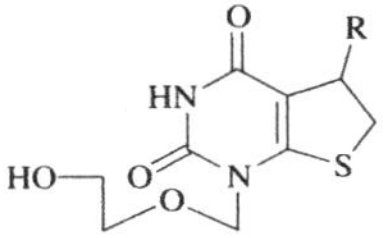 R: H (33); Ph (15)	658
	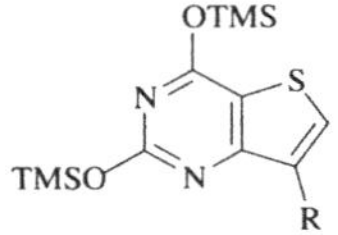 OTMS, TMSO, R	1. $SnCl_4$, $Cl(CH_2)_2Cl$, –30°, 8 h 2. NH_3, MeOH	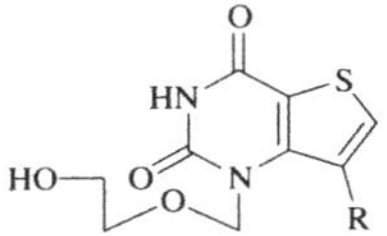 R: p-FC_6H_4 (24); p-ClC_6H_4 (31); p-BrC_6H_4 (27); Me (62) + N^3-isomer (23); Ph (35); p-MeC_6H_4 (27)	659
	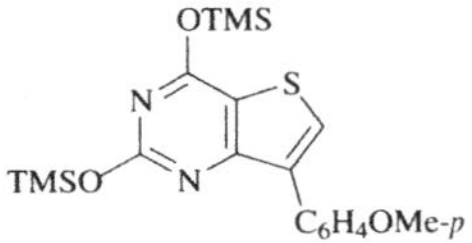 OTMS, TMSO, C_6H_4OMe-p	1. $SnCl_4$, $Cl(CH_2)_2Cl$, –30°, 8 h 2. NH_3, MeOH	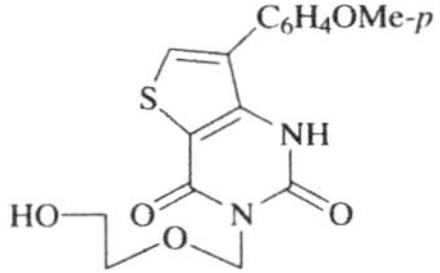 C_6H_4OMe-p (54) + N^1-isomer (48)	659
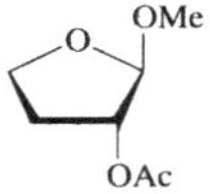 OMe, OAc	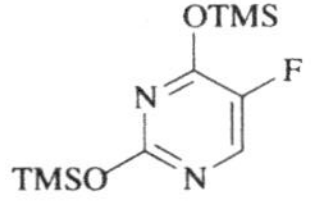 OTMS, F, TMSO	$SnCl_4$, $Cl(CH_2)_2Cl$, rt, 12 h	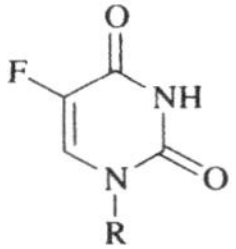 (33) R 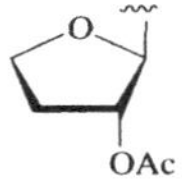 = (OAc)	660
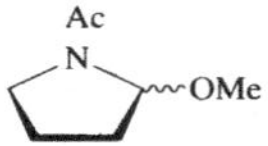 Ac, N, OMe	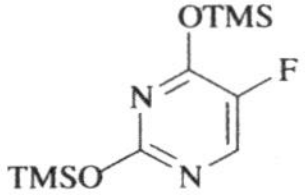 OTMS, F, TMSO	$SnCl_4$, CH_2Cl_2, –13 to –15°, 3 h 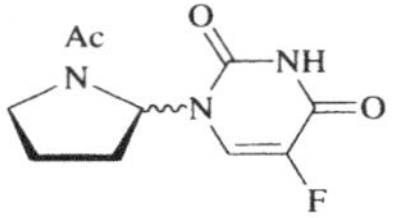	Ac, F (81)	642

TABLE III. REACTIONS OF SILYLATED HETEROCYCLIC BASES WITH PROTECTED SUGARS - $SnCl_4$ CATALYST (*Continued*)

Sugar	Base	Conditions	Product(s) and Yield(s) (%)	Refs.
		$SnCl_4$, MeCN, −20 to −25°, 24 h	(60)	642
		$SnCl_4$, CH_2Cl_2, MeCN −20 to −25°, 6 h	(36)	642
C_{7-8}		$SnCl_4$, CH_2Cl_2, 20°, 2 d	R: 2-furyl (48); 3-furyl (68); 2-pyridyl (48); 3-pyridyl (51)	661
C_8		$SnCl_4$, MeCN	(—)	662
		$SnCl_4$, CH_2Cl_2, rt, 4 h	(65)	640

Glycosyl donor	Heterocycle	Conditions	Product(s) and Yield(s) (%)		Refs.
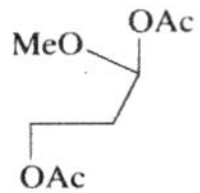 MeO, OAc, OAc	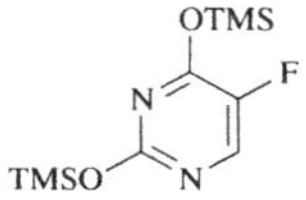 OTMS, F, TMSO, N	$SnCl_4$, MeCN, −78°, 15 min; 25°, 4 h	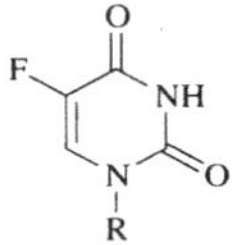 F, O, NH, O, N–R (47)	R = MeO, OAc 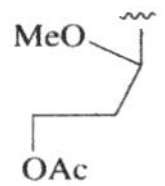	663
	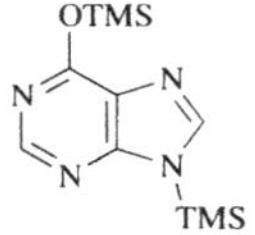 OTMS, N, TMS	$SnCl_4$, MeCN, −78°, 15 min; 25°, 20 h	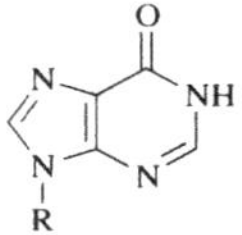 O, NH, N–R	(45) + N^7-isomer (—)	664
	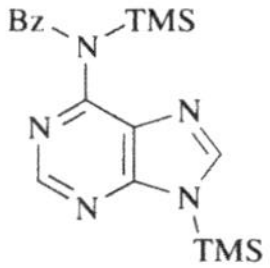 Bz–N–TMS, N, TMS	$SnCl_4$, MeCN, −78°, 15 min; 25°, 20 h	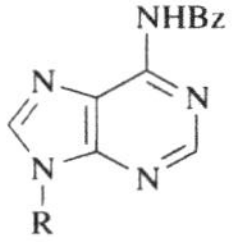 NHBz, N–R	(81)	664
	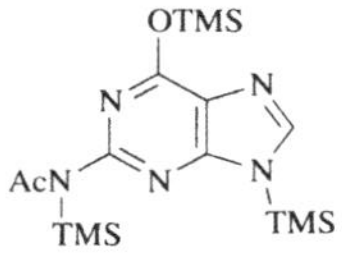 OTMS, AcN–TMS, N–TMS	$SnCl_4$, MeCN, −78°, 15 min; 25°, 20 h	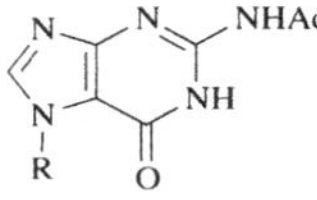 NHAc, NH, O, N–R	(21) + N^7-isomer (12)	664
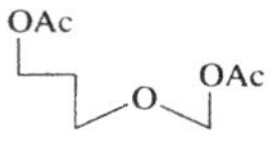 OAc, O, OAc	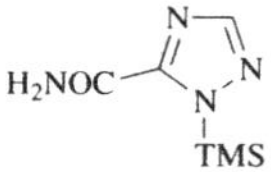 H_2NOC, N–TMS	1. $SnCl_4$ 2. NH_3, MeOH	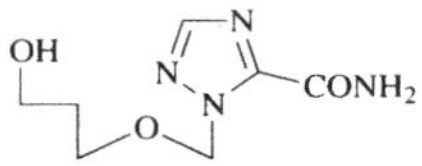 OH, O, $CONH_2$	(—)	653
	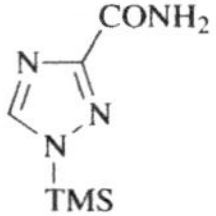 $CONH_2$, N–TMS	1. $SnCl_4$ 2. NH_3, MeOH	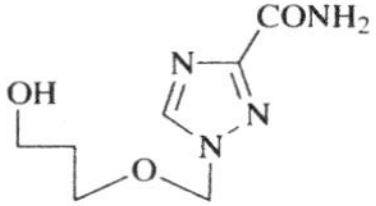 OH, O, $CONH_2$	(—)	653
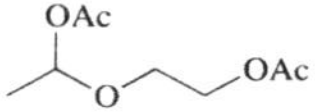 OAc, O, OAc	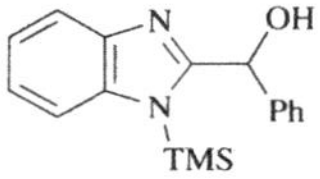 OH, Ph, N–TMS	$SnCl_4$, MeCN, reflux	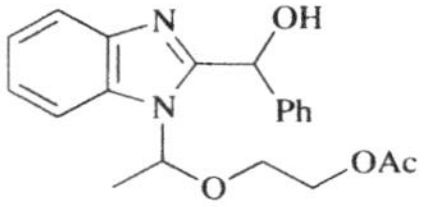 OH, Ph, O, OAc	(41)	665

TABLE III. REACTIONS OF SILYLATED HETEROCYCLIC BASES WITH PROTECTED SUGARS - $SnCl_4$ CATALYST (*Continued*)

Sugar	Base	Conditions	Product(s) and Yield(s) (%)	Refs.
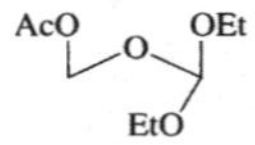 AcO, O, OEt, EtO	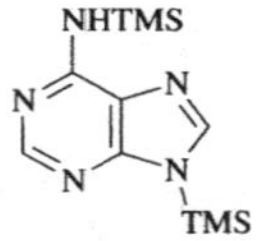 NHTMS, N, N, N, N, TMS	$SnCl_4$, MeCN, 30°, 4 h	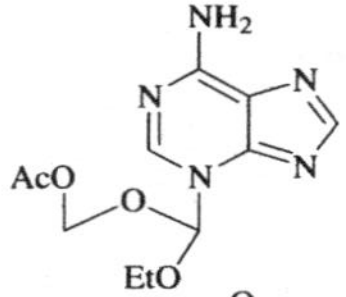 NH_2, N, N, N, N, AcO, O, EtO (25) + N^4-isomer (20)	666
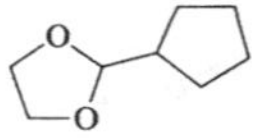 O, O	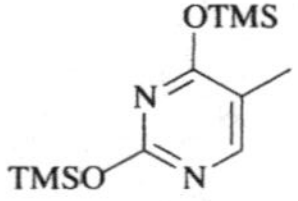 OTMS, N, N, TMSO	$SnCl_4$, CH_2Cl_2, 20°, 2 d	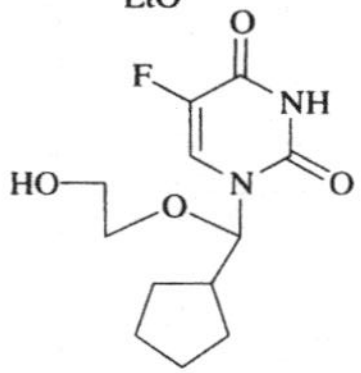 O, F, NH, O, N, HO, O (65)	661
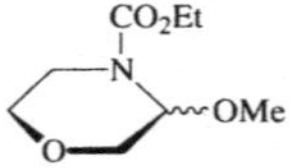 CO_2Et, N, OMe, O	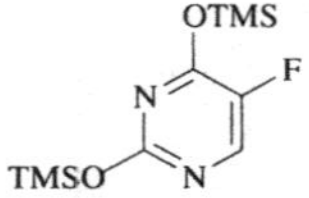 OTMS, F, N, N, TMSO	$SnCl_4$, MeCN, –20 to –25°, 24 h	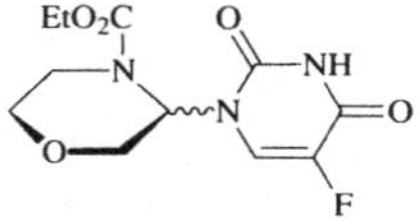 EtO_2C, N, O, NH, N, O, F, O (84)	642
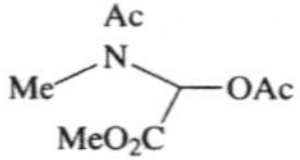 Ac, Me, N, OAc, MeO_2C	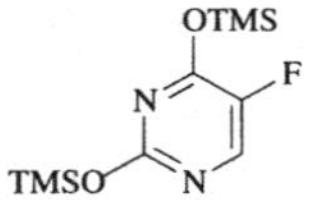 OTMS, F, N, N, TMSO	$SnCl_4$, MeCN, rt, 1 h	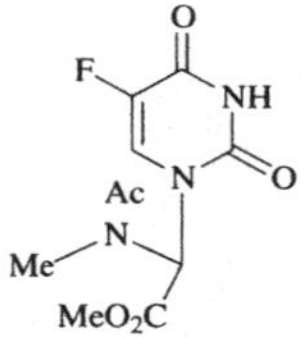 O, F, NH, N, O, Ac, N, Me, MeO_2C (—)	667
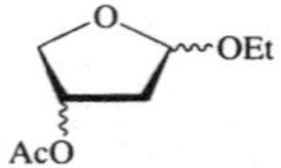 O, OEt, AcO	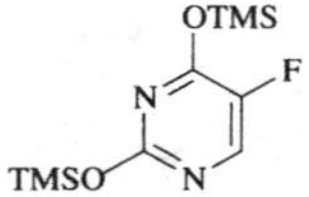 OTMS, F, N, N, TMSO	1. $SnCl_4$, $Cl(CH_2)_2Cl$, rt, 36 h 2. $NaHCO_3$ 3. NH_3, MeOH, 5°, 12 h	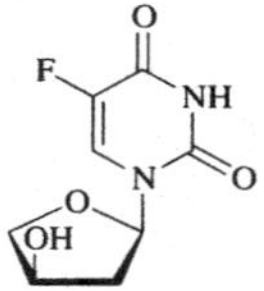 O, F, NH, N, O, O, OH (80)	660

C_9				
2-($C_6H_3Cl_2$-2,6)-1,3-dioxolane	5-F, 2,4-bis(TMSO)pyrimidine (TMSO, N, N, OTMS, F)	$SnCl_4$, CH_2Cl_2, rt, 12 h	HO–CH_2CH_2–O–CH(2,6-$Cl_2C_6H_3$)–N (5-F uracil: O, NH, O, F) (—)	637
2-(C_6H_4R)-1,3-dioxolane	5-F, 2,4-bis(TMSO)pyrimidine (TMSO, N, N, OTMS, F)	$SnCl_4$, CH_2Cl_2, rt, 12 h	HO–CH_2CH_2–O–CH(RC_6H_4)–N (5-Me uracil: O, NH, O) R: *p*-Cl (30); *m*-NO_2 (14)	637
2-Ph-1,3-dioxolane	5-F, 2,4-bis(TMSO)pyrimidine (TMSO, N, N, OTMS, F)	1. $SnCl_4$, $CHCl_3$, rt, 1 h 2. $NaHCO_3$, MeOH 3. Ac_2O, rt, 3 h	AcO–CH_2CH_2–O–CH(Ph)–N (5-F uracil: O, NH, O, F) (27)	637
	5-Me, 2,4-bis(TMSO)pyrimidine (TMSO, N, N, OTMS)	$SnCl_4$, CH_2Cl_2, 20°, 2 d	HO–CH_2CH_2–O–CH(Ph)–N (5-Me uracil: O, NH, O) (63)	661
2-(C_6H_{11})-1,3-dioxolane	5-R, 2,4-bis(TMSO)pyrimidine (TMSO, N, N, OTMS, R)	$SnCl_4$, CH_2Cl_2	HO–CH_2CH_2–O–CH(C_6H_{11})–N (5-R uracil: O, NH, O, R) R: F, rt, 12 h (37); Me, 20°, 2 d (73)	637 661

TABLE III. REACTIONS OF SILYLATED HETEROCYCLIC BASES WITH PROTECTED SUGARS - $SnCl_4$ CATALYST (*Continued*)

Sugar	Base	Conditions	Product(s) and Yield(s) (%)	Refs.
AcO, OAc, O	Bz, TMS, N, N, TMSO, N	$SnCl_4$, $Cl(CH_2)_2Cl$	AcO, O, N, O, N, NHBz (—)	668
OAc, O, AcO	R, TMS, N, N, TMSO, N	$SnCl_4$, $Cl(CH_2)_2Cl$	NHR, N, N, O, O, AcO; R: H (—), Bz (—)	668
	Bz, TMS, N, N, N, N, N, TMS	$SnCl_4$, $Cl(CH_2)_2Cl$	Bz, TMS, N, N, N, N, N, O, AcO (—) + α-anomer	668
AcO, O, Cl, AcO	OTMS, N, CF_2, TMSO, N	$SnCl_4$, $Cl(CH_2)_2Cl$, 4°, 16 h	AcO, O, N, O, NH, O, AcO, F_2C (—) α:β = 88:63	547
	OTMS, N, N, Et, TMSO, N	$SnCl_4$, $Cl(CH_2)_2Cl$, MeCN, -25°, 5 min; 20°, 25 min	O, Et, N, NH, AcO, O, N, O, AcO (—)	669

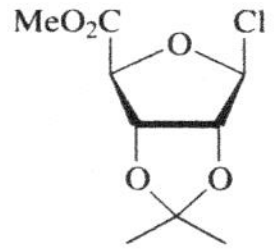 MeO$_2$C, O, Cl, O, O	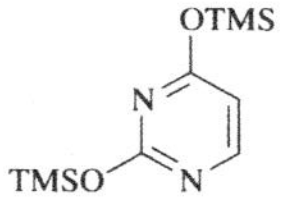 OTMS, N, TMSO, N	$SnCl_4$, $Cl(CH_2)_2Cl$, 60°, 3 h	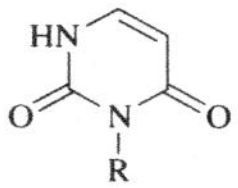 HN, O, N, O, R (32) R = MeO$_2$C, O, O, O	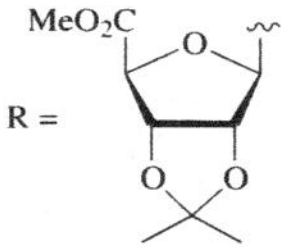 224
	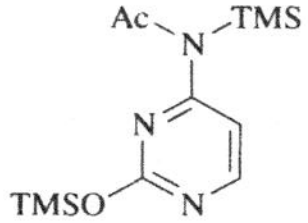 Ac, N, TMS, N, TMSO, N	$SnCl_4$, $Cl(CH_2)_2Cl$, 60°, 4 h	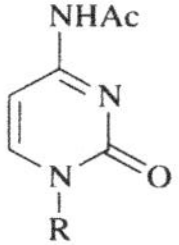 NHAc, N, N, O, R (33) + α-anomer (4) + N^3 isomer (7)	224
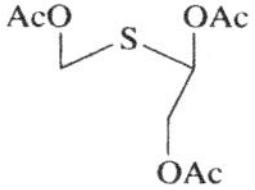 AcO, S, OAc, OAc	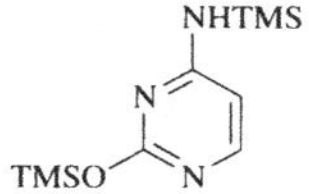 NHTMS, N, TMSO, N	1. $SnCl_4$ 2. NaOMe	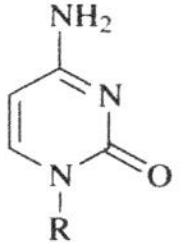 NH_2, N, N, O, R (63) R = HO, S, OH	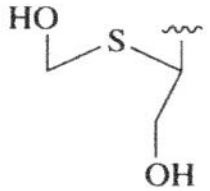 670
	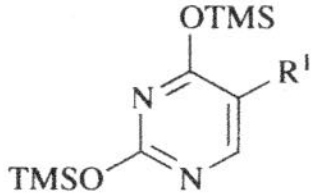 OTMS, R^1, N, TMSO, N	1. $SnCl_4$ 2. NaOMe	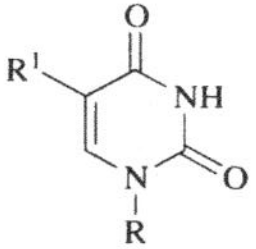 R^1, O, NH, N, O, R; R^1: F (49), I (63)	670
	Bz, N, TMS, N, N, N, N, TMS	$SnCl_4$	NHBz, N, N, N, N, R (32)	670
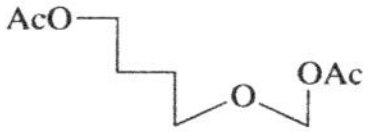 AcO, O, OAc	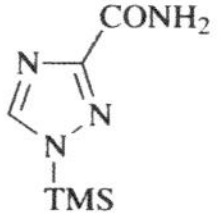 $CONH_2$, N, N, N, TMS	1. $SnCl_4$ 2. NH_3, MeOH 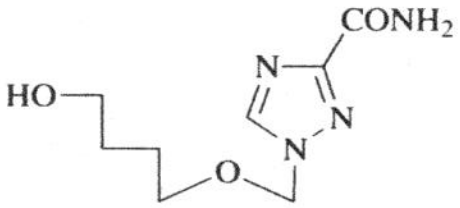	HO, O, N, N, N, $CONH_2$ (—)	653
	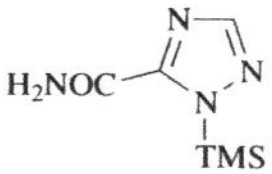 H_2NOC, N, N, N, TMS	1. $SnCl_4$ 2. NH_3, MeOH 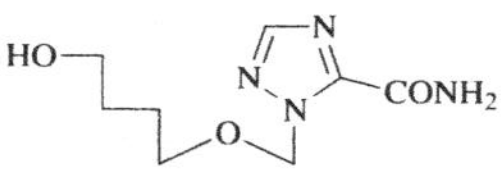	HO, O, N, N, N, $CONH_2$ (—)	653

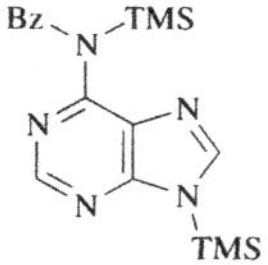

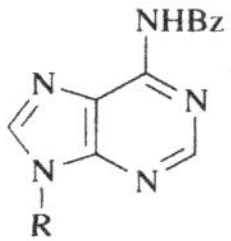

TABLE III. REACTIONS OF SILYLATED HETEROCYCLIC BASES WITH PROTECTED SUGARS - $SnCl_4$ CATALYST (*Continued*)

Sugar	Base	Conditions	Product(s) and Yield(s) (%)	Refs.
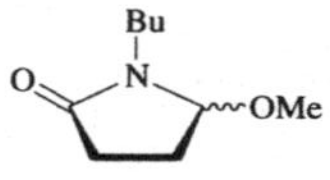 Bu, O, N, OMe	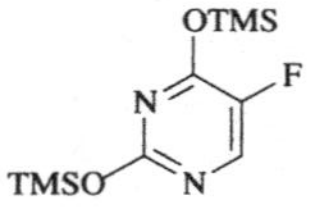 OTMS, F, N, TMSO, N	$SnCl_4$, MeCN, CH_2Cl_2, −20 to −25°, 6 h	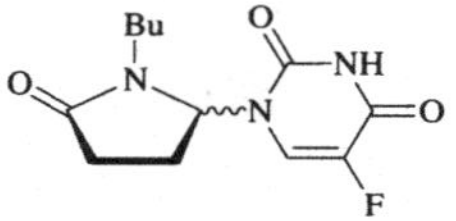 Bu, O, N, O, NH, N, O, F (38)	642
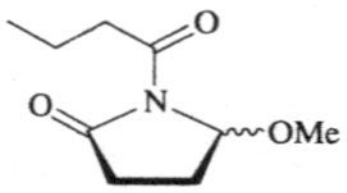 O, O, N, OMe	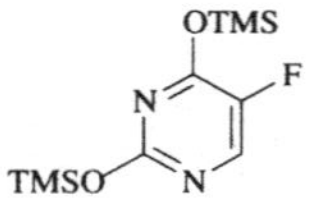 OTMS, F, N, TMSO, N	$SnCl_4$, CH_2Cl_2, −13 to −15°, 3 h	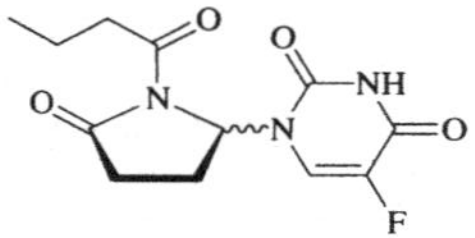 O, O, N, O, NH, N, O, F (78)	642
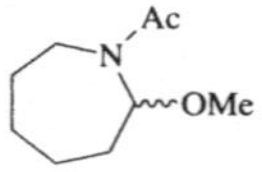 Ac, N, OMe	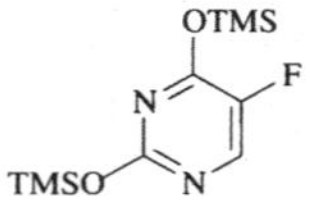 OTMS, F, N, TMSO, N	$SnCl_4$, MeCN, −20 to −25°, 24 h	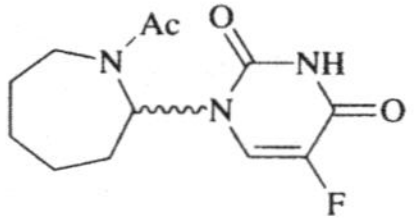 Ac, N, O, NH, N, O, F (70)	642
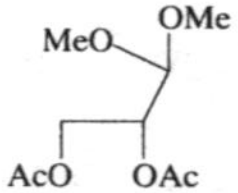 OMe, MeO, AcO, OAc	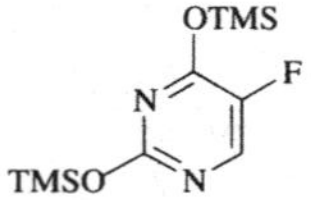 OTMS, F, N, TMSO, N	$SnCl_4$, MeCN, −78°, 15 min; 25°, 20 h	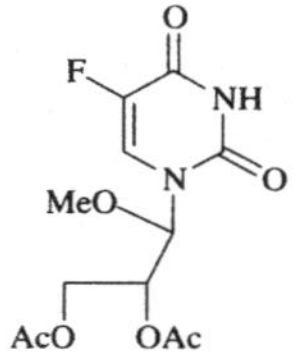 O, F, NH, N, O, MeO, AcO, OAc (60)	663
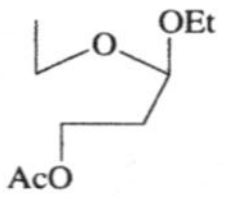 OEt, O, AcO	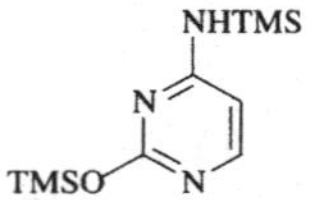 NHTMS, N, TMSO, N	1. $SnCl_4$, MeCN, 30°, 1 d 2. NH_3, MeOH	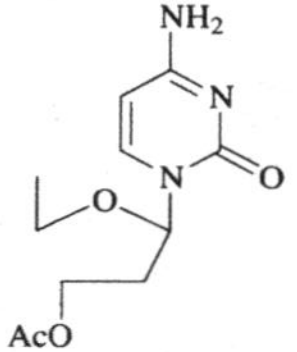 NH_2, N, N, O, O, AcO (88)	555

Substrate	Silylated base	Conditions	Product(s) and Yield(s) (%)	Refs.
	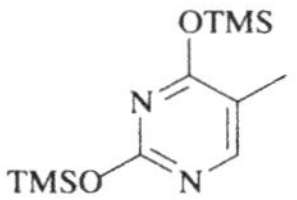 TMSO, N, OTMS	1. $SnCl_4$, MeCN, 30°, 1 d 2. NH_3, MeOH	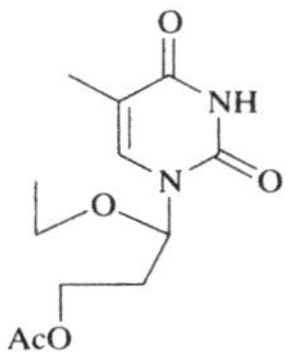 O, NH, N, O, AcO (89)	555
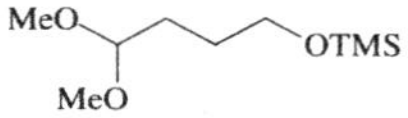 MeO, MeO, OTMS	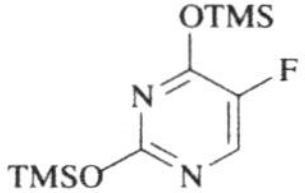 TMSO, N, OTMS, F	$SnCl_4$, $Cl(CH_2)_2Cl$, 23–25°, 1 h	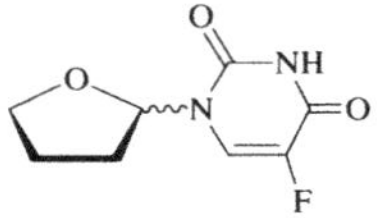 O, NH, N, O, F (84)	671
C_{10} 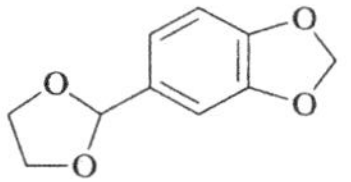 O, O, O	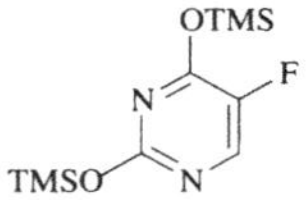 TMSO, N, OTMS, F	1. $SnCl_4$, CH_2Cl_2, rt, 12 h 2. $NaHCO_3$, MeOH	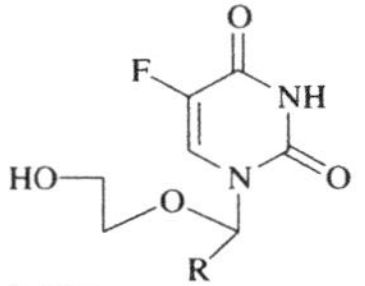 F, O, NH, N, O, HO, R **I** (29) R 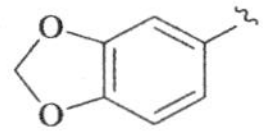 = O, O	637
		$SnCl_4$, CH_2Cl_2, rt, 3 h	**I** (51)	641
 BzO, O, Cl	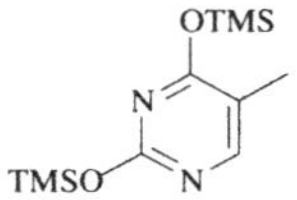 TMSO, N, OTMS	$SnCl_4$, $Cl(CH_2)_2Cl$, rt, 17 h	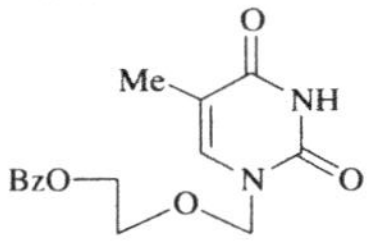 Me, O, NH, N, O, BzO (61)	643
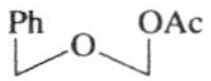 Ph, O, OAc	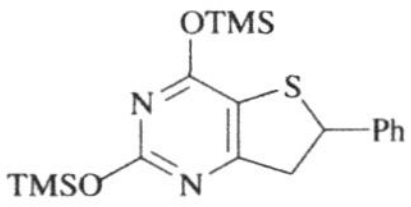 TMSO, N, OTMS, S, Ph	1. $SnCl_4$, $Cl(CH_2)_2Cl$, rt, 18 h 2. NH_3, MeOH	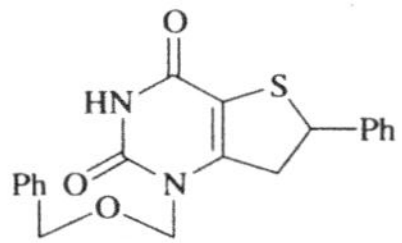 O, HN, S, Ph, Ph, O, N, O (20)	658
	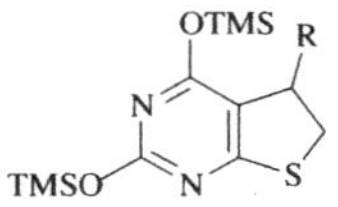 TMSO, N, OTMS, R, S	1. $SnCl_4$, $Cl(CH_2)_2Cl$, rt, 18 h 2. NH_3, MeOH	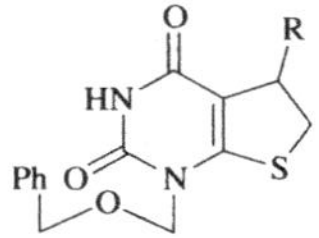 O, R, HN, S, Ph, O, N, O R: H (41); Ph (33)	658

TABLE III. REACTIONS OF SILYLATED HETEROCYCLIC BASES WITH PROTECTED SUGARS - $SnCl_4$ CATALYST (*Continued*)

Sugar	Base	Conditions	Product(s) and Yield(s) (%)	Refs.
	OTMS, TMSO, N, N, S, R	$SnCl_4$, $Cl(CH_2)_2Cl$, −30°, 8 h	O, HN, S, Ph O, N, O, R	659
	R			
	Me		(33)	
	Ph		(42)	
	p-FC_6H_4		(37)	
	p-ClC_6H_4		(34)	
	p-BrC_6H_4		(24)	
	p-MeC_6H_4		(32)	
	p-$MeOC_6H_4$		(32)	
	OTMS, TMSO, N, N, S	1. $SnCl_4$, $Cl(CH_2)_2Cl$, rt, 18 h 2. NH_3, MeOH	O, HN, S, Ph O, N, O (46)	658
O, O, O, O, OAc, OAc	OTMS, TMSO, N, N, R	$SnCl_4$, $Cl(CH_2)_2Cl$, rt, 20 h	R, O, NH, N, O, O, O, O, O, OAc R: Br (42); I (54)	672
MeO_2C, O, Cl, AcO, OAc	OTMS, TMSO, N, N	$SnCl_4$, $Cl(CH_2)_2Cl$, 25°, 20 h	O, NH, N, O, R (39) + N^3-isomer R = MeO_2C, O, AcO, OAc	224

Conditions	Product(s) and Yield(s) (%)	Refs.
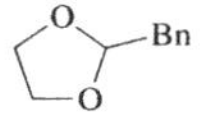 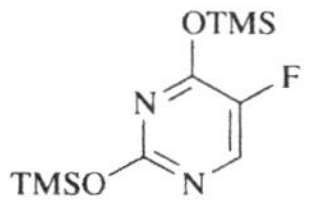 $SnCl_4$, CH_2Cl_2, rt, 12 h	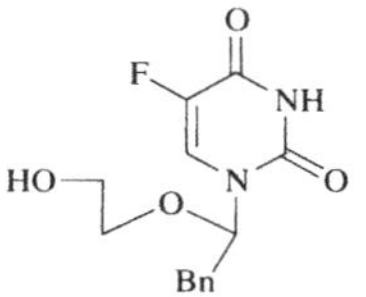 (45)	637
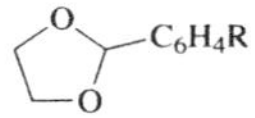 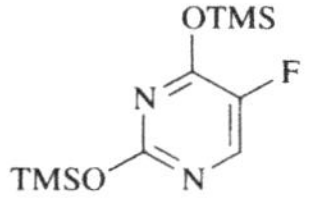 $SnCl_4$, CH_2Cl_2, rt, time	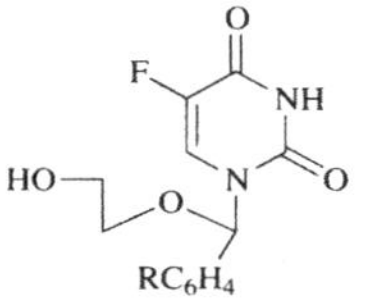 R / time: *o*-OMe, 12 h (23); *p*-OMe, 1 h (—); *o*-Me, 12 h (30)	637
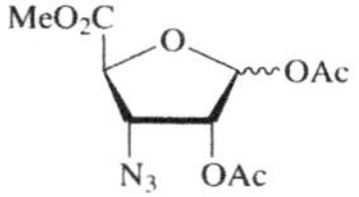 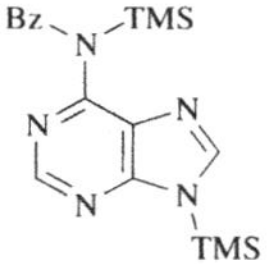 $SnCl_4$, $Cl(CH_2)_2Cl$, 60-70°, 4 h	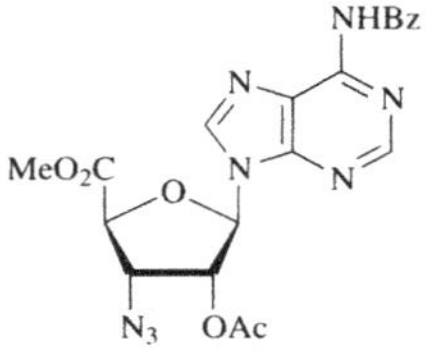 (57)	673
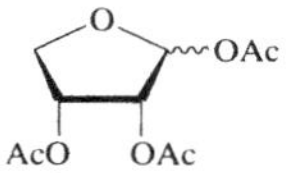 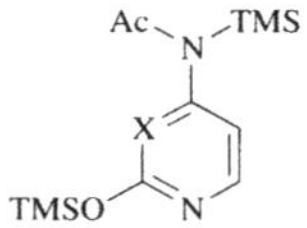 $SnCl_4$, $Cl(CH_2)_2Cl$, rt, time	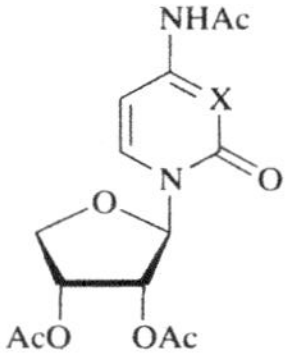 X / time: CH, 40 h (43); N, 48 h (29)	674
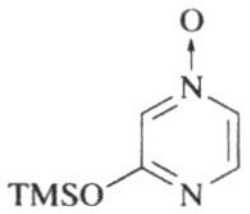 $SnCl_4$, $Cl(CH_2)_2Cl$, rt, 72 h	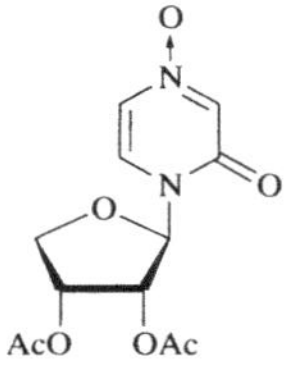 (69)	674

TABLE III. REACTIONS OF SILYLATED HETEROCYCLIC BASES WITH PROTECTED SUGARS - $SnCl_4$ CATALYST (*Continued*)

Sugar	Base	Conditions	Product(s) and Yield(s) (%)	Refs.
	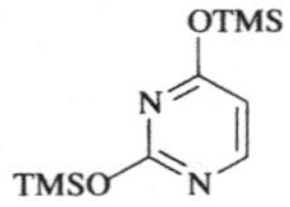	$SnCl_4$, $Cl(CH_2)_2Cl$, rt, 12 h	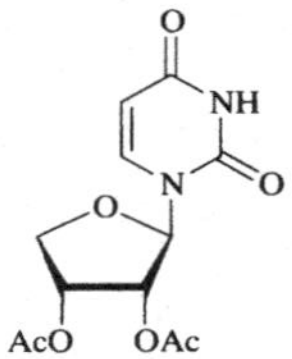 (52)	674
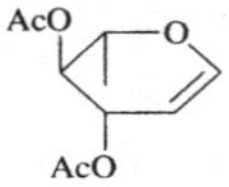	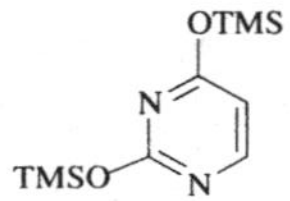	$SnCl_4$, $Cl(CH_2)_2Cl$, −15°, 2 h	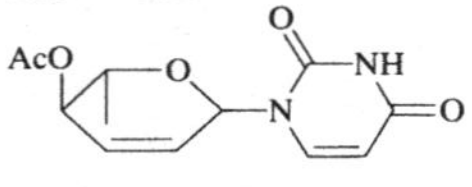 (60) α:β = 1.5:1	675
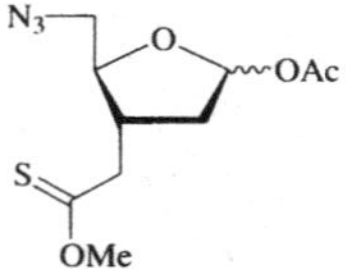	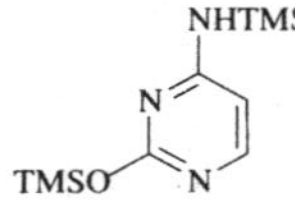	$SnCl_4$, $Cl(CH_2)_2Cl$, rt	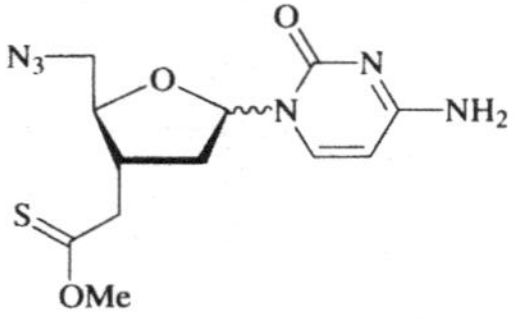 (—) α:β = 4:96	278
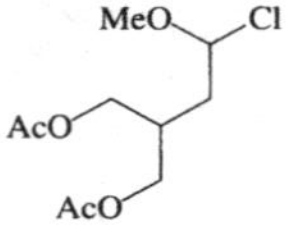	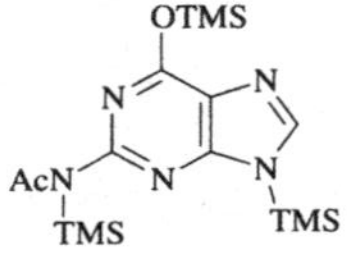	$SnCl_4$, MeCN, 25°, 44 h	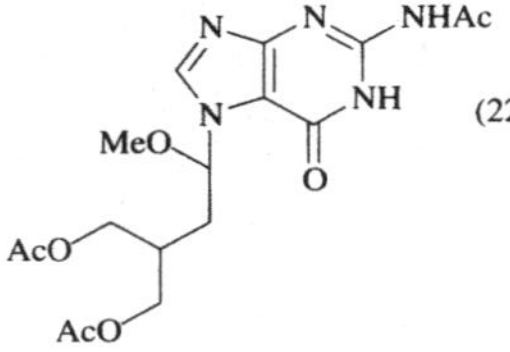 (22) + N^9-isomer (16)	676
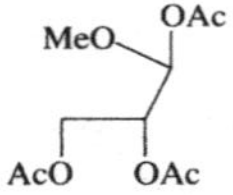	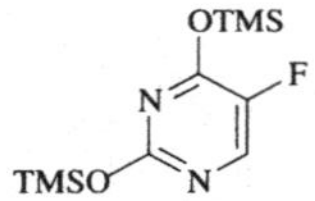	$SnCl_4$, MeCN, −78°, 15 min; 25°, 18 h	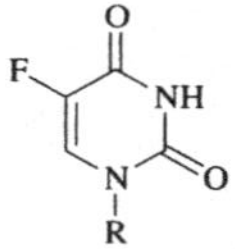 (57) R = 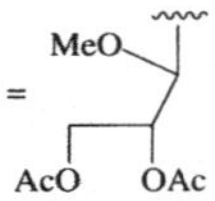	663

Reactant	Reactant	Conditions	Product(s) and Yield(s) (%)	Refs.
		$SnCl_4$, MeCN, −78°, 15 min; 25°, 20 h	(57) N^7, N^9 mixture	664
		$SnCl_4$, MeCN, −78°, 15 min; 25°, 20 h	(40) + N^9-isomer (28)	664
		$SnCl_4$, MeCN, −78°, 15 min; 25°, 20 h	(36) + N^9-isomer (9)	664
		$SnCl_4$, MeCN, −78°, 15 min; 25°, 20 h	(31)	664
		$SnCl_4$, MeCN, −78°, 15 min; 25°, 20 h	(27)	664
		$SnCl_4$, $Cl(CH_2)_2Cl$, 23-25°, 1 h	(72) $\alpha:\beta = 1:3$	671

TABLE III. REACTIONS OF SILYLATED HETEROCYCLIC BASES WITH PROTECTED SUGARS - $SnCl_4$ CATALYST (*Continued*)

Sugar	Base	Conditions	Product(s) and Yield(s) (%)	Refs.
C_{11}				
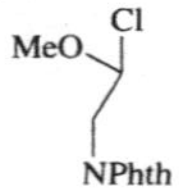 MeO, Cl, NPhth	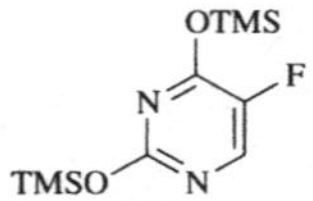 OTMS, F, N, N, TMSO	$SnCl_4$, CH_2Cl_2, rt, 12 h	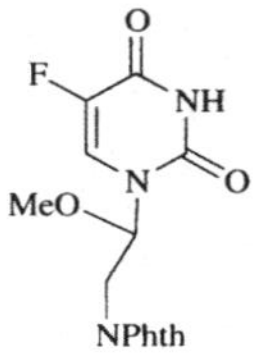 O, F, NH, N, O, MeO, NPhth (32) + N^3-isomer (5)	677
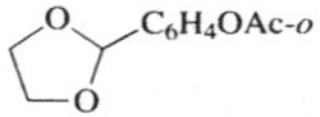 O, O, C_6H_4OAc-*o*	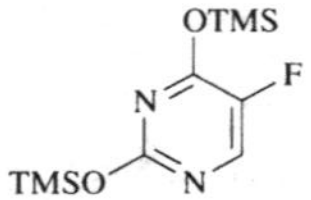 OTMS, F, N, N, TMSO	$SnCl_4$, CH_2Cl_2, rt, 1 h	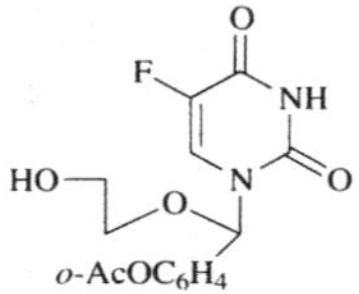 O, F, NH, N, O, HO, O, *o*-$AcOC_6H_4$ (47)	641
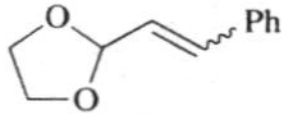 O, O, Ph	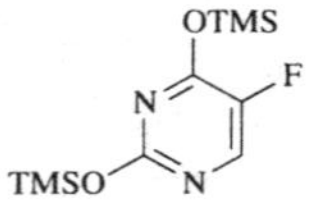 OTMS, F, N, N, TMSO	$SnCl_4$, CH_2Cl_2, rt, 1 h	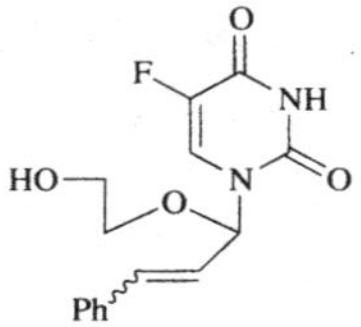 O, F, NH, N, O, HO, O, Ph (26)	637
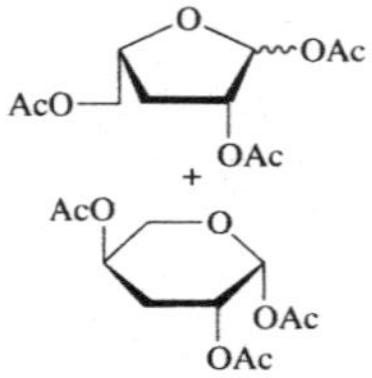 AcO, O, OAc, OAc + AcO, O, OAc, OAc	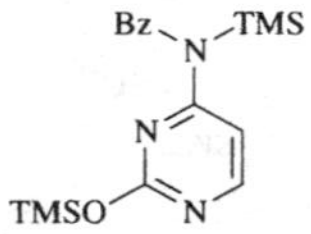 Bz, N, TMS, N, N, TMSO	$SnCl_4$, $Cl(CH_2)_2Cl$, rt, 5 h	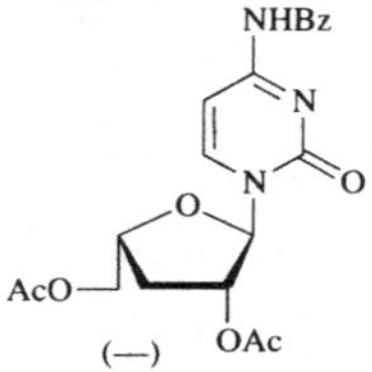 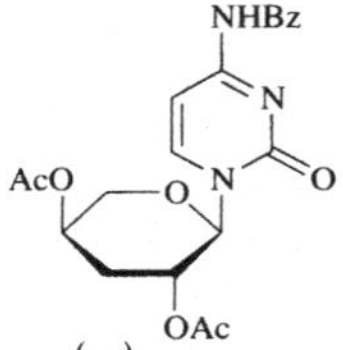 NHBz, N, N, O, O, AcO, OAc (—) + NHBz, N, N, O, AcO, O, OAc (—)	678

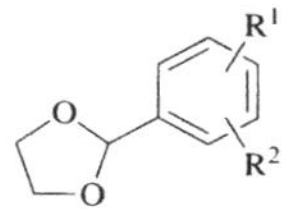

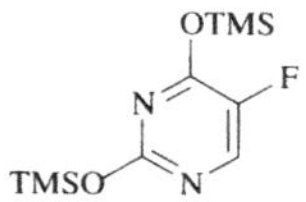

Substrate	Silylated base	Conditions	Product(s) and Yield(s) (%)	Refs.
2-aryl-1,3-dioxolane (R¹, R² on aryl ring)	TMSO / OTMS 5-fluoropyrimidine	1. $SnCl_4$, CH_2Cl_2, rt, 12 h 2. (see below) 3. (see below)	R^3O-ethoxy(aryl)methyl 5-fluorouracil (R^1, R^2, R^3)	637

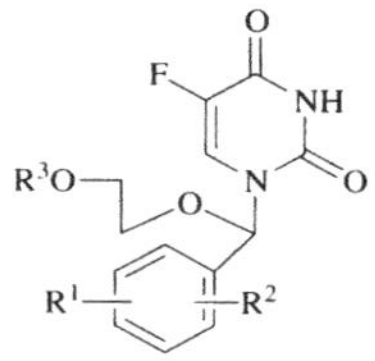

R^1	R^2	2	3	R^3	Yield
o-OEt	H	—	—	H	(20)
o-OAc	H	$NaHCO_3$,	Ac_2O, Py	Ac	(—)
p-OEt	H	—	—	H	(18)
o-OMe	*m*-OMe	—	—	H	(13)
p-OMe	*m*-OMe	—	—	H	(43)
p-OAc	*m*-OMe	—	—	H	(70)

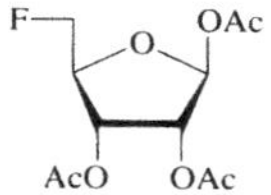

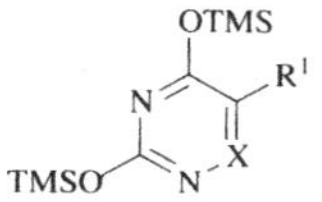

Substrate	Silylated base	Conditions	Product(s) and Yield(s) (%)	Refs.
5-fluoro-5-deoxy ribofuranose triacetate (F, OAc, AcO, OAc)	TMSO / OTMS, R^1, X (pyrimidine/triazine)	$SnCl_4$, $Cl(CH_2)_2Cl$, rt, 4.5 h	5'-fluoro nucleoside (F, AcO, OAc, R^1, X, NH)	679

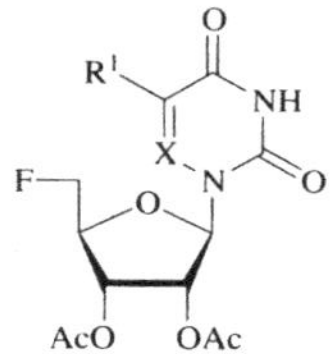

R^1	X	Yield
Me	CH	(70)
H	N	(31)

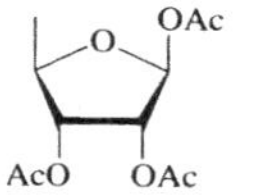

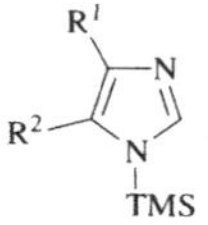

Substrate	Silylated base	Conditions	Product(s) and Yield(s) (%)	Refs.
5-deoxy ribofuranose triacetate (OAc, AcO, OAc)	1-TMS imidazole (R^1, R^2)	$SnCl_4$, $Cl(CH_2)_2Cl$, rt, time	imidazole nucleoside (R^1, R^2, AcO, OAc)	680

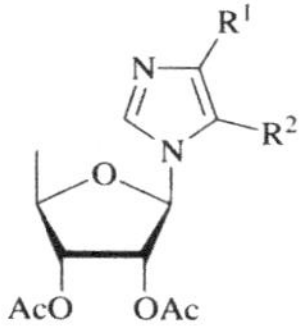

R^1	R^2	time	Yield
CO_2Me	CH_2CN	9 h	(68)
CH_2CN	CO_2Me	48 h	(48) + N^3-isomer (26)

TABLE III. REACTIONS OF SILYLATED HETEROCYCLIC BASES WITH PROTECTED SUGARS - $SnCl_4$ CATALYST (*Continued*)

Sugar	Base	Conditions	Product(s) and Yield(s) (%)	Refs.
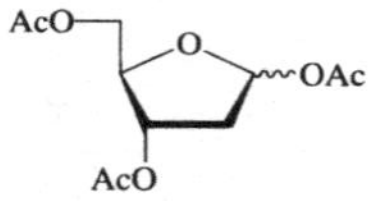 AcO, AcO, OAc	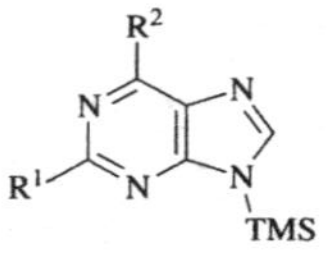 R^2, N, N, R^1, N, N, TMS		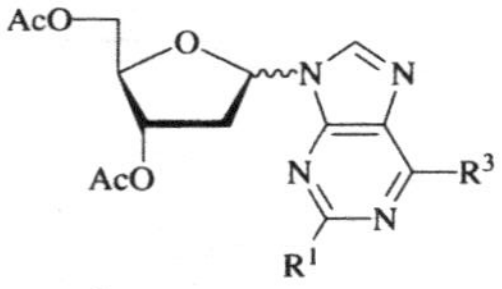 AcO, O, AcO, N, N, N, N, R^3, R^1	515
	R^1 / R^2		R^3 / / α:β	
	H / BzNTMS	$SnCl_4$, $Cl(CH_2)_2Cl$, rt, 12 h	NHBz (60) 68:32	
	H / BzNTMS	$SnCl_4$, MeCN, rt, 12 h	NHBz (95) 78:22	
	AcNTMS / O_2CNPh_2	$SnCl_4$, $Cl(CH_2)_2Cl$, rt, 16 h	O_2CNPh_2 (40) 75:25	
	AcNTMS / O_2CNPh_2	$SnCl_4$, MeCN, rt, 16 h	O_2CNPh_2 (80) 68:32	
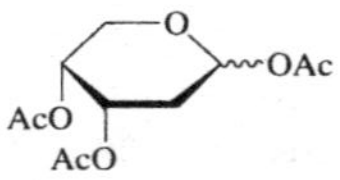 O, OAc, AcO, AcO	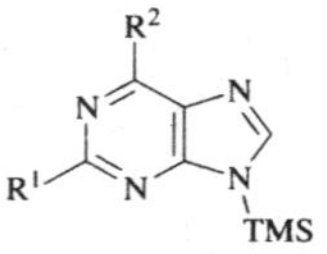 R^2, N, N, R^1, N, N, TMS	$SnCl_4$, MeCN, rt, 12 h	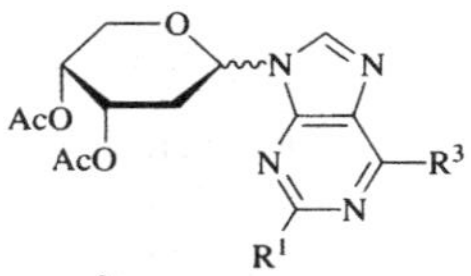 O, N, N, AcO, AcO, N, N, R^3, R^1	515
	R^1 / R^2		R^3 / / α:β	
	H / BzNTMS		NHBz (85) 52:48	
	AcNTMS / O_2CNPh_2		O_2CNPh_2 (80) 60:40	
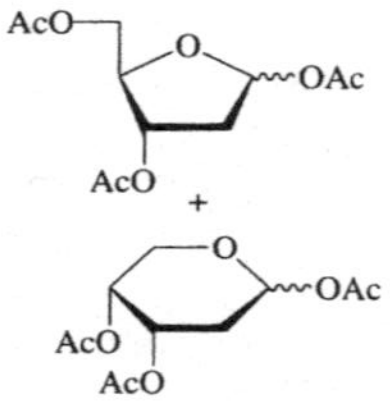 AcO, O, OAc, AcO + O, OAc, AcO, AcO	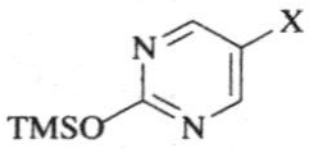 N, X, TMSO, N	$SnCl_4$, $Cl(CH_2)_2Cl$, rt, 5-12 h	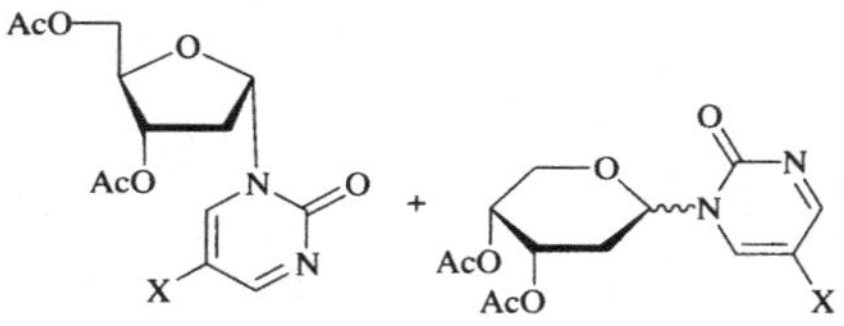 AcO, O, AcO, N, O, X, N + O, O, N, AcO, AcO, N, X	681
			X / /	
			Br (2) (1)	
			I (11) (15)	

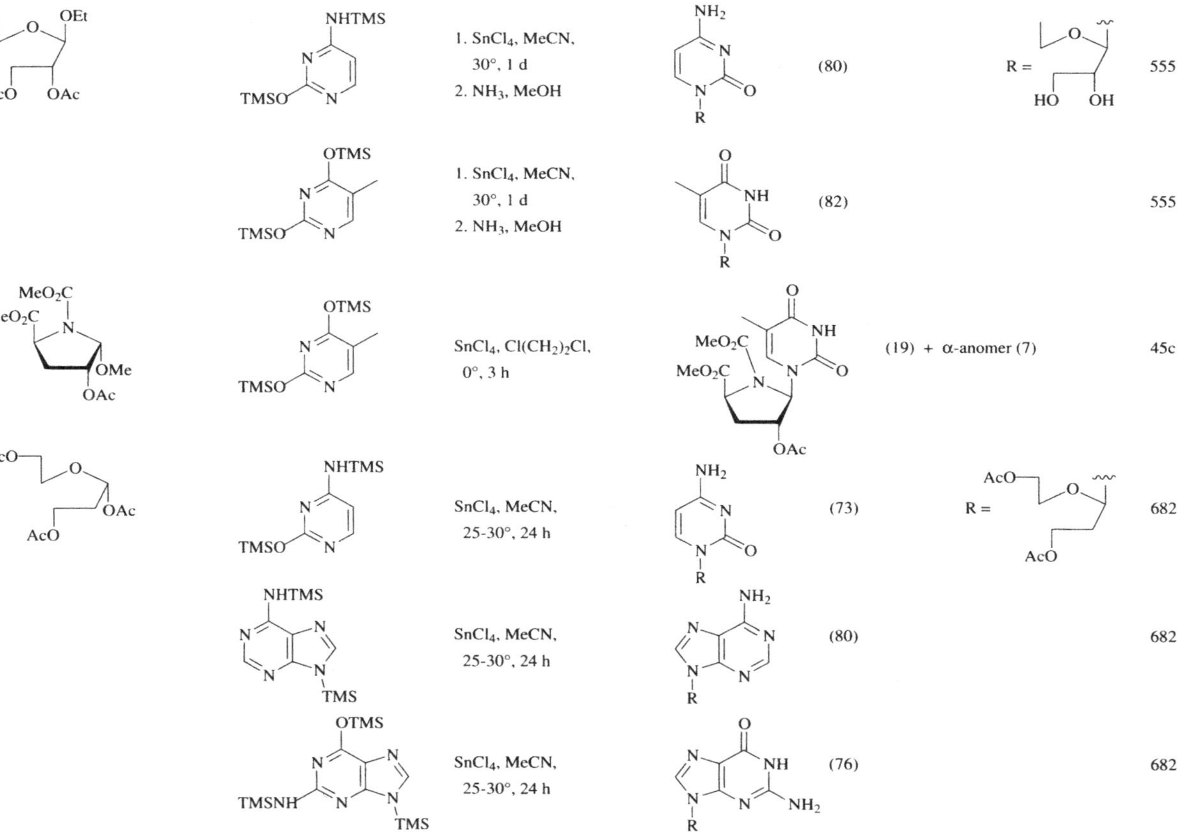

Reactant	Silylated base	Conditions	Product and Yield(s) (%)		Refs.
AcO, OAc, OEt	NHTMS, TMSO	1. $SnCl_4$, MeCN, 30°, 1 d 2. NH_3, MeOH	NH_2, R (80)	R = HO, OH	555
	OTMS, TMSO	1. $SnCl_4$, MeCN, 30°, 1 d 2. NH_3, MeOH	NH, R (82)		555
MeO_2C, MeO_2C, OMe, OAc	OTMS, TMSO	$SnCl_4$, $Cl(CH_2)_2Cl$, 0°, 3 h	MeO_2C, MeO_2C, NH, OAc (19) + α-anomer (7)		45c
AcO, OAc, AcO	NHTMS, TMSO	$SnCl_4$, MeCN, 25-30°, 24 h	NH_2, R (73)	R = AcO, AcO	682
	NHTMS, TMS	$SnCl_4$, MeCN, 25-30°, 24 h	NH_2, R (80)		682
	OTMS, TMSNH, TMS	$SnCl_4$, MeCN, 25-30°, 24 h	NH, NH_2, R (76)		682

TABLE III. REACTIONS OF SILYLATED HETEROCYCLIC BASES WITH PROTECTED SUGARS - $SnCl_4$ CATALYST (*Continued*)

Sugar	Base	Conditions	Product(s) and Yield(s) (%)	Refs.
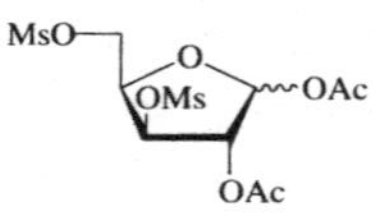 MsO, OMs, OAc, OAc	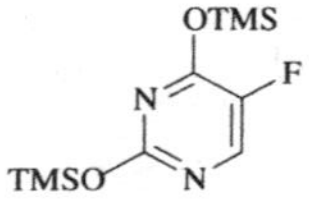 OTMS, F, N, N, TMSO	$SnCl_4$, MeCN, rt, 4 h	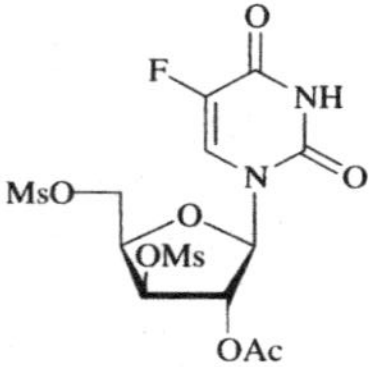 O, F, NH, N, O, MsO, O, OMs, OAc (59)	683
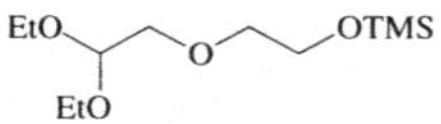 EtO, O, OTMS, EtO	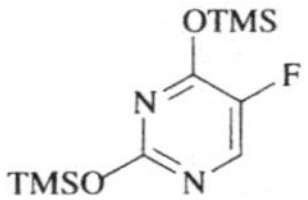 OTMS, F, N, N, TMSO	$SnCl_4$, $Cl(CH_2)_2Cl$, 23-25°, 1 h	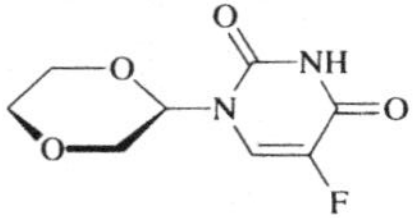 O, O, O, NH, N, O, F (75)	671
C_{12} 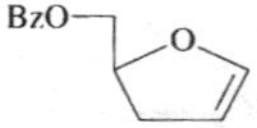 BzO, O	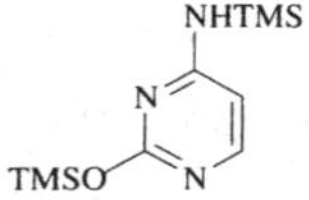 NHTMS, N, N, TMSO	$SnCl_4$, $Cl(CH_2)_2Cl$, rt, 5 h	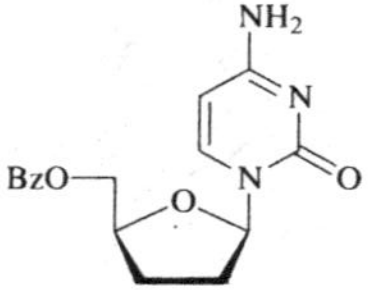 NH_2, N, N, O, BzO, O (—)	684
	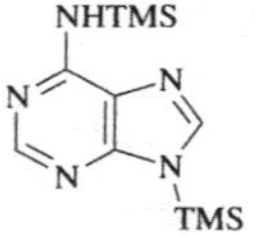 NHTMS, N, N, N, N, TMS	$SnCl_4$, $Cl(CH_2)_2Cl$, rt, 5 h	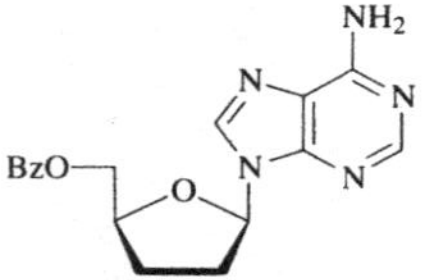 NH_2, N, N, N, N, BzO, O (—)	685
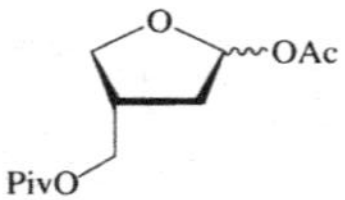 O, OAc, PivO	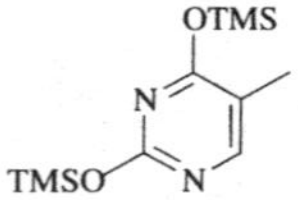 OTMS, N, N, TMSO	$SnCl_4$, $Cl(CH_2)_2Cl$, –50°, 2 h	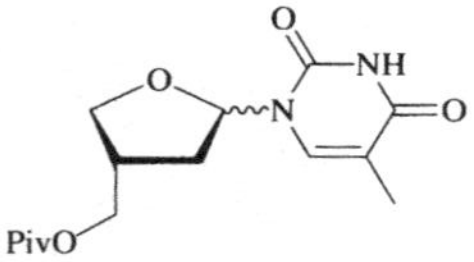 O, NH, O, N, O, PivO (40) α:β = 1:3	139a

Substrate	Silylated base	Conditions	Product(s) and Yield(s) (%)	Refs.
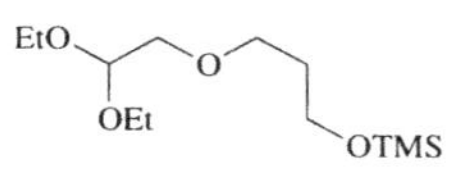 EtO, OEt, O, OTMS	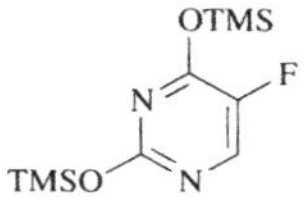 OTMS, F, TMSO, N	$SnCl_4$, $Cl(CH_2)_2Cl$, 23 to 25°, 1 h 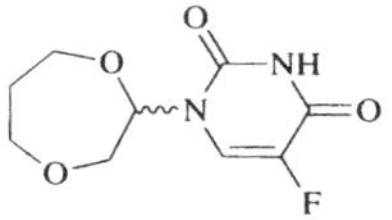	O, NH, N, O, F (28)	671
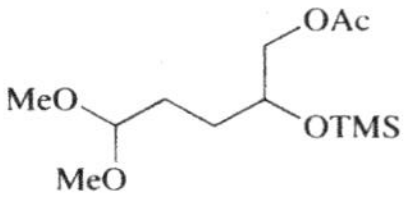 MeO, MeO, OAc, OTMS	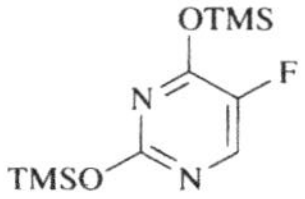 OTMS, F, TMSO, N	$SnCl_4$, $Cl(CH_2)_2Cl$, 23 to 25°, 1 h	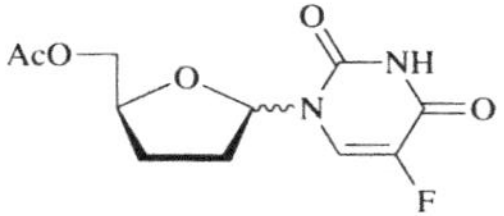 AcO, O, NH, N, O, F (67) α:β = 5:1	671
S, OAc, O, B, Ph	Bz, N, TMS, TMSO, N	$SnCl_4$	NHBz, N, O, N, R (57); R = S, O, B, Ph	686, 687
	OTMS, TMSO, N	$SnCl_4$	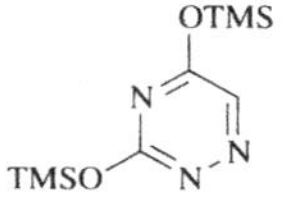 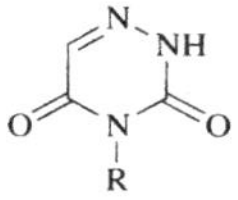 N, NH, O, N, R (42)	686, 687
	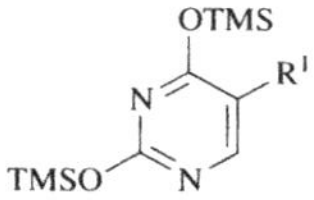 OTMS, R^1, TMSO, N	$SnCl_4$, CH_2Cl_2, rt, 12 h	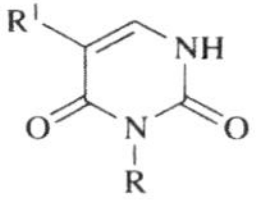 R^1, NH, O, N, R	687
	R^1 = H		(65)	
	F		(40)	
	Br		(60)	
	I		(57)	
	Me		(55)	

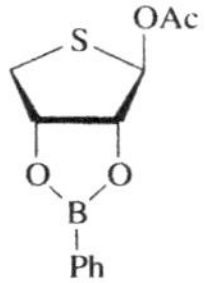

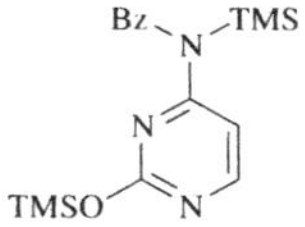

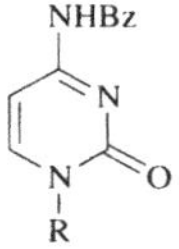

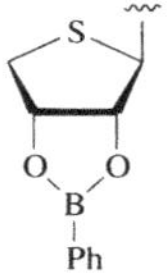

TABLE III. REACTIONS OF SILYLATED HETEROCYCLIC BASES WITH PROTECTED SUGARS - $SnCl_4$ CATALYST *(Continued)*

Sugar	Base	Conditions	Product(s) and Yield(s) (%)	Refs.
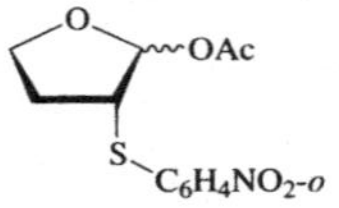 OAc; S–$C_6H_4NO_2$-*o*	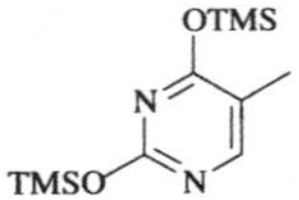 OTMS; TMSO	$SnCl_4$, CH_2Cl_2, –78 to 25°, 5 h	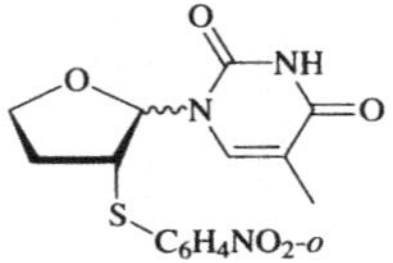 O; NH; O; S–$C_6H_4NO_2$-*o* (—) α:β = 1:16	227
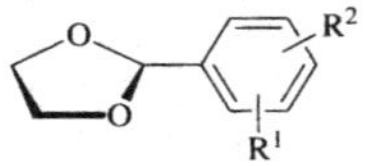 R^2; R^1	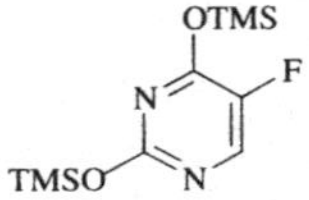 OTMS; F; TMSO	$SnCl_4$, CH_2Cl_2, rt, 12 h	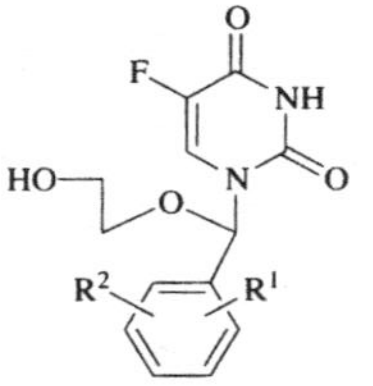 F; O; NH; O; HO; R^2; R^1	637

R^1	R^2	Product(s) and Yield(s) (%)
o-$OCH_2CH{=}CH_2$	H	(49)
p-$OCH_2CH{=}CH_2$	H	(47)
o-CO_2Et	H	(42)
p-OAc	*m*-OMe	(70)

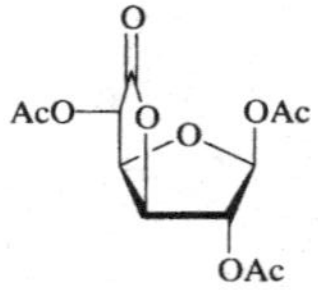

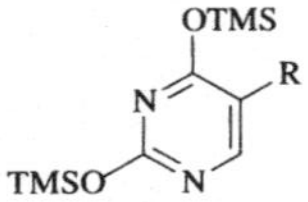

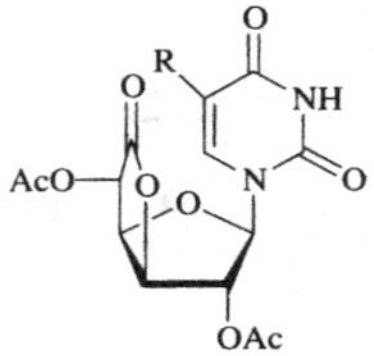

Sugar: AcO; O; OAc; OAc. Base: OTMS; R; TMSO. Product: O; R; NH; O; AcO; O; OAc.

R	Conditions	Product(s) and Yield(s) (%)	Refs.
F	$SnCl_4$, CH_2Cl_2, rt, 12 h	(35)	688
F	$SnCl_4$	(75)	688
Br	$SnCl_4$	(60)	688
I	$SnCl_4$	(56)	688

Conditions	Product(s) and Yield(s) (%)	Refs.
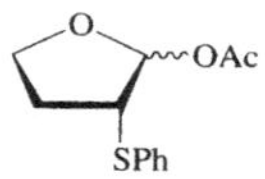 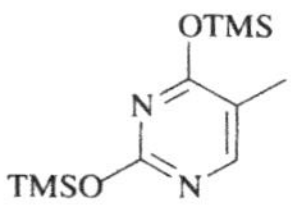 $SnCl_4$, CH_2Cl_2, –78 to 25°, 5 h	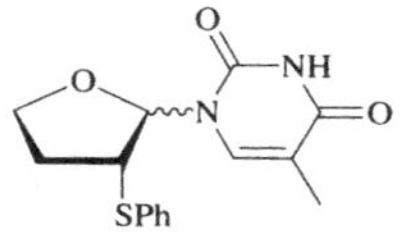 **I** (—) α:β = 1:17	227
$SnCl_4$, CH_2Cl_2, 25°	**I** (—) α:β = 1:30	227
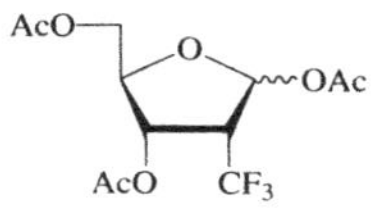 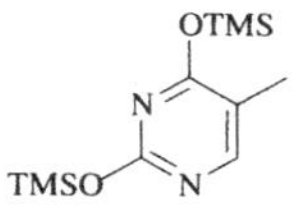 $SnCl_4$, MeCN, 50°	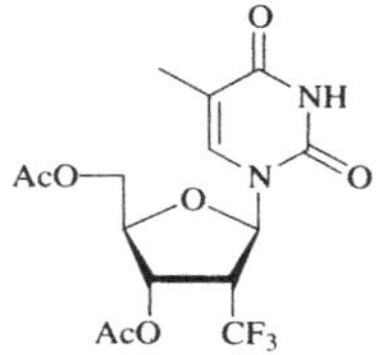 (40) α:β = 1:12	689
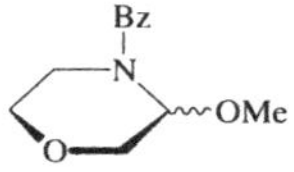 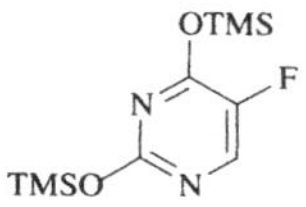 $SnCl_4$, MeCN, –20 to –25°, 24 h	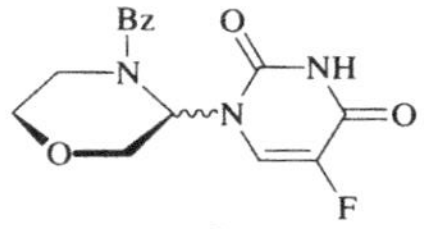 (71)	642
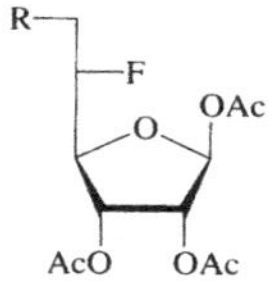 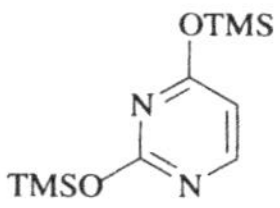 $SnCl_4$	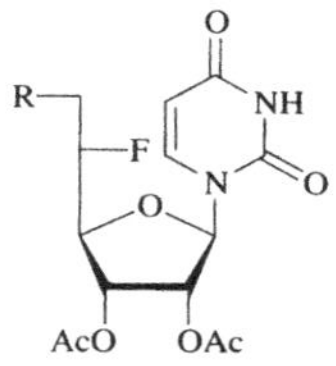 R: F (—); N_3 (—)	690
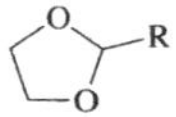 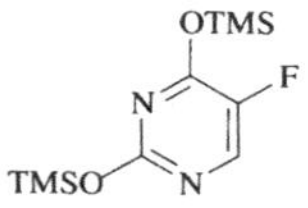 $SnCl_4$, CH_2Cl_2, rt, 12 h	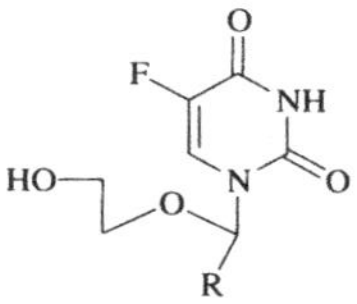 R: C_6H_4(Pr-*i*)-*p* (24); $C_6H_2(OMe)_3$-3,4,5 (—)	637

TABLE III. REACTIONS OF SILYLATED HETEROCYCLIC BASES WITH PROTECTED SUGARS - $SnCl_4$ CATALYST (*Continued*)

Sugar	Base	Conditions	Product(s) and Yield(s) (%)	Refs.
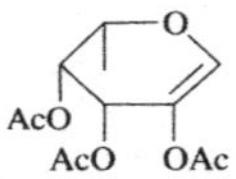 AcO, AcO, OAc	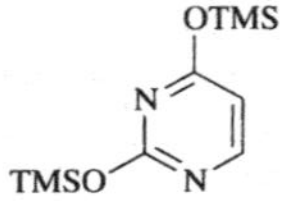 OTMS, N, N, TMSO	$SnCl_4$, CH_2Cl_2, 0°, 6 h	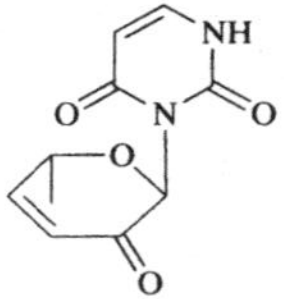 NH, O, N, O, O, O (50)	675
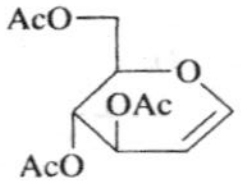 AcO, OAc, O, AcO	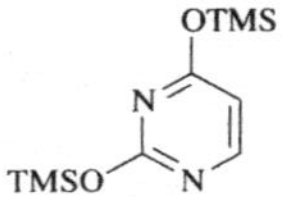 OTMS, N, N, TMSO	$SnCl_4$, MeCN	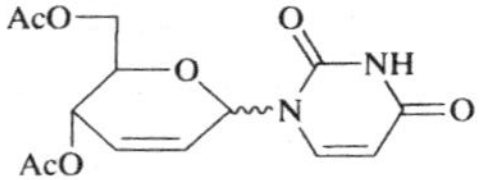 AcO, O, O, NH, N, O, AcO (78)	105
	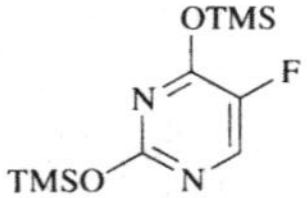 OTMS, F, N, N, TMSO	$SnCl_4$, MeCN, rt, 35 min	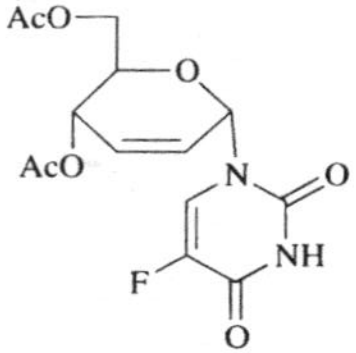 AcO, O, AcO, N, O, NH, F, O (36) + β-anomer (22)	306
	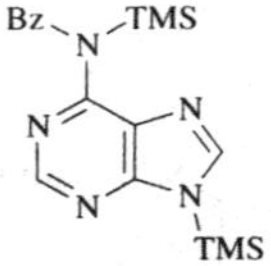 Bz, N, TMS, N, N, N, N, TMS	$SnCl_4$, MeCN, rt, 2 h	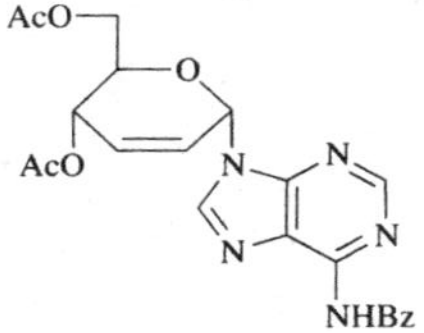 AcO, O, AcO, N, N, N, N, N, NHBz (18)	306
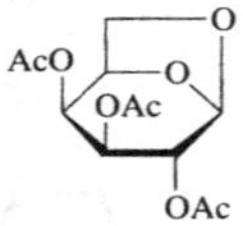 AcO, O, O, OAc, OAc	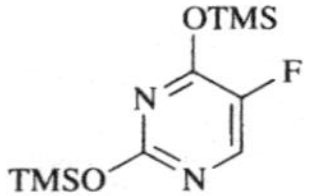 OTMS, F, N, N, TMSO	$SnCl_4$, MeCN, reflux, 10 h	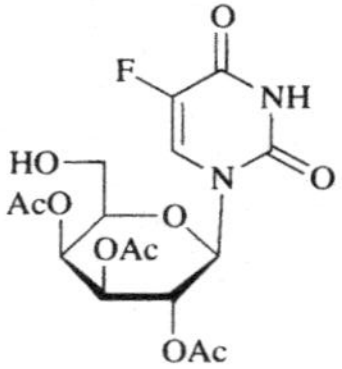 O, F, NH, HO, N, O, AcO, O, OAc, OAc (60)	692

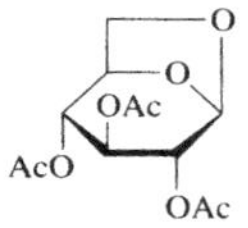	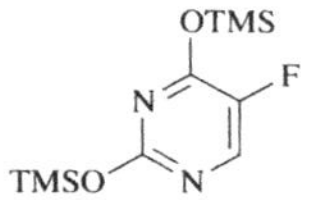	$SnCl_4$, MeCN, reflux, 10 h	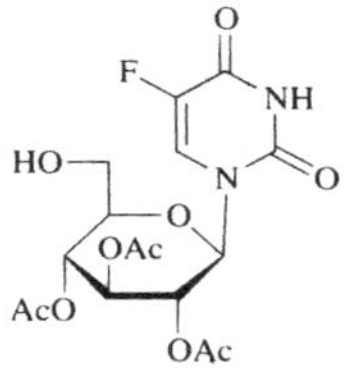	(80) 692
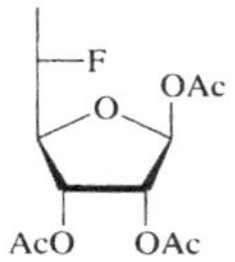	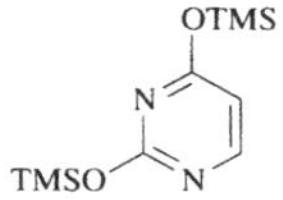	$SnCl_4$	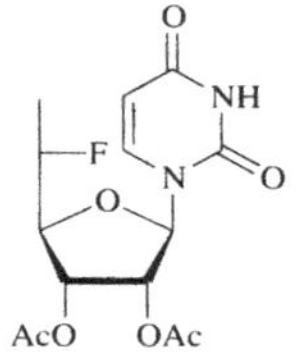	(—) 690
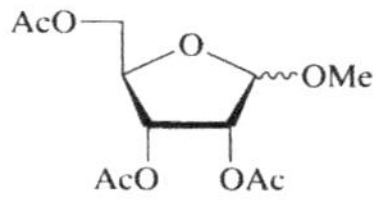	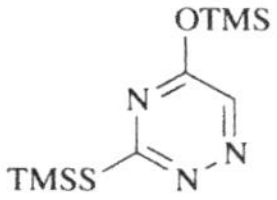	$SnCl_4$, $Cl(CH_2)_2Cl$, 22°, 5 h	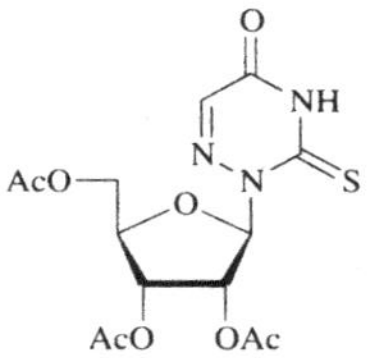	(73) 84
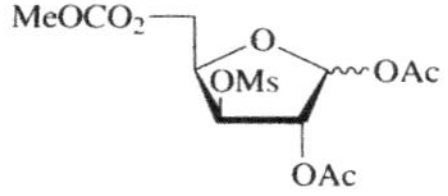	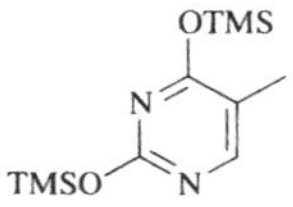	$SnCl_4$, CH_2Cl_2, rt, 18 h	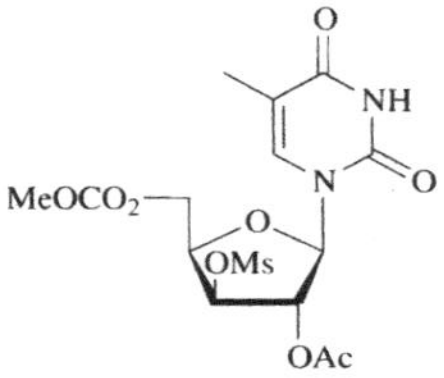	(66) 545

TABLE III. REACTIONS OF SILYLATED HETEROCYCLIC BASES WITH PROTECTED SUGARS - $SnCl_4$ CATALYST (*Continued*)

Sugar	Base	Conditions	Product(s) and Yield(s) (%)	Refs.
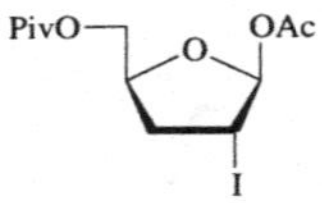	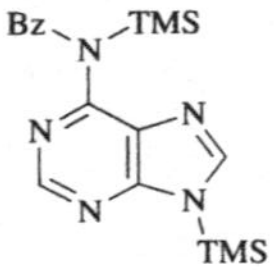	$SnCl_4$, $Cl(CH_2)_2Cl$, −20° to rt	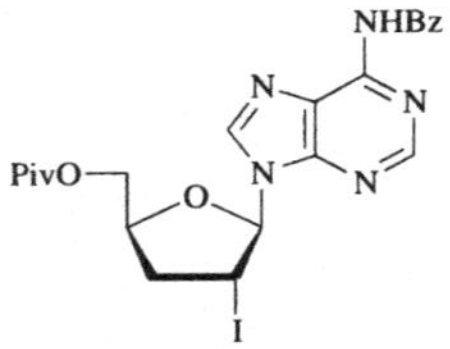 (45)	423
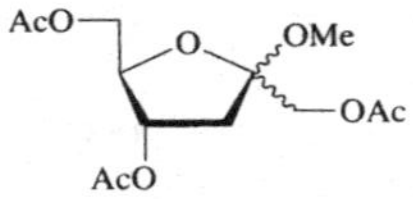	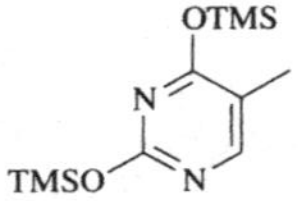	1. $SnCl_4$, MeCN, −20° to rt, 1 h 2. NH_3, MeOH	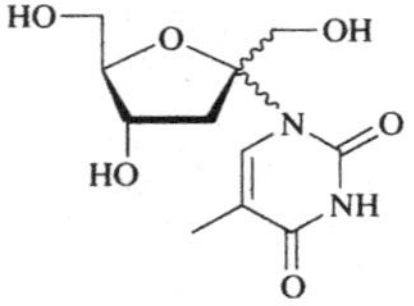 (41) $\alpha:\beta = 1:2$	693
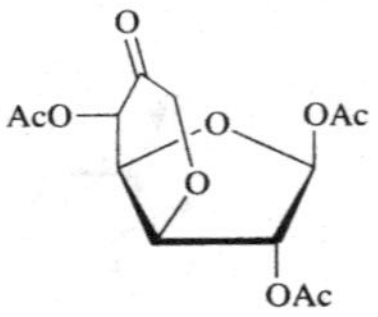	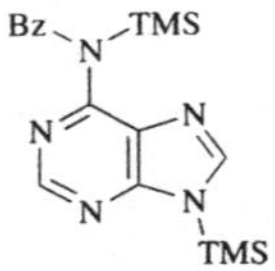	1. $SnCl_4$, $Cl(CH_2)_2Cl$, 48-72 h 2. NH_3, MeOH	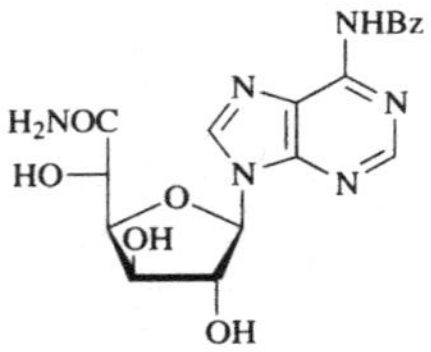 (64)	101
C_{13} 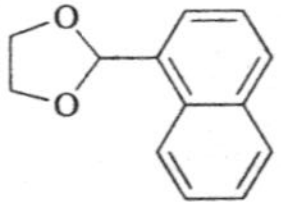	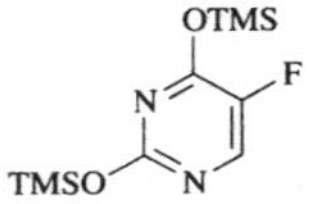	$SnCl_4$, CH_2Cl_2, rt, 12 h	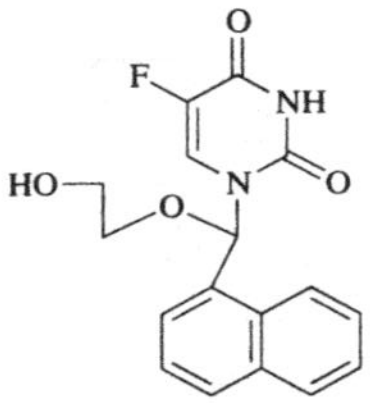 (44)	637

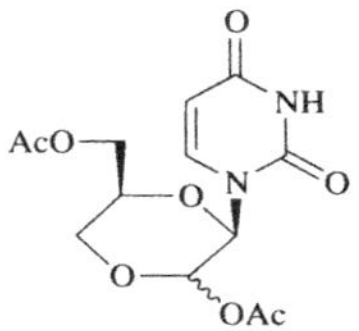

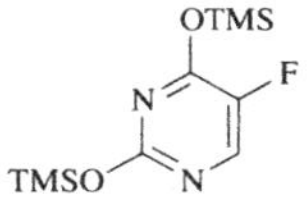

SnCl$_4$, MeCN, rt — (30) — 694

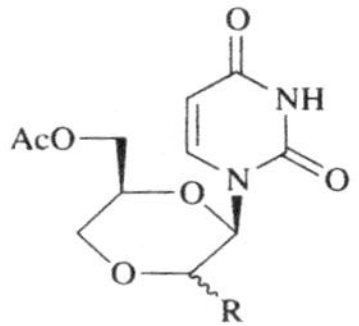

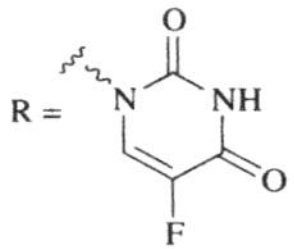

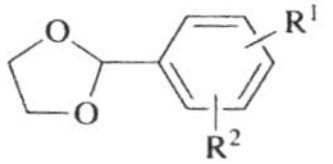

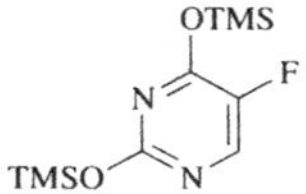

1. SnCl$_4$, CH$_2$Cl$_2$, rt, 12 h
2. (see below)
3. (see below)

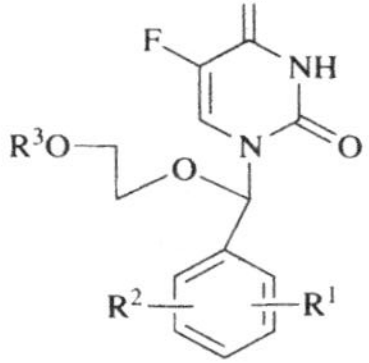

637

R^1	R^2	2	3	R^3	Yield
p-OAc	*m*-OAc	NaHCO$_3$, MeOH	Ac$_2$O, Py, rt, 3 h	Ac	(—)
o-OAc	*m*-OAc	—	—	H	(53)
p-OCH$_2$CH=CH$_2$	*m*-OMe	—	—	H	(51)
p-OAc	*m*-OEt	—	—	H	(41)
p-OPr-*n*	H	—	—	H	(19)

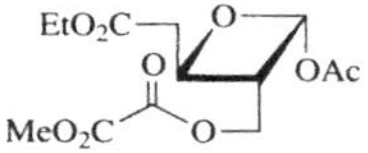

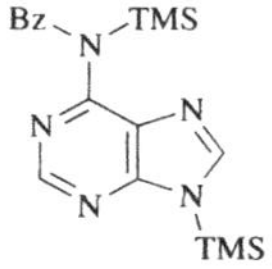

1. SnCl$_4$, Cl(CH$_2$)$_2$Cl
2. NaOMe
3. BzCl, Py

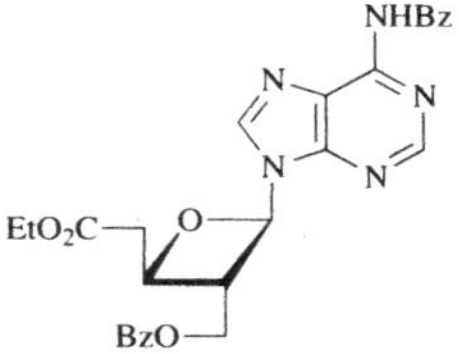

I (26) + α-anomer (9) — 233

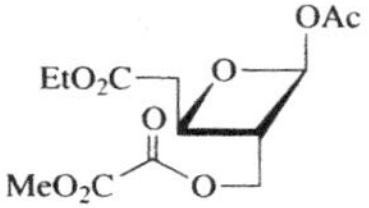

"

1. SnCl$_4$, Cl(CH$_2$)$_2$Cl, rt, 30 min
2. NaOMe, MeOH
3. BzCl, Py, CH$_2$Cl$_2$, rt, 12 h

I (16) + α-anomer (10) — 233

TABLE III. REACTIONS OF SILYLATED HETEROCYCLIC BASES WITH PROTECTED SUGARS - $SnCl_4$ CATALYST (*Continued*)

Sugar	Base	Conditions	Product(s) and Yield(s) (%)	Refs.
MeO_2C, OAc, OAc, N_3, OAc	Ac, N, TMS, N, TMSO, N	$SnCl_4$, $Cl(CH_2)_2Cl$, 60°, 12 h	NHAc, N, MeO_2C, O, N, O, OAc, N_3, OAc (67)	695, 696
AcO, O, OAc, AcO, OAc	O, N, O, TMS	$SnCl_4$, MeCN, rt, 24 h	O, N, O, R (86) R = AcO, O, AcO, OAc	548, 697
	Ph, O, N, O, TMS	$SnCl_4$, MeCN, 35-40°, 18 h	Ph, O, N, O, R (48)	698
	TMSO, O, N, N, OTMS	$SnCl_4$, $Cl(CH_2)_2Cl$, rt, 16 d	O, NH, O, N, O, R (70)	699
	TMSNH, S, N, N, OTMS	$SnCl_4$, $Cl(CH_2)_2Cl$, 25°, 5 h	H_2N, N, S, N, O, R (60)	700

Reactant	Conditions	Product(s) and Yield(s) (%)	Refs.
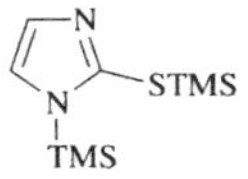 1-TMS-2-(TMS-thio)imidazole (STMS, N-TMS)	$SnCl_4$, $Cl(CH_2)_2Cl$, rt, 8 d	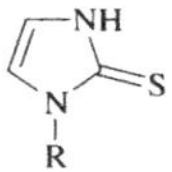 1-R-imidazole-2-thione (NH, =S, N–R) **I** (34) + bis(product) **II** (64)	701
	$SnCl_4$, MeCN, rt, 8 d	**I** (28) + bis(product) **II** (66)	701
	$SnCl_4$, $Cl(CH_2)_2Cl$, reflux, 45 min	**I** (42) + bis(product) **II** (52)	701
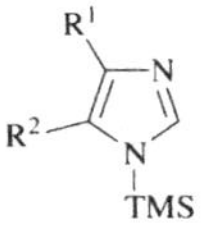 4-R^1-5-R^2-1-TMS-imidazole	$SnCl_4$, $Cl(CH_2)_2Cl$ 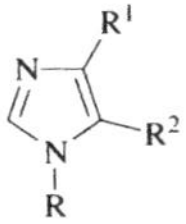	4-R^1-5-R^2-1-R-imidazole	

R^1	R^2	Conditions	R^1	R^2		N^3-isomer	Refs.
NO_2	Br	rt, 18 h	NO_2	Br	(33)	(14)	702
Me	CN	rt, 24 h	Me	CN	(38)	(16)	703
CO_2Me	CH_2CN	rt, 16-20 h	CO_2Me	CH_2CN	(95)	(—)	704
		rt, 6 h	CO_2Me	CH_2CN	(38)	(19) - β-isomer	705
		—	CH_2CN	CO_2Me	(35)	(30)	706
		rt, 6 h	CH_2CN	CO_2Me	(57)	(—)	707
CO_2Et	Me	reflux, 2 h	CO_2Et	Me	(43)	(6)	708
		reflux	CO_2Et	Me	(43)	(—)	709
TMSO	$CONH_2$	CH_2Cl_2, rt, 12 h	$CONH_2$	OH	(—)	(—)	710

Reactant	Conditions	Product(s) and Yield(s) (%)	Refs.
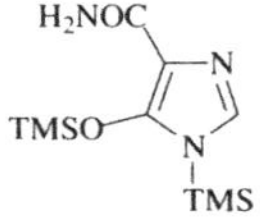 4-H_2NOC-5-TMSO-1-TMS-imidazole	$SnCl_4$, $Cl(CH_2)_2Cl$ 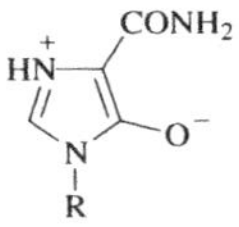	4-$CONH_2$-5-O^--1-R-imidazolium (HN^+) (14)	241
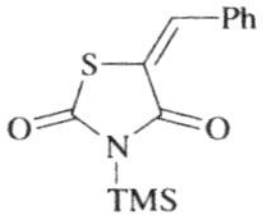 5-benzylidene-3-TMS-thiazolidine-2,4-dione (=CH–Ph, N–TMS)	$SnCl_4$, MeCN 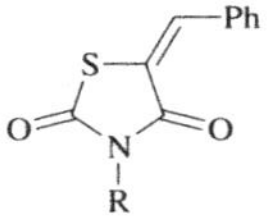	5-benzylidene-3-R-thiazolidine-2,4-dione (—)	711

TABLE III. REACTIONS OF SILYLATED HETEROCYCLIC BASES WITH PROTECTED SUGARS - $SnCl_4$ CATALYST (*Continued*)

Sugar	Base	Conditions	Product(s) and Yield(s) (%)	Refs.
	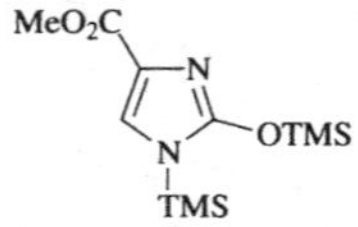 MeO_2C, N, OTMS, N–TMS	$SnCl_4$	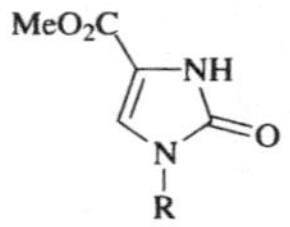 MeO_2C, NH, O, N–R (36) R = AcO, O, AcO, OAc 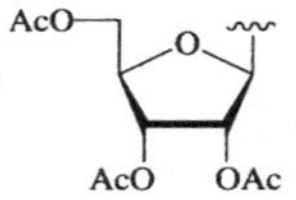	712
	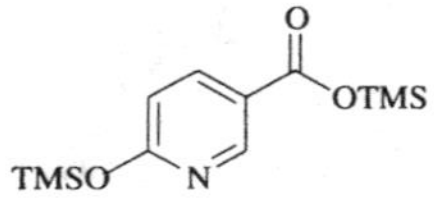 O, OTMS, TMSO, N	$SnCl_4$, MeCN, rt, 20 h	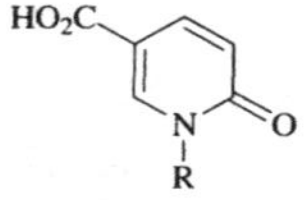 HO_2C, N–R, O (88)	713
	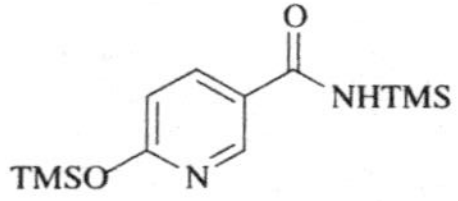 O, NHTMS, TMSO, N	$SnCl_4$, MeCN, $Cl(CH_2)_2Cl$, rt, 16 h	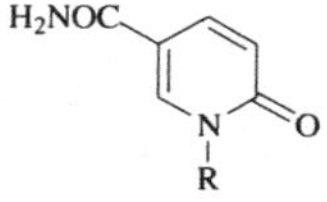 H_2NOC, N–R, O (85)	713
	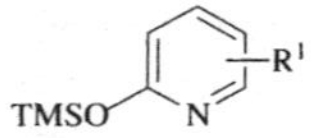 TMSO, N, R^1	1. $SnCl_4$, $Cl(CH_2)_2Cl$, reflux, 8 h; rt, 12 h 2. NH_3, MeOH, 0°, 4 d	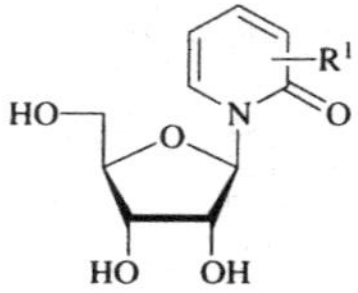 HO, O, N, O, R^1, HO, OH R^1: 3-NO_2 (51); 5-NO_2 (80)	714
	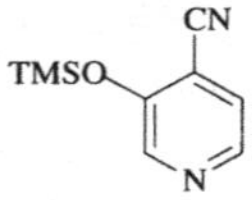 CN, TMSO, N	$SnCl_4$, $Cl(CH_2)_2Cl$, 0-5°, 1 h	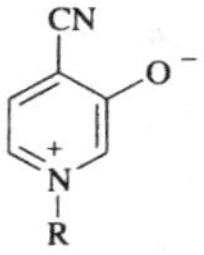 CN, O^-, N^+–R (57) R = AcO, O, AcO, OAc 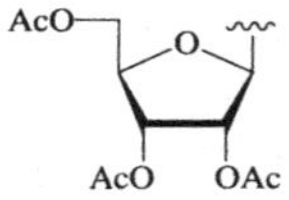	715
	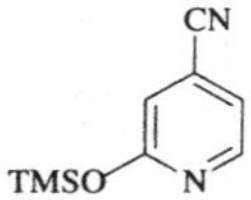 CN, TMSO, N	$SnCl_4$, $Cl(CH_2)_2Cl$, 0-5°, 30 min; rt, 15 h	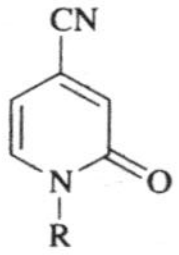 CN, N–R, O (86)	715

Substrate	Conditions	Product(s) and Yield(s) (%)	Refs.
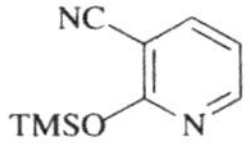 NC, TMSO (pyridine)	1. $SnCl_4$, $Cl(CH_2)_2Cl$, reflux, 8 h; rt, 12 ht 2. NH_3, MeOH, 0°, 4 d	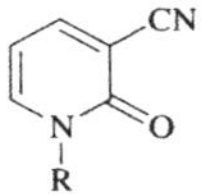 CN, N–R, O (48)	714
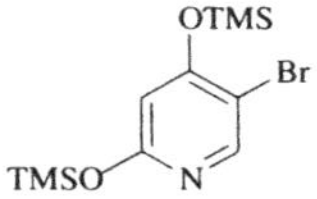 OTMS, Br, TMSO (pyridine)	$SnCl_4$, $Cl(CH_2)_2Cl$, rt, 18 h	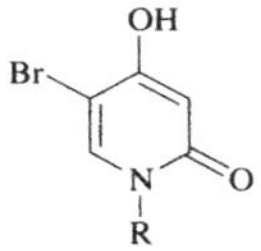 OH, Br, N–R, O (97)	716
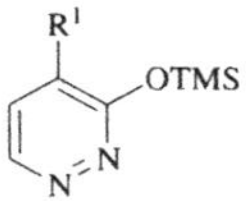 R^1, OTMS (pyridazine)	$SnCl_4$, $Cl(CH_2)_2Cl$, rt, 30 min	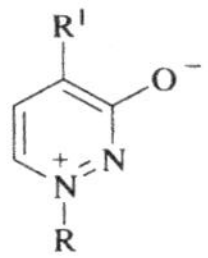 R^1, O^-, N^+–R R^1: Cl (77); CN (85); CO_2Et (45)	253
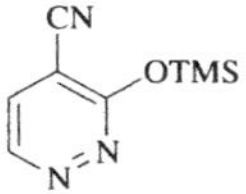 CN, OTMS (pyridazine)	$SnCl_4$, $Cl(CH_2)_2Cl$, reflux, 15 min	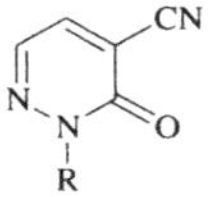 CN, N–R, O (51)	253
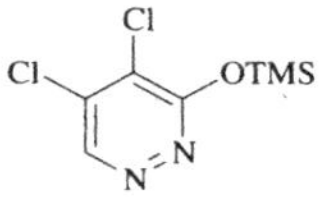 Cl, Cl, OTMS (pyridazine)	$SnCl_4$, $Cl(CH_2)_2Cl$, rt, 30 min	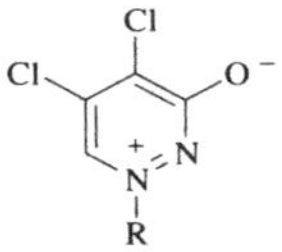 Cl, Cl, O^-, N^+–R (76)	253
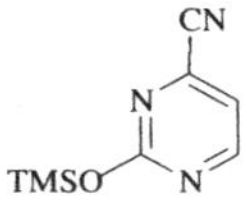 CN, TMSO (pyrimidine)	$SnCl_4$, $Cl(CH_2)_2Cl$, rt, 2 h	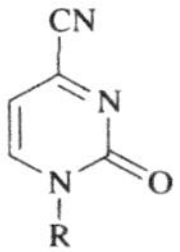 CN, N–R, O (73)	715
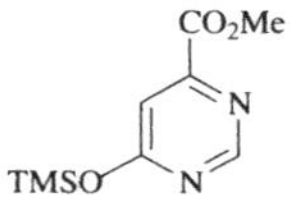 CO_2Me, TMSO (pyrimidine)	$SnCl_4$, $Cl(CH_2)_2Cl$, rt, 45 min	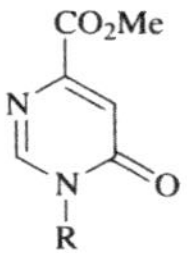 CO_2Me, N–R, O (80-90) R 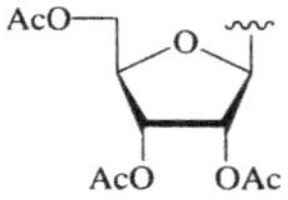 = 2,3,5-tri-O-acetylribofuranosyl (AcO, AcO, OAc)	715

TABLE III. REACTIONS OF SILYLATED HETEROCYCLIC BASES WITH PROTECTED SUGARS - $SnCl_4$ CATALYST (*Continued*)

Sugar	Base	Conditions	Product(s) and Yield(s) (%)	Refs.
	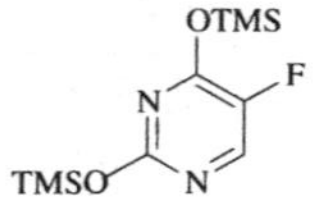 2,4-bis(OTMS)-5-F-pyrimidine	$SnCl_4$, MeCN, rt, 3 h	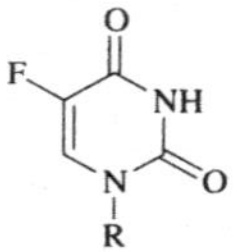 5-F-uracil, N-R (81); R = 2,3,5-tri-O-acetyl-ribofuranosyl (AcO, AcO, OAc) 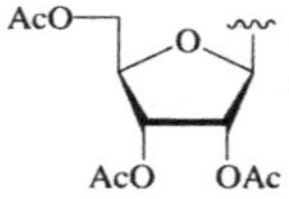	514b
	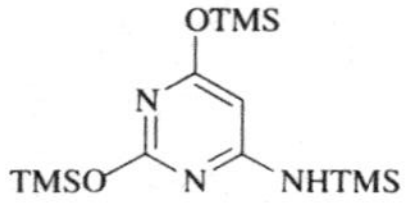 2,4-bis(OTMS)-6-NHTMS-pyrimidine	$SnCl_4$	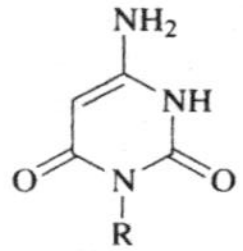 6-NH_2-uracil, N-R (—)	717
	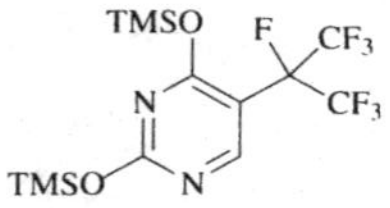 2,4-bis(TMSO)-5-($CF(CF_3)_2$)-pyrimidine	$SnCl_4$	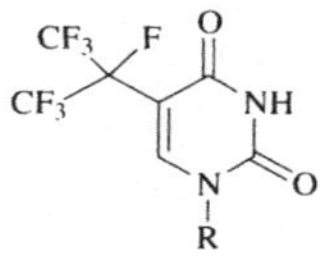 5-(CF_3)$_2$CF-uracil, N-R (—)	718
	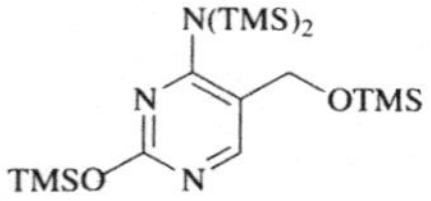 2-TMSO-4-$N(TMS)_2$-5-(CH_2OTMS)-pyrimidine	$SnCl_4$, MeCN	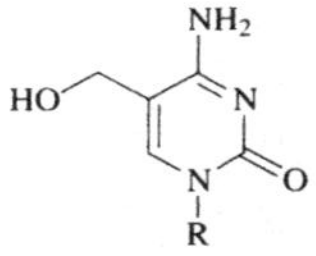 5-HOCH$_2$-cytosine, N-R (—)	719
	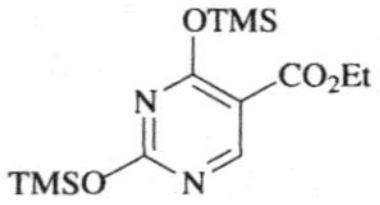 2,4-bis(OTMS)-5-CO_2Et-pyrimidine	$SnCl_4$, CH_2Cl_2, rt, 30 min	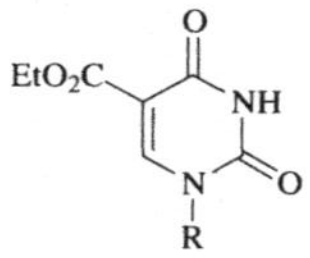 5-EtO_2C-uracil, N-R (55)	542
	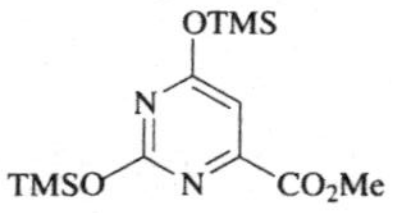 2,4-bis(OTMS)-6-CO_2Me-pyrimidine	$SnCl_4$, $Cl(CH_2)_2Cl$, rt, 72 h	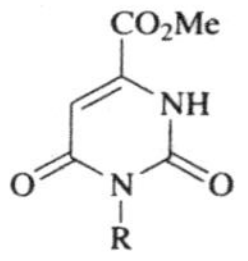 6-CO_2Me-uracil, N-R (40) + N^1,N^3-bis(isomer) (14)	720

Reactant	Conditions	Product (yield, %)		Ref.
	$SnCl_4$, $Cl(CH_2)_2Cl$, 20°, 8 h		(—)	721
	$SnCl_4$, $Cl(CH_2)_2Cl$, rt, 2.5 h		(91)	722
	$SnCl_4$		(94)	723
	$SnCl_4$, $Cl(CH_2)_2Cl$, rt, 16 h		(59)	724
	$SnCl_4$, $Cl(CH_2)_2Cl$, 23°, 4 h		(77)	725
	$SnCl_4$, $Cl(CH_2)_2Cl$, 25°, 72 h		(4)	239

TABLE III. REACTIONS OF SILYLATED HETEROCYCLIC BASES WITH PROTECTED SUGARS - $SnCl_4$ CATALYST (*Continued*)

Sugar	Base	Conditions	Product(s) and Yield(s) (%)	Refs.
	OTMS, N, TMSO, N, O	$SnCl_4$, $Cl(CH_2)_2Cl$, rt, 24 h	O, NH, O, N, O, R (86) R = AcO, O, AcO, OAc	726
	O, N, O, TMS	1. $SnCl_4$ 2. HCl, MeOH	O, N, O, R (16) R = HO, O, HO, OH	697
	O, N, O, TMS	1. $SnCl_4$ 2. HCl, MeOH	O, N, O, R (37)	697
	N, SH, N, TMS	$SnCl_4$, $Cl(CH_2)_2Cl$, rt, 7 d	HS, N, N, R (34) + bis(riboside) (12) R = AcO, O, AcO, OAc	727
	X, S, N, TMS	$SnCl_4$, $Cl(CH_2)_2Cl$, rt, 7 d	S, X, N, R; X: O (30), S (34)	727
	N, N, TMS	$SnCl_4$ (0.2 eq), MeCN, rt	N, N, R **I** (62) + N^1-isomer (4)	728

	$SnCl_4$ (0.2 eq), MeCN, reflux	**I** (44) + N^1-isomer (9)	728
	$SnCl_4$ (1 eq), MeCN, rt	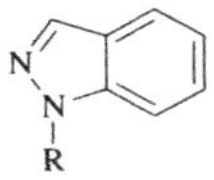 **I** (10)	728
	$SnCl_4$ (1 eq), MeCN, reflux	**I** (7)	728
	$SnCl_4$ (0.5 eq), MeCN, reflux	**I** (41)	728
	$SnCl_4$ (0.5 eq), MeCN, rt	**I** (39)	728
	$SnCl_4$ (1.5 eq), MeCN, rt	**I** (6)	728
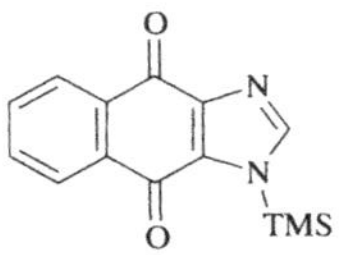	$SnCl_4$, $Cl(CH_2)_2Cl$, rt, 3 h	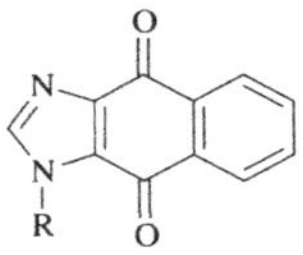 (54)	729
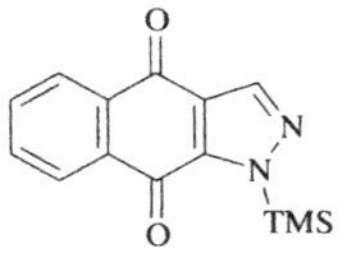	$SnCl_4$, $Cl(CH_2)_2Cl$, rt, 3 h	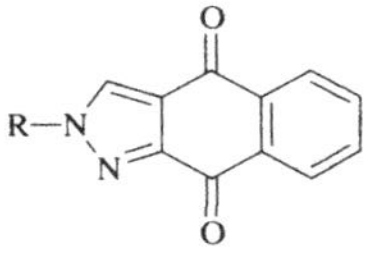 (76)	729
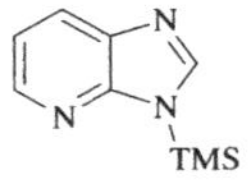	$SnCl_4$ (2.5 eq), MeCN, rt, 5 h	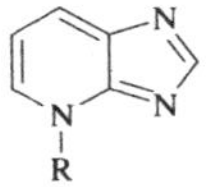 **I** (60) + N^3-isomer (24)	625
	$SnCl_4$ (1.1 eq), MeCN, rt, 5 h	**I** (70) + N^3-isomer (17)	625

TABLE III. REACTIONS OF SILYLATED HETEROCYCLIC BASES WITH PROTECTED SUGARS - $SnCl_4$ CATALYST (*Continued*)

Sugar	Base	Conditions	Product(s) and Yield(s) (%)	Refs.
		$SnCl_4$ (1.1 eq), MeCN, rt, 30 h	**I** (62) + N^3-isomer (16) + N^1-iscmer (11)	625
		$SnCl_4$ (3.6 eq), MeCN, rt, 30 h	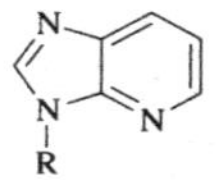 (42) + N^4-isomer (15) + N^1-isomer (26); R 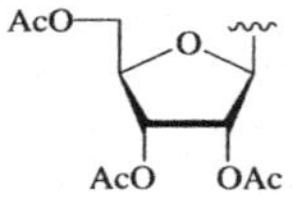 =	625
	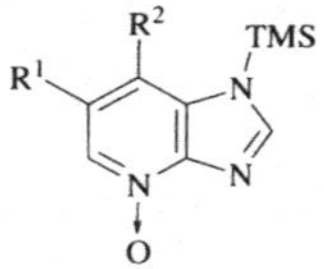	$SnCl_4$, MeCN, rt	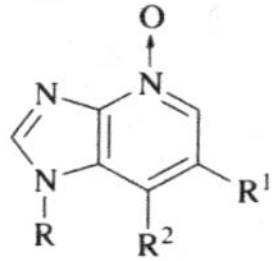	
	R¹ = H, R² = H	6 h	(83)	730-732
	R¹ = H, R² = Cl	6 h	(82)	730,731
	R¹ = Br, R² = H	18 h	(79)	732
	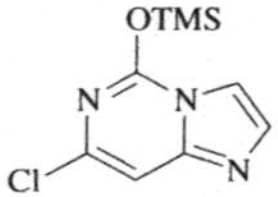	$SnCl_4$, $Cl(CH_2)_2Cl$, rt, 24 h	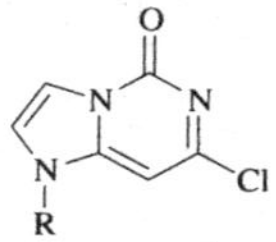 (64)	733
	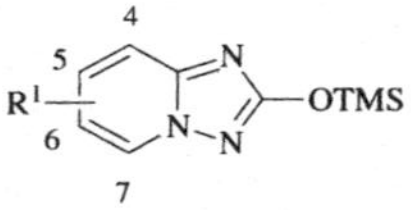	$SnCl_4$, $Cl(CH_2)_2Cl$, reflux	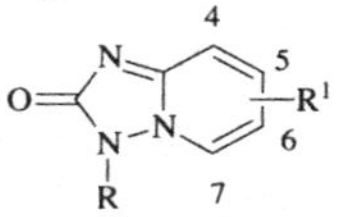	734
	R¹ = 4-Me	10 h	(47)	
	R¹ = 5-Me	8 h	(25)	
	R¹ = 6-Me	8 h	(50)	
	R¹ = 7-Me	10 h	(73)	
	R¹ = 5,7-Me_2	8 h	(57)	

Reactant	Conditions	Product(s) and Yield(s) (%)	Refs.
3-OTMS-[1,2,4]triazolo[4,3-a]pyridine (R¹ at 5–8)	$SnCl_4$, $Cl(CH_2)_2Cl$, reflux		734
R¹ = H	2 h	(96)	
R¹ = 4-Me	10 h	(48)	
R¹ = 5-Me	8 h	(37)	
R¹ = 6-Me	8 h	(21)	
R¹ = 5,7-Me_2	8 h	(2)	
1-TMS-naphtho[2,3-d]triazole-4,9-dione	$SnCl_4$, $Cl(CH_2)_2Cl$, rt, 3 h	(44)	729
TMSNH, TMSO, R¹ thieno[2,3-d]pyrimidine	1. $SnCl_4$, $Cl(CH_2)_2Cl$, rt, 4 h 2. NH_3, MeOH, rt, 20 h	R¹ = H (48); R¹ = Me (60); R = ribofuranosyl (HO, HO, OH)	525
4-OTMS-1-TMS-pyrazolo[3,4-d]pyrimidine	$SnCl_4$, $Cl(CH_2)_2Cl$, MS, 60°, 4.5 h	**I** (25) mixture of N^1-, N^2-, and N^5-isomers + N^1,N^5-bis(riboside) (23) + N^2,N^5-bis (riboside) (12)	510
	$SnCl_4$, $Cl(CH_2)_2Cl$, 60°, 4 h	**I** 3 monoriboside isomers + N^2,N^5- and N^1,N^5-bis(ribosides), 2:2:1	735

TABLE III. REACTIONS OF SILYLATED HETEROCYCLIC BASES WITH PROTECTED SUGARS - $SnCl_4$ CATALYST (*Continued*)

Sugar	Base	Conditions	Product(s) and Yield(s) (%)	Refs.
	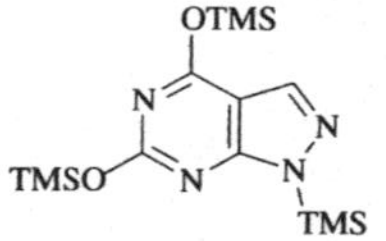 OTMS, TMSO, N, N, N, N, TMS	1. $SnCl_4$, $Cl(CH_2)_2Cl$, rt, 1 h 2. NH_3, MeOH	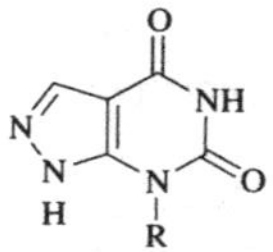 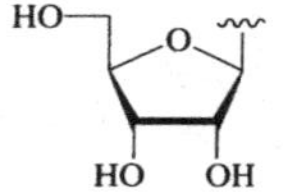 O, NH, N, N, H, N, R, O (—) R = HO, O, HO, OH	736
	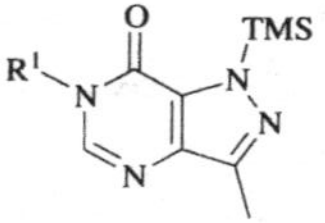 O, TMS, R^1, N, N, N, N	$SnCl_4$, MeCN, 25°, 18 h	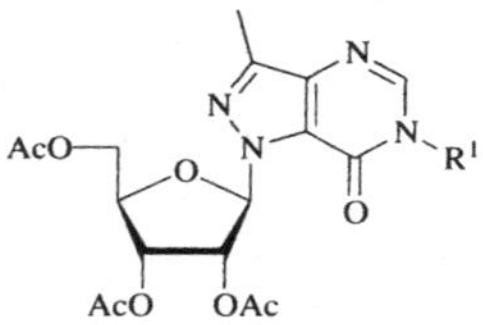 N, N, N, N, R^1, AcO, O, O, AcO, OAc	737
	R^1 = Bu		(75)	
	R^1 = i-C_5H_{11}		(80)	
	R^1 = Bn		(72)	
	R^1 = p-ClC_6H_4		(65)	
	R^1 = p-$ClC_6H_4CH_2$		(70)	
	R^1 = p-MeC_6H_4		(66)	
	R^1 = $BnCH_2$		(73)	
	R^1 = p-$MeOC_6H_4CH_2$		(75)	
	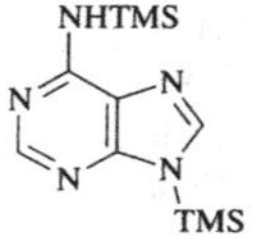 NHTMS, N, N, N, N, TMS	$SnCl_4$, MeCN, rt, 20 h	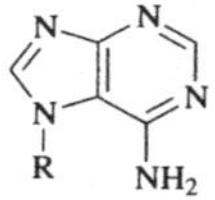 N, N, N, N, R, NH_2 (25) + α-anomer (18) + N^9-isomer (20) R 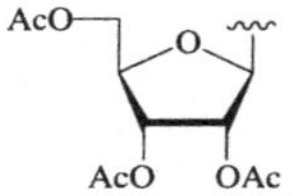 = AcO, O, AcO, OAc	625
	OTMS, N, N, N, N, AcN, TMS, TMS	$SnCl_4$, $Cl(CH_2)_2Cl$, rt, 12 h	N, N, NHAc, NH, N, R, O (70)	141

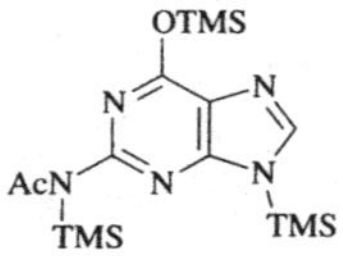

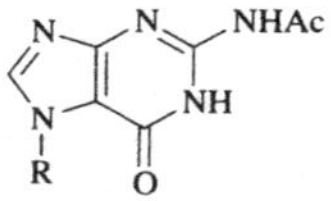

Reactant	Conditions	Product(s) and Yield(s) (%)	Refs.
	$SnCl_4$, MeCN, rt, 20 h	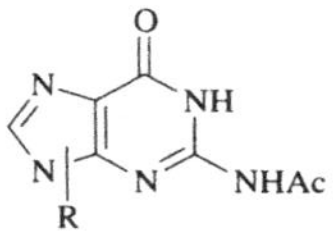 (78) N^7:N^9 = 95:1	142
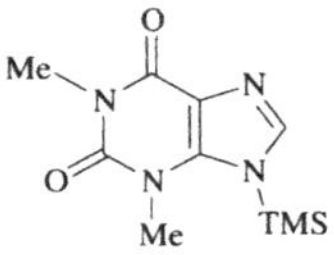 Me, N, O, N, Me, N, TMS	$SnCl_4$, $Cl(CH_2)_2Cl$, reflux, 30 min	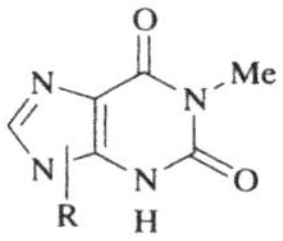 (—) N^7:N^9 = 4:1	738
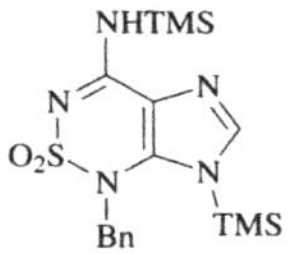 NHTMS, N, O_2S, N, Bn, N, TMS	$SnCl_4$, CH_2Cl_2, rt, 12 h	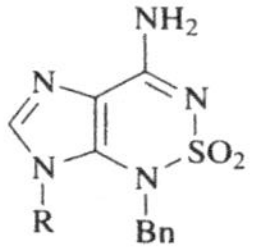 (15)	739
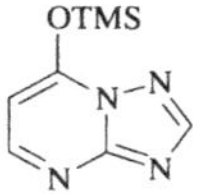 OTMS, N, N, N	$SnCl_4$, $Cl(CH_2)_2Cl$, reflux, 2 h	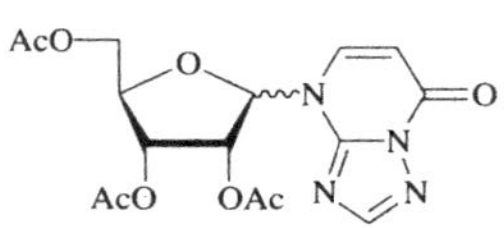 AcO, O, AcO, OAc, N, O, N, N (49) α:β = 1:2.4 + N^3-isomer (21), α:β = 1:1.5	257
	$SnCl_4$, $Cl(CH_2)_2Cl$, rt, 4 h	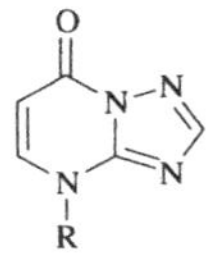 (46) + N^3-isomer (10) 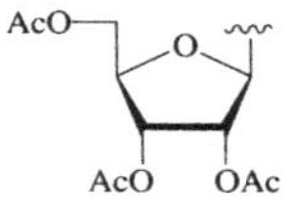	257

R = AcO, O, AcO, OAc

TABLE III. REACTIONS OF SILYLATED HETEROCYCLIC BASES WITH PROTECTED SUGARS - $SnCl_4$ CATALYST (*Continued*)

Sugar	Base	Conditions	Product(s) and Yield(s) (%)	Refs.
	OTMS, N, N, N, N	$SnCl_4$, $Cl(CH_2)_2Cl$, rt, 4 h	O, N, N, N, N–R (52) + N^1-isomer (11)	257
	TMS, N, N, N, O, N–Me	$SnCl_4$, $Cl(CH_2)_2Cl$, reflux, 6 h; rt, 18 h	O, Me—N, N, N, N–R (—)	714
	TMS, N, N, N, O, N–Bn	1. $SnCl_4$, $Cl(CH_2)_2Cl$, reflux, 8 h; 21°, 16 h 2. NaOMe, MeOH, rt, 16 h	O, Bn—N, N, N, N–R (78) R = HO—, O, HO, OH	714
	OTMS, F, CO_2Et, R^1, N	$SnCl_4$, $Cl(CH_2)_2Cl$, rt, 3 h	O, EtO_2C, F, R^1, N–R R = AcO—, O, AcO, OAc	740
	R^1 = H		(19)	
	R^1 = F		(25)	
	R^1 = Cl		(15)	

Sugar	Base	Conditions	Product(s) and Yield(s) (%)	Refs.
	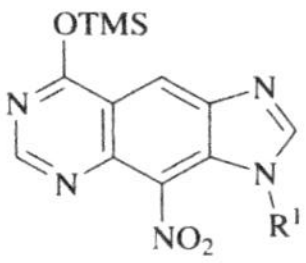 OTMS; N; N; N; N; NO_2; R^1	$SnCl_4$	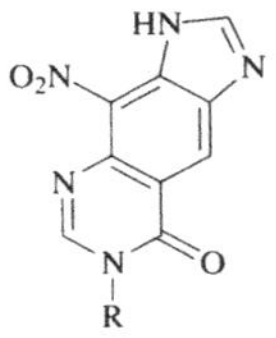 O_2N; HN; N; N; N; O; R	509
	R^1 = H	CH_2Cl_2, rt, 30 min	(22-35)	
	R^1 = TMS	MeCN, rt, 15 h	(19) + N^1-isomer (9) + N^3-isomer (7)	
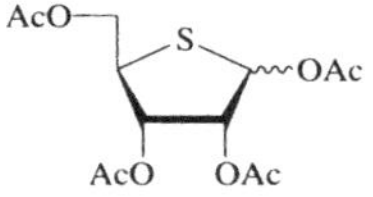 AcO; S; OAc; AcO; OAc	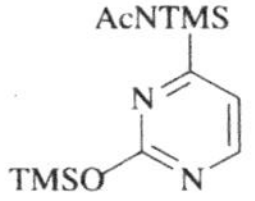 AcNTMS; N; TMSO; N	$SnCl_4$, $Cl(CH_2)_2Cl$, 25°, 20 h	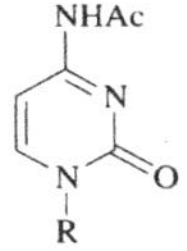 NHAc; N; N; O; R (65) + α-anomer (3) R 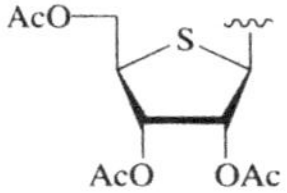 = AcO; S; AcO; OAc	215
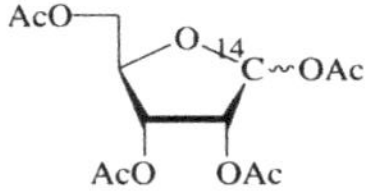 AcO; O; 14; C; OAc; AcO; OAc	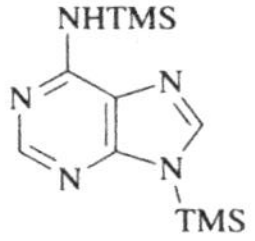 NHTMS; N; N; N; N; TMS	$SnCl_4$, MeCN	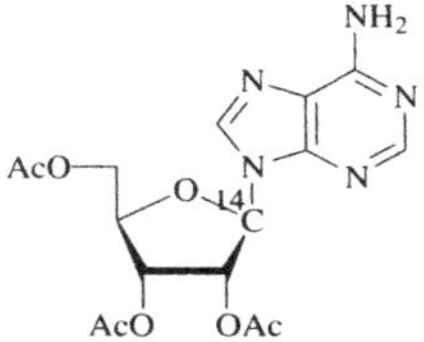 NH_2; N; N; N; N; AcO; O; 14; C; AcO; OAc (80)	741
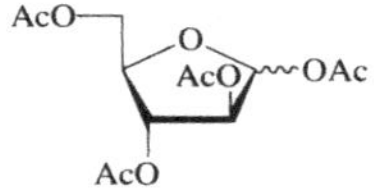 AcO; O; AcO; OAc; AcO	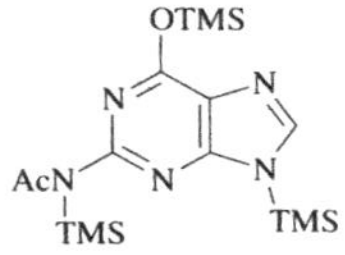 OTMS; N; N; AcN; TMS; N; N; TMS	$SnCl_4$, $Cl(CH_2)_2Cl$, 80°, 12 h	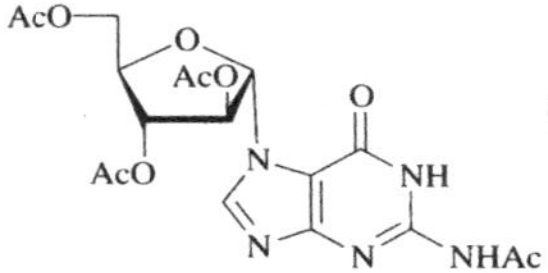 AcO; O; AcO; AcO; O; N; NH; N; N; NHAc (82)	141a

TABLE III. REACTIONS OF SILYLATED HETEROCYCLIC BASES WITH PROTECTED SUGARS - $SnCl_4$ CATALYST *(Continued)*

Sugar	Base	Conditions	Product(s) and Yield(s) (%)	Refs.
AcO, OAc, OAc, OAc	OTMS, N, N, TMSO	$SnCl_4$, CH_2Cl_2, rt, 18 h	AcO, O, OAc, OAc, N, NH, O, O (83)	742
	Bz, N, TMS, N, N, N, N, TMS	1. $SnCl_4$, $Cl(CH_2)_2Cl$, 48-72 h 2. NH_3, MeOH	HO, O, OH, OH, NHBz, N, N, N, N (50-70) + N^7-isomer	101
	OTMS, N, N, N, N, AcN, TMS, TMS	$SnCl_4$, $Cl(CH_2)_2Cl$, 24°, 24 h	AcO, O, OAc, OAc, N, N, N, NHAc, NH, O (76) + N^9-isomer (3)	141a
	N, N, N, N, TMSO	$SnCl_4$, $Cl(CH_2)_2Cl$, rt, 24 h	AcO, O, OAc, OAc, N, N, N, N, O (65)	743

Conditions	Product(s) and Yield(s) (%)	Refs.
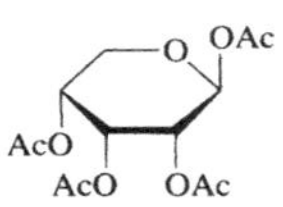 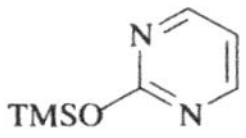 $SnCl_4$, $Cl(CH_2)_2Cl$, rt, 1 h	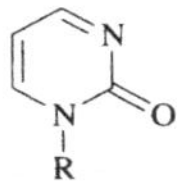 (69) R 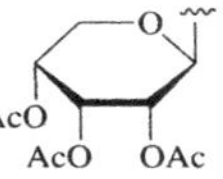 =	744
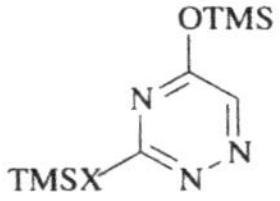 $SnCl_4$, $Cl(CH_2)_2Cl$	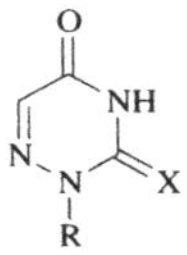 X: O, C_6H_6, 22°, 8 h (73); S, 50°, 3 h (82)	84
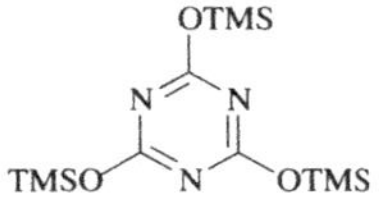 $SnCl_4$, $Cl(CH_2)_2Cl$, 23°, 55 h	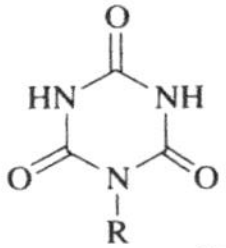 (59)	745
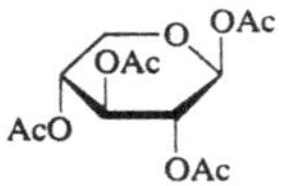 $SnCl_4$, $Cl(CH_2)_2Cl$, 23°, 6 h	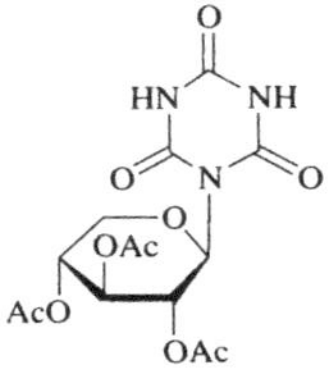 (63) + N^1,N^3-bis(isomer) (19)	745
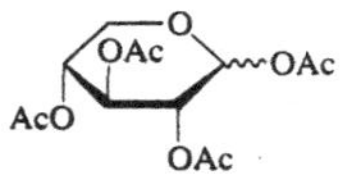 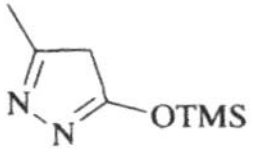 $SnCl_4$, $Cl(CH_2)_2Cl$, rt, 24 h	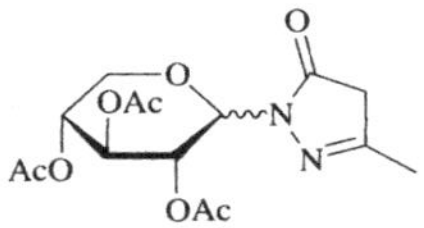 (26)	746
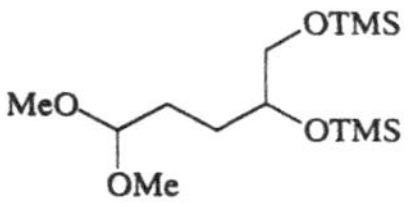 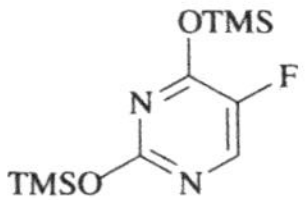 $SnCl_4$, $Cl(CH_2)_2Cl$, 23-25°, 1 h	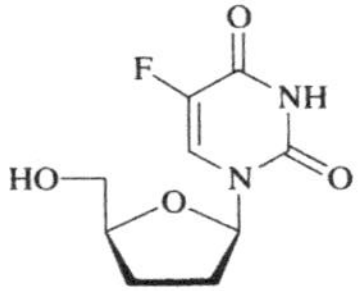 (62)	677

TABLE III. REACTIONS OF SILYLATED HETEROCYCLIC BASES WITH PROTECTED SUGARS - $SnCl_4$ CATALYST *(Continued)*

Sugar	Base	Conditions	Product(s) and Yield(s) (%)	Refs.
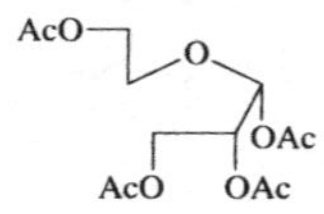 AcO, O, OAc, AcO, OAc	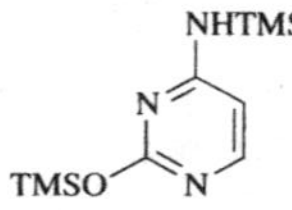 NHTMS, N, TMSO, N	1. $SnCl_4$, MeCN, 30°, 48 h 2. NH_3, MeOH	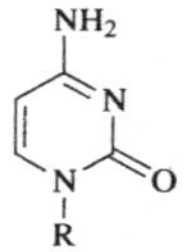 NH_2, N, O, N, R (66) R = HO, O, HO, OH 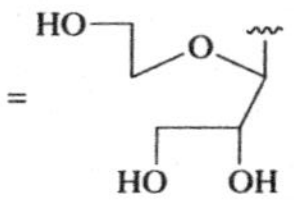	747
	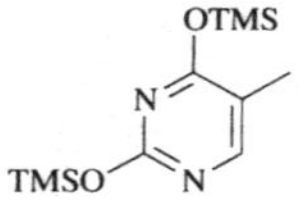 OTMS, N, TMSO, N	1. $SnCl_4$, MeCN, 30°, 48 h 2. NH_3, MeOH	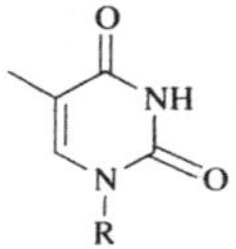 O, NH, O, N, R (67)	747
	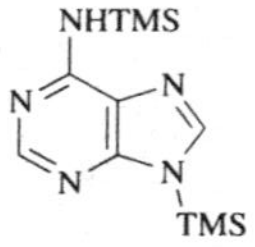 NHTMS, N, N, N, N, TMS	1. $SnCl_4$, MeCN, 30°, 48 h 2. NH_3, MeOH	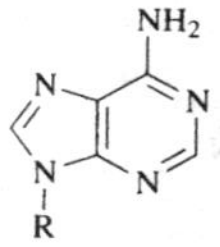 NH_2, N, N, N, N, R (71)	747
	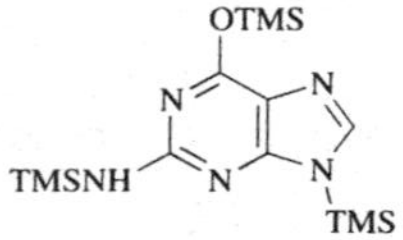 OTMS, N, N, TMSNH, N, N, TMS	1. $SnCl_4$, MeCN, 30°, 48 h 2. NH_3, MeOH	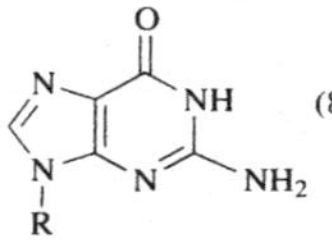 O, N, NH, N, N, NH_2, R (84)	747
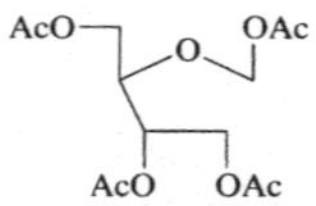 AcO, O, OAc, AcO, OAc	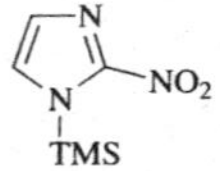 N, NO_2, N, TMS	$SnCl_4$, MeCN, 2-3 h	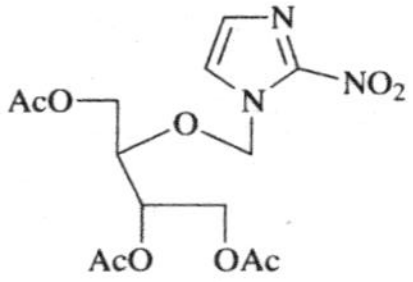 N, NO_2, AcO, O, N, AcO, OAc (81)	748
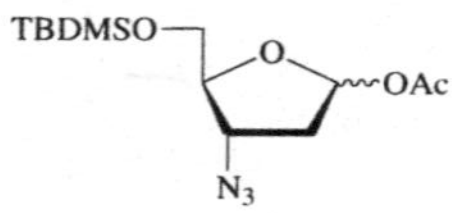 TBDMSO, O, OAc, N_3	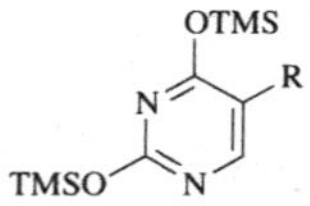 OTMS, R, N, TMSO, N	$SnCl_4$, MeCN, rt, 18 h	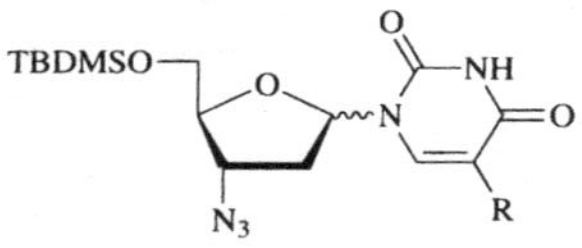 TBDMSO, O, O, NH, N, O, N_3, R R: H (82) α:β = 66:34; Me (68) α:β = 1:1	749

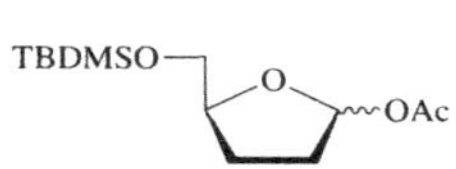 TBDMSO, OAc	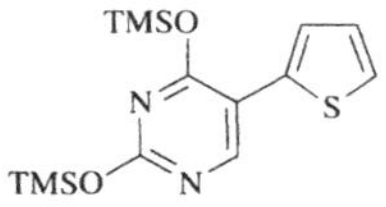 TMSO, TMSO, N, N, S	1. $SnCl_4$, MeCN, –15° 2. NH_3, MeOH, rt	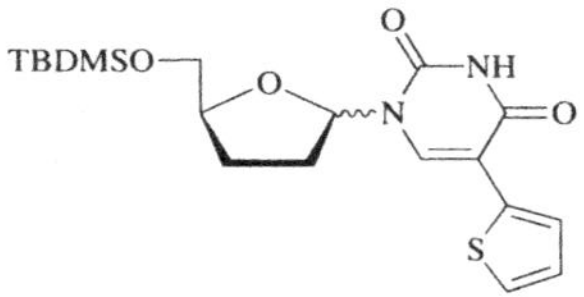 TBDMSO, O, NH, N, O, S (—) α:β = 1:2.1	750
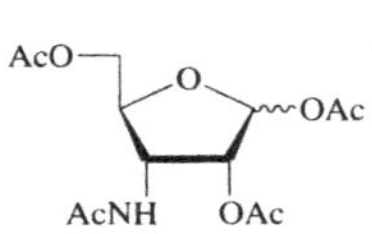 AcO, OAc, AcNH, OAc	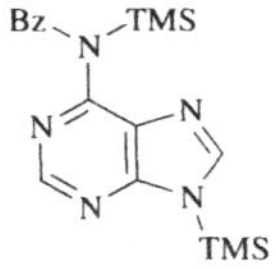 Bz, N, TMS, N, N, N, N, TMS	$SnCl_4$, $Cl(CH_2)_2Cl$, 80°, 18 h	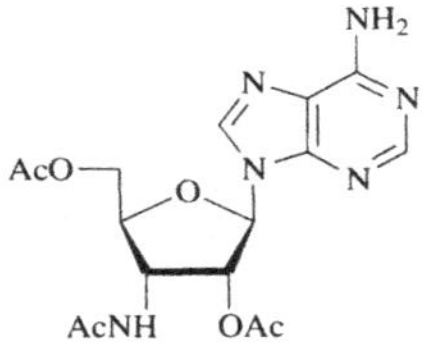 NH_2, N, N, N, N, AcO, AcNH, OAc (55)	751
C_{14} 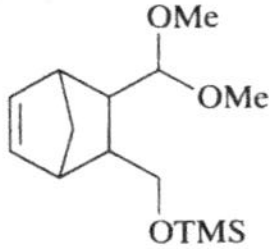 OMe, OMe, OTMS	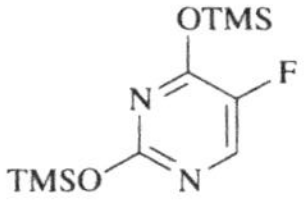 OTMS, F, N, N, TMSO	$SnCl_4$, $Cl(CH_2)_2Cl$, 23-25°, 1 h	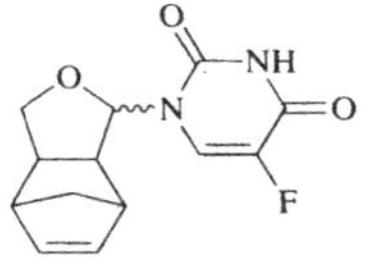 O, NH, O, N, F (61)	671
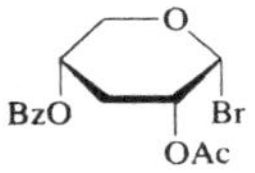 BzO, Br, OAc	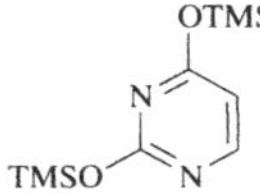 OTMS, N, N, TMSO	$SnCl_4$, MeCN, rt, 3 h	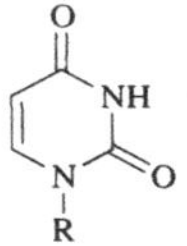 O, NH, N, O, R (50) R = BzO, OAc 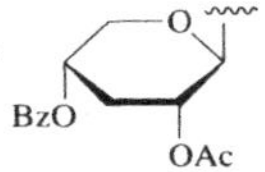	752
	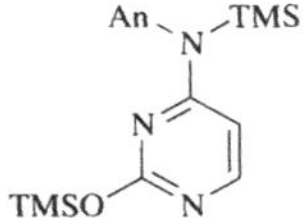 An, N, TMS, N, N, TMSO	$SnCl_4$, $Cl(CH_2)_2Cl$, rt, 1 h; 60°, 20 min	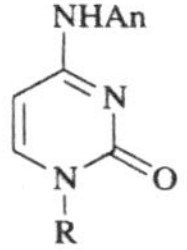 NHAn, N, N, O, R (43)	752
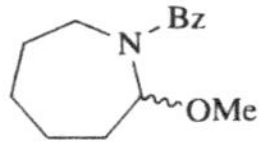 N, Bz, OMe	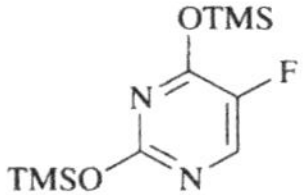 OTMS, F, N, N, TMSO	$SnCl_4$, MeCN, –20 to –25°, 24 h	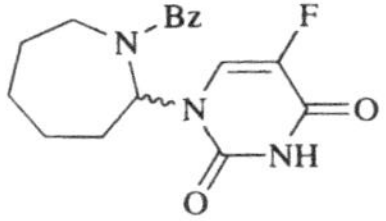 N, Bz, F, N, O, NH, O (73)	642

TABLE III. REACTIONS OF SILYLATED HETEROCYCLIC BASES WITH PROTECTED SUGARS - $SnCl_4$ CATALYST *(Continued)*

Sugar	Base	Conditions	Product(s) and Yield(s) (%)	Refs.
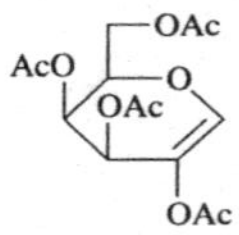	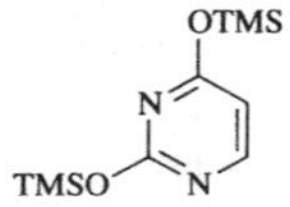	$SnCl_4$, CH_2Cl_2, 0°, 6 h	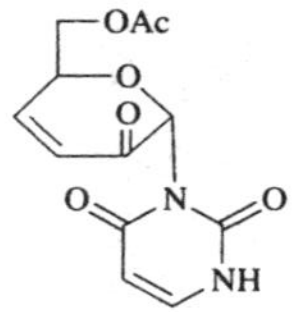 (65)	675
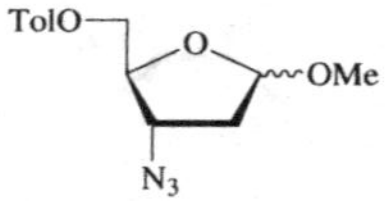	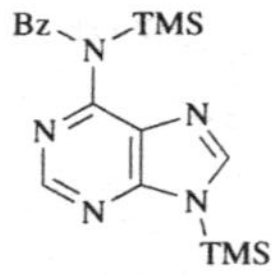	1. $SnCl_4$, MeCN, reflux, 3 h 2. NH_3, MeOH, rt, 24 h	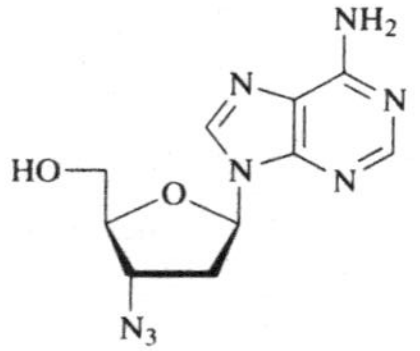 (34)	279
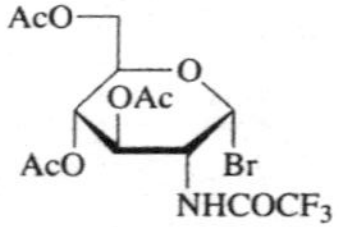	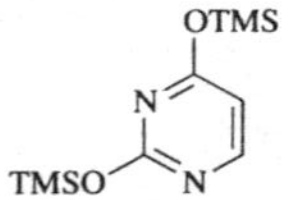	$SnCl_4$, $Cl(CH_2)_2Cl$, reflux, 4 h	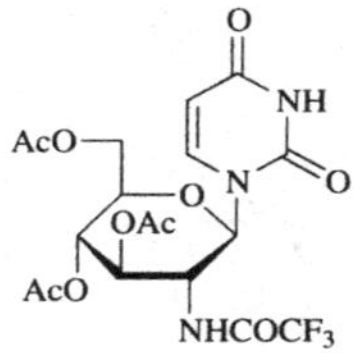 (80)	218
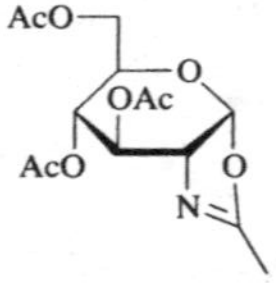	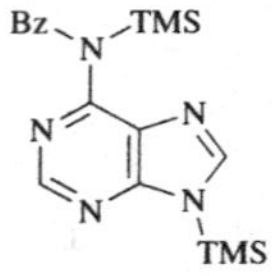	1. $SnCl_4$, $Cl(CH_2)_2Cl$, 50-60°, 5 h 2. *n*-$BuNH_2$, MeOH	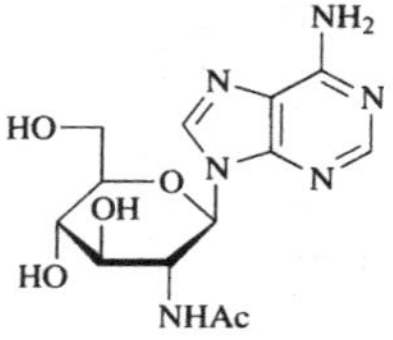 (58)	100

		1. $SnCl_4$, MeCN, –40 to –45°, 10 min 2. $NaHCO_3$	(54)	642
		$SnCl_4$, $Cl(CH_2)_2Cl$, rt, 3 h	(29)	753
		$SnCl_4$, $Cl(CH_2)_2Cl$, 40°, 2 h	R: I (75); Me (72)	754
		$SnCl_4$, $Cl(CH_2)_2Cl$, rt, 5 h	(75)	540
		$SnCl_4$	(—) R =	690

TABLE III. REACTIONS OF SILYLATED HETEROCYCLIC BASES WITH PROTECTED SUGARS - $SnCl_4$ CATALYST (*Continued*)

Sugar	Base	Conditions	Product(s) and Yield(s) (%)	Refs.
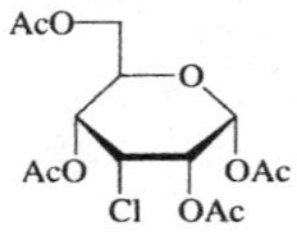 AcO, Cl, OAc, OAc	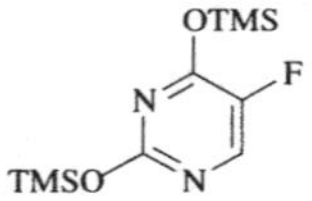 OTMS, F, TMSO	$SnCl_4$, $Cl(CH_2)_2Cl$, rt, 12 h	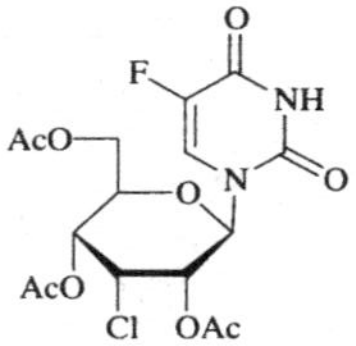 F, O, NH, O, AcO, AcO, Cl, OAc (75)	755
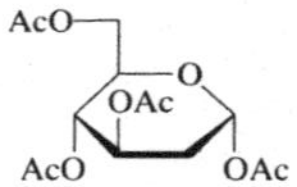 AcO, OAc, AcO, OAc	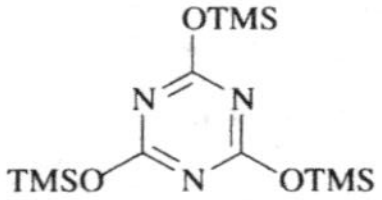 OTMS, TMSO, OTMS	$SnCl_4$, $Cl(CH_2)_2Cl$, 23°, 15 min	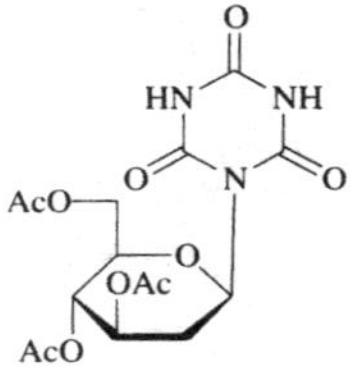 O, HN, NH, O, O, AcO, OAc, AcO (55) + N^1,N^3-bis(product) (12)	745
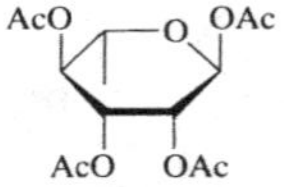 AcO, OAc, AcO, OAc	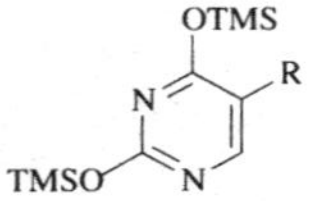 OTMS, R, TMSO	$SnCl_4$, $Cl(CH_2)_2Cl$	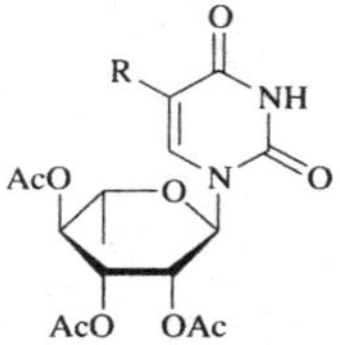 O, R, NH, O, AcO, AcO, OAc	
	R: F	rt, 5 h	(85)	756
	R: Cl	60°	(67)	757
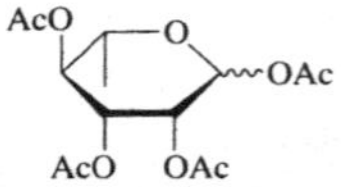 AcO, OAc, AcO, OAc	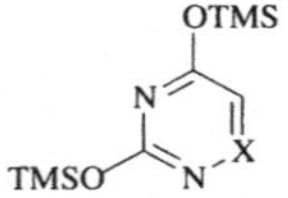 OTMS, X, TMSO	$SnCl_4$, $Cl(CH_2)_2Cl$	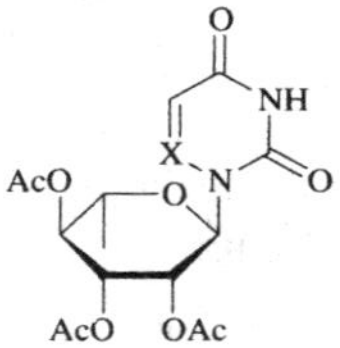 O, NH, X, O, AcO, AcO, OAc	

Substrate	Reagent	Conditions	Product(s) and Yield(s) (%)	Refs.
	X			
	CH	rt, 3 h	(—)	758
	N	50°, 2.5 h	(80)	759
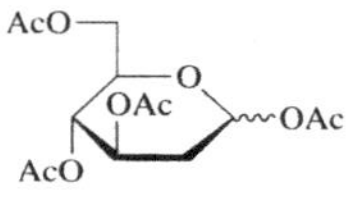 AcO, OAc, AcO, OAc (sugar structure)	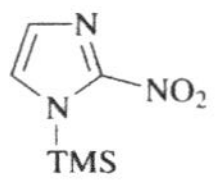 1-TMS-2-nitroimidazole (NO_2, TMS)	$SnCl_4$, $Cl(CH_2)_2Cl$, 60-70°, 45 min	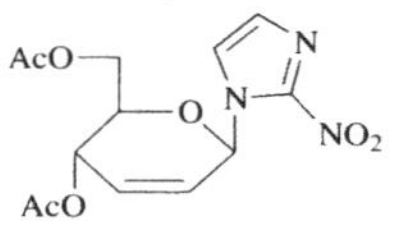 AcO, AcO, NO_2 (structure) (15) + α-anomer (5)	760
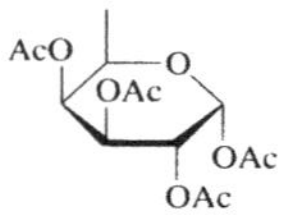 AcO, OAc, OAc, OAc (sugar structure)	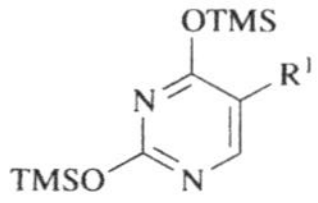 OTMS, R^1, TMSO (pyrimidine)	$SnCl_4$, MeCN, rt, 6 h	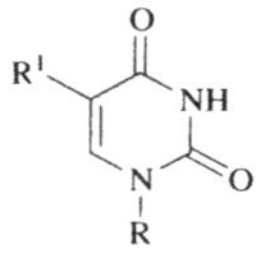 R^1, R (uracil); R^1: H (95), Me (98); R 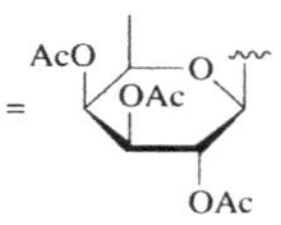 = AcO, OAc, OAc (sugar structure)	761
	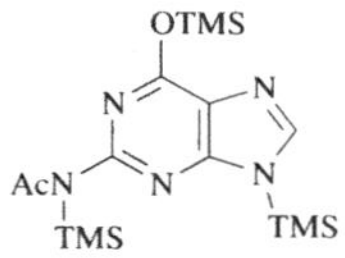 OTMS, AcN(TMS), TMS (purine)	$SnCl_4$, MeCN, reflux, 6 h	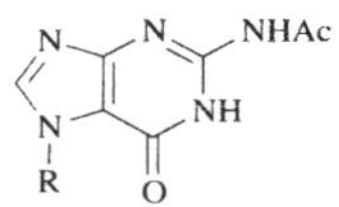 NHAc, R (guanine) (52)	761
	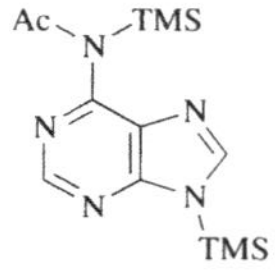 Ac–N–TMS, TMS (purine)	1. $SnCl_4$, $Cl(CH_2)_2Cl$, reflux, 2 h 2. NH_3, MeOH	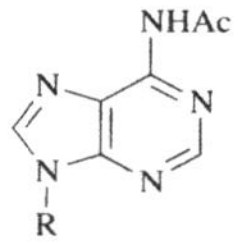 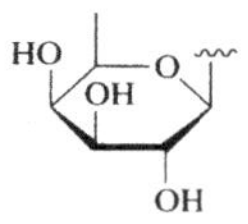 NHAc, R (adenine) (94); R = HO, OH, OH (sugar structure)	761
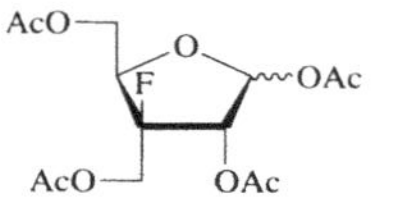 AcO, F, OAc, AcO, OAc (furanose)	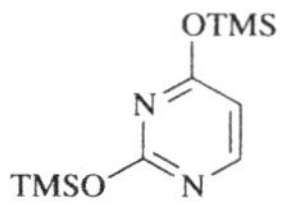 OTMS, TMSO (pyrimidine)	$SnCl_4$, $Cl(CH_2)_2Cl$, 20°, 24 h	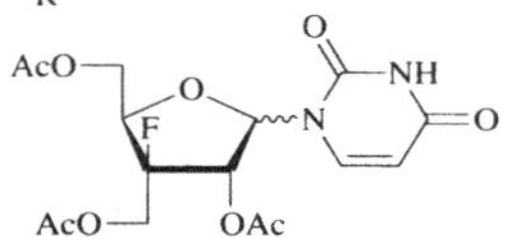 AcO, F, AcO, OAc (uracil nucleoside) (—) α:β = 1:1.8	762
C_{15} 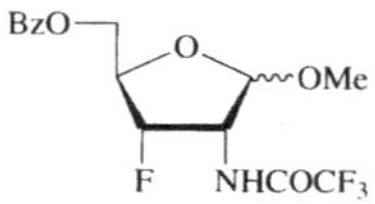 BzO, OMe, F, $NHCOCF_3$ (furanose)	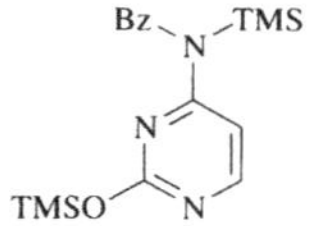 Bz–N–TMS, TMSO (pyrimidine)	$SnCl_4$, MeCN, reflux, 2 h	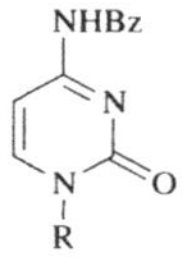 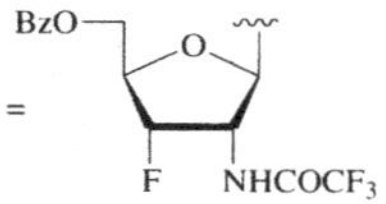 NHBz, R (cytosine) (82); R = BzO, F, $NHCOCF_3$ (furanose)	763

TABLE III. REACTIONS OF SILYLATED HETEROCYCLIC BASES WITH PROTECTED SUGARS - $SnCl_4$ CATALYST *(Continued)*

Sugar	Base	Conditions	Product(s) and Yield(s) (%)	Refs.
	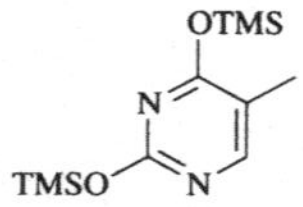 OTMS, N, TMSO, N	$SnCl_4$, MeCN, reflux, 45 min	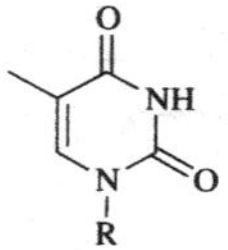 O, NH, O, N, R (63) R = BzO, O, F, $NHCOCF_3$ 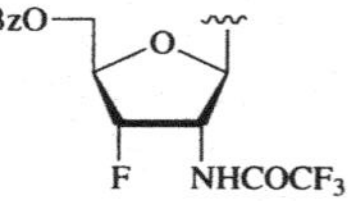	763
	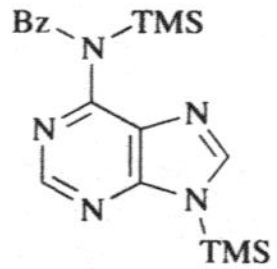 Bz, N, TMS, N, N, N, N, TMS	$SnCl_4$, MeCN, reflux, 30 min	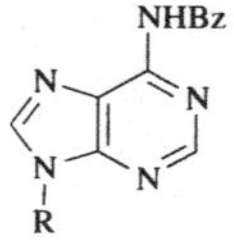 NHBz, N, N, N, N, R (53)	763
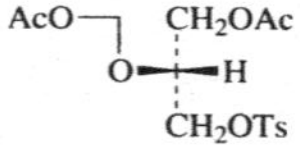 AcO, O, CH_2OAc, H, CH_2OTs	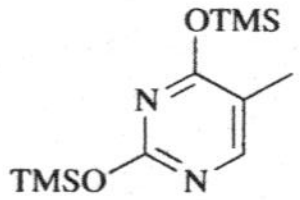 OTMS, N, TMSO, N	$SnCl_4$, $Cl(CH_2)_2Cl$, rt	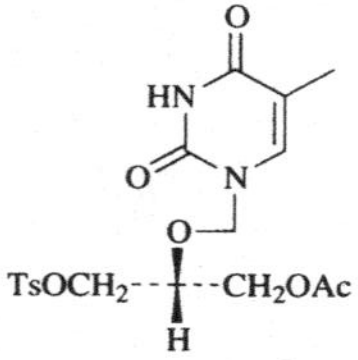 O, HN, O, N, O, $TsOCH_2$, CH_2OAc, H (42)	764
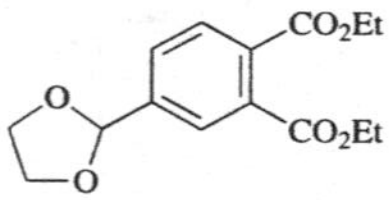 O, O, CO_2Et, CO_2Et	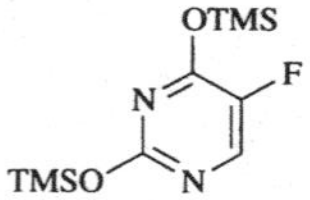 OTMS, F, N, TMSO, N	$SnCl_4$, CH_2Cl_2, rt, 12 h	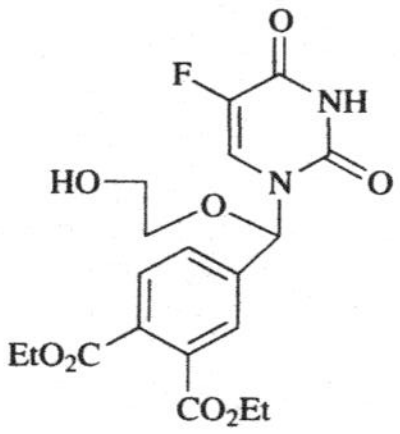 O, F, NH, HO, O, N, O, EtO_2C, CO_2Et (46)	637

Substrate	Reagent	Conditions	Product(s) and Yield(s) (%)	Refs.
OHC, N, CO_2Et, AcO, O, HO, O, O	OTMS, N, TMSO, N	$SnCl_4$, $Cl(CH_2)_2Cl$, rt, 24 h	OHC, N, CO_2Et, AcO, O, O, O; O, NH, O, N (72)	765
O, BnO, N H, O, N, OMe	OTMS, F, N, TMSO, N	$SnCl_4$, CH_2Cl_2, –13 to –15°, 3 h	O, BnO, N H, O, N; O, NH, O, N, F (72)	642
CO_2Me, O, OAc, OAc, AcO, OAc	OTMS, N, TMSO, N	$SnCl_4$, $Cl(CH_2)_2Cl$, 70°, 5 h	O, NH, MeO_2C, O, N, O, OAc, AcO, OAc **I** (15) + N^3-isomer (10)	766
		$SnCl_4$, MeCN, rt, 24 h	**I** (44) + N^3-isomer (2)	767
	Bz, N, TMS, N, TMSO, N	1. $SnCl_4$ (1.3 eq) 2. NH_3, MeOH	NH_2, N, MeO_2C, O, N, O, OH, HO, OH (70)	768

TABLE III. REACTIONS OF SILYLATED HETEROCYCLIC BASES WITH PROTECTED SUGARS - $SnCl_4$ CATALYST *(Continued)*

Sugar	Base	Conditions	Product(s) and Yield(s) (%)	Refs.
	Ac, TMS, N, N, N, TMSO	1. $SnCl_4$, $Cl(CH_2)_2Cl$, rt, 12 h 2. reflux, 1 h	NHAc, N, N, O, R (17) R = MeO$_2$C, O, OAc, AcO, OAc	769
	OTMS, X, N, N, TMSO	$SnCl_4$, MeCN, rt, 24 h	X, O, NH, N, O, R X: F (80); Cl (63); Br (61); I (55)	767
	Bz, TMS, N, N, N, N, N, TMS	$SnCl_4$, $Cl(CH_2)_2Cl$, 60-70°, 5 h	NHBz, N, N, N, N, R (63)	100
CO$_2$Me, O, OAc, OAc, AcO, OAc	OTMS, F, N, N, TMSO	$SnCl_4$	F, O, NH, N, O, R (—)	770
CO$_2$Me, AcO, O, OAc, OAc, OAc	OTMS, N, N, TMSO	$SnCl_4$, $Cl(CH_2)_2Cl$, 70°, 5 h	O, NH, N, O, R (58) R = MeO$_2$C, AcO, O, OAc, OAc	766
	Bz, TMS, N, N, N, N, N, TMS	$SnCl_4$, $Cl(CH_2)_2Cl$, 60-70°, 5 h	NHBz, N, N, N, N, R (41)	100

Substrate	Silylated base	Conditions	Product(s) and Yield(s) (%)	Refs.
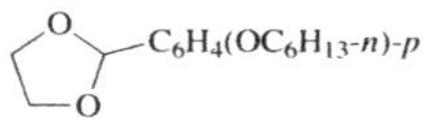 $C_6H_4(OC_6H_{13}\text{-}n)\text{-}p$ (dioxolane)	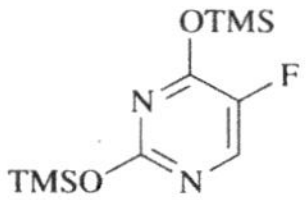 OTMS, F, N, TMSO	$SnCl_4$, CH_2Cl_2, rt, 12 h	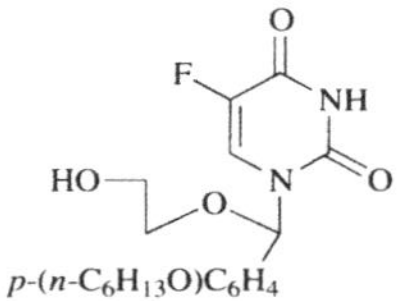 F, NH, O, HO, N, O, $p\text{-}(n\text{-}C_6H_{13}O)C_6H_4$ (19)	637
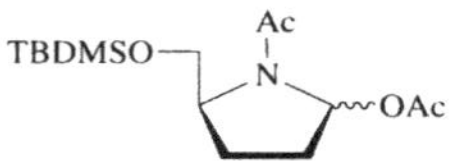 TBDMSO, Ac, N, OAc	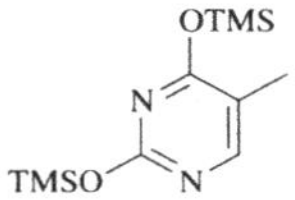 OTMS, N, TMSO	$SnCl_4$, MeCN, −15°, 3 h	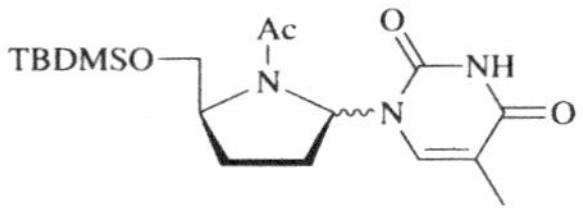 TBDMSO, Ac, N, O, NH, O (55) α:β = 3:2	45
C_{16} 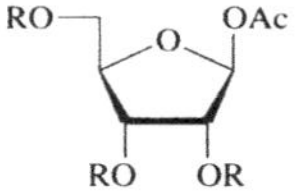 RO, OAc, O, RO, OR; R = $CO_2CH_2CCl_3$	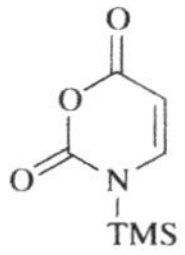 O, O, N, TMS	$SnCl_4$, $Cl(CH_2)_2Cl$, rt, time	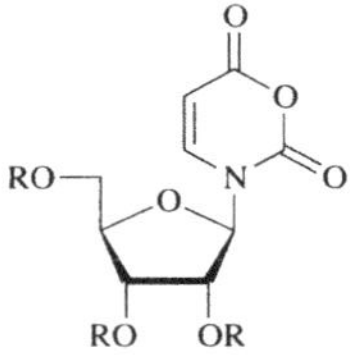 RO, O, N, O, RO, OR; time 24 h (90), 2 h (87)	771, 722
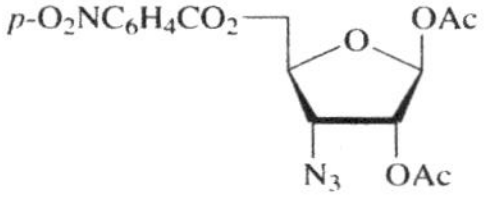 $p\text{-}O_2NC_6H_4CO_2$, O, OAc, N_3, OAc	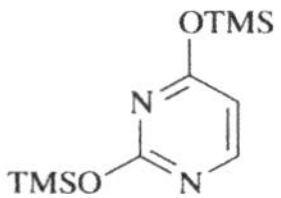 OTMS, N, TMSO	1. $SnCl_4$, $Cl(CH_2)_2Cl$, MeCN, reflux 1.5 h 2. NH_3, MeOH	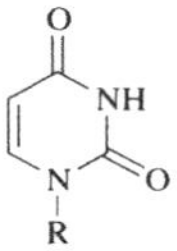 O, NH, N, O, R (97); R = HO, O, N_3, OH 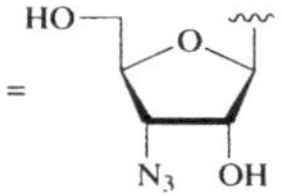	772
	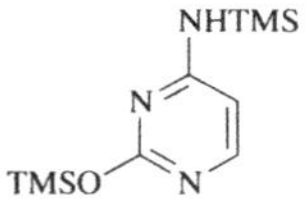 NHTMS, N, TMSO	1. $SnCl_4$, $Cl(CH_2)_2Cl$, reflux, 1.5 h 2. NH_3, MeOH	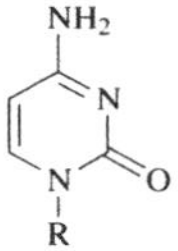 NH_2, N, N, O, R (73)	772
	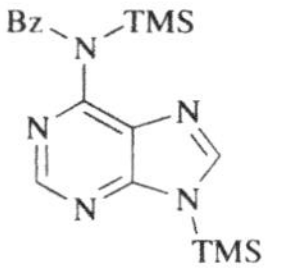 Bz, N, TMS, N, N, N, N, TMS	1. $SnCl_4$, $Cl(CH_2)_2Cl$, reflux, 2 h 2. NH_3, MeOH, rt, 3 h	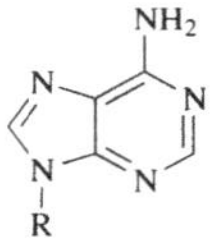 NH_2, N, N, N, N, R (68)	772

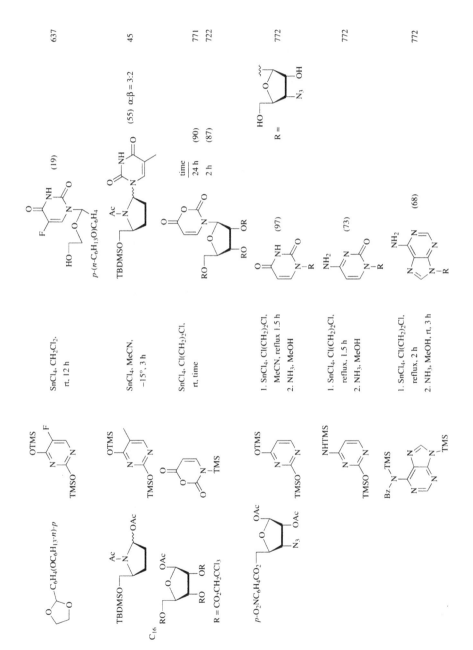

TABLE III. REACTIONS OF SILYLATED HETEROCYCLIC BASES WITH PROTECTED SUGARS - $SnCl_4$ CATALYST *(Continued)*

Sugar	Base	Conditions	Product(s) and Yield(s) (%)	Refs.
Ph, AcO, OAc, OAc	Bz, N, TMS, N, N, N, N, TMS	$SnCl_4$, $Cl(CH_2)_2Cl$, rt, 1 h	NHBz, N, N, N, N, Ph, O, AcO, OAc (80)	579
BzO, O, OAc, OAc	NHTMS, N, N, TMSO	$SnCl_4$, MeCN, rt, 45 min - 2 h	NH_2, N, N, O, R (60) R = BzO, O, OAc	773
	OTMS, N, N, TMSO	$SnCl_4$, MeCN, rt, 45 min - 2 h	O, NH, N, O, R (85)	773
	Bz, N, TMS, N, N, N, N, TMS	$SnCl_4$, MeCN, rt, 45 min - 2 h	NHBz, N, N, N, N, R (55)	773
	OTMS, N, N, N, N, TMSHN, TMS	$SnCl_4$, MeCN, rt, 45 min - 2 h	N^7 + N^9-β-isomers (57)	773

Sugar	Base	Conditions	Product (Yield, %)		Refs.
		$SnCl_4$, MeCN		(93)	774
		1. $SnCl_4$, (2-3 eq), $Cl(CH_2)_2Cl$, 48-72 h 2. NaOMe, MeOH		(60-70)	101
		1. $SnCl_4$, (1 eq), $Cl(CH_2)_2Cl$, 48-72 h 2. NaOMe, MeOH		(60-70)	101
		1. $SnCl_4$, $Hg(CN)_2$, MeCN, 50°, 2 h 2. NH_3, MeOH		(50)	629
		$SnCl_4$, $Cl(CH_2)_2Cl$, 60°, 1 h		(84)	629
		$SnCl_4$, MeCN, 55°, 1 h		(80)	629

TABLE III. REACTIONS OF SILYLATED HETEROCYCLIC BASES WITH PROTECTED SUGARS - $SnCl_4$ CATALYST (*Continued*)

Sugar	Base	Conditions	Product(s) and Yield(s) (%)	Refs.
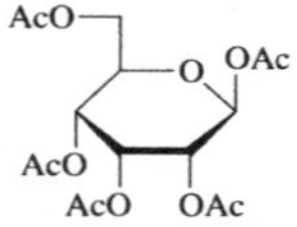 AcO, OAc, AcO, AcO, OAc (pyranose)	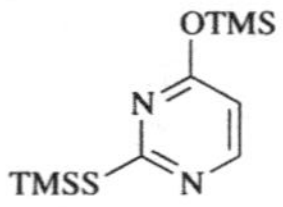 OTMS, TMSS, N, N (pyrimidine)	$SnCl_4$, $Cl(CH_2)_2Cl$, 23°, 12 h	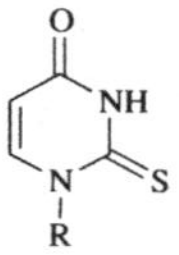 O, NH, S, N–R (structure) **I** (58) + N^3-isomer (34) + N^1,N^3-bis(isomer) (8); R 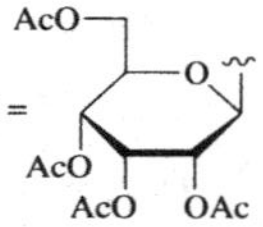 = AcO, O, AcO, AcO, OAc (pyranosyl)	90
		$SnCl_4$, MeCN, 23°, 12 h	**I** (90) + N^1,N^3-bis(isomer) (6)	90
	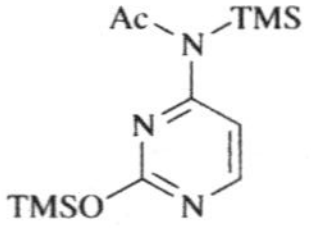 Ac, N, TMS, TMSO, N, N (pyrimidine)	$SnCl_4$, $Cl(CH_2)_2Cl$, 22°, 4 h	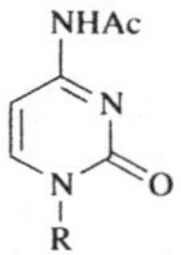 NHAc, N, O, N–R (structure) (83)	775
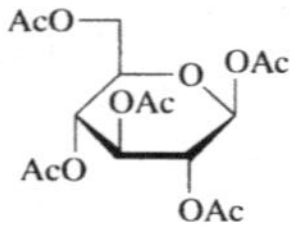 AcO, OAc, OAc, AcO, OAc (pyranose)	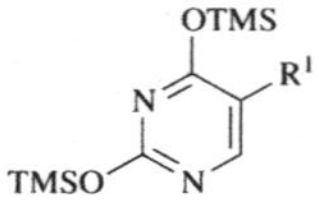 OTMS, R^1, TMSO, N, N (pyrimidine)	$SnCl_4$	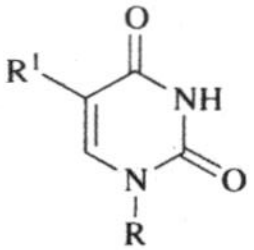 R^1, O, NH, O, N–R (structure); R = AcO, O, OAc, AcO, OAc (pyranosyl) 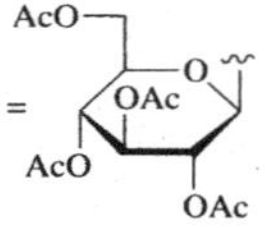	
	R^1			
	F	MeCN, <10°, 3 h	(80)	514b
	NO_2	$Cl(CH_2)_2Cl$, 23°, 5 h	(82)	90
		MeCN, 23°, 12 h	(77)	90
	OMe	$Cl(CH_2)_2Cl$, 23°, 3.5 h	(61) + N^3-isomer (25)	90
		MeCN, 23°, 12 h	(83) + N^3-isomer (15)	90
	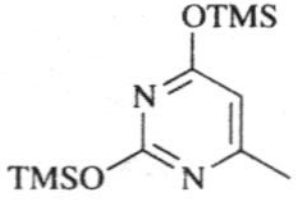 OTMS, TMSO, N, N (6-methylpyrimidine)	$SnCl_4$, MeCN, 22°, 3 h	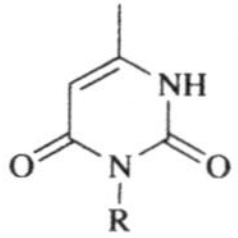 NH, O, O, N–R (structure) (42)	89a

Reactant	Conditions	Product(s) and Yield(s) (%)	Refs.
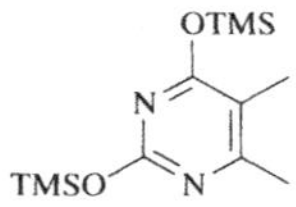 OTMS, N, N, TMSO	$SnCl_4$, MeCN, 22°, 15 h 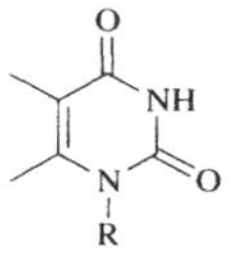	O, NH, O, N, R (22)	89a
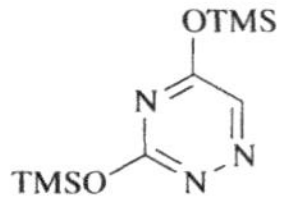 OTMS, N, N, N, N, TMSO	$SnCl_4$, $Cl(CH_2)_2Cl$, C_6H_6, 22°, 5 h	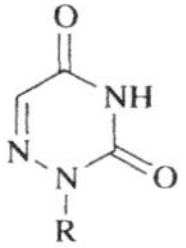 O, NH, O, N, N, R **I** (62)	84
	$SnCl_4$, $Cl(CH_2)_2Cl$, 23°, 12 h	**I** (82)	90
	$SnCl_4$, MeCN, 23°, 12 h	**I** (81)	90
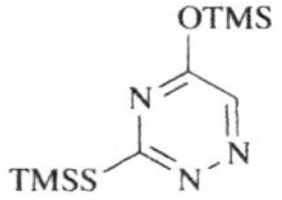 OTMS, N, N, N, TMSS	$SnCl_4$, $Cl(CH_2)_2Cl$, 50°, 3 h 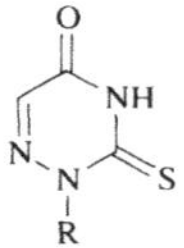	O, NH, S, N, N, R (79)	84
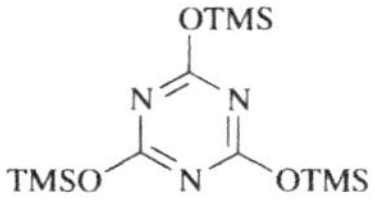 OTMS, N, N, N, TMSO, OTMS	$SnCl_4$, $Cl(CH_2)_2Cl$, 45°, 36 h	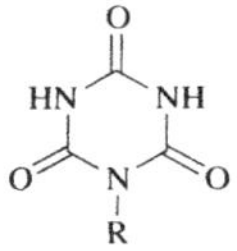 O, HN, NH, O, O, N, R **I** (32)	776
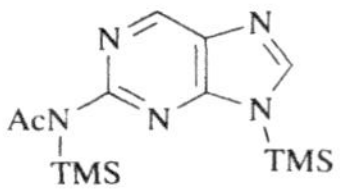 N, N, N, N, AcN, TMS, TMS	$SnCl_4$, MeCN, rt, 3 h	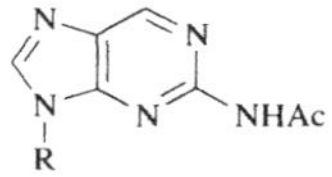 N, N, N, N, R, NHAc (11)	409
	$SnCl_4$, $Cl(CH_2)_2Cl$, MeCN, reflux, 0.5-1 h	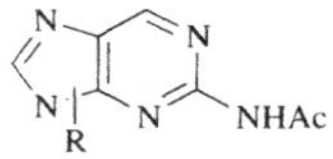 N, N, N, N, R, NHAc (78) $N^9:N^7 = 98:2$	409

TABLE III. REACTIONS OF SILYLATED HETEROCYCLIC BASES WITH PROTECTED SUGARS - $SnCl_4$ CATALYST (*Continued*)

Sugar	Base	Conditions	Product(s) and Yield(s) (%)	Refs.
	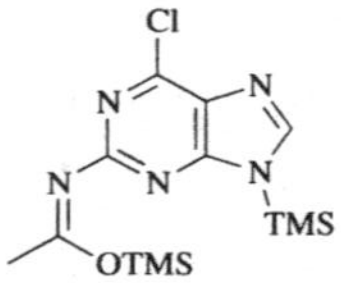	$SnCl_4$, MeCN, rt, 16 h	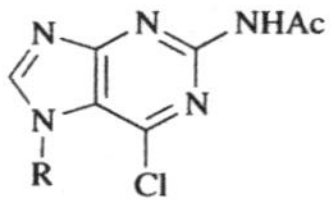 (59) + N^9-isomer (7)	409
		$SnCl_4$, $Cl(CH_2)_2Cl$, MeCN, reflux, 1 h	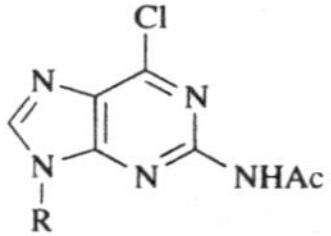 (33) + N^7-isomer (33)	409
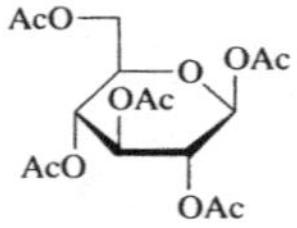	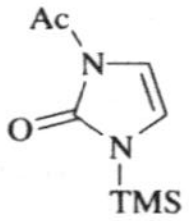	$SnCl_4$, $Cl(CH_2)_2Cl$, 20°, 18 h	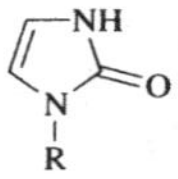 (32) + 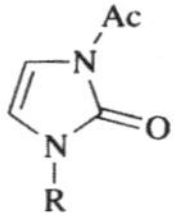 (12) R = 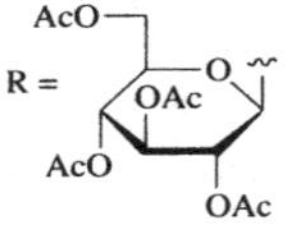	777
	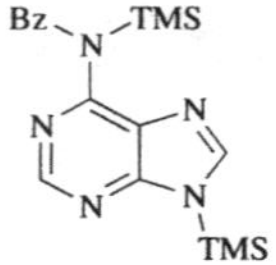	$SnCl_4$, $Cl(CH_2)_2Cl$, 60-70°, 5 h	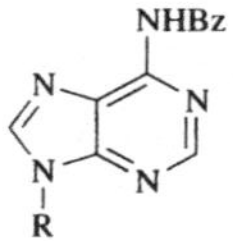 (68)	100
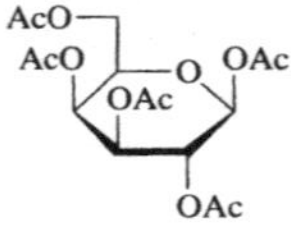	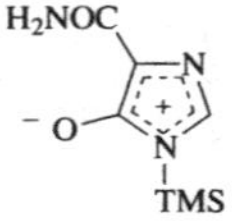	$SnCl_4$, TMSTf, CH_2Cl_2, reflux, 2 h	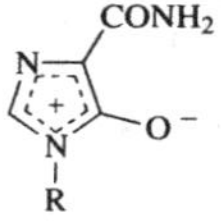 (—) R = 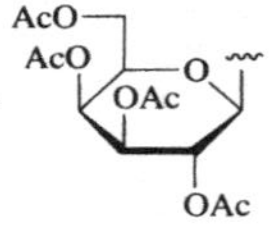	778
	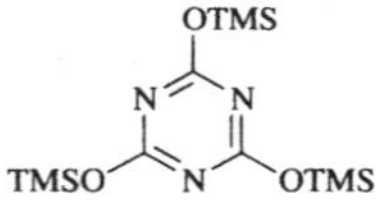	$SnCl_4$, $Cl(CH_2)_2Cl$, 23°, 40 h	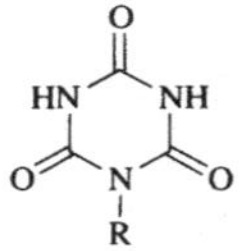 (60) + N^1,N^3-bis(riboside) (16)	745

Sugar	Base	Conditions	Product(s) and Yield(s) (%)		Refs.
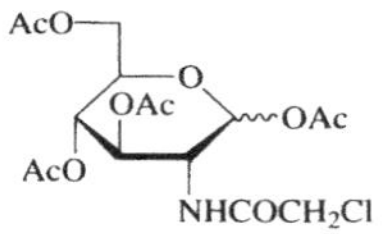	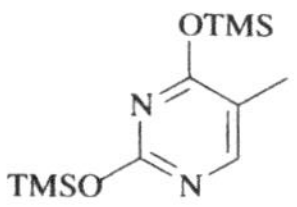	$SnCl_4$, $Cl(CH_2)_2Cl$, MS 4 Å		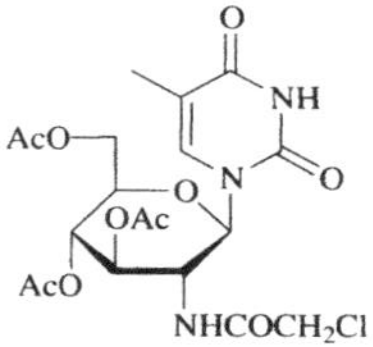 (64)	779
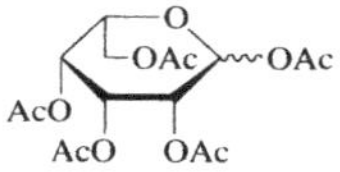	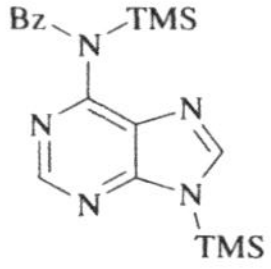	1. $SnCl_4$, $Cl(CH_2)_2Cl$, 60-70°, 5 h 2. NaOMe, MeOH		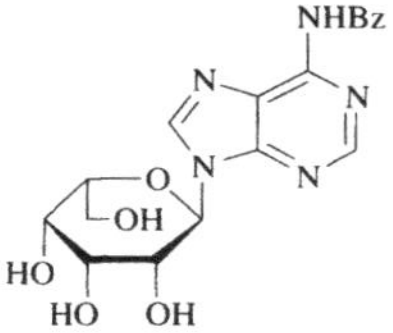 (68)	780
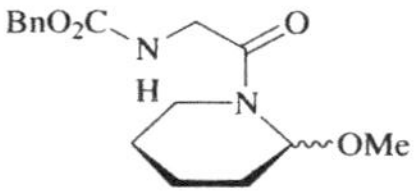	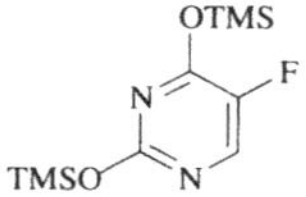	$SnCl_4$, MeCN, –40 to –45°, 10 min		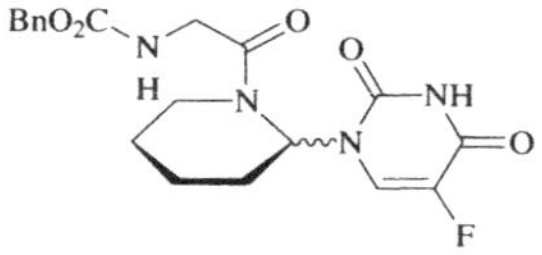 (40)	642
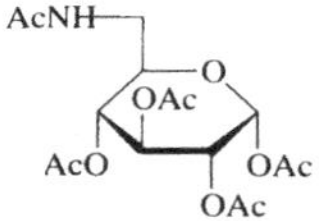	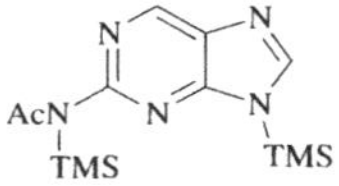	$SnCl_4$, $Cl(CH_2)_2Cl$, MeCN, 135°, 0.5-1 h		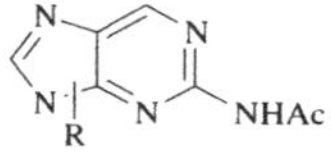 (67) N^9:N^7 = 98:2 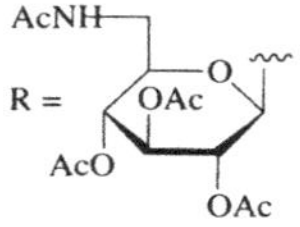	409
	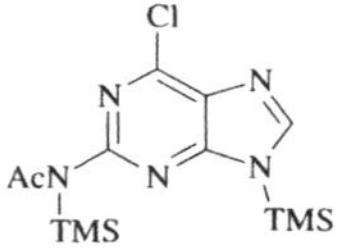	$SnCl_4$, MeCN, rt, 3 h		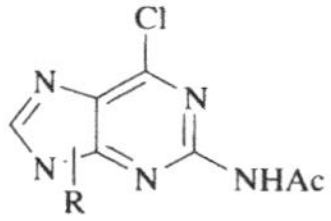 (22) N^9:N^7 = 59:41 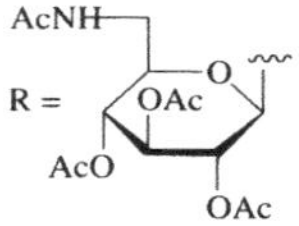	409
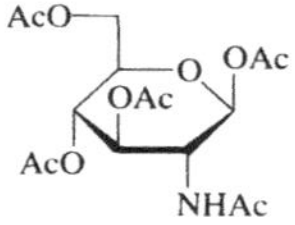	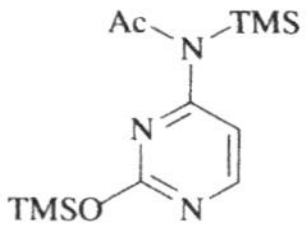	$SnCl_4$, $Cl(CH_2)_2Cl$, 70°, 5 h		(71)	766

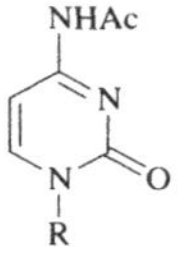

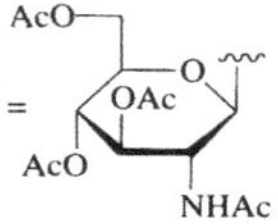

TABLE III. REACTIONS OF SILYLATED HETEROCYCLIC BASES WITH PROTECTED SUGARS - $SnCl_4$ CATALYST (*Continued*)

Sugar	Base	Conditions	Product(s) and Yield(s) (%)	Refs.
	OTMS, N, TMSO, N	$SnCl_4$, $Cl(CH_2)_2Cl$, 70°, 5 h	O, NH, O, N, R (50)	766
HO, EtO_2C, O, O, EtO_2C, O, O	OTMS, N, TMSO, N	1. $SnCl_4$ 2. Ac_2O	O, NH, AcO, EtO_2C, O, N, O, EtO_2C, O, OAc (—)	781
PivO, O, NO_2, O, O, CN	OTMS, N, TMSO, N	$SnCl_4$, MeCN, –30°, 20 min	O, NH, PivO, O, N, O, O, O, CN (40) α:β = 2:3	106
PivO, O, NO_2, OAc, O, O	OTMS, N, TMSO, N	$SnCl_4$, MeCN, 65°, 30 min	O, NH, PivO, O, N, O, OAc, O, O (15)	106

Substrate	Base	Conditions	Product(s) and Yield(s) (%)	Refs.
2-(n-$C_{13}H_{27}$)-1,3-dioxolane	5-fluoro-2,4-bis(trimethylsilyloxy)pyrimidine	$SnCl_4$, CH_2Cl_2, rt, 12 h	(14)	637
C_{17}				
BzO, AcO, CH_2NO_2, OAc furanose	OTMS, TMSO pyrimidine	$SnCl_4$, $Cl(CH_2)_2Cl$, 20°, 4 h	(—) α:β = 10:1	217
BnO, OAc, S, OMe furanose	NHTMS, TMSO pyrimidine	$SnCl_4$, $Cl(CH_2)_2Cl$, rt	**I** (—) α:β = 9:91	278
		$SnCl_4$, MeCN, rt	**I** (—) α:β = 67:33	278
PivO, NO_2, CO_2Me furanose	OTMS, TMSO pyrimidine	$SnCl_4$, MeCN, 0°, 1 h	(30) α:β = 1:1.2	106
MeO_2C, MeO_2C, OAc, OAc, OAc furanose	OTMS, TMSO pyrimidine	$SnCl_4$, $Cl(CH_2)_2Cl$, 0°, 12 h	(83)	782

TABLE III. REACTIONS OF SILYLATED HETEROCYCLIC BASES WITH PROTECTED SUGARS - $SnCl_4$ CATALYST (*Continued*)

Sugar	Base	Conditions	Product(s) and Yield(s) (%)	Refs.
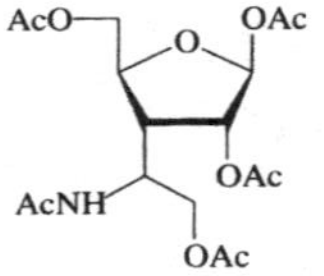 AcO, OAc, OAc, AcNH, OAc	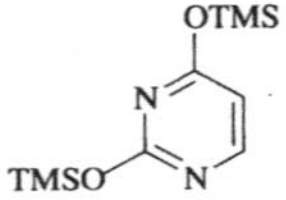 OTMS, N, TMSO, N	$SnCl_4$, $Cl(CH_2)_2Cl$, 20°, 4 h	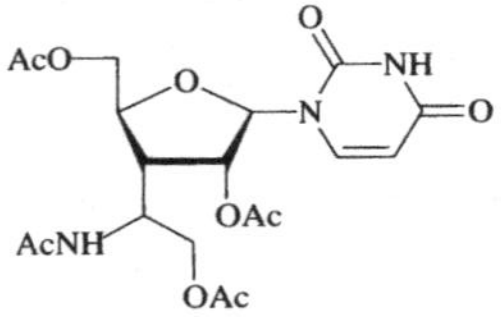 AcO, O, NH, O, N, OAc, AcNH, OAc (51) α:β = 2:1	216
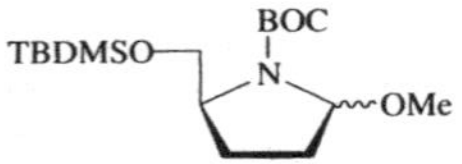 TBDMSO, BOC, N, OMe	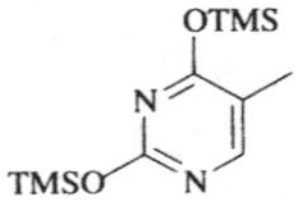 OTMS, N, TMSO, N	1. $SnCl_4$, $Cl(CH_2)_2Cl$, 0-20° 2. TBAF, THF, 0°	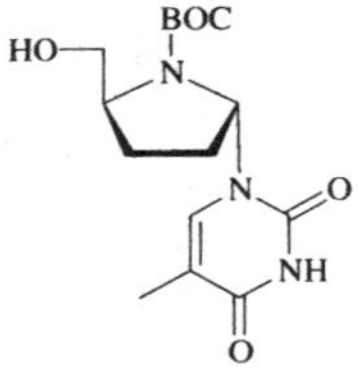 HO, BOC, N, N, O, NH, O (77) α:β = 9:1	783
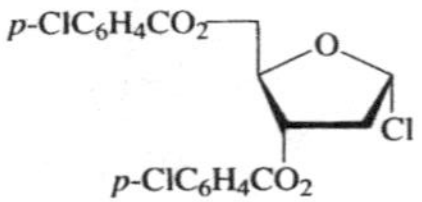 p-$ClC_6H_4CO_2$, O, Cl, p-$ClC_6H_4CO_2$	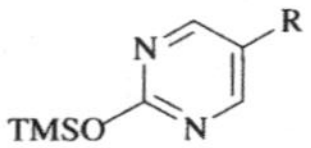 N, R, TMSO, N	$SnCl_4$, $Cl(CH_2)_2Cl$, 0°, 2-2.5 h	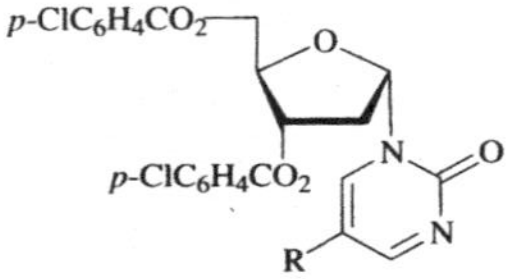 p-$ClC_6H_4CO_2$, O, p-$ClC_6H_4CO_2$, N, O, N, R R: I (20), β-anomer (20); C≡CH (14), β-anomer (12)	784
C_{18} 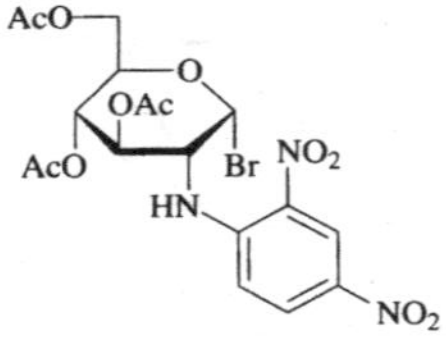 AcO, O, OAc, AcO, Br, NO_2, HN, NO_2	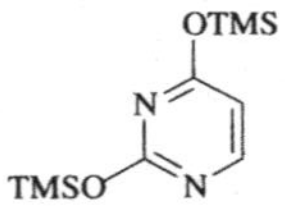 OTMS, N, TMSO, N	$SnCl_4$, $Cl(CH_2)_2Cl$, reflux, 4 h	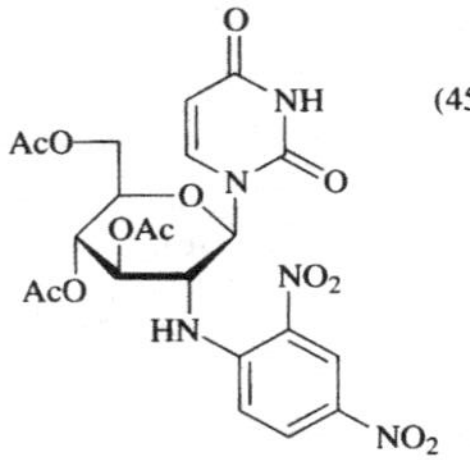 O, NH, AcO, O, N, O, OAc, AcO, NO_2, HN, NO_2 (45)	218

Sugar	Base	Conditions	Product and Yield(s) (%)	Refs.
CO_2Me, NHAc, OAc, OAc, OAc, OAc	OTMS, TMSO, N, N	$SnCl_4$, $Cl(CH_2)_2Cl$, rt, 21 h	MeO_2C, NHAc, OAc, OAc, OAc, O, NH, O (92)	785
MeO_2CO, OMs, OAc, $O_2CC_6H_4CF_3$-*m*	OTMS, TMSO, N, N	$SnCl_4$, CH_2Cl_2, rt, 20 h	MeO_2CO, OMs, $O_2CC_6H_4CF_3$-*m*, O, NH, O (52)	545
p-$O_2NC_6H_4CO_2$, OAc, AcNH, OAc	NHTMS, TMSO, N, N	1. $SnCl_4$, $Cl(CH_2)_2Cl$, MeCN, reflux, 1.5 h 2. NH_4OH (10%), rt, 12 h	NH_2, N, O, N–R (60) R = HO, O, N_3, OH	772
	Bz–N–TMS, N, N, N, N–TMS	1. $SnCl_4$, $Cl(CH_2)_2Cl$, MeCN, reflux, 8 h 2. NH_3, MeOH, rt, 30 h	NH_2, N, N, N, N, HO, O, AcNH, OH (64)	772

TABLE III. REACTIONS OF SILYLATED HETEROCYCLIC BASES WITH PROTECTED SUGARS - $SnCl_4$ CATALYST (*Continued*)

Sugar	Base	Conditions	Product(s) and Yield(s) (%)	Refs.
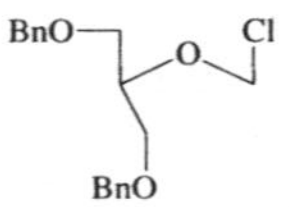 BnO, O, Cl, BnO	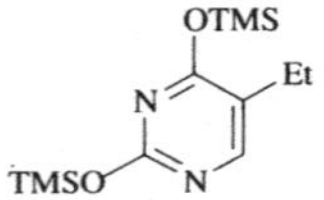 OTMS, Et, N, TMSO, N	$SnCl_4$, MeCN, rt, 48 h	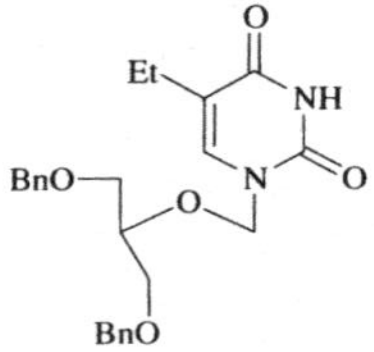 Et, O, NH, O, N, BnO, O, BnO (100) R = BnO, O, BnO 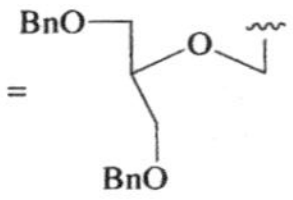	654
	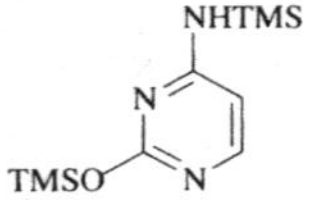 NHTMS, N, TMSO, N	$SnCl_4$, $Cl(CH_2)_2Cl$, rt, 12-14 h	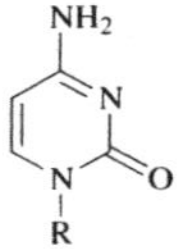 NH_2, N, N, O, R (34)	786, 787
	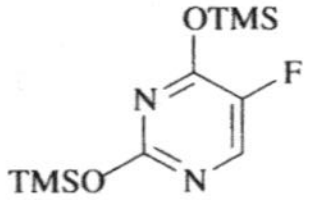 OTMS, F, N, TMSO, N	$SnCl_4$, $Cl(CH_2)_2Cl$, rt, 12 h	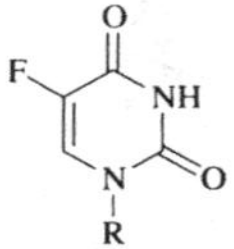 F, O, NH, N, O, R (88)	787
	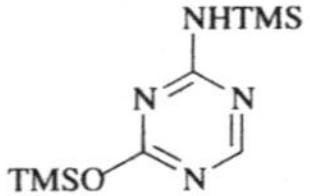 NHTMS, N, N, TMSO, N	$SnCl_4$, $Cl(CH_2)_2Cl$, rt, 12-14 h	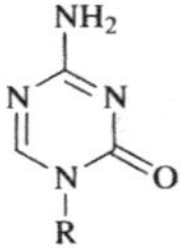 NH_2, N, N, N, O, R (39)	786, 787
	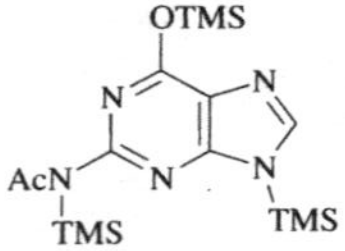 OTMS, N, N, AcN, TMS, N, N, TMS	$SnCl_4$, $Cl(CH_2)_2Cl$, rt, 12 h	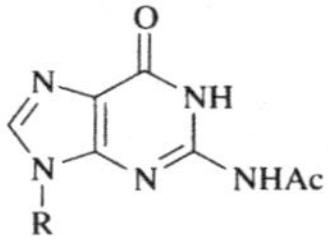 O, N, NH, N, R, N, NHAc (9) + N^7-isomer (3)	788

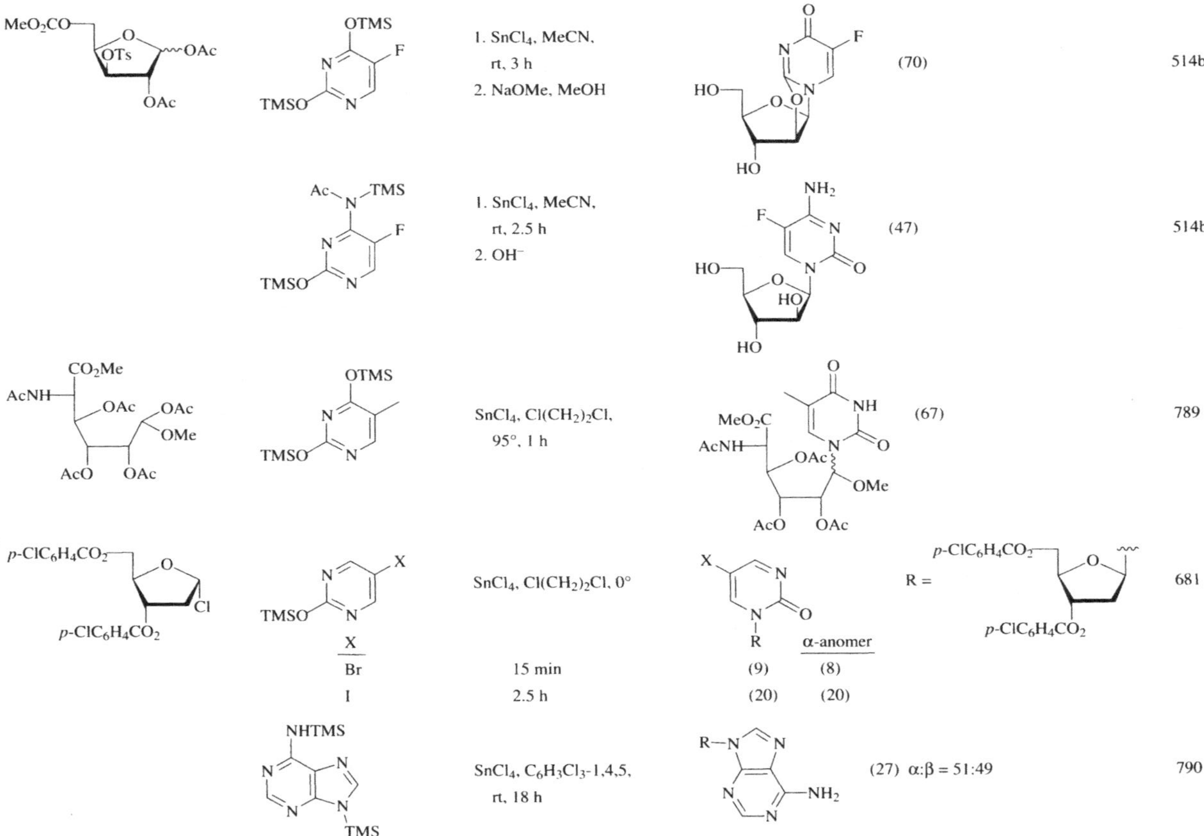

1. $SnCl_4$, MeCN, rt, 3 h
2. NaOMe, MeOH
(70)
514b
1. $SnCl_4$, MeCN, rt, 2.5 h
2. OH^-
(47)
514b
$SnCl_4$, $Cl(CH_2)_2Cl$, 95°, 1 h
(67)
789
C_{19}
$SnCl_4$, $Cl(CH_2)_2Cl$, 0°
X
Br
I
15 min
2.5 h
α-anomer
(9)
(8)
(20)
(20)
R =
681
$SnCl_4$, $C_6H_3Cl_3$-1,4,5, rt, 18 h
(27) α:β = 51:49
790

TABLE III. REACTIONS OF SILYLATED HETEROCYCLIC BASES WITH PROTECTED SUGARS - $SnCl_4$ CATALYST (*Continued*)

Sugar	Base	Conditions	Product(s) and Yield(s) (%)	Refs.
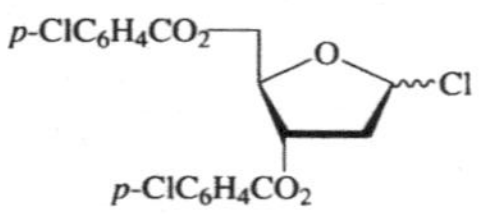 p-$ClC_6H_4CO_2$, p-$ClC_6H_4CO_2$ (Cl)	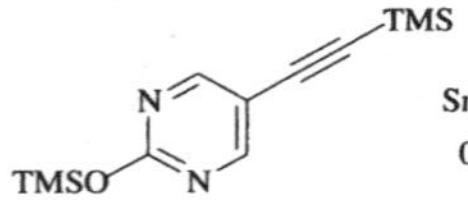 TMSO, TMS	$SnCl_4$, $Cl(CH_2)_2Cl$, 0°, 2 h	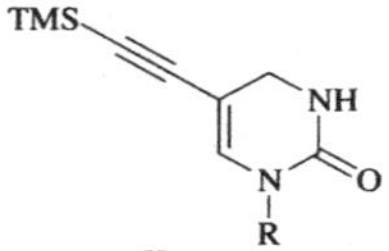 TMS, NH, N, R, O (14) + α-anomer (12)	791
	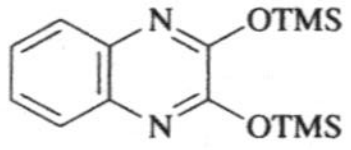 N, OTMS, OTMS	$SnCl_4$, CH_2Cl_2, rt, 16 h	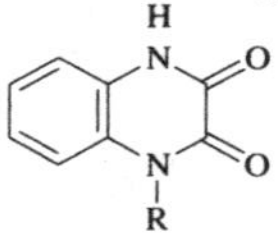 H, N, O, N, O, R (13) + α-anomer (5)	657
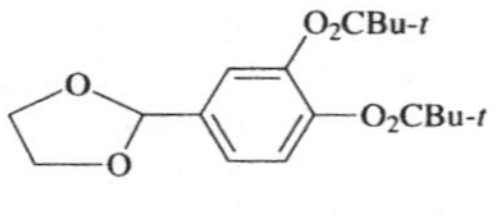 O, O, O_2CBu-t, O_2CBu-t	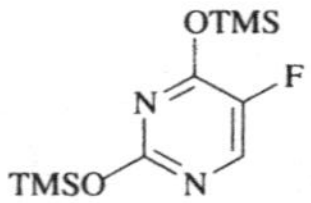 OTMS, F, TMSO	$SnCl_4$, CH_2Cl_2, rt, 12 h	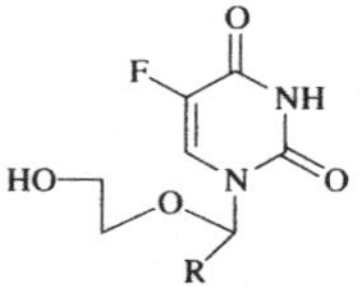 O, F, NH, HO, O, N, O, R (40) R = O_2CBu-t, O_2CBu-t 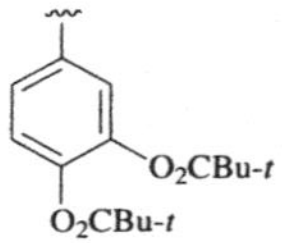	637
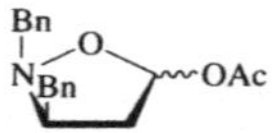 Bn, O, N, Bn, OAc	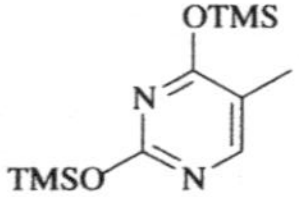 OTMS, TMSO	$SnCl_4$, CH_2Cl_2, 0°	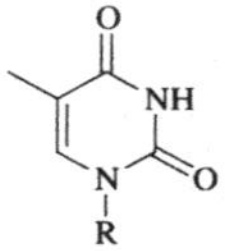 O, NH, N, O, R (90) α:β = 3:2 R 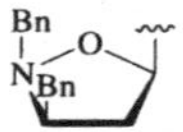 = Bn, O, N, Bn	792
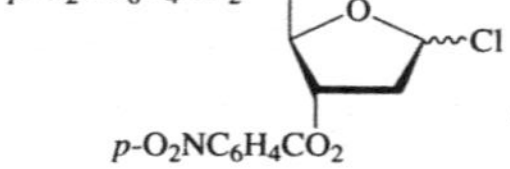 p-$O_2NC_6H_4CO_2$, p-$O_2NC_6H_4CO_2$ (Cl)	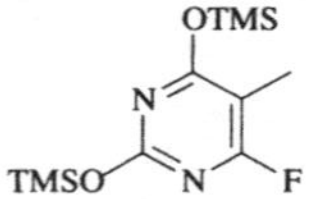 OTMS, TMSO, F	$SnCl_4$, $Cl(CH_2)_2Cl$, 0°, 24 h	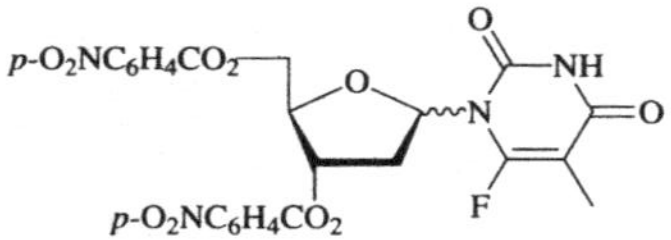 p-$O_2NC_6H_4CO_2$, O, NH, N, O, F, p-$O_2NC_6H_4CO_2$ (—)	721
	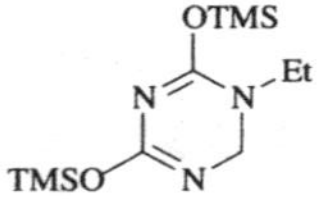 OTMS, N, Et, TMSO	1. $SnCl_4$, $Cl(CH_2)_2Cl$, MeCN, –25°, 5 min 2. 20°, 25 min	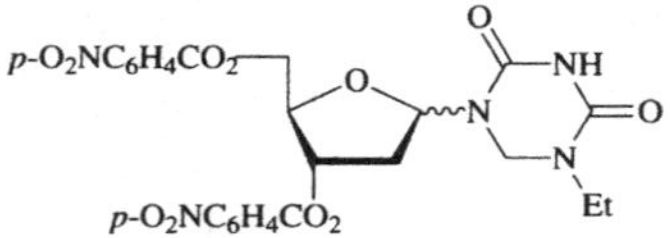 p-$O_2NC_6H_4CO_2$, O, NH, N, O, N, Et, p-$O_2NC_6H_4CO_2$ (—)	669

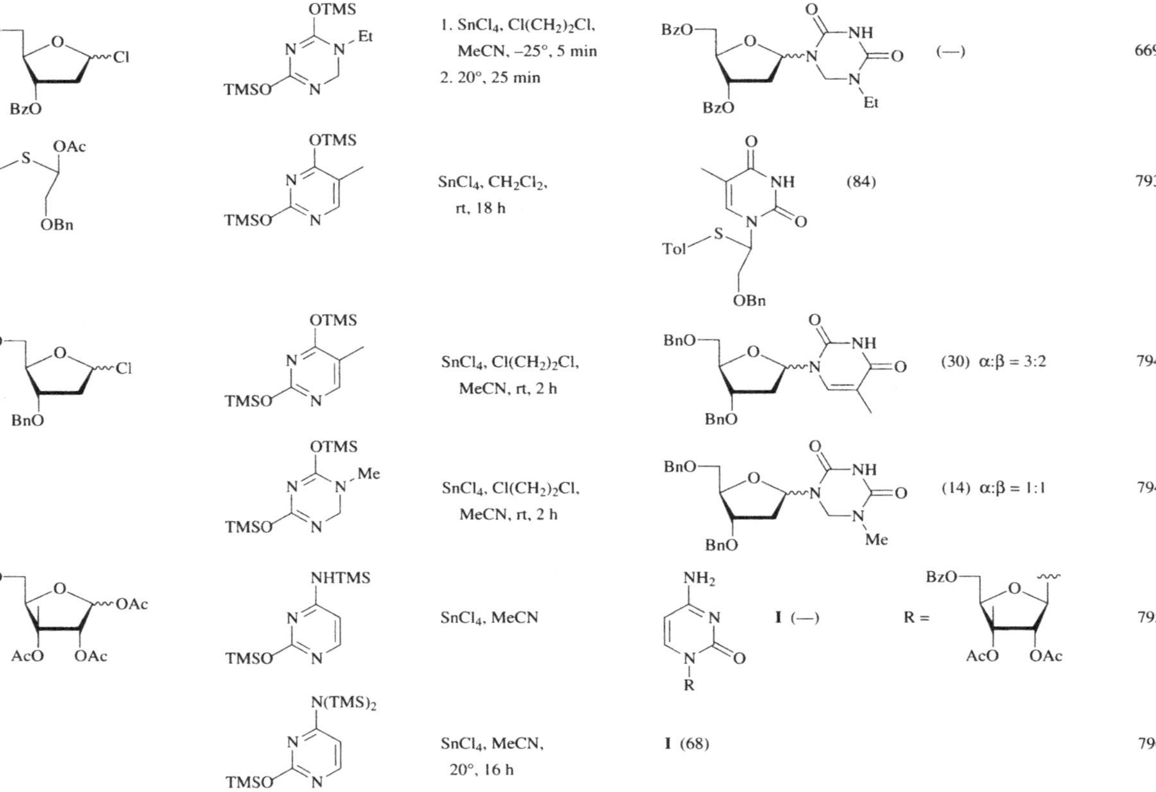
1. $SnCl_4$, $Cl(CH_2)_2Cl$, MeCN, −25°, 5 min
2. 20°, 25 min
(—)
669
$SnCl_4$, CH_2Cl_2, rt, 18 h
(84)
793
$SnCl_4$, $Cl(CH_2)_2Cl$, MeCN, rt, 2 h
(30) α:β = 3:2
794
$SnCl_4$, $Cl(CH_2)_2Cl$, MeCN, rt, 2 h
(14) α:β = 1:1
794
$SnCl_4$, MeCN
I (—)
R =
795
$SnCl_4$, MeCN, 20°, 16 h
I (68)
796

TABLE III. REACTIONS OF SILYLATED HETEROCYCLIC BASES WITH PROTECTED SUGARS - $SnCl_4$ CATALYST (*Continued*)

Sugar	Base	Conditions	Product(s) and Yield(s) (%)	Refs.
	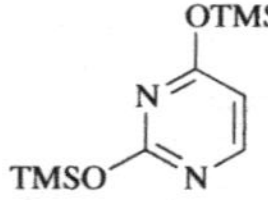	$SnCl_4$, MeCN, 20°, 16 h	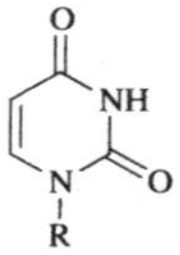 (80) + N^3-isomer (7)	796
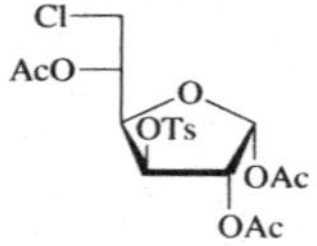	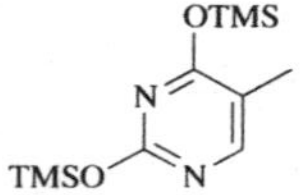	$SnCl_4$, $Cl(CH_2)_2Cl$, 20°, 3 h	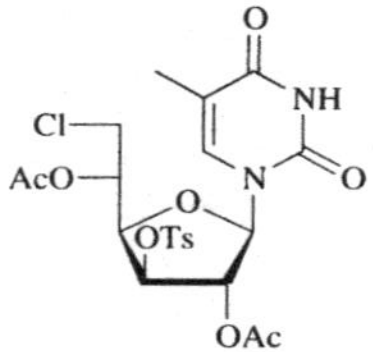 (37)	797
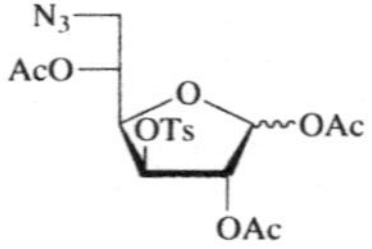	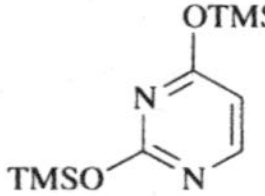	$SnCl_4$, $Cl(CH_2)_2Cl$, 20°, 6 h	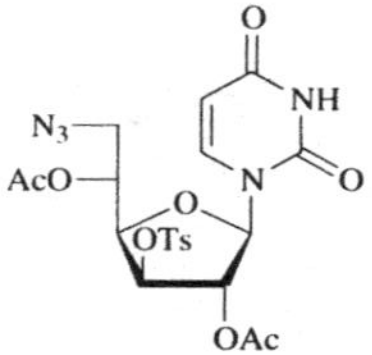 (42) + α-anomer (15)	797
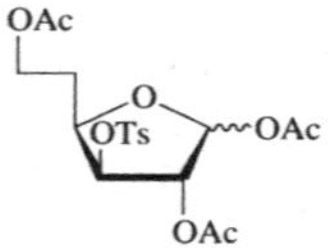	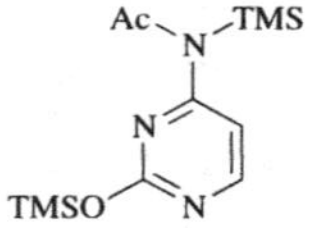	$SnCl_4$, $Cl(CH_2)_2Cl$, rt, 12 h	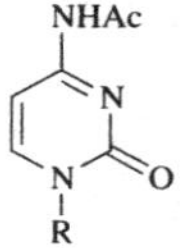 (46) R = 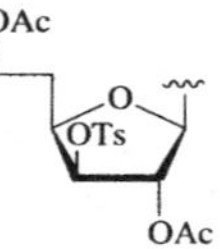	798
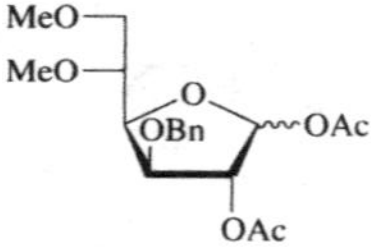	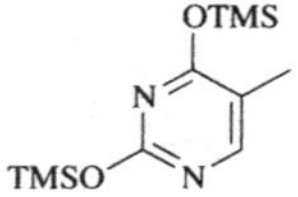	$SnCl_4$, CH_2Cl_2, rt, 18 h	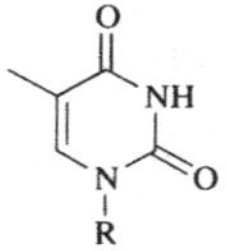 (85) R = 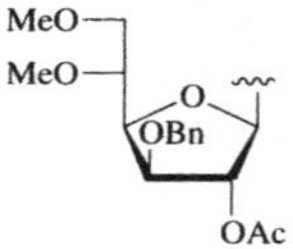	742

Substrate	Silylated base	Conditions	Product(s) and Yield(s) (%)	Refs.
C_{20} BzO, OBz, S, OAc 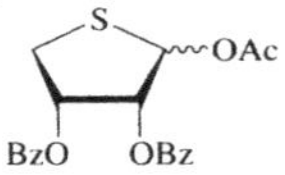	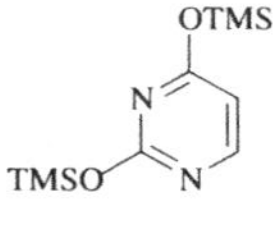 TMSO, N, N, OTMS	$SnCl_4$, CH_2Cl_2, rt, 12 h, or $SnCl_4$	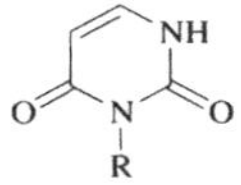 NH, O, N, O, R (35); R = S, BzO, OBz 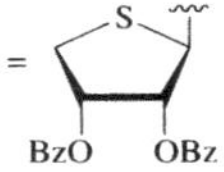	686, 687
	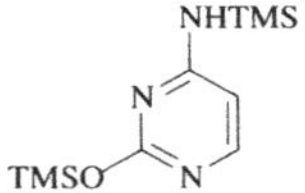 TMSO, N, N, NHTMS	$SnCl_4$	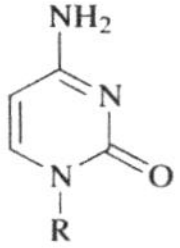 NH_2, N, O, N, R (61)	686
	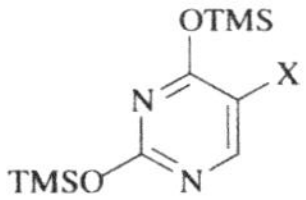 TMSO, N, N, OTMS, X	$SnCl_4$	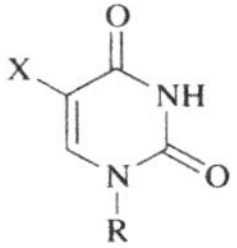 O, X, NH, O, N, R; X: F (—) + N^3-isomer (—); I (—)	686
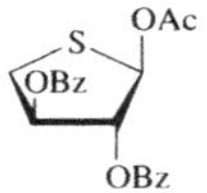 OAc, S, OBz, OBz	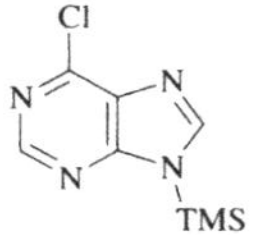 Cl, N, N, N, N, TMS	$SnCl_4$, CH_2Cl_2, rt, 12 h	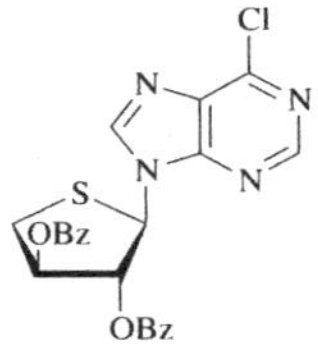 Cl, N, N, N, N, S, OBz, OBz (14)	687
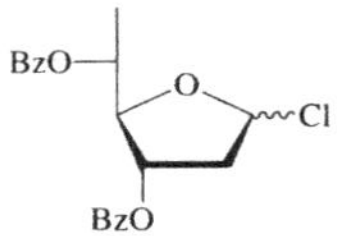 BzO, O, Cl, BzO	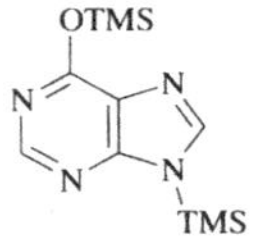 OTMS, N, N, N, N, TMS	$SnCl_4$, $Cl(CH_2)_2Cl$, rt, 12 h	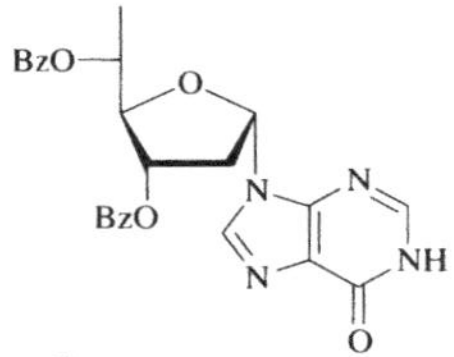 BzO, O, BzO, N, N, N, NH, O (6) + mixture of nucleoside isomers (38)	527
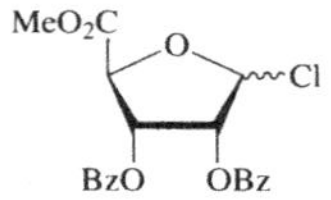 MeO_2C, O, Cl, BzO, OBz	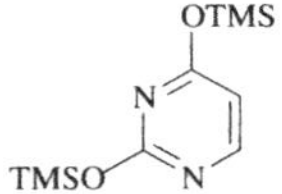 TMSO, N, N, OTMS	$SnCl_4$, $Cl(CH_2)_2Cl$, 60°, 4 h	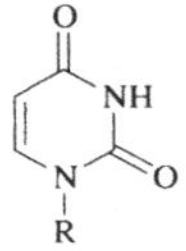 O, NH, O, N, R (47) + N^3-β-isomer (20); R 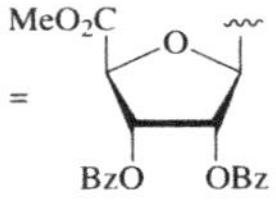 = MeO_2C, O, BzO, OBz	224

TABLE III. REACTIONS OF SILYLATED HETEROCYCLIC BASES WITH PROTECTED SUGARS - $SnCl_4$ CATALYST (*Continued*)

Sugar	Base	Conditions	Product(s) and Yield(s) (%)	Refs.
	Ac, N, TMS; N; TMSO; N	$SnCl_4$, $Cl(CH_2)_2Cl$, 25°, 20 h	NHAc; N; N; O; R (82)	224
BzO; O; OAc; BzO	OTMS; R[1]; N; TMSO; N; S	1. $SnCl_4$, CH_2Cl_2, rt, 18 h 2. NH_3, MeOH	O; R[1]; HN; O; N; R; S — R^1: H (35), Ph (15); R = HO; O; HO	658
	OTMS; N; S; TMSO; N; R[1]	1. $SnCl_4$,$Cl(CH_2)_2Cl$, –30°, 18 h 2. NH_3, MeOH	O; HN; S; O; N; R; R[1]	659
	R^1			
	p-FC_6H_4		(31)	
	p-ClC_6H_4		(24)	
	p-BrC_6H_4		(35) + N^3-isomer (15)	
	Me		(63)	
	Ph		(43)	
	Tol		(32)	
	p-$MeOC_6H_4$		(64)	
	OTMS; N; S; TMSO; N	1. $SnCl_4$, $Cl(CH_2)_2Cl$, 24°, 18 h 2. NH_3, MeOH	O; HN; S; O; N; R (20)	658

BzO, PhS, OAc	OTMS, R, TMSO	$SnCl_4$, CH_2Cl_2, rt, 12 h	BzO, PhS, O, NH, O, R; R: H (73) α:β 31:69; Me (34) α:β 52:48	229
	Ac, TMS, TMSO	$SnCl_4$, CH_2Cl_2, rt, 12 h	BzO, PhS, O, NHAc (89) α:β = 43:57	229
BnO, TolO, Cl	OTMS, TMSO	$SnCl_4$, $Cl(CH_2)_2Cl$, MeCN, rt, 2 h	BnO, TolO, O, NH, O (15) α:β = 1::2	794
	OTMS, Me, TMSO	$SnCl_4$, $Cl(CH_2)_2Cl$, MeCN, rt, 2 h	BnO, TolO, O, NH, O, Me (20) α:β = 1:1	794
TolO, BnO, Cl	OTMS, TMSO	$SnCl_4$, $Cl(CH_2)_2Cl$, MeCN, rt, 2 h	O, NH, O, R (3); R = TolO, BnO	794
	OTMS, Me, TMSO	$SnCl_4$, $Cl(CH_2)_2Cl$, MeCN, rt, 2 h	Me, O, NH, O, R (11)	794

TABLE III. REACTIONS OF SILYLATED HETEROCYCLIC BASES WITH PROTECTED SUGARS - $SnCl_4$ CATALYST *(Continued)*

Sugar	Base	Conditions	Product(s) and Yield(s) (%)	Refs.
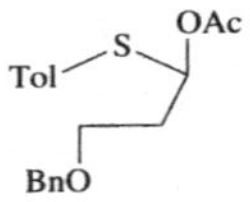 Tol–S, OAc, BnO	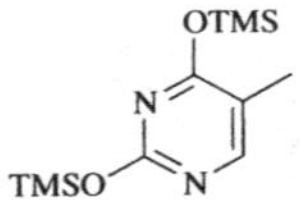 OTMS, TMSO	$SnCl_4$ (3 eq), CH_2Cl_2, rt, 3 h	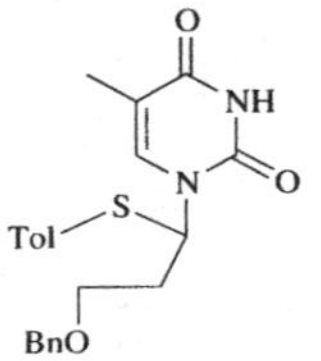 Tol–S, BnO (52)	793
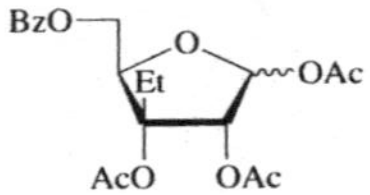 BzO, Et, OAc, AcO, OAc	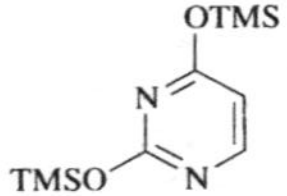 OTMS, TMSO	$SnCl_4$, MeCN	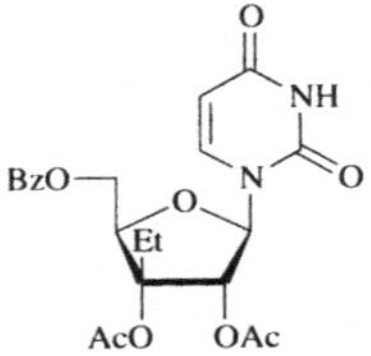 BzO, Et, AcO, OAc (—)	795
C_{21} 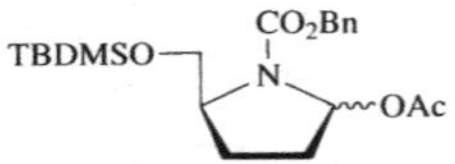 TBDMSO, CO_2Bn, OAc	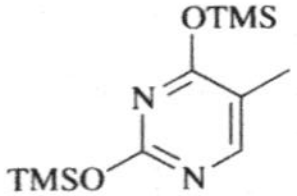 OTMS, TMSO	$SnCl_4$, CH_2Cl_2, –15°, 3 h	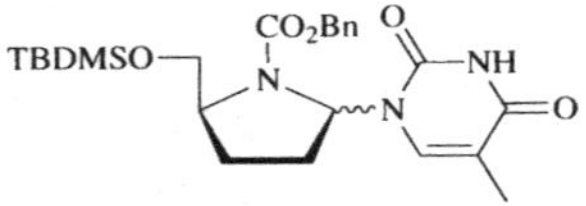 TBDMSO, CO_2Bn (64)	45
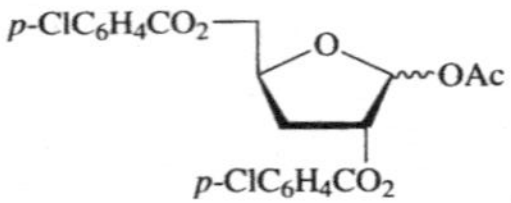 p-$ClC_6H_4CO_2$, OAc, p-$ClC_6H_4CO_2$	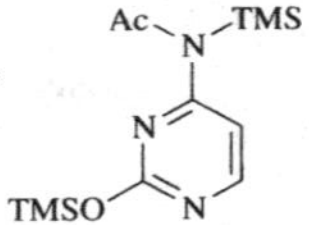 Ac, TMS, TMSO	$SnCl_4$, MeCN, rt, 12 h	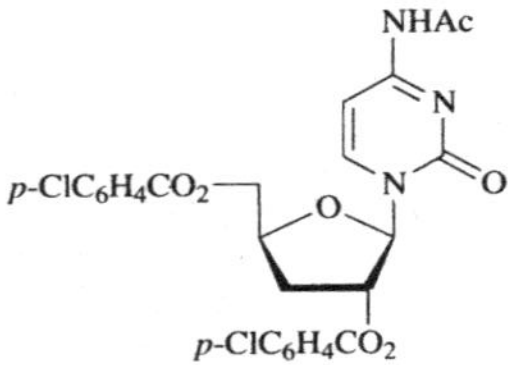 NHAc, p-$ClC_6H_4CO_2$, p-$ClC_6H_4CO_2$ (100)	800
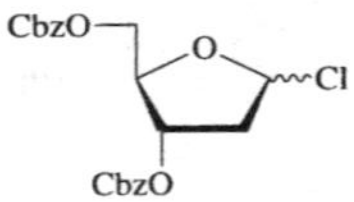 CbzO, Cl, CbzO	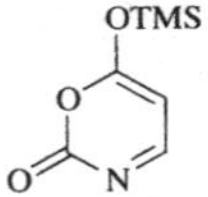 OTMS	$SnCl_4$, C_6H_6	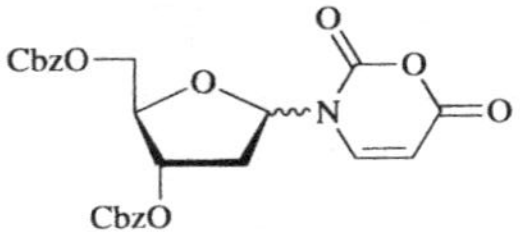 CbzO, CbzO (15) α:β = 1:1	801

BnO, OMe, MeO_2C, OTBDMS	OTMS, TMSO, N	$SnCl_4$, MeCN	BnO, MeO_2C, OTBDMS, NH, O (35) + α-anomer (35)	802
TolO, Cl, TolO	CN, CO_2Me, N, TMS	$SnCl_4$, $Cl(CH_2)_2Cl$	TolO, TolO, CN, CO_2Me; rt, 18 h (>70); 35°, 2 h (85) α:β = 80:20	803
	TMS, N, TMS, O	$SnCl_4$, Py, $Cl(CH_2)_2Cl$, MeCN, –35°, 45 min	R, NH, O (30) + α-anomer (32); R = TolO, TolO	557
	TMS, N, TMS, O	$SnCl_4$, $Cl(CH_2)_2Cl$, MeCN, Cond.	TolO, TolO, N, H, O Cond. α:β –20°, 1.5 h 60:40 (89) –35°, 0.75 h 50:50 (62)	804 805
	TMS, N, TMS, O	$Cl(CH_2)_2Cl$, MeCN 1. 50-60°, 2 h 2. $SnCl_4$, –20°, 1.5 h	TolO, TolO, N, H, O **I** (—) α:β = 40:60	804
		Py, MeCN 1. 50-60°, 2 h 2. $SnCl_4$, $Cl(CH_2)_2Cl$, MeCN, –35°, 45 min	**I** (89) α:β = 2:3	557

TABLE III. REACTIONS OF SILYLATED HETEROCYCLIC BASES WITH PROTECTED SUGARS - $SnCl_4$ CATALYST (*Continued*)

Sugar	Base	Conditions	Product(s) and Yield(s) (%)	Refs.
		$SnCl_4$, MeCN, $Cl(CH_2)_2Cl$, −35°, 45 min	(89) α:β = 2:3	557
		$SnCl_4$, $Cl(CH_2)_2Cl$, 0-5°, 2 h	(57) + α-anomer (20) R =	524
		$SnCl_4$, $Cl(CH_2)_2Cl$ −40°, 6.5 h	(94)	494
		$SnCl_4$, CH_2Cl_2, −78°, 5 h	(25)	494
		$SnCl_4$, $Cl(CH_2)_2Cl$		

Reactant	Conditions	Product(s) and Yield(s) (%)	Refs.
R^1			
TMS	20-22°, 4 h	(51)	806
Et	0°, 2 h	(57)	84
$CH{=}CH_2$	0-5°, 2-3 h, rt, 8-10 h	(30) + α-anomer (9)	807
$CH{=}CH_2$	C_6H_6, rt, 8-10 h	(30) + α-anomer (9)	808
CH_2OBn	rt, 2 h	(46)	809
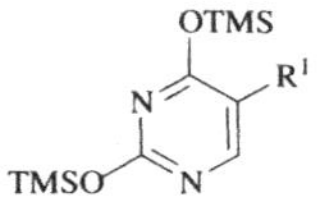 2,4-bis(TMSO)-5-R^1-pyrimidine	$SnCl_4$, $Cl(CH_2)_2Cl$	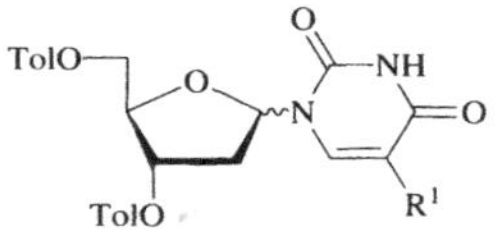 3,5-di-O-TolO-2-deoxyribosyl-5-R^1-uracil	
R^1			
Me	MeCN, rt, 2 h	(65) α:β = 1:3	794
i-Pr	MeCN (no $Cl(CH_2)_2Cl$), 0-4°, 18 h	(70) α:β = 1:10	810
s-Bu	—	(—)	718
i-Bu	—	(—)	718
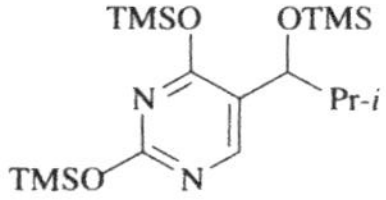 2,4-bis(TMSO)-5-[CH(OTMS)Pr-*i*]pyrimidine	$SnCl_4$, $Cl(CH_2)_2Cl$	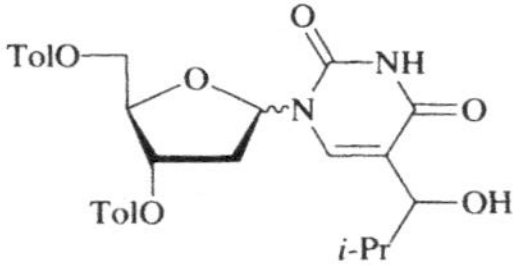 TolO-deoxyribosyl-5-[CH(OH)-*i*-Pr]uracil (—)	718
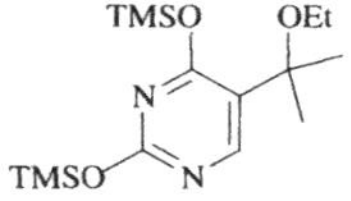 2,4-bis(TMSO)-5-[C(Me)$_2$OEt]pyrimidine	$SnCl_4$, $Cl(CH_2)_2Cl$	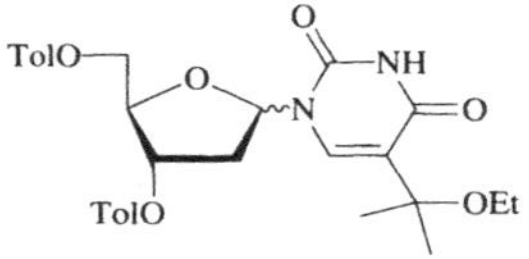 TolO-deoxyribosyl-5-[C(Me)$_2$OEt]uracil (—)	718
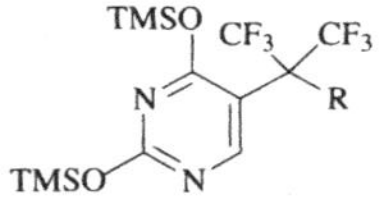 2,4-bis(TMSO)-5-[C(CF_3)$_2$R]pyrimidine	$SnCl_4$, $Cl(CH_2)_2Cl$	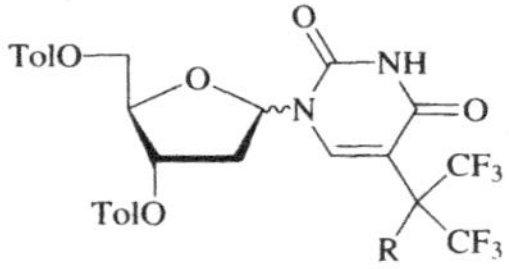 TolO-deoxyribosyl-5-[C(CF_3)$_2$R]uracil; R: F (—), OH (—)	718

TABLE III. REACTIONS OF SILYLATED HETEROCYCLIC BASES WITH PROTECTED SUGARS - $SnCl_4$ CATALYST (*Continued*)

Sugar	Base	Conditions	Product(s) and Yield(s) (%)	Refs.
	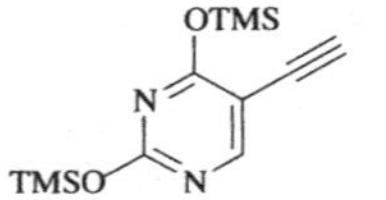	1. $SnCl_4$, CH_2Cl_2, 0°, 8 h 2. 20°, 18 h	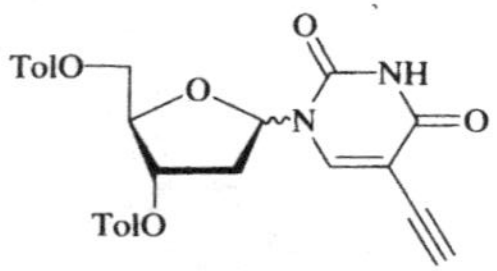 (85) α:β = 6:5	813
	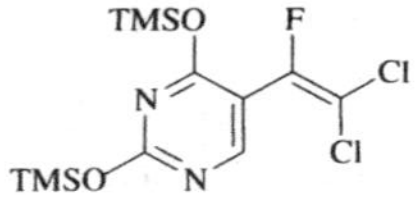	1. $SnCl_4$, CH_2Cl_2, 0°, 8 h 2. 20°, 24 h	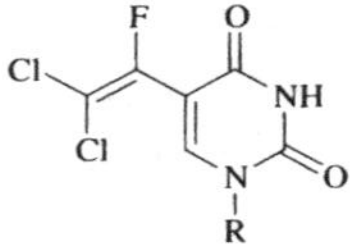 (30) + α-anomer (28) R 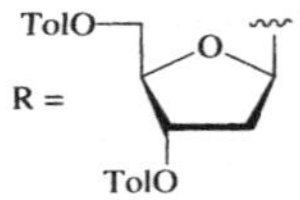 =	812
	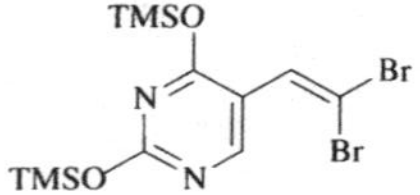	$SnCl_4$, $Cl(CH_2)_2Cl$, 0-5°	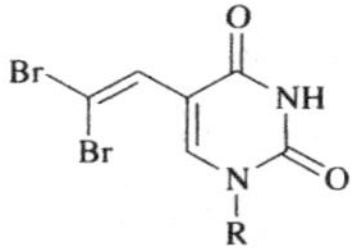 (71) α:β = 1:3	813
	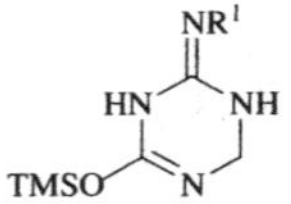	1. $SnCl_4$, $Cl(CH_2)_2Cl$, MeCN, –25°, 5 min 2. 20°, 25 min	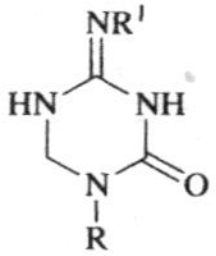 R^1: H (—); Ac (—)	669
	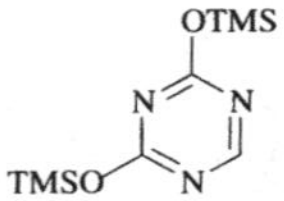	1. $SnCl_4$, $Cl(CH_2)_2Cl$, MeCN, –25°, 5 min 2. 20°, 25 min	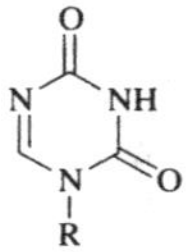 (—)	669

Reactant	Conditions	Product(s) and Yield(s) (%)	Refs.
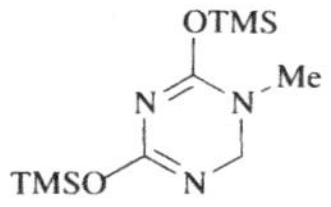 OTMS, N, N-Me, TMSO, N (structure)	1. $SnCl_4$, $Cl(CH_2)_2Cl$, MeCN, rt, 18 h 2. R^1MgBr	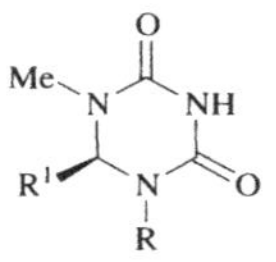 Me-N, NH, O, O, R^1, N-R (structure); R^1: Me (43), Ph (54)	512
	1. $SnCl_4$, $Cl(CH_2)_2Cl$, rt, 18 h 2. $NaBH_4$ 3. OH^-	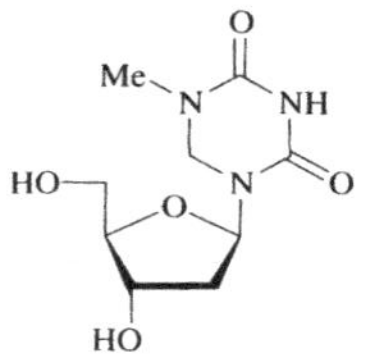 Me-N, NH, O, O, HO, HO (structure) (13) + α-anomer (12)	512
	$SnCl_4$, $Cl(CH_2)_2Cl$, MeCN, rt, 3-5 h 1. –20°, 5 min 2. 25°, 3-5 h	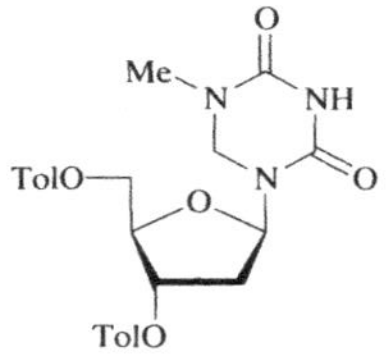 Me-N, NH, O, O, TolO, TolO (structure) **I** (23)	239
	1. $SnCl_4$, $Cl(CH_2)_2Cl$, MeCN, –25°, 5 min 2. 20°, 25 min	**I** (29)	669
	1. $SnCl_4$, MeCN, –20°, 5 min 2. 25°, 3-5 h	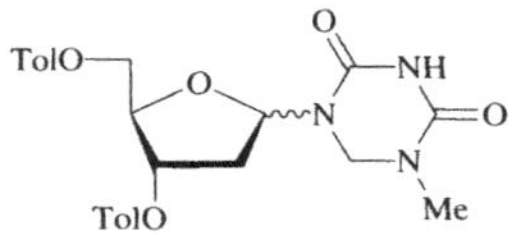 TolO, TolO, NH, O, O, N-Me (structure) **II** (57) α:β = 1:1	239
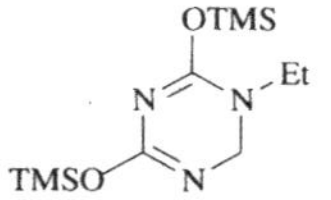 OTMS, N, N-Et, TMSO, N (structure)	1. $SnCl_4$, $Cl(CH_2)_2Cl$, MeCN, –25°, 5 min 2. 20°, 25 min	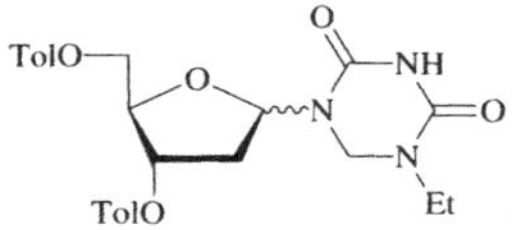 TolO, TolO, NH, O, O, N-Et (structure) (—)	669

TABLE III. REACTIONS OF SILYLATED HETEROCYCLIC BASES WITH PROTECTED SUGARS - $SnCl_4$ CATALYST (*Continued*)

Sugar	Base	Conditions	Product(s) and Yield(s) (%)	Refs.
	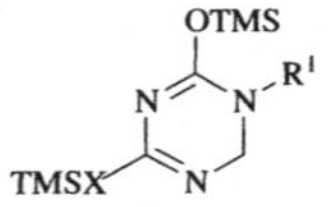 [structure: TMSX, OTMS, R[1]]	1. $SnCl_4$, $Cl(CH_2)_2Cl$, MeCN, –25°, 5 min 2. 20°, 25 min	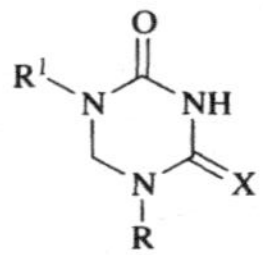 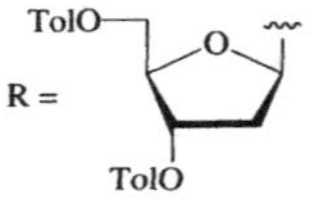 [structure: R[1], NH, X, R]; R = [structure: TolO, TolO]	669
	X = O, R^1 = *n*-Pr		(—)	
	X = O, R^1 = *n*-Bu		(—)	
	X = O, R^1 = *i*-Pr		(—)	
	X = O, R^1 = $MeOCH_2$		(—)	
	X = O, R^1 = $MeSCH_2$		(—)	
	X = O, R^1 = *c*-C_3H_5		(—)	
	X = S, R^1 = Me		(—)	
	X = S, R^1 = Et		(—)	
	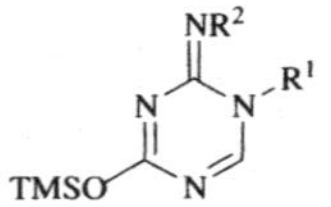 [structure: TMSO, NR^2, R[1]]	1. $SnCl_4$, $Cl(CH_2)_2Cl$, MeCN, –25°, 5 min 2. 20°, 25 min	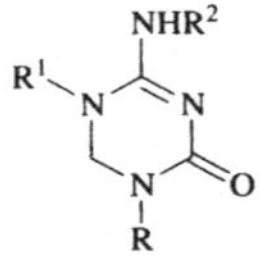 [structure: R[1], NHR^2, O, R]	669
	R^1 = Me, R^2 = H		(—)	
	R^1 = Et, R^2 = H		(—)	
	R^1 = *n*-Pr, R^2 = H		(—)	
	R^1 = *i*-Pr, R^2 = H		(—)	
	R^1 = *c*-C_3H_5, R^2 = H		(—)	
	R^1 = Me, R^2 = Ac		(—)	
	R^1 = Et, R^2 = Ac		(—)	
	R^1 = *n*-Pr, R^2 = Ac		(—)	
	R^1 = *i*-Pr, R^2 = Ac		(—)	
	R^1 = *c*-C_3H_5, R^2 = Ac		(—)	

	$SnCl_4$ (1.37 eq), MeCN, 0°, 40 min	**I** (40) α:β = 1:1.3	814
	$SnCl_4$ (1.46 eq), MeCN/$Cl(CH_2)_2Cl$ (2:5), 20°, 135 min	**I** (25) α:β = 1:1.8	814
	$SnCl_4$ (0.68 eq), MeCN, 20°, 125 min	**I** (30) α:β = 1:2.4	814
	$SnCl_4$ (0.68 eq), $Cl(CH_2)_2Cl$, 20°, 1 h	**I** (46) α:β = 1:2.5	814
	$SnCl_4$ (0.68 eq), $Cl(CH_2)_2Cl$, 20°, 85 min	**I** (45) α:β = 1:2.6	814
	$SnCl_4$, $Cl(CH_2)_2Cl$, MeCN, rt, 4 h	R^1 = Me: (21) + α-anomer (9) R^1 = Ph: (21) + α-anomer (9) + N^3-α- and N^3-β-isomers (20)	512
	$SnCl_4$, $Cl(CH_2)_2Cl$, 25°	(34) α:β = 3:3.2	815, 546
	1. $SnCl_4$, $Cl(CH_2)_2Cl$, 0°, 105 min 2. NaOMe, MeOH	(33) + β-anomer (10)	816

TABLE III. REACTIONS OF SILYLATED HETEROCYCLIC BASES WITH PROTECTED SUGARS - $SnCl_4$ CATALYST (*Continued*)

Sugar	Base	Conditions	Product(s) and Yield(s) (%)	Refs.
	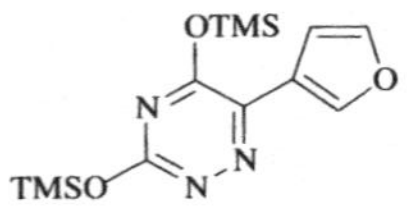 OTMS, TMSO, N, N, N	$SnCl_4$, $Cl(CH_2)_2Cl$, rt, 2.5 h	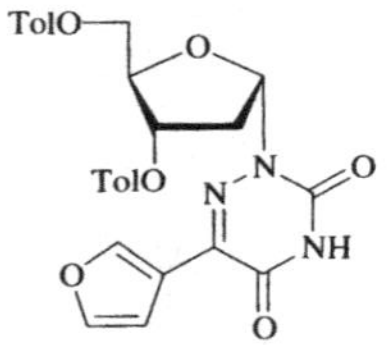 TolO, TolO, N, N, O, NH, O (33)	546
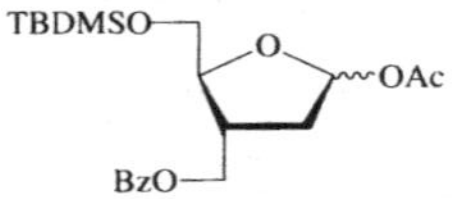 TBDMSO, OAc, BzO	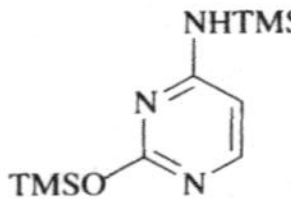 NHTMS, TMSO, N, N	$SnCl_4$	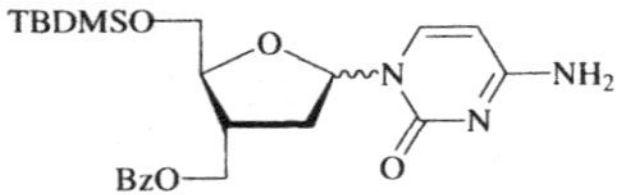 TBDMSO, BzO, N, NH$_2$, N, O (—) α:β = 1:1	817
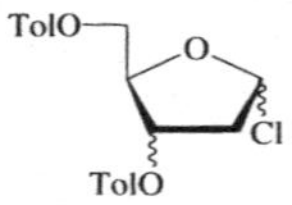 TolO, Cl, TolO	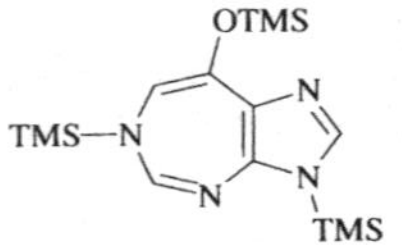 OTMS, TMS—N, N, N, TMS	$SnCl_4$, $Cl(CH_2)_2Cl$, MeCN, –35°, 45 min	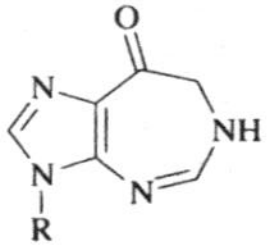 O, N, NH, N, R (35) + α-anomer (17); R 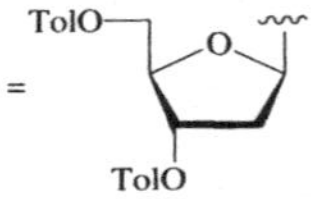 = TolO, TolO	818
	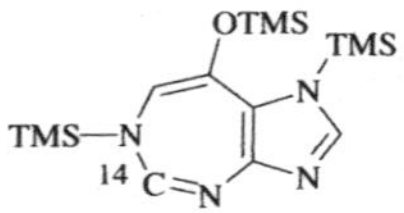 OTMS, TMS, TMS—N, N, N, 14 C=N	$SnCl_4$, $Cl(CH_2)_2Cl$/MeCN, (1:1), –35°	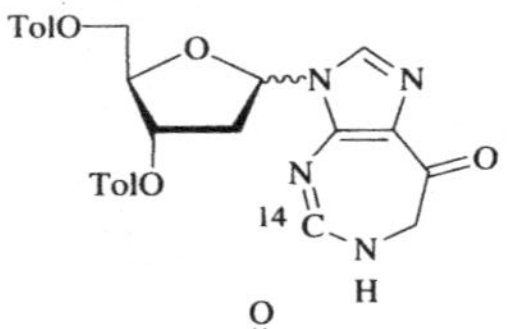 TolO, TolO, N, N, O, N, 14 C, N, H (26)	819
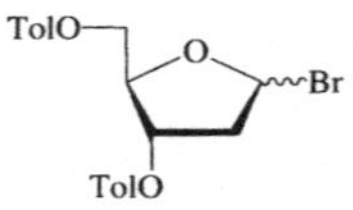 TolO, Br, TolO	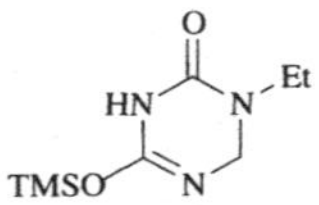 O, HN, N, Et, TMSO, N	1. $SnCl_4$, $Cl(CH_2)_2Cl$, MeCN, –25°, 5 min 2. 20°, 25 min	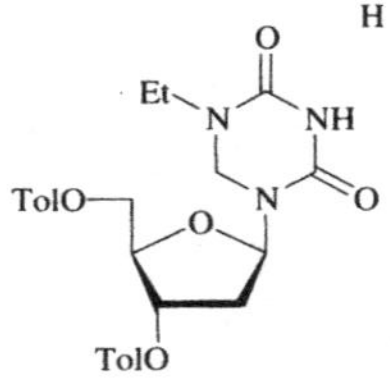 O, Et, N, NH, N, O, TolO, TolO (—)	669

		$SnCl_4$, $Cl(CH_2)_2Cl$, 20°, 12 h	(28)	820
		$SnCl_4$, $Cl(CH_2)_2Cl$, –5 to 0°	(25) α:β = 1:2	801
		$SnCl_4$, $Cl(CH_2)_2Cl$, rt	(44)	789
		$SnCl_4$, RSCl, CH_2Cl_2, –78 to 25°, 2 h R: Ph; Tipp	(65) α:β = 1:18 (R = Ph) (70) α:β = 1:23 (R = Tipp)	421a
		$SnCl_4$, $Cl(CH_2)_2Cl$, MeCN, 20°, 16 h	(83)	821

TABLE III. REACTIONS OF SILYLATED HETEROCYCLIC BASES WITH PROTECTED SUGARS - $SnCl_4$ CATALYST *(Continued)*

Sugar	Base	Conditions	Product(s) and Yield(s) (%)	Refs.
	OTMS, TMSO, N, N, R^1	$SnCl_4$, RSCl, CH_2Cl_2, –78 to 25°, 2 h	TBDPSO, O, N, O, NH, O, SR, R^1	421a
	R^1: H	R: Tipp	(52) α:β = 1:99	
	Me	Ph	(68) α:β = 1:42	
	Me	Tipp	(60) α:β = 1:44	
C_{22} MeO, O, OAc, BzO, OBz	Bz, N, TMS, N, N, N, N, TMS	$SnCl_4$, $Cl(CH_2)_2Cl$, 20°, 16 h	NHBz, N, N, N, N, R (80)	821
	N, TMSO, N	$SnCl_4$, MeCN, rt, 2 d	N, N, O, R (45) R = MeO, O, BzO, OBz	822
	OTMS, N, TMSO, N	$SnCl_4$, $Cl(CH_2)_2Cl$, 20°, 16 h	O, NH, N, O, R (86)	822
BzO, O, OAc, BzO, OMe	OTMS, N, TMSO, N	1. $SnCl_4$, $Cl(CH_2)_2Cl$, 20°, 16 h 2. NaOMe, MeOH, reflux, 24 h	O, NH, N, O, HO, O, HO, OMe (24) + N^3-isomer (16)	225

		$SnCl_4$, MeCN, 40°, 1.5 h	(88)	823
		$SnCl_4$, MeCN, rt	R: H (95) α:β = 46:54; Me (97) α:β = 55:45	370b
		$SnCl_4$, MeCN, rt	(95) α:β = 42:58	370b
		$SnCl_4$, MeCN, rt	(92) α:β = 75:25	370b
		$SnCl_4$, MeCN, rt	(75) α:β = 65:35	370b
		$SnCl_4$, MeCN	(—)	795

TABLE III. REACTIONS OF SILYLATED HETEROCYCLIC BASES WITH PROTECTED SUGARS - $SnCl_4$ CATALYST *(Continued)*

Sugar	Base	Conditions	Product(s) and Yield(s) (%)	Refs.
OMe	OTMS, F, N, TMSO, N	1. $SnCl_4$, MeCN, rt, 1 h 2. MeOH	F, O, NH, O, N, O, O (11)	824
TBDPSO, O, OAc, S	NHTMS, N, TMSO, N	$SnCl_4$, MeCN, rt	NH_2, N, O, N, TBDPSO, O, S (80)	235
C_{23} RO, O, OAc, OR, OAc R = p-$O_2NC_6H_4CO$	Bz, N, TMS, N, N, N, N, TMS	$SnCl_4$, $Cl(CH_2)_2Cl$, rt, 18 h	NHBz, N, N, N, N, R^1 (90) R^1 = RO, O, OR, OAc	825
	OTMS, N, N, R^2, N, N, N, TMS, TMS R^2 = n-$C_{11}H_{23}CO$	$SnCl_4$, $Cl(CH_2)_2Cl$, rt, 24 h	H, N, R^2, N, N, NH, N, R^1, O **I** (37) + N^9-β-isomer (6)	825
		$SnCl_4$, $HgBr_2$, C_6H_6, reflux, 6 h	**I** (44) + N^9-β-isomer (20)	825

Sugar	Base	Conditions	Product(s) and Yield(s) (%)	Refs.
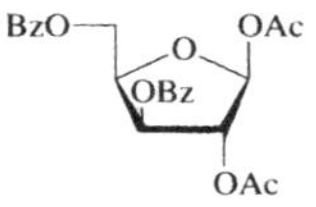 BzO, OAc, OBz, OAc	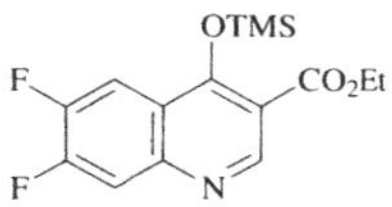 OTMS, F, F, CO_2Et	$SnCl_4$, $Cl(CH_2)_2Cl$, 20°	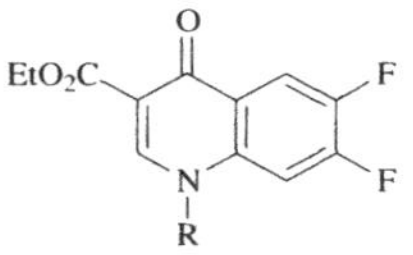 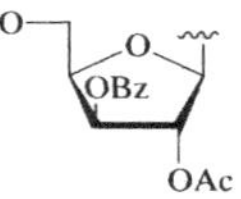 EtO_2C, O, F, F, N–R (61); R = BzO, OBz, OAc	826
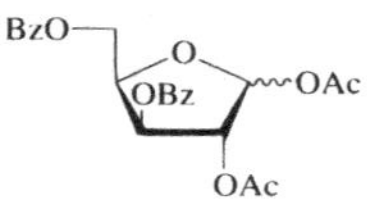 BzO, OBz, OAc, OAc	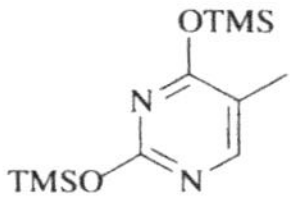 OTMS, TMSO	1. $SnCl_4$, $Cl(CH_2)_2Cl$, rt, 12 h 2. $SOCl_2$, MeCN, reflux, 90 min	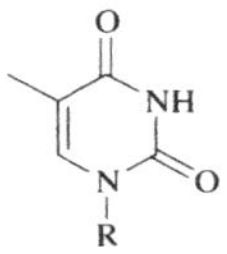 O, NH, O, N–R **I** (53)	827
		$SnCl_4$, MeCN, reflux, 5 h	**I** (32)	828
BnO, Ac, N, OAc, RO	OTMS, TMSO	$SnCl_4$, MeCN	O, NH, O, Ac, N, BnO, RO; R: Bz –15°, 1 h (93)	45b
			Bn –30°, 3 h (72) α:β = 1:2	45a
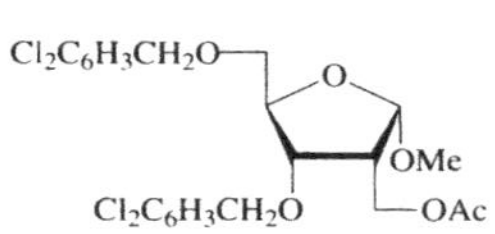 $Cl_2C_6H_3CH_2O$, OMe, $Cl_2C_6H_3CH_2O$, OAc	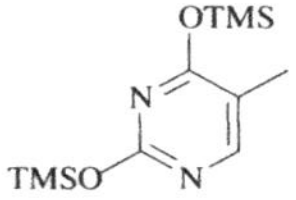 OTMS, TMSO	$SnCl_4$, MeCN, 50°	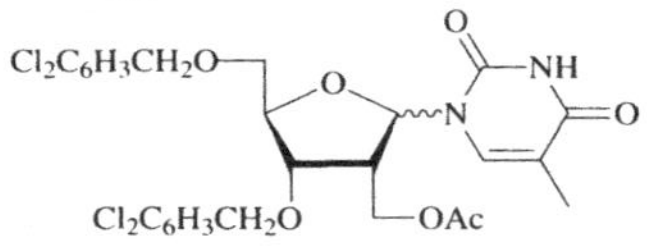 $Cl_2C_6H_3CH_2O$, O, NH, O, $Cl_2C_6H_3CH_2O$, OAc (66) α:β = 2:98	829
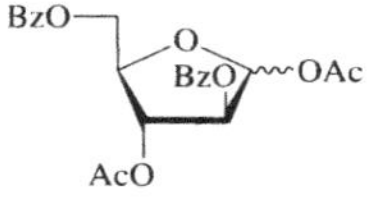 BzO, BzO, OAc, AcO	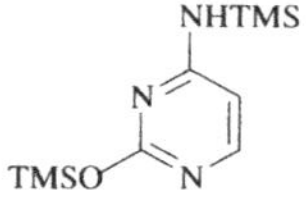 NHTMS, TMSO	$SnCl_4$, MeCN, rt	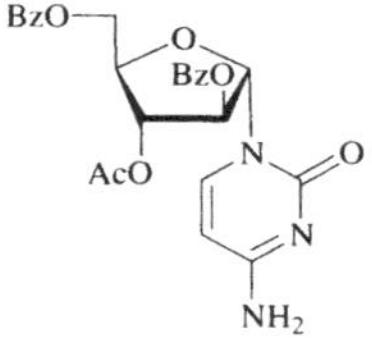 BzO, BzO, AcO, O, N, NH_2 (93)	830

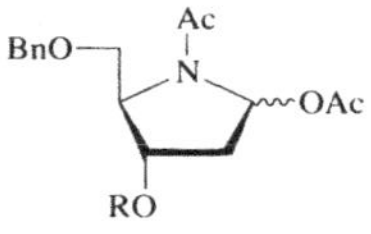

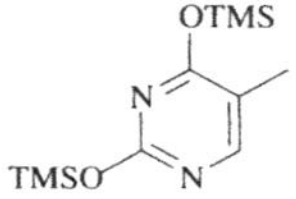

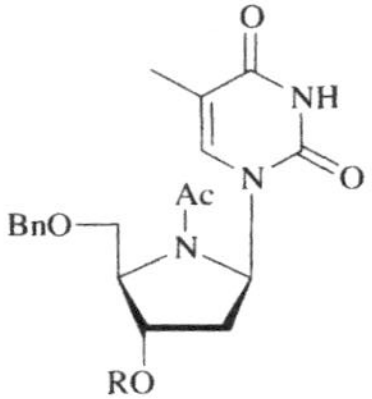

TABLE III. REACTIONS OF SILYLATED HETEROCYCLIC BASES WITH PROTECTED SUGARS - $SnCl_4$ CATALYST (*Continued*)

Sugar	Base	Conditions	Product(s) and Yield(s) (%)	Refs.
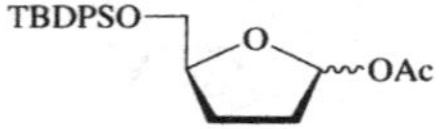 TBDPSO– OAc	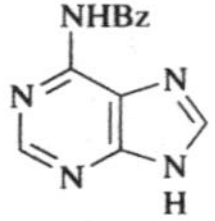 NHBz	$SnCl_4$, MeCN, –5°	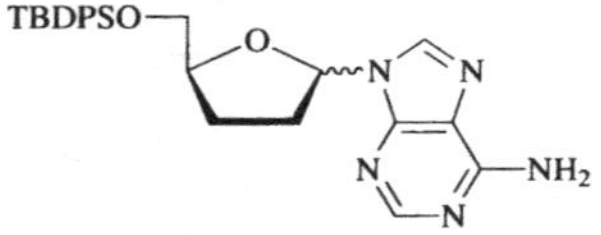 TBDPSO– NH$_2$ (—)	831
C_{24} 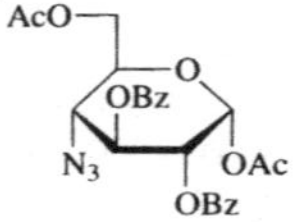 AcO– OBz N$_3$ OAc OBz	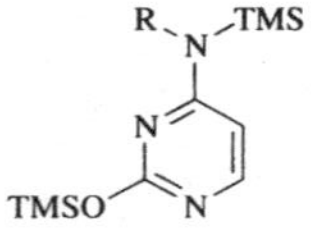 R–N–TMS TMSO	$SnCl_4$, $Cl(CH_2)_2Cl$, 50°, 1 h	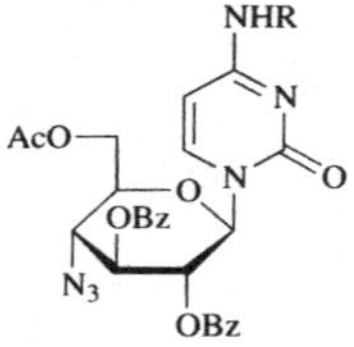 NHR AcO– OBz N$_3$ OBz; R: Ac (85), Bz (76)	832
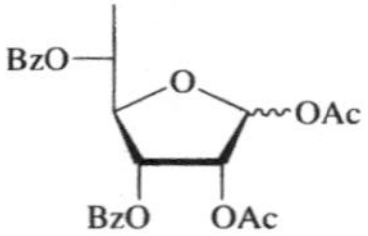 BzO– OAc BzO OAc	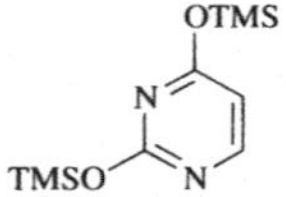 OTMS TMSO	1. $SnCl_4$, MeCN, rt, 24 h 2. NH_3, MeOH, rt, 5 d	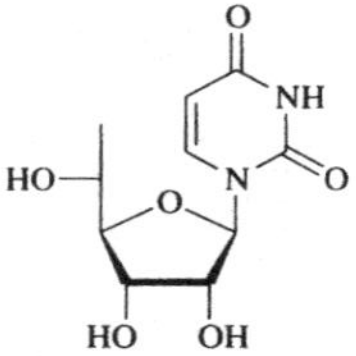 O NH HO– HO OH (16)	833
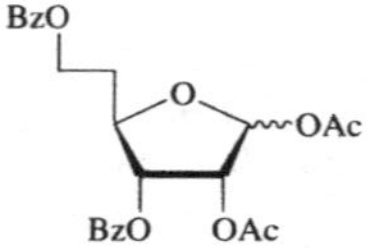 BzO OAc BzO OAc	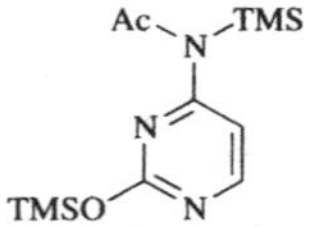 Ac–N–TMS TMSO	$SnCl_4$, $Cl(CH_2)_2Cl$, rt, 12 h	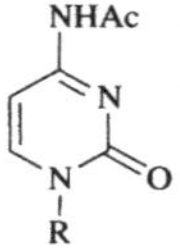 NHAc N–R (49); R = BzO BzO OAc 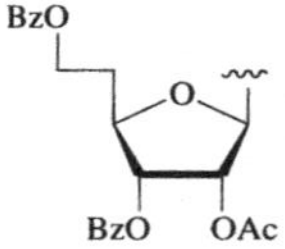	820, 834
	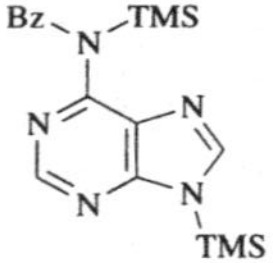 Bz–N–TMS N–TMS	$SnCl_4$, $Cl(CH_2)_2Cl$, rt, 16 h	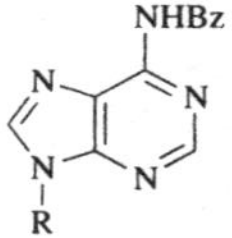 NHBz N–R (72)	821

		Conditions	Product(s) and Yield(s) (%)	Refs.
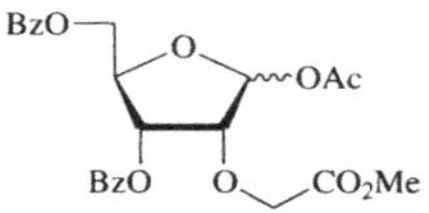	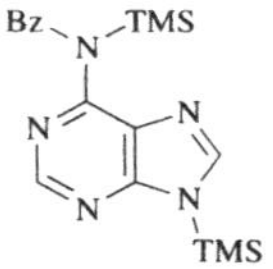	$SnCl_4$, MeCN, rt, 3 d	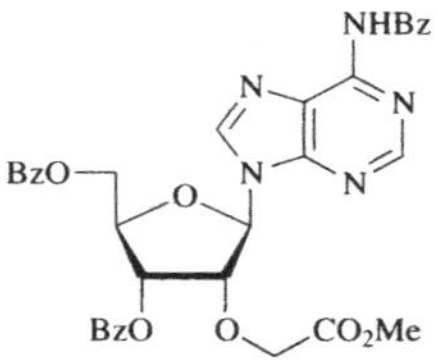 (63) + α-anomer (14)	835
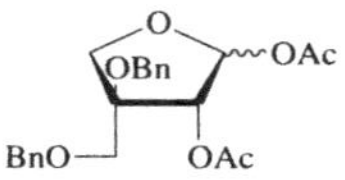	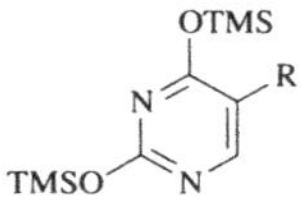	$SnCl_4$, $Cl(CH_2)_2Cl$, rt	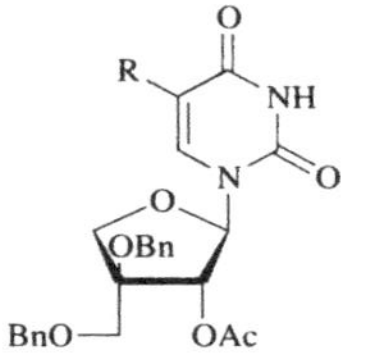 R: H, 18 h (85); Me, 13 h (81)	672
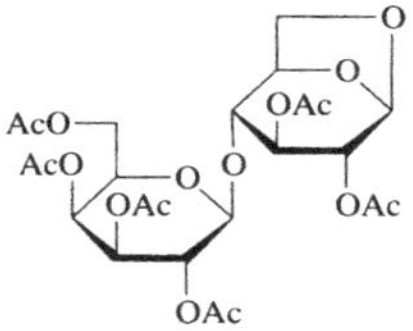	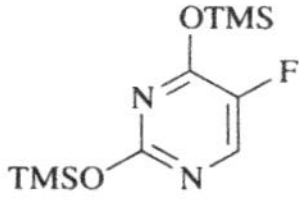	$SnCl_4$, reflux, 5 h	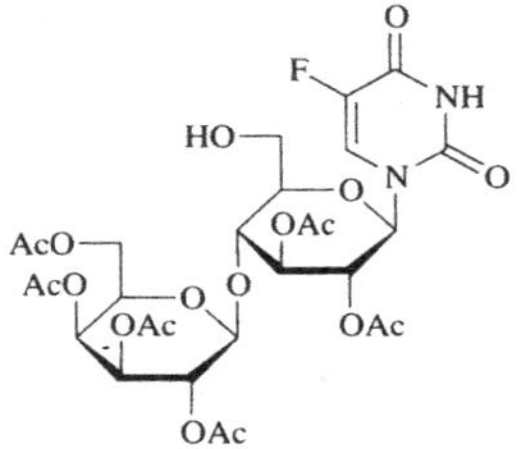 **I** (—)	836
		$SnCl_4$, MeCN, reflux, 10 h	**I** (35)	692
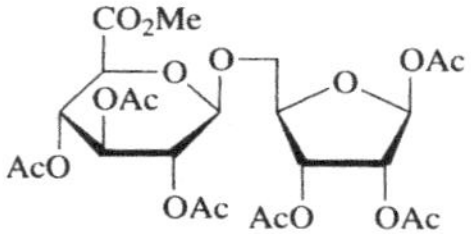	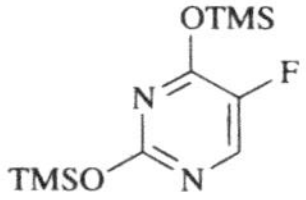	$SnCl_4$, $Cl(CH_2)_2Cl$, reflux, 30 min	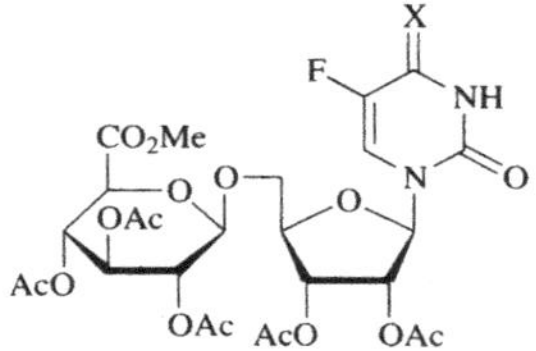 **I**, X = O (92)	837

TABLE III. REACTIONS OF SILYLATED HETEROCYCLIC BASES WITH PROTECTED SUGARS - $SnCl_4$ CATALYST (*Continued*)

Sugar	Base	Conditions	Product(s) and Yield(s) (%)	Refs.
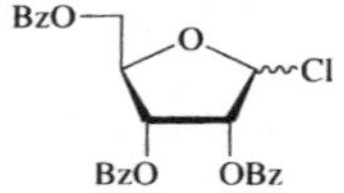	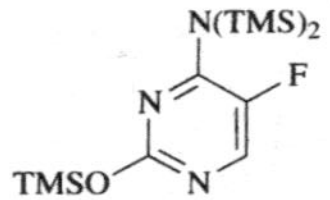 [structure]	$SnCl_4$, $Cl(CH_2)_2Cl$, reflux, 35 min	**I**, X = NH (98)	837
C_{26} [structure]	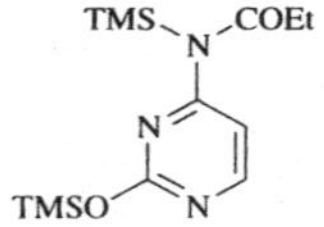 [structure]	$SnCl_4$, $Cl(CH_2)_2Cl$, 50°, 5 h	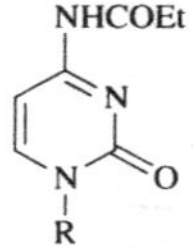 [structure] (—) R = [structure] 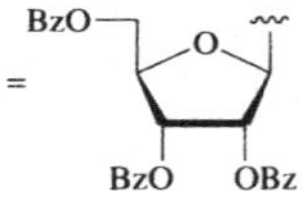	424a
	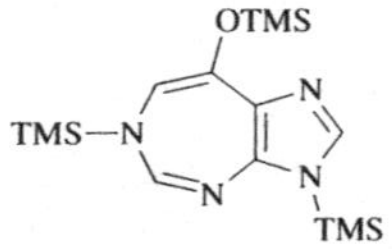 [structure]	$SnCl_4$, $Cl(CH_2)_2Cl$, MeCN, –35°, 45 min	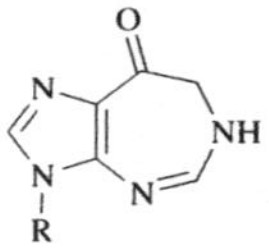 [structure] **I** (51)	557, 818
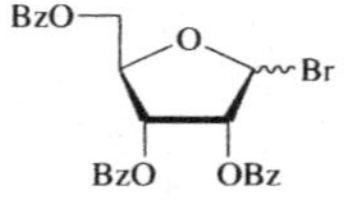 [structure]	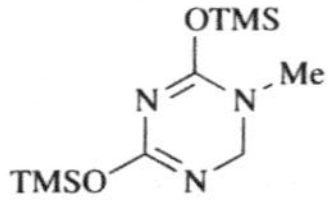 [structure]	$SnCl_4$, $Cl(CH_2)_2Cl$, rt, 48 h	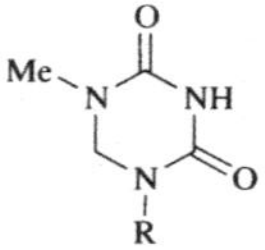 [structure] (21)	239
	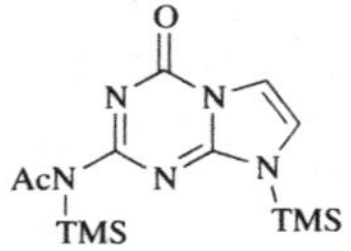 [structure]	$SnCl_4$, MeCN	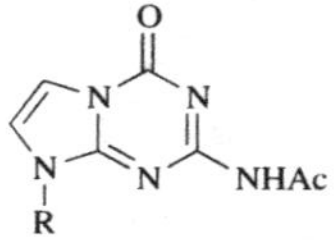 [structure] (—)	838
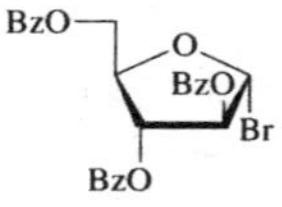 [structure]	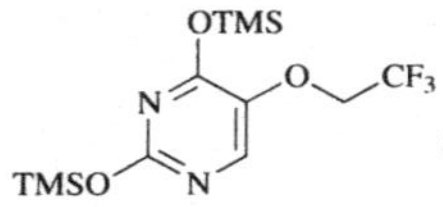 [structure]	$SnCl_4$, CH_2Cl_2, rt, 12 h	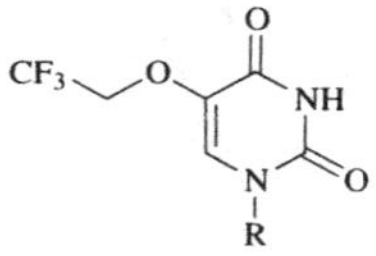 [structure] (57) R = [structure] 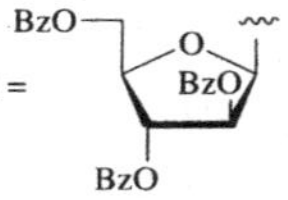	369

	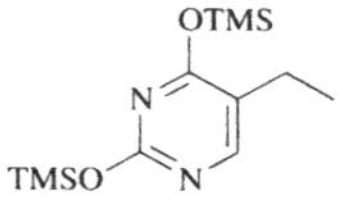 5-ethyl-2,4-bis(TMSO)pyrimidine	$SnCl_4$, $Cl(CH_2)_2Cl$, rt, 24 h	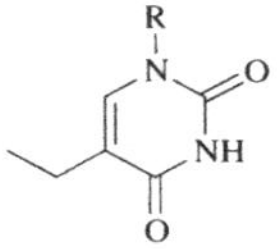 (82) R 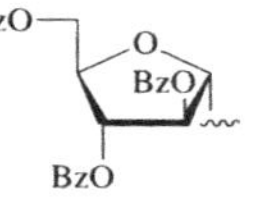 = 2,3,5-tri-O-Bz-ribofuranosyl	358
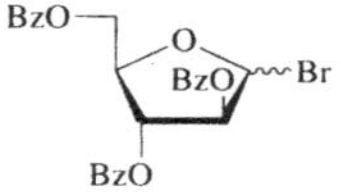 2,3,5-tri-O-Bz-ribofuranosyl bromide	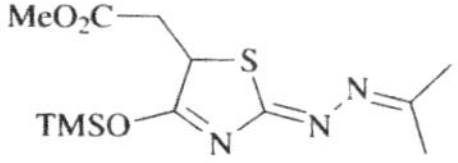 MeO_2C, TMSO, S, N, N, N	1. $SnCl_4$, MeCN, 0°, 10 min 2. 90°, 2 h 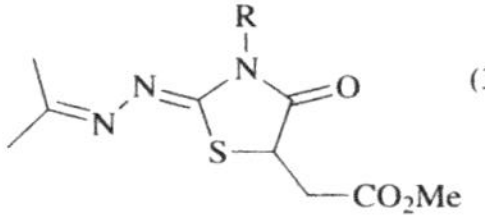	R, N, O, S, CO_2Me (31)	839
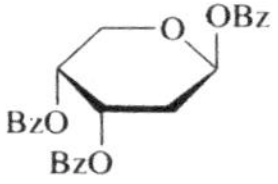 BzO, BzO, OBz	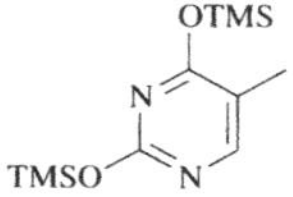 OTMS, TMSO, N, N	$SnCl_4$, $Cl(CH_2)_2Cl$, 20°	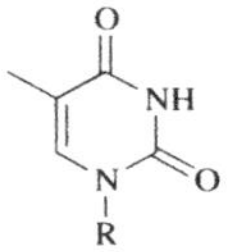 O, NH, O, N–R (36) R = BzO, BzO, O	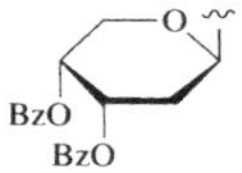 840
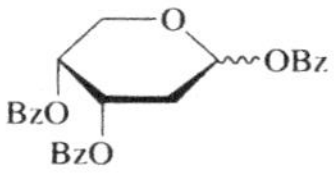 BzO, BzO, OBz	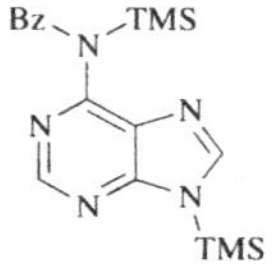 Bz, N, TMS, N, N, N, N, TMS	$SnCl_4$, $Cl(CH_2)_2Cl$, rt, 6 h	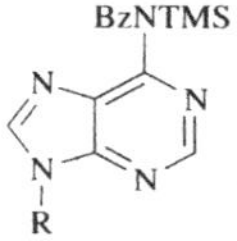 BzNTMS, N, N, N, N–R (28) + α-anomer (24)	841
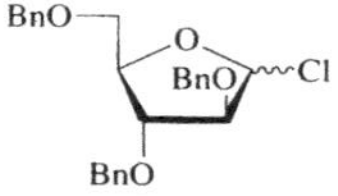 BnO, BnO, BnO, Cl	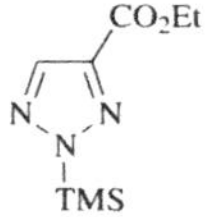 CO_2Et, N, N, N–TMS	$SnCl_4$, $Cl(CH_2)_2Cl$, rt, 12 h	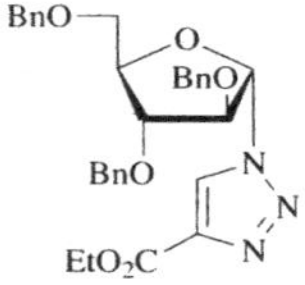 BnO, BnO, BnO, N, N, N, EtO_2C (20) + β-anomer (17) + N^3-α-isomer (15) + N^2-β-isomer (10) + N^2-α-isomer (17)	842
	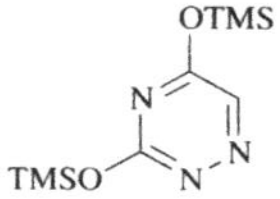 OTMS, TMSO, N, N, N	$SnCl_4$, $Cl(CH_2)_2Cl$, 22°, 16 h 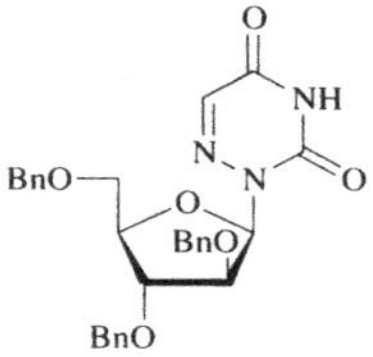	O, NH, O, N, N, BnO, BnO, BnO (42)	84

TABLE III. REACTIONS OF SILYLATED HETEROCYCLIC BASES WITH PROTECTED SUGARS - $SnCl_4$ CATALYST (*Continued*)

Sugar	Base	Conditions	Product(s) and Yield(s) (%)	Refs.
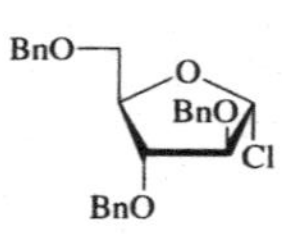 BnO—, O, BnO, Cl, BnO	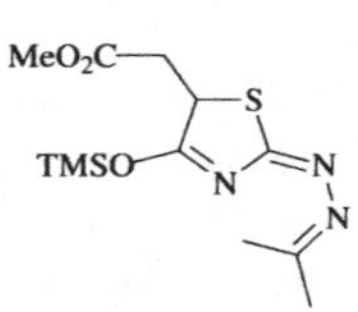 MeO_2C, S, TMSO, N, N, N	$SnCl_4$, rt, 12 h	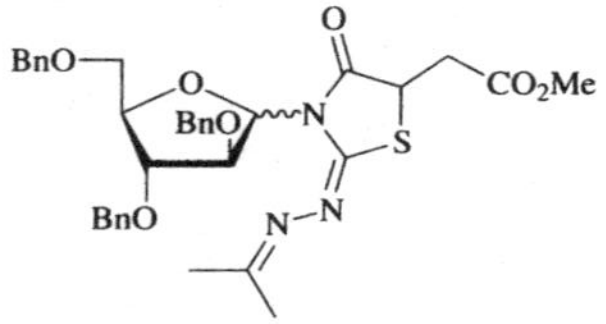 BnO—, O, BnO, N, O, S, CO_2Me, BnO, N—N	839
		C_6H_6	(40) α:β = 4:1	
		$Cl(CH_2)_2Cl$	(45) α:β = 6:1	
BnO—, O, BnO, Cl, BnO	MeO_2C, S, TMSO, N, N, N, R^1	$SnCl_4$ A. CH_2Cl_2, rt, 12 h B. $Cl(CH_2)_2Cl$, 0°, 30 min; rt, 12 h	BnO—, O, BnO, BnO, R^1, N, N, N, O, S, CO_2Me R^1: Me — A. (35) + β-anomer (6); Ph — B. (53) + β-anomer (6)	839
	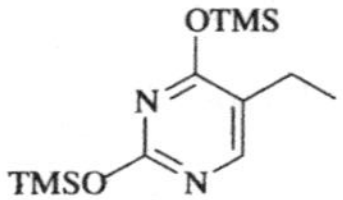 OTMS, N, TMSO, N	$SnCl_4$, $Cl(CH_2)_2Cl$, rt, 16 h	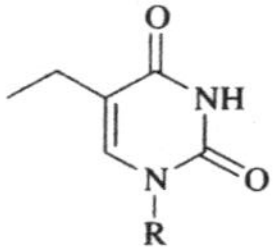 O, NH, O, N, R (85) α,β mixture R 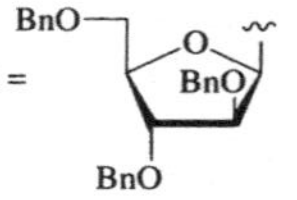 = BnO—, O, BnO, BnO	358
	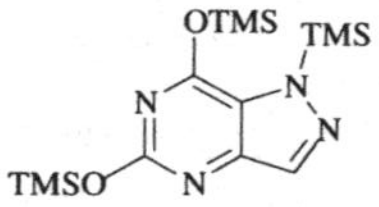 OTMS, TMS, N, N, N, TMSO, N	$SnCl_4$, $Cl(CH_2)_2Cl$, rt, 12 h	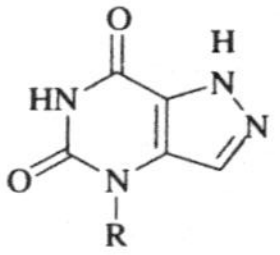 O, H, HN, N, N, O, N, R (29) + N^1-isomer (7)	362

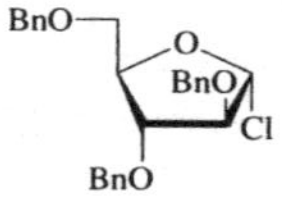

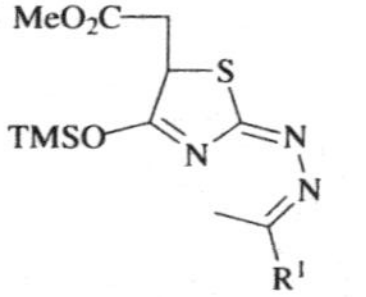

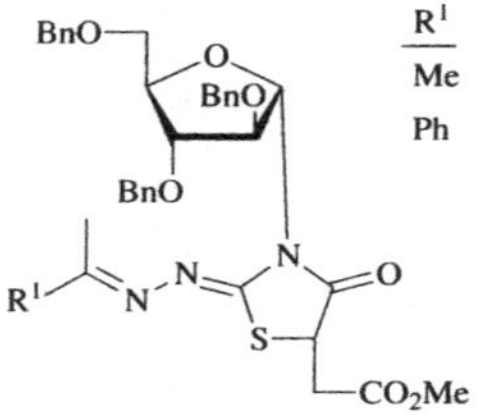

			$SnCl_4$, $Cl(CH_2)_2Cl$, rt, 12 h	(42) + N^5-isomer (15)	362
			$SnCl_4$, $Cl(CH_2)_2Cl$, MeCN, 22°, 16 h	(50)	192g
	R = $CO(CH_2)_2COPh$		$SnCl_4$, MeCN, rt, 24 h	(80)	843
C_{27}	R = p-$O_2NC_6H_4CO$		$SnCl_4$, $Cl(CH_2)_2Cl$, rt, 12 h	(20)	844
			$SnCl_4$, $Hg(CN)_2$, C_6H_6, 60°, 1.5 h	(56)	632

TABLE III. REACTIONS OF SILYLATED HETEROCYCLIC BASES WITH PROTECTED SUGARS - $SnCl_4$ CATALYST (*Continued*)

Sugar	Base	Conditions	Product(s) and Yield(s) (%)	Refs.
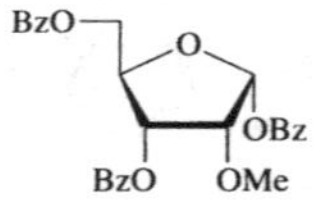 BzO, OBz, BzO, OMe	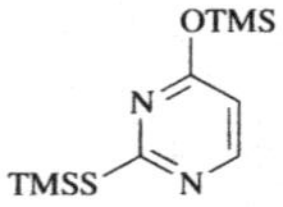 OTMS, N, TMSS, N	$SnCl_4$, MeCN, rt, 3 h	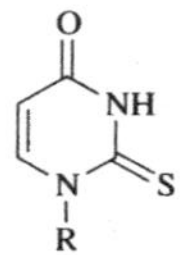 O, NH, N, S, R (39) R = BzO, O, BzO, OMe 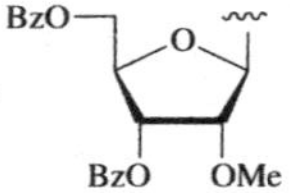	845
	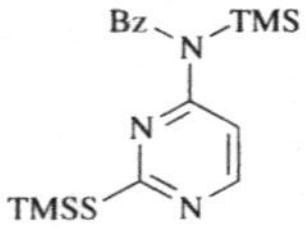 Bz, N, TMS, N, TMSS, N	$SnCl_4$, $Cl(CH_2)_2Cl$, reflux	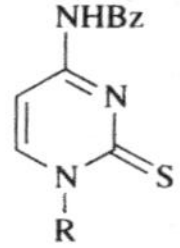 NHBz, N, N, S, R (70)	846, 847
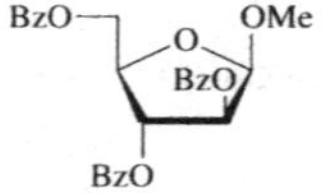 BzO, O, OMe, BzO, BzO	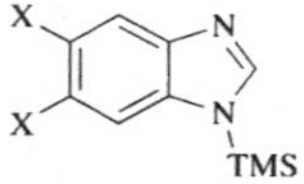 X, X, N, N, TMS	$SnCl_4$, $Cl(CH_2)_2Cl$, reflux, 30 min	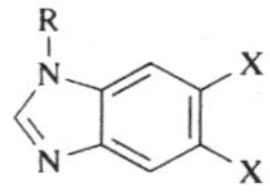 R, N, N, X, X **I** X = H (18) R 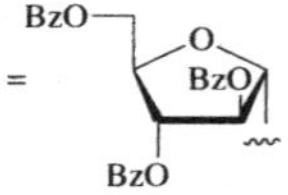 = BzO, O, BzO, BzO	848
		$SnCl_4$, $Cl(CH_2)_2Cl$, rt, 26 h, reflux, 15 min	**I** X = Cl (24)	848
	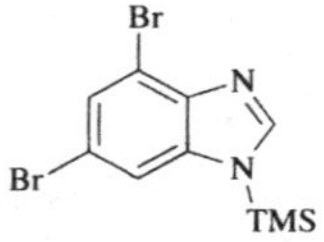 Br, Br, N, N, TMS	$SnCl_4$, $Cl(CH_2)_2Cl$, reflux, 6 h, rt, 12 h	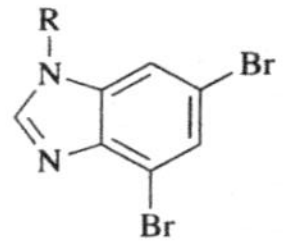 R, N, N, Br, Br (35) + isomer	849
C_{28} 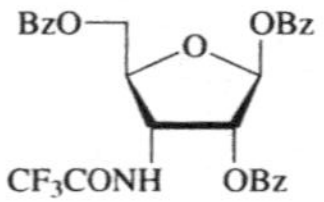 BzO, O, OBz, CF_3CONH, OBz	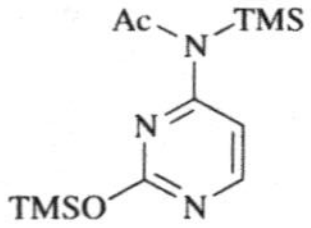 Ac, N, TMS, N, TMSO, N	$SnCl_4$, $Cl(CH_2)_2Cl$, 25°, 16 h	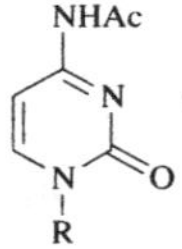 NHAc, N, N, O, R (69) R = BzO, O, CF_3CONH, OBz 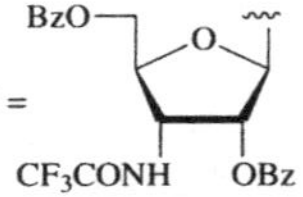	766

Sugar	Heterocycle	Conditions	Product(s) and Yield(s) (%)		Refs.
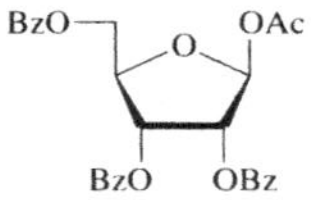	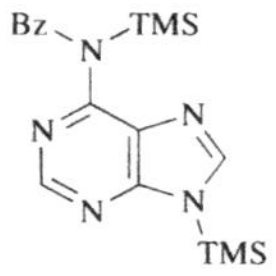 Bz–N(TMS) / N / N / N / N–TMS	$SnCl_4$, $Cl(CH_2)_2Cl$, 60-70°, 5 h	NHBz / N / N / N / N–R	(61)	100
	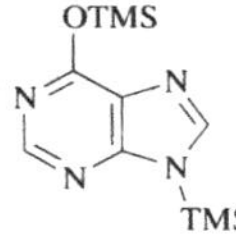 OTMS / N / N / N / N–TMS	$SnCl_4$, $Cl(CH_2)_2Cl$, 60-70°, 5 h	O / NH / N / N / N–R	(67)	100
BzO / O / OAc / BzO / OBz	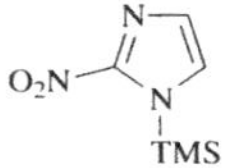 O_2N / N / N–TMS	$SnCl_4$, MeCN, 60°, 1 h	BzO / O / BzO / BzO / N / NO_2 / N	(38)	213
	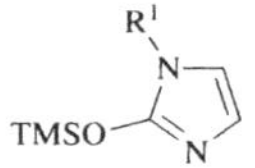 R^1 / N / TMSO / N	$SnCl_4$, $Cl(CH_2)_2Cl$	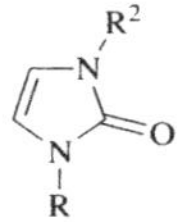 R^2 / N / O / N–R 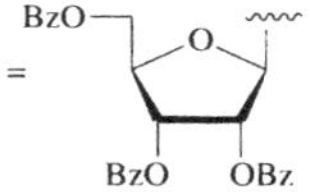	R = BzO / O / BzO / OBz	
	R^1		R^2		
	TMS	0°, 5 h	H	(19)	850
	Ac	20°, 6 h	Ac	(51)	777
	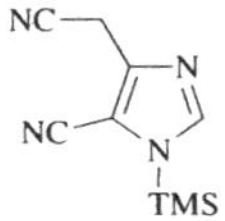 NC / N / NC / N–TMS	$SnCl_4$, $Cl(CH_2)_2Cl$, rt, 5.5 h	CN / N / CN / N–R	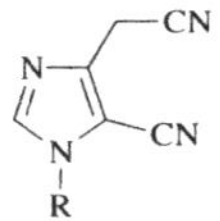 (72) + N^3-isomer (18)	851
	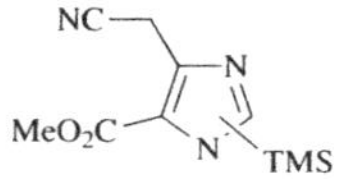 NC / N / MeO_2C / N / TMS	$SnCl_4$ (0.72 eq), $Cl(CH_2)_2Cl$, rt, 9 h	CN / N / CO_2Me / N–R	(35)	704

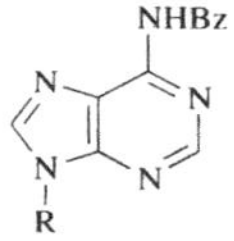

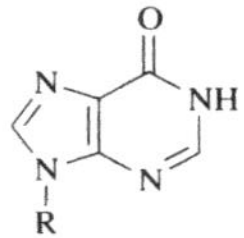

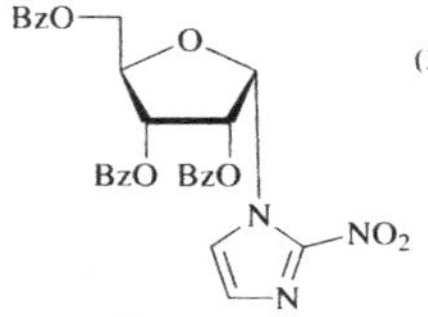

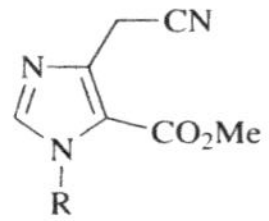

TABLE III. REACTIONS OF SILYLATED HETEROCYCLIC BASES WITH PROTECTED SUGARS - $SnCl_4$ CATALYST (*Continued*)

Sugar	Base	Conditions	Product(s) and Yield(s) (%)	Refs.
		$SnCl_4$, $Cl(CH_2)_2Cl$, rt, 24 h	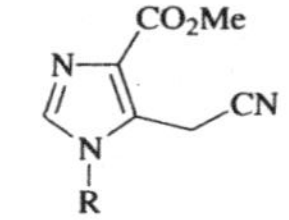 [imidazole: N-R, CO_2Me, CH_2CN] **I** (100) R 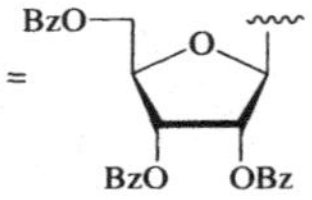 = [2,3,5-tri-O-benzoyl-ribofuranosyl: BzO, BzO, OBz]	704
		$SnCl_4$, $Cl(CH_2)_2Cl$, rt, 6 h	**I** (50) + N^3-β-isomer (10)	704
	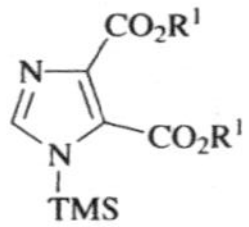 [imidazole: N-TMS, CO_2R^1, CO_2R^1]	$SnCl_4$, $Cl(CH_2)_2Cl$, rt, 48 h	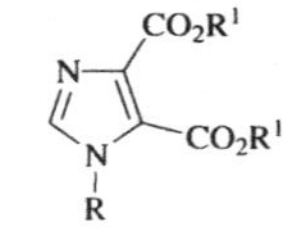 [imidazole: N-R, CO_2R^1, CO_2R^1] **I**, R^1 = Me, (82)	852
		$SnCl_4$, $Cl(CH_2)_2Cl$, rt, 2 d	**I**, R^1 = Me, (78)	853
		$SnCl_4$, $Cl(CH_2)_2Cl$, rt,15 min	**I**, R^1 = Me, (90)	543
		$SnCl_4$, $Cl(CH_2)_2Cl$, rt, 48 h	**I**, R^1 = Et, (84)	852
		$SnCl_4$, $Cl(CH_2)_2Cl$, rt, 2 d	**I**, R^1 = Et, (—)	853
	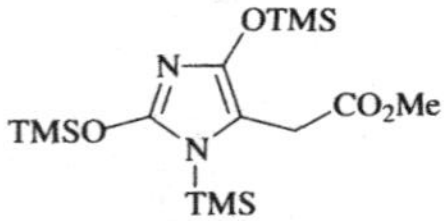 [imidazole: TMSO, OTMS, CH_2CO_2Me, N-TMS]	$SnCl_4$, $Cl(CH_2)_2Cl$, rt, 2 d	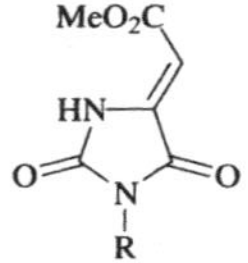 [hydantoin: MeO_2CCH=, HN, O, O, N-R] (80)	720
	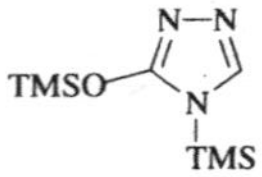 [1,2,4-triazole: TMSO, N–N, N-TMS]	$SnCl_4$, $Cl(CH_2)_2Cl$, MeCN, rt, 5 h	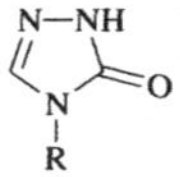 [triazolone: N–NH, =O, N-R] (21) + α-anomer (1) + N^2,N^4-bis(riboside) (29)	850

Substrate	Conditions	Product(s) and Yield(s) (%)		Refs.
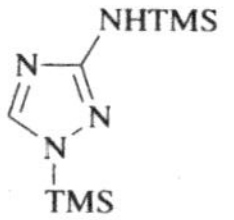 NHTMS, N, N, N, TMS	1. $SnCl_4$, $Cl(CH_2)_2Cl$, rt, 3.5 h 2. NH_3, MeOH, rt, 4 d	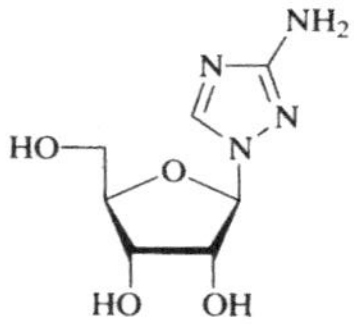 NH_2, N, N, N, HO, O, HO, OH	(—) + N^2-isomer (—) + N^4-isomer (—)	854
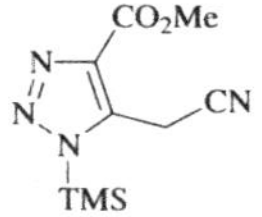 CO_2Me, N, N, N, CN, TMS	$SnCl_4$ (0.72 eq), $Cl(CH_2)_2Cl$, rt, 25 h	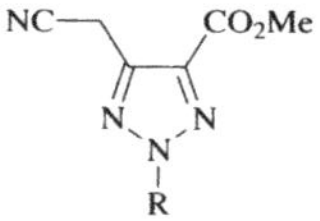 NC, CO_2Me, N, N, N, R	(94) R = BzO, O, BzO, OBz 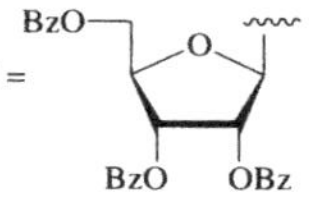	855
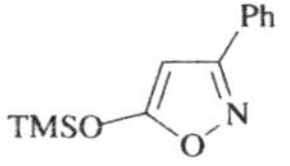 Ph, TMSO, N, O	$SnCl_4$	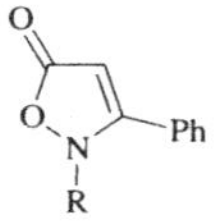 O, O, N, Ph, R	(—)	856
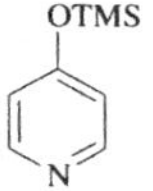 OTMS, N	$SnCl_4$, $Cl(CH_2)_2Cl$, reflux, 1 h	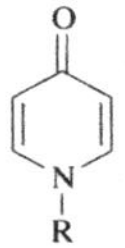 O, N, R	(63) + *O*-riboside (2)	99
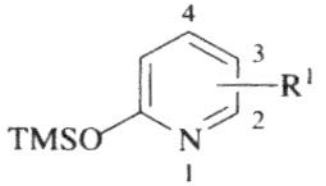 4, 3, R^1, 2, TMSO, N, 1	$SnCl_4$, $Cl(CH_2)_2Cl$	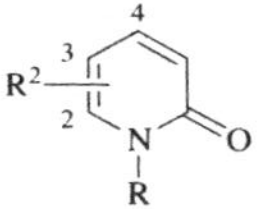 4, R^2, 3, 2, N, O, R		
R^1		R^2		
H	rt, 12 h	H	(85)	99, 388
4-NHTMS	rt, 8 h	4-NH_2	(97)	856
3-NO_2	reflux, 8 h	3-NO_2	(72)	714

TABLE III. REACTIONS OF SILYLATED HETEROCYCLIC BASES WITH PROTECTED SUGARS - $SnCl_4$ CATALYST (*Continued*)

Sugar	Base	Conditions	Product(s) and Yield(s) (%)	Refs.
	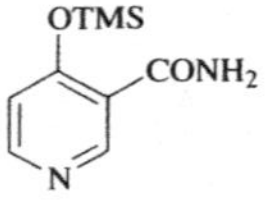 OTMS, CONH2 (pyridine)	1. $SnCl_4$, $Cl(CH_2)_2Cl$, reflux, 1.5 h 2. NH_3, MeOH, rt, 16 h	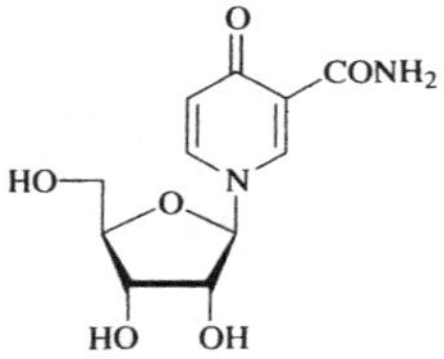 O, CONH2, HO, N, O, HO, OH (69)	858
	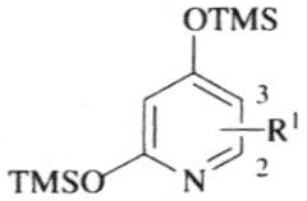 OTMS, TMSO, N, 3, 2, R[1]	$SnCl_4$, $Cl(CH_2)_2Cl$, rt, 18 h	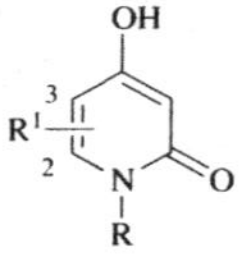 OH, R[1], 3, 2, N, O, R; R = BzO, O, BzO, OBz 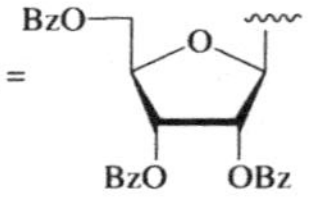	859
	R[1]			
	3-Me		(40)	
	2-Me		(37)	
	3-Et		(55)	
	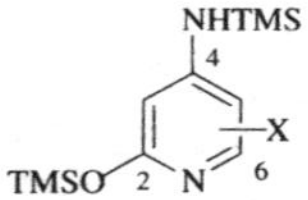 NHTMS, TMSO, 2, 4, N, 6, X	$SnCl_4$, $Cl(CH_2)_2Cl$	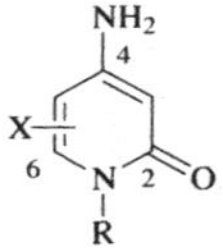 NH2, 4, X, 6, N, 2, O, R	
	X			
	5-F	0°, 30 min; rt, 24 h	(65)	860
	5-F	0°, 15 min; rt, 12 h	(61)	861
	3-F	0°, 30 min; rt, 24 h	(98)	860
	3-Br	rt, 24 h	(56)	861
	5-Me	rt, 18 h	(40)	859
	5-Et	rt, 18 h	(55)	859
	3-F,5-F	0°, 30 min; rt, 24 h	(90)	860

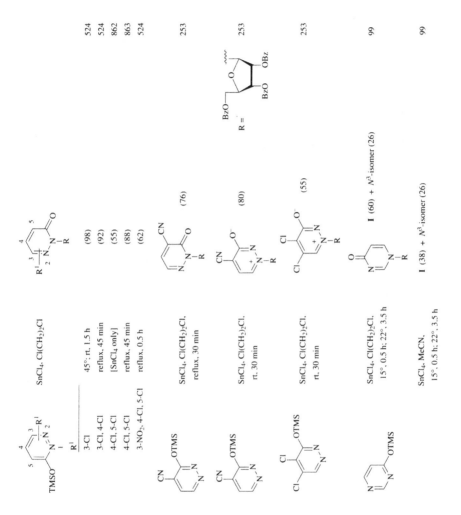

Reactant	Conditions	Product(s) and Yield(s) (%)	Refs.
TMSO-pyridazine, R^1	$SnCl_4$, $Cl(CH_2)_2Cl$	pyridazin-3-one, N-R	
R^1			
3-Cl	45°; rt, 1.5 h	(98)	524
3-Cl, 4-Cl	reflux, 45 min	(92)	524
4-Cl, 5-Cl	[$SnCl_4$ only]	(55)	862
4-Cl, 5-Cl	reflux, 45 min	(88)	863
3-NO_2, 4-Cl, 5-Cl	reflux, 0.5 h	(62)	524
CN, OTMS pyridazine	$SnCl_4$, $Cl(CH_2)_2Cl$, reflux, 30 min	CN, O, N-R (76)	253
CN, OTMS pyridazine	$SnCl_4$, $Cl(CH_2)_2Cl$, rt, 30 min	CN, O^-, N^+-R (80) R = BzO, O, BzO, OBz	253
Cl, Cl, OTMS pyridazine	$SnCl_4$, $Cl(CH_2)_2Cl$, rt, 30 min	Cl, Cl, O^-, N^+-R (55)	253
N, N, OTMS pyrimidine	$SnCl_4$, $Cl(CH_2)_2Cl$, 15°, 0.5 h; 22°, 3.5 h	O, N, N-R **I** (60) + N^3-isomer (26)	99
	$SnCl_4$, MeCN, 15°, 0.5 h; 22°, 3.5 h	**I** (38) + N^3-isomer (26)	99

TABLE III. REACTIONS OF SILYLATED HETEROCYCLIC BASES WITH PROTECTED SUGARS - $SnCl_4$ CATALYST (*Continued*)

Sugar	Base	Conditions	Product(s) and Yield(s) (%)	Refs.
	TMSS, N, N	$SnCl_4$, MeCN, 22°, 5 min	N, N–R, S (97)	99
	5 N, 4, 3, R[1], 2, TMSO, N 1	$SnCl_4$	4, N 5, R[1], 3, 2, N–R, O	
	R[1]			
	H	MeCN, 22°	(73)	99
	5-F	$Cl(CH_2)_2Cl$, rt, 18 h	(26)	388
	4-Me	MeCN, rt, 12 h	(62)	822
	OTMS, 13C, N, TMSO, N	$SnCl_4$, MeCN, rt, 16 h	O, 13C, NH, N–R, O (90)	864
	OTMS, N, TMSO, N	$SnCl_4$, $Cl(CH_2)_2Cl$, 23°, 12 h	O, NH, N–R, O **I** (84) + N^3-isomer (9)	90
		$SnCl_4$, MeCN, 23°, 12 h	**I** (89) + N^3-isomer (4)	90

Reactant	Conditions	Product(s) and Yield(s) (%)		Refs.
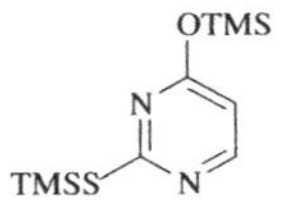 2-TMSS-4-OTMS-pyrimidine	$SnCl_4$, MeCN	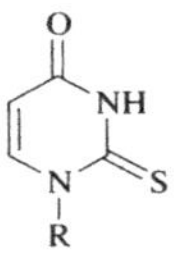 **I** (85)		865
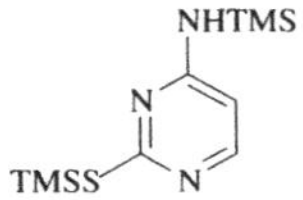 2-TMSS-4-NHTMS-pyrimidine	$SnCl_4$, $Cl(CH_2)_2Cl$, 22°, 2 h	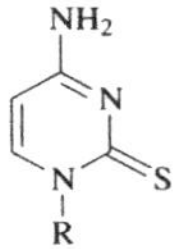 (95)		84
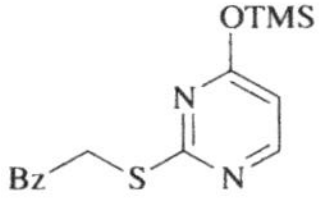 2-(BzCH$_2$S)-4-OTMS-pyrimidine	$SnCl_4$, $Cl(CH_2)_2Cl$, rt, 30 min	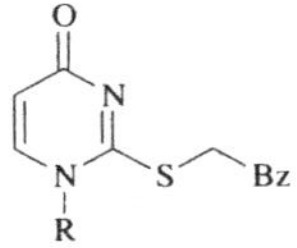 (45)		845
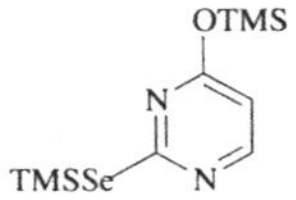 2-TMSSe-4-OTMS-pyrimidine	1. $SnCl_4$, $Cl(CH_2)_2Cl$, 2. NaOMe, MeOH, rt	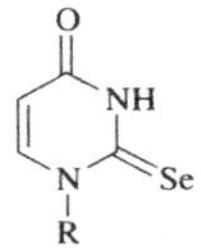 (30)	R 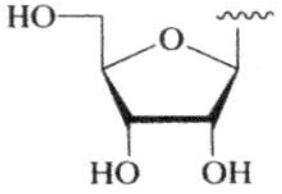 =	866
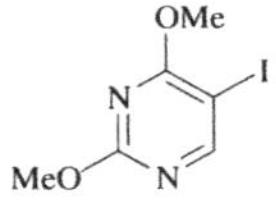 2,4-(MeO)$_2$-5-I-pyrimidine	$SnCl_4$, $Cl(CH_2)_2Cl$, 22°, 2 h	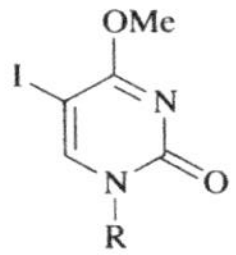 (67)		84
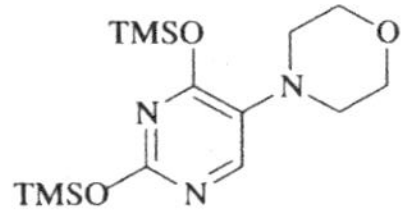 2,4-(TMSO)$_2$-5-morpholino-pyrimidine	$SnCl_4$, MeCN 23°, 1 h	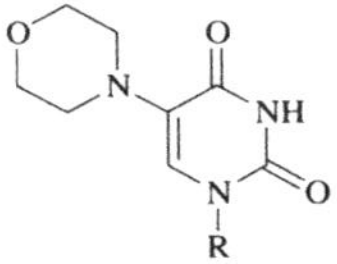 **I** (53) + N^3-isomer (32) + N^1,N^3-bis(riboside) (12)		90, 193
	$SnCl_4$, $Cl(CH_2)_2Cl$, 23°, 2 h	**I** (39) + N^3-isomer (18) + N^1,N^3-bis(riboside) (42)		90, 193

TABLE III. REACTIONS OF SILYLATED HETEROCYCLIC BASES WITH PROTECTED SUGARS - $SnCl_4$ CATALYST (*Continued*)

Sugar	Base	Conditions	Product(s) and Yield(s) (%)	Refs.
	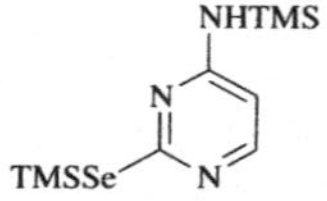 2-TMSSe-4-NHTMS-pyrimidine	1. $SnCl_4$, $Cl(CH_2)_2Cl$, rt, 3 h 2. NH_3, MeOH, rt	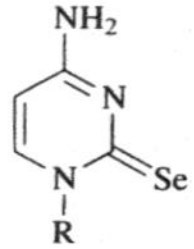 4-NH_2-1-R-pyrimidine-2-selone (31)	811
	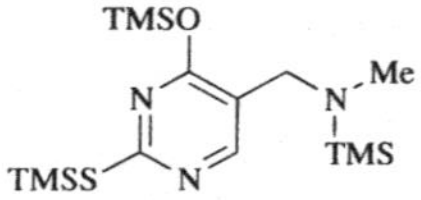 2-TMSS-4-TMSO-5-[$CH_2N(Me)TMS$]-pyrimidine	1. $SnCl_4$, $Cl(CH_2)_2Cl$, 22°, 1 h 2. NH_3, MeOH, 24°, 1 h	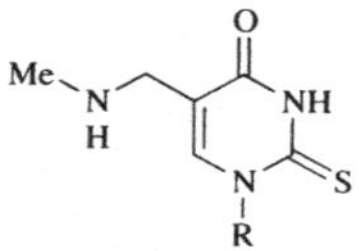 5-(MeNHCH$_2$)-1-R-2-thiouracil (20)	126
	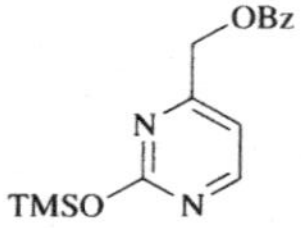 2-TMSO-4-(CH_2OBz)-pyrimidine	$SnCl_4$, $Cl(CH_2)_2Cl$, rt, 2 d	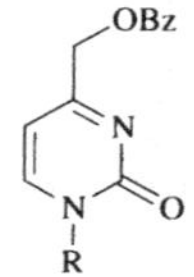 4-(CH_2OBz)-1-R-pyrimidin-2-one (18) R = 2,3,5-tri-O-benzoyl-ribofuranosyl (BzO, BzO, OBz) 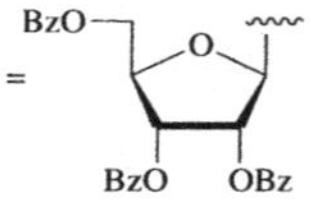	867
	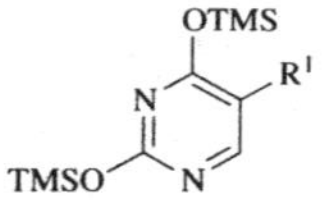 2,4-bis(TMSO)-5-R^1-pyrimidine	$SnCl_4$	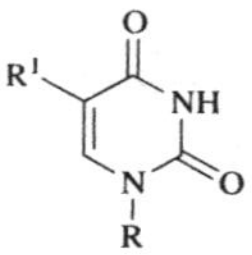 5-R^1-1-R-uracil	
	R^1			
	F	$MeNO_2$, rt, 3 h	(90)	514b
	F	C_6H_6, rt, 3 h	(34)	514b
	F	$Cl(CH_2)_2Cl$, rt, 3 h	(48)	514b
	F	$Cl(CH_2)_2Cl$, rt, 13 h	(99)	868
	F	MeCN, <10°, 1 h	(98)	514b
	F	CH_2Cl_2, rt, 4 h	(86)	869
	Me	—	(—)	870
	Et	$Cl(CH_2)_2Cl$, 22°, 20 h	(95)	84
	Bu-*n*	$Cl(CH_2)_2Cl$, 22°, 5 h	(95)	84

Reactant	Conditions	Product(s) and Yield(s) (%)	Refs.
Bu-*n*	MeCN, 22°, 5 h	(84)	84
$CH{=}CH_2$	$Cl(CH_2)_2Cl$, rt, 16 h	(58)	808
$C{\equiv}CH$	$Cl(CH_2)_2Cl$, rt, 4 h	(75)	872
c-C_3H_5	$Cl(CH_2)_2Cl$, rt, 2 h	(53)	873
NO_2	$Cl(CH_2)_2Cl$, MeCN, 22°, 0.5 h	(98)	84
NO_2	$Cl(CH_2)_2Cl$, 23°, 0.1 h	(97)	90, 193
NO_2	MeCN, 23°, 0.1 h	(98)	90
OMe	MeCN, rt, 16 h	(62)	874
OMe	$Cl(CH_2)_2Cl$, 23°, 2.5 h	(53) + N^3-isomer (27) + N^1,N^3-bis(riboside) (13)	90, 193
OMe	MeCN, 24°, 12 h	(90) + N^3-isomer (3)	193, 90
OCH_2CF_3	MeCN, 25°, 12 h	(54)	369
OCH_2CO_2Et	MeCN, rt, 16 h	(67)	874

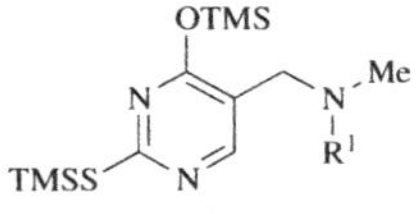

2-TMSS-4-OTMS-5-[$CH_2N(Me)R^1$]pyrimidine; $SnCl_4$ → 5-[$CH_2N(Me)R^2$]-2-thiouracil, N1-R; R = 2,3,5-tri-O-benzoyl-ribofuranosyl (BzO, BzO, OBz) — 126

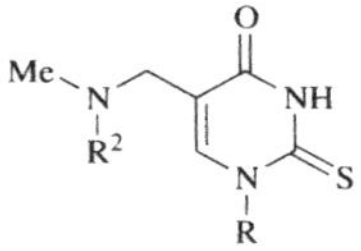

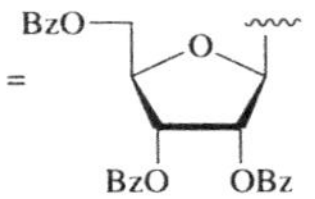

R^1	Conditions	Product(s) and Yield(s) (%)	R^2	Refs.
Ac	$Cl(CH_2)_2Cl$, 24°, 3 h	(20) + N^3-isomer (19)	Ac	126
$COCF_3$	MeCN	(70-80)	$COCF_3$	875
BOC	$Cl(CH_2)_2Cl$, 24°, 1.5 h	(—)	H	126
$CO_2CH_2CCl_3$	MeCN	(70-80)	$CO_2CH_2CCl_3$	875

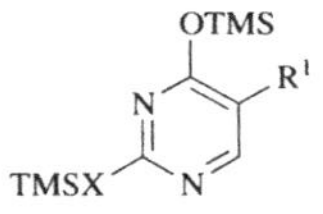

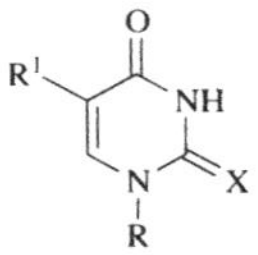

2-TMSX-4-OTMS-5-R^1-pyrimidine → 5-R^1-1-R-uracil (2-X)

X	R^1	Conditions	Product(s) and Yield(s) (%)	Refs.
O	SO_2Me	$SnCl_4$	(—)	876
S	OMe	$SnCl_4$, MeCN	(62) + N^3-anomer (15)	865

TABLE III. REACTIONS OF SILYLATED HETEROCYCLIC BASES WITH PROTECTED SUGARS - $SnCl_4$ CATALYST (*Continued*)

Sugar	Base	Conditions	Product(s) and Yield(s) (%)	Refs.
	NHTMS, R^1, TMSO	$SnCl_4$, $Cl(CH_2)_2Cl$,	NH_2, R^1, N–R	
	R^1 = F	MeCN, rt, 3 h	(45)	877
	R^1 = C≡CH	rt, 2 h	(90)	872
	NHTMS, Pr-*i*, TMSO	1. $SnCl_4$ or $SnCl_4$/$SnBr_4$ 2. NH_3, MeOH	NH_2, *i*-Pr, N–R (72) R = HO, HO, OH	878
	Ac, N, TMS, F, TMSO	1. $SnCl_4$, MeCN, <10°, 15 min; rt, 2 h 2. NaOMe, MeOH	NH_2, F, HO, HO, OH (67)	514b
	OTMS, N(Me)Ac, TMSS	$SnCl_4$, $Cl(CH_2)_2Cl$	Me–N(Ac), O, NH, S, N–R (70) R = BzO, BzO, OBz	125, 126

Substrate	Conditions	Product(s) and Yield(s) (%)	Refs.
		(~70)	125, 126
X = O	$SnCl_4$, MeCN, 24°	(52) + N^1-isomer (41) + N^1,N^3-bis(riboside) (3)	193,89a
X = O	$SnCl_4$, $Cl(CH_2)_2Cl$, 24°	(68) + N^1-isomer (13)	193,89a
X = S	$SnCl_4$, MeCN, 22°, 3 h	(39) + *S*-riboside (20)	89a
X = S	$SnCl_4$, $Cl(CH_2)_2Cl$, 22°, 4 h	(26)	89a
	$SnCl_4$, $Cl(CH_2)_2Cl$, rt, 12 h	(95)	879
	1. $SnCl_4$, MeCN, rt, 6 h 2. NaOMe, MeOH, rt, 3 h	(70)	544
	$SnCl_4$, $Cl(CH_2)_2Cl$, rt, 72 h	(67)	720

TABLE III. REACTIONS OF SILYLATED HETEROCYCLIC BASES WITH PROTECTED SUGARS - $SnCl_4$ CATALYST (*Continued*)

Sugar	Base	Conditions	Product(s) and Yield(s) (%)	Refs.
	OTMS, N, O, N, Cl, Bn	$SnCl_4$, MeCN, 10°, 1 h; rt, 40 h	Cl, Bn, N, O, N, O, R (90) R = BzO, O, BzO, OBz	255
	OTMS, N, TMSX, N	$SnCl_4$	HN, X, N, O, R	89a
	X			
	O	$Cl(CH_2)_2Cl$, 22°, 6 h	(60) + N^1-isomer (10)	
	O	MeCN, 20°, 2 h	(17) + N^1-isomer (66)	
	S	$Cl(CH_2)_2Cl$, 22°, 5 h	(8)	
	S	MeCN, 20°, 2 h	(20) + *S*-riboside (38)	
	OTMS, R^1, N, TMSO, N	$SnCl_4$	O, R^1, NH, N, O, R	
	R^1			
	Me	MeCN, 22-24°, 12 h	(66) + N^3-isomer (17)	89a, 193
	i-Pr	MeCN, 23°, 24 h	(53) + N^3-isomer (25)	90
	i-Pr	$Cl(CH_2)_2Cl$, rt, 72 h	(18) + N^3-isomer (41)	90
	NO_2	MeCN, 23°, 0.6 h	(73) + bis(riboside) (19)	90
	NO_2	MeCN, 2 h	(73) + N^3-isomer (19)	193
	NO_2	$Cl(CH_2)_2Cl$, 23°, 0.9-1 h	(84) + N^3-isomer (14)	193, 90

Reactant	Conditions	Product(s) and Yield(s) (%)	Refs.
TMSO, OTMS, OTMS, Me-substituted pyrimidine	$SnCl_4$	HO, NH, N–R uracil (40-43)	880
TMSO, OTMS, OTMS, R^1 pyrimidine	1. $SnCl_4$ 2. NaOMe, MeOH, rt, time	R^1, NH, N–R barbiturate; R = ribofuranosyl (HO, HO, OH)	544
R^1 = Me	MeCN, rt, 24 h	time 3 h (88)	
R^1 = Et	$Cl(CH_2)_2Cl$, rt, 20 h	time 2 h (93)	
R^1 = $CH_2CH{=}CH_2$	MeCN, rt, 20 h	time 2 h (76)	
R^1 = Bn	MeCN, rt, 27 h	time 2 h (70)	
TMSO, OTMS, OTMS, R^1 pyrimidine	$SnCl_4$, MeCN, rt, time	R^1, NH, N–R barbiturate; R = ribofuranosyl (BzO, BzO, OBz)	364
R^1 = Me	time 90 h	(26)	
R^1 = n-Pr	time 40 h	(29)	
TMSO, TMSO, $SC_6H_4R^1$-p, CO_2Et pyrimidine	$SnCl_4$, MeCN, 60°, 3 h	p-$R^1C_6H_4S$, CO_2Et, NH, N–R uracil	809
R^1 = F		(17) + N^1,N^3-bis(riboside) (18)	
R^1 = Br		(28) + N^1,N^3-bis(riboside) (17)	
R^1 = OBu-n		(15) + N^1,N^3-bis(riboside) (22)	
R^1 = Bu-t		(20) + N^1,N^3-bis(riboside) (27)	
R^1 = SMe		(22) + N^1,N^3-bis(riboside) (21)	

TABLE III. REACTIONS OF SILYLATED HETEROCYCLIC BASES WITH PROTECTED SUGARS - $SnCl_4$ CATALYST (*Continued*)

Sugar	Base	Conditions	Product(s) and Yield(s) (%)	Refs.
	OTMS, TMSO, N, N, S, S, Bu-*t*	$SnCl_4$, MeCN, rt, 1 h	O, HN, O, N, R, S, S, Bu-*t* (75-93) R = BzO, O, BzO, OBz	881
	N, NHTMS, N, OTMS	$SnCl_4$, $Cl(CH_2)_2Cl$, 25°, 3 h	N, NH_2, N, O, R; R^1 NHTMS (87)	882
			H (60)	883
			Me (55)	883
			$C_{10}H_{21}$ (62)	883
	OTMS, O, O, N	$SnCl_4$, $Cl(CH_2)_2Cl$, rt, 2.5 h	O, O, N, O, R (90)	722
	OTMS, N, TMSO, N, O	$SnCl_4$, $Cl(CH_2)_2Cl$, rt, 22 h	O, NH, O, N, O, R (68)	726
	NMe_2, N, TMSO, N, O	$SnCl_4$, $Cl(CH_2)_2Cl$, rt, 18 h	NMe_2, N, O, N, O, R (61)	726
	OTMS, N, N, N	$SnCl_4$, MeCN, 22°, 24 h	O, N, N, N, R (67)	99

Reactant	Conditions	Product(s) and Yield(s) (%)	Refs.
OTMS, TMSO (2,4-bis(trimethylsilyloxy)-1,2,4-triazine)	$SnCl_4$, MeCN, 23°, 12 h	O, NH, O, N–R; **I** (93)	90
	$SnCl_4$, $Cl(CH_2)_2Cl$, 23°, 4 h	**I** (97)	90, 84
OTMS, TMSO	1. $SnCl_4$, $Cl(CH_2)_2Cl$, 22°, 4 h 2. Pyrrolidine, rt, 3 h	NC_4H_8, O, N–R (57)	79
OTMS, R^1, TMSO	$SnCl_4$, $Cl(CH_2)_2Cl$	R^1, O, NH, O, N–R	884
R^1 = $NHCH_2CO_2H$		(26) + bis(riboside) (9)	
R^1 = $NHCH_2CO_2Me$		(58) + R^1 = $NHCH_2CO_2H$ (5)	
OTMS, R^1, TMSO	$SnCl_4$, $Cl(CH_2)_2Cl$	R^1, O, NH, O, N–R	
R^1 = Br	rt, 6 h	(87)	885
R^1 = c-C_3H_5	rt, 27 h	(30)	873
R^1 = *N*-morpholinyl	rt, 6 h	(71)	885
R^1 = 3-furyl	rt, 2 h	(88)	546
R^1 = SMe	rt, 8 h	(70)	885
R^1 = SBn	25°, 12 h	(82)	885

TABLE III. REACTIONS OF SILYLATED HETEROCYCLIC BASES WITH PROTECTED SUGARS - $SnCl_4$ CATALYST (*Continued*)

Sugar	Base	Conditions	Product(s) and Yield(s) (%)	Refs.
	STMS, R[1], N, N, N, TMSO	$SnCl_4$, $Cl(CH_2)_2Cl$	S, R[1], NH, N, N, O, R	
	R[1]			
	SMe	rt, 6 h	(80)	886
	SBn	rt, 8 h	(84)	886
	OTMS, N, N, Me, TMSO, N	$SnCl_4$, C_6H_6, 25°, 18 h	O, Me, N, NH, N, O, R (18)	239
	O, N, N, Me, TMSO, N, Ac	$SnCl_4$, MeCN, rt, 94 h	Ac, N, N, Me, O, N, O, R (41)	239
	R[1], X, N, R[2], OTMS	$SnCl_4$, $Cl(CH_2)_2Cl$ reflux, 40 min; rt, 30 min	R[2], R[1], X, N, O, R	887
	R[1] R[2] X			
	H H O		(74)	
	Me H O		(63)	
	H Me S		(69)	

Reactant	Conditions	Product(s) and Yield(s) (%)	Refs.
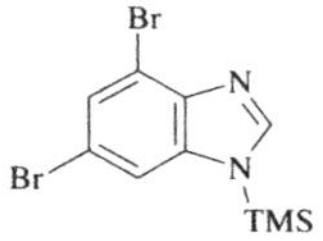 [structure]	$SnCl_4$, $Cl(CH_2)_2Cl$, rt, 24 h	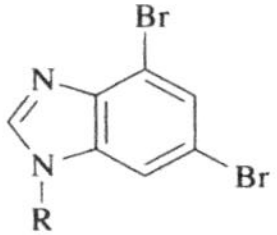 [structure] (25) + N^3-isomer (11)	849
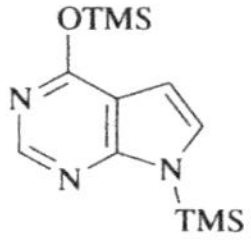 [structure]	1. $SnCl_4$, $Cl(CH_2)_2Cl$, rt, 18 h 2. NH_3, MeOH, rt, 24 h 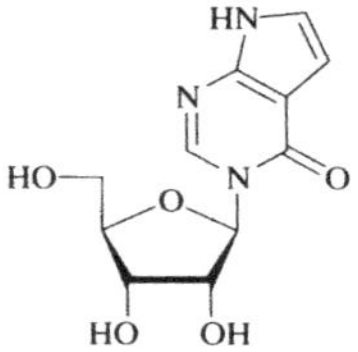	[structure] (25)	888
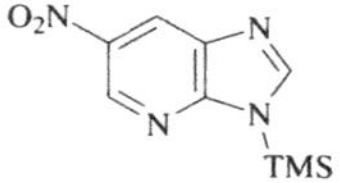 [structure]	$SnCl_4$, MeCN, rt, 12 h	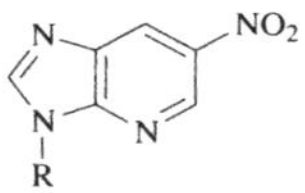 [structure] (35) + α-anomer (11) R 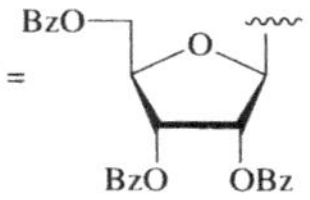 = [structure: 2,3,5-tri-O-benzoylribofuranosyl]	889
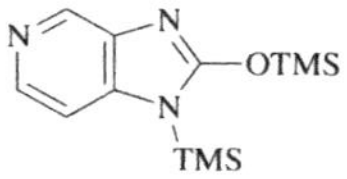 [structure]	$SnCl_4$, $Cl(CH_2)_2Cl$/MeCN (1:1), 0°, 7 h	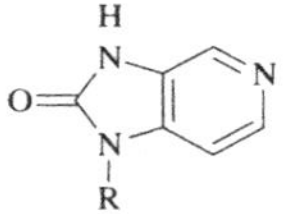 [structure] (23) + N^3-isomer (19)	890
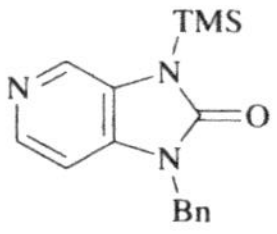 [structure]	$SnCl_4$, $Cl(CH_2)_2Cl$/MeCN (1:1), 0°, 7 h 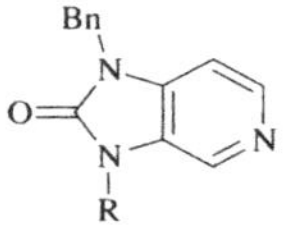	[structure] (31)	890
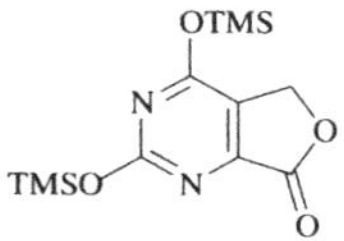 [structure]	$SnCl_4$, $Cl(CH_2)_2Cl$, rt, 5 h	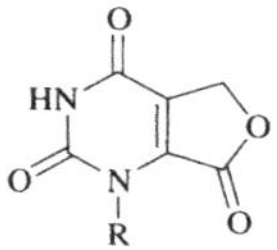 **I** (18) + bis(riboside) (81)	891
	$SnCl_4$, MeCN, 0-5°, 0.5 h	**I** (81)	891

TABLE III. REACTIONS OF SILYLATED HETEROCYCLIC BASES WITH PROTECTED SUGARS - $SnCl_4$ CATALYST (*Continued*)

Sugar	Base	Conditions	Product(s) and Yield(s) (%)	Refs.
	OTMS, N, N, S	1. $SnCl_4$, $Cl(CH_2)_2Cl$, rt, 18 h 2. NH_3, MeOH, rt, 24 h	S, N, N, O, R (25) R = HO, O, HO, OH	888
	OTMS, R[1], N, TMSO, N, S	1. $SnCl_4$, $Cl(CH_2)_2Cl$, rt, 18-20 h 2. NH_3, MeOH, rt, 20 h	O, R[1], HN, O, N, S, R R[1]: H (30); Me (51)	525
	OTMS, N, S, TMSO, N	1. $SnCl_4$, $Cl(CH_2)_2Cl$, rt, 18 h 2. NH_3, MeOH, rt, 3 d	O, HN, S, O, N, R (67)	892
	OTMS, N, S, TMSO, N, R[1]	1. $SnCl_4$, $Cl(CH_2)_2Cl$, −30°, 8 h 2. NH_3, MeOH	R[1], S, NH, O, N, O, R	659

R^1		N^1-isomer
FC_6H_4	(34)	(24)
ClC_6H_4	(23)	(18)
BrC_6H_4	(24)	(20)
Me	(41)	(—)
Ph	(36)	(—)

Reactant	Conditions	Product(s) and Yield(s) (%)	Refs.
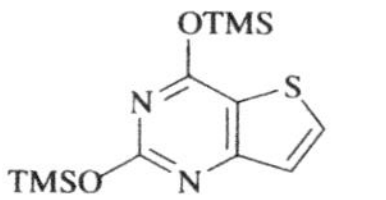 OTMS, TMSO (thieno[3,2-d]pyrimidine)	$SnCl_4$, $Cl(CH_2)_2Cl$, rt, 18 h	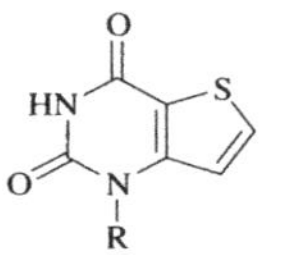 HN, O, S, N–R (98) R = BzO, O, BzO, OBz 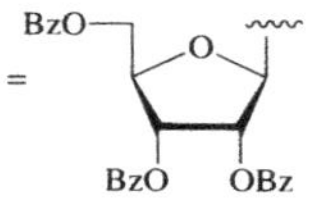	893
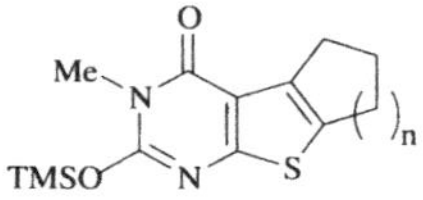 Me, O, TMSO, N, S, ( )n	$SnCl_4$, $Cl(CH_2)_2Cl$, rt, 12 h	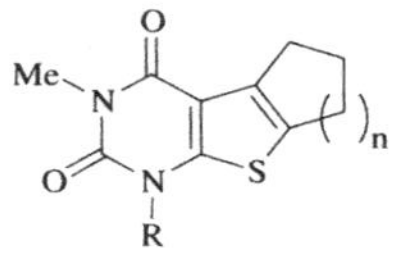 Me, O, N–R, S, ( )n; n = 1 (70); n = 2 (23)	894
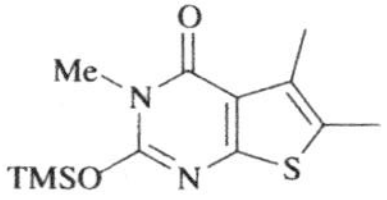 Me, O, TMSO, N, S	$SnCl_4$, $Cl(CH_2)_2Cl$, rt, 12 h	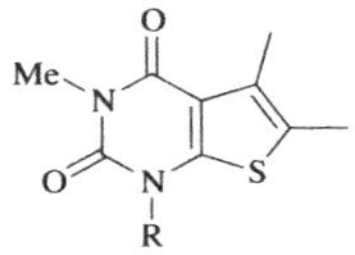 Me, O, N–R, S (26)	894
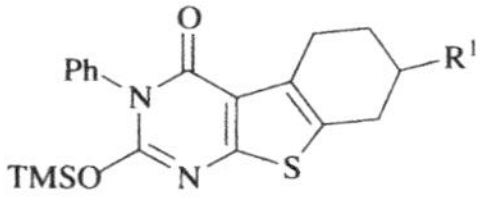 Ph, O, TMSO, N, S, R[1]	$SnCl_4$, $Cl(CH_2)_2Cl$, rt, 12 h	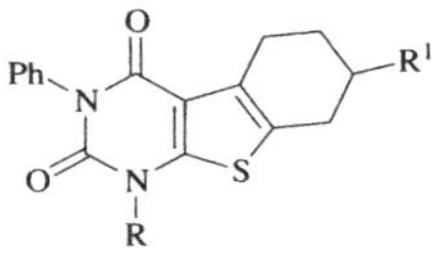 Ph, O, N–R, S, R[1]; R[1] = H (41); Me (45)	894
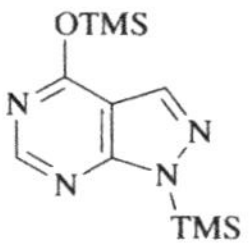 OTMS, N, N–TMS	$SnCl_4$, $Cl(CH_2)_2Cl$, 60°, 6 h	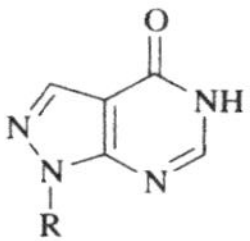 NH, O, N–R **I** (17) + N^2-isomer (7) + N^5-isomer (12) + N^1,N^5-bis(riboside) (38) + N^2,N^5-bis(riboside) (9)	735, 510
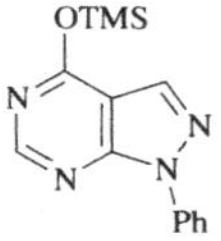 OTMS, N, N–Ph	1. $SnCl_4$, $Cl(CH_2)_2Cl$, rt, 18 h 2. NH_3, MeOH, rt, 24 h	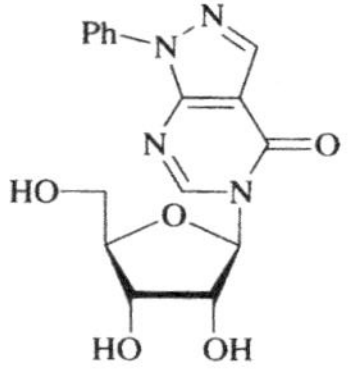 Ph–N, O, N, HO, HO, OH (30)	888

TABLE III. REACTIONS OF SILYLATED HETEROCYCLIC BASES WITH PROTECTED SUGARS - $SnCl_4$ CATALYST (*Continued*)

Sugar	Base	Conditions	Product(s) and Yield(s) (%)	Refs.
	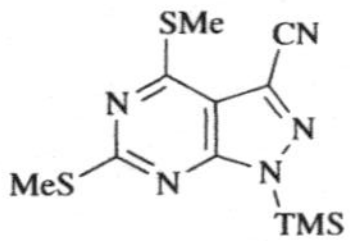 [structure: SMe, CN, MeS, N, TMS]	$SnCl_4$, $Cl(CH_2)_2Cl$, reflux, 2 h	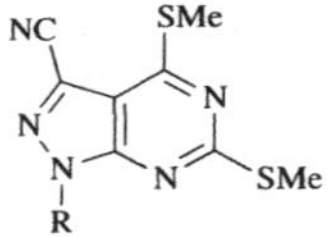 [structure: NC, SMe, SMe, R] (96) R = [structure: BzO, O, BzO, OBz] 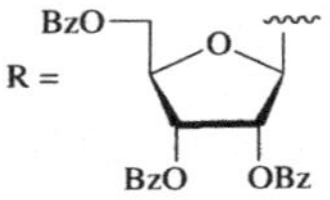	895
	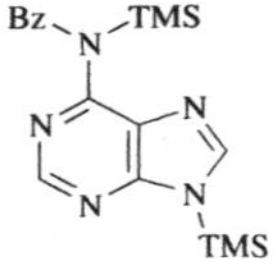 [structure: Bz, N, TMS, N, TMS]	1. $SnCl_4$, $Cl(CH_2)_2Cl$, 48-72 h 2. NH_3, MeOH	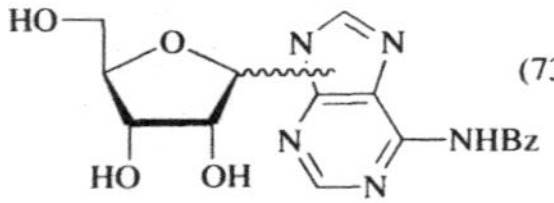 [structure: HO, O, HO, OH, N, NHBz] (73) N^7:N^9 = 1:1	101
	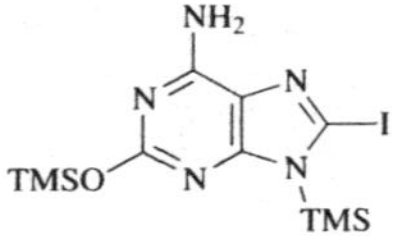 [structure: NH_2, TMSO, I, TMS]	$SnCl_4$, $Cl(CH_2)_2Cl$ rt, 18 h	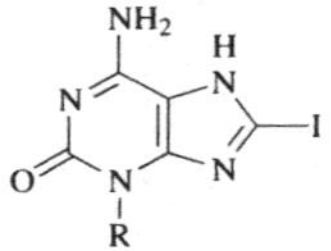 [structure: NH_2, H, I, O, R] (—) R = [structure: BzO, O, BzO, OBz] 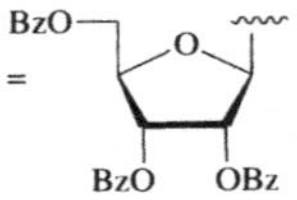	896
	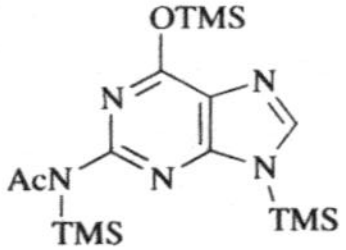 [structure: OTMS, AcN, TMS, TMS]	$SnCl_4$, MeCN, rt, 4 h	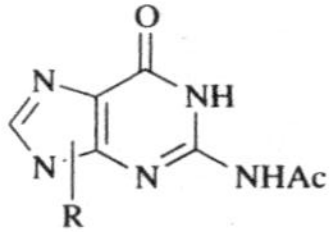 [structure: O, NH, NHAc, R] (81) N^7:N^9 = 3:1	142
	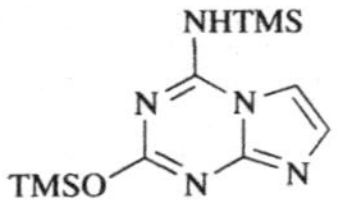 [structure: NHTMS, TMSO]	$SnCl_4$, $Cl(CH_2)_2Cl$, rt, 4 h	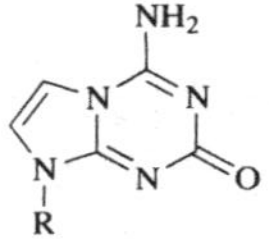 [structure: NH_2, O, R] (60)	897
	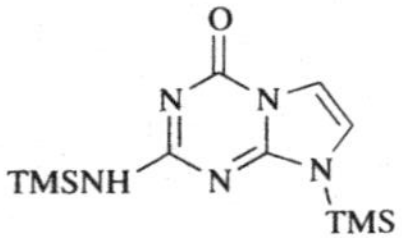 [structure: O, TMSNH, TMS]	$SnCl_4$, $Cl(CH_2)_2Cl$, rt, 30 h	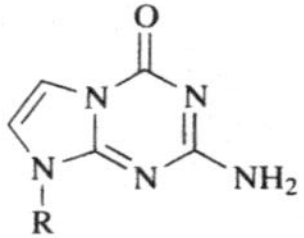 [structure: O, NH_2, R] (56)	898

Substrate	Conditions	Product(s) and Yield(s) (%)	Refs.
(imidazotriazinone; AcN(TMS), N-TMS)	$SnCl_4$, MeCN, rt, 30 h	(NHAc product, N–R) (85)	898
(OTMS, N-TMS imidazopyridazine)	$SnCl_4$, $Cl(CH_2)_2Cl$, rt, 15 min	sugar:base 1:1 (74), N^3,N^6-bis(riboside) (20); 1.5:1 (40), (40); 2:1 (50), (44)	543
(BnO, OTMS imidazopyridazine)	$SnCl_4$, $Cl(CH_2)_2Cl$, rt, 1 h	(OBn product) **I** (56) + N^3-isomer **II** (13)	899
	$SnCl_4$, $Cl(CH_2)_2Cl$, rt, 3 d	(OBn product) **II** (33) + N^1-isomer **I** (18)	900
(OTMS, OTMS, N-TMS imidazopyridazine)	$SnCl_4$, $Cl(CH_2)_2Cl$, rt, 12 h	(NH, NH dione product) (38) + N^1,N^5 and N^1,N^6 bis(ribosides) (58)	543
(TMSO triazolopyrimidine)	$SnCl_4$, $Cl(CH_2)_2Cl$, rt, 24 h	(triazolopyrimidinone, N–R) (31)	743

TABLE III. REACTIONS OF SILYLATED HETEROCYCLIC BASES WITH PROTECTED SUGARS - $SnCl_4$ CATALYST (*Continued*)

Sugar	Base	Conditions	Product(s) and Yield(s) (%)	Refs.
	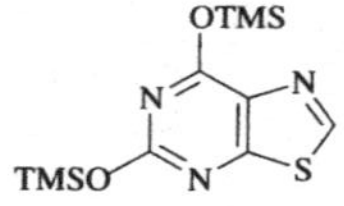 OTMS, TMSO, N, N, N, S	$SnCl_4$, $Cl(CH_2)_2Cl$, rt, 4 h	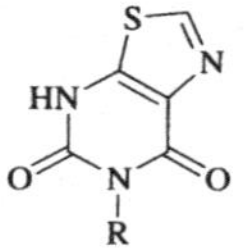 HN, S, N, O, N, O, R (56)	901
	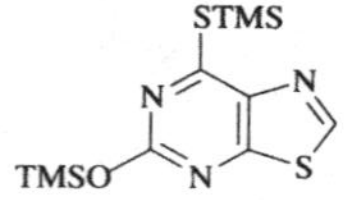 STMS, TMSO, N, N, N, S	$SnCl_4$, $Cl(CH_2)_2Cl$, rt, 15 h	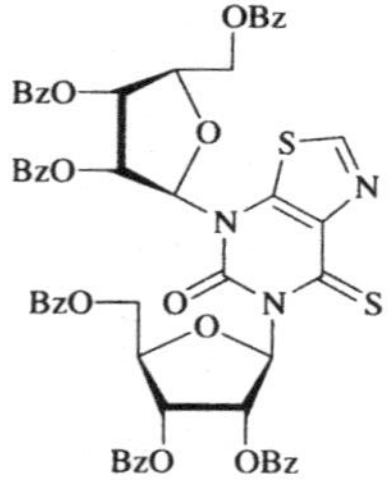 OBz, BzO, BzO, O, S, N, N, BzO, O, N, S, O, BzO, OBz (—)	901
	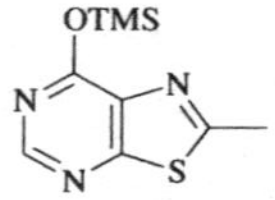 OTMS, N, N, N, S	1. $SnCl_4$, $Cl(CH_2)_2Cl$, rt, 18 h 2. NH_3, MeOH, rt, 24 h	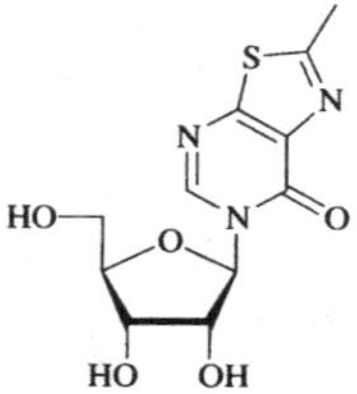 S, N, N, HO, N, O, O, HO, OH (25)	888
	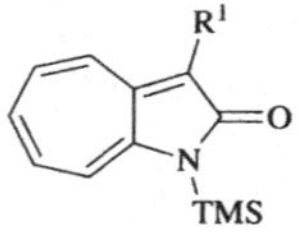 R^1, O, N, TMS	$SnCl_4$, $Cl(CH_2)_2Cl$, reflux, 1 h	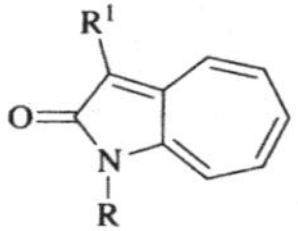 R^1, O, N, R R^1: H (56); CO_2Me (69) R 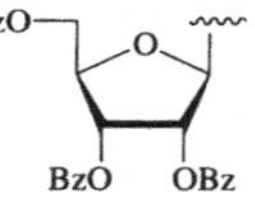 = BzO, O, BzO, OBz	902

Substrate	Conditions	Product(s) and Yield(s) (%)	Refs.
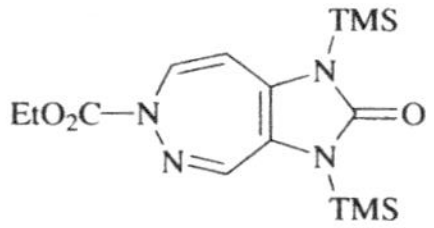	$SnCl_4$, $Cl(CH_2)_2Cl$, MeCN, 0°, 7 h	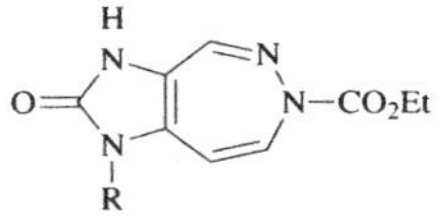 (20) + N^3-isomer (15) + N^1,N^3-bis(riboside) (13)	890
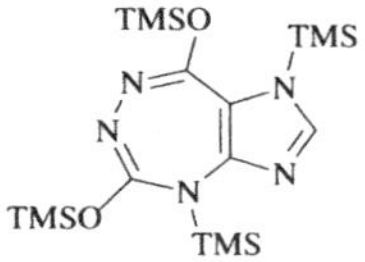	$SnCl_4$, CH_2Cl_2, MeCN	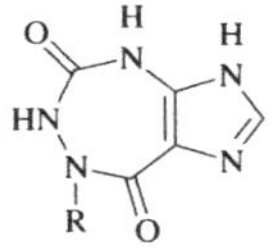 (39) + N^3-isomer (2)	507
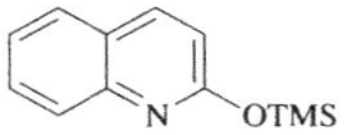	$SnCl_4$, MeCN, 22°, 42 h	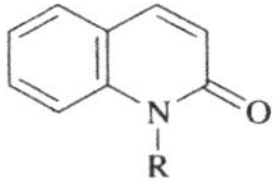 (79)	99
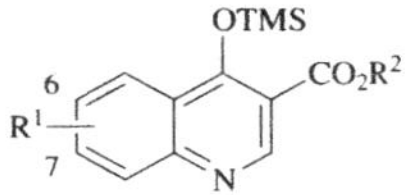	$SnCl_4$, $Cl(CH_2)_2Cl$	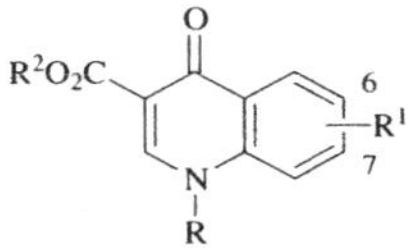	

R^1	R^2	Conditions	Yield (%)	Refs.
H	Et	22°, 2 h	(82)	99
7-Me	Et	rt, 3 h	(82)	740
6-F	Et	rt, 3 h	(83)	740
6,7-F_2	Et	rt, 3 h	(94)	740
6-F,7-Cl	H	rt, 3 h	(18)	740
6-F,7-Cl	Et	rt, 3 h	(60)	740

Substrate	Conditions	Product(s) and Yield(s) (%)	Refs.
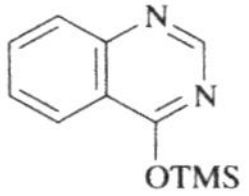	1. $SnCl_4$, $Cl(CH_2)_2Cl$, rt, 18 h 2. NH_3, MeOH, rt, 24 h	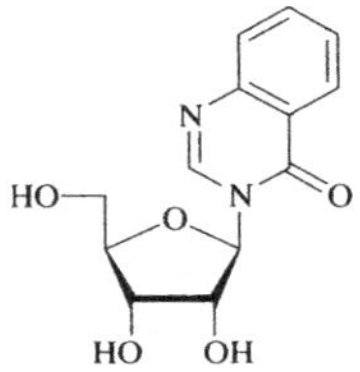 (27)	888

TABLE III. REACTIONS OF SILYLATED HETEROCYCLIC BASES WITH PROTECTED SUGARS - $SnCl_4$ CATALYST (*Continued*)

Sugar	Base	Conditions	Product(s) and Yield(s) (%)	Refs.
	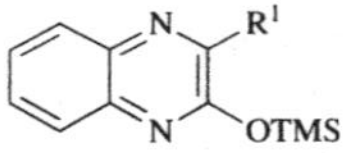 [structure] R[1]	$SnCl_4$, MeCN	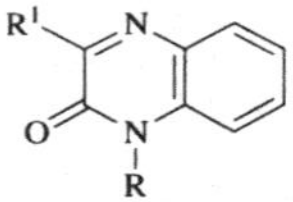 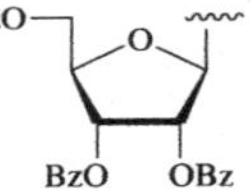 [structure] R = [structure] BzO, BzO, OBz	903
	H	rt, 3 d	(8)	
	Me	50°, 20 h	(5)	
	[structure] OTMS, TMSO, OTMS, R[1]	$SnCl_4$, C_6H_6, reflux, 4 h	[structure] R[1]	
			H (19)	904
			Me (23)	904
			Ph (52)	904
	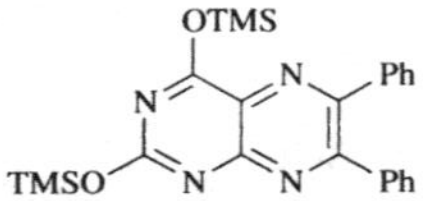 [structure] OTMS, TMSO, Ph, Ph	$SnCl_4$, $Cl(CH_2)_2Cl$, rt, 4 h	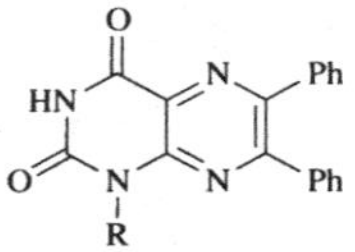 [structure] (20) + N^3-isomer (8) + bis(riboside) (21)	905
	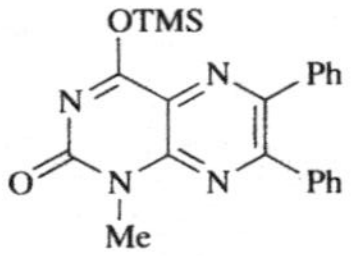 [structure] OTMS, Me, Ph, Ph	$SnCl_4$, CH_2Cl_2, rt, 1 d	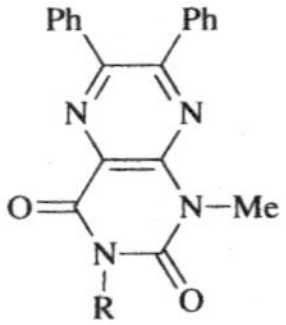 [structure] (48)	906

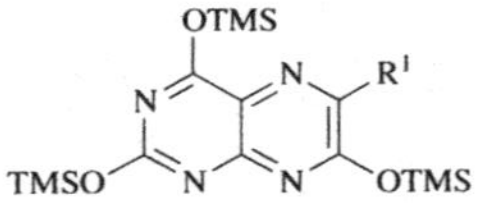

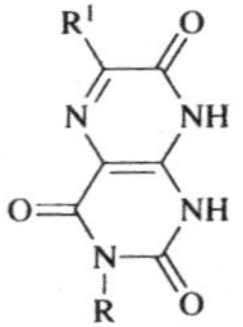

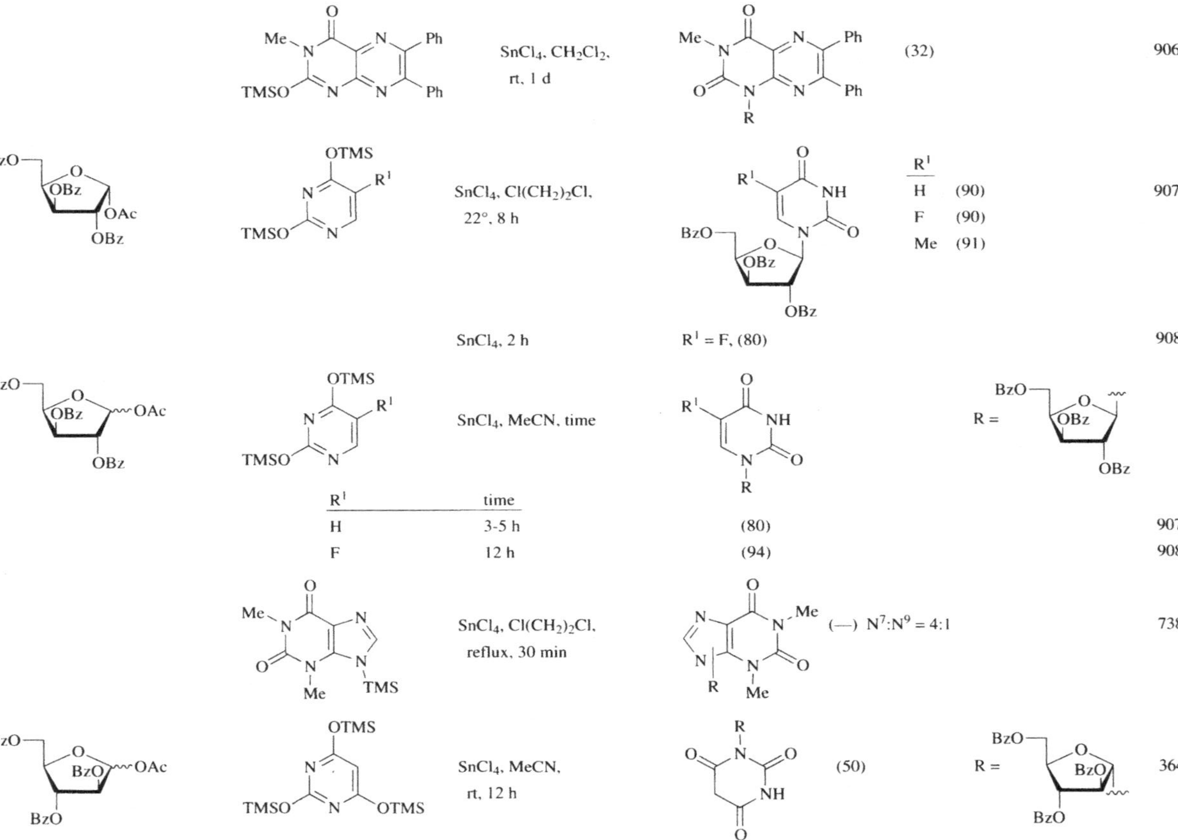
SnCl$_4$, CH$_2$Cl$_2$, rt, 1 d
(32)
906
SnCl$_4$, Cl(CH$_2$)$_2$Cl, 22°, 8 h
R^1
H (90)
F (90)
Me (91)
907
SnCl$_4$, 2 h
R^1 = F, (80)
908
SnCl$_4$, MeCN, time
R =
R^1 time
H 3-5 h (80) 907
F 12 h (94) 908
SnCl$_4$, Cl(CH$_2$)$_2$Cl, reflux, 30 min
(—) N^7:N^9 = 4:1
738
SnCl$_4$, MeCN, rt, 12 h
(50)
R =
364

TABLE III. REACTIONS OF SILYLATED HETEROCYCLIC BASES WITH PROTECTED SUGARS - $SnCl_4$ CATALYST (*Continued*)

Sugar	Base	Conditions	Product(s) and Yield(s) (%)	Refs.
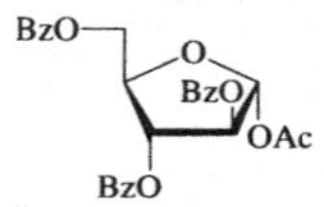 BzO, BzO, OAc (structure)	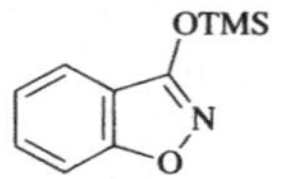 OTMS (structure)	$SnCl_4$, $Cl(CH_2)_2Cl$, reflux, 40 min; rt, 30 min	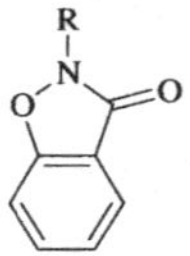 R (structure) (74)	887
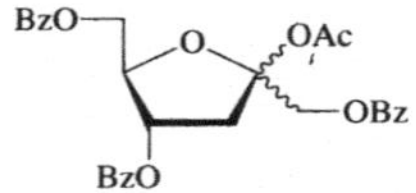 BzO, OAc, OBz, BzO (structure)	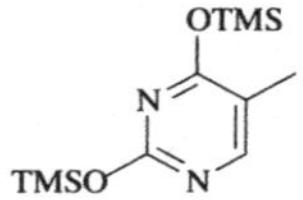 OTMS, TMSO (structure)	1. $SnCl_4$, $Cl(CH_2)_2Cl$, −20° to rt, 1 h 2. NH_3, MeOH	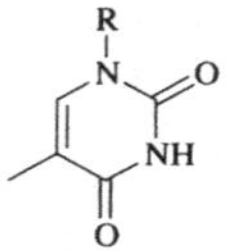 R, NH (structure) **I** (58) α:β = 1:5; R 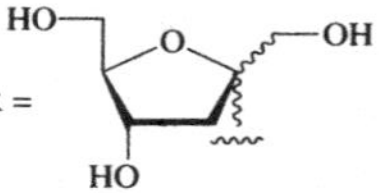 = HO, OH, HO (structure)	693
		1. $SnCl_4$, MeCN, −20° to rt, 1 h 2. NH_3, MeOH	**I** (62) α:β = 1:3	693
	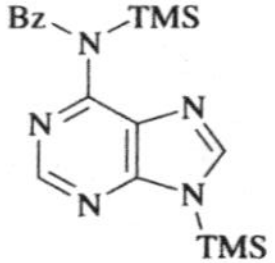 Bz, TMS, TMS (structure)	1. $SnCl_4$, $Cl(CH_2)_2Cl$, −20° to rt, 1 h 2. NH_3, MeOH	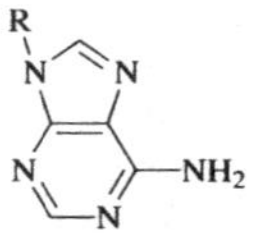 R, NH_2 (structure) **I** (64) α:β = 30:1	693
		1. $SnCl_4$, MeCN, −20° to rt, 1 h 2. NH_3, MeOH	**I** (61) α:β = 11:1	693
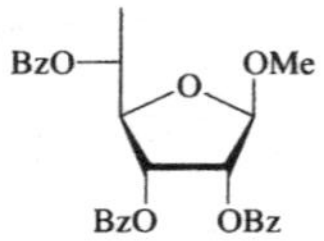 BzO, OMe, BzO, OBz (structure)	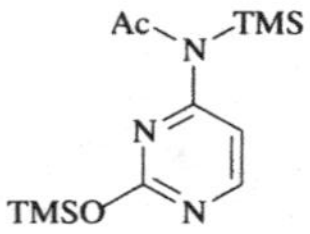 Ac, TMS, TMSO (structure)	$SnCl_4$, $Cl(CH_2)_2Cl$, 80°, 2 h	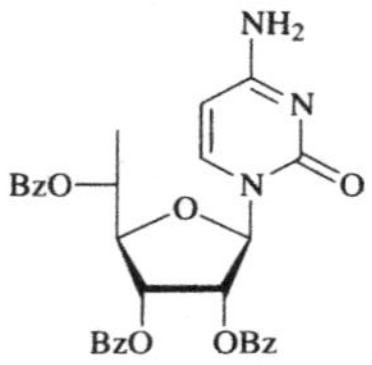 NH_2, BzO, BzO, OBz (structure) (83)	911

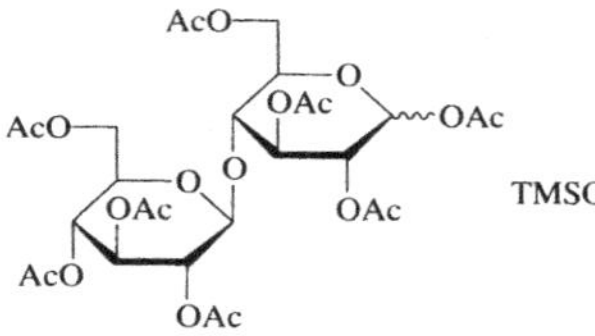 AcO, AcO, OAc, OAc, O, O, OAc, OAc, AcO, OAc	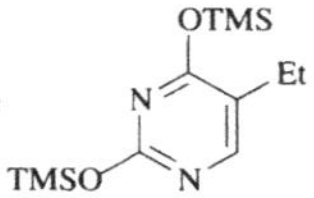 OTMS, Et, N, N, TMSO	$SnCl_4$, $Cl(CH_2)_2Cl$, 80°, 2 h	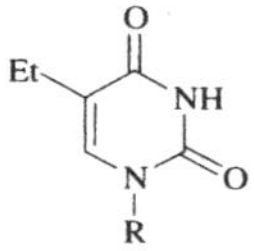 Et, O, NH, O, N–R (78); R = AcO, AcO, OAc, O, O, O, OAc, AcO, OAc, OAc 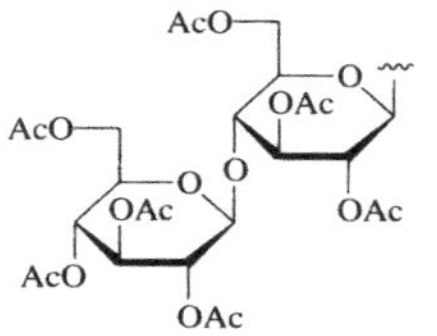	98
	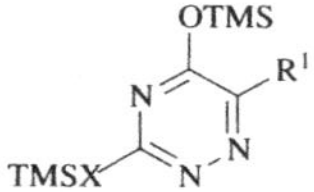 OTMS, R[1], N, N, N, TMSX	$SnCl_4$, $Cl(CH_2)_2Cl$, reflux, 2 h	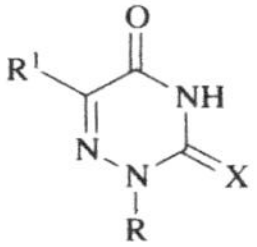 R[1], O, NH, N, N–R, X	98
	X O, R[1] H		(80)	
	X O, R[1] Me		(63)	
	X S, R[1] Me		(95)	
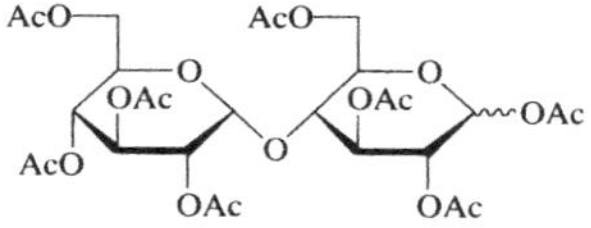 AcO, AcO, O, O, OAc, OAc, O, OAc, AcO, OAc, OAc	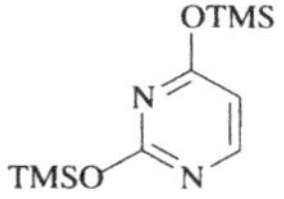 OTMS, N, N, TMSO	$SnCl_4$, $Cl(CH_2)_2Cl$, reflux, 2 h	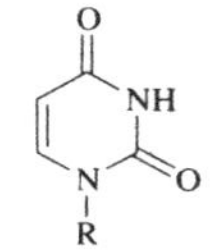 O, NH, O, N–R (71); R = AcO, AcO, O, O, OAc, OAc, O, OAc, AcO, OAc, OAc 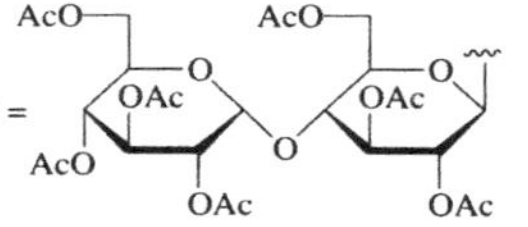	98
	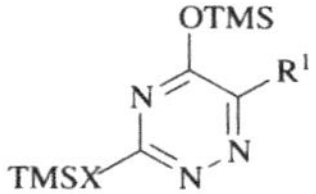 OTMS, R[1], N, N, N, TMSX	$SnCl_4$, $Cl(CH_2)_2Cl$	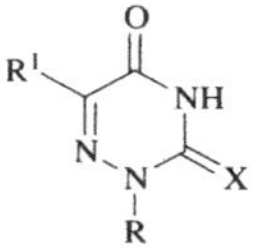 R[1], O, NH, N, N–R, X	98
	X O, R[1] H	reflux, 2 h	(72)	
	X O, R[1] Me	60°, 4 h	(63)	
	X S, R[1] Me	60°, 4 h	(70)	

TABLE III. REACTIONS OF SILYLATED HETEROCYCLIC BASES WITH PROTECTED SUGARS - $SnCl_4$ CATALYST (*Continued*)

Sugar	Base	Conditions	Product(s) and Yield(s) (%)	Refs.
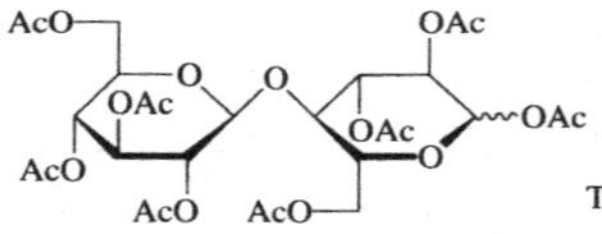	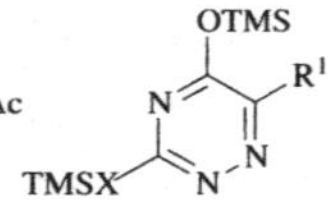 X / R[1]: O / H; O / Me; S / Me	$SnCl_4$, $Cl(CH_2)_2Cl$, reflux, 2 h	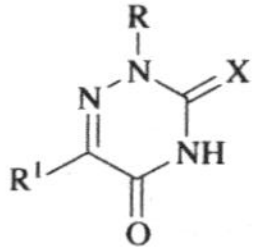 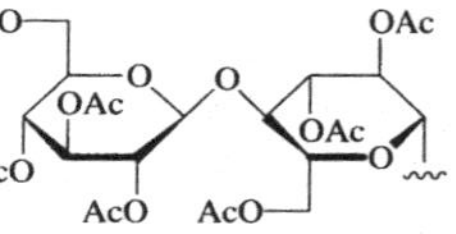 (80); (69); (66)	98
C_{29} 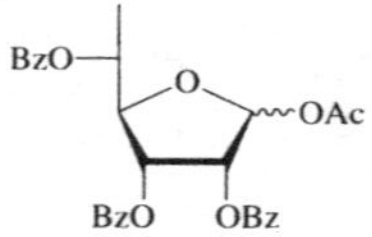	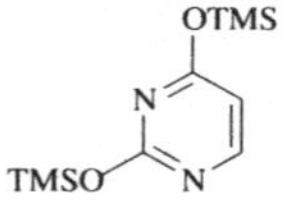	$SnCl_4$, MeCN	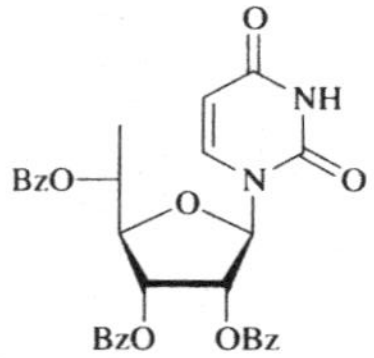 (—)	912
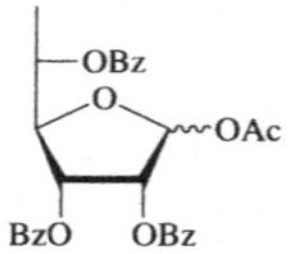	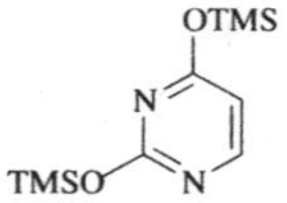	$SnCl_4$, MeCN	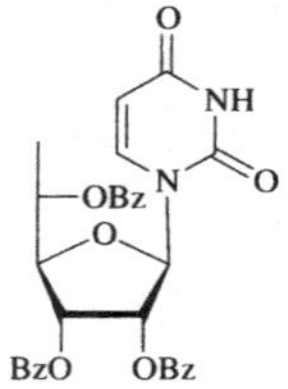 (—)	912
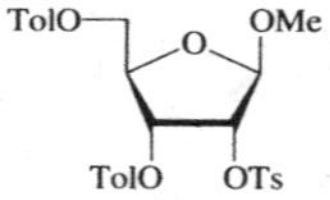	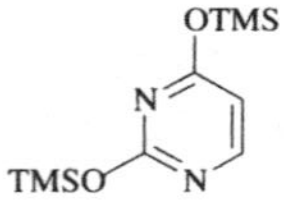	$SnCl_4$	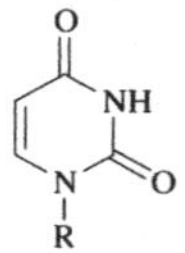 (—) 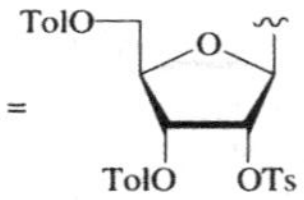	192k

Substrate	Reagent	Conditions	Product(s) and Yield(s) (%)	Refs.
	NHTMS, N, N, N, N, TMS	$SnCl_4$	NH_2, N, N, N, N, R (—)	192k
TBDPSO, O, OAc, SR^1	OTMS, R^2, N, N, TMSO	$SnCl_4$	TBDPSO, O, N, O, NH, O, SR^1, R^2	
R^1	R^2			
$C_6H_4NO_2$-*o*	Me	CH_2Cl_2, –78 to 25°, 5 h	(—) α:β = 1:9	227
Ph	H	$Cl(CH_2)_2Cl$, rt, 12 h	(96) α:β = 3:97	229
Ph	H	$Cl(CH_2)_2Cl$, 0°, 12 h	(91) α:β = 7:93	229
Ph	H	$Cl(CH_2)_2Cl$ (6 eq), rt	(96) α:β = 3:97	228
Ph	H	$Cl(CH_2)_2Cl$ (1.8 eq), 0°, 3.5 h	(91) α:β = 7:93	228
Ph	Me	$Cl(CH_2)_2Cl$, 0°, 12 h	(87) α:β = 8:92	229
Ph	Me	$Cl(CH_2)_2Cl$, 0°, 4 h	(87) α:β = 8:92	228
Ph	Me	CH_2Cl_2, –78 to 25°	(—) α:β = 1:14	227
	Ac, N, TMS, N, N, TMSO	$SnCl_4$, $Cl(CH_2)_2Cl$, 0°, 4-12 h	TBDPSO, O, N, O, N, NHAc, SPh **I** (93) α:β = 11:89	228, 229
TBDPSO, O, OMe, BnO	OTMS, CF_3, N, N, TMSO	$SnCl_4$, CH_2Cl_2, 5°, 1.5 h	TBDPSO, O, BnO, N, O, NH, O, CF_3 (5) + β-anomer (4)	913

TABLE III. REACTIONS OF SILYLATED HETEROCYCLIC BASES WITH PROTECTED SUGARS - $SnCl_4$ CATALYST (*Continued*)

Sugar	Base	Conditions	Product(s) and Yield(s) (%)	Refs.
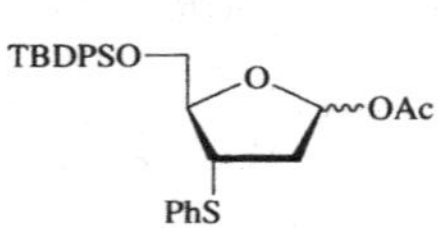 TBDPSO, OAc, PhS	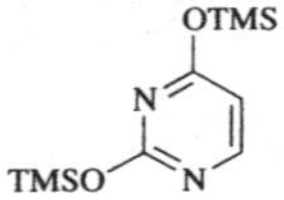 OTMS, TMSO, N, N	$SnCl_4$, CH_2Cl_2, rt, 12 h	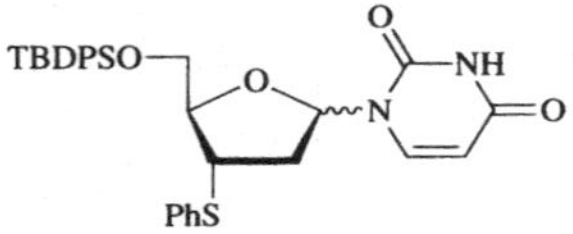 TBDPSO, PhS, O, NH, O (70) α:β = 53:47	229
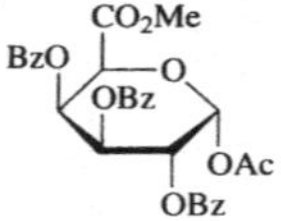 C_{30} CO_2Me, BzO, OBz, OAc, OBz	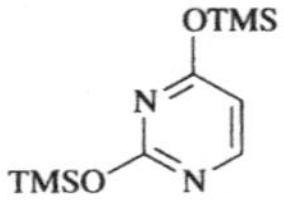 OTMS, TMSO, N, N	$SnCl_4$, $Cl(CH_2)_2Cl$, 70°, 5 h	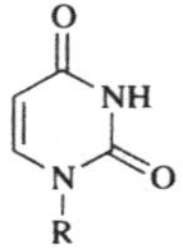 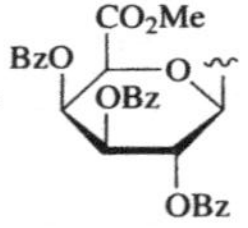 O, NH, O, N–R (75) R = CO_2Me, BzO, OBz, OBz	766
	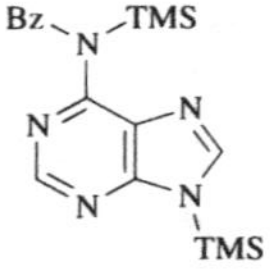 Bz, N, TMS, N–TMS	$SnCl_4$, $Cl(CH_2)_2Cl$, 60-70°, 5 h	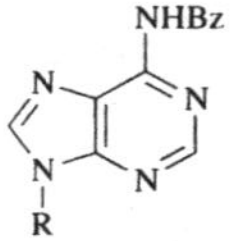 NHBz, N–R (65)	100
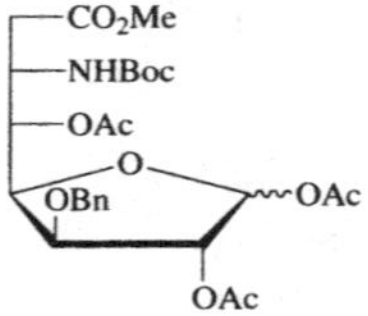 CO_2Me, NHBoc, OAc, OBn, OAc, OAc	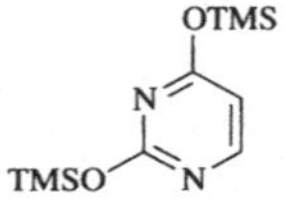 OTMS, TMSO, N, N	$SnCl_4$, $Cl(CH_2)_2Cl$, rt, 28 h	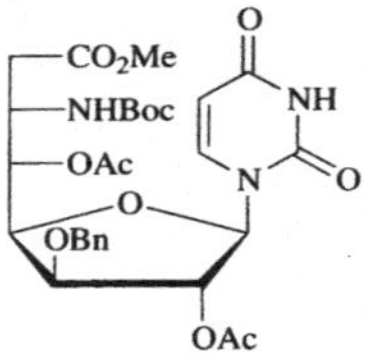 CO_2Me, NHBoc, OAc, OBn, OAc, O, NH, O (85)	785
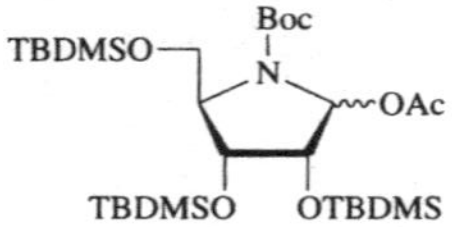 TBDMSO, Boc, N, OAc, TBDMSO, OTBDMS	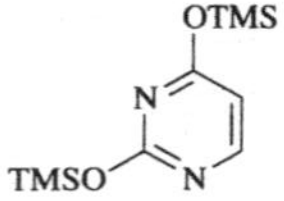 OTMS, TMSO, N, N	$SnCl_4$, $Cl(CH_2)_2Cl$, 0 to 20°	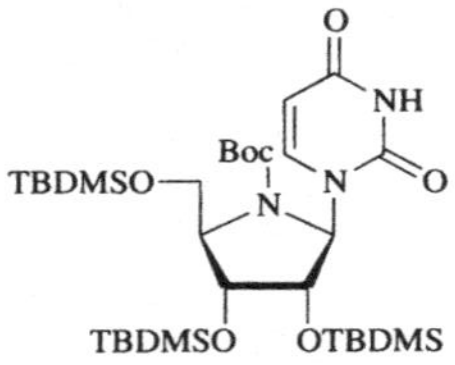 TBDMSO, Boc, N, TBDMSO, OTBDMS, O, NH, O (73) α:β = 8:92	783

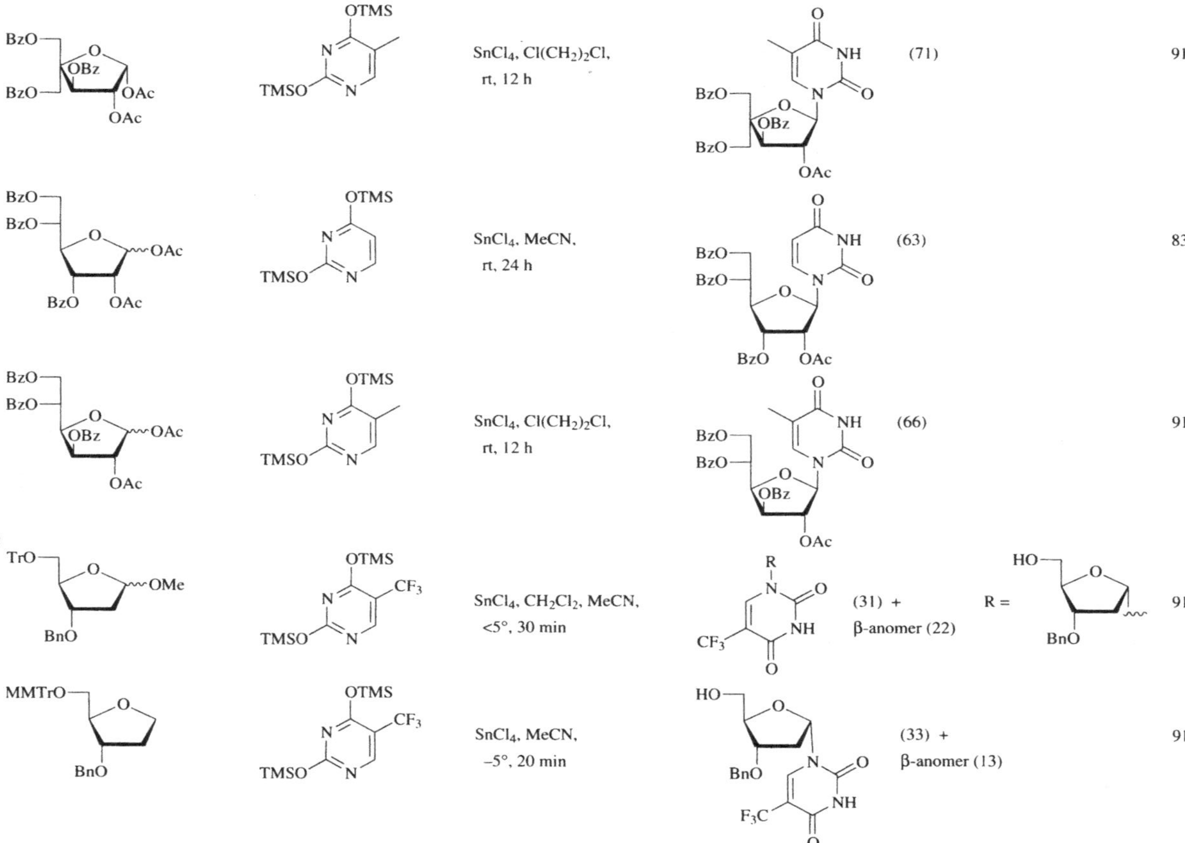

Substrate	Silylated base	Conditions	Product(s) and Yield(s) (%)	Refs.
C_{31}		$SnCl_4$, $Cl(CH_2)_2Cl$, rt, 12 h	(71)	914
		$SnCl_4$, MeCN, rt, 24 h	(63)	833
		$SnCl_4$, $Cl(CH_2)_2Cl$, rt, 12 h	(66)	915
C_{32}		$SnCl_4$, CH_2Cl_2, MeCN, <5°, 30 min	(31) + β-anomer (22) R =	913
		$SnCl_4$, MeCN, –5°, 20 min	(33) + β-anomer (13)	913

TABLE III. REACTIONS OF SILYLATED HETEROCYCLIC BASES WITH PROTECTED SUGARS - $SnCl_4$ CATALYST (*Continued*)

Sugar	Base	Conditions	Product(s) and Yield(s) (%)	Refs.
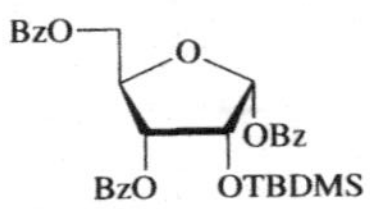	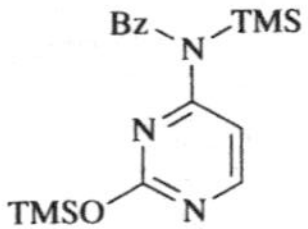	$SnCl_4$, $Cl(CH_2)_2Cl$, reflux	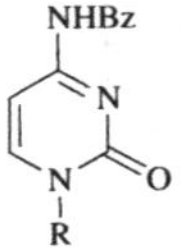 (95) R = 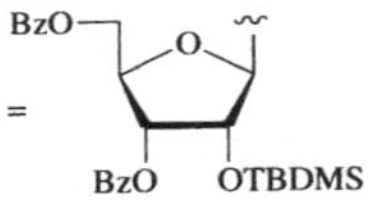	846
	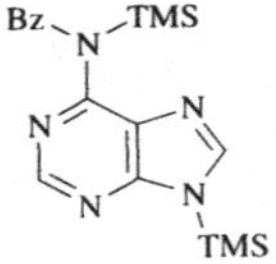	$SnCl_4$, $Cl(CH_2)_2Cl$, reflux	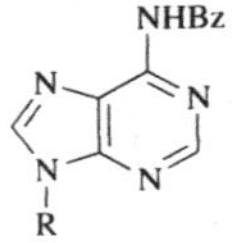 (93)	846
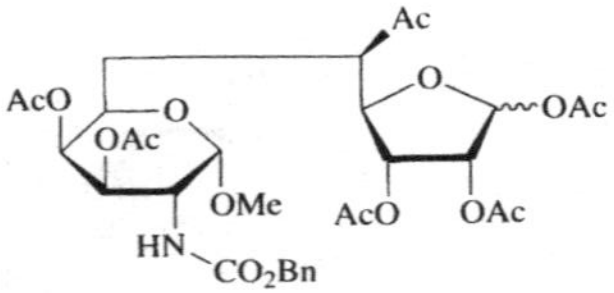	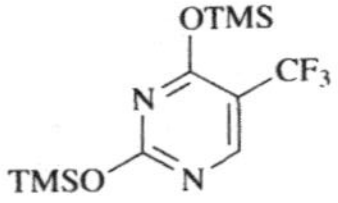	$SnCl_4$, $Cl(CH_2)_2Cl$	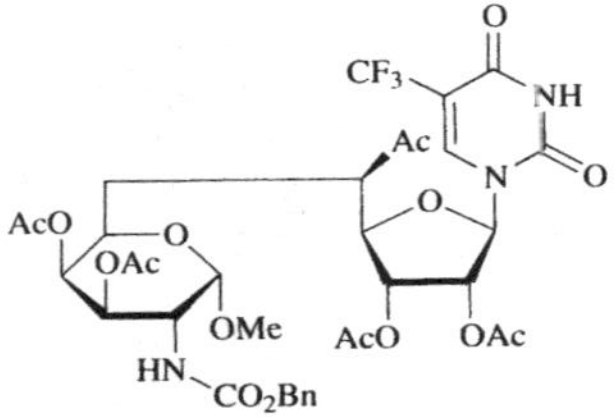 (75)	916
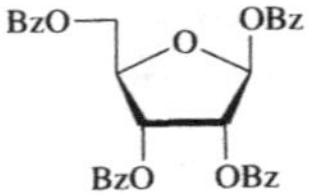	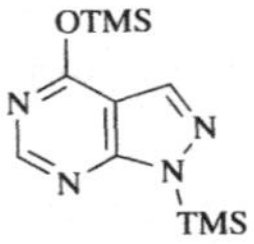	$SnCl_4$, $Cl(CH_2)_2Cl$, 60-70°, 5 h	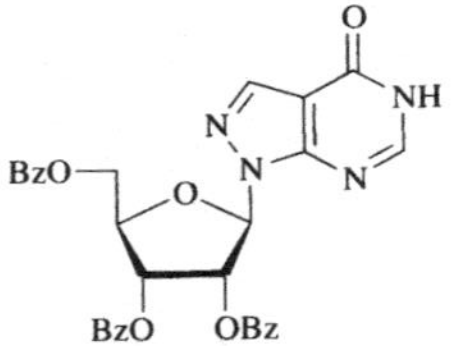 (39)	100
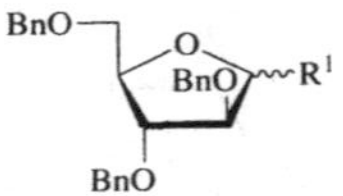 $R^1 = O_2CC_6H_4NO_2$-*p*	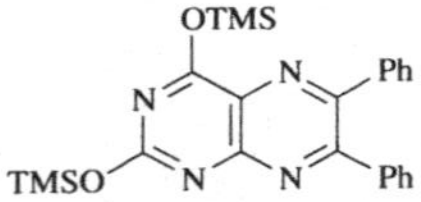	$SnCl_4$, CH_2Cl_2, rt, 5 d	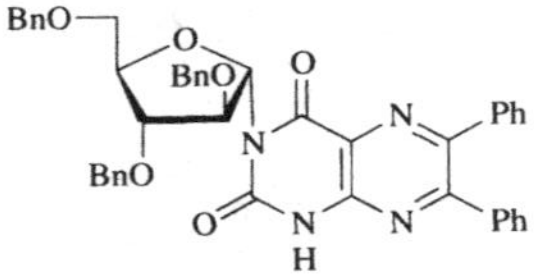 (33)	917

Reactant	Silylated base	Conditions	Product(s) and Yield(s) (%)	Refs.
		$SnCl_4$	(76)	508
C_{34}		$SnCl_4$, $Hg(CN)_2$, C_6H_6, 60°, 1.5 h	(56)	632, 918
C_{35}		$SnCl_4$, $Cl(CH_2)_2Cl$, rt, 2 h	(45) $\alpha:\beta = 1:0.9$	919
		$SnCl_4$, $Cl(CH_2)_2Cl$, 70°, 5 h	(75) R =	766
		$SnCl_4$, $Cl(CH_2)_2Cl$, 60-70°, 5 h	(65)	100

TABLE III. REACTIONS OF SILYLATED HETEROCYCLIC BASES WITH PROTECTED SUGARS - $SnCl_4$ CATALYST (*Continued*)

Sugar	Base	Conditions	Product(s) and Yield(s) (%)	Refs.
TBDPSO, PhS, OAc, PhS	OTMS, N, TMSO, N	$SnCl_4$, MeCN, 30 min	TBDPSO, PhS, N, O, NH, O, PhS (64) α:β = 23:77	305a
BzO, O, OBz, BzO, OAc, OBz	Cl, N, N, AcN, TMS, N, N, TMS	$SnCl_4$, MeCN, rt, 24 h	Cl, N, N, N, N, NHAc, BzO, O, OBz, BzO, OBz (65) N^9:N^7 = 10:90	409
C_{36} BzO, OBz, O, OAc, BzO, OBz	Bz, N, TMS, N, N, N, N, TMS	1. $SnCl_4$, $Cl(CH_2)_2Cl$, reflux, 2 h 2. NaOMe, MeOH, reflux, 1 h	NH_2, N, N, HO, OH, O, N, N, HO, OH (68)	780
BzO, O, OAc, BzO, OBz, BzO	OTMS, N, TMSO, N	$SnCl_4$, MeCN, rt, 1 h	BzO, O, OBz, BzO, BzO, N, O, NH, O (88)	920

		SnCl$_4$, MeCN, rt, 1 h	(59) + N^9-isomer (7)	409
		SnCl$_4$, MeCN, rt, 4 h	(61)	142
C$_{37}$		SnCl$_4$, Cl(CH$_2$)$_2$Cl, 60°, 12 h	(71)	766
		SnCl$_4$, Cl(CH$_2$)$_2$Cl, 60°, 12 h	(85)	766
		SnCl$_4$, Cl(CH$_2$)$_2$Cl, MS, 60-70°, 5 h	(59)	100
		SnCl$_4$, CH$_2$Cl$_2$, 60°, 12 h	(81)	921

TABLE III. REACTIONS OF SILYLATED HETEROCYCLIC BASES WITH PROTECTED SUGARS - $SnCl_4$ CATALYST (*Continued*)

Sugar	Base	Conditions	Product(s) and Yield(s) (%)	Refs.
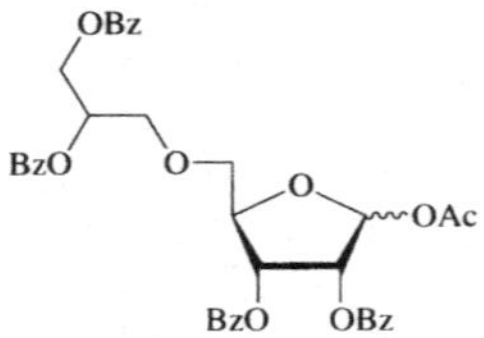 OBz, BzO, O, OAc, BzO, OBz	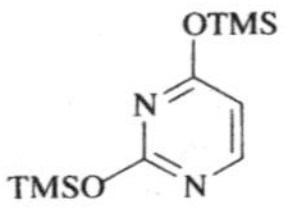 OTMS, N, TMSO, N	$SnCl_4$, MeCN, rt, 2 d	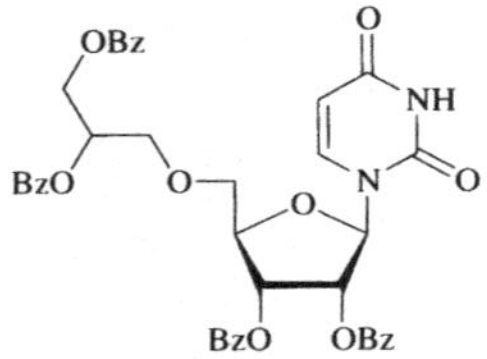 OBz, BzO, O, O, NH, N, O, BzO, OBz (72)	871
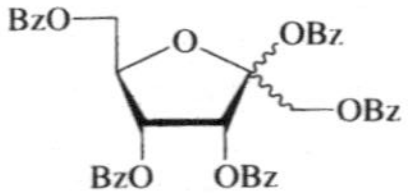 BzO, O, OBz, OBz, BzO, OBz	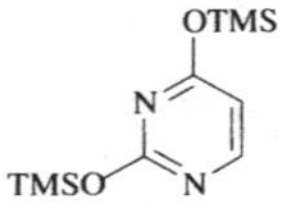 OTMS, N, TMSO, N	$SnCl_4$, MeCN	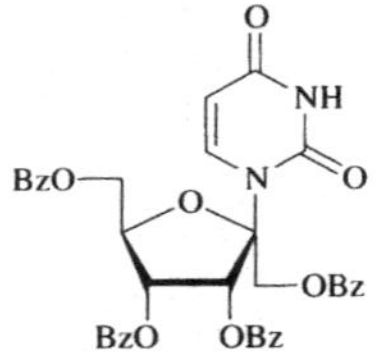 O, NH, N, O, BzO, O, OBz, BzO, OBz (81)	918
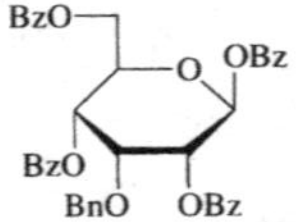 BzO, O, OBz, BzO, BnO, OBz	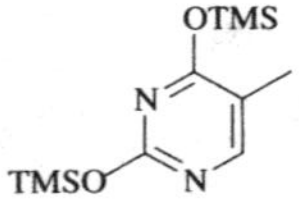 OTMS, N, TMSO, N	$SnCl_4$, $Cl(CH_2)_2Cl$, 40°, 4 h	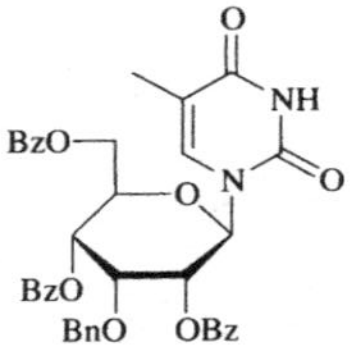 O, NH, N, O, BzO, O, BzO, BnO, OBz (80)	922

TABLE IV. REACTIONS WITH TRIMETHYLSILYL AND SILVER TRIFLATES AND PERCHLORATES

Sugar	Base	Conditions	Product(s) and Yield(s) (%)	Refs.
C_3 MeO⁀OMe	2,4-bis(OTMS)pyrimidine	TMSOTf, CH_2Cl_2, –30°, 2-12 h	1-(MeOCH$_2$)uracil (74)	923
C_4 tetrahydrothiophene S-oxide	Bz–N(Me)–TMS	TMSOTf, Et_3N, ZnI_2, PhMe, rt, 48 h	Me–N(R)–Bz (63); R = tetrahydrothiophen-2-yl	414
	2-TMSO-5-R[1]-pyridine	TMSOTf, Et_3N, ZnI_2, PhMe, rt, 48 h	5-R[1]-1-R-pyridin-2-one; R[1] = H (47), Cl (65)	414
	2-TMSO-4-NHTMS-pyrimidine	TMSOTf, Et_3N, ZnI_2, PhMe, rt, 48 h	1-R-cytosine (61)	414
	2-TMSO-4-OTMS-5-R[1]-pyrimidine	TMSOTf, Et_3N, ZnI_2, PhMe, rt, 48 h	5-R[1]-1-R-uracil; R[1] = H (67), F (53), Me (84)	414
	6-Cl-9-TMS-purine	TMSOTf, Et_3N, ZnI_2, PhMe, rt, 48 h	6-Cl-9-R-purine (59)	414

TABLE IV. REACTIONS WITH TRIMETHYLSILYL AND SILVER TRIFLATES AND PERCHLORATES (*Continued*)

Sugar	Base	Conditions	Product(s) and Yield(s) (%)	Refs.
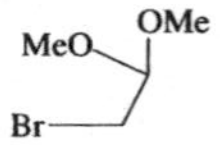 MeO, OMe, Br	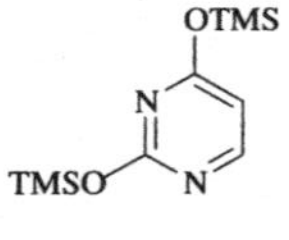 OTMS, N, N, TMSO	TMSOTf, CH_2Cl_2, –30°, 2-12 h	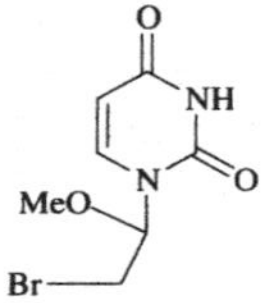 O, NH, O, N, MeO, Br (40)	923
C_5				
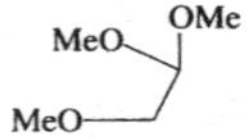 MeO, OMe, MeO	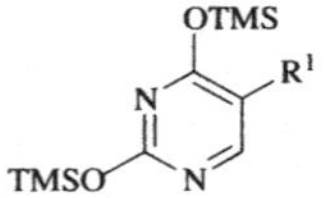 OTMS, R^1, N, N, TMSO	TMSOTf, CH_2Cl_2, –30°, 2-12 h	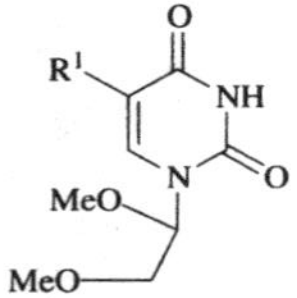 R^1, O, NH, O, N, MeO, MeO; R^1: H (81), F (83)	923
C_6				
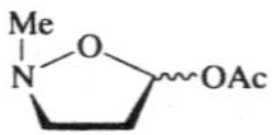 Me, N, O, OAc	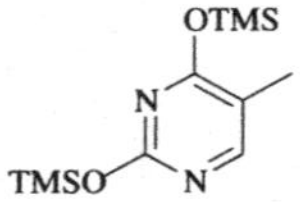 OTMS, N, N, TMSO	TMSOTf,CH_2Cl_2, 25°, 3 h	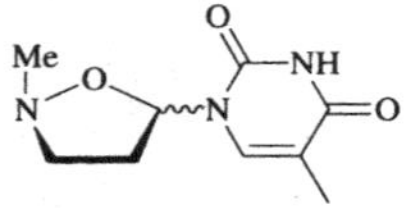 Me, N, O, N, O, NH, O (70)	924
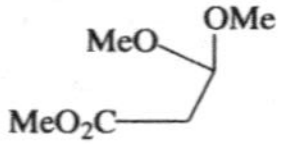 MeO, OMe, MeO_2C	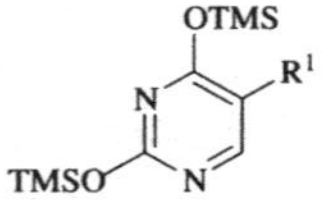 OTMS, R^1, N, N, TMSO	TMSOTf, CH_2Cl_2, –30°, 2-12 h	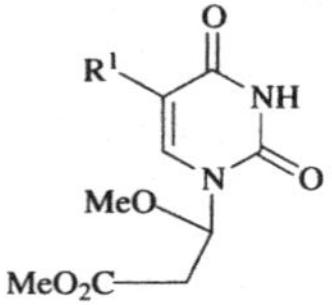 R^1, O, NH, O, N, MeO, MeO_2C; R^1: H (56), F (81)	923
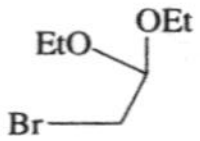 EtO, OEt, Br	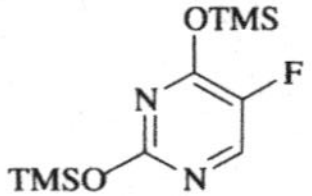 OTMS, F, N, N, TMSO	TMSOTf, CH_2Cl_2, –30°, 2-12 h	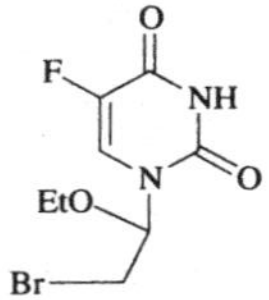 F, O, NH, O, N, EtO, Br (71)	923

Substrate	Nucleobase	Conditions	Product(s) and Yield(s) (%)	Refs.
C_7 EtO, OEt, Cl	OTMS, R^1, TMSO, N	TMSOTf, CH_2Cl_2, –30°, 2-12 h	R^1, O, NH, O, N, EtO, Cl; R^1: H (56); F (91)	923
C_8 N_3, O, OMe, AcO	OTMS, TMSO, N	TMSOTf, MeCN, –40 to 0°, 6 h; rt, 16 h	O, NH, O, N–R (35) + α-anomer (13); R = N_3, O, AcO	925
	i-PrCO, N, TMS, TMSO, N	TMSOTf, MeCN, –40 to 0°, 4 h; rt, 24 h	NHCOPr-i, N, O, N–R (24) + α-anomer (10)	925
	OTMS, TMSO, N	TMSOTf, MeCN, –40 to 0°, 8.5 h rt, 16 h	O, NH, O, N–R (18) + α-anomer (7)	925
	i-PrCO, N, TMS, N, N–TMS	TMSOTf, MeCN, –40 to 0°, 4 h	NHCOPr-i, N, N–R (5) + α-anomer (2)	925
AcO, O, OMe	NHTMS, TMSO, N	TBDMSOTf, MeCN, 25°, 12 h	AcO, O, N, O, NH_2 (80) α:β = 1.4:1	926

TABLE IV. REACTIONS WITH TRIMETHYLSILYL AND SILVER TRIFLATES AND PERCHLORATES (*Continued*)

Sugar	Base	Conditions	Product(s) and Yield(s) (%)	Refs.
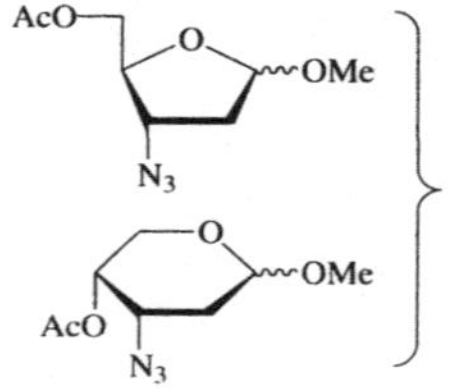 AcO, OMe, N_3 / AcO, OMe, N_3	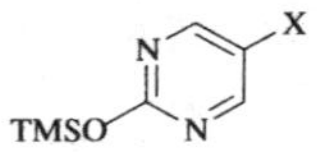 TMSO, X; X = Br, I	1. TMSOTf 2. NH_3, MeOH	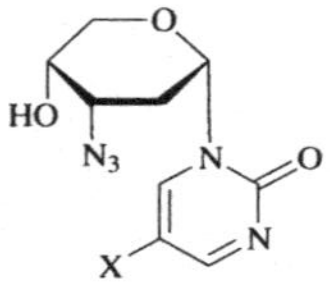 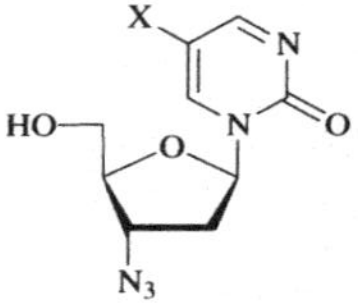 HO, N_3, X + HO, X, N_3 X = Br: (17) + (12) X = I: (15) + (15)	149
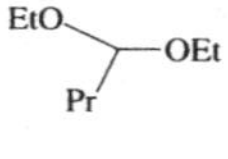 EtO, OEt, Pr	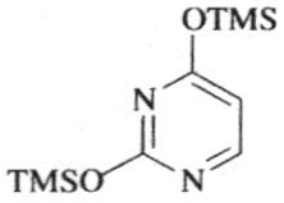 TMSO, OTMS	TMSOTf, CH_2Cl_2, –30°, 2-12 h	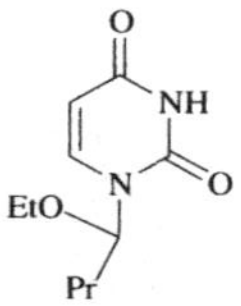 EtO, Pr (77)	923
C_9				
MeO, OMe, o-BrC_6H_4	TMSO, OTMS	TMSOTf, CH_2Cl_2, –30°, 2-12 h	MeO, o-BrC_6H_4 (74)	923
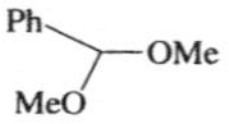 Ph, OMe, MeO	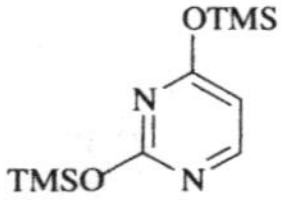 TMSO, OTMS	TMSOTf, CH_2Cl_2, –30°, 2-12 h	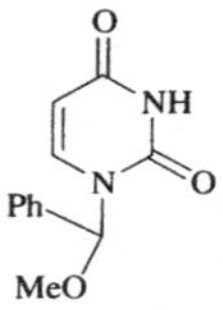 Ph, MeO (91)	923

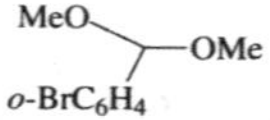

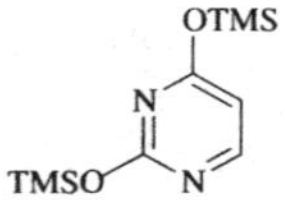

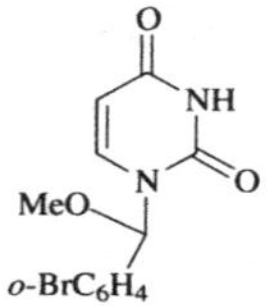

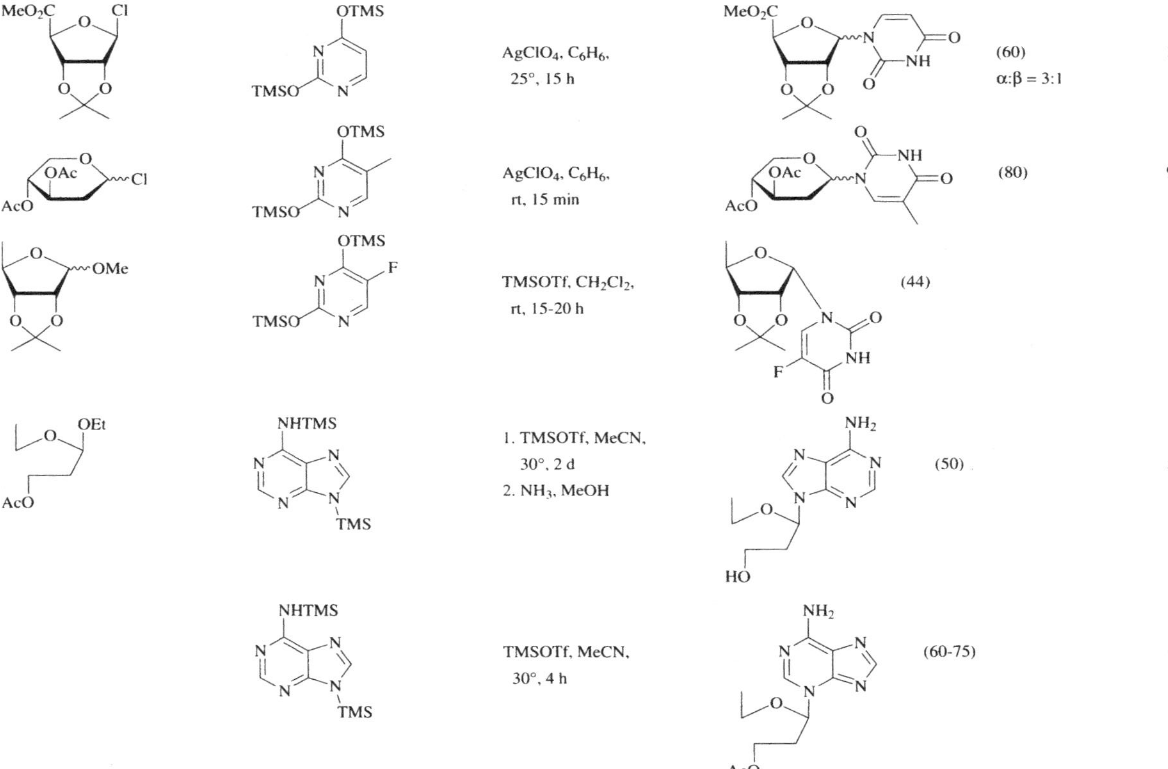

Conditions	Yield (%)	Refs.
$AgClO_4$, C_6H_6, 25°, 15 h	(60) α:β = 3:1	224
$AgClO_4$, C_6H_6, rt, 15 min	(80)	927
TMSOTf, CH_2Cl_2, rt, 15-20 h	(44)	220
1. TMSOTf, MeCN, 30°, 2 d 2. NH_3, MeOH	(50)	555
TMSOTf, MeCN, 30°, 4 h	(60-75)	666

TABLE IV. REACTIONS WITH TRIMETHYLSILYL AND SILVER TRIFLATES AND PERCHLORATES (*Continued*)

Sugar	Base	Conditions	Product(s) and Yield(s) (%)	Refs.
C_{10} 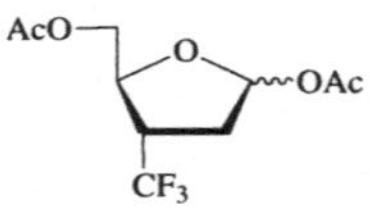	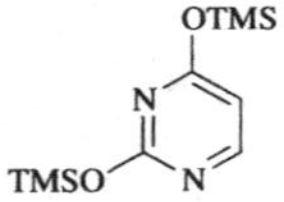	TMSOTf, $Cl(CH_2)_2Cl$, reflux, 40 min	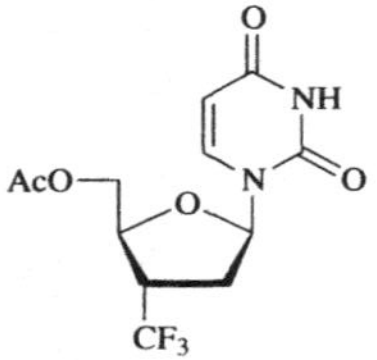 (—)	296
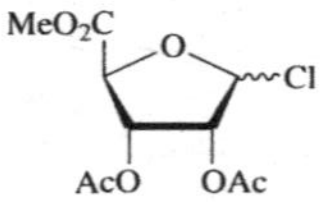	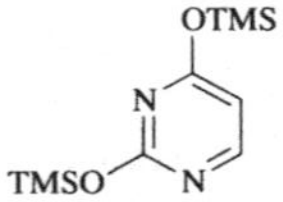	$AgClO_4$, C_6H_6, 25°, 18 h	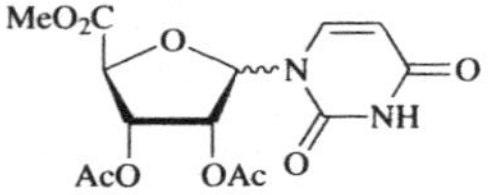 (34) α:β = 1:4	224
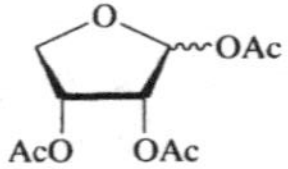	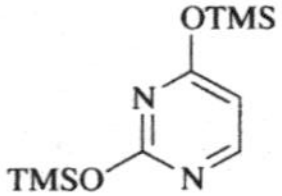	1. TMSOTf, MeCN, reflux, 2 h 2. NH_3, MeOH, 25°, 24 h	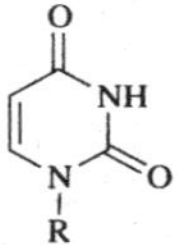 (43) R = 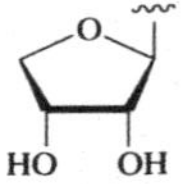	928
	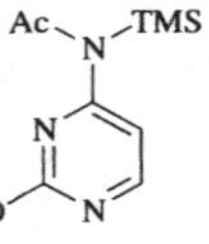	1. TMSOTf, MeCN, reflux, 2 h 2. NH_3, MeOH, 25°, 24 h	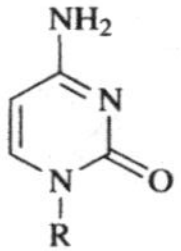 (55)	928
	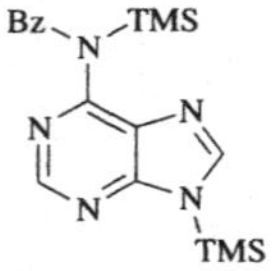	1. TMSOTf, MeCN, reflux, 5 h 2. NH_3, MeOH, 25°, 24 h	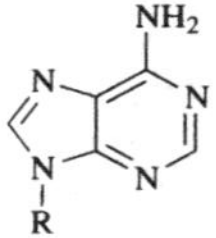 (63)	928

	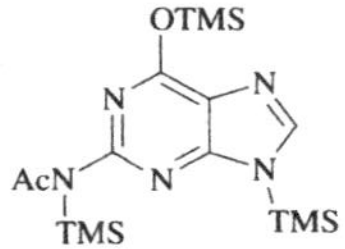	1. TMSOTf, MeCN, reflux, 3 h 2. NH_3, MeOH, 25°, 24 h	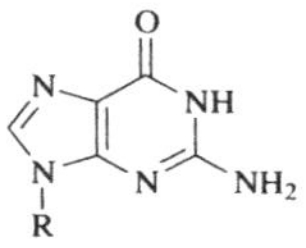 (25)	928
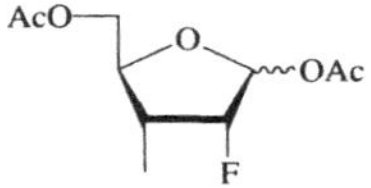	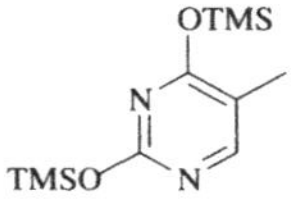	TMSOTf, MeCN, rt, 12 h	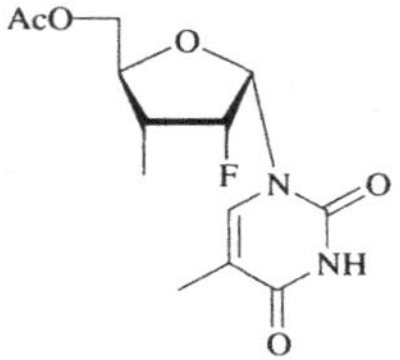 (55) + β-anomer (28)	392
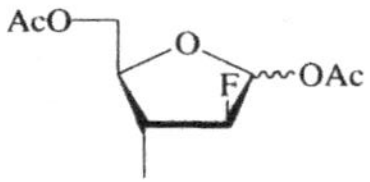	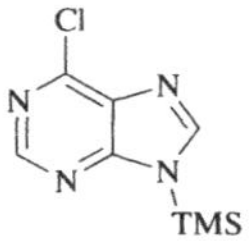	TMSOTf, MeCN, 0°, 2 d	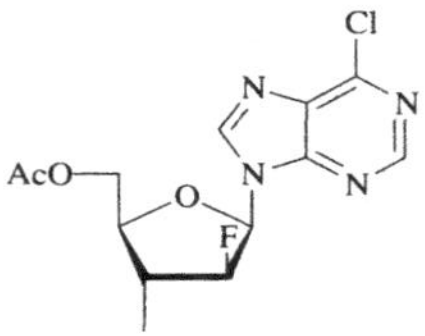 (31) + N^9-α-isomer (17) + N^7-α-isomer (17) + N^7-β-isomer (11)	392
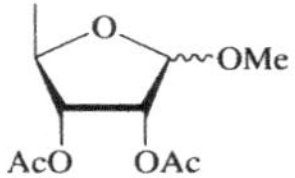	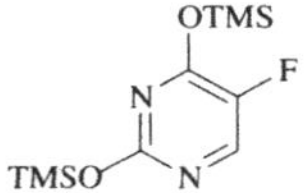	TMSOTf, CH_2Cl_2, rt, 20 h	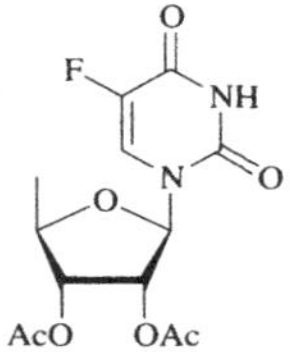 (58)	221
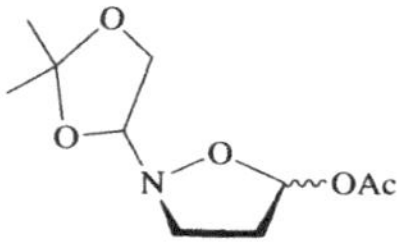	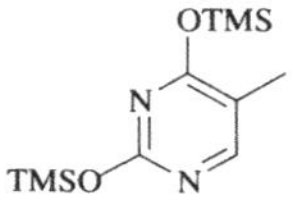	TMSOTf, CH_2Cl_2, 25°, 3 h	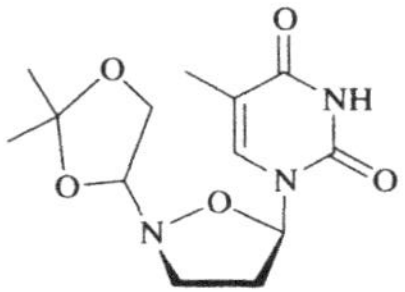 (44)	924

TABLE IV. REACTIONS WITH TRIMETHYLSILYL AND SILVER TRIFLATES AND PERCHLORATES (*Continued*)

Sugar	Base	Conditions	Product(s) and Yield(s) (%)	Refs.
C_{11}				
MeO, OMe, BzO	OTMS, TMSO (pyrimidine)	TMSOTf, CH_2Cl_2, −30°, 2-12 h	MeO, BzO (uracil) (66)	923
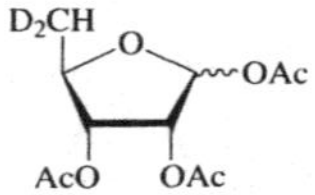 D_2CH, AcO, OAc, OAc	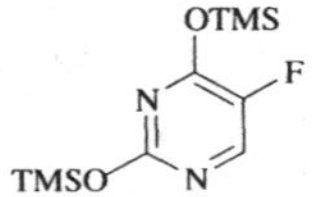 OTMS, F, TMSO (pyrimidine)	TMSOTf, CH_2Cl_2, rt, 5 h	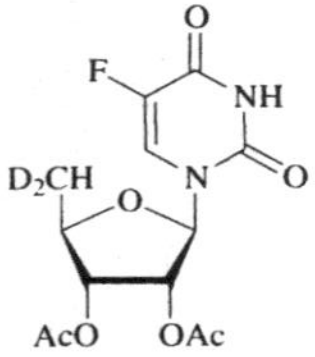 F, D_2CH, AcO, OAc (58)	221
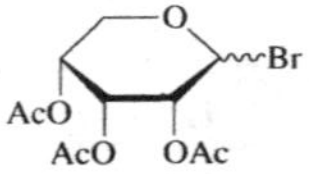 AcO, AcO, OAc, Br	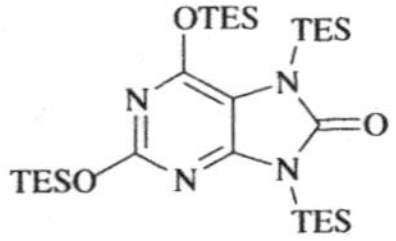 OTES, TES, TESO, TES	$AgClO_4$, PhMe, 60°	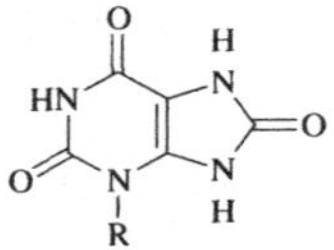 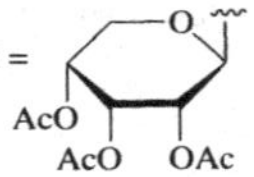 HN, N–R, H, H (56); R = AcO, AcO, OAc	74
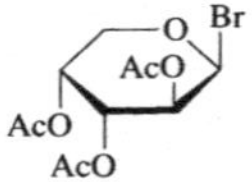 Br, AcO, AcO, AcO	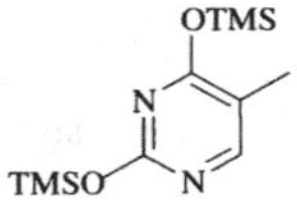 OTMS, TMSO (pyrimidine)	$AgClO_4$, C_6H_6, rt, 16 h	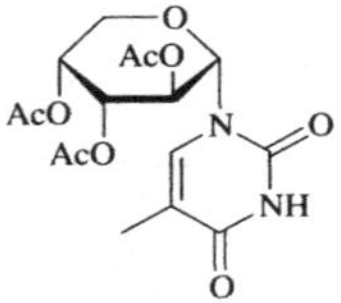 AcO, AcO, AcO, NH (78)	80
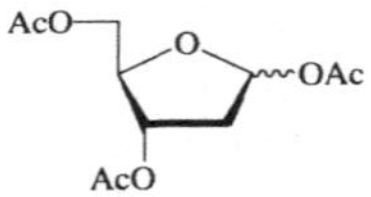 AcO, AcO, OAc	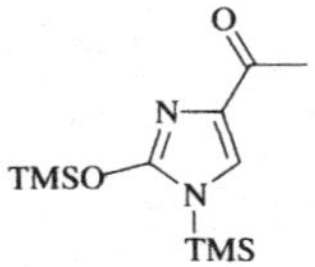 TMSO, N–TMS (imidazole)	TMSOTf, $Cl(CH_2)_2Cl$, −40° to rt, 0.5 h	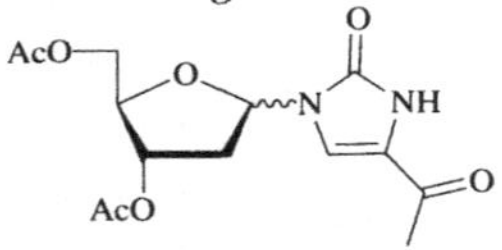 AcO, AcO, NH (49)	929

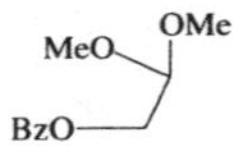

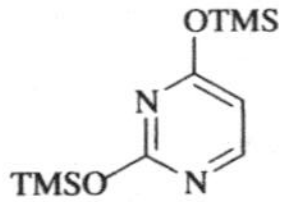

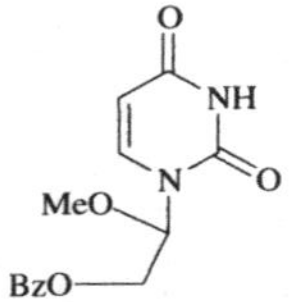

Glycosyl donor	Base	Conditions	Product(s) and Yield(s) (%)	Refs.
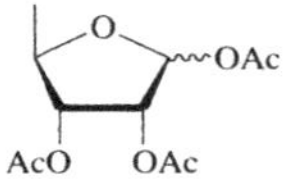	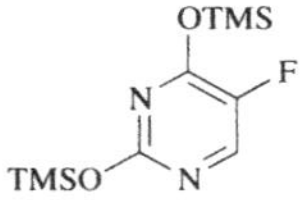	TMSOTf	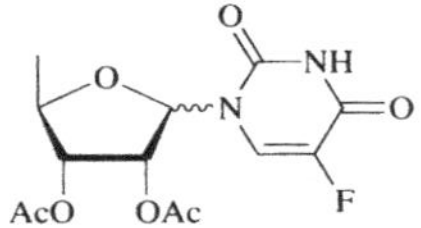 (—) α:β = 1:9	220
		TMSOTf, CH_2Cl_2, rt, 20 h	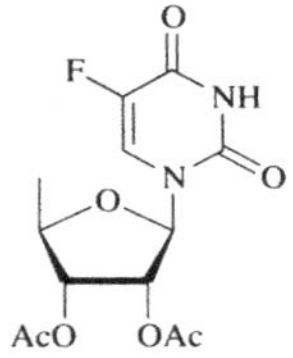 (65) + α-anomer (6)	221, 930
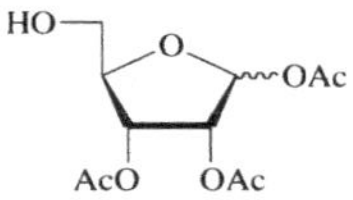	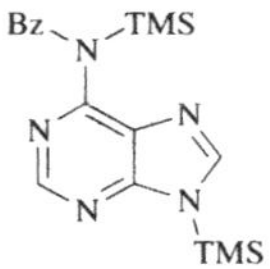	TMSOTf, $Cl(CH_2)_2Cl$, 20°, 16 h	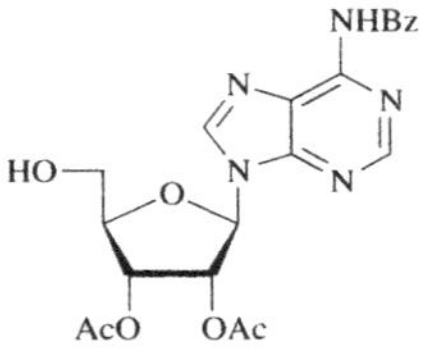 (55)	931
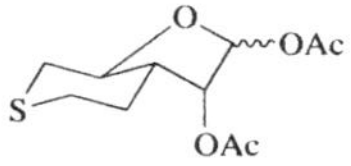	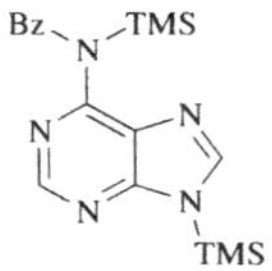	TMSOTf, $Cl(CH_2)_2Cl$, reflux, 15 h	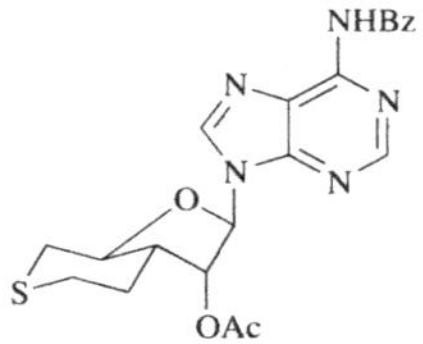 (88)	932
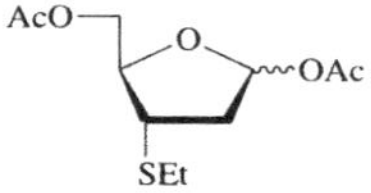	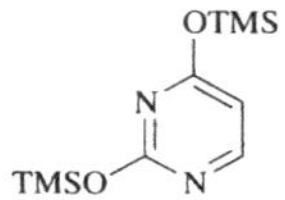	TMSOTf, MeCN, –30°, 2 h	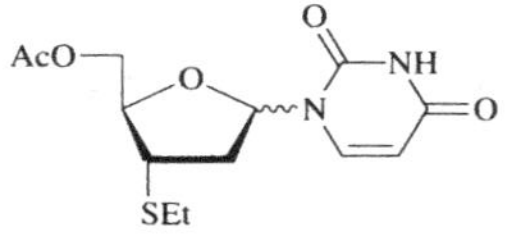 (57)	933

TABLE IV. REACTIONS WITH TRIMETHYLSILYL AND SILVER TRIFLATES AND PERCHLORATES (*Continued*)

Sugar	Base	Conditions	Product(s) and Yield(s) (%)	Refs.
	NHCOPr-*i*; TMSO; N	TMSOTf, MeCN, –30°, 2 h	AcO; O; SEt; N; O; N; NHCOPr-*i* (21) + β-anomer (13)	933
PivO; O; OMe; N_3	OTMS; N; TMSO; N	1. TMSOTf, MeCN, 0°, 30 min 2. NaOMe, MeOH, rt, 1 h	O; NH; HO; O; N; O; N_3 (33)	934
i-PrO; O; O; OPr-*i*	Cl; N; N; AcN; TMS; N; N; TMS	TMSOTf, $Cl(CH_2)_2Cl$, –30°, 2 h	N; N; NHAc; N; N; O; Cl; OPr-*i* (58) N^7:N^9 = 10:1	521
C_{12} O; OAc; SPh	OTMS; N; TMSO; N	TMSOTf, CH_2Cl_2, 25°	O; NH; O; N; O; SPh (—) α:β = 1:4	227
O; AcO; O; O; OAc; OAc	OTMS; N; N; TMSO; N; N; TMS	TMSOTf, $Cl(CH_2)_2Cl$, 70°, 16 h	O; O; N; NH; AcO; O; O; N; N; O; H; OAc (42) + N^7-isomer (22)	935

Substrate	Reagent	Conditions	Product(s) and Yield(s) (%)	Refs.
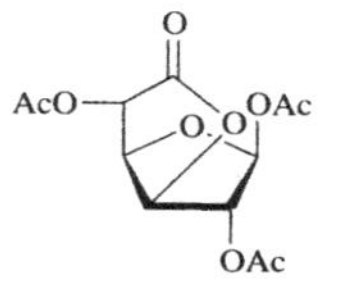	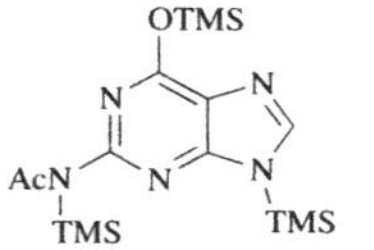	TMSOTf, $Cl(CH_2)_2Cl$, 70°, 16 h	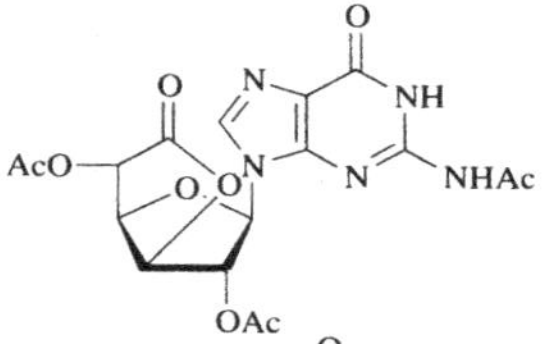 (49) + N^7-isomer (35)	935
	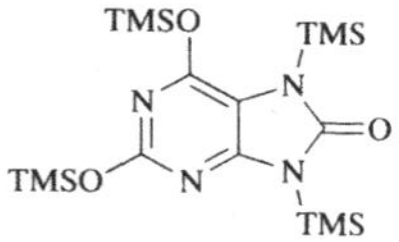	TMSOTf, $Cl(CH_2)_2Cl$, 60°, 15 h	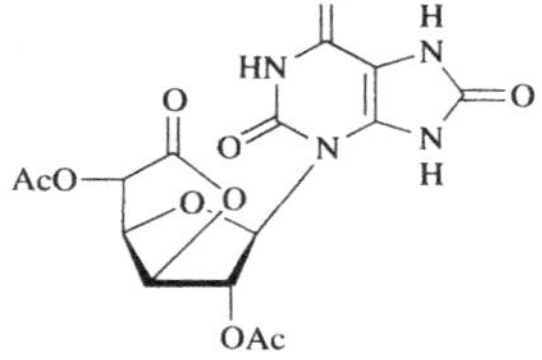 (54)	935
	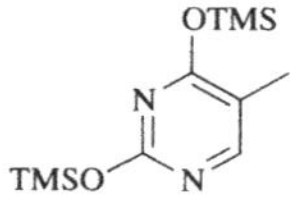	TMSOTf, CH_2Cl_2, 25°, 3 h	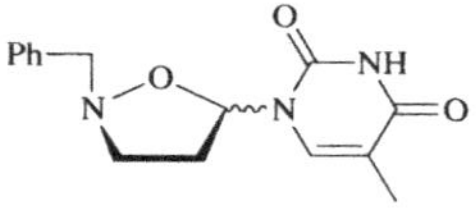 (75)	924
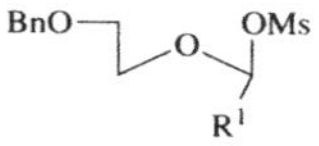	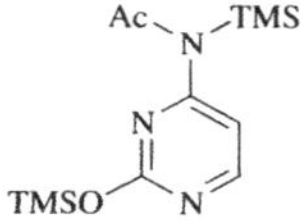	TMSOTf, CH_2Cl_2	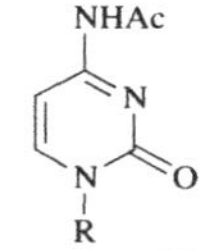 R^1: CF_3 (29); CF_2H (42); R 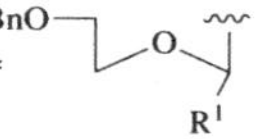 =	936
	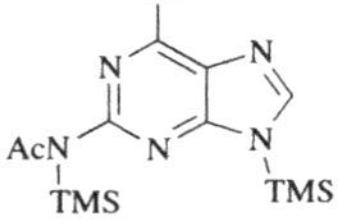	TMSOTf, CH_2Cl_2	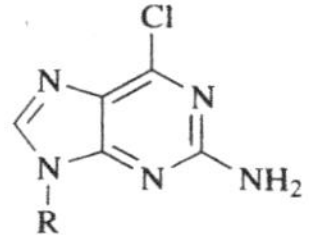 R^1: CF_3 (8); CF_2H (31); N^7:N^9 = 1:1	936
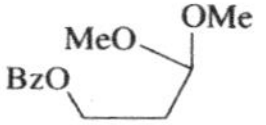	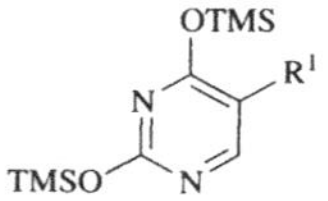	TMSOTf, CH_2Cl_2, –30°, 2-12 h	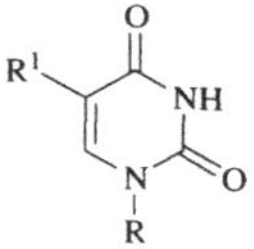 R^1: H (80); F (71); R 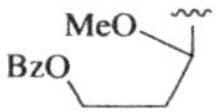 =	923

TABLE IV. REACTIONS WITH TRIMETHYLSILYL AND SILVER TRIFLATES AND PERCHLORATES (*Continued*)

Sugar	Base	Conditions	Product(s) and Yield(s) (%)	Refs.
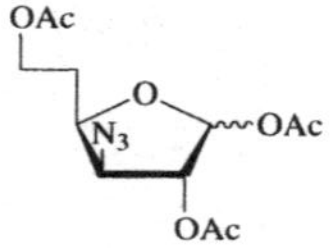 OAc; N_3; OAc; OAc	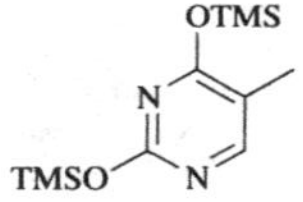 OTMS; TMSO	TMSOTf, MeCN, rt	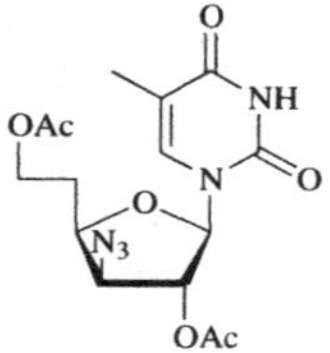 OAc; N_3; OAc; O; NH; O (41)	937
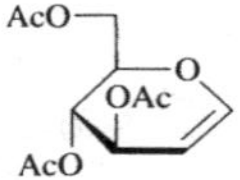 AcO; OAc; AcO	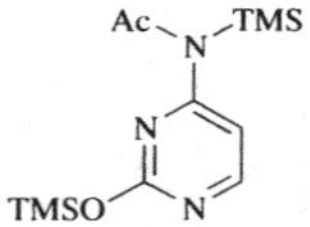 Ac, TMS; TMSO	TMSOTf, CH_2Cl_2, rt, 30 min	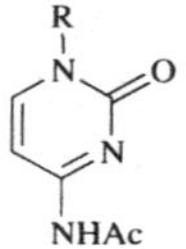 R; O; NHAc (46) + β-anomer (25); R 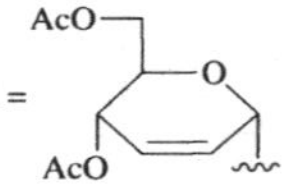 = AcO; AcO	306
	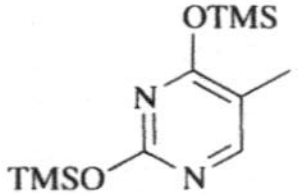 OTMS; TMSO	TMSOTf, CH_2Cl_2, rt, 55 min	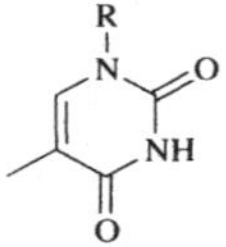 R; O; NH; O (21) + β-anomer (25)	306
	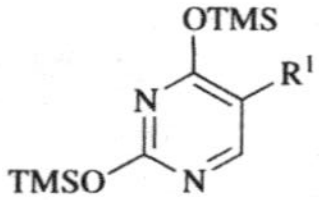 OTMS; R^1; TMSO	TMSOTf, MeCN, 0°	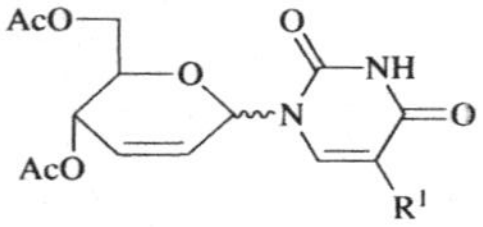 AcO; AcO; O; NH; O; R^1	310

R^1	Yield (%)	α:β
N(Me)Bn	(67)	1:1
1-pyrrolidinyl	(74)	1:1
NHC_6H_4OEt-*p*	(39)	1:1
N-4-methylpiperidinyl	(62)	1:1
N-4-methylpiperazinyl	(39)	1:1
N-4-benzylpiperazinyl	(63)	1:1

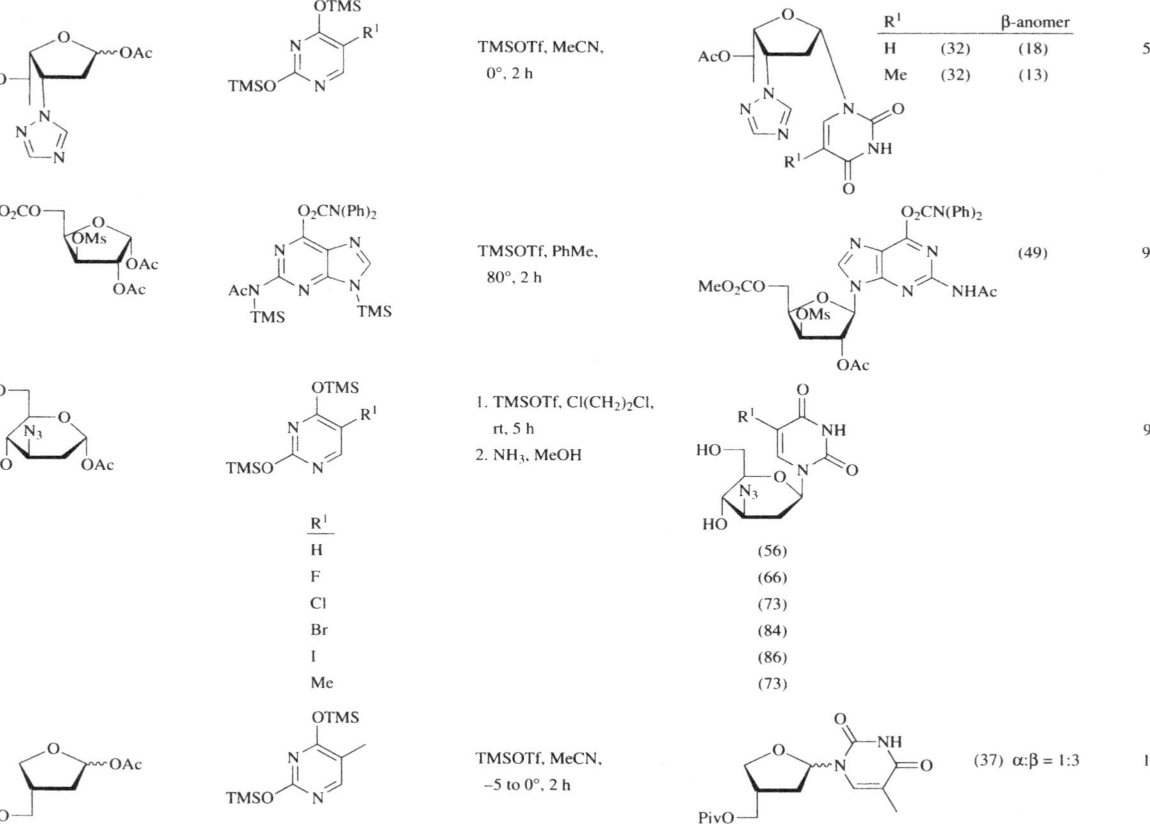

Conditions	Product(s) and Yield(s) (%)	Refs.
TMSOTf, MeCN, 0°, 2 h	R^1 = H (32), β-anomer (18); R^1 = Me (32), β-anomer (13)	531
TMSOTf, PhMe, 80°, 2 h	(49)	938
1. TMSOTf, $Cl(CH_2)_2Cl$, rt, 5 h; 2. NH_3, MeOH	R^1 = H (56); F (66); Cl (73); Br (84); I (86); Me (73)	939
TMSOTf, MeCN, −5 to 0°, 2 h	(37) α:β = 1:3	139a

TABLE IV. REACTIONS WITH TRIMETHYLSILYL AND SILVER TRIFLATES AND PERCHLORATES (*Continued*)

Sugar	Base	Conditions	Product(s) and Yield(s) (%)	Refs.
TBDMSO; F; F; OMs	Ac, N, TMS; N; TMSO; N	TMSOTf, $Cl(CH_2)_2Cl$, reflux, 4 h	TBDMSO; F; F; N; O; N; NHAc (44) α:β = 1:1	506
C_{13} BzO; O; O; OAc	NHTMS; F; N; TMSO; N	1. TMSOTf, CH_2Cl_2, 0°, 20 min 2. rt, 2 h 3. Ac_2O, DMAP	BzO; O; O; N; O; N; F; NH_2 (46) + β–anomer (30)	940
	OTMS; N; TMSO; N	TMSOTf, MeCN, 25°, 66 h	BzO; O; O; N; O; NH; O (76) α:β = 1:1	942
BzO; O; S; OAc	NHTMS; F; N; TMSO; N	1. TMSOTf, CH_2Cl_2, MS, 0°, 20 min 2. 25°, 18 h 3. Ac_2O, DMAP	BzO; O; S; N; O; N; F; NHAc (46) + *cis*-β–anomer (30)	940, 942
BzO; O; OMe	OTMS; N; TMSO; N	TMSOTf, MeCN, 0°, 10 min	BzO; O; N; O; NH; O (80) α:β = 1:1	943

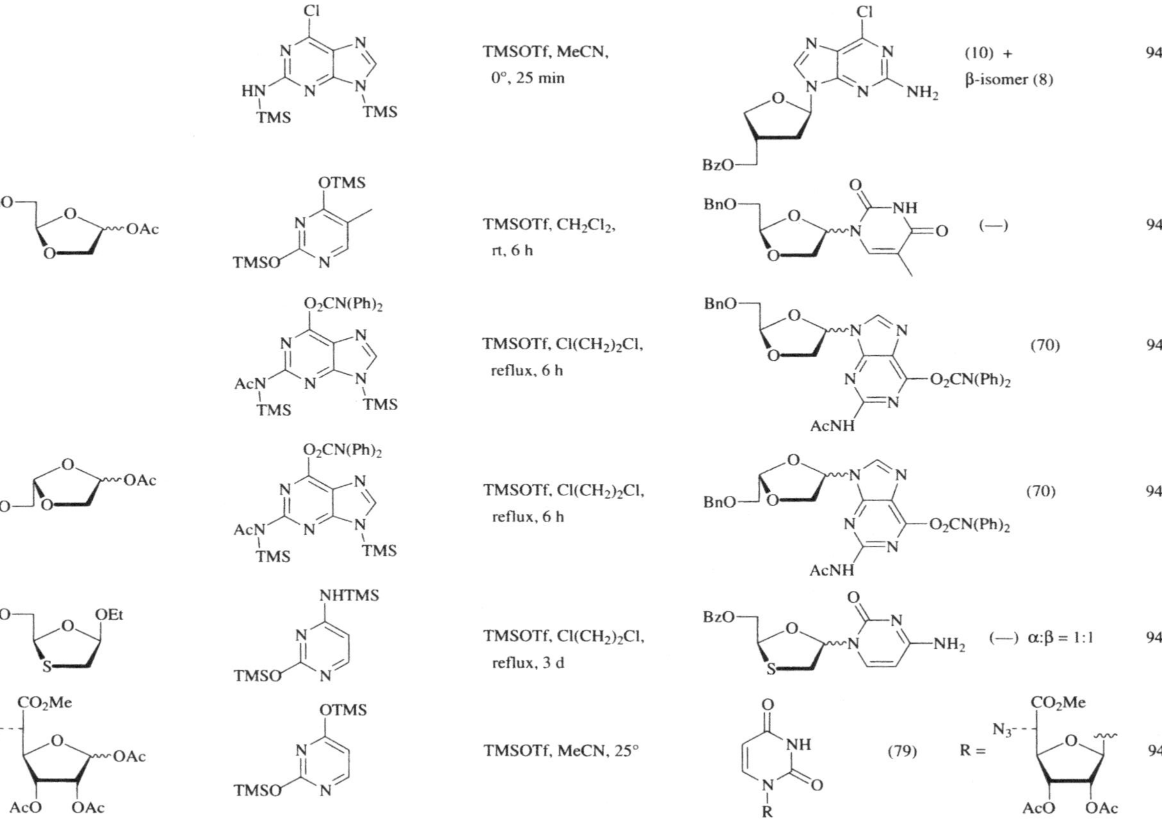

Conditions	Product(s) and Yield(s) (%)	Refs.
TMSOTf, MeCN, 0°, 25 min	(10) + β-isomer (8)	943
TMSOTf, CH_2Cl_2, rt, 6 h	(—)	944
TMSOTf, $Cl(CH_2)_2Cl$, reflux, 6 h	(70)	945
TMSOTf, $Cl(CH_2)_2Cl$, reflux, 6 h	(70)	945
TMSOTf, $Cl(CH_2)_2Cl$, reflux, 3 d	(—) α:β = 1:1	946
TMSOTf, MeCN, 25°	(79) R =	947

TABLE IV. REACTIONS WITH TRIMETHYLSILYL AND SILVER TRIFLATES AND PERCHLORATES (*Continued*)

Sugar	Base	Conditions	Product(s) and Yield(s) (%)	Refs.
	OTMS, OTMS, TMSO, N	TMSOTf, MeCN, 25°	HO, O, NH, N, O, R (66)	947
BnO, MeO, OMe	OTMS, TMSO, N	TMSOTf, CH_2Cl_2, –30°, 2-12 h	O, NH, N, O, MeO, BnO (70)	923
AcO, O, OAc, AcO, OAc	OTMS, N, Cl, N, TMS	TMSOTf, MeCN, rt, 75 h	Cl, HN, O, N, R (50) R = AcO, O, AcO, OAc	627
AcO, O, OAc, AcO, OAc	TMSO, N, TMSNH, O, N, TMS	TMSOTf, $Cl(CH_2)_2Cl$, $SnCl_4$, reflux, 1 h	$CONH_2$, HN^+, O^-, N, R (83)	242
		TMSOTf, $Cl(CH_2)_2Cl$, reflux, 1 h	O^-, HN^+, $CONH_2$, N, R (61) + N^1-isomer (17) + bis(riboside) (9)	242
	TMSO, N, TMSNH, O, N, $C_6H_4NO_2$-*p*	TMSOTf, $Cl(CH_2)_2Cl$, $SnCl_4$, 60°, 1 h	*p*-$O_2NC_6H_4$, $CONH_2$, N^+, O^-, N, R (76)	243

Substrate	Conditions	Product(s) and Yield(s) (%)	Refs.
3-R[1]-2-OTMS-pyridine	TMSOTf, MeCN, rt, 15 h	R^1 = $CONH_2$ (50); CO_2Et (75)	948
NHBz, BzO pyrimidine	1. TMSOTf, $Cl(CH_2)_2Cl$, 0° 2. NH_4OH, MeOH, rt, 3 d	NH$_2$, HO, HO, OH (90)	424
OTMS, TMSO, NH, R^1: Me; *n*-Pr; $(Me)_2$	TMSOTf, $Cl(CH_2)_2Cl$, 60°, 15 min	O, R^1, NH, HN, N–R, O; R = AcO, O, AcO, OAc N^1- + N^6-isomers (15), N^1:N^6 = 2:75 N^1- + N^6-isomers (6), N^1:N^6 = 1:1 N^1- + N^6-isomers (40), N^1:N^6 = 0.7:1	949a
Cl, R^1, N–TMS; R^1:	TMSOTf, MeCN, rt, 16 h	Cl, N–R, R^1	
Cl		(74)	627
Me		(69)	949
MeS		(87)	627
NHTMS, H_2NO_2S, N–TMS	TMSOTf, MeCN, rt, 5 h	NH_2, N–R, SO_2NH_2 (53)	950

TABLE IV. REACTIONS WITH TRIMETHYLSILYL AND SILVER TRIFLATES AND PERCHLORATES (*Continued*)

Sugar	Base	Conditions	Product(s) and Yield(s) (%)	Refs.
	Cl; H_2NO_2S; N; TMS	TMSOTf, MeCN, rt, 2 h	Cl; N; SO_2NH_2; R (71) R = AcO; O; AcO; OAc	950
	TMSNH; SO_2; N; TMS	TMSOTf, MeCN, rt, 1 h	H_2N; SO_2; N; R (46)	951
	$O_2CN(Ph)_2$; AcN; TMS; N; TMS	TMSOTf, PhMe, 80°, 1 h	$O_2CN(Ph)_2$; N; NHAc; R (91)	140
	OTMS; AcN; TMS; N; TMS	TMSOTf, $Cl(CH_2)_2Cl$, reflux, 1.5 h	O; NH; NHAc; N; R (81) N^7:N^9 = 1:3	142
	O; $C_6H_4NO_2$-*p*; Br; N; TMS	TMSOTf, MeCN, reflux, 1 h	*p*-$O_2NC_6H_4$; O; N; Br; R (95) N^7:N^9 = 1:20	143
	TMSO; TMS; TMSO; N; O; TMS	TMSOTf, $Cl(CH_2)_2Cl$, 5°, 72 h	H; H; N; O; NH; O; R (81)	952

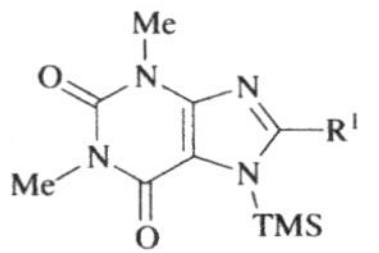

Sugar	Base	Conditions	Product(s) and Yield(s) (%)	Refs.
		TMSOTf, $Cl(CH_2)_2Cl$,	R^1 H 50°, 1 h (91)	953
			R^1 Br 60°, 2 h (75)	953
	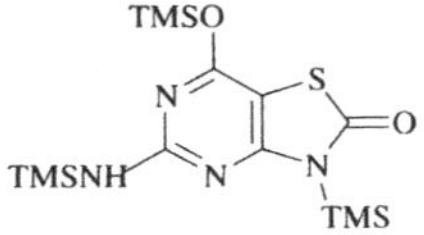	TMSOTf, MeCN, rt, 12-16 h	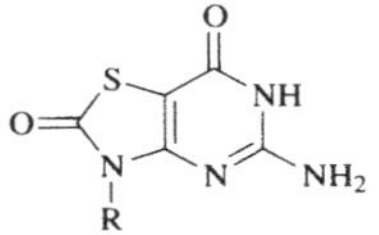 (29)	954
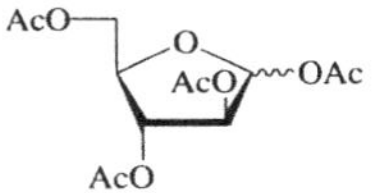	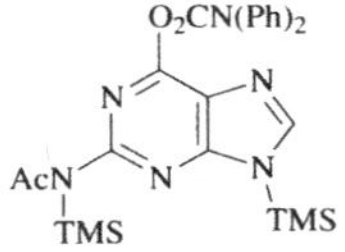	TMSOTf, PhMe, 80°, 1 h	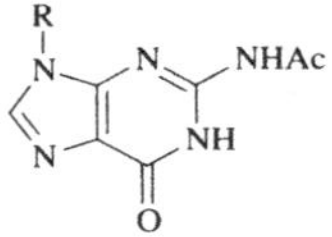 (84) N^7:N^9 = 1:2 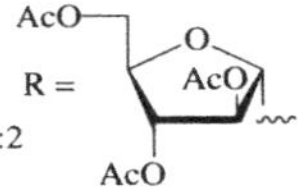	141
		TMSOTf, $Cl(CH_2)_2Cl$, 50°, 1 h	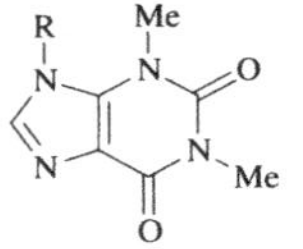 (62)	368
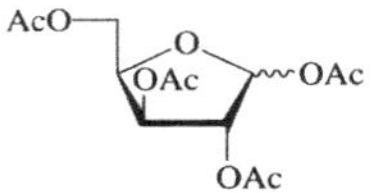	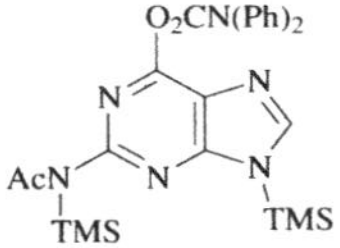	TMSOTf, PhMe, 80°, 1 h	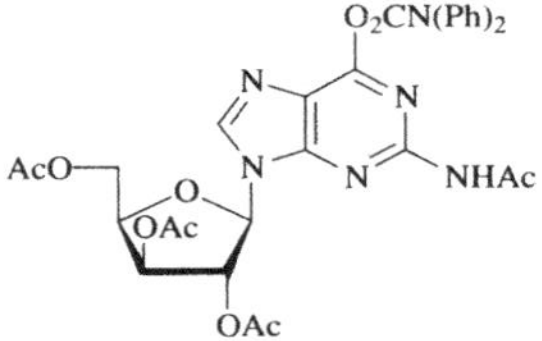 (67) N^7:N^9 = 1:2	141
		1. TMSOTf, $Cl(CH_2)_2Cl$, rt, 3 d 2. NH_4OH, MeOH	(52)	424

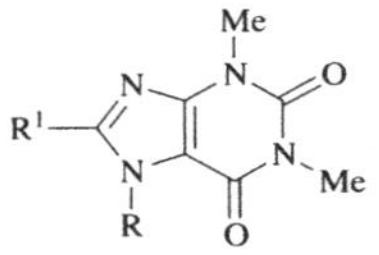

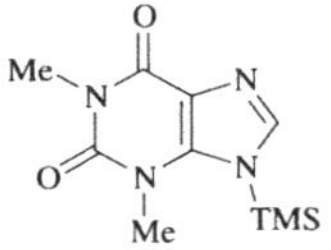

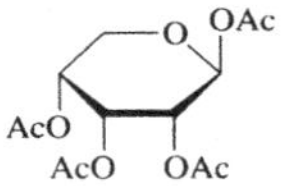

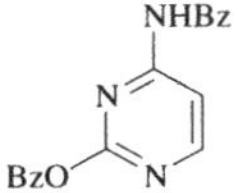

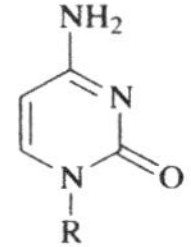

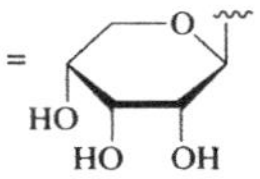

TABLE IV. REACTIONS WITH TRIMETHYLSILYL AND SILVER TRIFLATES AND PERCHLORATES (*Continued*)

Sugar	Base	Conditions	Product(s) and Yield(s) (%)	Refs.
AcO, OAc, AcO, O, OAc	OTMS, N, N, AcN, TMS, N, N, TMS	TMSOTf, $Cl(CH_2)_2Cl$, reflux, 1 h	AcO, O, OAc, AcO, N, N, NHAc, N, NH, O; N^9 (59), N^7 (7)	537
S, OAc, OAc, AcO, OAc	OTMS, N, R[1], TMSO, N	TMSOTf, $Cl(CH_2)_2Cl$, reflux, time	O, R[1], NH, N, O, S, OAc, AcO, OAc	955
			R[1] F, time 5 h (46)	
			R[1] I, time 6 h (52)	
			R[1] Me, time 3.5 h (50)	
	OTMS, N, N, 3, R[1], TMSO, N, N, 2	TMSOTf	O, HN, N, 3, R[1], O, N, N, 2, S, OAc, AcO, OAc	955
	R[1] 3-C_6H_4Cl-*p*	CH_2Cl_2, rt, 6 h	(40)	
	R[1] 2-C_6H_4Cl-*p*	CH_2Cl_2, rt, 7 h	(40)	
	R[1] 3-C_6H_4Br-*p*	$Cl(CH_2)_2Cl$, rt, 48 h	(26)	
	R[1] 2-C_6H_4Br-*p*	$Cl(CH_2)_2Cl$, rt, 24 h	(30)	

Sugar	Base	Conditions	Product(s) and Yield(s) (%)		Refs.
HO(CH$_2$)$_3$O, O(CH$_2$)$_3$OH, Ph	TMSO, OTMS, N, N	TMSOTf, CH$_2$Cl$_2$, –30°, 2-12 h	HO(CH$_2$)$_3$O, Ph, N, NH, O, O (51)		923
p-O$_2$NC$_6$H$_4$CO$_2$, N$_3$, O, OMe	TMSO, OTMS, N, N	1. TMSOTf, MeCN, 90°, 3 h 2. NaOMe, MeOH	HO, N$_3$, O, N, NH, O, O (63)		956
TBDMSO, O, OAc, N$_3$	TMSO, OTMS, N, N, R^1	TMSOTf	TBDMSO, O, N$_3$, N, NH, O, O, R^1		
	R^1			α:β	
	H	Cl(CH$_2$)$_2$Cl, rt, 18 h	(55)	1:2	280
	H	MeCN, rt, 24 h	(76)	1:1	749
	Me	Cl(CH$_2$)$_2$Cl, rt, 24 h	(66)	55:45	749
	Me	Cl(CH$_2$)$_2$Cl	(66)	1:1	280, 749
	Me	MeCN, rt, 24 h	(66)	53:47	749
TBDMSO, O, OAc	TMSO, OTMS, N, N, CF$_3$	TMSOTf, –10°, 30 min	CF$_3$, NH, O, O, N, TBDMSO, O (—)		292

TABLE IV. REACTIONS WITH TRIMETHYLSILYL AND SILVER TRIFLATES AND PERCHLORATES (*Continued*)

Sugar	Base	Conditions	Product(s) and Yield(s) (%)	Refs.
	NHTMS, N, TMSO, N	TMSOTf, $Cl(CH_2)_2Cl$	TBDMSO, O, N, O, N, NH_2 (—) α:β = 2:1	282
	Cl, N, N, N, N, TMS	TMSOTf, $Cl(CH_2)_2Cl$, rt, 30 min	TBDMSO, O, N, N, N, Cl, N (82) α:β = 1:1	298
	O, N, N, AcN, TMS, N, N, TMS	TMSOTf, MeCN, 25°, 4.5 h	TBDMSO, O, N, N, N, O, N, H_2N (56) α:β = 1:1	957
CO_2Me, N_3, O, OAc, AcO, OAc	OTMS, N, TMSO, N	TMSOTf, CH_2Cl_2, rt TMSOTf, $Cl(CH_2)_2Cl$, rt	MeO_2C, N_3, O, NH, N, O, O, AcO, OAc (93) (84)	958 959
C_{14} BzO, O, F, Br, AcO	OTMS, N, TMSO, N, R^1	TMSOTf, MeCN, reflux, time	R^1, O, NH, BzO, N, O, O, F, AcO R^1 F, time 1.5 h (25) R^1 Cl, time 5 h (21)	360

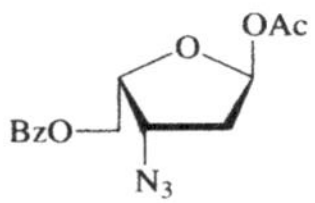	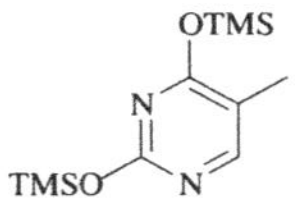	1. TMSOTf, MeCN, $-35°$, 45 min 2. rt, 30 min	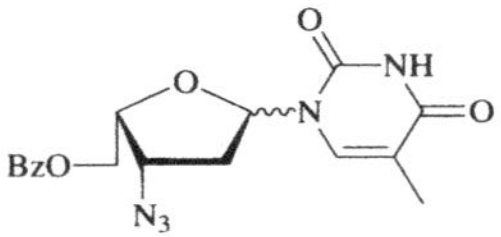 (58) $\alpha:\beta = 5:3$	960, 961
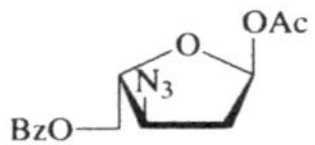	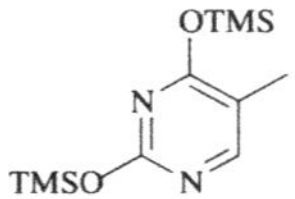	1. TMSOTf, MeCN, $-30°$, 20 min 2. rt, 30 min	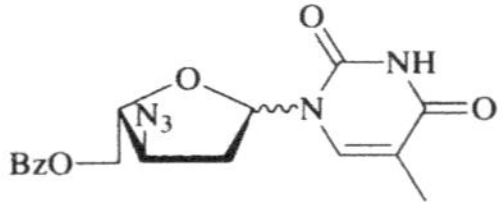 (72) $\alpha:\beta = 2:3$	960
	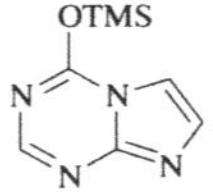	TMSOTf, MeCN, 25°, 4 h	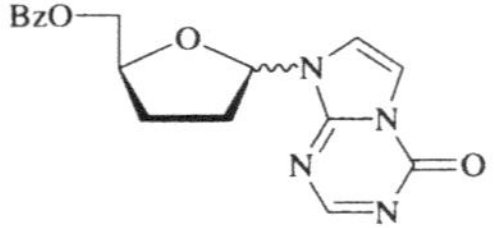 (76) $\alpha:\beta = 1:1$	957
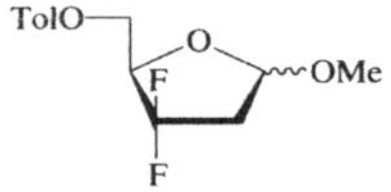	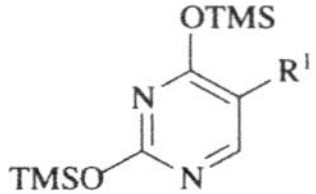	TMSOTf, MeCN, reflux, 3 h	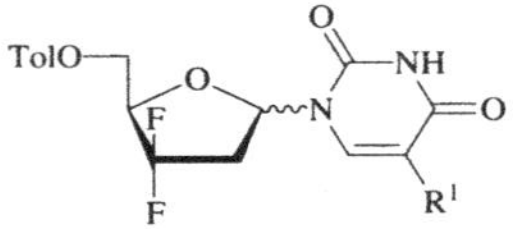 R^1 / $\alpha:\beta$: H (26) 1:1 Me (31) 1:1	506
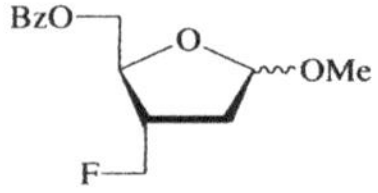	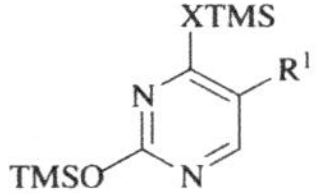	1. TBDMSOTf, CH_2Cl_2, rt, 24 h 2. NH_3, MeOH, rt, 24 h	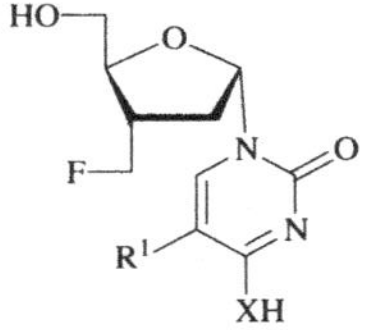 R^1 / X / β-anomer: H NH (38) (26) Me O (36) (28)	962, 963
	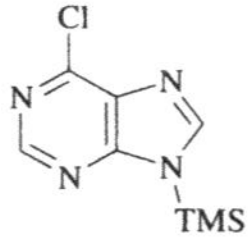	1. TBDMSOTf, CH_2Cl_2, rt or 40°, 24 h 2. NH_3, MeOH, rt, 24 h or 100°, 20 h	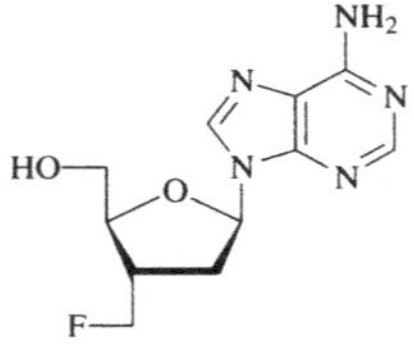 (39) + α-anomer (26)	962, 963

TABLE IV. REACTIONS WITH TRIMETHYLSILYL AND SILVER TRIFLATES AND PERCHLORATES (*Continued*)

Sugar	Base	Conditions	Product(s) and Yield(s) (%)	Refs.
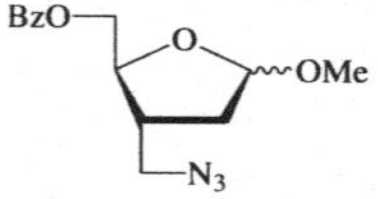 BzO, OMe, N_3	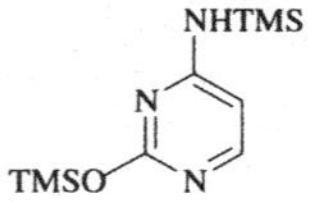 NHTMS, N, TMSO, N	1. TBDMSOTf, CH_2Cl_2, rt, 24 h 2. NH_3, MeOH, rt, 24 h	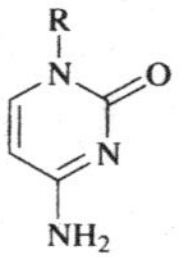 R, N, O, N, NH_2 (43) + β-anomer (23) R 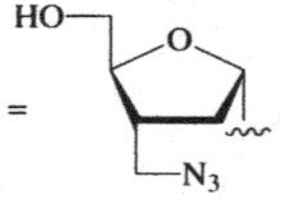 = HO, O, N_3	962, 963
	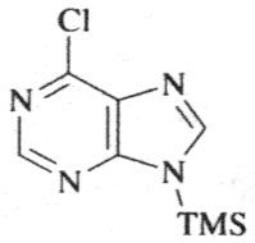 Cl, N, N, N, N, TMS	1. TBDMSOTf, CH_2Cl_2, rt, 24 h 2. NH_3, MeOH, rt, 24 h	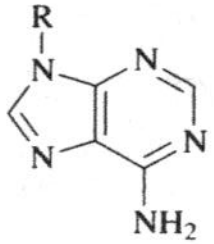 R, N, N, N, N, NH_2 (17) + β-anomer (17)	962
	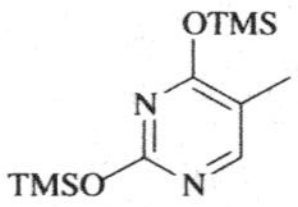 OTMS, N, TMSO, N	1. TBDMSOTf, CH_2Cl_2, rt, 24 h 2. NH_3, MeOH, rt, 24 h	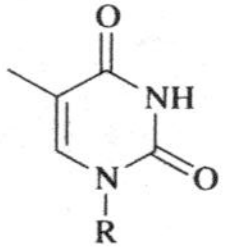 O, NH, N, O, R (30) + α-anomer (29) R 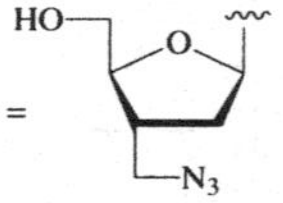 = HO, O, N_3	962, 963
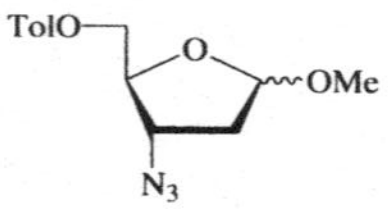 TolO, OMe, N_3	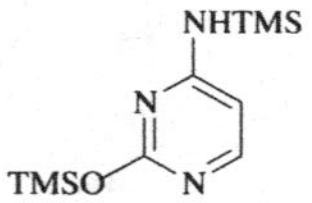 NHTMS, N, TMSO, N	1. TBDMSOTf, CH_2Cl_2, reflux, 6 h 2. NH_3, MeOH, rt, 24 h	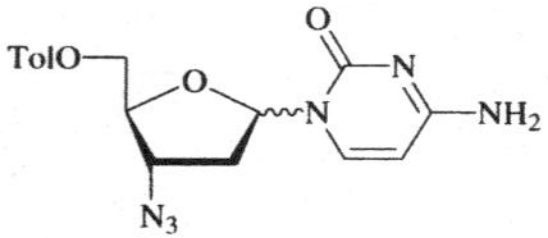 TolO, O, N_3, N, O, N, NH_2 (37)	279
	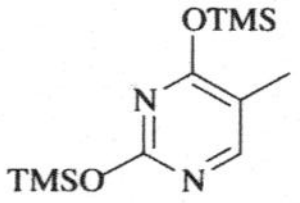 OTMS, N, TMSO, N	1. TBDMSOTf, CH_2Cl_2, reflux, 4 h 2. NH_3, MeOH, 20°, 18 h	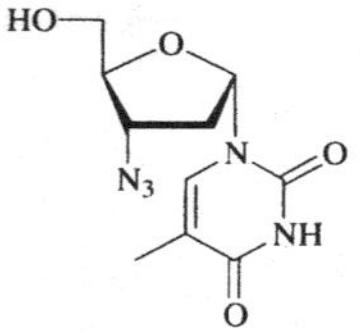 HO, O, N_3, N, O, NH, O (32) + β-anomer (20)	279

Sugar	Base	Conditions	Product(s) and Yield(s) (%)	Refs.
	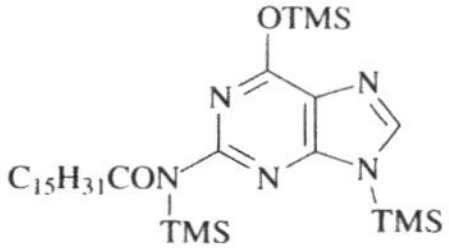 $C_{15}H_{31}CON(TMS)$–purine, OTMS, N-TMS	1. TBDMSOTf, CH_2Cl_2, reflux, 3 h 2. NH_3, MeOH, rt, 24 h	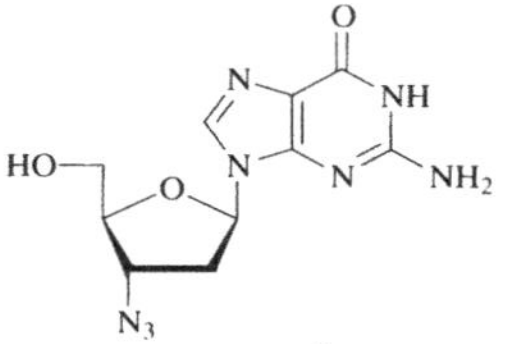 HO–, N_3 nucleoside (guanine: O, NH, NH_2) (29)	279
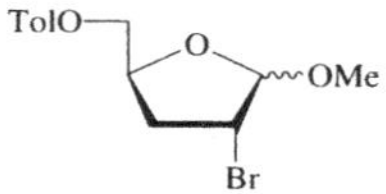 TolO–, Br, OMe	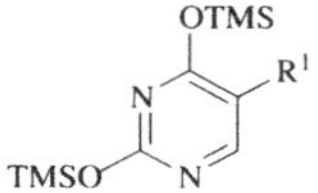 TMSO, OTMS, R^1 pyrimidine	1. TMSOTf, MeCN, 0° 2. rt, time	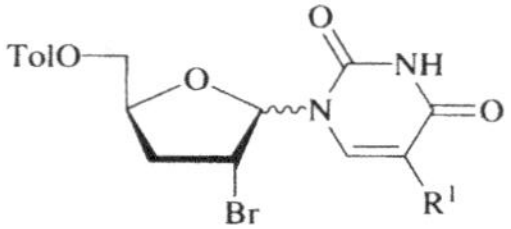 TolO–, Br, R^1 uracil nucleoside R^1 / time: H 12 h (63-68); F 4 h (63-68); Me 24 h (63-68)	964
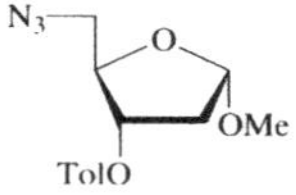 N_3–, TolO, OMe	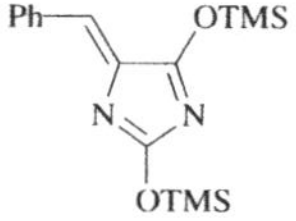 Ph, OTMS, OTMS imidazole	1. TMSOTf, MeCN, –50°, 5 min 2. –30°, 12 h	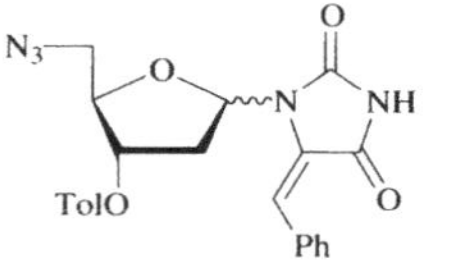 N_3–, TolO, NH, O, Ph (42) α:β = 1:1	965
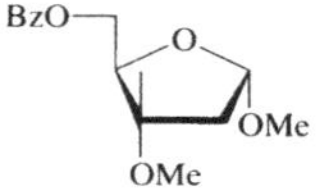 BzO–, OMe, OMe	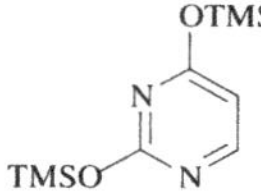 TMSO, OTMS pyrimidine	TMSOTf, $Cl(CH_2)_2Cl$	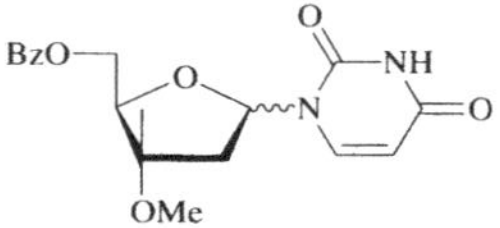 BzO–, OMe, NH, O (—) α:β = 1:1	966
 TolO–, OMe	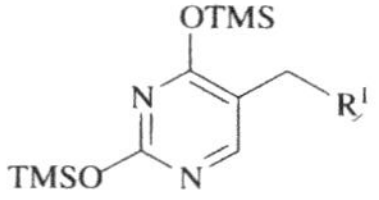 TMSO, OTMS, R^1 pyrimidine R^1: OTMS, OMe, OBn	TMSOTf, CH_2Cl_2, 0°, 2-3 h	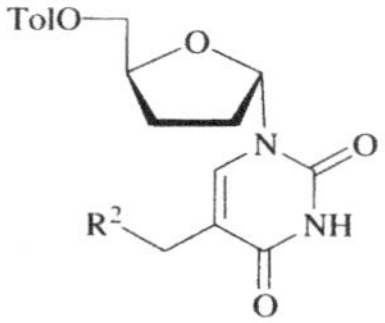 TolO–, R^2 nucleoside R^2 / (α+β) / β-anomer: OH (18) (12); OMe (24) (16); OBn (24) (16)	967

TABLE IV. REACTIONS WITH TRIMETHYLSILYL AND SILVER TRIFLATES AND PERCHLORATES (*Continued*)

Sugar	Base	Conditions	Product(s) and Yield(s) (%)	Refs.
	OTMS, R[1], N, TMSO, N	1. TMSOTf, MeCN, 0°, 15 min 2. rt, 0.5-3 h	TolO, O, N, O, NH, R[1], O	968
	R[1]			**β-anomer**
	2-pyrrolidinyl		(37)	(16)
	N(Me)Bn		(—)	(—)
	N(Me)C_6H_{11}		(22)	(12)
	NHC_6H_4OEt-*o*		(—)	(—)
AcO, O, OMe, PhX	OTMS, N, TMSO, N	TBDMSOTf, CH_2Cl_2	AcO, O, N, O, NH, O, PhX X: S (30); Se (50)	969
BzO, O, $O(CH_2)_2OMe$, S	N—N, N, N, CO_2Et, TMS	1. TMSOTf, MeCN, 0°, 5 h 2. rt, 15 h	BzO, O, N, N=N, N, S, EtO_2C (50) α:β = 4:1	970

		$AgClO_4$, $MeNO_2$, 20°	(70)	R = 74
		$AgClO_4$, C_6H_6, 20°, 0.25 h	(61)	80
		$AgClO_4$, PhMe, 60°, 75 min	(85)	971
		$AgClO_4$, PhMe, 70°	(45)	74
		$AgClO_4$, PhMe, 20°	(70)	74
		$AgClO_4$, C_6H_6, 20°, 16 h	(50)	R = 80

TABLE IV. REACTIONS WITH TRIMETHYLSILYL AND SILVER TRIFLATES AND PERCHLORATES (*Continued*)

Sugar	Base	Conditions	Product(s) and Yield(s) (%)	Refs.
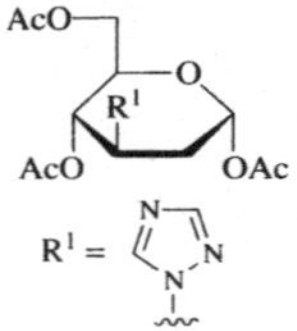 AcO, R[1], AcO, OAc; R^1 = 1,2,4-triazol-1-yl	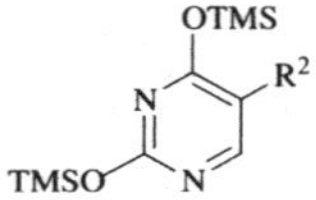 TMSO, N, N, OTMS, R^2	1. TMSOTf, MeCN, 0°, 1 h 2. rt, 23 h	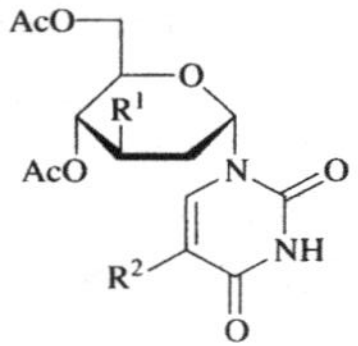 AcO, R[1], AcO, N, O, NH, O, R^2 R^2 / / β-anomer H (14) (8) Me (21) (9)	972
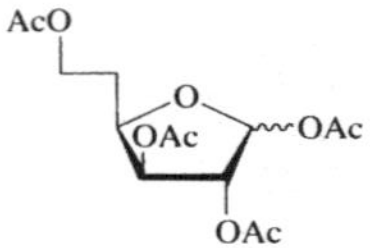 AcO, OAc, OAc, OAc	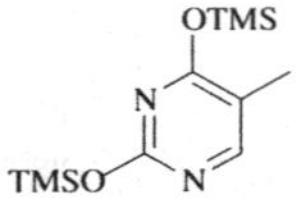 TMSO, N, N, OTMS	1. TMSOTf, MeCN, rt, 4 h 2. NaOMe, MeOH, rt, 1 h	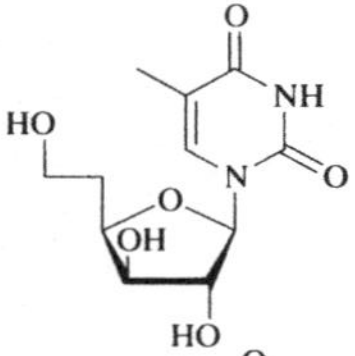 HO, OH, HO, N, O, NH, O (79)	937
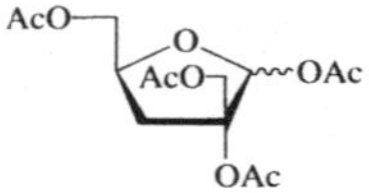 AcO, AcO, OAc, OAc	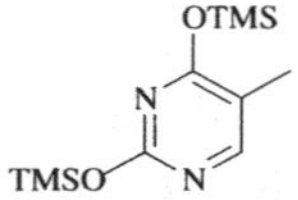 TMSO, N, N, OTMS	TMSOTf, MeCN, rt	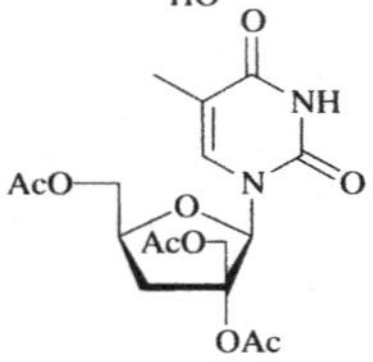 AcO, AcO, OAc, N, O, NH, O (50)	973
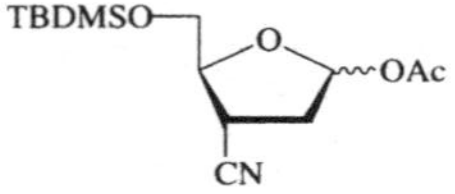 TBDMSO, CN, OAc	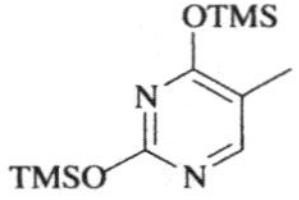 TMSO, N, N, OTMS	TMSOTf, CH_2Cl_2, rt, 1 h	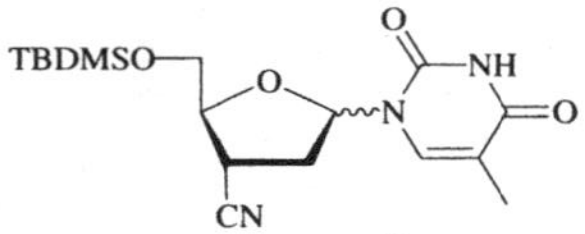 TBDMSO, CN, N, O, NH, O (91) α:β = 5:4	114
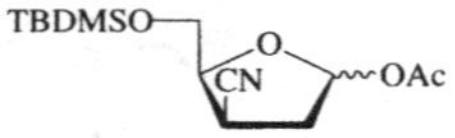 TBDMSO, CN, OAc	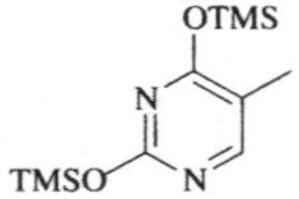 TMSO, N, N, OTMS	TMSOTf, CH_2Cl_2, rt, 1 h	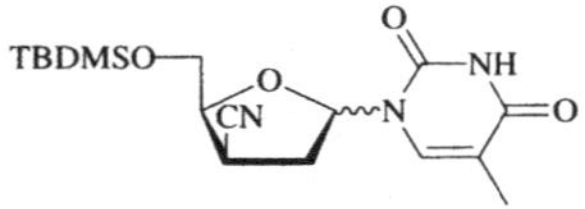 TBDMSO, CN, N, O, NH, O (50) α:β = 3:7	114

Carbohydrate	Heterocycle	Conditions	Product(s) and Yield(s) (%)	Refs.
C_{15} BzO, F, NHCOCF₃, OMe 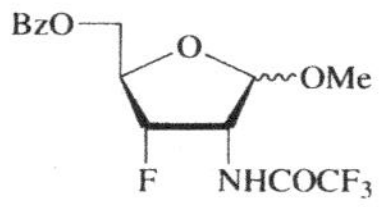	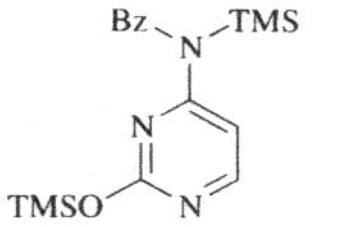 Bz(TMS)N, TMSO pyrimidine	TMSOTf, MeCN, reflux, 1.5 h	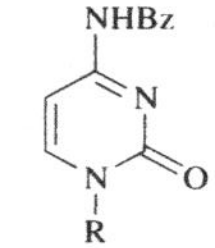 NHBz, N–R (82) R = BzO, F, NHCOCF₃ 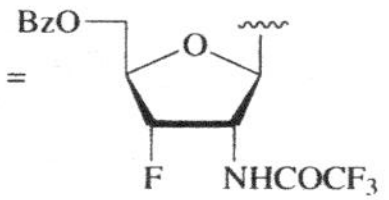	763
	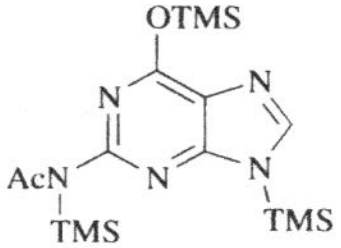 OTMS, AcN(TMS), N-TMS purine	TMSOTf, MeCN, reflux, 30 min	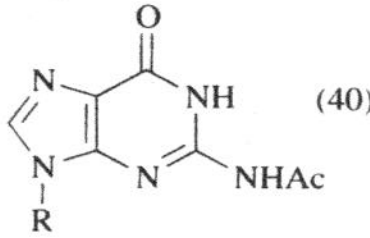 NH, NHAc, N–R (40) + N^7-isomer (20)	763
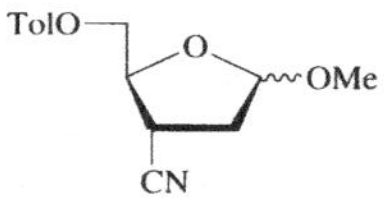 TolO, CN, OMe	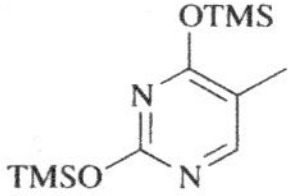 OTMS, TMSO pyrimidine	TMSOTf, MeCN, reflux, 3 h	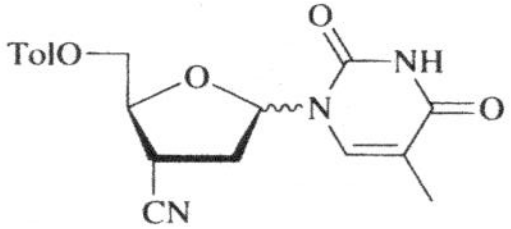 TolO, CN, NH (62) α:β = 1:1	287
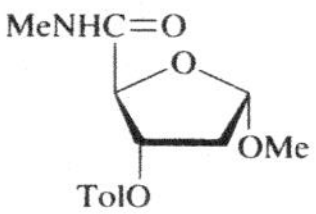 MeNHC=O, TolO, OMe	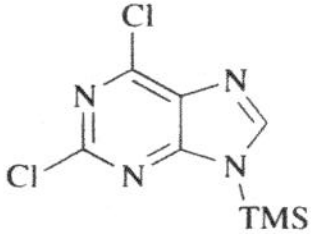 Cl, Cl, N-TMS purine	TMSOTf, CH_2Cl_2, rt, 16 h	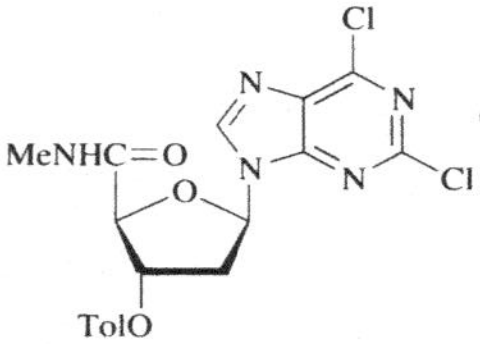 MeNHC=O, TolO, Cl, Cl (64) α:β = 1:1	909[a]
 BzO, OAc	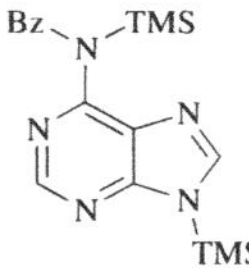 Bz(TMS)N, N-TMS purine	TMSOTf, MeCN, 4 h	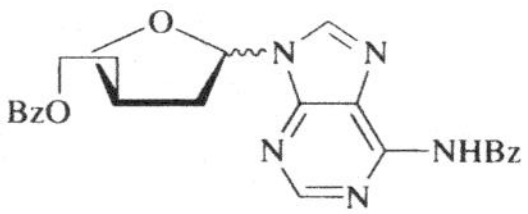 BzO, NHBz (35) α:β = 3:2	974
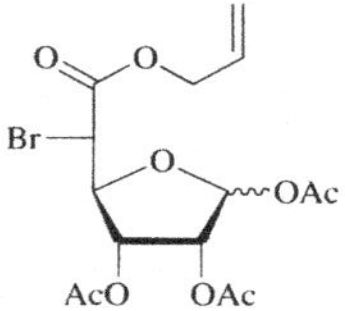 Br, AcO, OAc, OAc	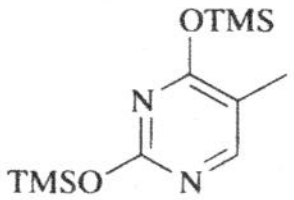 OTMS, TMSO pyrimidine	TMSOTf, MeCN, 60°, 18 h	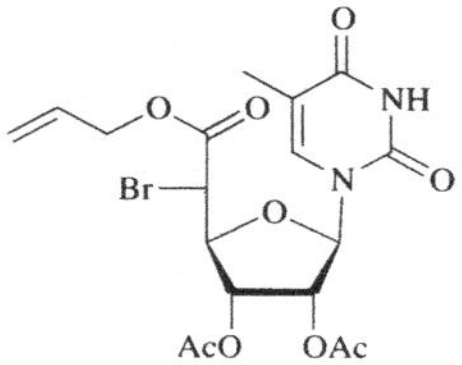 Br, AcO, OAc, NH (75)	975

TABLE IV. REACTIONS WITH TRIMETHYLSILYL AND SILVER TRIFLATES AND PERCHLORATES (*Continued*)

Sugar	Base	Conditions	Product(s) and Yield(s) (%)	Refs.
AcNH, CO_2Me, OAc, AcO, OAc	OTMS, N, TMSO, N	TMSOTf, $Cl(CH_2)_2Cl$, reflux	MeO_2C, AcNH, O, NH, N, O, AcO, OAc (23)	789
CO_2Me, AcNH, O, OAc AcO, OAc	OTMS, N, TMSO, N	TMSOTf, $Cl(CH_2)_2Cl$, 95°	CO_2Me, AcNH, O, OAc AcO, N, O, NH, O (79)	789
O, $(EtO)_2P$, O, OMe, t-BuO_2C	OTMS, N, TMSO, N	1. TMSOTf, MeCN, 10°, 0.5 h 2. rt, 24 h	O, NH, O, $(EtO)_2P$, N, O, t-BuO_2C (19) + α-anomer (12)	976
C_{16}				
BzO, O, OBz	OTMS, N, TMSO, N	TMSOTf, CH_2Cl_2, 0 or 25°, 8 h	O, NH, BzO, O, N, O (45-50)	650, 977
	R^1, N, N, N, N, TMS	TMSOTf, CH_2Cl_2, 0°	BzO, O, N, N, N, N, R^2 R^1 / R^2 Cl / Cl (42) NHTMS / NH_2 (39)	977

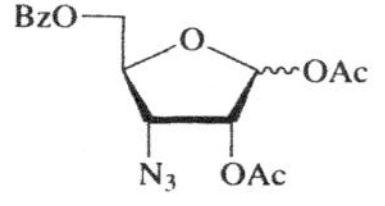 BzO, O, OAc, N_3, OAc	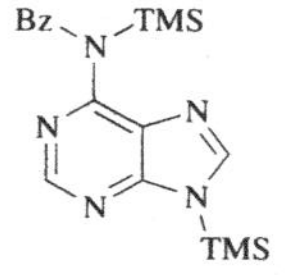 Bz, N, TMS, N, N, N, N, TMS	TMSOTf, $Cl(CH_2)_2Cl$, reflux, 5 h	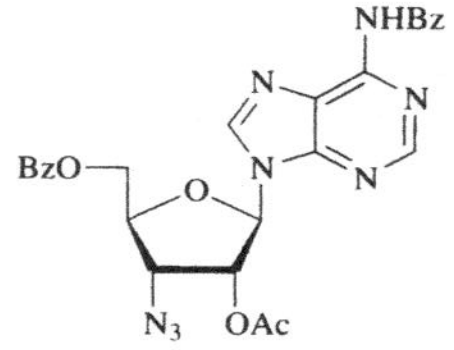 NHBz, N, N, N, N, BzO, O, N_3, OAc (77)	979
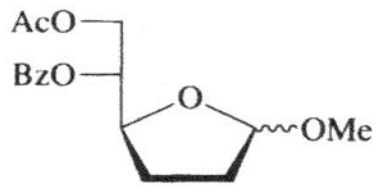 AcO, BzO, O, OMe	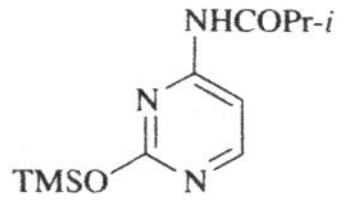 NHCOPr-*i*, N, TMSO, N	1. TMSOTf, MeCN, 0°, 20 min 2. 20°, 30 min	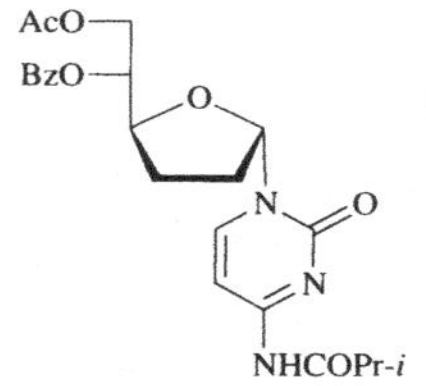 AcO, BzO, O, N, O, N, NHCOPr-*i* (45) + β-anomer (18)	980
	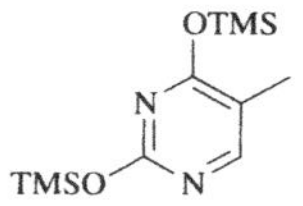 OTMS, N, TMSO, N	TMSOTf, MeCN, 0°, 2 h	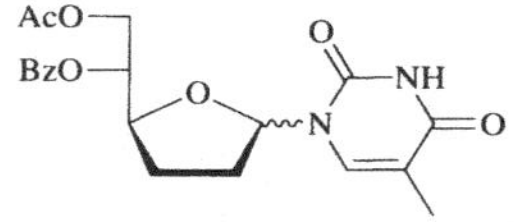 AcO, BzO, O, O, NH, N, O (37) α:β = 3:2	980
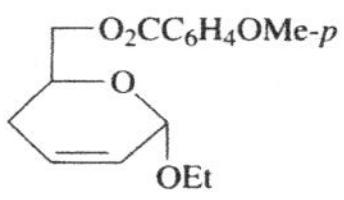 $O_2CC_6H_4OMe$-*p*, O, OEt	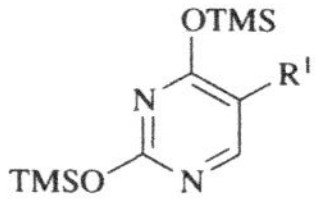 OTMS, N, R^1, TMSO, N	TMSOTf, MeCN	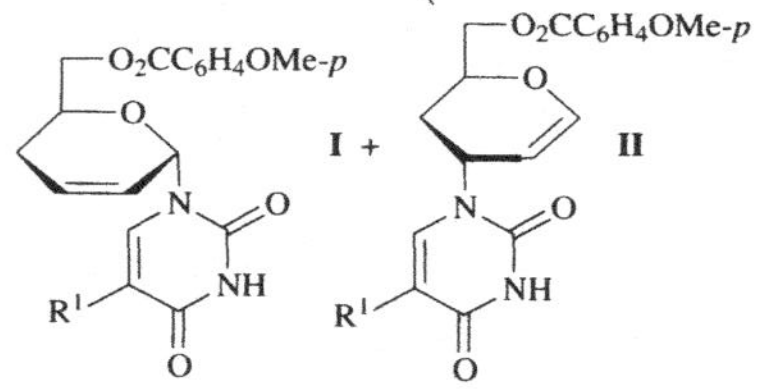 $O_2CC_6H_4OMe$-*p*, O, N, O, NH, R^1, O **I** + $O_2CC_6H_4OMe$-*p*, O, N, O, NH, R^1, O **II**	311
	R^1			
	H	15 min	**I** (36) + β-anomer (8) + **II** (24)	
	Me	0°, 30 min	**I** (23) + β-anomer (13) + **II** (11)	

TABLE IV. REACTIONS WITH TRIMETHYLSILYL AND SILVER TRIFLATES AND PERCHLORATES (*Continued*)

Sugar	Base	Conditions	Product(s) and Yield(s) (%)	Refs.
	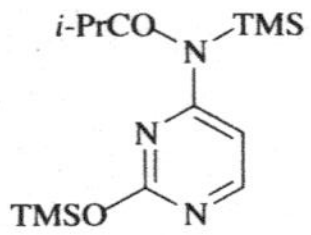 *i*-PrCO, TMS, N, N, TMSO, N	TMSOTf, MeCN, 0°, 2 h	 $O_2CC_6H_4OMe$-*p*, O, N, O, N, NHCOPr-*i* **I** + $O_2CC_6H_4OMe$-*p*, O, N, O, N, NHCOPr-*i* 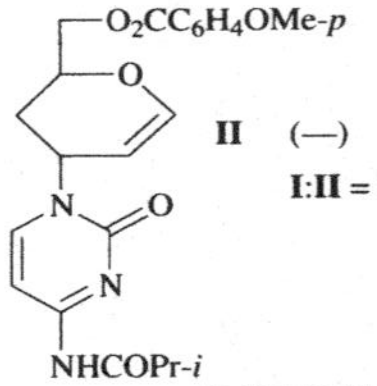 **II** (—) **I**:**II** = 59:41	311
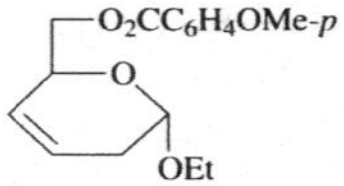 $O_2CC_6H_4OMe$-*p*, O, OEt	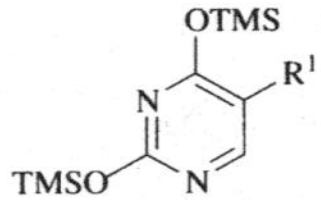 OTMS, N, R^1, TMSO, N	TMSOTf, MeCN, rt, time	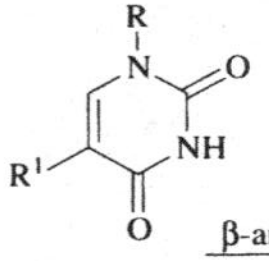 R, 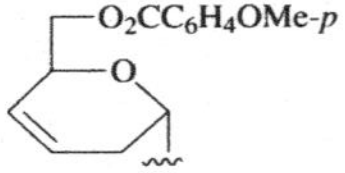 N, O, NH, R^1, O R = $O_2CC_6H_4OMe$-*p*, O	311
	R^1	time	β-anomer	
	H	5 h	(37) (36)	
	Me	2 h	(22) (15)	
	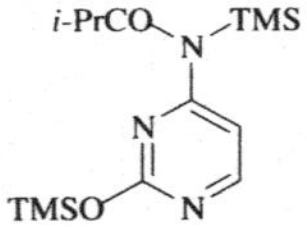 *i*-PrCO, TMS, N, N, TMSO, N	TMSOTf, MeCN, rt, 3 h	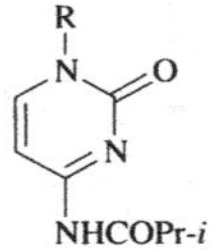 R, N, O, N, NHCOPr-*i* (23) + β-anomer (19)	311
	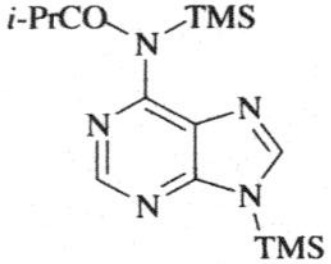 *i*-PrCO, TMS, N, N, N, N, N, TMS	TMSOTf, MeCN, 0°, 4 h	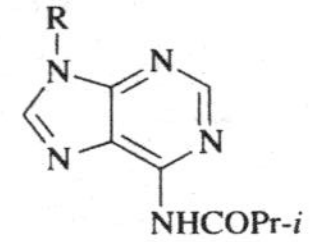 R, N, N, N, N, NHCOPr-*i* (13) + β-anomer (8)	311
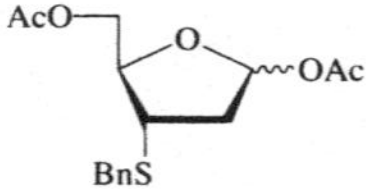 AcO, O, OAc, BnS	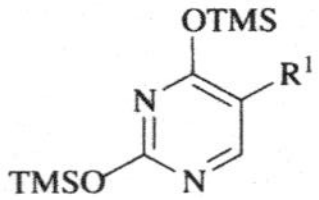 OTMS, N, R^1, TMSO, N	TMSOTf, MeCN, –30°, 2 h	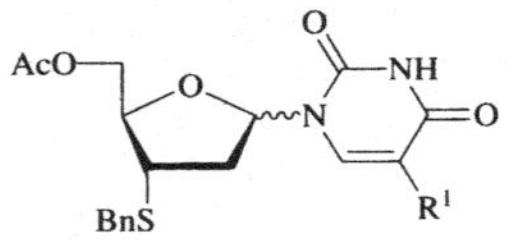 AcO, O, BnS, N, O, NH, O, R^1 — R^1: H (57); Me (33)	933

Substrate	Silylated base	Conditions	Product(s) and Yield(s) (%)	Refs.
	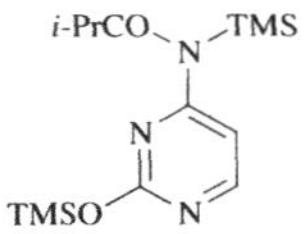 i-PrCO–N(TMS), TMSO, N	TMSOTf, MeCN, –30°, 2 h	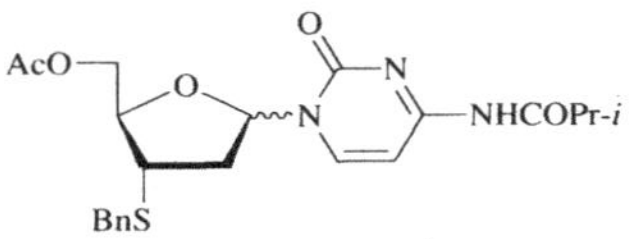 AcO, O, BnS, N, NHCOPr-i (30)	933
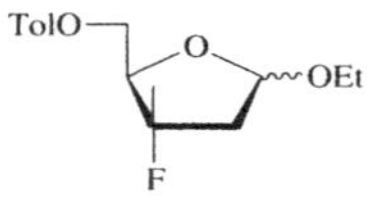 TolO, O, OEt, F	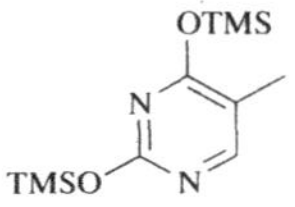 OTMS, N, TMSO	TMSOTf, MeCN, rt, 2.5 h	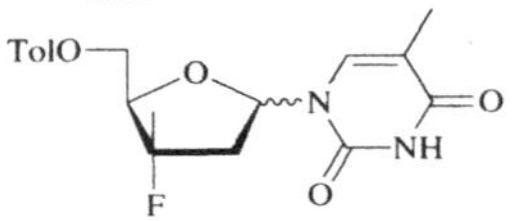 TolO, O, F, N, NH, O (63)	981
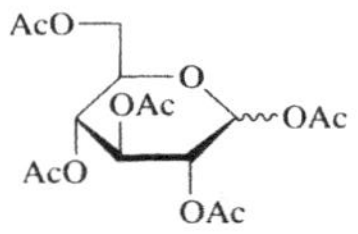 AcO, O, OAc, AcO, OAc, OAc	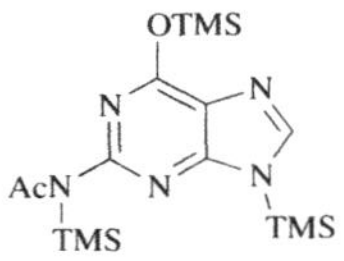 OTMS, N, AcN, TMS, TMS	TMSOTf, $Cl(CH_2)_2Cl$, reflux, 16 h	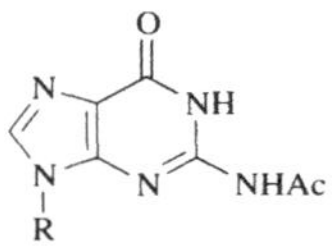 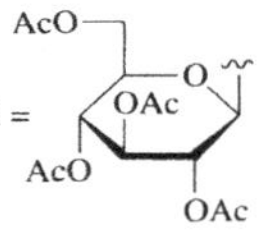 O, N, NH, NHAc, R; R = AcO, OAc, O, AcO, OAc (72) N^7:N^9 = 1:2	142
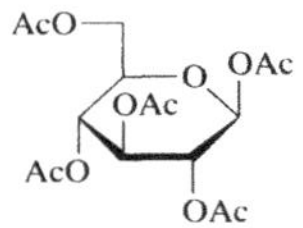 AcO, O, OAc, OAc, AcO, OAc	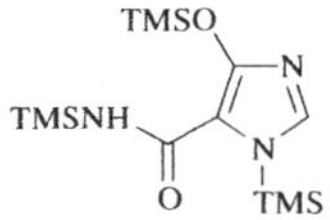 TMSO, N, TMSNH, O, TMS	TMSOTf, $Cl(CH_2)_2Cl$, $SnCl_4$, reflux, 1 h	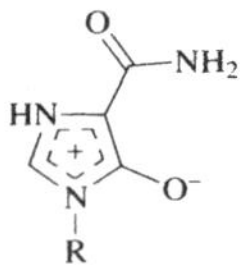 O, NH_2, HN, (+), N, R, O^- **I** (65)	242
		TMSOTf, $Cl(CH_2)_2Cl$, reflux, 1 h	**I** (22) + N^3-isomer (49) + bis(glucoside) (14)	242
	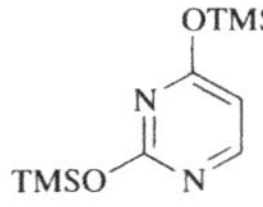 OTMS, N, TMSO	TMSOTf, $Cl(CH_2)_2Cl$, reflux, 2.5 h	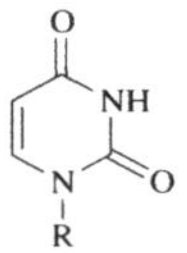 O, NH, N, O, R (92)	133

TABLE IV. REACTIONS WITH TRIMETHYLSILYL AND SILVER TRIFLATES AND PERCHLORATES (*Continued*)

Sugar	Base	Conditions	Product(s) and Yield(s) (%)	Refs.
	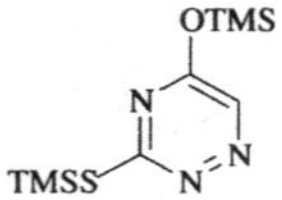 OTMS, TMSS, N, N, N	TMSOTf, $Cl(CH_2)_2Cl$, C_6H_6, reflux, 3.5 h	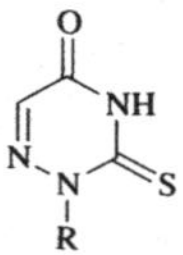 O, NH, S, N, N, R (65)	133
	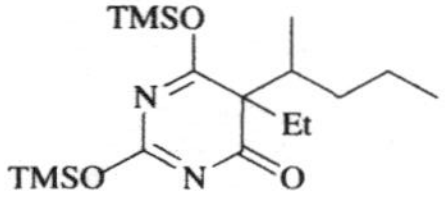 TMSO, N, Et, TMSO, N, O	TMSOTf, $Cl(CH_2)_2Cl$, CH_2Cl_2, rt, 18 h	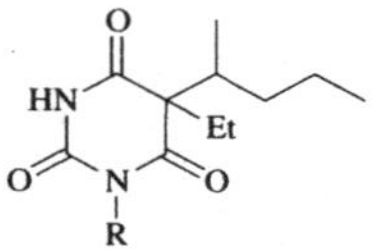 O, HN, Et, O, N, O, R (59)	982
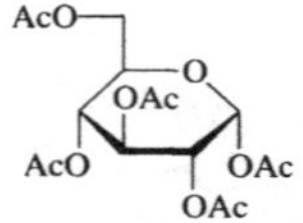 AcO, O, OAc, AcO, OAc, OAc	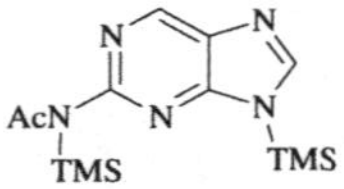 N, N, AcN, TMS, N, N, TMS	TMSOTf, $Cl(CH_2)_2Cl$, reflux, 16 h	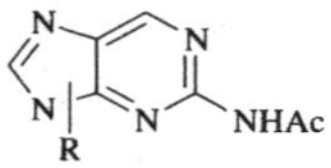 N, N, N, R, N, NHAc (51) N^9:N^7 = 96:4	409
	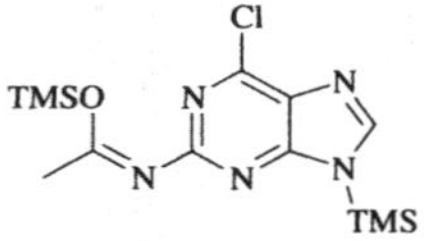 Cl, TMSO, N, N, N, N, N, TMS	TMSOTf, $Cl(CH_2)_2Cl$, reflux, 1 h	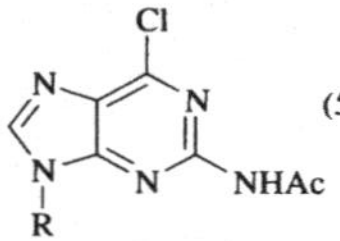 Cl, N, N, N, R, N, NHAc (56) + N^7-isomer (11)	409
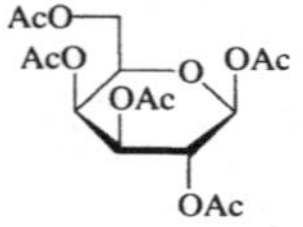 AcO, AcO, O, OAc, OAc, OAc	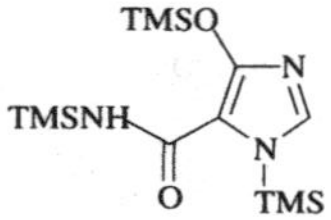 TMSO, N, TMSNH, N, O, TMS	TMSOTf, $Cl(CH_2)_2Cl$, $SnCl_4$, reflux, 1 h	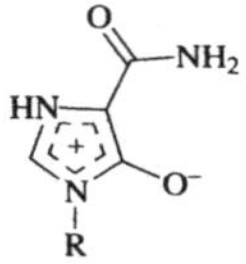 O, NH_2, HN, +, O^-, N, R **I** (67) R 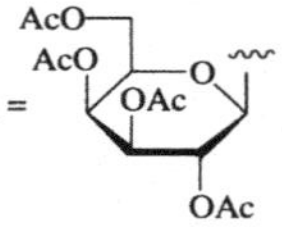 = AcO, AcO, O, OAc, OAc	242
		TMSOTf, $Cl(CH_2)_2Cl$, reflux, 1 h	**I** (6) + N^3-isomer (47) + bis(glucoside) (10)	242

		TMSOTf, PhMe, 80°	(52) + (35)	143a
		TMSOTf, $Cl(CH_2)_2Cl$, reflux, 5-16 h	**I** (48) + N^7-isomer (16)	409
C_{17} (R^1 = pyrrolidin-1-yl)		1. TMSOTf, $Cl(CH_2)_2Cl$, −20°, 1 h 2. $NaHCO_3$, H_2O	(37) + β-anomer (34)	520
		1. TMSOTf, MeCN, 0°, 15 min 2. rt, 2 h	(54)	983

TABLE IV. REACTIONS WITH TRIMETHYLSILYL AND SILVER TRIFLATES AND PERCHLORATES (*Continued*)

Sugar	Base	Conditions	Product(s) and Yield(s) (%)	Refs.
AcO, OAc, O, N, O, O	Ph, OTMS, N, N, OTMS	1. TMSOTf, MeCN, –50°, 5 min 2. –30°, 12 h	HN, O, O, N, R, Ph; R = AcO, O, N, O, O (45) + α-anomer (44)	965
	OTMS, N, R[1], TMSO, N	TMSOTf, MeCN, –30°, time	O, R[1], NH, N, O, R	
	R[1]	**time**		
	Cl	2 h	(40) + α-anomer (11)	984
	Me	2 h	(17)	984
	Me	2 h	(39) + α-anomer (24)	985
	CH_2OTMS	2 h	(30) α:β = 2:3	986
	CH_2OMe	2 h	(30)	986
	CH_2OBu-*s*	2-3 h	(26)	986
	$CH_2O(CH_2)_2OBu$-*n*	2-3 h	(22)	986
	$CH_2OC_{10}H_{21}$-*n*	2-3 h	(35)	986
	CH_2OBn	2-3 h	(28) + α-anomer (16)	986
O, N, O, O, OAc, AcO	OTMS, N, R[1], TMSO, N	TMSOTf, MeCN, –30 to 0°, 3 h	O, N, O, O, AcO, R[1], N, O, NH, O R[1] / α:β H (92) 1:2 Cl (81) 1:1 Me (92) 2:3	987

Sugar	Base	Conditions	Product(s) and Yield(s) (%)	Refs.
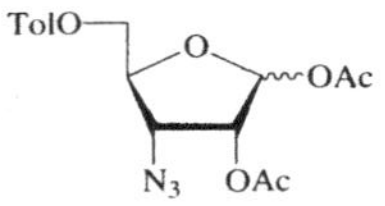 TolO, N_3, OAc, OAc (furanose)	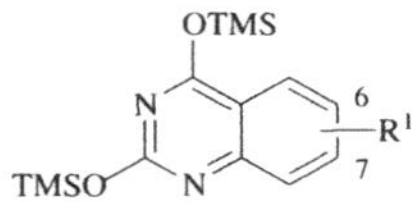 OTMS, TMSO, N, N, 6, 7, R^1 (quinazoline)	TMSOTf, $Cl(CH_2)_2Cl$, 75°, 3 h	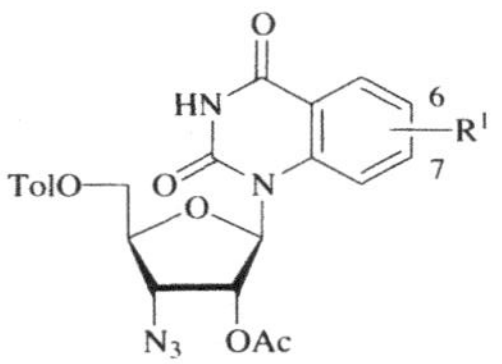 HN, O, O, N, 6, 7, R^1, TolO, N_3, OAc	988
	R^1			
	H		(73)	
	6-Cl		(64)	
	6-Me		(73)	
	7-Me		(23)	
	6,7-Me_2		(39)	
	6,7-OMe_2		(26)	
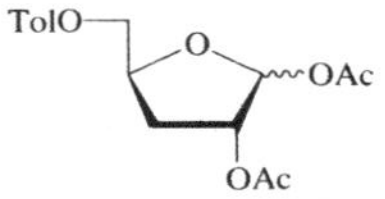 TolO, OAc, OAc (furanose)	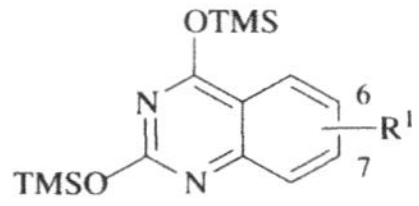 OTMS, TMSO, N, N, 6, 7, R^1 (quinazoline)	TMSOTf, $Cl(CH_2)_2Cl$, rt	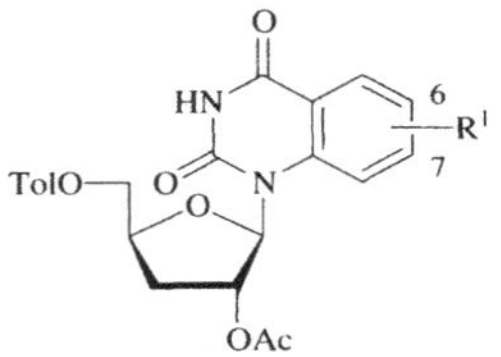 HN, O, O, N, 6, 7, R^1, TolO, OAc	989
	R^1			
	H		(59)	
	6-Me		(62)	
	7-Me		(67)	
	6,7-Me_2		(73)	
	6,7-OMe_2		(68)	
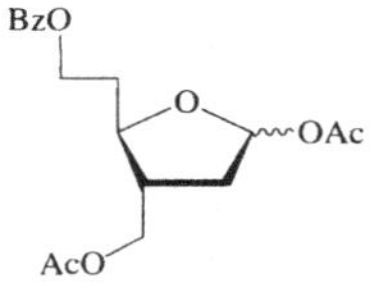 BzO, AcO, OAc (furanose)	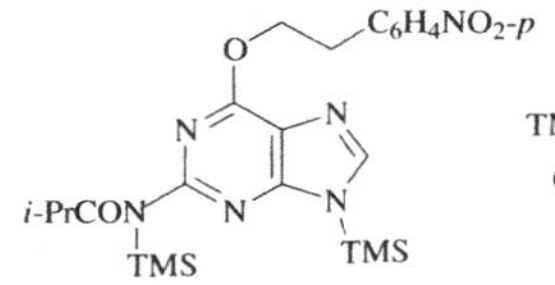 O, $C_6H_4NO_2$-*p*, N, N, N, N, *i*-PrCON, TMS, TMS (purine)	TMSOTf, MeCN, 60°, 18 h	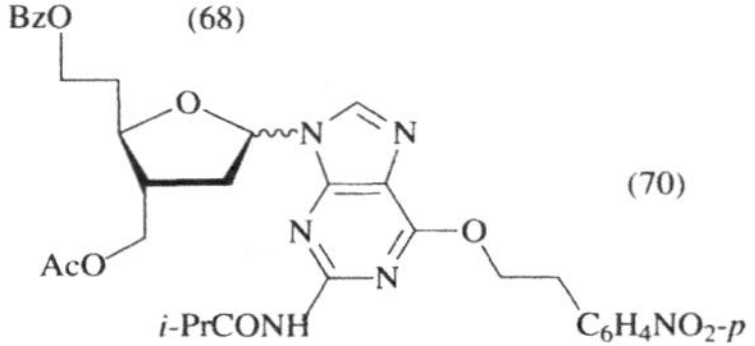 BzO, AcO, N, N, N, N, O, $C_6H_4NO_2$-*p*, *i*-PrCONH (70)	990

TABLE IV. REACTIONS WITH TRIMETHYLSILYL AND SILVER TRIFLATES AND PERCHLORATES (*Continued*)

Sugar	Base	Conditions	Product(s) and Yield(s) (%)	Refs.
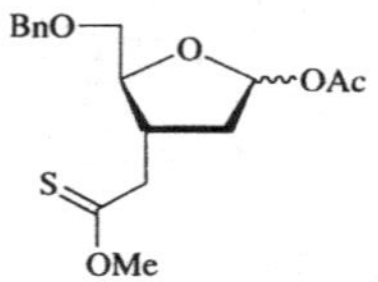 BnO, OAc, S, OMe	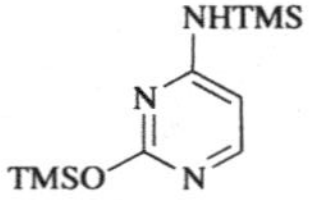 NHTMS, N, TMSO, N	TMSOTf, MeCN, rt	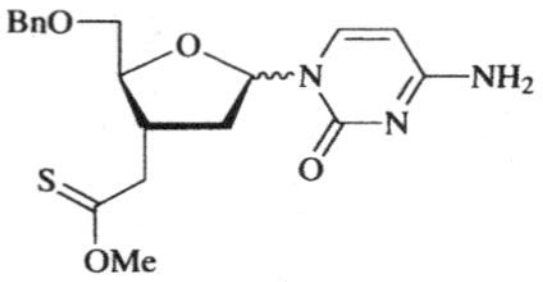 BnO, N, NH_2, N, O, S, OMe; **I** (—) α:β = 24:76	278
		TMSOTf, $Cl(CH_2)_2Cl$, 40-50°	**I** (—) α:β = 23:77	278
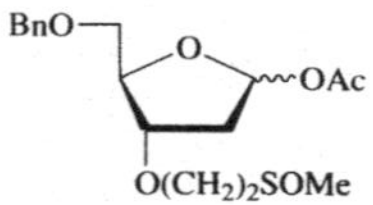 BnO, OAc, $O(CH_2)_2SOMe$	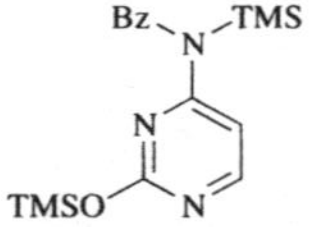 Bz, N, TMS, N, TMSO, N	TMSOTf, CH_2Cl_2, −23°	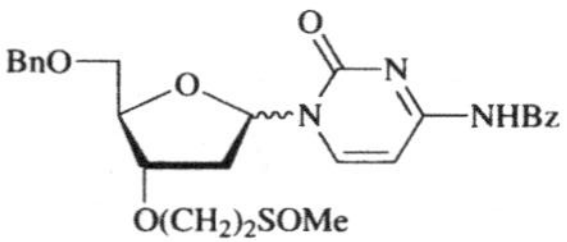 BnO, O, N, N, NHBz, $O(CH_2)_2SOMe$; (89) α:β = 25:75	277
	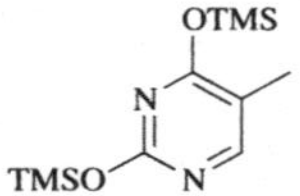 OTMS, N, TMSO, N	TMSOTf, CH_2Cl_2, −23°	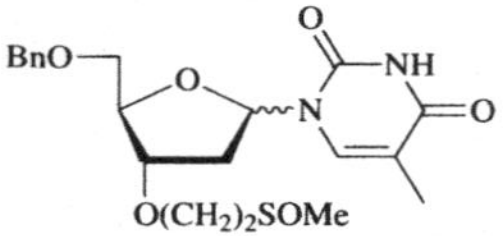 BnO, O, NH, N, O, $O(CH_2)_2SOMe$; **I** (89) α:β = 11:89	277
		TMSOTf, CH_2Cl_2, 0°, 15 min	**I** (96)	991
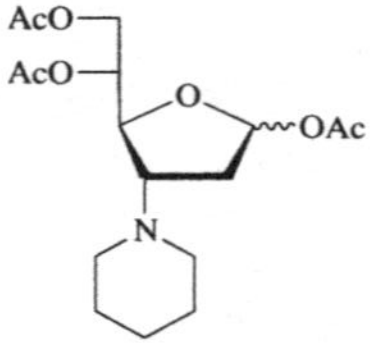 AcO, AcO, OAc, N	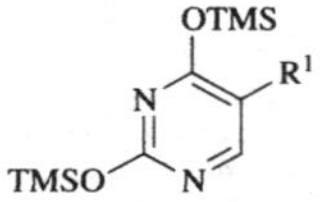 OTMS, R^1, N, TMSO, N	TMSOTf, $Cl(CH_2)_2Cl$	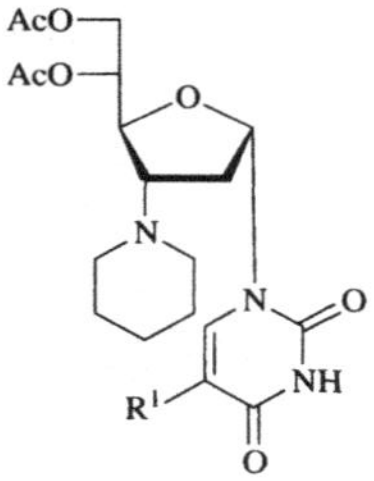 AcO, AcO, N, N, O, NH, R^1, O	520
	R^1			β-anomer	
	H	0°, 1 h	(40)	(34)	
	Me	−20°, 1 h; 0°, 1 h	(38)	(31)	

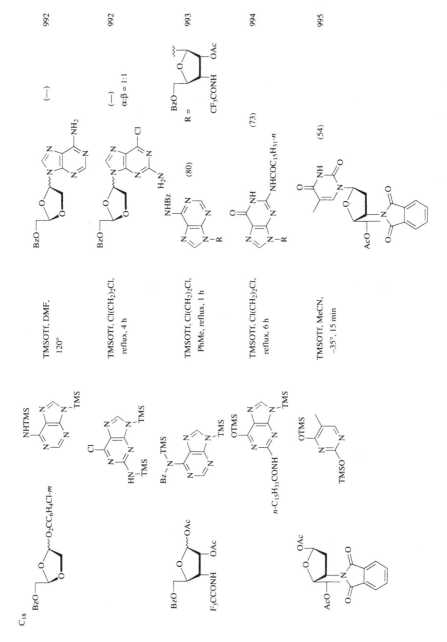

C_{18}

Substrate	Silylated base	Conditions	Product(s) and Yield(s) (%)		Refs.
BzO, $O_2CC_6H_4Cl$-*m*	NHTMS, TMS	TMSOTf, DMF, 120°	BzO, NH_2	(—)	992
	Cl, HN, TMS, TMS	TMSOTf, $Cl(CH_2)_2Cl$, reflux, 4 h	BzO, Cl, H_2N	(—) α:β = 1:1	992
BzO, OAc, F_3CCONH, OAc	Bz, TMS, TMS	TMSOTf, $Cl(CH_2)_2Cl$, PhMe, reflux, 1 h	NHBz, R (80)	R = BzO, CF_3CONH, OAc	993
	OTMS, *n*-$C_{15}H_{31}$CONH, TMS	TMSOTf, $Cl(CH_2)_2Cl$, reflux, 6 h	O, NH, R, NHCO$C_{15}H_{31}$-*n*	(73)	994
OAc, AcO, phthalimido	OTMS, TMSO	TMSOTf, MeCN, –35°, 15 min	O, NH, O, AcO, phthalimido	(54)	995

TABLE IV. REACTIONS WITH TRIMETHYLSILYL AND SILVER TRIFLATES AND PERCHLORATES (*Continued*)

Sugar	Base	Conditions	Product(s) and Yield(s) (%)	Refs.
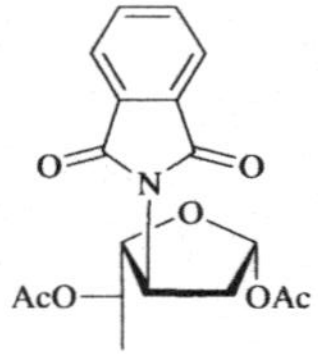 AcO, OAc	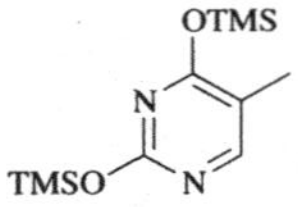 OTMS, TMSO, N	TMSOTf, MeCN, −15°, 30 min	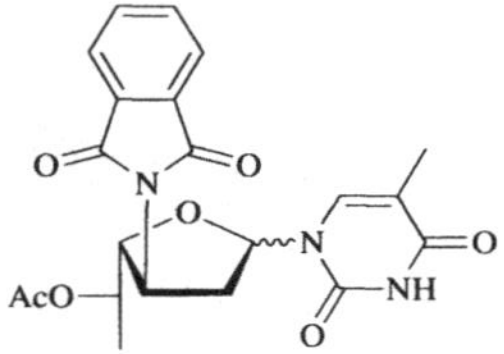 AcO, NH, O (75) α:β = 1.76:2	995
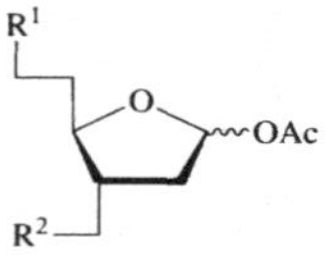 R^1, R^2, OAc	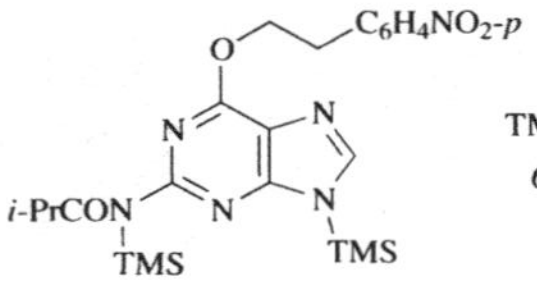 O, $C_6H_4NO_2$-*p*, *i*-PrCON, TMS, TMS	TMSOTf, MeCN, 60°, 18 h	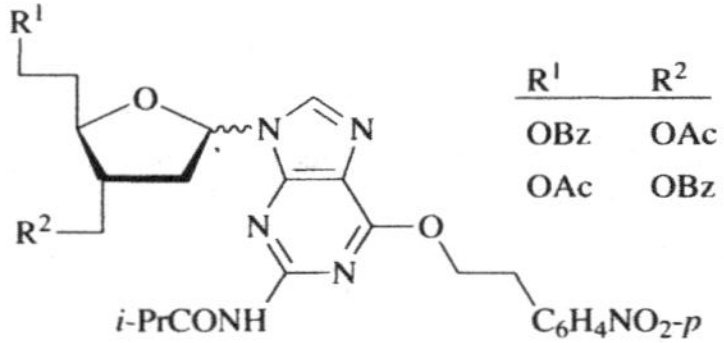 R^1, R^2, *i*-PrCONH, $C_6H_4NO_2$-*p* R^1 / R^2: OBz / OAc (—); OAc / OBz (—)	990
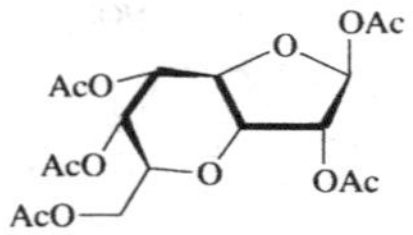 AcO, AcO, AcO, OAc, OAc	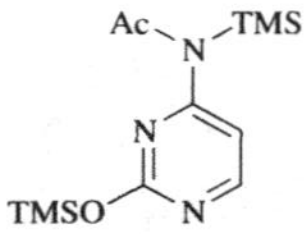 Ac, N, TMS, TMSO	TMSOTf, MeCN, rt, 5 h	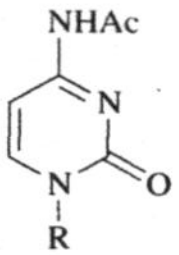 NHAc, N, O, R (53) R = AcO, AcO, AcO, OAc 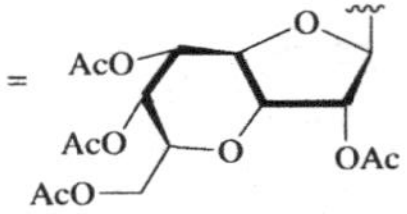	158
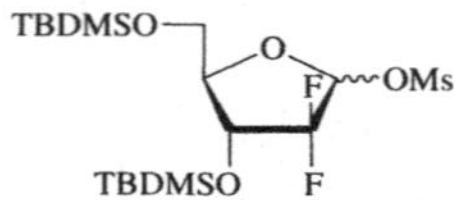 TBDMSO, TBDMSO, F, F, OMs	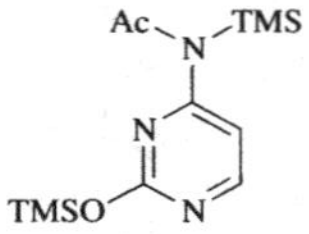 Ac, N, TMS, TMSO	1. TMSOTf, $Cl(CH_2)_2Cl$, reflux, 3-15 h 2. NH_3, MeOH	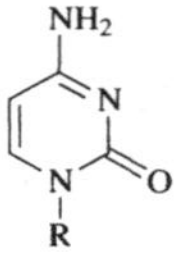 NH_2, N, O, R (10) + α-anomer (40) R 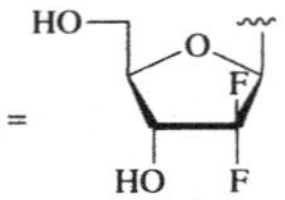 = HO, HO, F, F	302, 996
	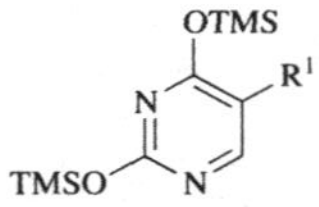 OTMS, R^1, TMSO	1. TMSOTf, $Cl(CH_2)_2Cl$, reflux, 2-15 h 2. HBr, CH_2Cl_2	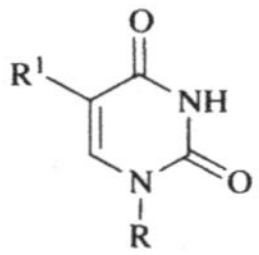 R^1, O, NH, O, R R^1: F (—); Me (—)	302, 996

Substrate	Silylated base	Conditions	Product(s) and Yield(s) (%)	Refs.
	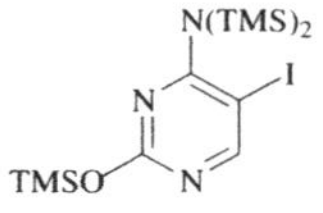 $N(TMS)_2$, I, TMSO	1. TMSOTf, $Cl(CH_2)_2Cl$, reflux, 3-15 h 2. HBr, CH_2Cl_2 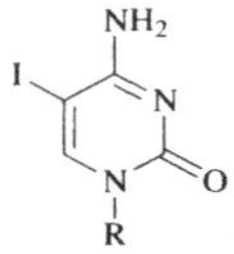	NH_2, I, N–R (—)	996
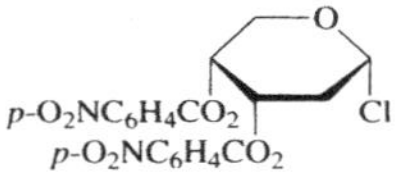 p-$O_2NC_6H_4CO_2$, p-$O_2NC_6H_4CO_2$, Cl	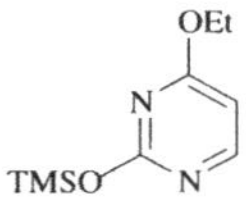 OEt, TMSO	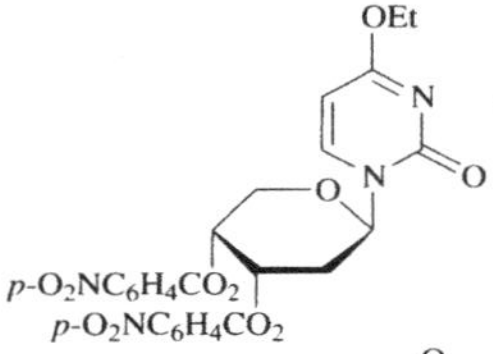 $AgClO_4$	OEt, p-$O_2NC_6H_4CO_2$, p-$O_2NC_6H_4CO_2$ (35) + α-anomer (55)	997
	OEt, TMSO	$AgClO_4$	OEt, p-$O_2NC_6H_4CO_2$, p-$O_2NC_6H_4CO_2$ (78)	998
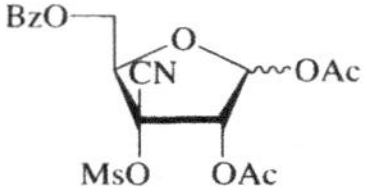 BzO, CN, OAc, MsO, OAc	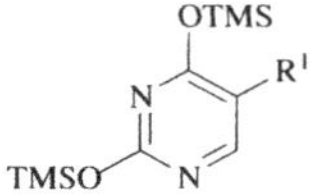 OTMS, R^1, TMSO	TMSOTf, MeCN, reflux, 3 h	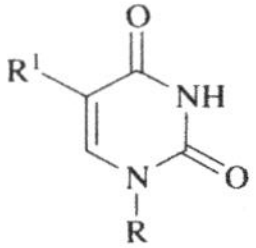 R^1, NH, N–R; R^1: H (78), Me (77), Et (93); R 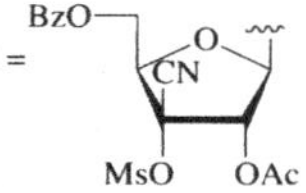 = BzO, CN, MsO, OAc	999
C_{19} 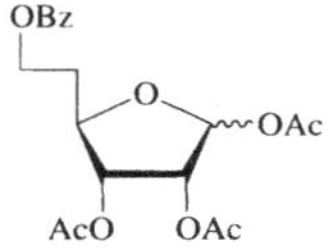 OBz, OAc, AcO, OAc	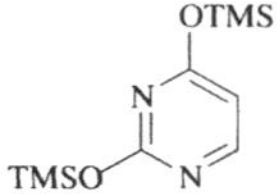 OTMS, TMSO	TMSOTf, $Cl(CH_2)_2Cl$, 20°, 16 h	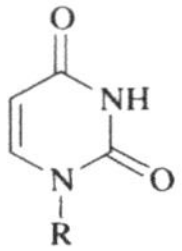 NH, N–R (91); R = OBz, AcO, OAc	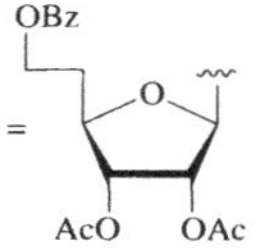 821
BnO, OAc, AcO, OAc	OTMS, TMSO	TMSOTf	NH, N–R (50); R = BnO, AcO, OAc	973

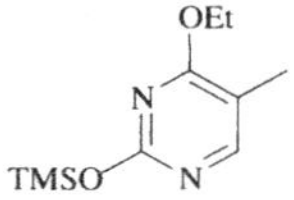

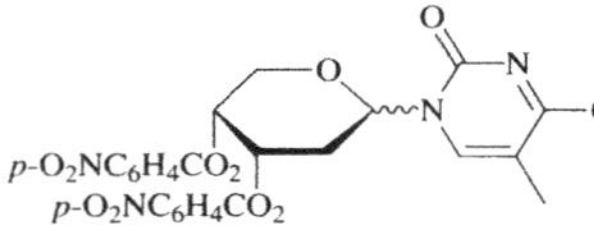

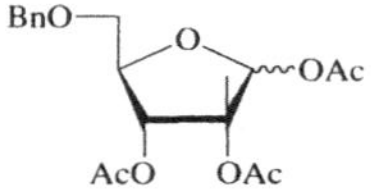

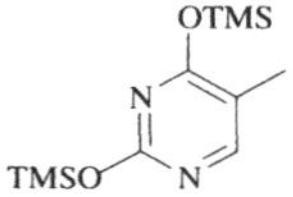

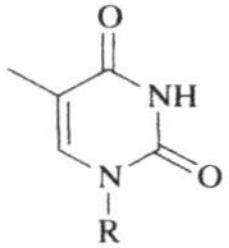

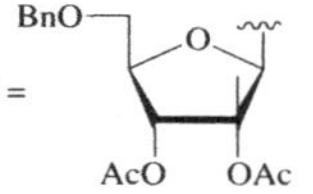

TABLE IV. REACTIONS WITH TRIMETHYLSILYL AND SILVER TRIFLATES AND PERCHLORATES (*Continued*)

Sugar	Base	Conditions	Product(s) and Yield(s) (%)	Refs.
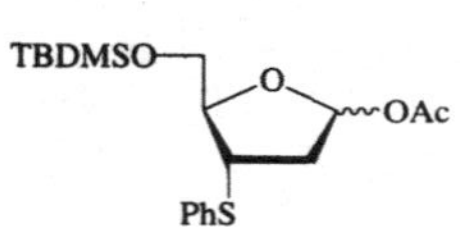 TBDMSO, OAc, PhS	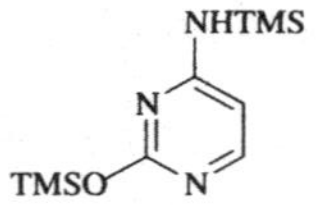 NHTMS, N, TMSO, N	TMSOTf, $Cl(CH_2)_2Cl$	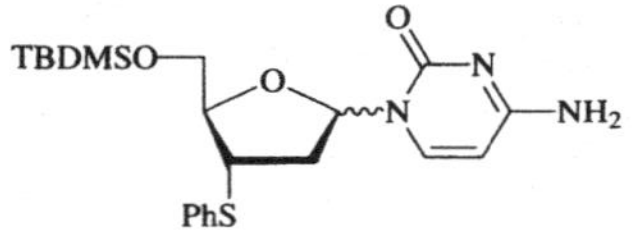 TBDMSO, O, N, N, NH_2, PhS (—) α:β = 1:1	282
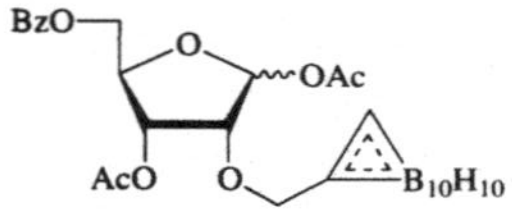 BzO, O, OAc, AcO, O, $B_{10}H_{10}$	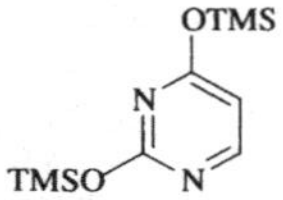 OTMS, N, TMSO, N	TMSOTf, MeCN, rt, 4 h	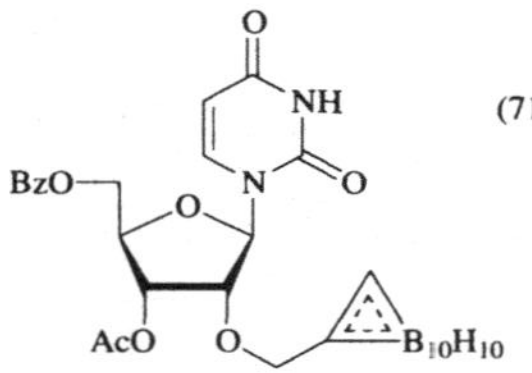 O, NH, BzO, N, O, AcO, O, $B_{10}H_{10}$ (71)	1000
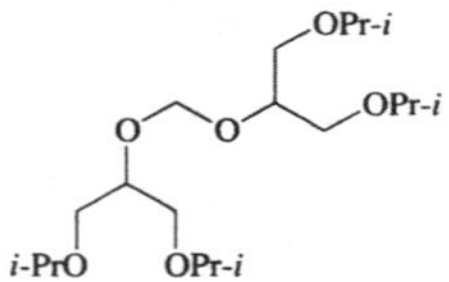 OPr-*i*, OPr-*i*, O, O, *i*-PrO, OPr-*i*	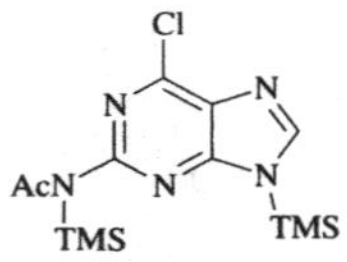 Cl, N, N, AcN, TMS, N, N, TMS	TMSOTf, $Cl(CH_2)_2Cl$, –30°, 2 h	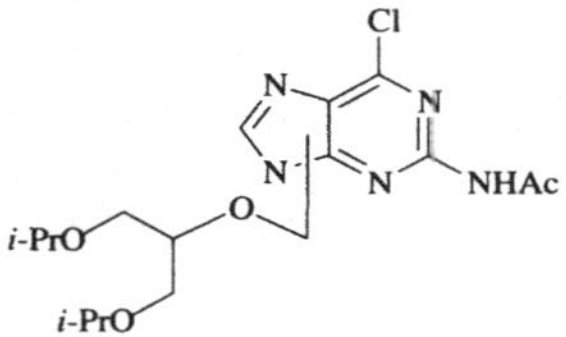 Cl, N, N, N, N, NHAc, *i*-PrO, O, *i*-PrO (58) N^7:N^9 = 10:1	521
C_{20} MeO_2C, O, Cl, BzO, OBz	OTMS, N, TMSO, N	$AgClO_4$, C_6H_6, 25°, 18 h	O, NH, MeO_2C, N, O, BzO, OBz (64)	224

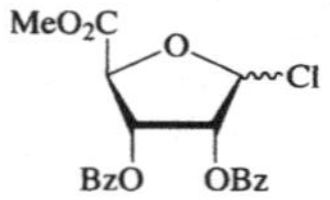

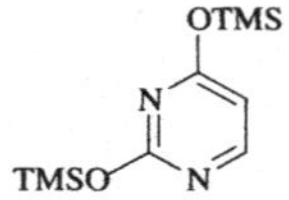

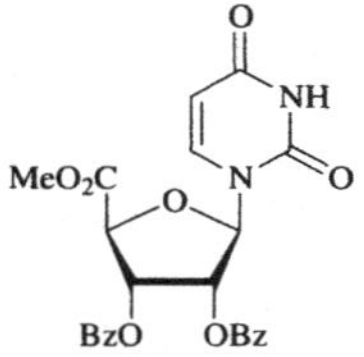

Sugar	Base	Conditions	Product(s) and Yield(s) (%)	Refs.
	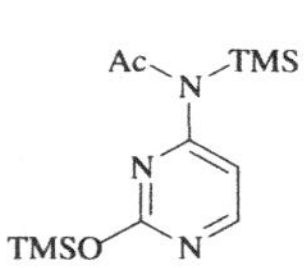	$AgClO_4$, C_6H_6, 25°, 17 h	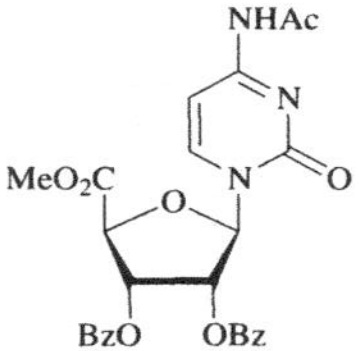 (71)	224
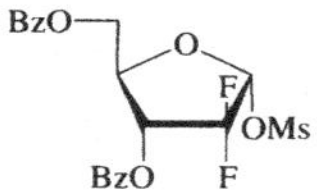	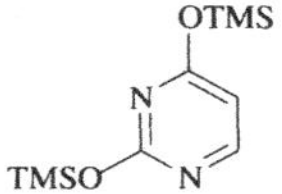	TMSOTf, $Cl(CH_2)_2Cl$, 83°, 5 h	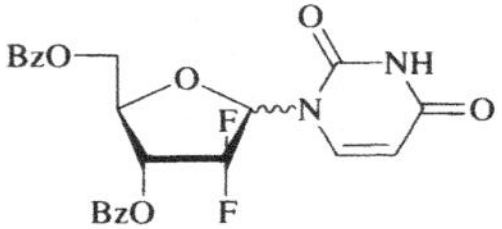 (—) α:β = 1.43:1	303
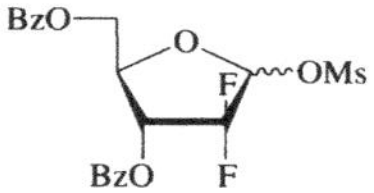	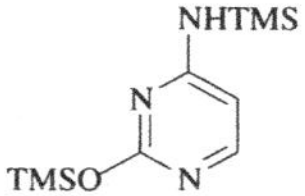	TMSOTf, $CHCl_2CH_2Cl$ 113°, 18 h	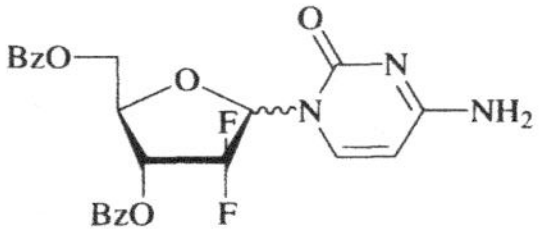 (87) α:β = 56:44	303
	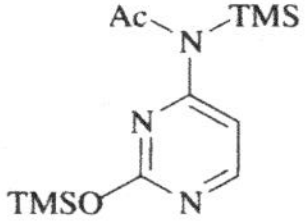	1. TMSOTf, $Cl(CH_2)_2Cl$, reflux, 8 h 2. NH_3, MeOH	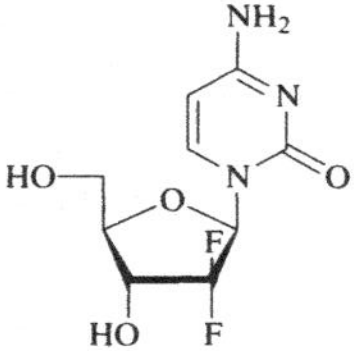 (—) α:β = 1:1	1001
		1. TMSOTf, $Cl(CH_2)_2Cl$, reflux, 12 h 2. NH_3, MeOH	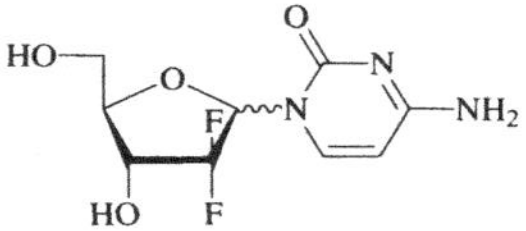 (49) α:β = 53:47	303
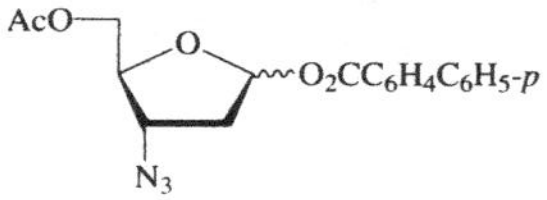	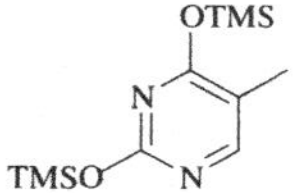	1. TMSOTf, MeCN, 2. $MeNH_2$, MeOH	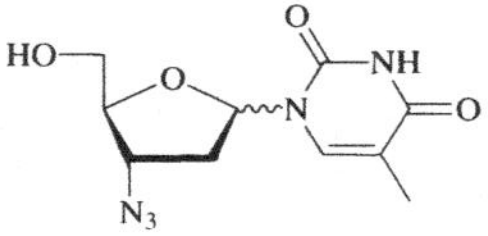 (—)	961

TABLE IV. REACTIONS WITH TRIMETHYLSILYL AND SILVER TRIFLATES AND PERCHLORATES (*Continued*)

Sugar	Base	Conditions	Product(s) and Yield(s) (%)	Refs.
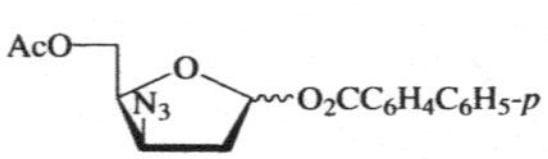 AcO– N$_3$ ~O$_2$CC$_6$H$_4$C$_6$H$_5$-*p*	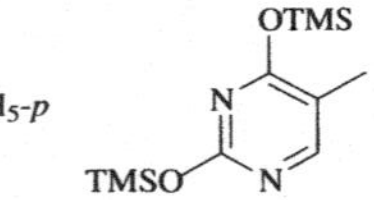 OTMS, TMSO	1. TMSOTf, MeCN 2. $MeNH_2$, MeOH	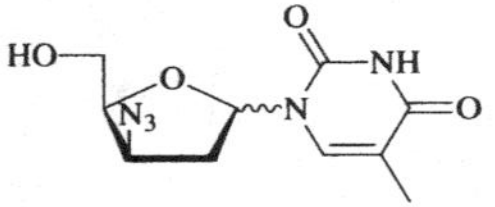 HO– N$_3$ (—)	961
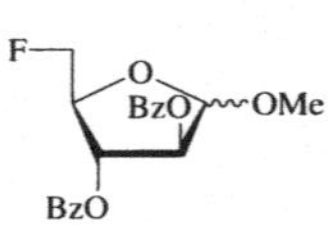 F– BzO ~OMe, BzO	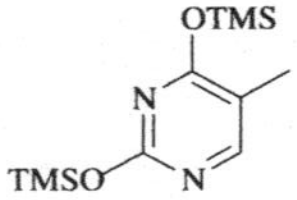 OTMS, TMSO	TMSOTf, MeCN, rt, 5 d	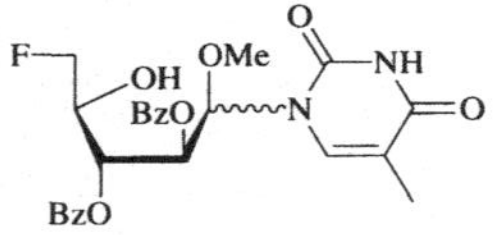 F– OH OMe BzO, BzO (44)	925
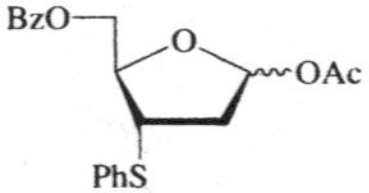 BzO– ~OAc, PhS	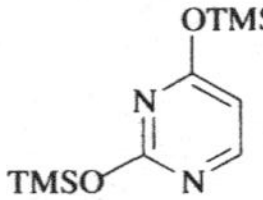 OTMS, TMSO	TMSOTf, CH_2Cl_2, $Sn(OTf)_2$, rt, 12 h	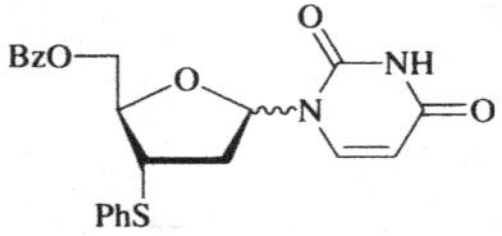 BzO– PhS **I** (78) α:β = 63:37	229
		TMSOTf, CH_2Cl_2, $HgBr_2$, rt, 12 h	**I** (61) α:β = 69:31	229
		TMSOTf, CH_2Cl_2, rt, 12 h	**I** (87) α:β = 60:40	229
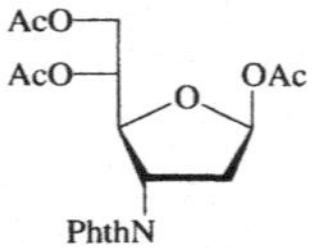 AcO– AcO– OAc, PhthN	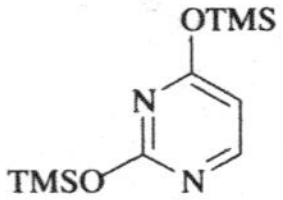 OTMS, TMSO	1. TMSOTf, MeCN 2. $MeNH_2$, EtOH	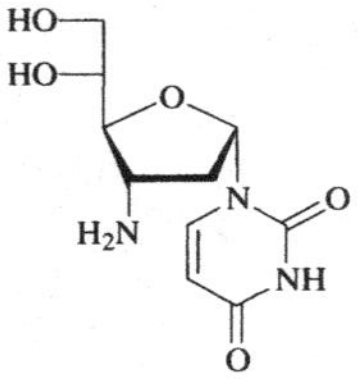 HO– HO– H_2N (—)	530

Sugar	Base	Conditions	Product(s) and Yield(s) (%)	Refs.
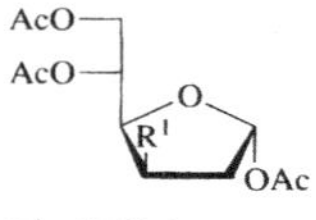 AcO, AcO, R^1, OAc; R^1 = N(Phth)	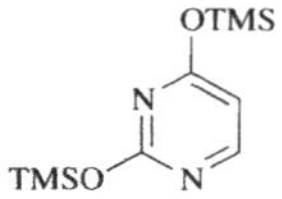 OTMS, TMSO	1. TMSOTf, MeCN 2. $MeNH_2$, EtOH	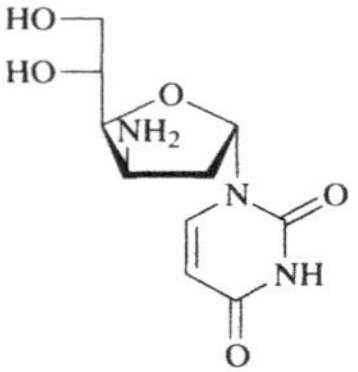 HO, HO, NH_2, NH, O (—)	530
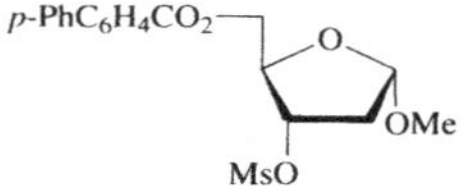 p-$PhC_6H_4CO_2$, OMe, MsO	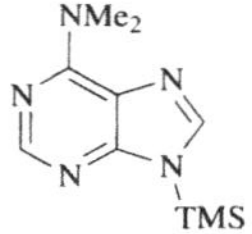 NMe_2, TMS	TMSOTf, MeCN, −30°, 2 h	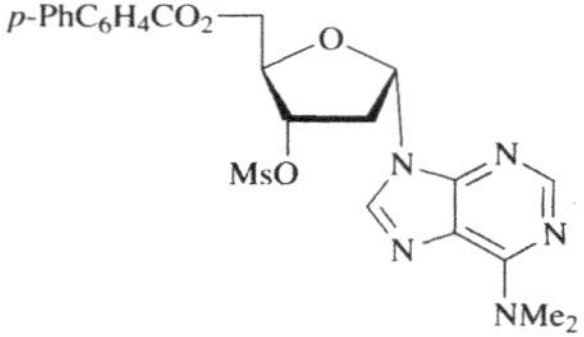 p-$PhC_6H_4CO_2$, MsO, NMe_2 (18) + α-anomer (21)	1002
	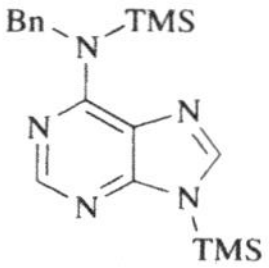 Bn, TMS, TMS	TMSOTf, MeCN, −30 to 20°, 18 h	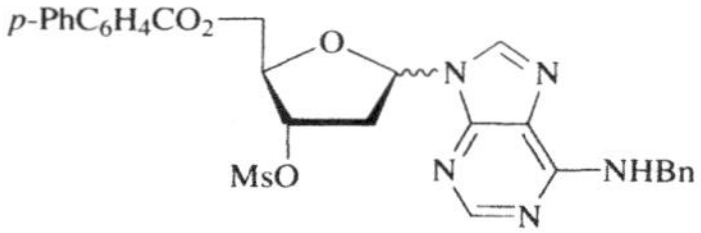 p-$PhC_6H_4CO_2$, MsO, NHBn (15) α:β = 3:1	1003
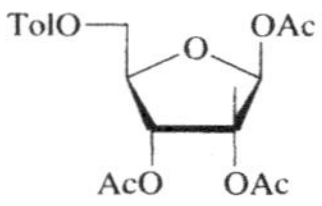 TolO, OAc, AcO, OAc	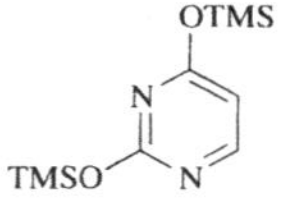 OTMS, TMSO	TMSOTf, $Cl(CH_2)_2Cl$, 20°, 16 h	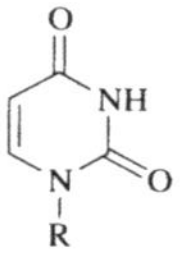 O, NH, O, R (89); R = TolO, AcO, OAc 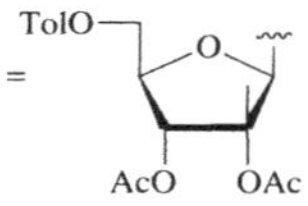	1004
	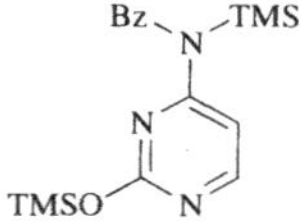 Bz, TMS, TMSO	TMSOTf, $Cl(CH_2)_2Cl$, 20°, 16 h	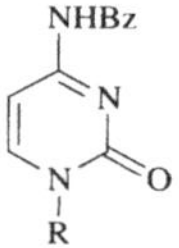 NHBz, O, R (65)	1004

TABLE IV. REACTIONS WITH TRIMETHYLSILYL AND SILVER TRIFLATES AND PERCHLORATES (*Continued*)

Sugar	Base	Conditions	Product(s) and Yield(s) (%)	Refs.
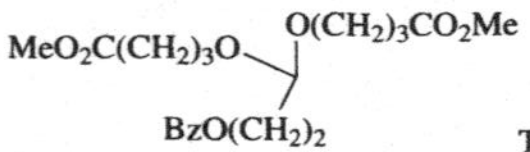 $MeO_2C(CH_2)_3O$, $O(CH_2)_3CO_2Me$, $BzO(CH_2)_2$	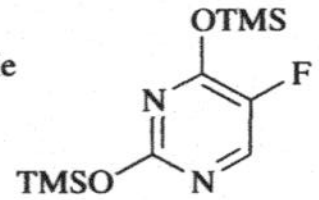 OTMS, F, TMSO	TMSOTf, CH_2Cl_2, –30°, 2-12 h	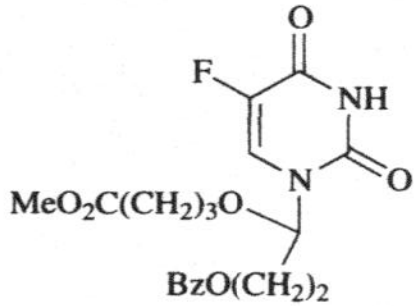 F, NH, O, $MeO_2C(CH_2)_3O$, $BzO(CH_2)_2$ (62)	923
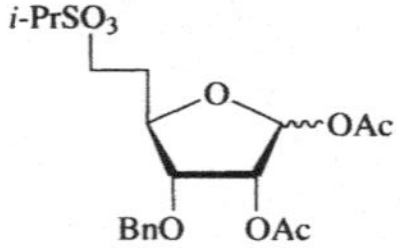 *i*-$PrSO_3$, OAc, BnO, OAc	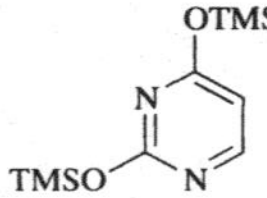 OTMS, TMSO	TMSOTf, $Cl(CH_2)_2Cl$, rt, 20 h	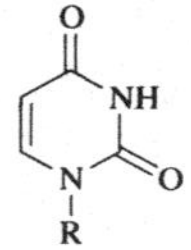 O, NH, N–R, O (75) R = *i*-$PrSO_3$, BnO, OAc 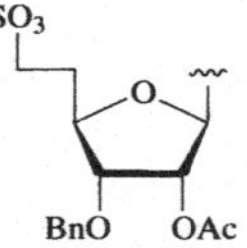	508
	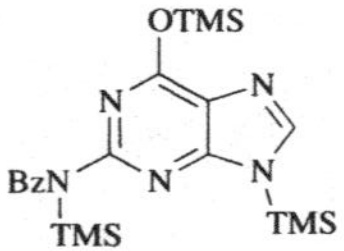 OTMS, BzN(TMS), N–TMS	TMSOTf, $Cl(CH_2)_2Cl$, 40°, 4.5 h	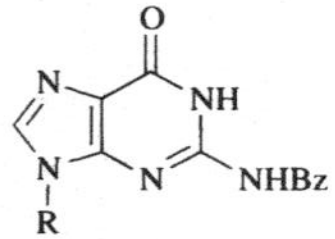 O, NH, N–R, NHBz (35) + N^7-isomer (13)	508
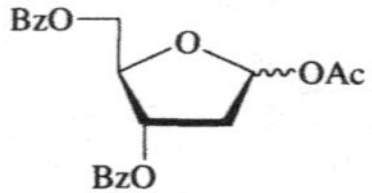 BzO, OAc, BzO	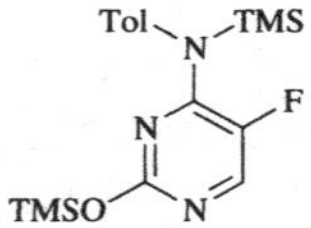 Tol–N–TMS, F, TMSO	TMSOTf, MeCN, rt, 1.3 h	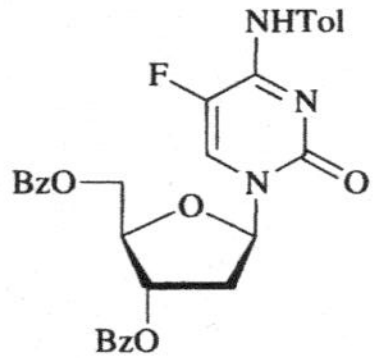 NHTol, F, BzO, BzO (31) + α-anomer (18)	151
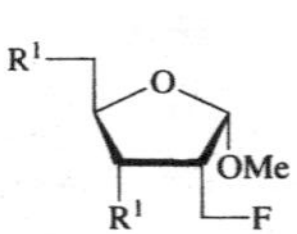 R^1, OMe, R^1, F; $R^1 = Cl_2C_6H_3CH_2O$	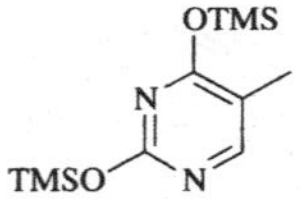 OTMS, TMSO	TMSOTf, MeCN, 50°	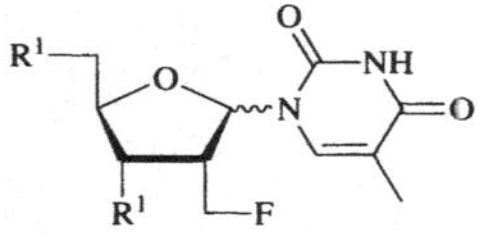 R^1, NH, O, R^1, F (76) α:β = 1:2.7	689

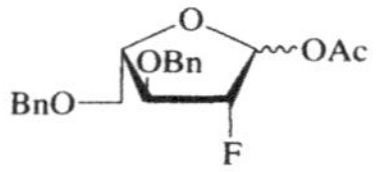

Substrate	Silylated base	Conditions	Product(s) and Yield(s) (%)	Refs.
BnO, OBn, OAc, F (sugar)	OTMS, TMSO (pyrimidine, 5-methyl)	TMSOTf	BnO, OBn, F; O, NH, O (nucleoside) (—)	978
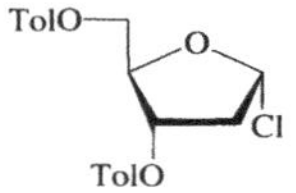 TolO, TolO, Cl (sugar)	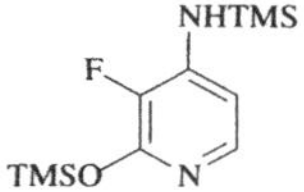 NHTMS, F, TMSO (pyridine)	1. TMSOTf, $Cl(CH_2)_2Cl$, –25°, 30 min 2. rt, 24 h	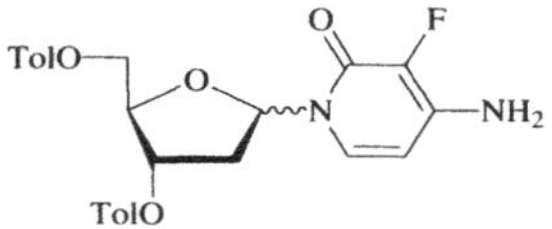 TolO, TolO; O, F, NH_2 (83)	860, 861
	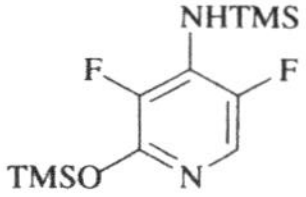 NHTMS, F, F, TMSO (pyridine)	1. TMSOTf, $Cl(CH_2)_2Cl$, 0°, 30 min 2. rt, 24 h	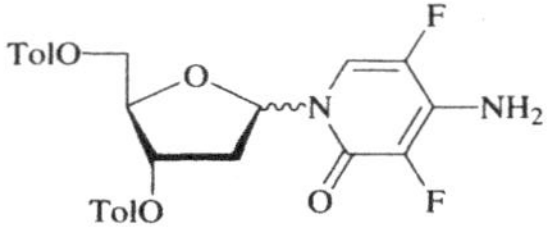 TolO, TolO; F, NH_2, O, F (82) α:β = 1:1	860
	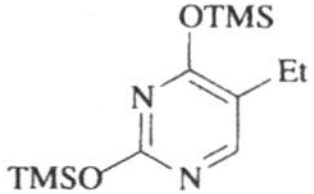 OTMS, Et, TMSO (pyrimidine)	TMSOTf, $Cl(CH_2)_2Cl$, 24°, 3 h	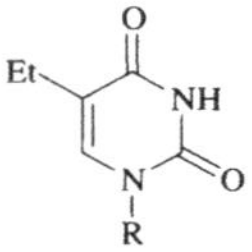 Et, O, NH, O, N–R (58) + α-anomer (32)	133
	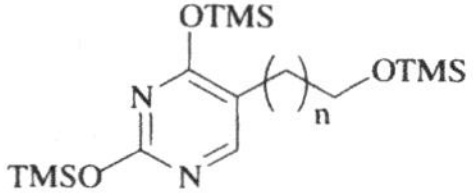 OTMS, $(\ )_n$OTMS, TMSO (pyrimidine)	TMSOTf, $CHCl_3$, rt, 1 h	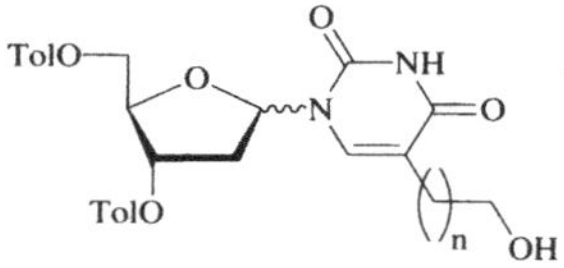 TolO, TolO; O, NH, O, $(\ )_n$OH n = 1: (84), β:α 5:1 n = 2: (—), β:α —	1005
	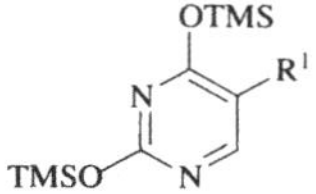 OTMS, R^1, TMSO (pyrimidine)	TMSOTf, $CHCl_3$, rt, 1 h	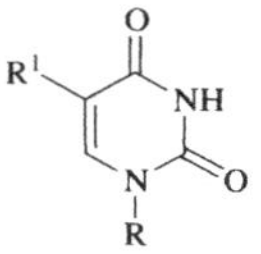 R^1, O, NH, O, N–R; R = TolO, TolO (sugar) 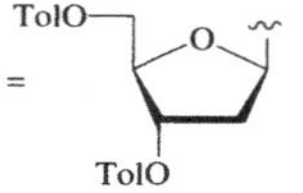	
	R^1 = $(CH_2)_3F$		(72) + α-anomer (10)	360
	R^1 = $(CH_2)_3Cl$		(65) + α-anomer (15)	1005
	R^1 = $(CH_2)_3Br$		(71) + α-anomer (15)	1005

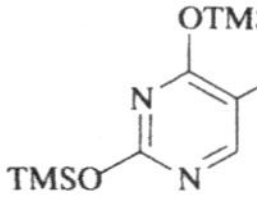

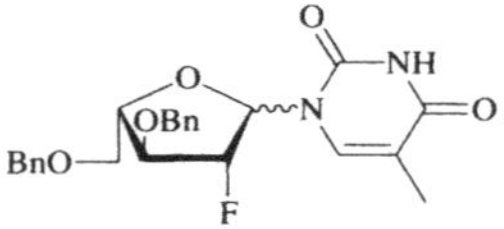

TABLE IV. REACTIONS WITH TRIMETHYLSILYL AND SILVER TRIFLATES AND PERCHLORATES (*Continued*)

Sugar	Base R^1	Base R^2	Conditions	Product(s) and Yield(s) (%)	α-anomer	Refs.
	2,4-bis(TMSO)pyrimidine (OTMS, TMSO, R^1, R^2)		1. TMSOTf, $CHCl_3$, rt 2. NaOEt, EtOH	1-R-5-R^1-6-R^2-uracil, R = 3,5-di-O-Tol-2-deoxyribofuranosyl (TolO)		
	$CH_3CH_2CH(CH_3)$–CO_2Et	H	1 h	(17)	(18)	1006
	$CH_3CH{=}CH$–CO_2Et	H	—	(61)	(37)	1006
	$CH_3CH{=}C(CH_3)$–CO_2Et	H	—	(74)	(—)	1006
	$CH_3CH{=}CH$–CO_2Et (Z)	H	—	(61)	(—)	1006
	$CH_3CH{=}CH$–Br	Me	—	(10)	(—)	1006
	$CH_3CH{=}C(CH_3)$–Br	H	—	(31)	(—)	1006
	$CH_3CH{=}CH$–CO_2Et	Me	—	(25)	(—)	1006
	$CH_3CH{=}C(CH_3)$–CO_2Et	H	—	(74)	(—)	1006
	SCF_3	H	62°, 4 h	(85)	(—)	1007, 1008
	SO_2CF_3	H	62°, 3 h	(89)	(—)	1007
	$SeCF_3$	H	62°, 4 h	(84)	(—)	1007

Reactant	Conditions	Product(s) and Yield(s) (%)	Refs.
	TMSOTf, dioxane, 90°, 4 h	(67)	1007
	TMSOTf, MeCN, rt, 2 h	(67) + α-anomer (12)	1009
	TMSOTf, $Cl(CH_2)_2Cl$, C_6H_6, rt to reflux, 2 h	(18) + N^7 β-isomer (16) + N^7 α-isomer (6)	505
	1. TMSOTf, $CHCl_3$, rt, 3 h 2. NH_3, MeOH	(65) + α-anomer (35) R =	893
	1. TMSOTf, $Cl(CH_2)_2Cl$, reflux, 3 h 2. NaOMe	(39)	1010
	TMSOTf, 110°, 30 min	(14) α,β mixture	954

TABLE IV. REACTIONS WITH TRIMETHYLSILYL AND SILVER TRIFLATES AND PERCHLORATES (*Continued*)

Sugar	Base	Conditions	Product(s) and Yield(s) (%)	Refs.
BzO, OMe, OBz	NHTMS, N, TMSO, N	1. TBDMSOTf, CH_2Cl_2, rt, 24 h 2. NH_3, MeOH	HO, O, HO, N, O, N, NH_2 (30) + β-anomer (24)	963, 1011
	OTMS, N, TMSO, N	1. TBDMSOTf, CH_2Cl_2, rt, 24 h 2. NH_3, MeOH, reflux, 24 h	O, NH, N, O, R (33) + α-anomer (33) R = HO, O, HO	1011
	Cl, N, N, N, N, TMS	1. TBDMSOTf, MeCN or CH_2Cl_2, rt, 24 h 2. NH_3, MeOH 100°, 20 h	NH_2, N, N, N, N, R (20) + α-anomer (10)	1011, 963
BzO, BzO, O, OMe	NHTMS, N, TMSO, N	1. TBDMSOTf, CH_2Cl_2 2. NH_3, MeOH	R, N, O, N, NH_2 (24) + β-anomer (20) R = HO, O, HO	1012
	OTMS, N, TMSO, N	TBDMSOTf, CH_2Cl_2	R, N, O, NH, O (41) + β-anomer (36)	1012

Glycosyl donor	Base	Conditions	Product(s) and Yield(s) (%)	Refs.
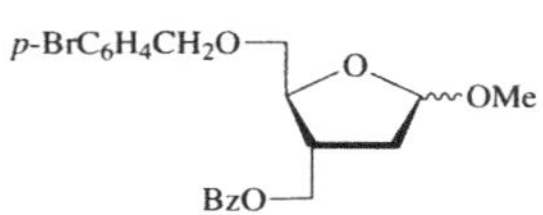 p-$BrC_6H_4CH_2O$ / BzO furanoside, ~OMe	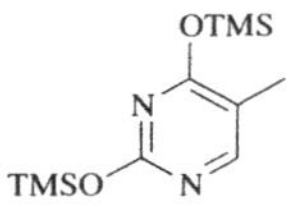 TMSO / OTMS pyrimidine	1. TBDMSOTf, CH_2Cl_2, rt, 24 h 2. H_2, Pd/C 3. NH_3, MeOH	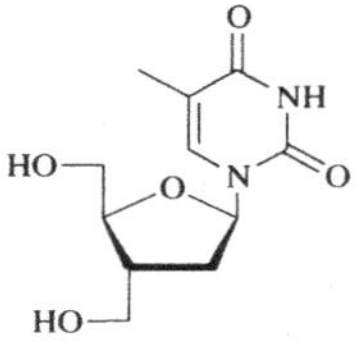 HO / HO nucleoside (33) + α-anomer (33)	1011
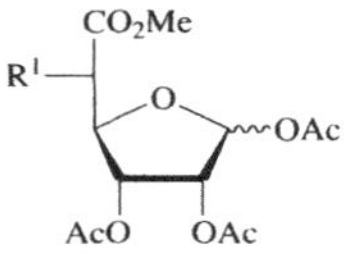 R^1, CO_2Me, AcO, OAc, ~OAc	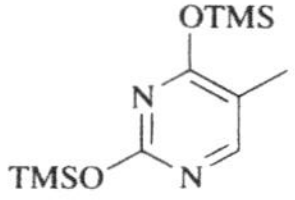 TMSO / OTMS pyrimidine	TMSOTf, $Cl(CH_2)_2Cl$, reflux	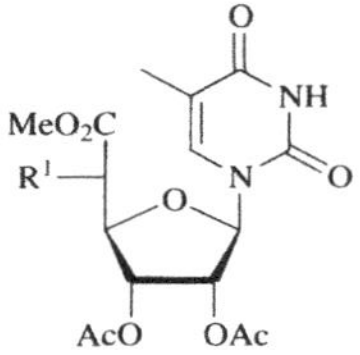 MeO_2C, R^1, AcO, OAc nucleoside R^1 = CbzNH (82) R^1 = CbzNH (75)	789, 1013 1014
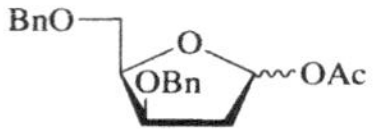 BnO, OBn, ~OAc	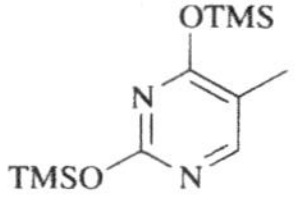 TMSO / OTMS pyrimidine	TMSOTf, CH_2Cl_2, rt, 1 h	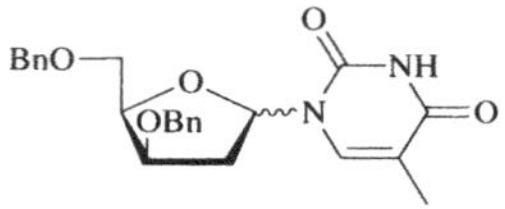 BnO, OBn nucleoside (87) α:β = 1:1	405
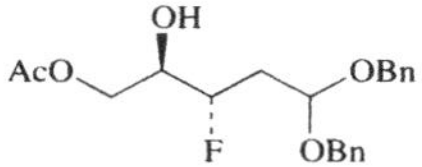 AcO, OH, F, OBn, OBn	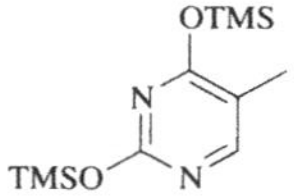 TMSO / OTMS pyrimidine	TMSOTf, MeCN, 25°, 2.75 h	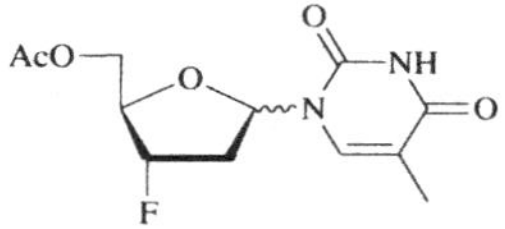 AcO, F nucleoside (71) α:β = 1:2.8	300
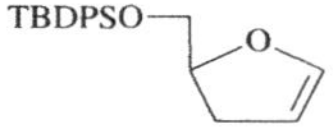 TBDPSO glycal	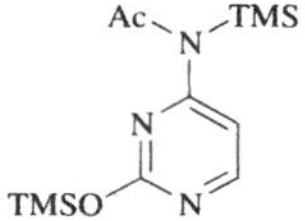 TMSO / Ac, N, TMS pyrimidine	TMSOTf, PhSCl, CH_2Cl_2, −78 to 0°, 1.5 h	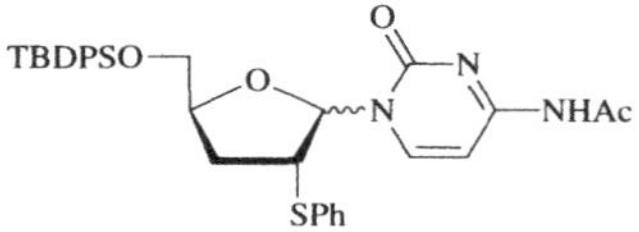 TBDPSO, SPh, NHAc nucleoside (60) α:β = 1:5	421a
		TMSOTf, TippSCl, CH_2Cl_2, −78 to 25°, 2 h	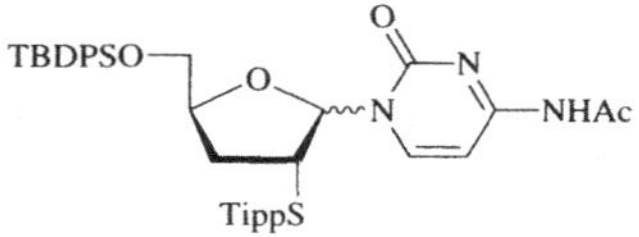 TBDPSO, TippS, NHAc nucleoside (75) α:β = 1:6	421a

TABLE IV. REACTIONS WITH TRIMETHYLSILYL AND SILVER TRIFLATES AND PERCHLORATES (*Continued*)

Sugar	Base	Conditions	Product(s) and Yield(s) (%)	Refs.
	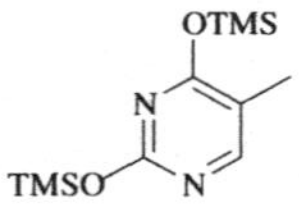 OTMS, N, TMSO, N	TMSOTf, PhSCl, CH_2Cl_2, –78 to 0°, 1.5 h	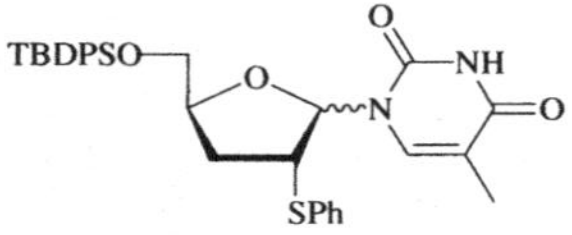 TBDPSO, O, N, O, NH, O, SPh (50) $\alpha:\beta$ = 1:4	421a
	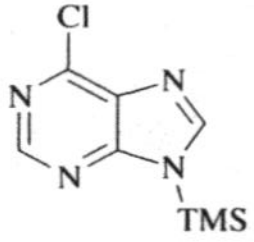 Cl, N, N, N, N, TMS	TMSOTf, TippSCl, Et_2O, –78 to 25°, 2 h	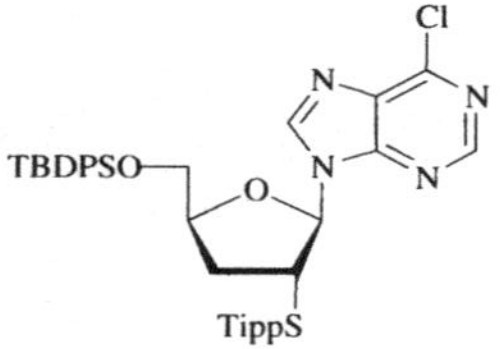 TBDPSO, O, N, N, Cl, N, N, TippS (80) $\alpha:\beta$ = 1:5	421a
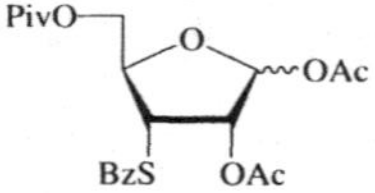 PivO, O, OAc, BzS, OAc	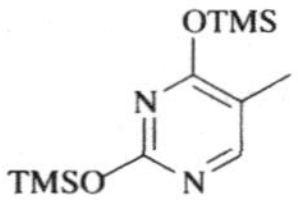 OTMS, N, TMSO, N	TMSOTf, CH_2Cl_2, or MeCN, 70°	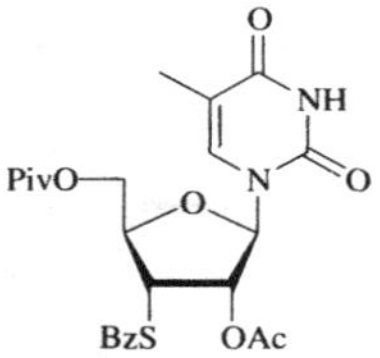 O, NH, N, O, PivO, O, BzS, OAc (85)	1015
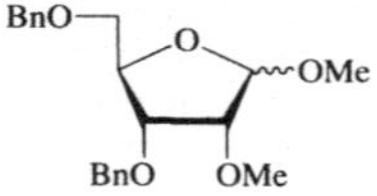 BnO, O, OMe, BnO, OMe	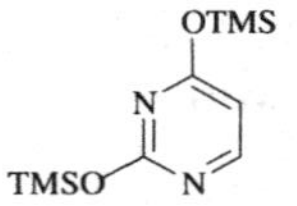 OTMS, N, TMSO, N	TMSOTf, MeCN, 25°, 2-3 h	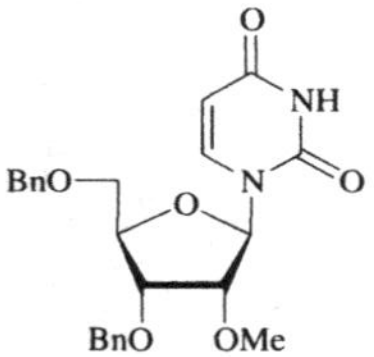 O, NH, N, O, BnO, O, BnO, OMe (65) + α-anomer (11)	226b

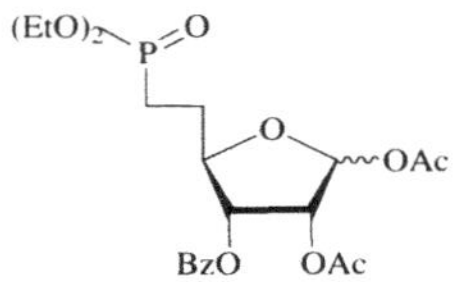	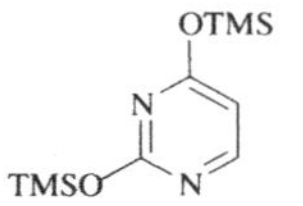	TMSOTf, Cl(CH$_2$)$_2$Cl, rt, 16 h	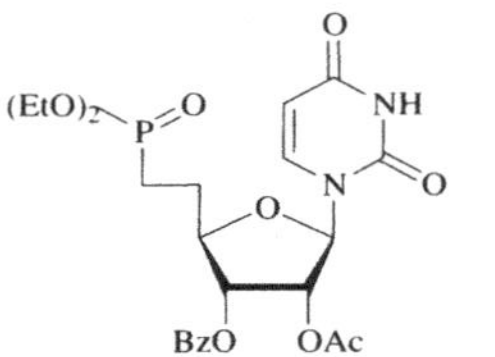 (30) + N^7-ethyluracil (36)	821
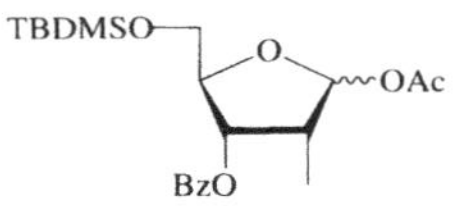	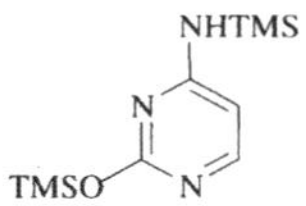	TMSOTf, MeCN, rt	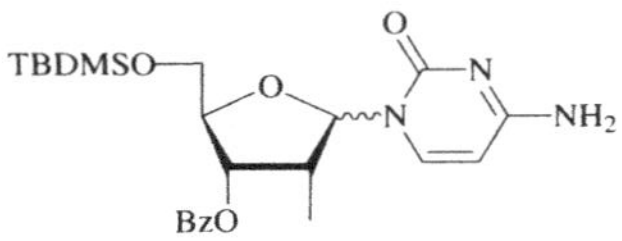 (72) α:β = 1:9	829
	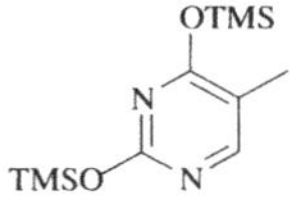	TMSOTf, Cl(CH$_2$)$_2$Cl, rt	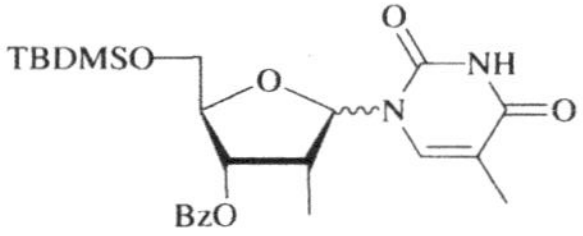 (77) α:β = 1:6.7	829
C$_{22}$ 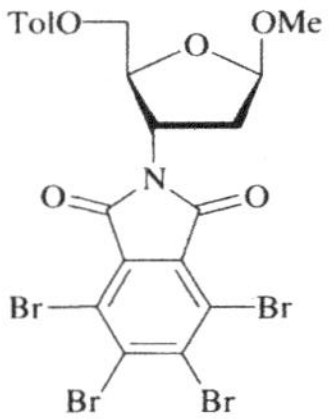	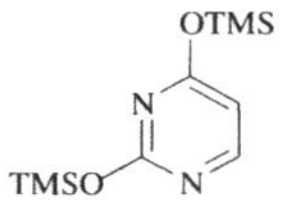	TMSOTf, MeCN, 0°, 15 min; rt, 2 h	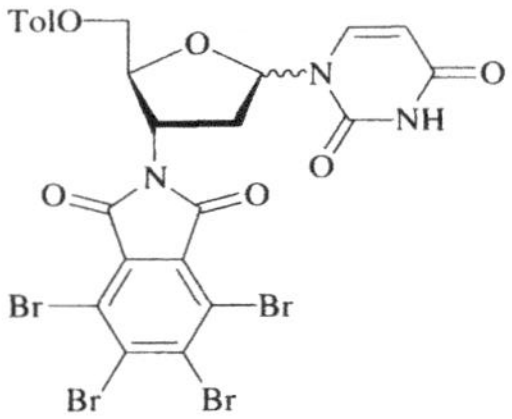 (49)	983
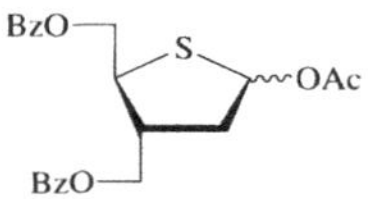	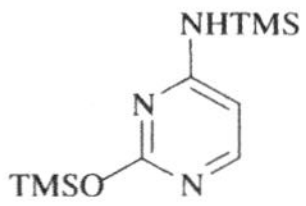	TBDMSOTf, MeCN, 0°, 1 h	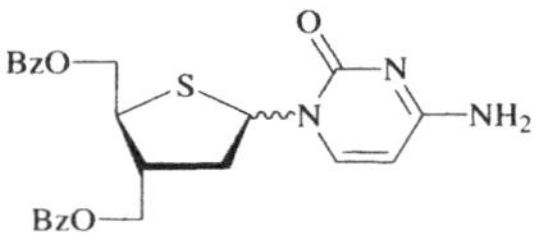 (37)	963

TABLE IV. REACTIONS WITH TRIMETHYLSILYL AND SILVER TRIFLATES AND PERCHLORATES (*Continued*)

Sugar	Base	Conditions	Product(s) and Yield(s) (%)	Refs.
	Cl, N, N, N, N, TMS	1. TMSOTf, CH_2Cl_2, rt, 24 h 2. NH_3, MeOH	HO, S, HO, N, N, N, N, NH_2 (33) + β-anomer (28)	1016
BzO, O, OAc, OBz	OTMS, N, TMSO, N	TMSOTf	O, NH, BzO, O, N, O, OBz (55) + α-anomer (22)	922
BzO, O, OMe, BzO	OTMS, R^1, N, TMSO, N	TBDMSOTf, MeCN, CH_2Cl_2, rt, 15 h	O, R^1, NH, BzO, O, N, O, BzO R^1: H (68); Me (72) α:β = 1:12	236
	Cl, N, N, N, N, TMS	TBDMSOTf, MeCN, CH_2Cl_2, rt, 15 h	Cl, N, N, BzO, O, N, N, BzO (68)	236

Substrate	Silylated base	Conditions	Product(s) and Yield(s) (%)	Refs.
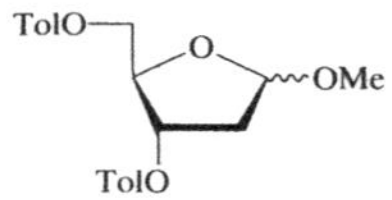 TolO, TolO, OMe (2-deoxyribofuranoside)	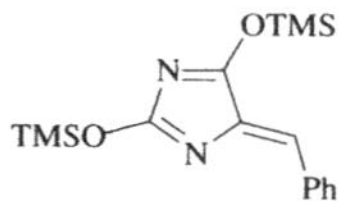 TMSO, N, N, OTMS, Ph	TMSOTf, MeCN, –30°, 4 h; –10°, 3 d	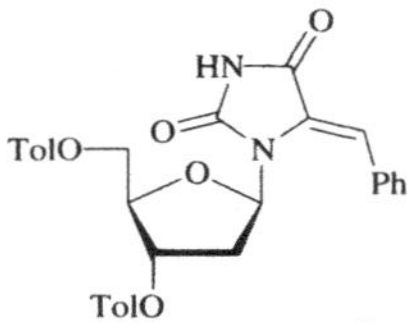 TolO, TolO, HN, O, O, N, Ph (55) + α-anomer (19)	1017
	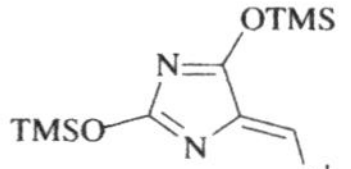 TMSO, N, N, OTMS, R^1	TMSOTf, MeCN, –30°, 4 h; –10°, 3 d	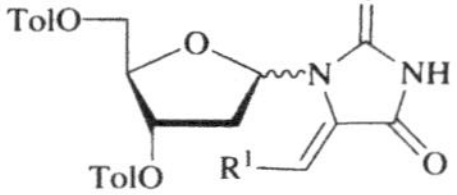 TolO, TolO, N, NH, O, O, R^1	1017
	R^1			
	Me		(63)	
	2-thienyl		(80) α:β = 1:2	
	2-naphthyl		(78)	
	1-naphthyl		(20)	
	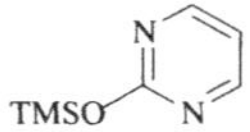 TMSO, N, N	TMSOTf, MeCN, 20°, 3 h 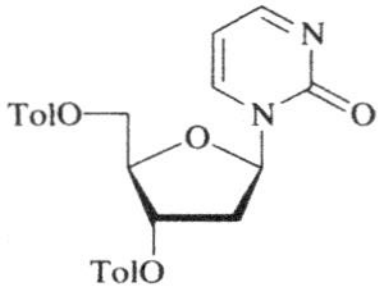	TolO, TolO, N, N, O (30)	147
	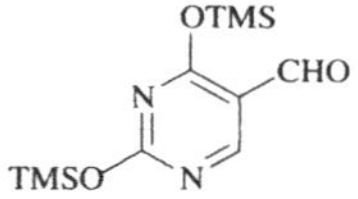 TMSO, N, N, OTMS, CHO	TMSOTf, MeCN, –50°, 2 h; –10°, 12 h; rt, 4 h	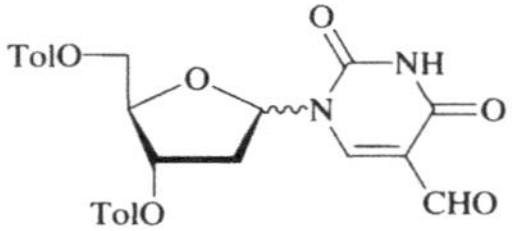 TolO, TolO, N, NH, O, O, CHO (63) α:β = 1:2	1018
	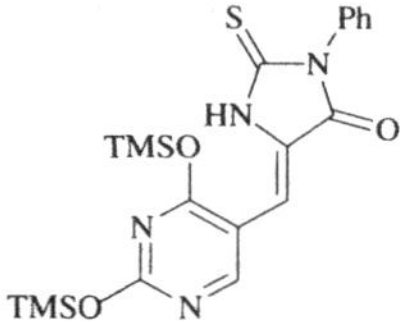 TMSO, N, N, TMSO, HN, N, S, Ph, O	TMSOTf, MeCN, –50°, 2 h; –10°, 12 h; rt, 12 h	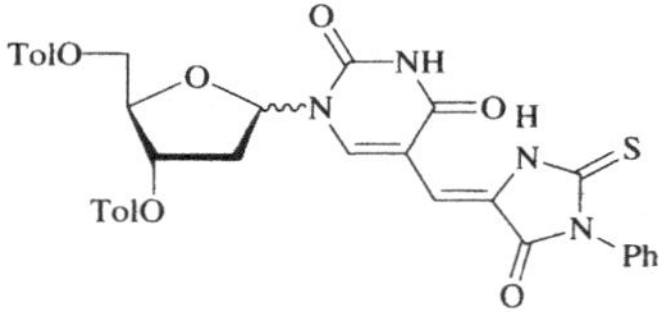 TolO, TolO, N, NH, O, O, H, N, S, N, Ph, O (46) α:β = 1:4	1018

TABLE IV. REACTIONS WITH TRIMETHYLSILYL AND SILVER TRIFLATES AND PERCHLORATES (*Continued*)

Sugar	Base	Conditions	Product(s) and Yield(s) (%)	Refs.
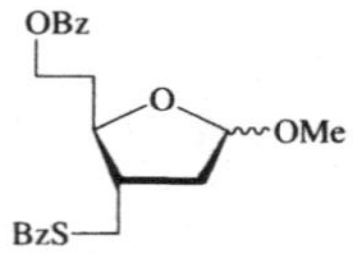 OBz; BzS; OMe	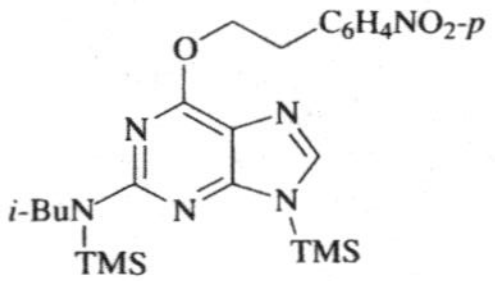 i-BuN; TMS; TMS; $C_6H_4NO_2$-*p*	TMSOTf, MeCN, 60°, 18 h	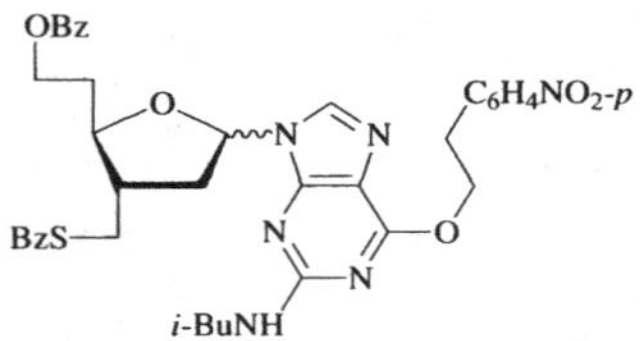 OBz; BzS; i-BuNH; $C_6H_4NO_2$-*p* (—)	990
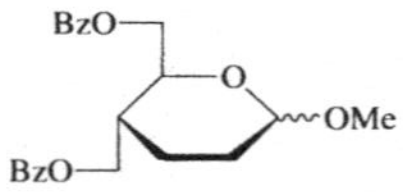 BzO; BzO; OMe	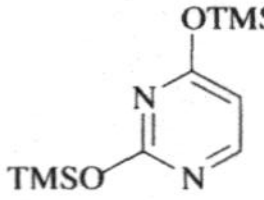 OTMS; TMSO	1. TBDMSOTf, MeCN, CH_2Cl_2 2. NaOMe, MeOH	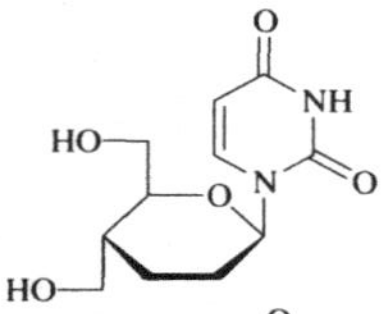 HO; HO; NH (47)	236
	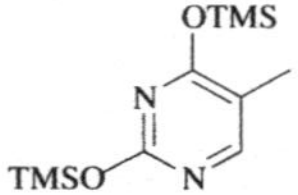 OTMS; TMSO	TBDMSOTf, MeCN, CH_2Cl_2, rt, 15 h	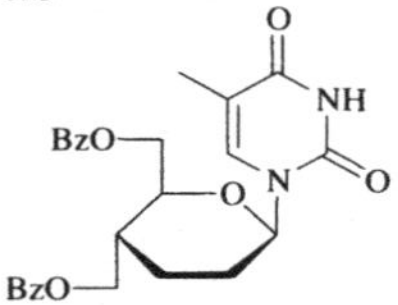 BzO; BzO; NH (66) + α-anomer (6)	236
	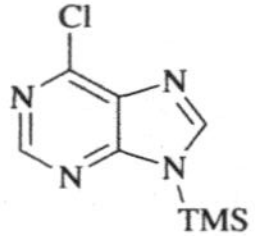 Cl; TMS	TBDMSOTf, MeCN, CH_2Cl_2	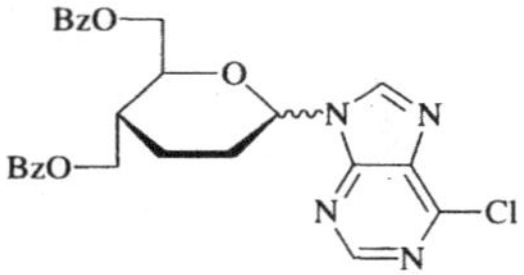 BzO; BzO; Cl (68) α:β = 1:13	236
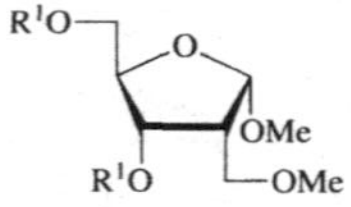 R^1O; R^1O; OMe; OMe; $R^1 = Cl_2C_6H_3CH_2$	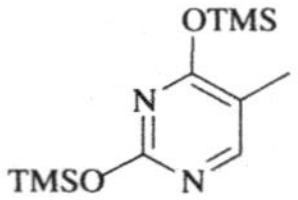 OTMS; TMSO	TMSOTf, $Cl(CH_2)_2Cl$, 50°	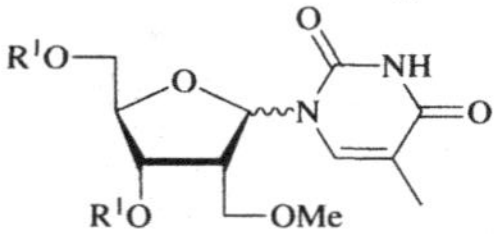 R^1O; R^1O; OMe; NH (76) α:β = 1:3	829

Glycosyl donor	Base	Conditions	Product(s) and Yield(s) (%)	Refs.
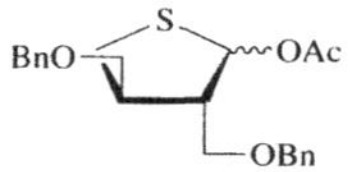 BnO, S, OAc, OBn	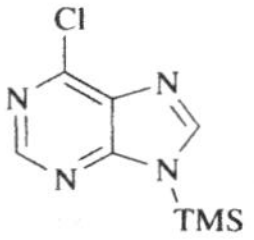 Cl, N, N, N, N, TMS	TMSOTf, CH_2Cl_2, rt, 3 h	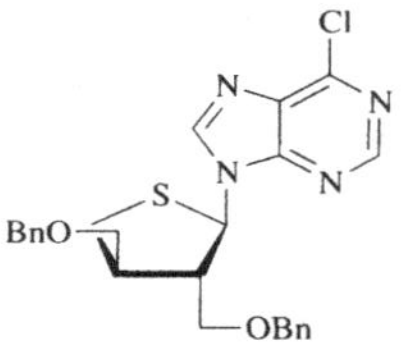 Cl, N, N, N, N, BnO, S, OBn (85) α:β = 2.7:1	1019
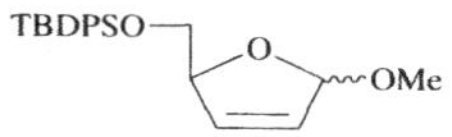 TBDPSO, O, OMe	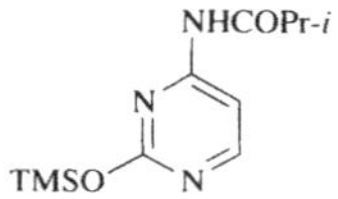 NHCOPr-*i*, N, TMSO, N	TMSOTf, MeCN, –20°, 2.5 h	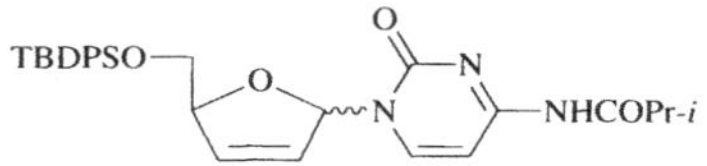 TBDPSO, O, N, O, N, NHCOPr-*i* (49)	1020
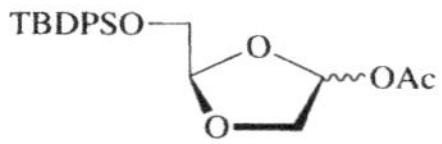 TBDPSO, O, O, OAc	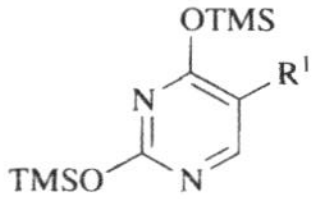 OTMS, R^1, N, TMSO, N	TMSOTf, $Cl(CH_2)_2Cl$, rt	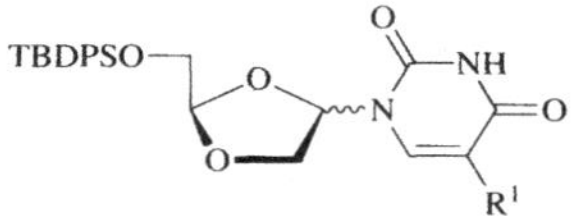 TBDPSO, O, O, N, O, NH, O, R^1	
	R^1 = H	2 h	(60)	1021
	R^1 = I	1 h	(35) α:β = 1:1	
	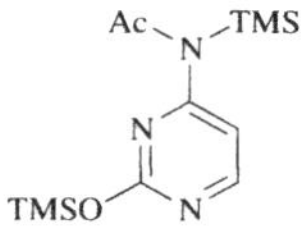 Ac, N, TMS, N, TMSO, N	TMSOTf, $Cl(CH_2)_2Cl$, rt, 45 min	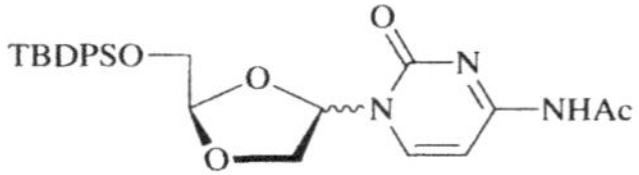 TBDPSO, O, O, N, O, N, NHAc (78) α:β = 1:1	1021
	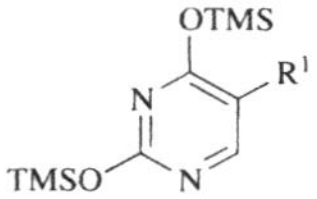 OTMS, R^1, N, TMSO, N	TMSOTf	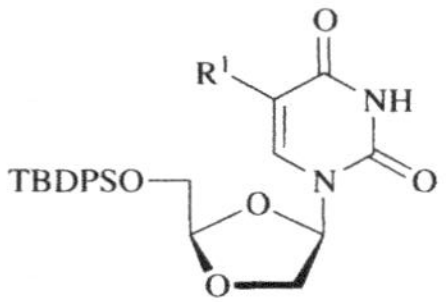 R^1, O, NH, O, N, TBDPSO, O, O (β) + α-anomer	
	R^1 = H	$Cl(CH_2)_2Cl$	(45) + α-anomer (29)	1021
	R^1 = Cl	$Cl(CH_2)_2Cl$, rt, 1 h	(30) + α-anomer (27)	1021
	R^1 = Br	$Cl(CH_2)_2Cl$, rt, 1 h	(35) + α-anomer (30)	1021
	R^1 = Me	$Cl(CH_2)_2Cl$, rt, 1 h	(45) + α-anomer (29)	1022

TABLE IV. REACTIONS WITH TRIMETHYLSILYL AND SILVER TRIFLATES AND PERCHLORATES (*Continued*)

Sugar	Base	Conditions	Product(s) and Yield(s) (%)	Refs.
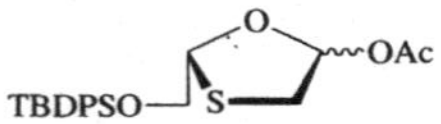 TBDPSO–(oxathiolane)–OAc	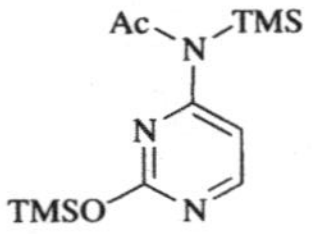 Ac(TMS)N–pyrimidine–OTMS (TMSO)	TMSOTf, $Cl(CH_2)_2Cl$	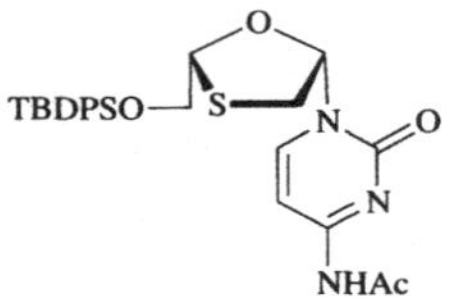 TBDPSO … N, NHAc	
		0° to rt	(64)	249
		rt, 1.5 h	(45) + β-anomer (22)	1023
	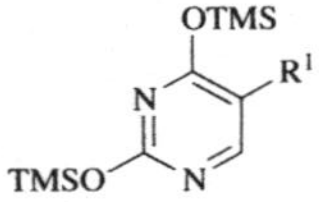 OTMS, R^1, TMSO	TMSOTf, $Cl(CH_2)_2Cl$, rt, 2 h	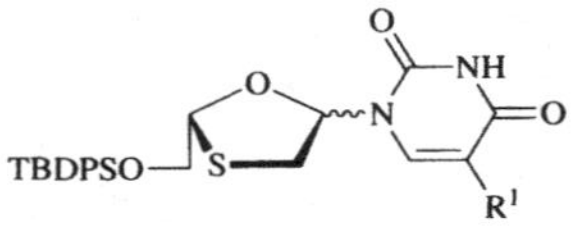 TBDPSO … NH, O, R^1	1023
	R^1			
	F		(93)	
	Cl		(58)	
	Br		(53)	
	I		(88)	
	Me		(72)	
	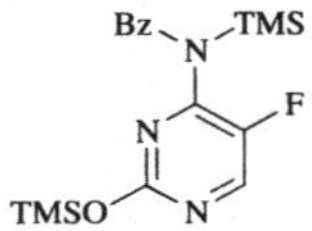 Bz(TMS)N, F, TMSO	TMSOTf, $Cl(CH_2)_2Cl$, rt, 2 h	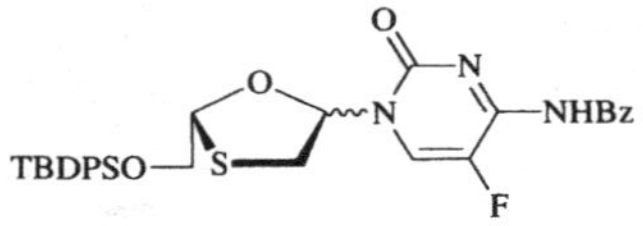 TBDPSO … NHBz, F (42) + β-anomer (32)	1023
	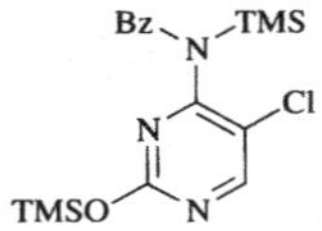 Bz(TMS)N, Cl, TMSO	TMSOTf, $Cl(CH_2)_2Cl$, rt, 2 h	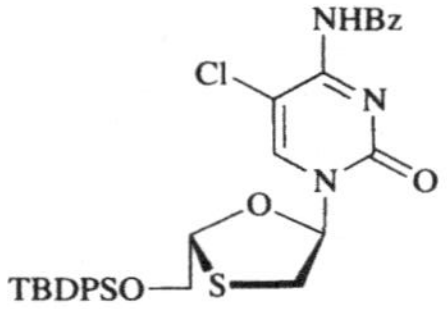 NHBz, Cl, TBDPSO … (44) + α-anomer (30)	1023

Substrate	Reagent	Conditions	Product(s) and Yield(s) (%)	Refs.
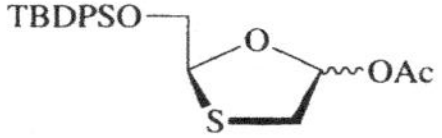 TBDPSO–(oxathiolane)–OAc	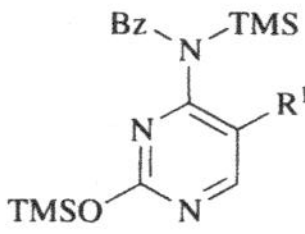 Bz(TMS)N–pyrimidine (2-TMSO, 5-R^1)	TMSOTf, $Cl(CH_2)_2Cl$, rt, time	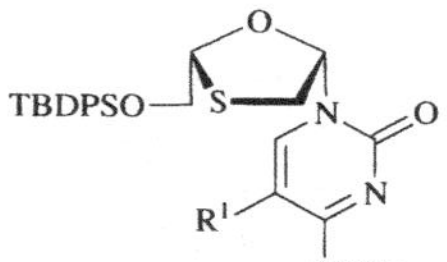 TBDPSO–oxathiolanyl–5-R^1-4-NHBz-cytosine + β-anomer	1023
	R^1	time	(α) / β-anomer	
	Br	1.5 h	(45) / (30)	
	I	1 h	(36) / (24)	
	Me	2 h	(35) / (28)	
	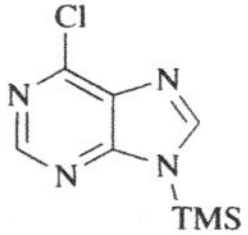 6-Cl-9-TMS-purine	TMSOTf, CH_2Cl_2, rt, 30 min; reflux, 14 h	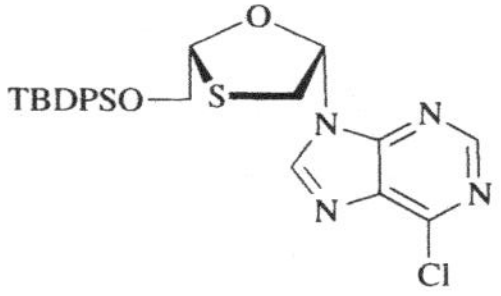 (35) + β-anomer (30)	1023
	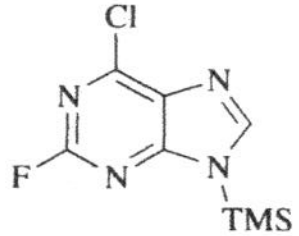 6-Cl-2-F-9-TMS-purine	TMSOTf, CH_2Cl_2, rt, 16 h	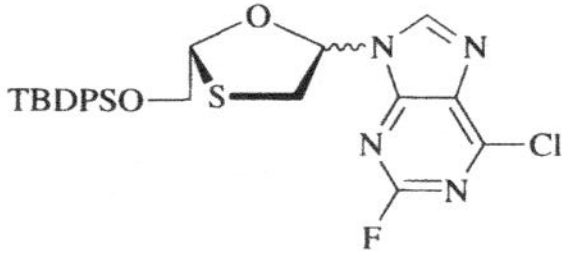 (60)	1023
	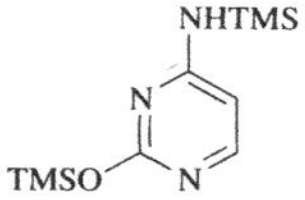 NHTMS, TMSO-pyrimidine	TMSOTf, $Cl(CH_2)_2Cl$, rt, 45 min	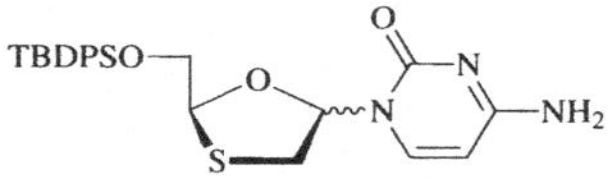 –NH_2 (—) α:β = 1:1	235
	Ac(TMS)N, TMSO-pyrimidine	TMSOTf, $Cl(CH_2)_2Cl$, 0° to rt	NHAc (60)	1024

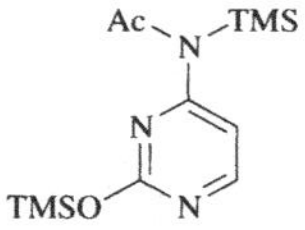

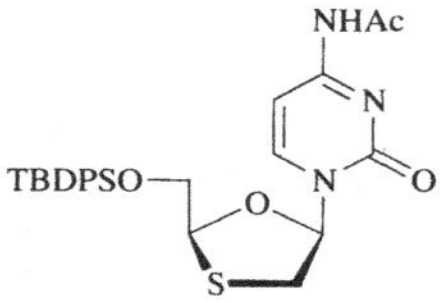

TABLE IV. REACTIONS WITH TRIMETHYLSILYL AND SILVER TRIFLATES AND PERCHLORATES (*Continued*)

Sugar	Base	Conditions	Product(s) and Yield(s) (%)	Refs.
	Ac, N, TMS; N, N; TMSO	TMSOTf, $Cl(CH_2)_2Cl$	TBDPSO, O, S, N, O, N, NHAc (50) $\alpha:\beta = 1:2$	1025
TBDPSO, O, OMe, F	OTMS; N, N; TMSO	1. TMSOTf, MeCN, rt, 2.5 h 2. n-Bu_4NF, THF, rt, 1 h	HO, O, F, N, NH, O, O **I** (35) + α-anomer (35)	281
TBDPSO, O, OMe, F	OTMS; N, N; TMSO	1. TMSOTf, MeCN, rt, 40 min 2. n-Bu_4NF, THF, rt, 30 min	**I** (44) + α-anomer (20)	281
TBDPSO, O, I, OMe	OTMS; N, N; TMSO; R[1]	TMSOTf, MeCN, –10°, 30 min; rt, 1-4 h	TBDPSO, O, I, N, NH, O, O, R[1] R[1] / α:β H (70) 1:1 OMe (67) 1:1	1026
	OTMS; N, N; TMSO; n	TMSOTf, MeCN, –17°, 10-12 h	n, TBDPSO, O, I, N, NH, O, O n / / α-anomer 3 (16) (5) 4 (13) (9) 6 (32) (21)	1027
TBDPSO, O, OMe, N_3	OTMS; N, N; TMSS	TMSOTf, MeCN, 20°, 30 min	TBDPSO, O, N_3, N, NH, S, O (—) $\alpha:\beta = 2:1$ α-anomer (16)	286

Reactant	Conditions	Product(s) and Yield(s) (%)	Refs.
OTMS, TMSO, N, N (pyrimidine)	1. TBDMSOTf, MeCN, rt, 1 h 2. F_3CCO_2H (50% aq), rt, 20 h	HO, O, N_3, N, NH, O, O — **I** (33) + α-anomer (29)	281
OTMS, OTMS, TMSO, N, N	1. TMSOTf, MeCN, 5°, 3 h 2. $NaHCO_3$	TBDPSO, O, N_3, N, NH, O, O, OH (74)	1028
OTMS, OR^1, TMSO, N, N	TMSOTf, MeCN, 5°, 3 h	TBDPSO, O, N_3, N, NH, O, O, OR^1	
R^1			
Me		(56)	1028
Et		(83)	1028
Pr-*n*		(79)	1028
Bu-*n*		(69)	1028
C_5H_{11}-*n*		(67)	1028
C_6H_{13}-*n*		(36)	1028
OTMS, OEt, TMSS, N, N	TMSOTf, MeCN, 5°, 3 h	TBDPSO, O, N_3, N, S, NH, EtO, O — (15) α:β = 5:3	286

TABLE IV. REACTIONS WITH TRIMETHYLSILYL AND SILVER TRIFLATES AND PERCHLORATES (*Continued*)

Sugar	Base	Conditions	Product(s) and Yield(s) (%)	Refs.
	NMe_2, N, N, N, N, TMS	TMSOTf, MeCN, –30 to 0°, 3 h	TBDPSO, O, N, N, N_3, N, N, NMe_2 (35) α:β = 4:1	1002
	TMSN, N, N, N, N, TMS	1. TMSOTf, MeCN, –30 to 20°, 72 h 2. Bu_4NF, THF	HO, O, N, N, N_3, N, N, NH (14) + α-anomer (25)	1003
	Bn, N, TMS, N, N, N, N, TMS	TMSOTf, MeCN, –30 to 20°, 72 h	TBDPSO, O, N, N, N_3, N, N, NHBn (43) α:β = 1:1	1003
TBDPSO, O, OMe	R^1, N, TMS, N, N, N, N, TMS $R^1 = 3,4-Cl_2C_6H_3CH_2$	1. TMSOTf, MeCN, –30 to –10°, 4 h 2. Bu_4NF, THF	HO, O, N, N, N, N, NHR^1 (28) α:β = 2:1	1003
	Bn, N, TMS, N, N, N, N, TMS	TMSOTf, MeCN, –30 to –10°, 4 h	TBDPSO, O, N, N, N, N, NHBn (41) α:β = 1:1	1003

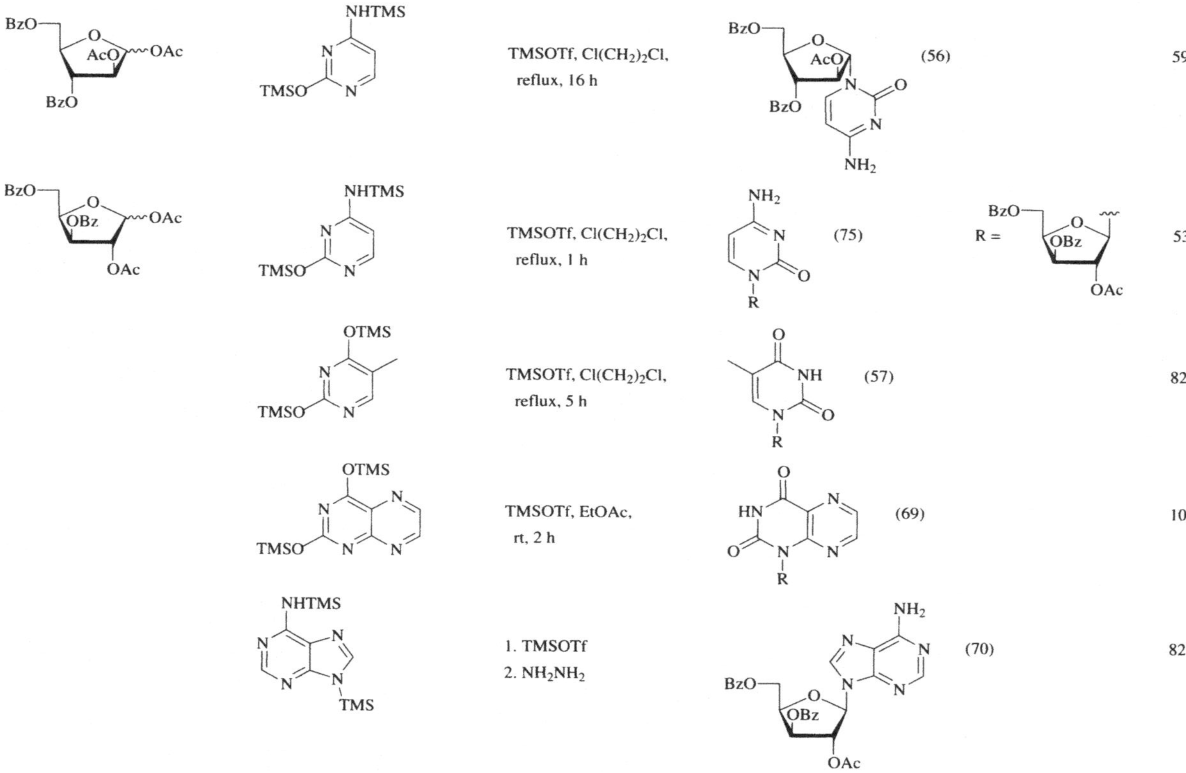
C23
TMSOTf, Cl(CH2)2Cl, reflux, 16 h
(56)
595
TMSOTf, Cl(CH2)2Cl, reflux, 1 h
(75)
R =
537
TMSOTf, Cl(CH2)2Cl, reflux, 5 h
(57)
828
TMSOTf, EtOAc, rt, 2 h
(69)
1029
1. TMSOTf
2. NH2NH2
(70)
828

TABLE IV. REACTIONS WITH TRIMETHYLSILYL AND SILVER TRIFLATES AND PERCHLORATES (*Continued*)

Sugar	Base	Conditions	Product(s) and Yield(s) (%)	Refs.
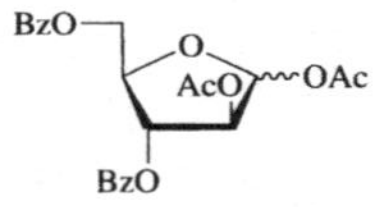 BzO, AcO, OAc, BzO	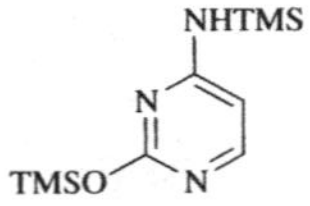 NHTMS, N, TMSO, N	1. TMSOTf, $Cl(CH_2)_2Cl$, reflux, 2 h 2. NH_3, MeOH, rt, 12 h	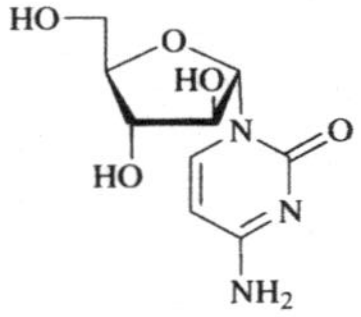 HO, O, HO, N, O, HO, N, NH_2 (56)	595
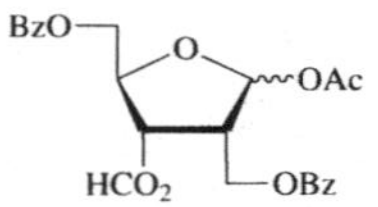 BzO, O, OAc, HCO_2, OBz	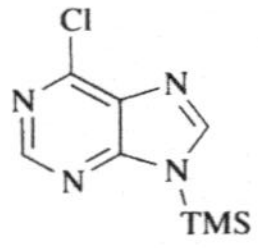 Cl, N, N, N, N, TMS	TMSOTf, MeCN, rt, 20 min	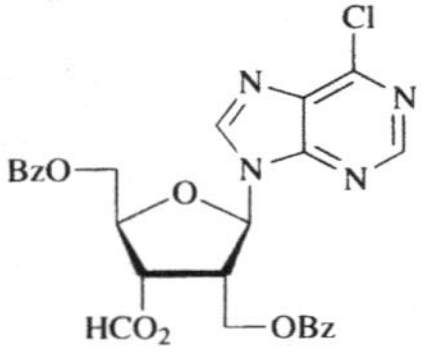 Cl, N, N, BzO, N, N, O, HCO_2, OBz (66)	1030
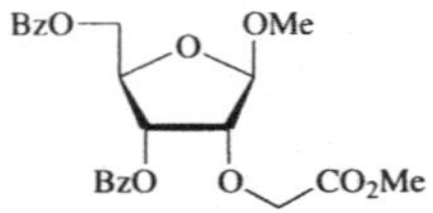 BzO, O, OMe, BzO, O, CO_2Me	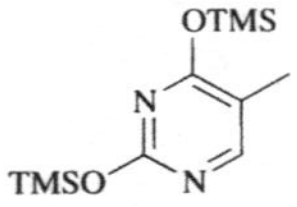 OTMS, N, TMSO, N	TMSOTf, MeCN, 50°, 3 h	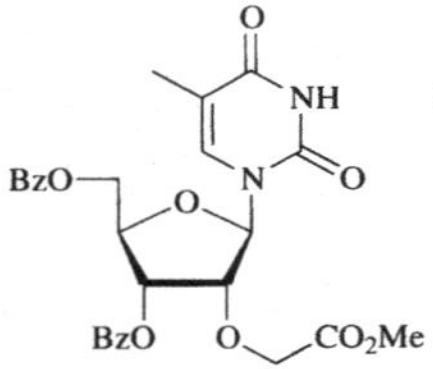 O, NH, BzO, N, O, O, BzO, O, CO_2Me (78)	835
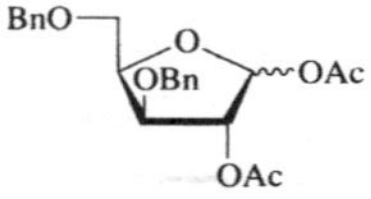 BnO, O, OBn, OAc, OAc	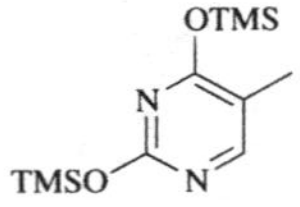 OTMS, N, TMSO, N	TMSOTf, $Cl(CH_2)_2Cl$, 60°, 8 h	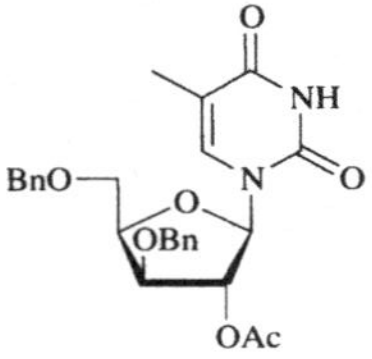 O, NH, BnO, N, O, O, OBn, OAc (36)	1031

Glycosyl donor	Base	Conditions	Product(s) and Yield(s) (%)	Refs.
TBDPSO, OAc	OTMS, TMSO	TMSOTf, $Cl(CH_2)_2Cl$, rt, 12-16 h	TBDPSO, O, NH, O (78)	1032
OMs, OBn, MeO_2C, OAc, AcO, OAc	OTMS, CO_2Me, TMSO	TMSOTf, MeCN, 0°, 3 h; rt, 5 h	O, MeO_2C, NH, OMs, OBn, O, MeO_2C, AcO, OAc (91)	155, 154
TBDPSO, OMe, MeS	OTMS, TMSO	TMSOTf, MeCN, –20°, 4 h	TBDPSO, O, NH, O, MeS (20)	1020
	OTMS, TMSO	1. TMSOTf, MeCN, –20°, 4 h 2. NH_3, MeOH	TBDPSO, O, NH, O, MeS (42) α:β = 1.4:1	1020
	NHCOPr-*i*, TMSO	TMSOTf, MeCN, –20°, 4 h; rt, 17 h	TBDPSO, O, N, NHCOPr-*i*, MeS (28)	1020
TBDPSO, OMe, MsO	OTMS, TMSO	TMSOTf, MeCN, –30°, 15 min; rt, 4 h	TBDPSO, O, NH, O, MsO (82) α:β = 1:2	1026

TABLE IV. REACTIONS WITH TRIMETHYLSILYL AND SILVER TRIFLATES AND PERCHLORATES (*Continued*)

Sugar	Base	Conditions	Product(s) and Yield(s) (%)	Refs.
C_{23} 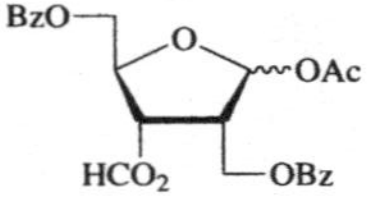 BzO, OAc, HCO_2, OBz	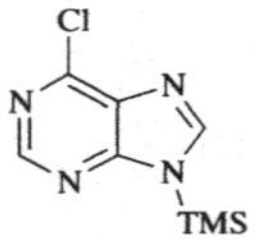 Cl, N, N, N, N, TMS	TMSOTf, MeCN, rt, 15 min	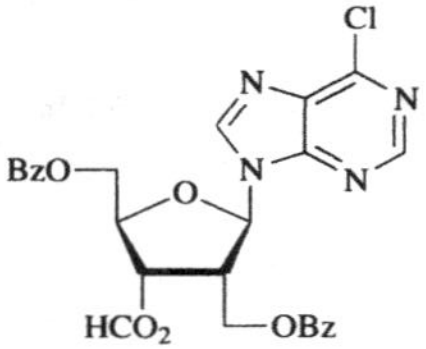 Cl, N, N, N, N, BzO, O, HCO_2, OBz (91)	1030
C_{24} 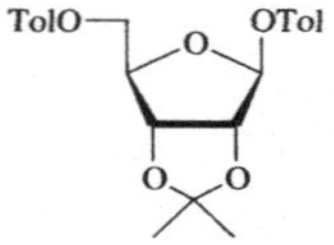 TolO, OTol, O, O, O	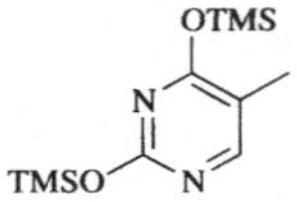 OTMS, N, N, TMSO	TMSOTf, $Cl(CH_2)_2Cl$, rt, 2 h	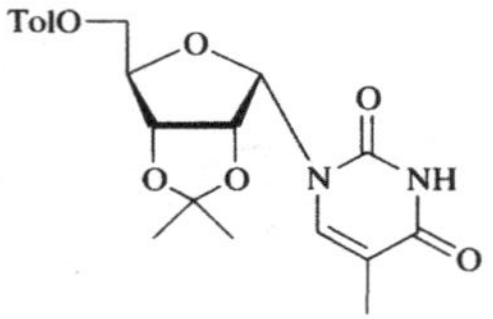 TolO, O, O, O, N, NH, O, O (65) + β-anomer (10-15)	223
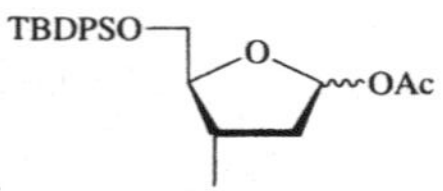 TBDPSO, O, OAc	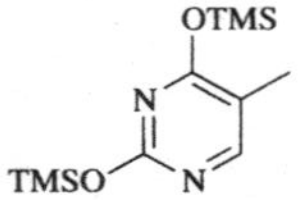 OTMS, N, N, TMSO	TMSOTf, $Cl(CH_2)_2Cl$, rt, 12-16 h	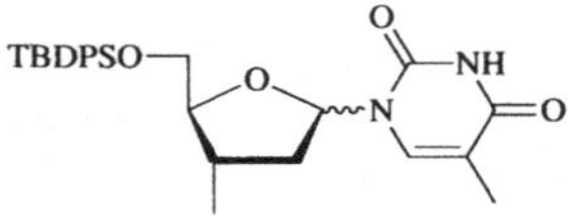 TBDPSO, O, N, NH, O, O (91) α:β = 1:1.6	1032
C_{25} 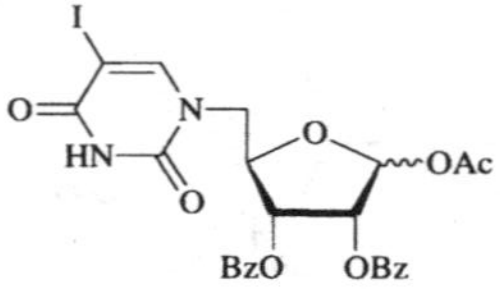 I, O, HN, N, O, O, OAc, BzO, OBz	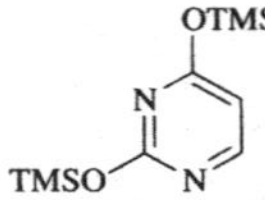 OTMS, N, N, TMSO	TMSOTf, MeCN	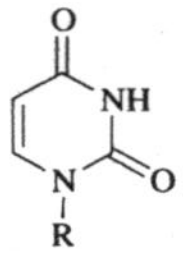 O, NH, O, N, R (89) R = I, O, HN, N, O, O, BzO, OBz 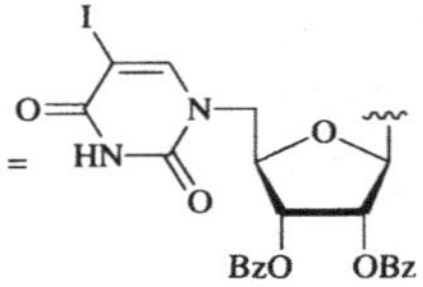	1033
	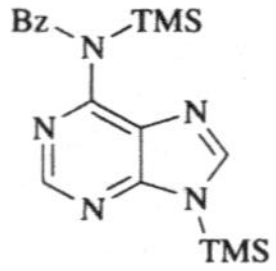 Bz, N, TMS, N, N, N, N, TMS	TMSOTf, MeCN	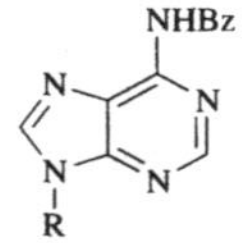 NHBz, N, N, N, N, R (56)	1033

		TMSOTf, MeCN, rt, 24 h	(59) R =	1034
		TMSOTf, MeCN, rt, 24 h	(50)	1034
		TMSOTf, MeCN, rt, 24 h	(60)	1034
		TMSOTf, MeCN, rt, 48 h	(30)	1034
		TMSOTf, MeCN, rt, 16 h	(75) N^7:N^9 = 1:19	143
		TMSOTf, $Cl(CH_2)_2Cl$	(79) α:β = 2:1	297

TABLE IV. REACTIONS WITH TRIMETHYLSILYL AND SILVER TRIFLATES AND PERCHLORATES (*Continued*)

Sugar	Base	Conditions	Product(s) and Yield(s) (%)	Refs.
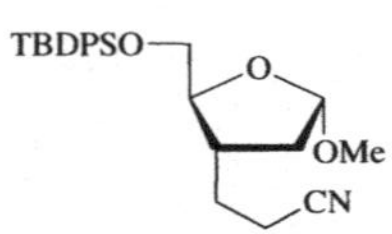 TBDPSO, OMe, CN	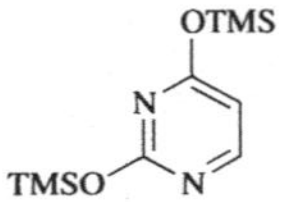 OTMS, N, TMSO, N	TMSOTf, MeCN, –30°, 30 min; 20°, 2 h	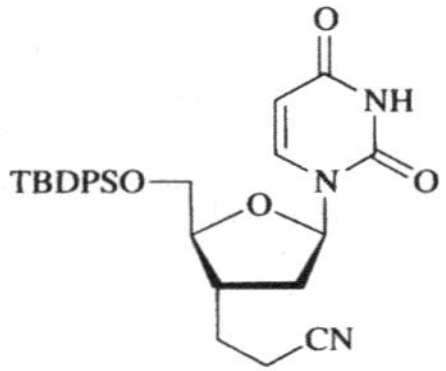 O, NH, O, N, TBDPSO, O, CN (33) + α-anomer (15)	1035
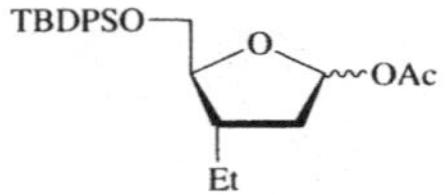 TBDPSO, O, OAc, Et	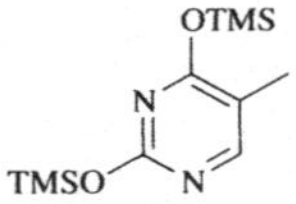 OTMS, N, TMSO, N	TMSOTf, $Cl(CH_2)_2Cl$, rt, 12-16 h	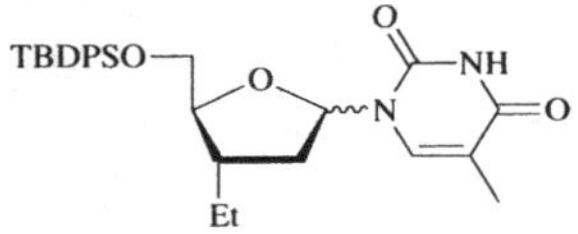 TBDPSO, O, NH, O, N, O, Et (85) α:β = 1:1.8	1032
C_{26} 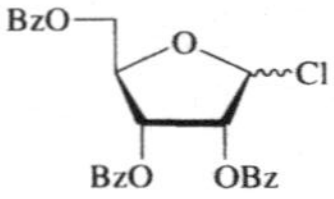 BzO, O, Cl, BzO, OBz	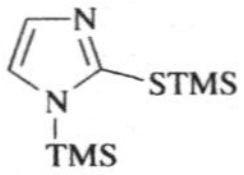 N, STMS, N, TMS	$AgClO_4$, C_6H_6, rt, 24 h	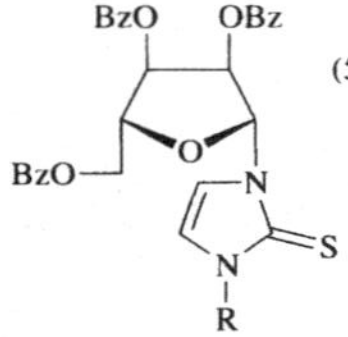 BzO, OBz, BzO, O, N, S, N, R (53) R = BzO, O, BzO, OBz 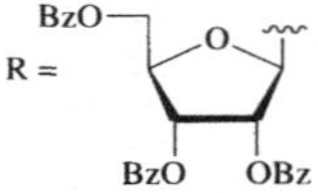	701
	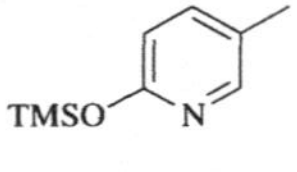 TMSO, N	AgOTf, CH_2Cl_2, –70°	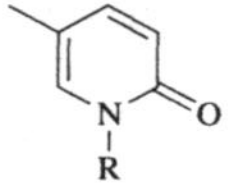 N, R, O (75)	212
	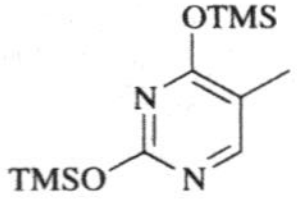 OTMS, N, TMSO, N	$AgClO_4$, C_6H_6, 20°, 16 h	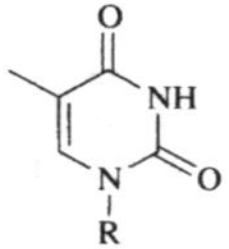 O, NH, N, O, R (75)	80

Reactant	Conditions	Product(s) and Yield(s) (%)	Refs.
OTMS; MeO, N	1. AgOTf, $MeNO_2$, –18° 2. MeOH, Amberlite IR 45	O; N, OMe; R (51)	212a
OMe; TMSO, N	1. AgOTf, CH_2Cl_2, –70° 2. MeOH, Amberlite IR 45	OMe; N, O; R (75)	212
OTMS; N, CO_2Et; TMSS, N	$AgClO_4$, C_6H_6, rt, 15 min	O; EtO_2C, NH; N, S; R (86)	81, 971
OTMS; N, N(Me)BOC; TMSS, N	1. $AgClO_4$, PhMe, 24°, 1 h 2. NH_3, MeOH, 24°, 30 h	O; Me–NH, NH; HO, O, N, S; HO, OH (50)	126
	$AgClO_4$, C_6H_6	O; Me–NH, NH; N, S; R (~70) R = BzO, O; BzO, OBz	125
Bn; O, N; TMSO, N	$AgClO_4$, C_6H_6, rt, 12 h	Bn; N, O; N, O; R (83)	1036

TABLE IV. REACTIONS WITH TRIMETHYLSILYL AND SILVER TRIFLATES AND PERCHLORATES (*Continued*)

Sugar	Base	Conditions	Product(s) and Yield(s) (%)	Refs.
	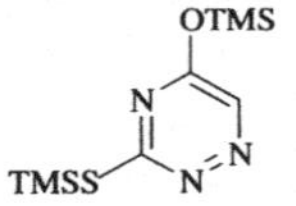 OTMS, TMSS, N, N, N	$AgClO_4$, C_6H_6, rt, 12 h	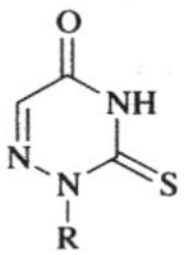 O, NH, N, N, S, R (78)	971
	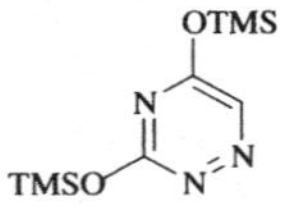 OTMS, TMSO, N, N, N	1. $AgClO_4$, C_6H_6, 22°, 1 h 2. pyrrolidine, rt, 23 h	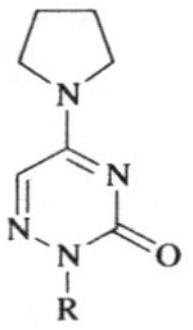 N, N, N, N, O, R (46)	79
	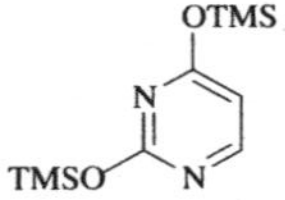 OTMS, TMSO, N, N	AgOTf, $MeNO_2$, –18°	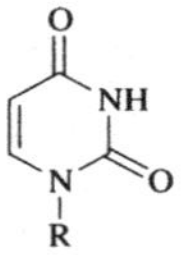 O, NH, N, O, R (62)	212
	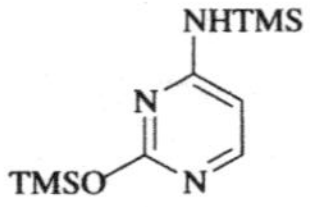 NHTMS, TMSO, N, N	AgOTf, $MeNO_2$, –18°	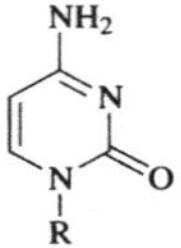 NH_2, N, N, O, R (50)	212
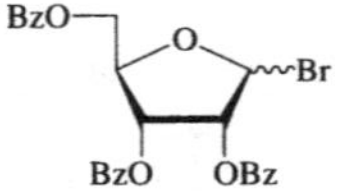 BzO, O, Br, BzO, OBz	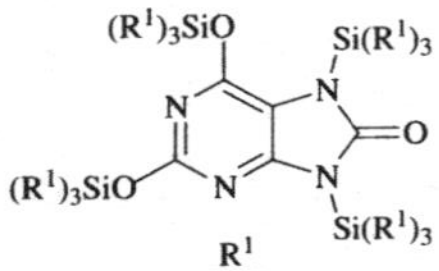 $(R^1)_3SiO$, $Si(R^1)_3$, N, N, O, $(R^1)_3SiO$, N, N, $Si(R^1)_3$		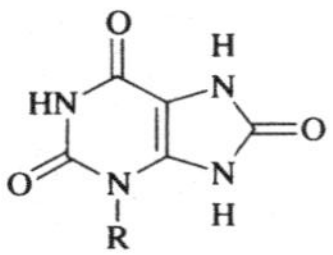 O, H, HN, N, O, O, N, N, R, H	
	R^1			
	Me	$AgClO_4$	(90)	73
	Et	$AgClO_4$, PhMe, 50°	(90)	74

Substrate	Silylated base	Conditions	Product(s) and Yield(s) (%)	Refs.
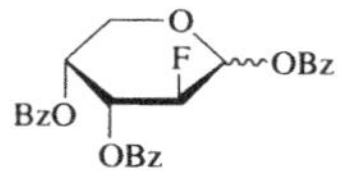 BzO, OBz, F, OBz	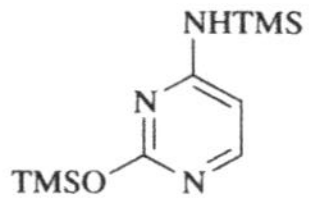 NHTMS; TMSO, N	1. TMSOTf, $Cl(CH_2)_2Cl$, reflux, 24 h 2. NH_3, MeOH, 12 h	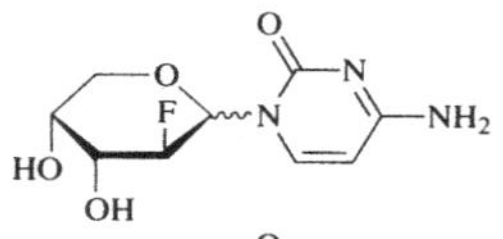 HO, OH, F, NH_2 (63) α:β = 65:35	1037
	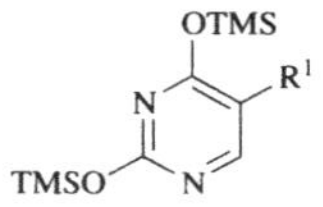 OTMS, R^1; TMSO, N	1. TMSOTf, $Cl(CH_2)_2Cl$, reflux, 24 h 2. NH_3, MeOH, 12 h	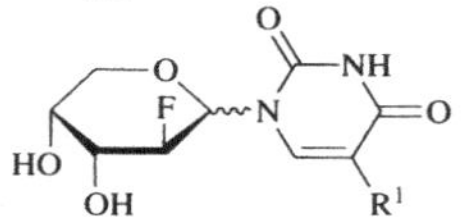 HO, OH, F, NH, R^1	1037
	R^1 H		(88) α:β 58:42	
	F		(89) 59:41	
	I	step 1 - 60 h	(72) 58:42	
	Me		(85) 59:41	
	Et		(67) 55:45	
	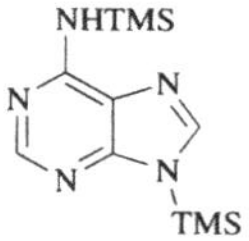 NHTMS; N-TMS	1. TMSOTf, $Cl(CH_2)_2Cl$, reflux, 24 h 2. NH_3, MeOH, 12 h	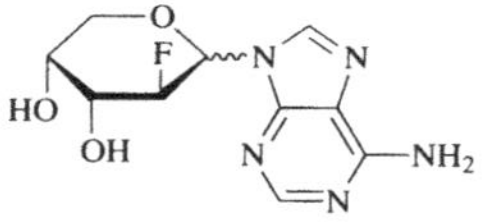 HO, OH, F, NH_2 (83) α:β = 58:42	1037
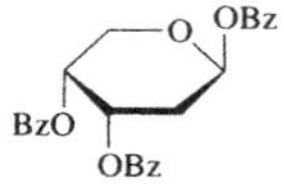 BzO, OBz, OBz	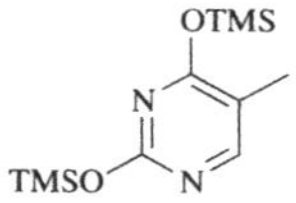 OTMS; TMSO, N	TMSOTf, $Cl(CH_2)_2Cl$	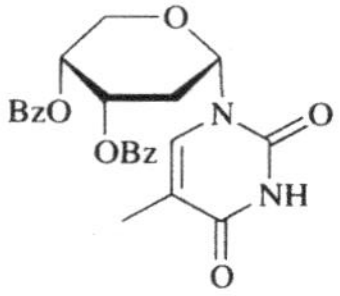 BzO, OBz, NH (63) + β anomer (25)	840
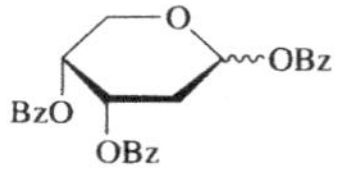 BzO, OBz, OBz	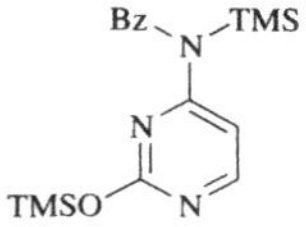 Bz-N-TMS; TMSO, N	TMSOTf, $Cl(CH_2)_2Cl$ rt, 4 h	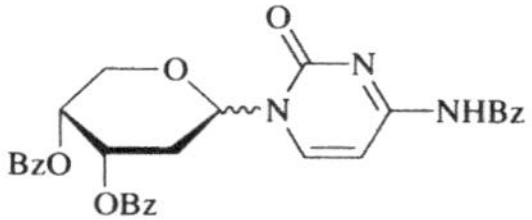 BzO, OBz, NHBz (94) α:β = 80:20	1037, 1038

TABLE IV. REACTIONS WITH TRIMETHYLSILYL AND SILVER TRIFLATES AND PERCHLORATES (*Continued*)

Sugar	Base	Conditions	Product(s) and Yield(s) (%)	Refs.
	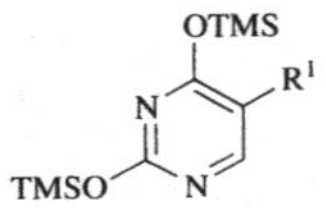 OTMS, R^1, N, N, TMSO	TMSOTf, $Cl(CH_2)_2Cl$ rt, 4 h	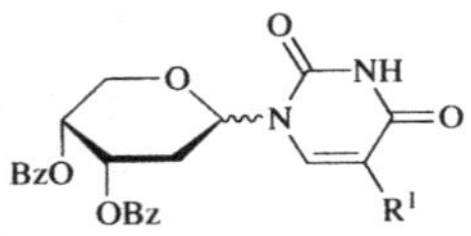 O, NH, N, O, BzO, OBz, R^1	1037, 1038
	R^1		α:β	
	H		(91) 66:34	
	F		(90) 68:32	
	I		(90) 76:24	
	Me		(93) 80:20	
	Et		(93) 73:27	
	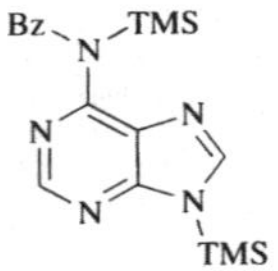 Bz, N, TMS, N, N, N, N, TMS	$TMSClO_4$, $Cl(CH_2)_2Cl$, reflux 6 h	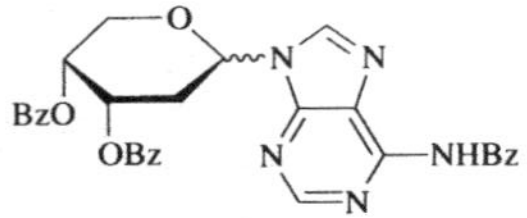 O, N, N, BzO, OBz, N, N, NHBz (62) α:β = 55:45	1037
	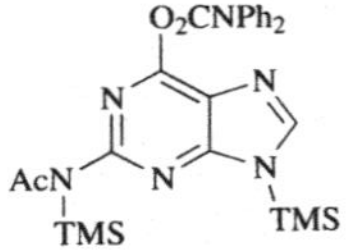 O_2CNPh_2, N, N, AcN, N, N, TMS, TMS	$TMSClO_4$, $Cl(CH_2)_2Cl$, reflux 6 h	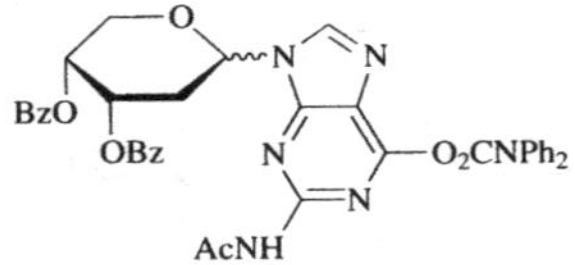 O, N, N, BzO, OBz, N, N, O_2CNPh_2, AcNH (55) α:β = 65:35	1037
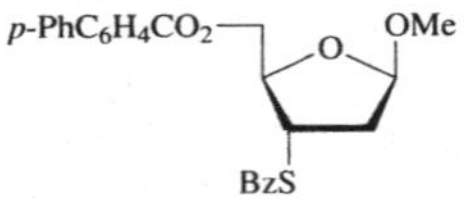 p-$PhC_6H_4CO_2$, O, OMe, BzS	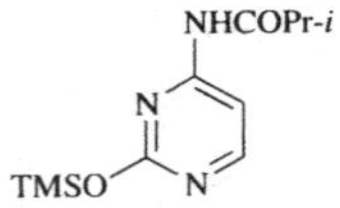 NHCOPr-*i*, N, TMSO, N	TMSOTf, MeCN, 25°, 5-6 h	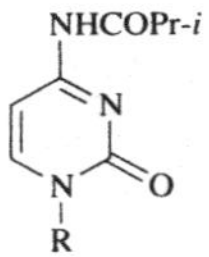 NHCOPr-*i*, N, N, O, R (70) R = p-$PhC_6H_4CO_2$, O, BzS 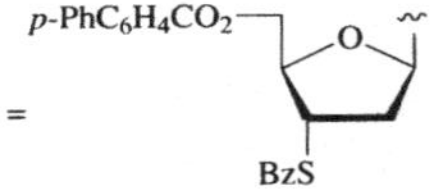	1020

Reactants	Conditions	Product(s) and Yield(s) (%)	Refs.
TMSO, N, OTMS, R^1 (silylated pyrimidine)	TMSOTf, MeCN, 25°, 5-6 h	R^1, O, NH, O, N, R; R^1: H (64), Me (54)	1020
O, OAc, OBn, O, OAc, AcO, OAc; TMSO, N, OTMS, CO_2Me	TMSOTf, $Cl(CH_2)_2Cl$, rt, 2 h	MeO_2C, O, NH, O, N, O, OAc, OBn, O, AcO, OAc (80)	156
TBDPSO, O, OAc, OAc; TMSO, N, NHTMS	TBDMSOTf, MeCN, CH_2Cl_2, rt, 15 h	NH_2, N, O, N, R (84); R = TBDPSO, O, OAc	1039
TMSO, N, OTMS	TBDMSOTf, MeCN, CH_2Cl_2, rt, 15 h	O, NH, O, N, R **I** (—)	962
	TBDMSOTf, MeCN, CH_2Cl_2, 0°, 30 min rt. 15 h	**I** (97)	1039
Cl, N, N, TMS (silylated 6-chloropurine)	TBDMSOTf, MeCN, CH_2Cl_2, 0°, 30 min rt. 15 h	Cl, N, N, R (87)	1039

TABLE IV. REACTIONS WITH TRIMETHYLSILYL AND SILVER TRIFLATES AND PERCHLORATES (*Continued*)

Sugar	Base	Conditions	Product(s) and Yield(s) (%)	Refs.
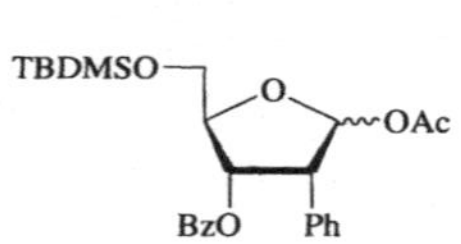 TBDMSO, OAc, BzO, Ph	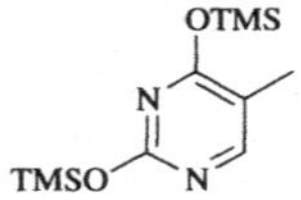 OTMS, N, TMSO, N	TMSOTf, $Cl(CH_2)_2Cl$, rt	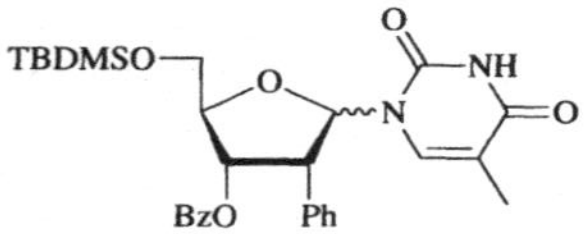 TBDMSO, O, NH, N, O, BzO, Ph (68) α:β = 1:3.5	829
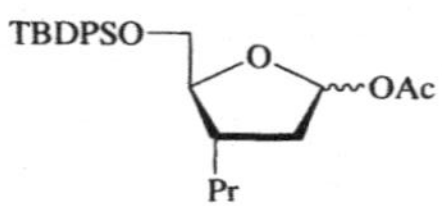 TBDPSO, OAc, Pr	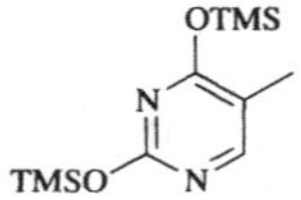 OTMS, N, TMSO, N	TMSOTf, $Cl(CH_2)_2Cl$, rt, 12-16 h	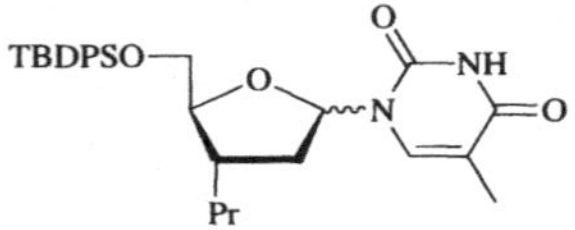 TBDPSO, O, NH, N, O, Pr (88) α:β = 1:1.7	1032
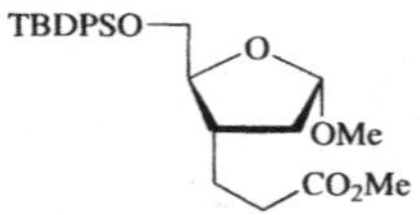 TBDPSO, OMe, CO_2Me	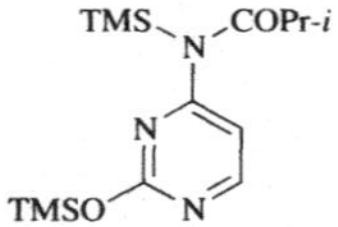 TMS, N, COPr-*i*, N, TMSO, N	TMSOTf, MeCN, –35°, 1.5 h	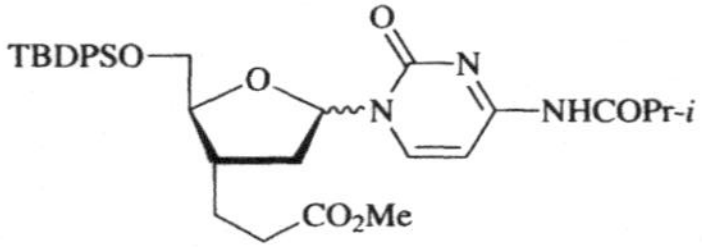 TBDPSO, O, N, N, NHCOPr-*i*, CO_2Me (59)	1035
C_{27} 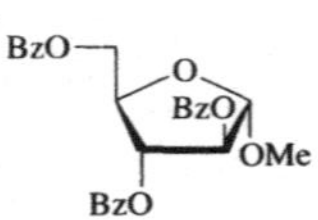 BzO, BzO, OMe, BzO	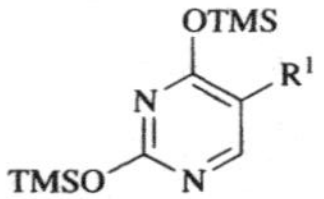 OTMS, R^1, N, TMSO, N	TMSOTf, MeCN, rt, 5 d	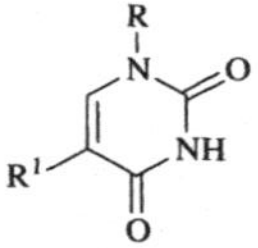 R, 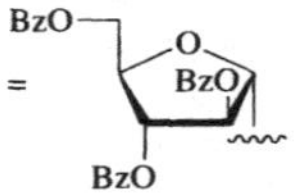 N, O, NH, R^1, O; R = BzO, BzO, BzO	370
	R^1			
	Cl		(81)	
	Me		(63)	
	n-C_6H_{13}		(6)	
	N-4-benzylpiperidinyl		(68)	
	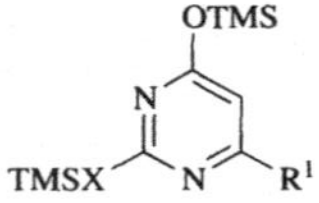 OTMS, N, TMSX, N, R^1	TMSOTf, MeCN, rt, 5 d	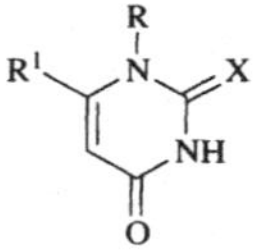 R, R^1, N, X, NH, O	366

Substrate	Base	Conditions	Product(s) and Yield(s) (%)	Refs.
	X / R^1: O / Me; O / Et; S / Me; S / Et		(2) bis(riboside) (81); (38) (46); (24) (—) N^3 products; (66) (—) N^3 products	
BzO, BzO, OMe, OBz (ribofuranose)	Cl, N, TMSO, N	TMSOTf, MeCN, rt, 30 min	BzO, O, BzO, OMe, O, N, N, Cl (96) $\alpha:\beta$ = 1:14	538a
	OTMS, R^1, N, TMSO, N	TMSOTf, MeCN, rt, 30 min	R^1, O, NH, O, N, R; R = BzO, O, BzO, OMe	538a
	R^1: H		(75)	
	F		(68)	
	NO_2		(56)	
	Me		(55)	
	CF_3		(62)	
BzO, S, SBn, BzO	NHTMS, N, TMSO, N	1. TMSOTf, MeCN, –20°, 1 h 2. NH_3, MeOH	HO, S, HO, N, O, N, NH_2 (36) + β-anomer (28)	1016
	OTMS, N, TMSO, N	1. TMSOTf, $Hg(OAc)_2$, CH_2Cl_2, rt, 24 h 2. NH_3, MeOH	HO, S, HO, N, O, NH, O (35) + β-anomer (24)	1016

TABLE IV. REACTIONS WITH TRIMETHYLSILYL AND SILVER TRIFLATES AND PERCHLORATES (*Continued*)

Sugar	Base	Conditions	Product(s) and Yield(s) (%)	Refs.
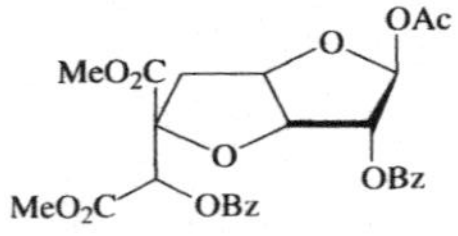 MeO$_2$C, OAc, OBz, MeO$_2$C, OBz	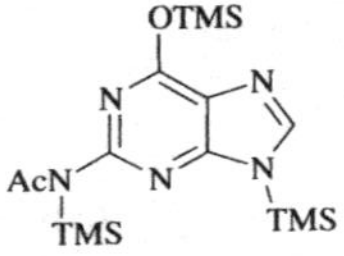 OTMS, AcN(TMS), N-TMS	TMSOTf, Cl(CH$_2$)$_2$Cl, rt, 4 d	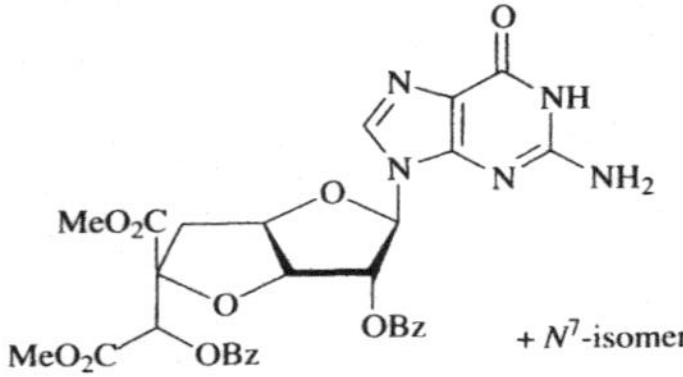 MeO$_2$C, MeO$_2$C, OBz, OBz, NH, NH$_2$ (22) + N^7-isomer (41)	782
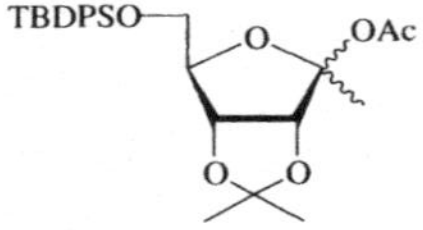 TBDPSO, OAc	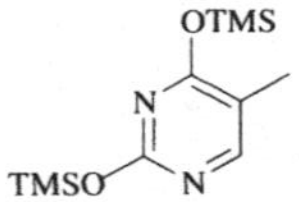 OTMS, TMSO	TMSOTf, MeCN, 0°, 1 h	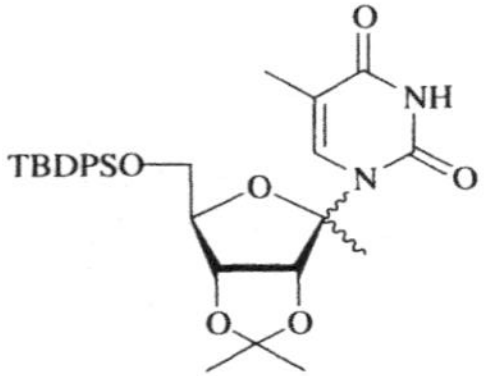 TBDPSO, NH, O: 0° (66) α:β 1.2:1; −30° (45) α:β 1:2	1040
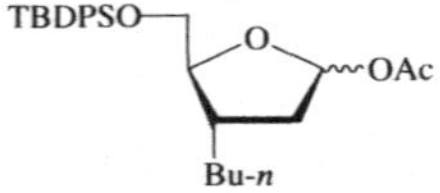 TBDPSO, OAc, Bu-*n*	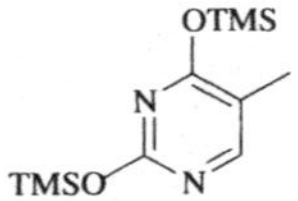 OTMS, TMSO	TMSOTf, Cl(CH$_2$)$_2$Cl, rt, 12-16 h	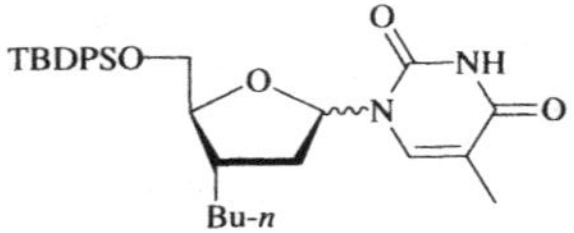 TBDPSO, NH, Bu-*n* (86) α:β = 1:1.7	1032
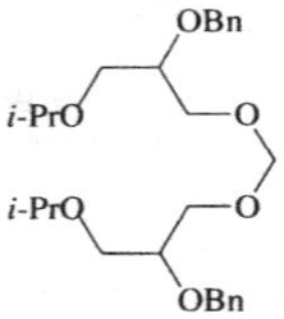 *i*-PrO, OBn, *i*-PrO, OBn	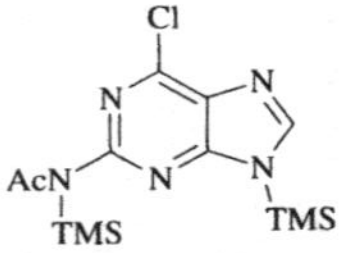 Cl, AcN(TMS), N-TMS	TMSOTf, Cl(CH$_2$)$_2$Cl, −30°, 2 h	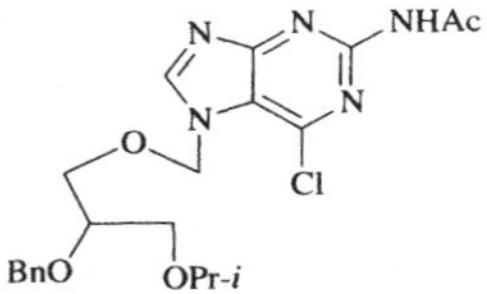 NHAc, Cl, BnO, OPr-*i* (51)	521

C_{28} BzO–(ribofuranose)–OAc, BzO, OBz | 4,5-dicyano-2-formamido(HN, OHC)-1-TMS imidazole (HN–TMS) | TMSOTf, CH_2Cl_2, rt, 2.5 h | OHC–HN, CN, CN, HN–R (60) | R = BzO–(ribofuranosyl), BzO, OBz | 161

Imidazole (R^3, R^2, R^1, N–TMS) | TMSOTf | Imidazole (R^3, R^2, R^1, N–R)

R^1	R^2	R^3	Conditions		N^7-isomer	Ref.
NO_2	$SCH_2P(O)(OEt)_2$	H	MeCN, rt, 42 min	(19)	(7)	1041
NO_2	SMe	H	MeCN, 0°, 3 min	(80)	(—)	1041
SMe	NO_2	H	$Cl(CH_2)_2Cl$, rt, 16 h	(51)	(—)	1041
NO_2	SCH_2CO_2Me	H	MeCN, 0°, 90 s	(72)	(0.8)	1041
NO_2	$SCH_2CH(CO_2Me)CH_2CO_2Me$	H	MeCN, 0°, 90 s	(63)	(—)	1041
CO_2Me	CO_2Me	Me	$Cl(CH_2)_2Cl$, rt, 3 h	(60)	(—)	502
CO_2Me	CO_2Me	Me	$Cl(CH_2)_2Cl$, rt, 18 h	(60)	(—)	1042
CO_2Me	CO_2Me	Et	$Cl(CH_2)_2Cl$, rt, 3 h	(52)	(—)	502
CO_2Me	CO_2Me	*i*-Pr	$Cl(CH_2)_2Cl$, rt, 3 h	(83)	(—)	502
CO_2Me	CO_2Me	Bu	$Cl(CH_2)_2Cl$, rt, 3 h	(55)	(—)	502
CO_2Et	CO_2Et	Ph	$Cl(CH_2)_2Cl$, rt, 3 h	(45)	(—)	502
CO_2Et	CO_2Et	Bn	$Cl(CH_2)_2Cl$, rt, 3 h	(61)	(—)	502

1,2,4-Triazole (R^1, N–TMS) | TMSOTf | 1,2,4-Triazole (R^1, N–R)

R^1	Conditions	Product	Ref.
H	$Cl(CH_2)_2Cl$, MeCN, 24°, 3 h	(61)	133
CO_2Me	$Cl(CH_2)_2Cl$, MeCN, 4°, 4 h	(47) + N^2-isomer (20)	133
CH_2CO_2H	$Cl(CH_2)_2Cl$, C_6H_6, 100°, 4 h	(86)	1043

TABLE IV. REACTIONS WITH TRIMETHYLSILYL AND SILVER TRIFLATES AND PERCHLORATES (*Continued*)

Sugar	Base	Conditions	Product(s) and Yield(s) (%)	Refs.
BzO– (furanose) OAc, BzO, OBz	MeO_2C, CO_2Me triazole, N–H	HMDS, TMSOTf, MeCN, rt, 12 h	MeO_2C, CO_2Me triazole, N–R (94)	1044
	P, N, N–TMS	1. TMSOTf, MeCN, rt, 8 h 2. NH_3, MeOH	HO– (furanose) HO, HO, N, N, P (16)	214
	CO_2Et, P, N, N–TMS	1. TMSOTf, MeCN, rt, 16 h 2. NH_3, MeOH	CO_2Et, P, N, N–R (44) R = HO– (furanose) HO, OH	214
	CO_2Et, P, EtO_2C, N, N–TMS	1. TMSOTf, MeCN, rt, 24 h 2. NH_3, MeOH	$CONH_2$, P, H_2NOC, N, N–R (—)	214
	TMSO, N (pyridine)	TMSOTf, $Cl(CH_2)_2Cl$	N–R, O (pyridone) R = BzO– (furanose) BzO, OBz	
		82°, 2 h	(86)	132
		C_6H_6, 100°, 1.5 h	(85)	1043
		reflux, 1.5 h	(86)	133

Substrate	Conditions	Product(s) and Yield(s) (%)	Refs.
R^1-pyridine (4-substituted)	TMSOTf, $Cl(CH_2)_2Cl$, reflux,	N-R-pyridin-4-ylidene (R^2)	
R^1 = OTMS, R^2 = O	3.5 h	(83)	133
R^1 = NHTMS, R^2 = NH	2.5 h	(80)	133
3-AcO-2-TMSO-pyridine	TMSOTf, CH_2Cl_2, rt, 24 h	3-OH-1-R-pyridin-2-one (81)	1045
4-NHTMS-3-R^1-2-TMSX-pyridine	TMSOTf, $Cl(CH_2)_2Cl$, rt	4-NH_2-3-R^1-1-R-pyridin-2-(=X)	
R^1 = Cl, X = O	24 h	(99)	861, 1046
R^1 = Br, X = O	18 h	(41)	1046
R^1 = NO_2, X = NH	—	(—)	1047
4-OTMS-3-R^1-2-TMSO-pyridine	TMSOTf, $Cl(CH_2)_2Cl$, rt, 18 h	4-OH-3-R^1-1-R-pyridin-2-one	1046
R^1 = Cl		(56)	
R^1 = Br		(84)	
R^1 = NO_2		(82)	
4-OTMS-3-MeO-2-methylpyridine	TMSOTf, CH_2Cl_2, rt, 24 h	3-OMe-2-methyl-1-R-pyridin-4-one (88)	1045

TABLE IV. REACTIONS WITH TRIMETHYLSILYL AND SILVER TRIFLATES AND PERCHLORATES (*Continued*)

Sugar	Base	Conditions	Product(s) and Yield(s) (%)	Refs.
	OTMS	TMSOTf, $Cl(CH_2)_2Cl$, rt, 2 h	O, N, N, R (98)	845
BzO, O, OAc, *, *, BzO, OBz	OTMS, N, TMSO, N	1. TMSOTf, $Cl(CH_2)_2Cl$, reflux, 4 h 2. NH_3, MeOH	O, NH, N, O, HO, O, *, *, HO, OH (89) * ^{13}C-enriched	36
BzO, O, OAc, BzO, OBz	OTMS, N, TMSO, N	$TMSClO_4$, $Cl(CH_2)_2Cl$	O, NH, N, O, R (56) R = BzO, O, BzO, OBz	
		22°, 7 d	(76)	132
		82°, 6 h	(80)	132
		24°, 1 week	(81)	133
		C_6H_6, 100°, 4 h	(80)	1043
		C_6H_6, 24°, 1 week	(76)	1043
	NHTMS, N, TMSO, N	TMSOTf, $Cl(CH_2)_2Cl$, C_6H_6, 100°, 1 h	NH_2, N, N, O, R **I** (85)	1043

Substrate	Silylated base	Conditions	Product(s) and Yield(s) (%)	Refs.
		TMSOTf, $Cl(CH_2)_2Cl$		
		reflux, 1 h	**I** (98)	133
		82°, 1 h	**I** (90)	132
		20°, 1 h	**I** (90)	1043
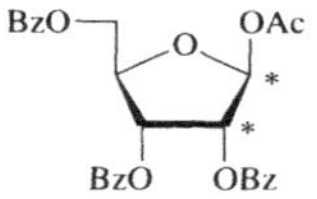 BzO, OAc, BzO, OBz (* labelled)	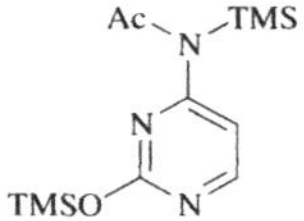 Ac–N–TMS, TMSO, N	1. TMSOTf, $Cl(CH_2)_2Cl$, reflux, 2 h 2. NH_3, MeOH, 16 h	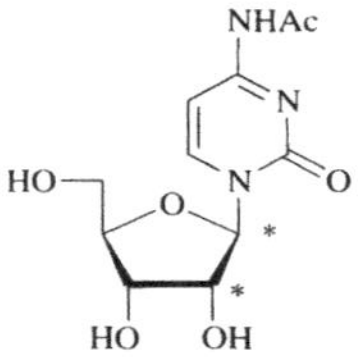 NHAc, HO, HO, OH (95) * ^{13}C-enriched	36
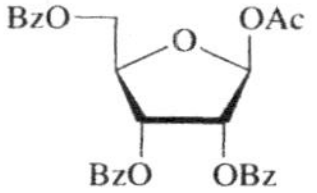 BzO, OAc, BzO, OBz	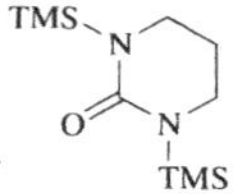 TMS–N, N–TMS, O	TMSOTf, $Cl(CH_2)_2Cl$, 60°, 1 h	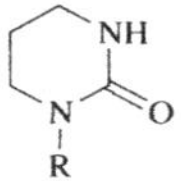 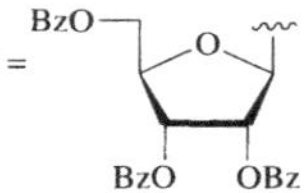 NH, N–R, O (16) R = BzO, BzO, OBz	162
	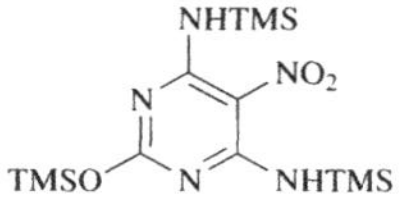 NHTMS, NO_2, TMSO, NHTMS	TMSOTf, MeCN, rt, 18 h	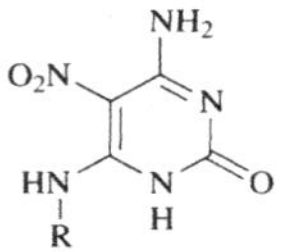 NH_2, O_2N, HN–R, N–H, O (72)	455
	OTMS, R^1, TMSO	TMSOTf, $Cl(CH_2)_2Cl$, 24°, time	R^1, O, NH, N–R, O R^1 = NO_2, time 2 h (93) R^1 = OMe, time 4-7 h (86-90)	133 193, 133
	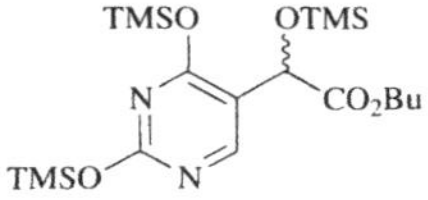 TMSO, OTMS, CO_2Bu, TMSO	TMSOTf	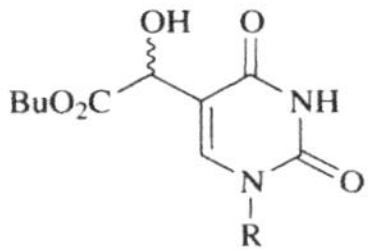 BuO_2C, OH, O, NH, N–R, O (76) α:β = 1:1	611

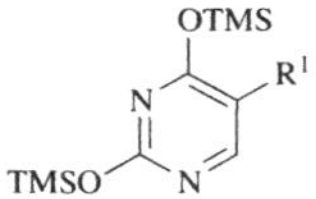

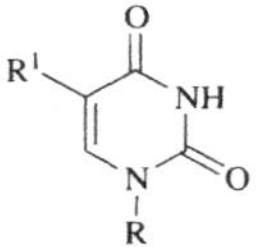

TABLE IV. REACTIONS WITH TRIMETHYLSILYL AND SILVER TRIFLATES AND PERCHLORATES (*Continued*)

Sugar	Base	Conditions	Product(s) and Yield(s) (%)	Refs.
	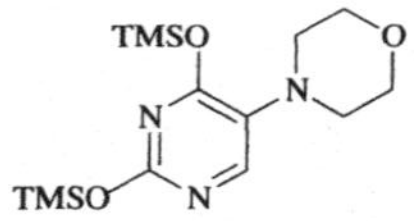 TMSO, TMSO, N, N, N, O	TMSOTf, $Cl(CH_2)_2Cl$, 24°, 10 or 24 h	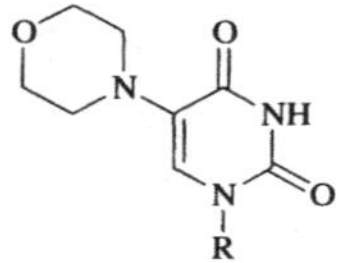 O, N, O, NH, N, O, R (95-99) R 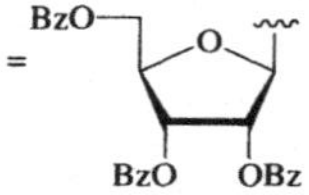 = BzO, O, BzO, OBz	133, 193
	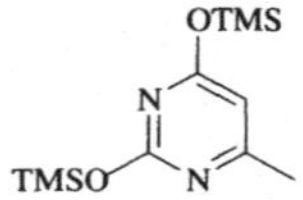 OTMS, N, TMSO, N	TMSOTf, $Cl(CH_2)_2Cl$, 24°	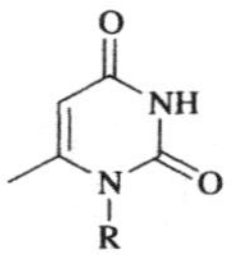 O, NH, N, O, R **I** (47) + N^3-isomer (18) + N^1,N^3-bis(riboside) (12)	193
		TMSOTf, MeCN, 24°	**I** (71) + N^3-isomer (2) + N^1,N^3-bis(riboside) (11)	193
		TMSOTf, $Cl(CH_2)_2Cl$, MeCN, 4 to 24°, 2 h	**I** (75) + N^3-isomer (4) + N^1,N^3-bis(riboside) (11)	133
	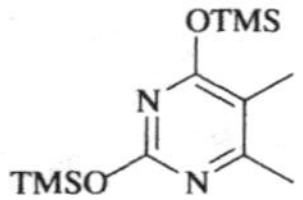 OTMS, N, TMSO, N	TMSOTf, $Cl(CH_2)_2Cl$, 24°, 3 h	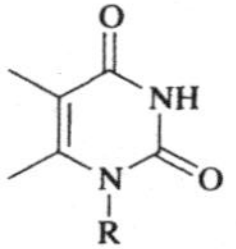 O, NH, N, O, R (82) + N^3-isomer (9)	133, 193
	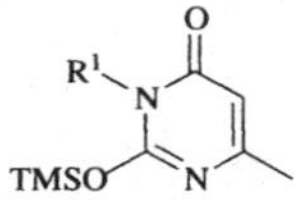 O, R^1, N, TMSO, N	TMSOTf, MeCN	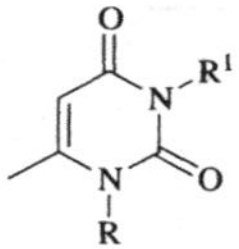 O, N, R^1, N, O, R; R^1: Me (—); Bn (85)	256
	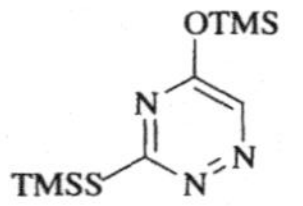 OTMS, N, TMSS, N, N	$TMSClO_4$, $Cl(CH_2)_2Cl$, 82°, 3 h	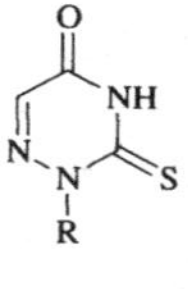 O, NH, N, N, S, R (67)	132

Substrate	Conditions	Product(s) and Yield(s) (%)		Refs.
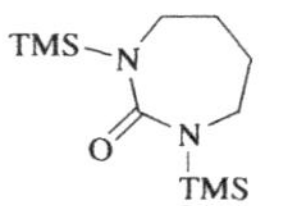 TMS–N ... N–TMS (structure)	TMSOTf, $Cl(CH_2)_2Cl$, 25°, 18 h	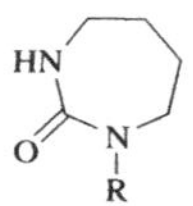 **I** (14)	R 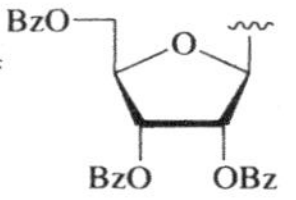 = (BzO, BzO, OBz ribofuranosyl structure)	162
	TMSOTf, $Cl(CH_2)_2Cl$, 60°, 1 h	**I** (10)		162
	TMSOTf, C_6H_6, 25°, 18 h	**I** (7)		162
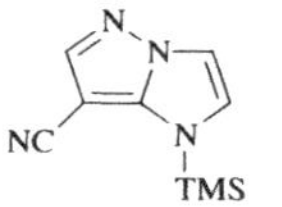 NC, N–TMS (structure)	TMSOTf, MeCN	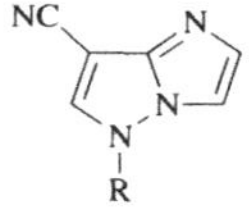 NC, N–R (structure) **I**		
	0-rt, 6 d	**I** (53) + N^1-isomer (47)		1049
	0-5°, 1 h	**I** (96) + N^1-isomer (4)		1049
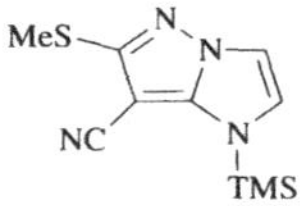 MeS, NC, N–TMS (structure)	TMSOTf, MeCN, 0-5°, 2 d	SMe, CN, N–R (structure) (89)		1049
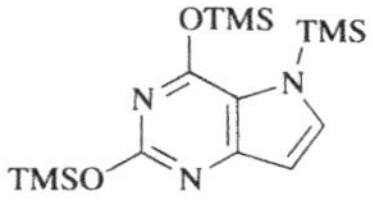 OTMS, TMS, TMSO (structure)	TMSOTf, MeCN, rt, 12 h	H, O, NH, O, N–R (structure) (82)		362
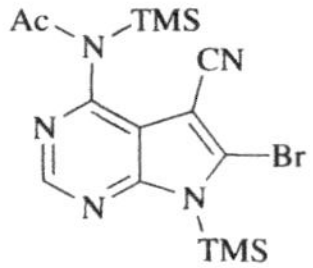 Ac, TMS, CN, Br, N–TMS (structure)	TMSOTf, $Cl(CH_2)_2Cl$, rt, 1 h; 80-85°, 20 h	NC, NHAc, Br, N–R (structure) (90)		1050

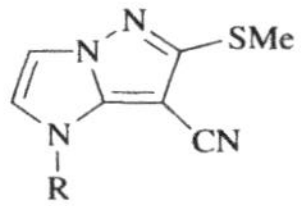

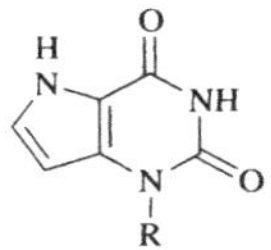

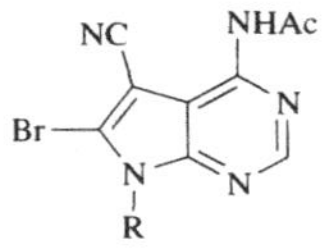

TABLE IV. REACTIONS WITH TRIMETHYLSILYL AND SILVER TRIFLATES AND PERCHLORATES (*Continued*)

Sugar	Base	Conditions	Product(s) and Yield(s) (%)	Refs.
	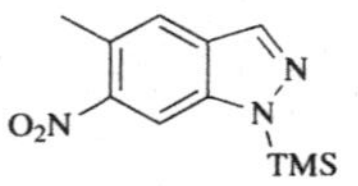	TMSOTf, $Cl(CH_2)_2Cl$, rt, 6 h	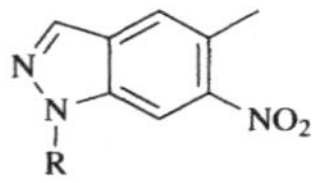 (64)	1051
	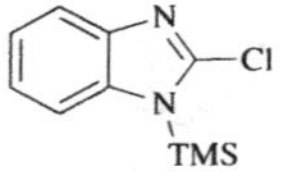	1. TMSOTf, $Cl(CH_2)_2Cl$, reflux, 1 h 2. NaOH, EtOH	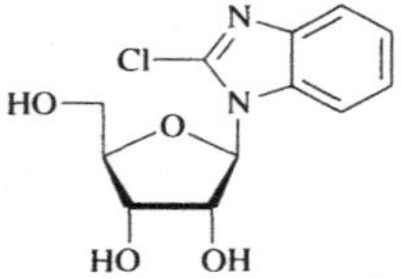 (81)	1052
		TMSOTf	R^1: CF_3 (—)	1053
			CN (—)	1053
	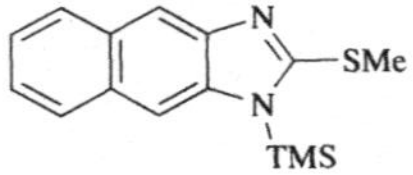	TMSOTf, $Cl(CH_2)_2Cl$, 25°, 24 h	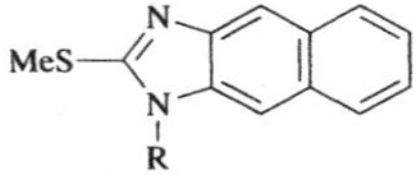 (76) R = 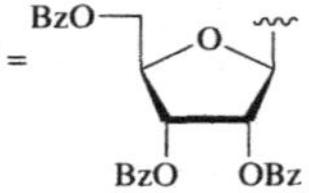	1054
	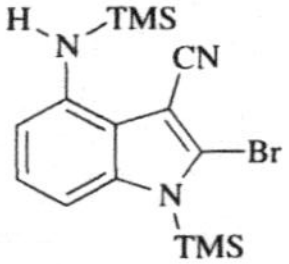	TMSOTf, $Cl(CH_2)_2Cl$, rt, 1 h; 75-80°, 18 h; reflux, 3 h	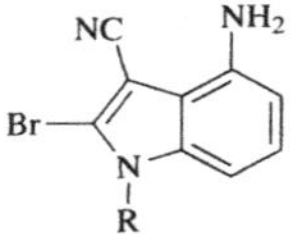 (88)	1050

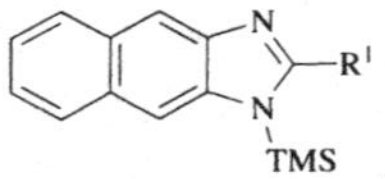

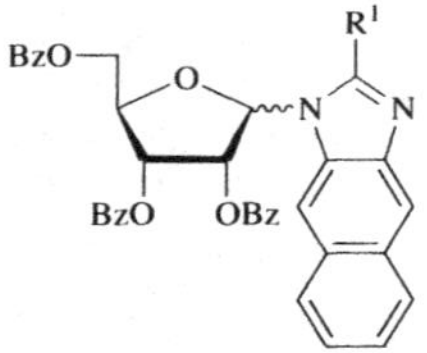

Reactant	Conditions	Product(s) and Yield(s) (%)	Refs.
XTMS, TMSO (pyrazolo[1,5-a]pyrimidine)	TMSOTf, $Cl(CH_2)_2Cl$, rt, 6 h	XH, N–R, =O; X: NH (78); O (55)	361; 361
OTMS, TMSO, N-TMS (pyrazolo[3,4-d]pyrimidine)	TMSOTf, $Cl(CH_2)_2Cl$, rt, 6 h	O, NH, H, N–R, =O (66); R = BzO, O, BzO, OBz (ribofuranosyl)	361
OTMS, R[1], R[2], N-TMS	TMSOTf, MeCN	R[1], O, NH, R[2], N–R	
R[1] = Br, R[2] = H	reflux, 1 week	(29)	1055
R[1] = CN, R[2] = H	−20° to rt, 4 h	(46)	1055
R[1] = Ph, R[2] = Br	rt, 16 h	(67)	1056
OTMS, TMS, TMSO	TMSOTf, MeCN, rt, 12 h	H, O, NH, N–R, =O (56)	362
Bz–N–TMS, N-TMS (purine)	TMSOTf, $Cl(CH_2)_2Cl$, PhMe, rt, 1 h	NHBz, N–R (92)	1048

TABLE IV. REACTIONS WITH TRIMETHYLSILYL AND SILVER TRIFLATES AND PERCHLORATES (*Continued*)

Sugar	Base	Conditions	Product(s) and Yield(s) (%)	Refs.
	Bz–N(TMS), N–TMS	1. $TMSClO_4$, $Cl(CH_2)_2Cl$, 82°, 7 h 2. NH_3, MeOH	NH_2, N–R **I** (81) R = HO, O, HO OH	132
		1. $TMSClO_4$, $Cl(CH_2)_2Cl$, 82°, 7 h 2. NH_3, MeOH	**I** (81)	133
		1. $TMSClO_4$, $Cl(CH_2)_2Cl$, C_6H_6, 22°, 16 h 2. NH_3, MeOH	**I** (86)	1043
	OTMS, TMSO, N–TMS	1. TMSOTf, $Cl(CH_2)_2Cl$, reflux, 1 h 2. NH_3, MeOH	O, NH, O, N–R, N H (49)	133
	OTMS, AcN(TMS), N–TMS	1. TMSOTf, $Cl(CH_2)_2Cl$, reflux, 1.5 h 2. NH_3, MeOH	O, NH, NH_2, N–R (66)	133
		TMSOTf, $Cl(CH_2)_2Cl$, reflux, 1.5 h	O, NH, NHAc, N, R (79) N^7:N^9 = 1:6 R = BzO, O, BzO OBz	142

Reactant	Conditions	Product(s) and Yield(s) (%)	Refs.
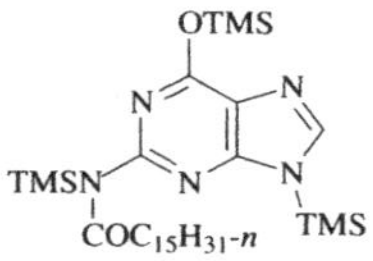 OTMS, TMSN, $COC_{15}H_{31}$-*n*, TMS	TMSOTf, $Cl(CH_2)_2Cl$, PhMe, rt, 4 h	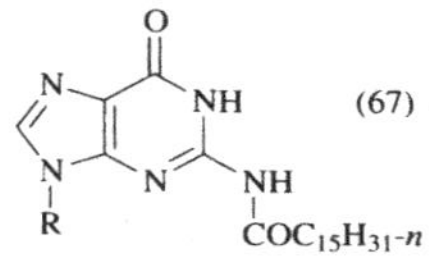 O, NH, NH, $COC_{15}H_{31}$-*n*, R (67) + N^7-β-isomer (10)	1048
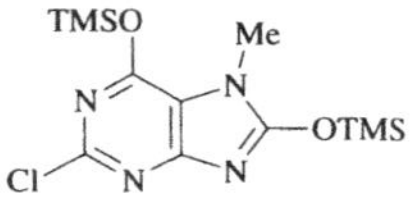 TMSO, Me, OTMS, Cl	TMSOTf, MeCN, reflux, 12 h	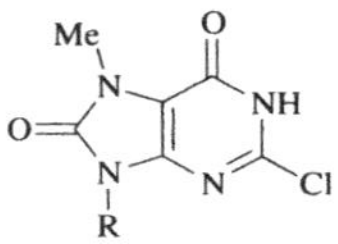 Me, O, NH, Cl, R (85)	1057
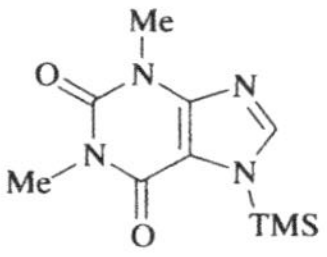 Me, O, Me, O, TMS	1. TMSOTf, $Cl(CH_2)_2Cl$, 24°, 1 h 2. NH_3, MeOH	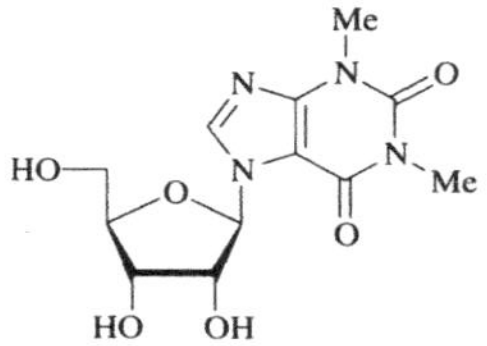 Me, O, Me, O, HO, HO, OH (82)	133
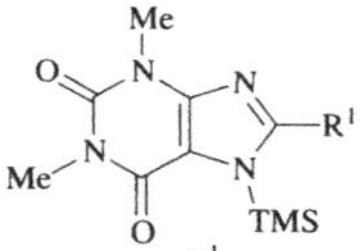 Me, O, Me, O, TMS, R^1	TMSOTf, $Cl(CH_2)_2Cl$, rt	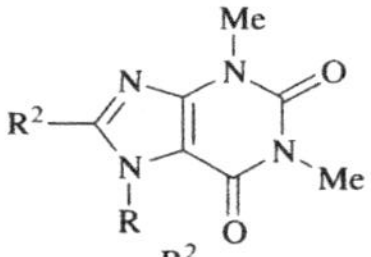 Me, O, Me, O, R^2, R 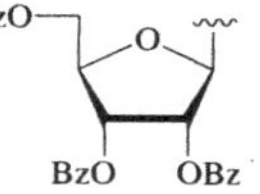	1058
R^1: Cl	10 h	R^2: Cl (72)	
R^1: N(Bn)TMS	6 h	R^2: NHBn (61)	
		R = BzO, O, BzO, OBz	
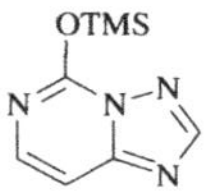 OTMS	TMSOTf, $Cl(CH_2)_2Cl$, reflux, 3.5 h	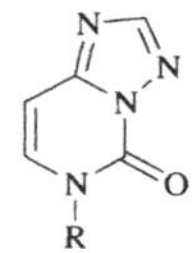 O, R (82)	957

TABLE IV. REACTIONS WITH TRIMETHYLSILYL AND SILVER TRIFLATES AND PERCHLORATES (*Continued*)

Sugar	Base	Conditions	Product(s) and Yield(s) (%)	Refs.
	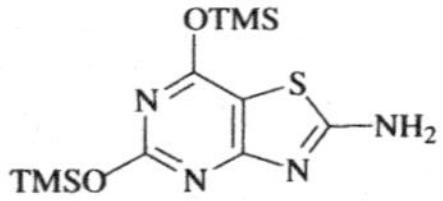 OTMS, N, S, NH$_2$, TMSO, N, N	TMSOTf, MeCN, rt, 12 h	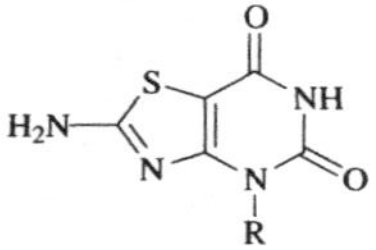 O, S, NH, H$_2$N, N, N, O, R (25)	361, 1058
	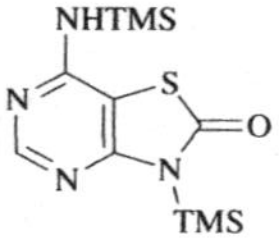 NHTMS, N, S, O, N, N, TMS	TMSOTf, MeCN, rt, 12 h; reflux, 2 d	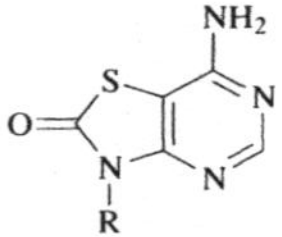 NH$_2$, S, N, O, N, N, R (24)	1058
		TMSOTf, MeCN, rt, 12 h	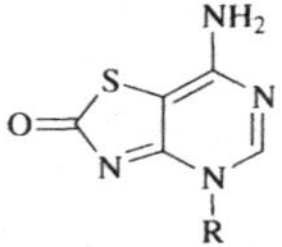 NH$_2$, S, N, O, N, N, R (95)	1058
	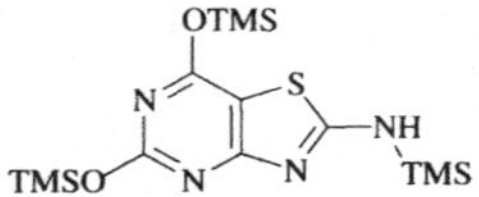 OTMS, N, S, NH, TMSO, N, N, TMS	TMSOTf, MeCN, rt, 12 h	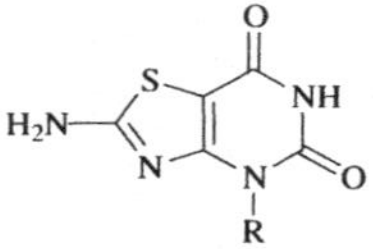 O, S, NH, H$_2$N, N, N, O, R (82)	361
	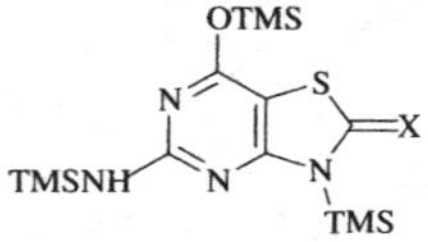 OTMS, N, S, X, TMSNH, N, N, TMS	TMSOTf, MeCN, rt, 16 h	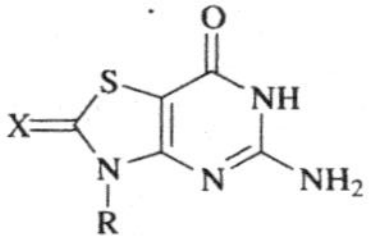 O, S, NH, X, N, N, NH$_2$, R X: O (77); S (59)	1058
	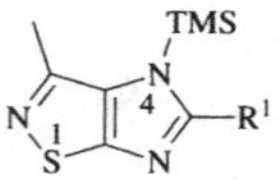 TMS, N, 4, R^1, N, 1, S, N	TMSOTf, MeCN, 80°, 0.5-1.5 h	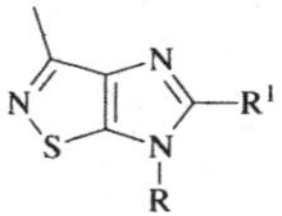 N, R^1, N, S, N, R R^1: H (37) + N^4-isomer (26); SMe (61)	350

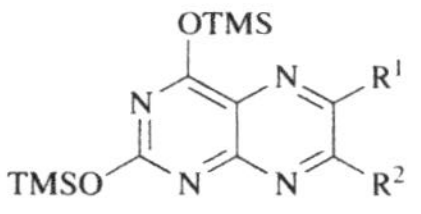

Sugar	Silylated base	Conditions	Product(s) and Yield(s) (%)		Refs.
	TMSO, OTMS pteridine (R^1, R^2)	TMSOTf, $Cl(CH_2)_2Cl$	HN, O, N–R pteridinedione (R^1, R^2)		
	R^1 / R^2				
	H / H	24°, 1.3 h	(93)		133
	H / H	100°, 4 h	(84)		1043
	H / H	24°, 24 h	(64)		132
	p-ClC_6H_4 / H	rt, 3 h	(84)		1059
	H / p-ClC_6H_4	rt, 3 h	(71)		1059
	p-BrC_6H_4 / H	rt, 2 h	(90)		1060
	H / p-BrC_6H_4	rt, 2 h	(84)		1060
BzO, BzO, BzO, OAc	TMSO, OTMS	TMSOTf, MeCN, rt, 18 h	BzO, BzO, BzO, N, O, NH, O	(44)	365
BzO, O_2COMe, BzO, OBz	TMSO, OTMS, R^1	AgOTf, $Ph_2Sn{=}S$, MeCN	R^1, O, NH, O, N–R	R = BzO, BzO, OBz	124
	R^1 = H	80°, 2 h	(99)		
	R^1 = Me	60°, 5 h	(96)		
	Bz–N–TMS, TMSO	AgOTf, $Ph_2Sn{=}S$, MeCN, 60°, 4.5 h	NHBz, N, O, N–R	(57)	124

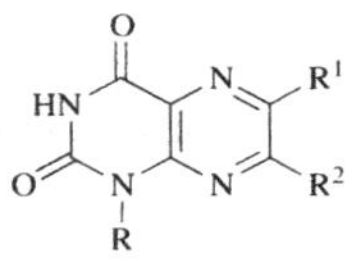

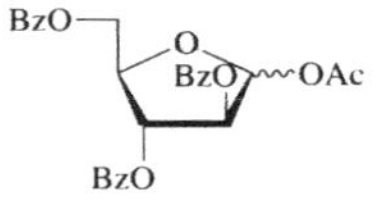

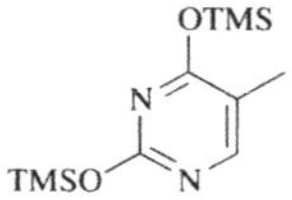

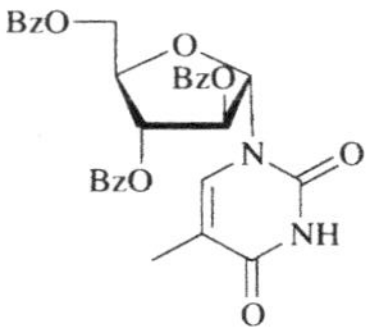

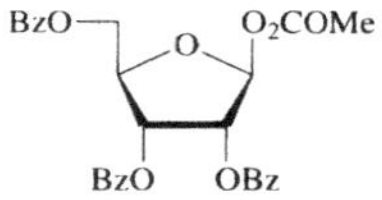

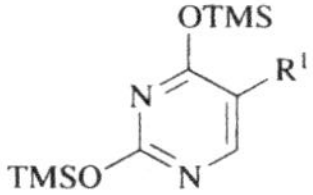

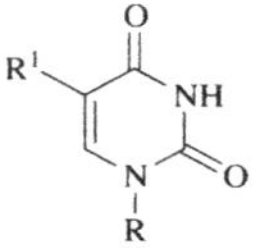

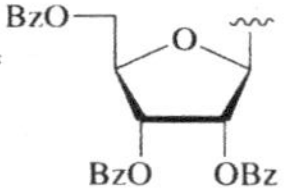

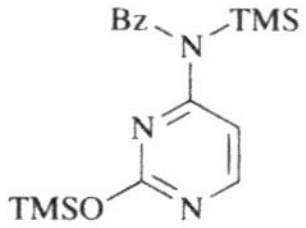

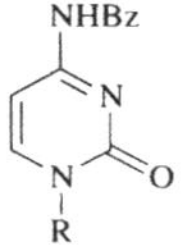

TABLE IV. REACTIONS WITH TRIMETHYLSILYL AND SILVER TRIFLATES AND PERCHLORATES (*Continued*)

Sugar	Base	Conditions	Product(s) and Yield(s) (%)	Refs.
	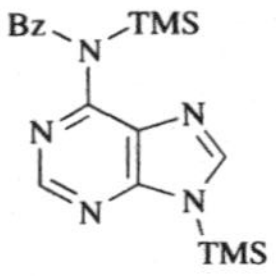 Bz(TMS)N, N-TMS	AgOTf, $Ph_2Sn{=}S$, MeCN, reflux, 14 h	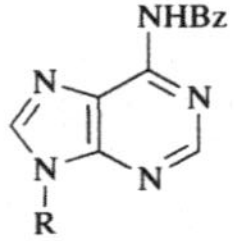 NHBz, N–R (49); R = BzO, BzO, OBz 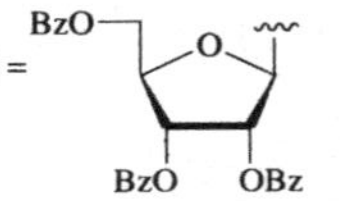	124
	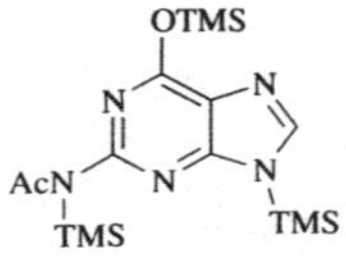 OTMS, AcN(TMS), N-TMS	AgOTf, $Ph_2Sn{=}S$, MeCN, reflux, 16h	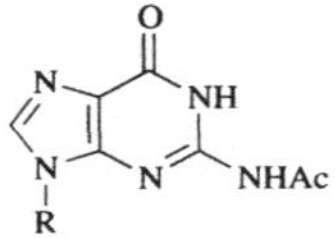 O, NH, NHAc, N–R (77)	124
	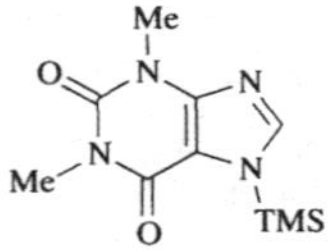 Me, O, Me, O, N-TMS	AgOTf, $Ph_2Sn{=}S$, MeCN, 60°, 1.5 h	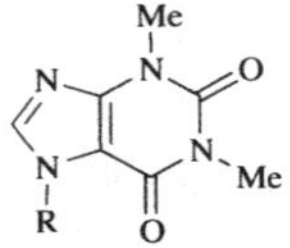 Me, O, Me, O, N–R (96)	124
C_{29} 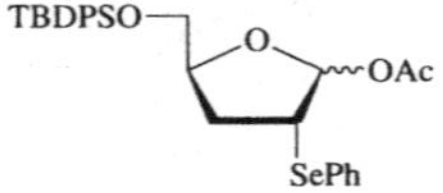 TBDPSO, OAc, SePh	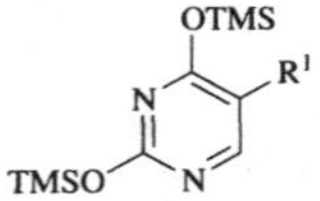 OTMS, R^1, TMSO	TMSOTf, $Cl(CH_2)_2Cl$, 5 to 24°, 15 min	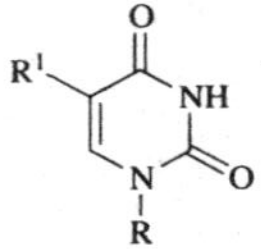 R^1, O, NH, O, N–R; R^1: H (67), Me (78); R 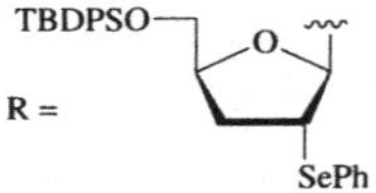 = TBDPSO, SePh	230a
	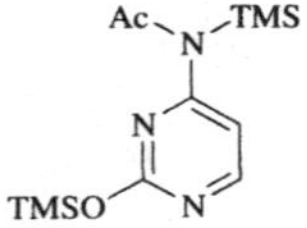 Ac(TMS)N, TMSO	TMSOTf	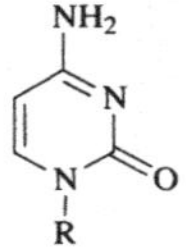 NH_2, O, N–R (72)	230a
	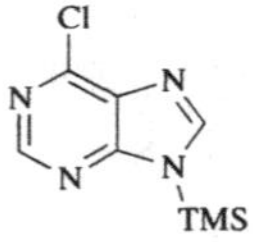 Cl, N-TMS	TMSOTf	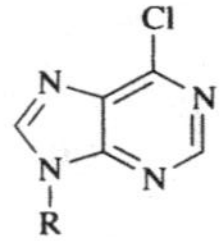 Cl, N–R (89)	230a

Substrate	Base	Conditions	Product(s) and Yield(s) (%)	Refs.
		TMSOTf, $Cl(CH_2)_2Cl$, rt	**I** (90) $\alpha:\beta = 21:79$	229, 228
		TMSOTf, MeCN, rt, 12 h	**I** (96) $\alpha:\beta = 11:89$	229, 228
		TMSOTf, $Cl(CH_2)_2Cl$,	(—) $\alpha:\beta = 20:80$	229
		TMSOTf, $Cl(CH_2)_2Cl$, reflux, 2.5 h	(86)	932
		TMSOTf, $Cl(CH_2)_2Cl$, reflux, 15 h	(90)	932
		TMSOTf, $Cl(CH_2)_2Cl$, reflux, 30 min	X: O (90); S (78)	1061

TABLE IV. REACTIONS WITH TRIMETHYLSILYL AND SILVER TRIFLATES AND PERCHLORATES (*Continued*)

Sugar	Base	Conditions	Product(s) and Yield(s) (%)	Refs.
C_{30} 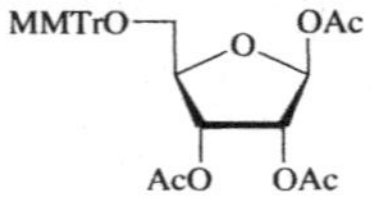 MMTrO, AcO, OAc, OAc	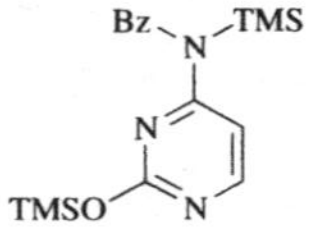 Bz, N, TMS, N, TMSO, N	TMSOTf, $Cl(CH_2)_2Cl$, 20°, 16 h	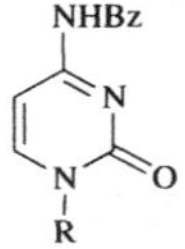 NHBz, N, O, N, R (70); R = HO, O, AcO, OAc 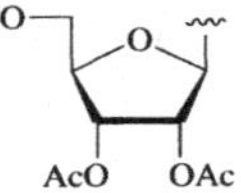	931
	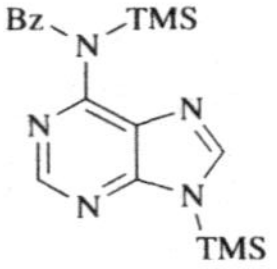 Bz, N, TMS, N, N, N, N, TMS	TMSOTf, $Cl(CH_2)_2Cl$, 20°, 16 h	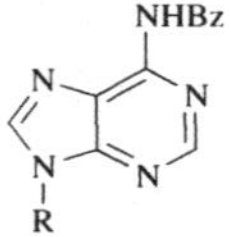 NHBz, N, N, N, N, R (66)	931
C_{31} 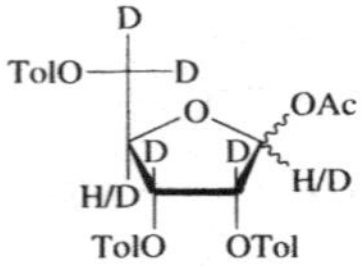 TolO, D, D, O, OAc, D, D, H/D, H/D, TolO, OTol	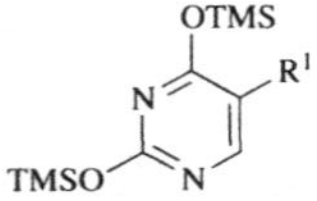 OTMS, R[1], N, TMSO, N	TMSOTf, $Cl(CH_2)_2Cl$, 70°, 4 h	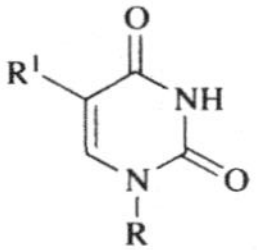 O, R[1], NH, O, N, R; R[1]: H (75), Me (91); R 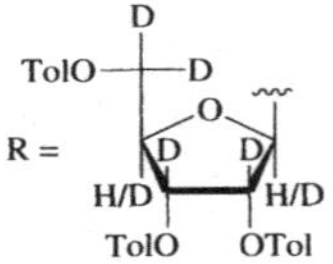 = TolO, D, D, O, D, D, H/D, H/D, TolO, OTol	1062
	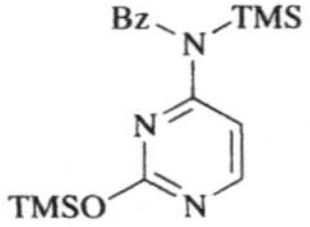 Bz, N, TMS, N, TMSO, N	TMSOTf, $Cl(CH_2)_2Cl$, 70°, 4 h	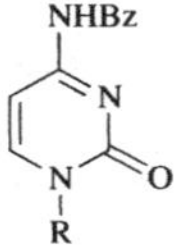 NHBz, N, O, N, R (85)	1062
	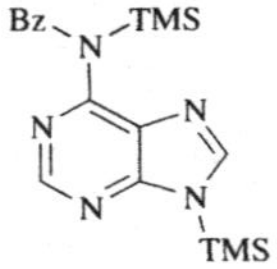 Bz, N, TMS, N, N, N, N, TMS	TMSOTf, $Cl(CH_2)_2Cl$, 70°, 4 h	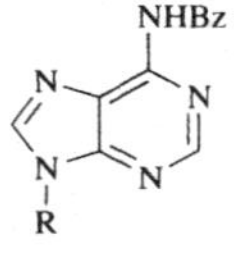 NHBz, N, N, N, N, R (60)	1062

Substrate	Reagent	Conditions	Product(s) and Yield(s) (%)		Refs.
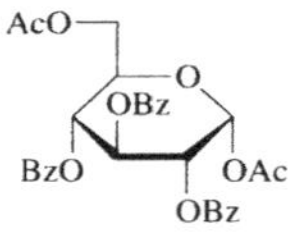 (AcO, OBz, BzO, OAc, OBz pyranose)	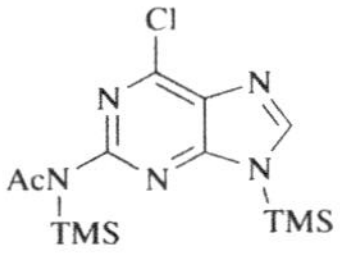 (Cl, AcN(TMS), N-TMS purine)	TMSOTf, $Cl(CH_2)_2Cl$, 135°, 16 h	(Cl, NHAc purine nucleoside; AcO, OBz, BzO, OBz)	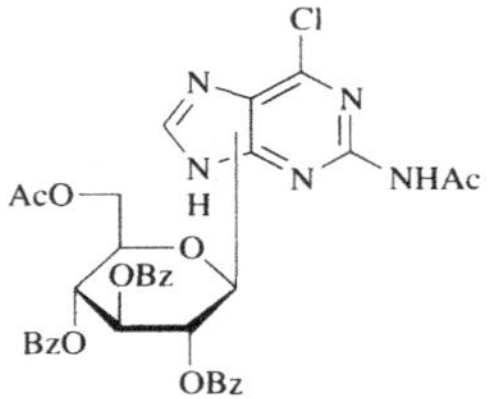 (69) N^9:N^7 = 98:2	409
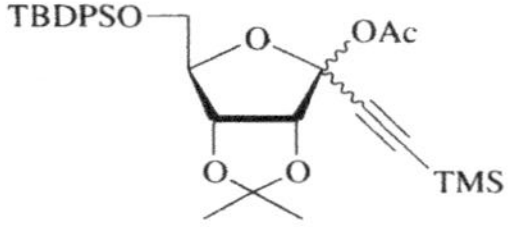 (TBDPSO, OAc, TMS-alkynyl, isopropylidene furanose)	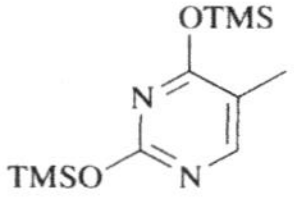 (OTMS, TMSO pyrimidine)	TMSOTf, MeCN, 0°, 4 h	(TBDPSO, NH, O, TMS nucleoside)	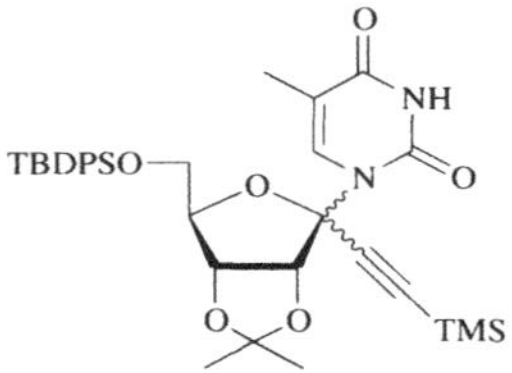 (97)	1040
C_{32} 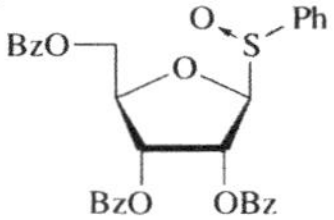 (BzO, O←S–Ph, BzO, OBz furanose)	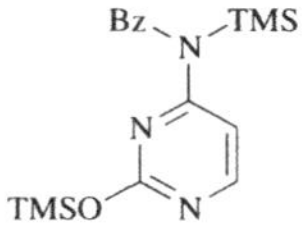 (Bz–N–TMS, TMSO pyrimidine)	TMSOTf, $Cl(CH_2)_2Cl$, rt, 1 min	(NHBz cytosine, N–R) (95)	R = (BzO, BzO, OBz furanosyl)	412
	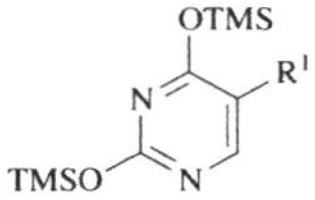 (OTMS, R^1, TMSO pyrimidine)	TMSOTf, $Cl(CH_2)_2Cl$, rt, 1 min	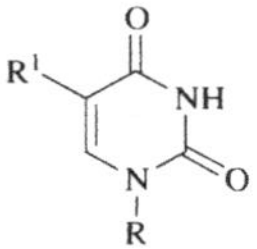 (R^1, NH, O, N–R)	R^1: H (90); Me (92)	412
	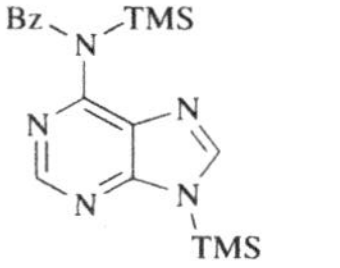 (Bz–N–TMS, N-TMS purine)	TMSOTf, $Cl(CH_2)_2Cl$, rt, 1 min	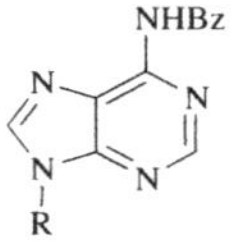 (NHBz purine, N–R) **I** (40)		412
		TMSOTf, $Cl(CH_2)_2Cl$, rt, 0.5 h	**I** (90)		412

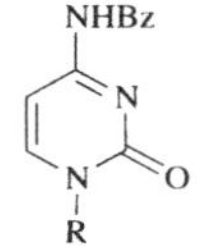

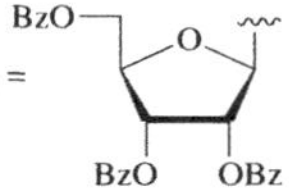

TABLE IV. REACTIONS WITH TRIMETHYLSILYL AND SILVER TRIFLATES AND PERCHLORATES (*Continued*)

Sugar	Base	Conditions	Product(s) and Yield(s) (%)	Refs.
	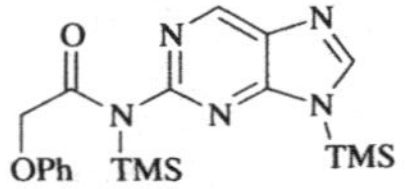 [structure]	TMSOTf, $Cl(CH_2)_2Cl$, rt, 0.5 h	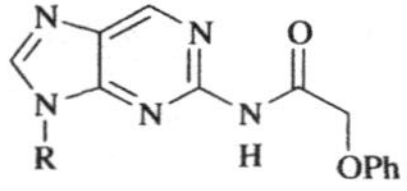 [structure] (60)	412
	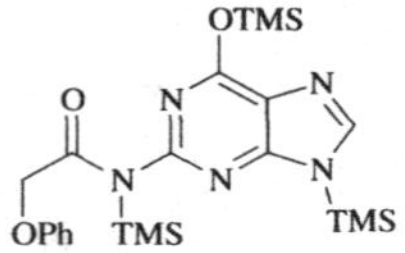 [structure]	TMSOTf, $Cl(CH_2)_2Cl$, rt, 0.5 h	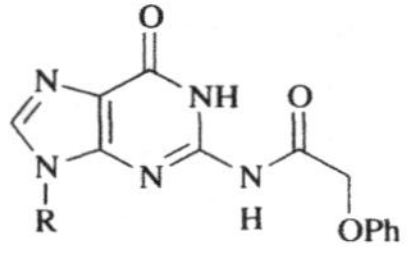 [structure] (65) + N^7-isomer (22)	412
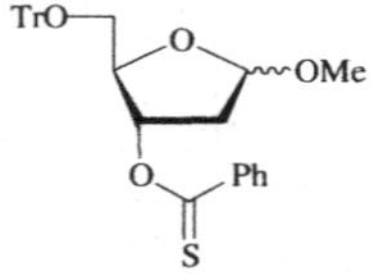 [structure]	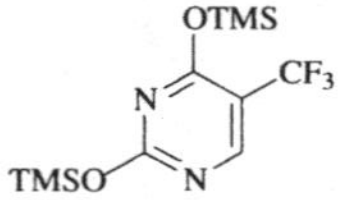 [structure]	TMSOTf, MeCN, –40 to –45°, 0.5 h	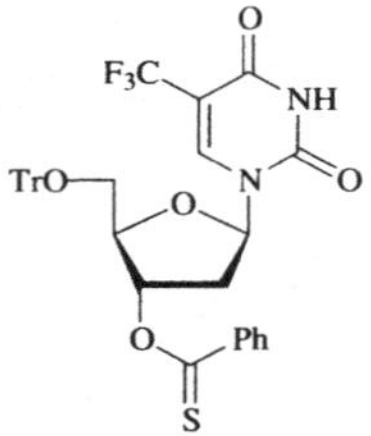 [structure] (89)	1063
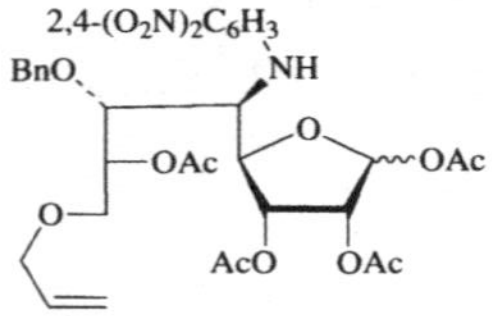 [structure]	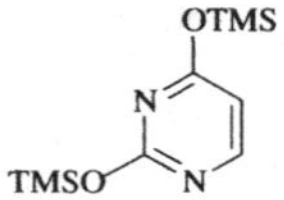 [structure]	TMSOTf, MeCN, reflux, 45 min	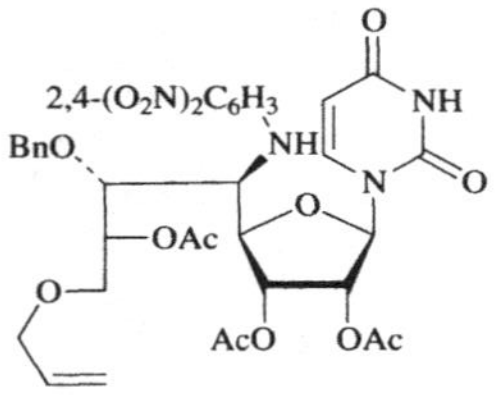 [structure] (92)	157

			TMSOTf, CH_2Cl_2, < 5°, 2.5 h		(31) + β-anomer (22)	913
			TMSOTf, MeCN, 0°, 3.5 h		(27)	1040
C_{33}			TMSOTf, CH_2Cl_2, rt, 24 h		(51) + α-anomer (27)	1045
			1. TMSOTf, CH_2Cl_2, rt, 24 h 2. MeCOCl, Et_3N		(86) α:β = 1:2	1045
			TMSOTf, MeCN, rt, 24 h		(12)	893
			1. TMSOTf, $Cl(CH_2)_2Cl$ 2. NaOH, MeOH 3. H_2SO_4, MeOH		(67)	299

TABLE IV. REACTIONS WITH TRIMETHYLSILYL AND SILVER TRIFLATES AND PERCHLORATES (*Continued*)

Sugar	Base	Conditions	Product(s) and Yield(s) (%)	Refs.
C_{34}				
BzO, OBz, BzO, OBz	Cl, N, N, N, N, TMS	TMSOTf, MeCN, 80°, 1 h	Cl, N, N, N, N, BzO, O, BzO, OBz (78)	1064
R^1O, O, OR^1, R^1O, R^1O; $R^1 = p\text{-}O_2NC_6H_4CO_2$	OTMS, N, TMSO, N	1. TMSOTf, MeCN, reflux, 2 h 2. Na_2CO_3	O, NH, N, O, R (75); R = R^1O, O, R^1O, R^1O	1065
	Ac, N, TMS, N, TMSO, N	1. TMSOTf, MeCN, 25°, 10 d 2. Na_2CO_3	NHAc, N, N, O, R (82)	1065
	Cl, N, N, R^2, N, N, TMS	TMSOTf, MeCN	Cl, N, N, N, N, R^2, R	1065
	R^2			
	H	25°, 12 h, $NaHCO_3$	(58) + N^7-β-isomer (30)	
	Cl	rt, 16 h	(66) + α-anomer (8)	

Substrate	Reagent	Conditions	Product(s) and Yield(s) (%)	Refs.
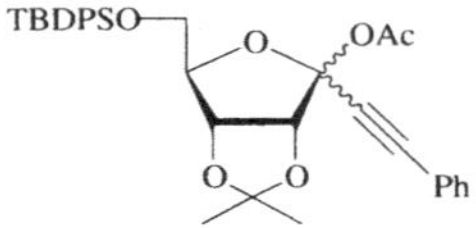 TBDPSO	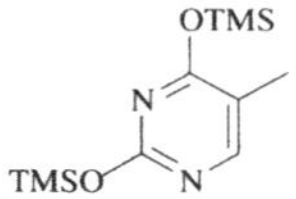 OTMS, TMSO	TMSOTf, MeCN, 0°, 4.5 h	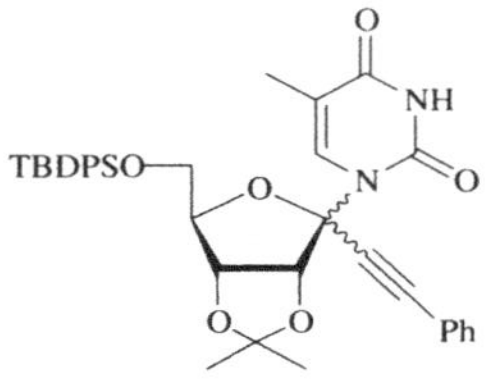 (94) $\alpha:\beta = 1:2.4$	1040
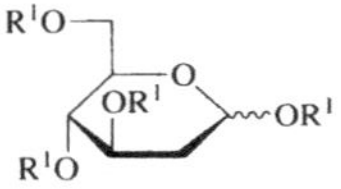 $R^1 = p\text{-}O_2NC_6H_4CO_2$	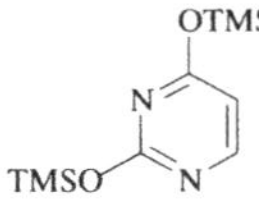 OTMS, TMSO	TMSOTf, $Cl(CH_2)_2Cl$	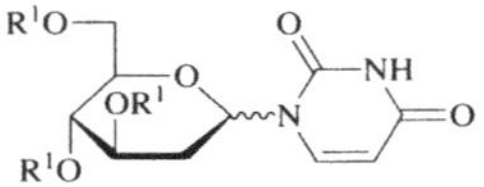 (—) $\alpha:\beta = 3:2$	1038
C_{35} TBDPSO, PhS, OAc	OTMS, H (Me), TMSO	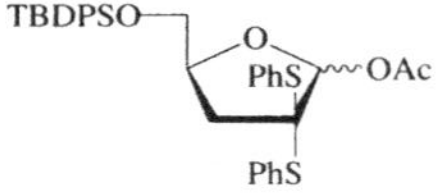 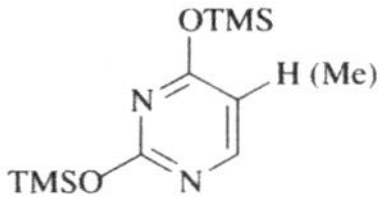 TMSOTf, MeCN, 0°, 4 h	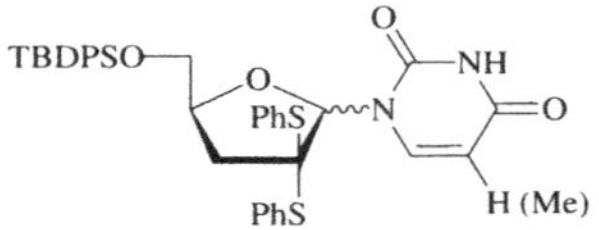 (83) $\alpha:\beta = 19:81$	305a, 305
	NHTMS, TMSO	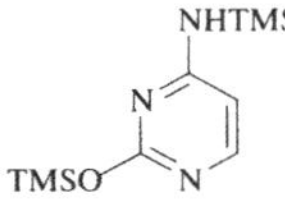 TMSOTf, MeCN, 0°, 4 h	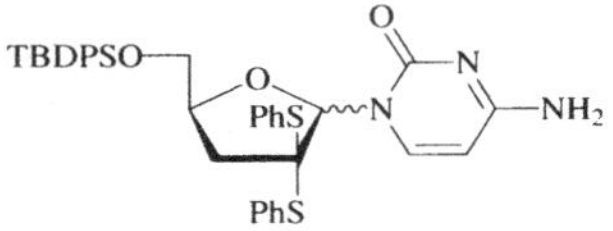 (79) $\alpha:\beta = 20:80$	305a
C_{36} BzO, OBz, OAc	Cl, AcN(TMS), N-TMS	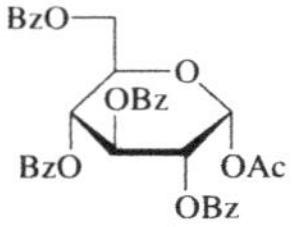 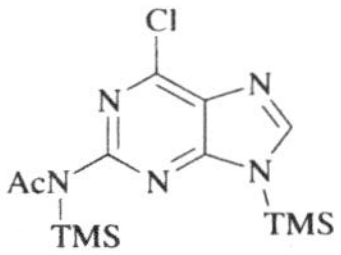 TMSOTf, $Cl(CH_2)_2Cl$, reflux, 1 h	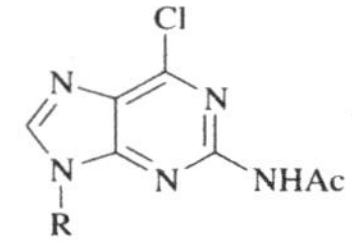 (70) + N^7-isomer (2) 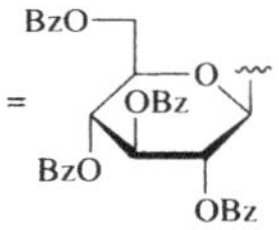	409
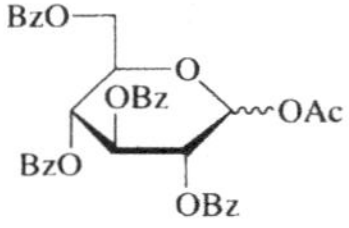 BzO, OBz, OAc	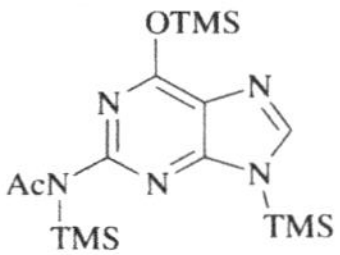 OTMS, AcN(TMS), N-TMS	TMSOTf, $Cl(CH_2)_2Cl$, reflux, 16 h	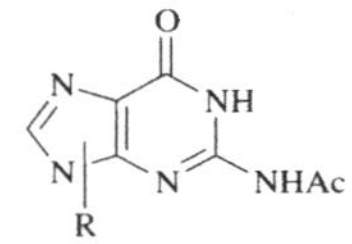 (56) $N^7:N^9 = 1:8$	142

TABLE IV. REACTIONS WITH TRIMETHYLSILYL AND SILVER TRIFLATES AND PERCHLORATES (*Continued*)

Sugar	Base	Conditions	Product(s) and Yield(s) (%)	Refs.
C_{38}				
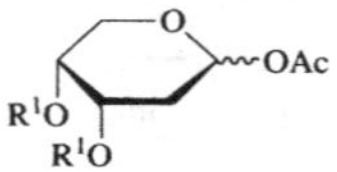 $R^1 = p\text{-}t\text{-}BuC_6H_4CO_2$	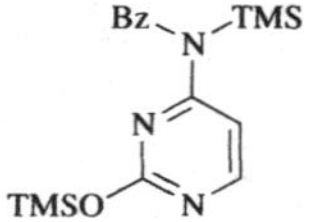	TMSOTf, $Cl(CH_2)_2Cl$, rt, 4 h	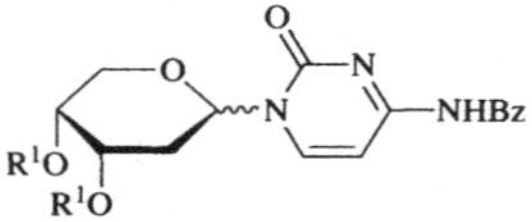 (95)	1037
	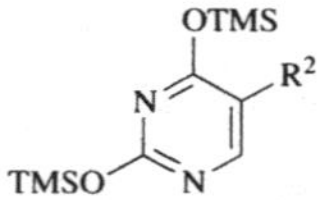	TMSOTf, $Cl(CH_2)_2Cl$, rt, 4 h	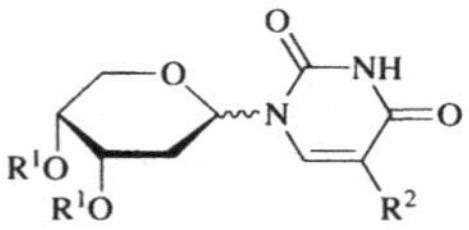 R^2 / α:β F (94) 58:42 I (93) 63:37 Me (94) 75:25	1037
C_{39}				
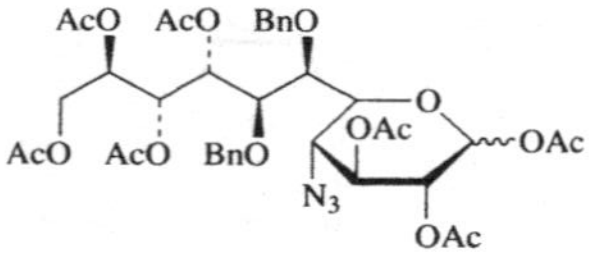	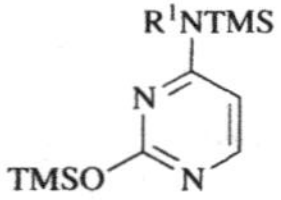	TMSOTf, $PhNO_2$, 127°, 3.5 h	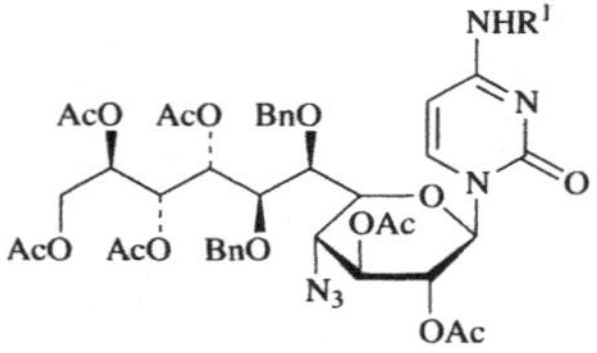 R^1 H (76) Ac (76)	160 159
C_{40}				
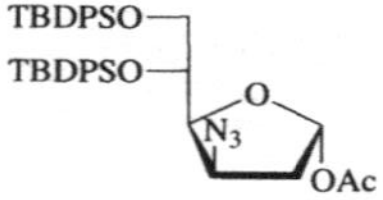	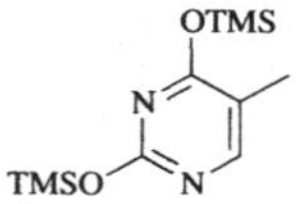	1. TMSOTf, MeCN, –20°, 90 min 2. Bu_4NF, THF, rt, 24 h	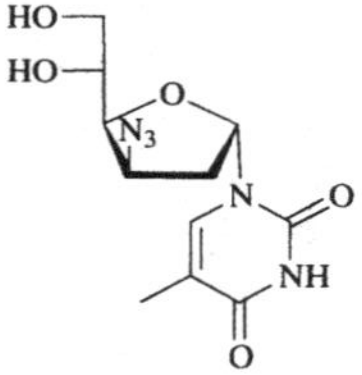 (31) + β-anomer (24)	980, 961

Sugar	Base	Conditions	Product(s) and Yield(s) (%)	Refs.
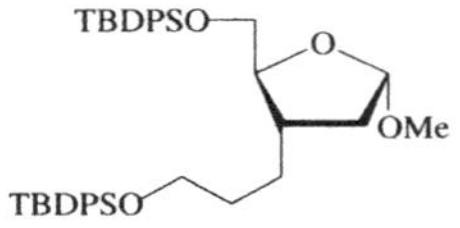 TBDPSO, TBDPSO, OMe	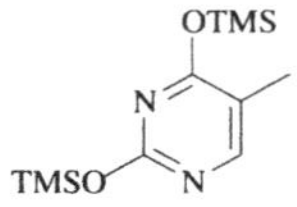 OTMS, N, TMSO, N	TMSOTf, MeCN, −10°, 1 h; rt, 12 h	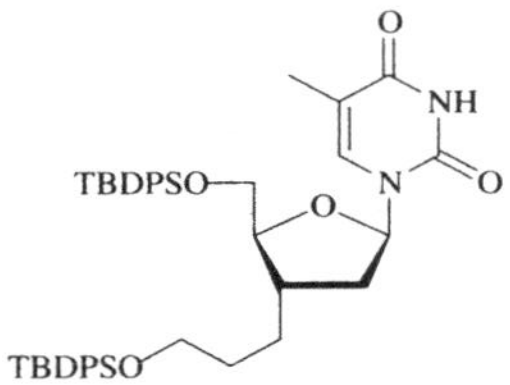 O, NH, O, N, TBDPSO, O, TBDPSO (42) + α-anomer (38)	1066
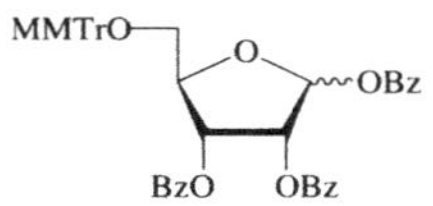 MMTrO, O, OBz, BzO, OBz	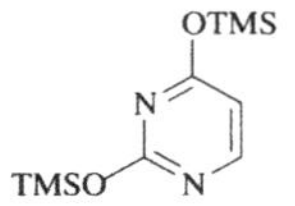 OTMS, N, TMSO, N	TMSOTf, $Cl(CH_2)_2Cl$, 20°, 16 h	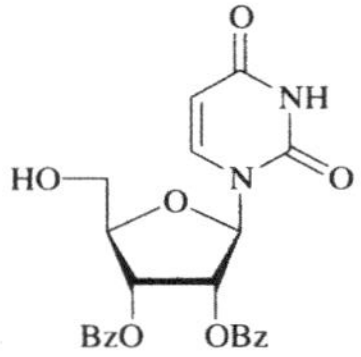 O, NH, O, N, HO, O, BzO, OBz (70)	931

[a] The 1-β-methoxy sugar does not react under these conditions.

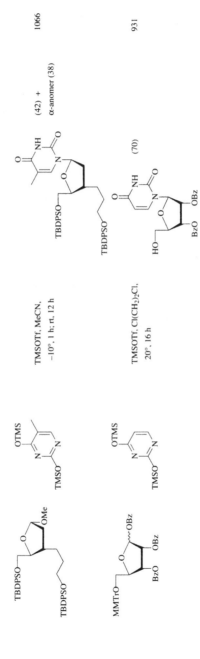

TABLE V. REACTIONS WITH TITANIUM TETRACHLORIDE AS CATALYST

Sugar	Base	Conditions	Product(s) and Yield(s) (%)	Refs.
C_4 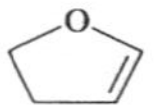	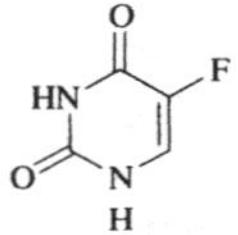	$TiCl_4$, Py, 80°, 4 h	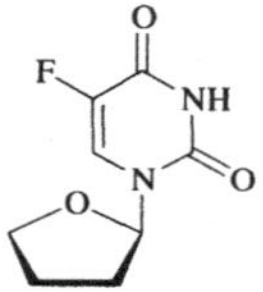 (69)	1067
C_6 	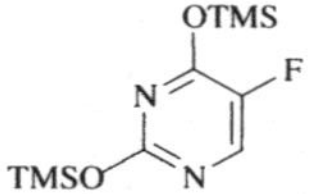 TMSO, OTMS, F	$TiCl_4$, CH_2Cl_2, rt, 0.5 h	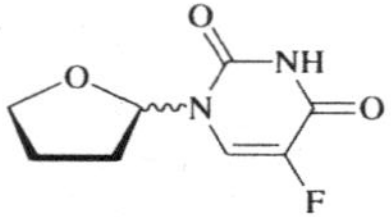 (60)	528
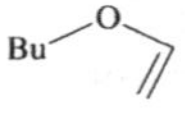 Bu–O	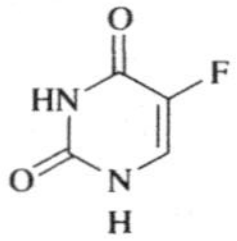	$TiCl_4$, Py, reflux, 2 h	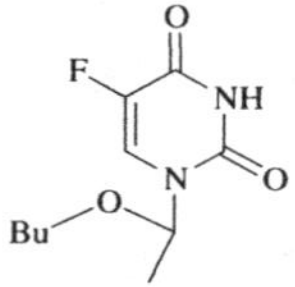 (49) + N^1,N^3-bis(isomer) (17)	638
C_{10} 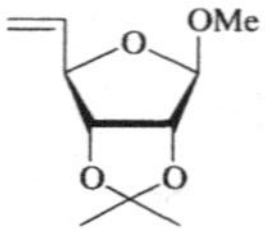 OMe	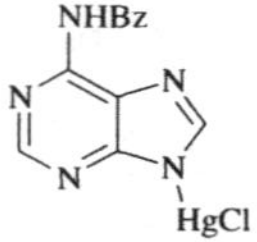 NHBz, HgCl	1. $TiCl_4$, $Cl(CH_2)_2Cl$, reflux, 22 h 2. NaOMe, MeOH	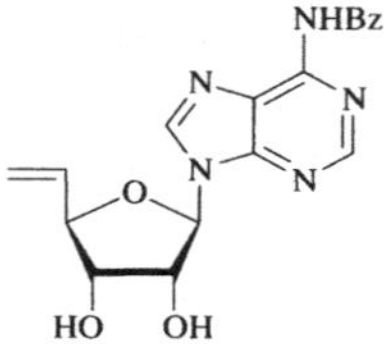 (44)	1068
C_{13} 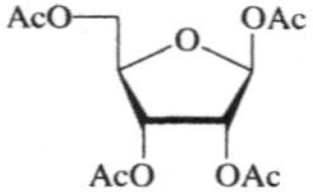 AcO, OAc, AcO, OAc	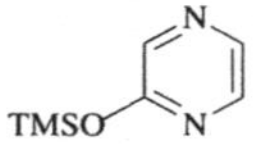 TMSO	1. $TiCl_4$, $Cl(CH_2)_2Cl$, reflux, 5 h 2. NaOMe, MeOH	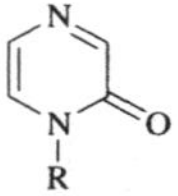 (40) R 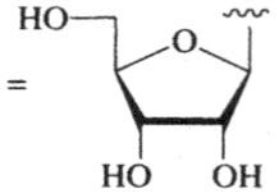 =	1069

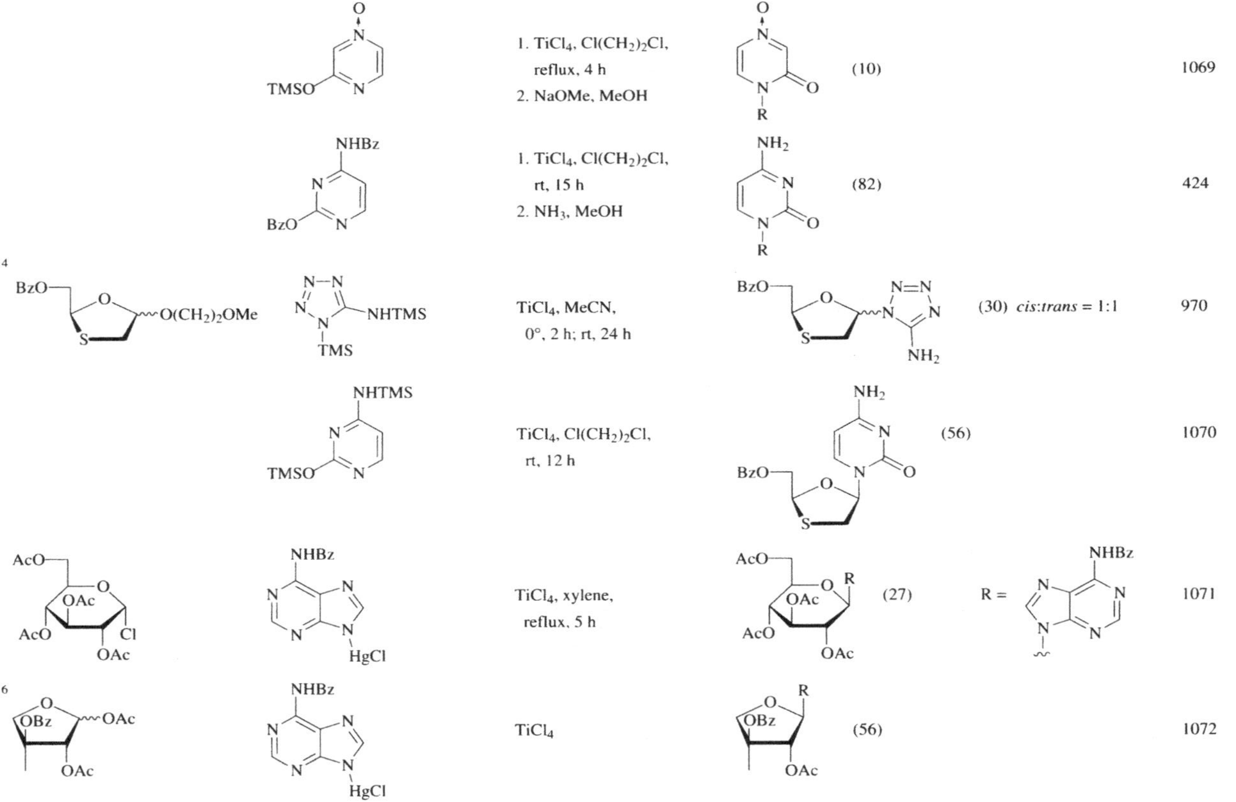

TMSO
1. $TiCl_4$, $Cl(CH_2)_2Cl$, reflux, 4 h
2. NaOMe, MeOH
R
(10)
1069
NHBz
BzO
1. $TiCl_4$, $Cl(CH_2)_2Cl$, rt, 15 h
2. NH_3, MeOH
NH_2
R
(82)
424
C_{14}
BzO
$O(CH_2)_2OMe$
NHTMS
TMS
$TiCl_4$, MeCN, 0°, 2 h; rt, 24 h
BzO
NH_2
(30) *cis*:*trans* = 1:1
970
NHTMS
TMSO
$TiCl_4$, $Cl(CH_2)_2Cl$, rt, 12 h
NH_2
BzO
(56)
1070
AcO
OAc
AcO
Cl
OAc
NHBz
HgCl
$TiCl_4$, xylene, reflux, 5 h
AcO
R
OAc
AcO
OAc
(27)
R =
NHBz
1071
C_{16}
OBz
OAc
OAc
NHBz
HgCl
$TiCl_4$
R
OBz
OAc
(56)
1072

TABLE V. REACTIONS WITH TITANIUM TETRACHLORIDE AS CATALYST (*Continued*)

Sugar	Base	Conditions	Product(s) and Yield(s) (%)	Refs.
AcO, OAc, AcO, OAc, NHAc	NHBz, N, N, N, N, HgCl	$TiCl_4$, $CdCO_3$, xylene, reflux, 1.5 h	AcO, OAc, R, AcO, NHAc (30)	1071
C_{17-20} BzO, OAc, R^1, OAc	NHBz, N, N, N, N, HgCl	1. $TiCl_4$, $Cl(CH_2)_2Cl$, Celite, reflux, 24 h 2. NaOMe, MeOH	NH_2, HO, R^1, OH; R^1: Me (63), Et (46), Bu-*n* (58)	1173
C_{19} R^1, Cl, R^1; $R^1 = p\text{-}ClC_6H_4CO_2$	TMSO, N, N	$TiCl_4$, $Cl(CH_2)_2Cl$, MS, rt, 24 h	R^1, O, N, N, R^1 (86) $\alpha:\beta = 2.1:1$	1073
BzO, OAc, AcO, OAc	NHBz, N, N, N, N, HgCl	1. $TiCl_4$, $Cl(CH_2)_2Cl$, reflux, 21 h 2. NaOMe, MeOH	NH_2, OH, HO, OH **I** + OH, HO, HO, NH_2 **II** (—) 1:1	1074

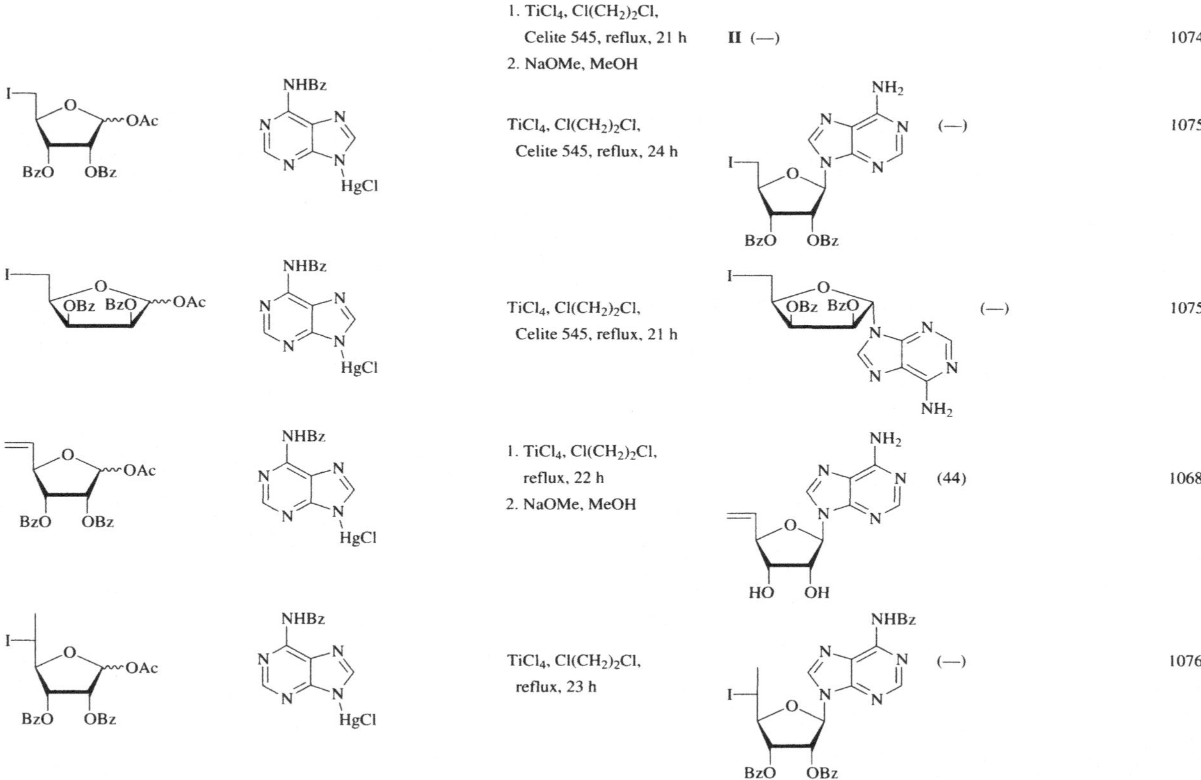
C21
NHBz
HgCl
OAc
BzO
OBz
1. TiCl4, Cl(CH2)2Cl, Celite 545, reflux, 21 h
2. NaOMe, MeOH
II (—)
1074
TiCl4, Cl(CH2)2Cl, Celite 545, reflux, 24 h
NH2
(—)
1075
TiCl4, Cl(CH2)2Cl, Celite 545, reflux, 21 h
(—)
1075
C22
1. TiCl4, Cl(CH2)2Cl, reflux, 22 h
2. NaOMe, MeOH
HO
OH
(44)
1068
TiCl4, Cl(CH2)2Cl, reflux, 23 h
(—)
1076

TABLE V. REACTIONS WITH TITANIUM TETRACHLORIDE AS CATALYST (*Continued*)

Sugar	Base	Conditions	Product(s) and Yield(s) (%)	Refs.
I; OAc; BzO; OBz	NHBz; N; N; N; N; HgCl	$TiCl_4$, $Cl(CH_2)_2Cl$, reflux, 23 h	NHBz; N; N; N; N; I; O; BzO; OBz (—)	1076
C_{23} BzO; O; OAc; AcNH; OBz	NHBz; N; N; N; N; HgCl	$TiCl_4$, $Cl(CH_2)_2Cl$, reflux, 19 h	BzO; O; N; N; AcNH; OBz; N; N; NHBz (57)	85
C_{24} OBz; O; OBz; OAc; OAc	NHBz; N; N; N; N; HgCl	1. $TiCl_4$, $Cl(CH_2)_2Cl$, reflux, 7 h 2. NaOMe, MeOH	NH_2; N; N; N; N; OH; O; OH; OH (30)	1077
BzO; S; OBz; OBz	NHBz; N; N; N; N; HgCl	$TiCl_4$, $Cl(CH_2)_2Cl$, reflux, 7 h	BzO; S; R; OBz **I** (78) R = NHBz; N; N; N; N	1078
BzO; S; OBz; OBz	NHBz; N; N; N; N; HgCl	$TiCl_4$, $Cl(CH_2)_2Cl$, reflux, 7 h	**I** (82)	1078

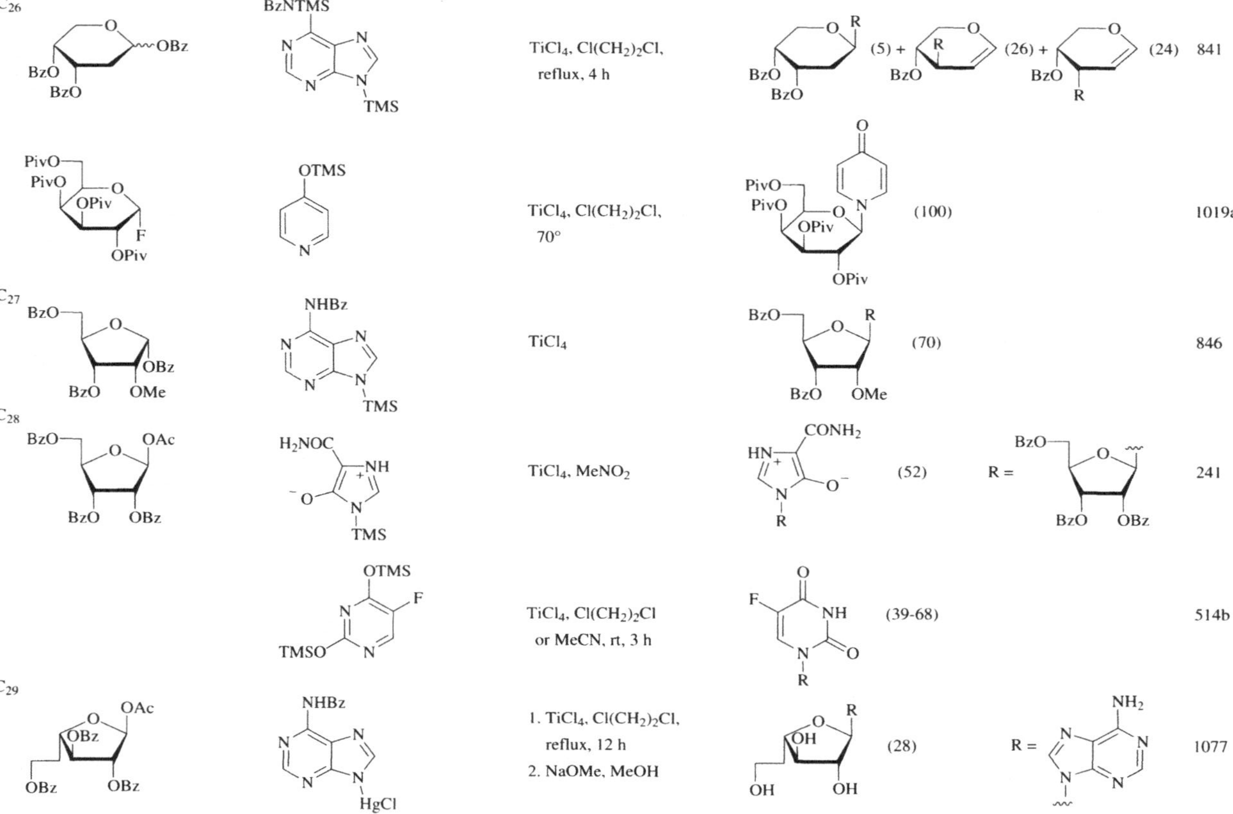

C_{26}
TiCl4, Cl(CH2)2Cl, reflux, 4 h
(5) + (26) + (24)
841
TiCl4, Cl(CH2)2Cl, 70°
(100)
1019a
C_{27}
TiCl4
(70)
846
C_{28}
TiCl4, MeNO2
(52)
241
TiCl4, Cl(CH2)2Cl or MeCN, rt, 3 h
(39-68)
514b
C_{29}
1. TiCl4, Cl(CH2)2Cl, reflux, 12 h
2. NaOMe, MeOH
(28)
1077

TABLE V. REACTIONS WITH TITANIUM TETRACHLORIDE AS CATALYST (*Continued*)

Sugar	Base	Conditions	Product(s) and Yield(s) (%)	Refs.
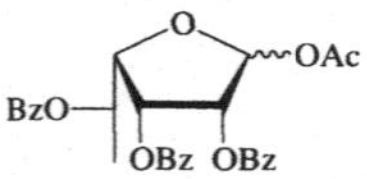 BzO, OAc, OBz, OBz	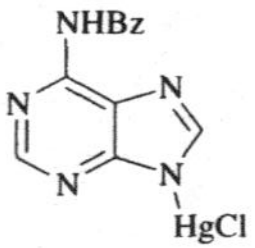 NHBz, N, N, N, N, HgCl	1. $TiCl_4$, $Cl(CH_2)_2Cl$, reflux, 2 h 2. NaOMe, MeOH, reflux, 1 h	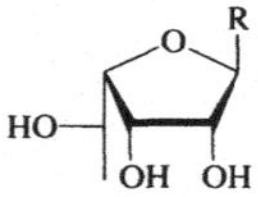 HO, R, OH, OH (31-39)	1079
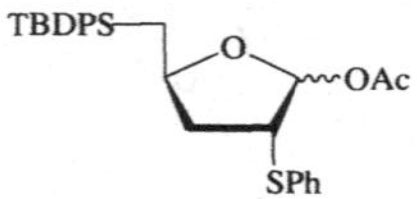 TBDPS, OAc, SPh	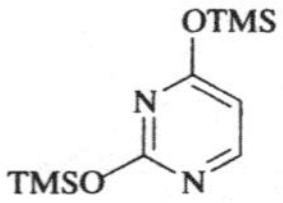 OTMS, N, N, TMSO	$TiCl_4$ (1-6 eq), $Cl(CH_2)_2Cl$, rt	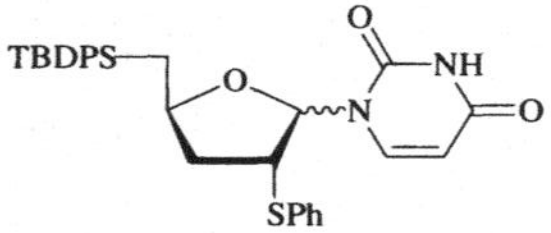 TBDPS, O, NH, O, N, SPh (65) α:β = 15:85	228, 229

TABLE VI. REACTIONS WITH BORON TRIFLUORIDE ETHERATE AS CATALYST

Sugar	Base	Conditions	Product(s) and Yield(s) (%)	Refs.
C_4 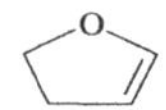	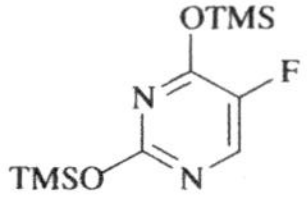	1. H^+ 2. BF_3•OEt_2, dioxane, rt, 3 h	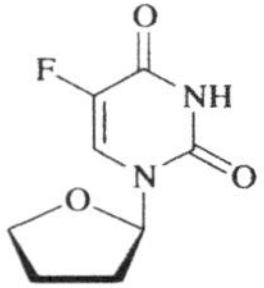 (93)	639
C_6 	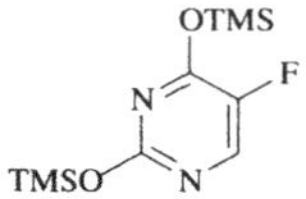	BF_3•OEt_2 (1 eq), CH_2Cl_2, rt, 1 h	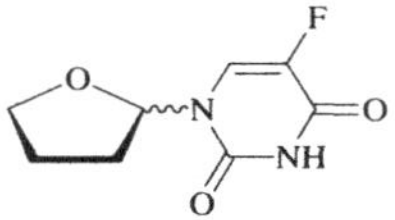 (70)	528
		BF_3•OEt_2 (0.02 eq), CH_2Cl_2, rt, 1 h	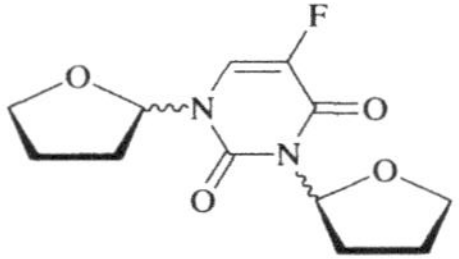 (67)	528
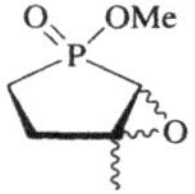	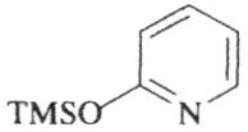	BF_3•OEt_2, MeCN, reflux, 48 h	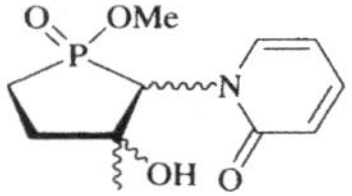 **I** (15)	652
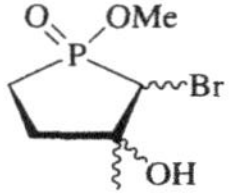	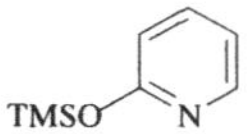	BF_3•OEt_2 (0.5 eq), MeCN, reflux, 24 h	**I** (62)	652
C_{12} 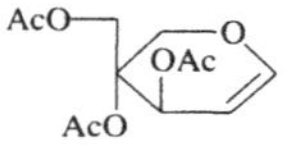	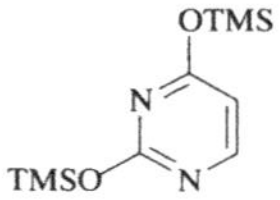	BF_3•OEt_2, EtOAc, −20°, 1 h	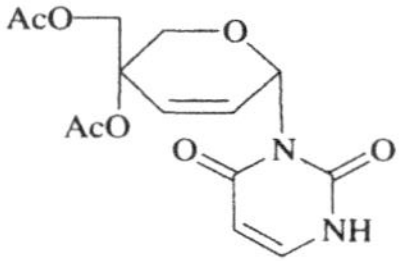 (35) + β-anomer (32)	107

TABLE VI. REACTIONS WITH BORON TRIFLUORIDE ETHERATE AS CATALYST (*Continued*)

Sugar	Base	Conditions	Product(s) and Yield(s) (%)	Refs.
C_{13} (AcO, OAc, AcO, OAc structure)	(OTMS, N, N, N, N, TMS structure)	1. BF_3•OEt_2, dioxane, reflux, 30 min 2. NH_3, MeOH	(structure) (44) + α-anomer (44) + N^7-isomer (5)	1080
	(OTMS, N, N, N, N, TMS structure)	BF_3•OEt_2, dioxane, reflux, 30 min	**I** (38) R = (AcO, AcO, OAc structure)	1081
		BF_3•OEt_2, dioxane, 60°, 24 h	**I** (—) N^2 and N^5 ribosides + N^1,N^5-bis(riboside) 2:2:1	735
	(TMS-NH, O_2S, TMS structure)	BF_3•OEt_2, CH_2Cl_2, rt, 4 h	(structure) (36)	1082
	(TMS-NH, O_2S, Me structure)	1. BF_3•OEt_2, CH_2Cl_2, rt, 6 h 2. Zn, AcOH, reflux, 4 h	(structure) (40)	1083
	(Me, MeS, OTMS structure)	BF_3•OEt_2, CH_2Cl_2, rt, 2 h	(structure) (53)	1084

Sugar	Base	Conditions	Product(s) and Yield(s) (%)	Refs.
C_{16}				
BzO, OAc, OAc (furanose)	NHTMS, O_2S, TMS, R^1	BF_3•OEt_2, CH_2Cl_2, rt, 4 h	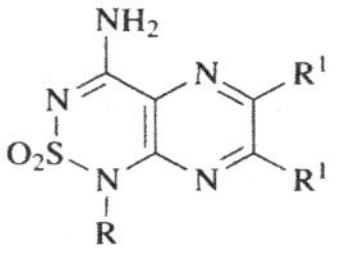 NH_2, O_2S, R, R^1; R^1: H (12) + N^8-isomer (7); Ph (40)	1085
	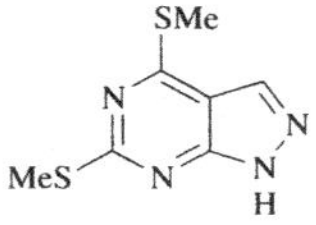 SMe, MeS, H	BF_3•OEt_2, MeCN	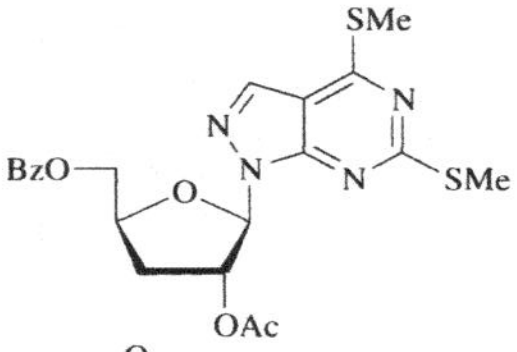 SMe, BzO, OAc, SMe (60) + N^2-β-isomer (2)	1086, 1087
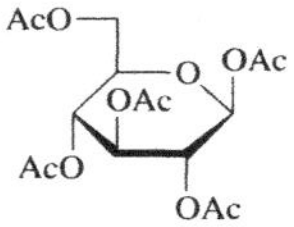 AcO, OAc, OAc, AcO, OAc (pyranose)	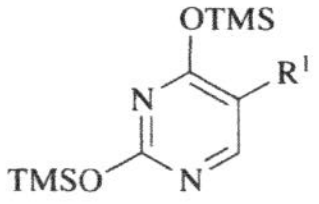 OTMS, TMSO, R^1	BF_3•OEt_2, HMPA, 190°, 15 min	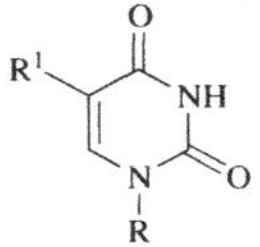 O, NH, O, R, R^1; R^1: H (26); Me (14); R 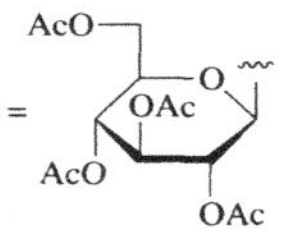 = AcO, OAc, AcO, OAc	102
	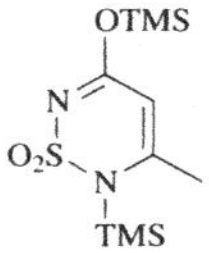 OTMS, O_2S, TMS	BF_3•OEt_2, CH_2Cl_2, rt, 4 h	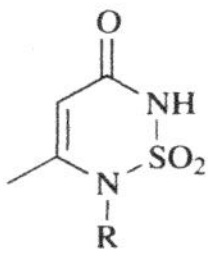 O, NH, SO_2, R (45) + bis(riboside) (32)	1088
	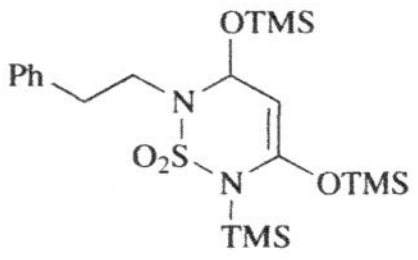 Ph, OTMS, O_2S, OTMS, TMS	BF_3•OEt_2, CH_2Cl_2, rt, 4 h	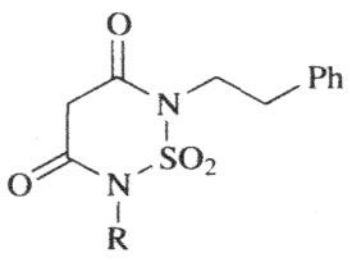 O, Ph, SO_2, O, R (79)	1089
	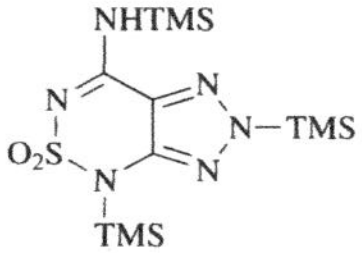 NHTMS, TMS, O_2S, TMS	BF_3•OEt_2, MeCN, rt, 12 h	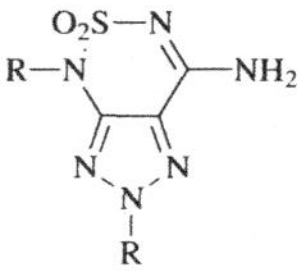 O_2S, R, NH_2, R (35)	1090

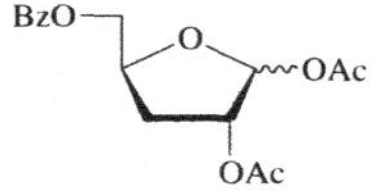

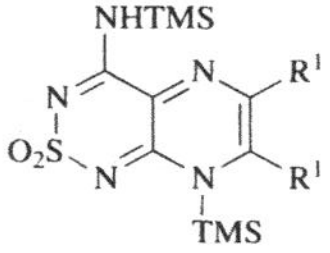

TABLE VI. REACTIONS WITH BORON TRIFLUORIDE ETHERATE AS CATALYST (*Continued*)

Sugar	Base	Conditions	Product(s) and Yield(s) (%)	Refs.
	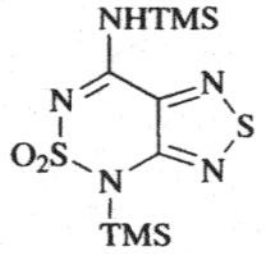	$BF_3 \cdot OEt_2$, MeCN, rt, 12 h	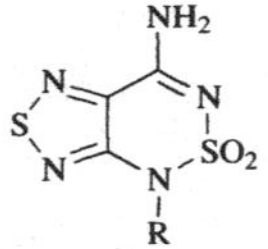 (64) R 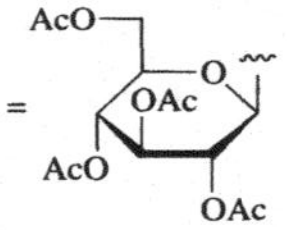 =	1090
	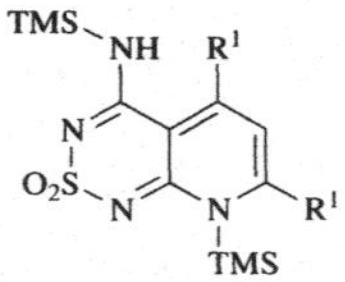	$BF_3 \cdot OEt_2$, CH_2Cl_2, rt, 4 h	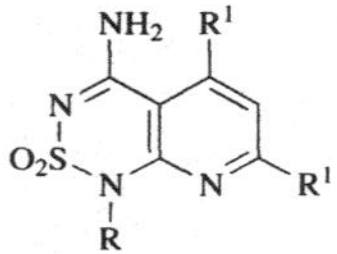 R^1: Me (4); Ph (66)	1082
	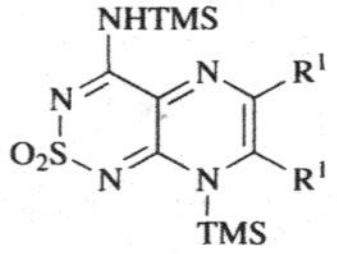	$BF_3 \cdot OEt_2$, CH_2Cl_2, rt, 4 h	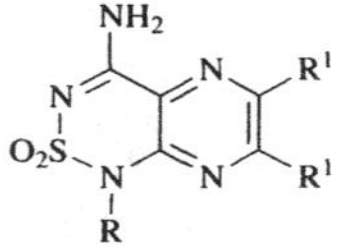 R^1: H (16); Me (81); Ph (81)	1091
C_{17} 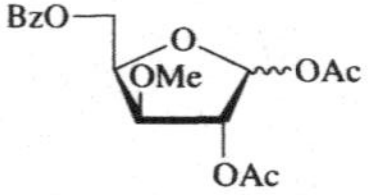	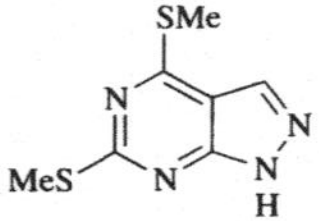	$BF_3 \cdot OEt_2$, MeCN	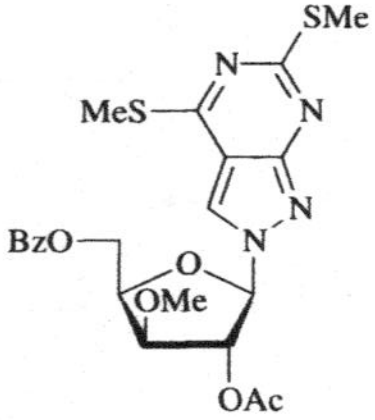 (56)	1086
C_{19} 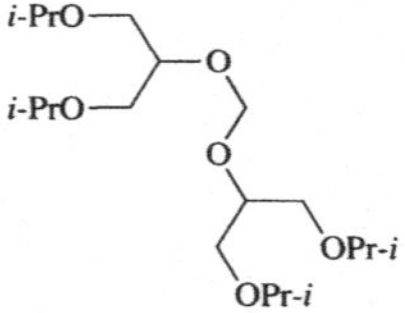	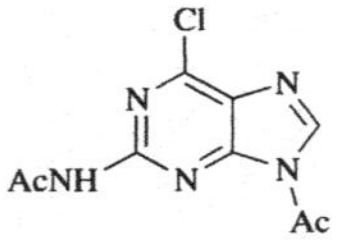	$BF_3 \cdot OEt_2$, *N*-methylpyrrolidone, 100°, 4 h	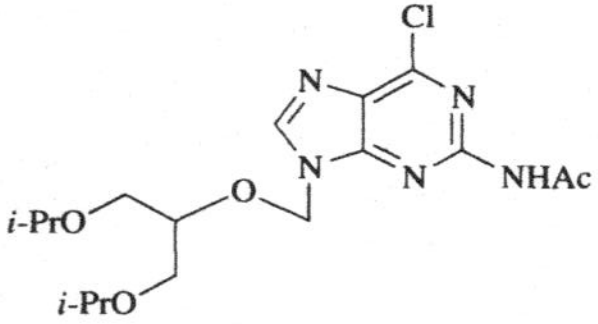 (63)	521

	Glycosyl Donor	Base	Conditions	Product(s) and Yield(s) (%)	Refs.
C_{19}			BF_3•OEt_2, EtOAc, rt, 2 h	(72)	1029
C_{20-21}			BF_3•OEt_2, $MeNO_2$, reflux, 2 h	R^1: F (52); Cl (82); N_3 (32); H (43)	519
C_{23}			BF_3•OEt_2, MeCN	(60) N^1:N^3 = 6:1	1086
			BF_3•OEt_2, MeCN	(40) + N^1-α-isomer (20)	1086

TABLE VI. REACTIONS WITH BORON TRIFLUORIDE ETHERATE AS CATALYST (*Continued*)

Sugar	Base	Conditions	Product(s) and Yield(s) (%)	Refs.
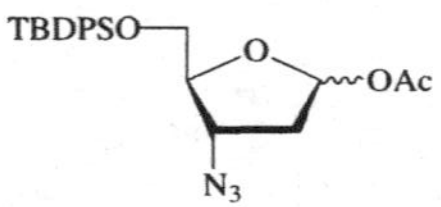 TBDPSO, OAc, N_3	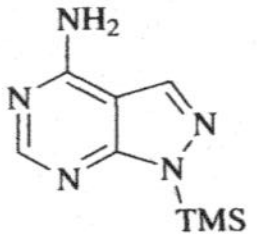 NH_2, TMS	BF_3•OEt_2, dioxane, rt, 12 h	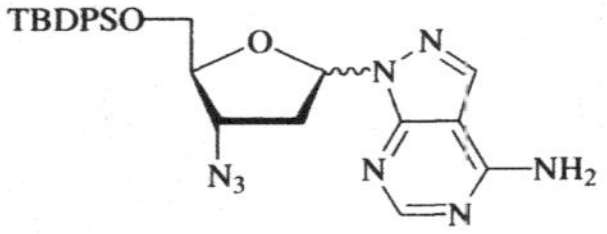 TBDPSO, N_3, NH_2 (65) α:β = 1:3	749
C_{25} 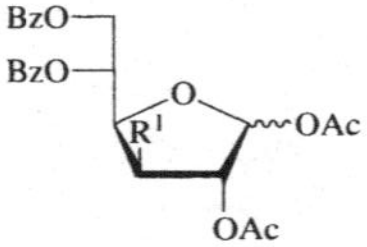 BzO, BzO, R^1, OAc, OAc	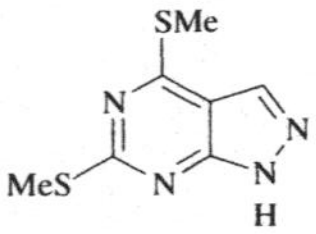 SMe, MeS, H	BF_3•OEt_2, MeCN	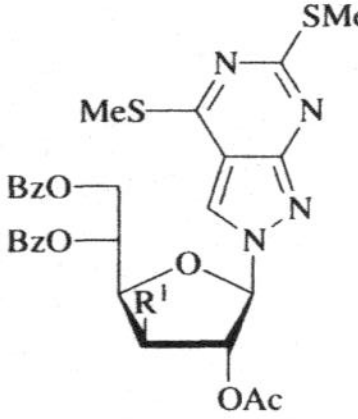 SMe, MeS, BzO, BzO, R^1, OAc; R^1: OMe (52) + α-isomer (10); OMs (30) + N^1-α-isomer (10)	1086
C_{28} 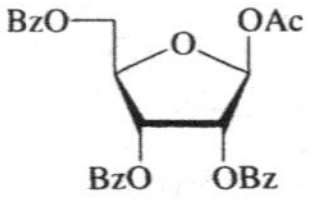 BzO, OAc, BzO, OBz	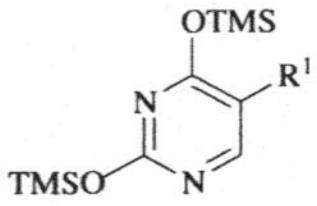 OTMS, R^1, TMSO	BF_3•OEt_2, HMPA, 190°, 15 min	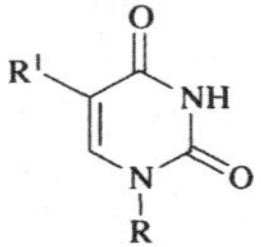 R^1, O, NH, O, R; R^1: H (14); Me (67); R 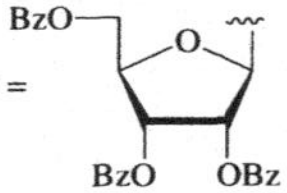 = BzO, BzO, OBz	102
	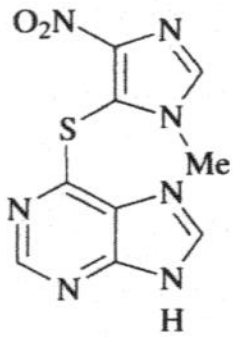 O_2N, S, Me, H	BF_3•OEt_2, $MeNO_2$, reflux, 45 min	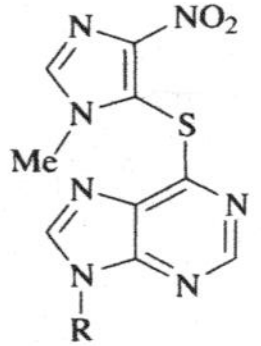 NO_2, S, Me, R (30)	1092

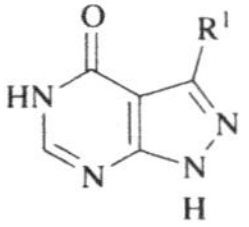

	BF_3•OEt_2, $MeNO_2$, reflux	R^1: Br, 90 min (45)	1055
		R^1: CN, 90 min (45)	1055
	BF_3•OEt_2, PhCN, 160-180°, 15 min	R^1 = CN (52)	1055
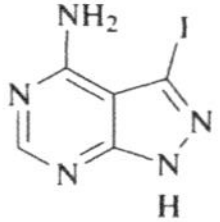	BF_3•OEt_2, $MeNO_2$, reflux, 2 h	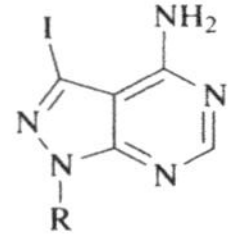 (86)	519
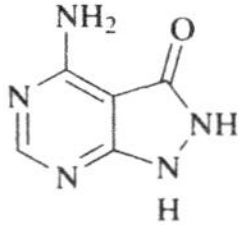	BF_3•OEt_2, $MeNO_2$, reflux, 15 min	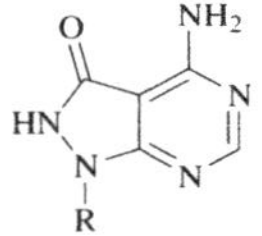 (37)	1095
	BF_3•OEt_2, solvent, reflux, 1 h	Solvent: $MeNO_2$ (60), α-anomer (10) MeCN (58), α-anomer (12) CH_2Cl_2 (40), α-anomer (5)	1094
	BF_3•OEt_2, $MeNO_2$, reflux, time	R^1 = Me, R^2 = Br, 5 min (62)	1056
		R^1 = OMe, R^2 = SMe, 40 min (78)	1093
		R^1 = OEt, R^2 = SMe, 20 min (61)	1095
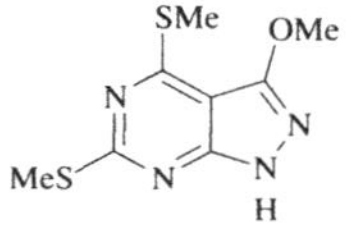	BF_3•OEt_2, solvent, reflux, 20 min	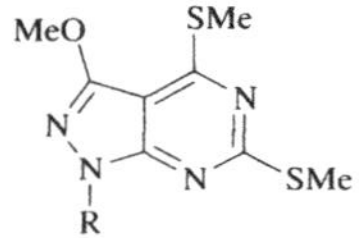 (45)	1095

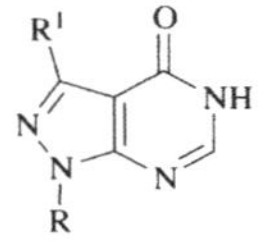

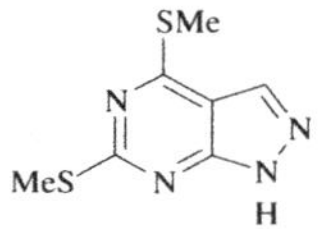

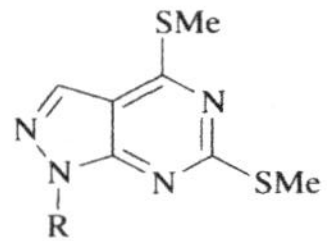

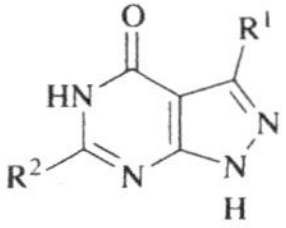

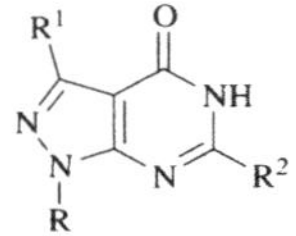

TABLE VI. REACTIONS WITH BORON TRIFLUORIDE ETHERATE AS CATALYST (*Continued*)

Sugar	Base	Conditions	Product(s) and Yield(s) (%)	Refs.
	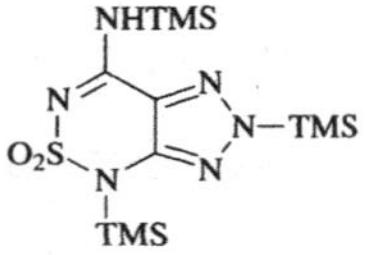 NHTMS, O_2S, N—TMS, TMS	BF_3•OEt_2, MeCN, rt, 12 h	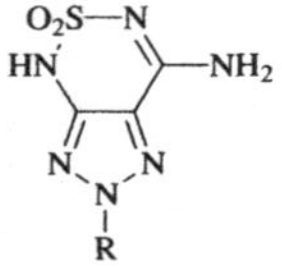 O_2S—N, HN, NH_2, R (35); R = BzO, O, BzO, OBz 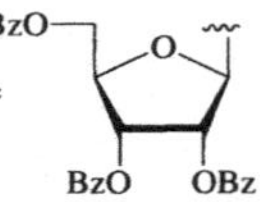	1090
	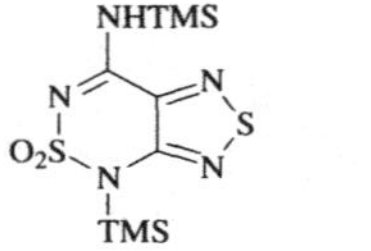 NHTMS, O_2S, S, TMS	BF_3•OEt_2, MeCN, rt, 12 h	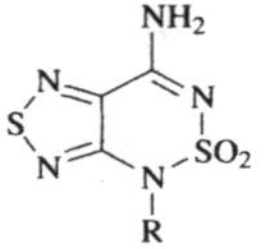 NH_2, S, SO_2, R (59)	1090
	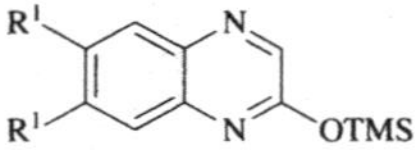 R[1], R[1], OTMS	BF_3•OEt_2, CH_2Cl_2, rt, time	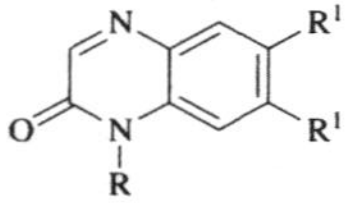 O, N, R, R[1], R[1]; R[1] / time: H / 1 h (19); Cl / 1 h (13); Me / 2.5 h (36)	1096
	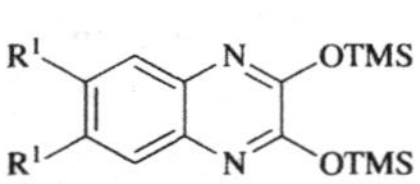 R[1], R[1], OTMS, OTMS	BF_3•OEt_2, CH_2Cl_2, rt, time	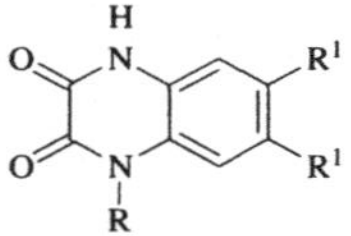 H, O, O, N, R, R[1], R[1]; R[1] / time: H / 1 h (40); Cl / 1 h (36); Me / 1 h (25)	1096
	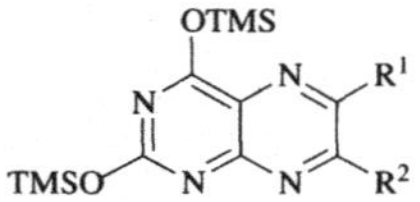 OTMS, TMSO, R[1], R[2]	BF_3•OEt_2	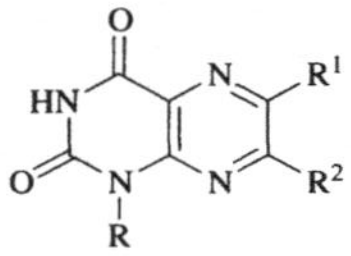 O, HN, O, R, R[1], R[2]	
	R[1] = H, R[2] = H	C_6H_6, rt, 1 h	(30) + N^3-isomer (19) + N^1,N^3-bis(riboside) (20)	1097
	R[1] = p-ClC_6H_4, R[2] = H	EtOAc, 20°, 30 min	(53)	104
	R[1] = H, R[2] = p-ClC_6H_4	EtOAc, 20°, 30 min	(68) + N^3-β-isomer (5)	104

Conditions	Product (yield, %)	Ref.
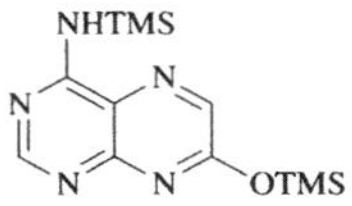 BF_3•OEt_2, $CHCl_3$, rt, 3 h	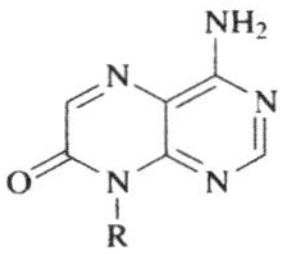 (53)	1098
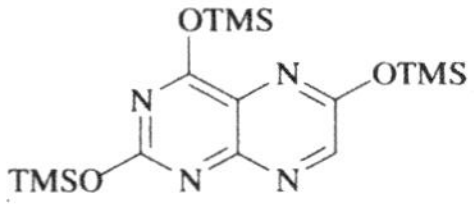 BF_3•OEt_2, CH_2Cl_2, rt, 2 h	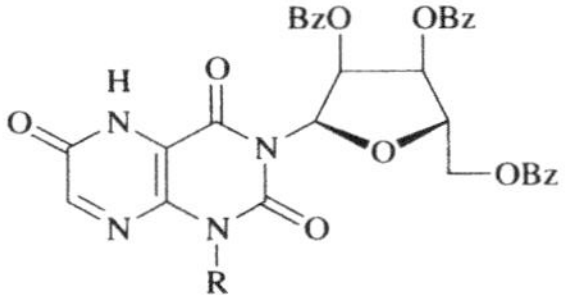 (65)	1099
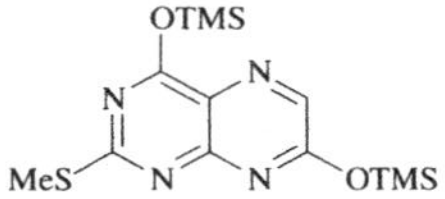 BF_3•OEt_2, CH_2Cl_2, rt, 4 h	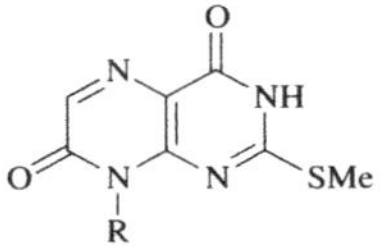 (82)	1084
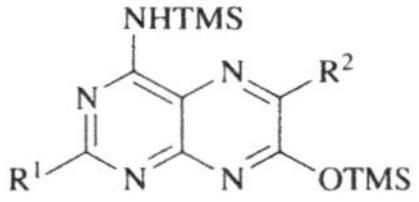 BF_3•OEt_2, CH_2Cl_2, rt, time	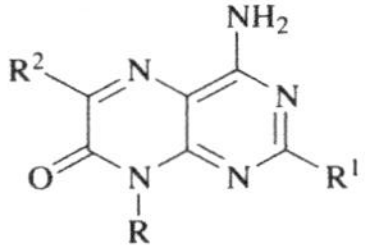 R^1 = Ph, R^2 = H, time 3 h (75); R^1 = H, R^2 = Ph, time 8 h (40)	1098
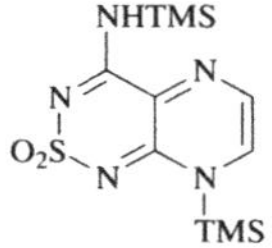 BF_3•OEt_2, CH_2Cl_2, rt, 4 h	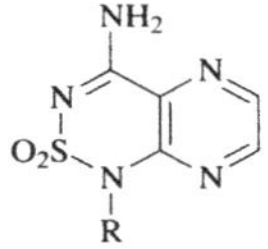 (31) + N^8-isomer (9)	1085

TABLE VI. REACTIONS WITH BORON TRIFLUORIDE ETHERATE AS CATALYST (*Continued*)

Sugar	Base	Conditions	Product(s) and Yield(s) (%)	Refs.
	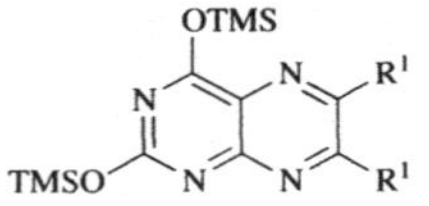 [structure: 2,4-bis(OTMS)pteridine, 6,7-R¹]	BF_3•OEt_2	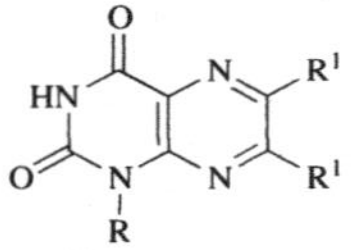 [structure: N-R pteridinedione, 6,7-R¹]; R = 2,3,5-tri-O-benzoylribofuranosyl (BzO, BzO, OBz) 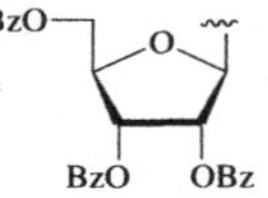	
	R^1			
	Me	C_6H_6, rt, 1 h	(25) + N^3-isomer (19) + N^1,N^3-bis(riboside) (26)	1097
	Ph	C_6H_6, rt, 1 h	(20) + N^3-isomer (17) + N^1,N^3-bis(riboside) (56)	1097
	Ph	C_6H_6, rt, 4 h	(9) + N^3-isomer (64) + N^1,N^3-bis(riboside) (9)	1100
	Ph	EtOAc, rt, 3 h	(71)	1100
	p-ClC_6H_4	CCl_4, 20°, 1 h	(45) + N^3-β-isomer (20) + N^3-α-isomer (11)	104
	p-ClC_6H_4	EtOAc, 20°, 1 h	(69) + N^3-isomer (2)	104
	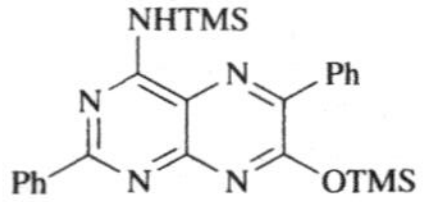 [structure: NHTMS, Ph, Ph, OTMS pteridine]	BF_3•OEt_2, CH_2Cl_2, rt, 24 h	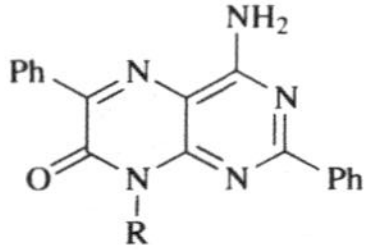 [structure: NH₂, Ph, Ph, N-R pteridinone] (59)	1098
	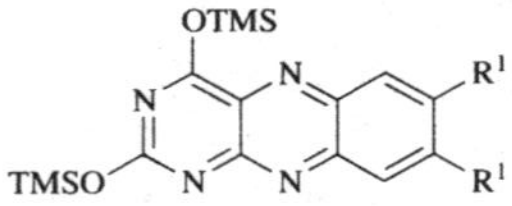 [structure: 2,4-bis(OTMS)benzo[g]pteridine, R¹, R¹]	BF_3•OEt_2, EtOAc, rt, 2 h	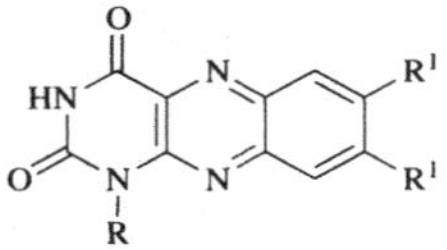 [structure: N-R benzo[g]pteridinedione, R¹, R¹] R^1: H (55); Me (54)	103
		BF_3•OEt_2, C_6H_6, rt, 4 h	R^1 = Me, (34) + N^3-isomer (19) + N^1,N^3-bis(isomer) (23)	103

Glycosyl Donor	Heterocycle	Conditions	Product(s) and Yield(s) (%)	Refs.
	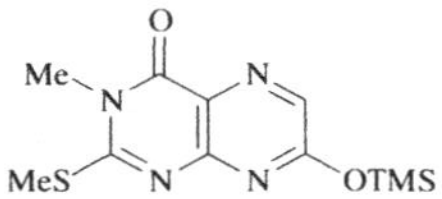 Me-N, MeS, OTMS (pteridine)	BF_3•OEt_2, CH_2Cl_2, rt, 2 h	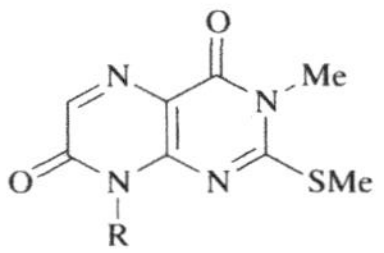 N-Me, SMe, N-R (75)	1084
	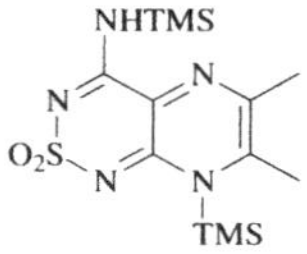 NHTMS, O_2S, N-TMS	BF_3•OEt_2, CH_2Cl_2, rt, 4 h	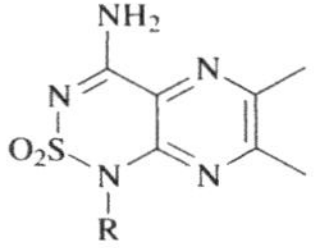 NH_2, O_2S, N-R (81)	1085
C_{31} 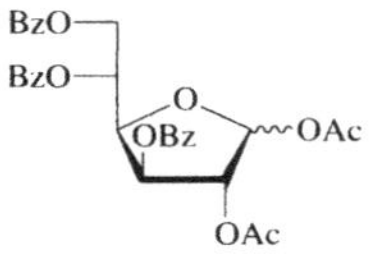 BzO, BzO, OBz, OAc, OAc	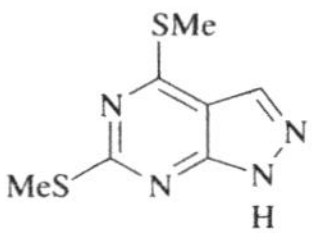 SMe, MeS, N-H	BF_3•OEt_2, MeCN	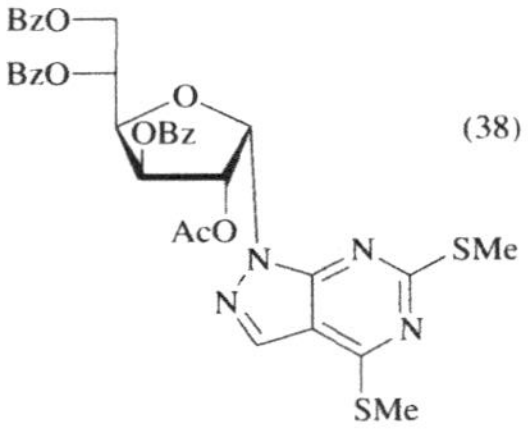 BzO, BzO, OBz, AcO, SMe, SMe (38) + N^2-β-isomer (30)	1086
C_{36} 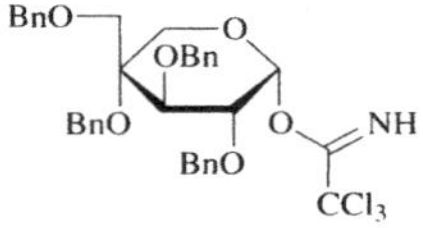 BnO, OBn, BnO, BnO, O-C(=NH)CCl_3	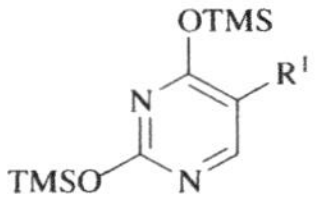 OTMS, R^1, TMSO	BF_3•OEt_2, CH_2Cl_2, rt, 2 h	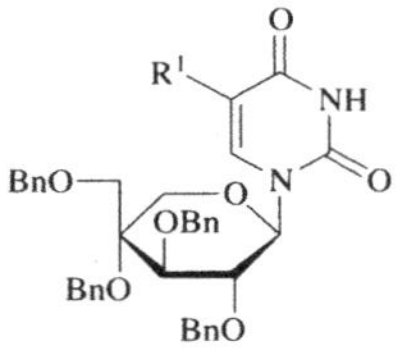 R^1, NH, BnO, OBn, BnO, BnO; R^1: H (62), Me (61)	1101
	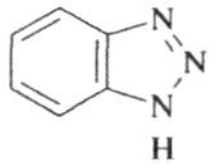 benzotriazole (N-H)	BF_3•OEt_2, CH_2Cl_2, –22°, 2 h	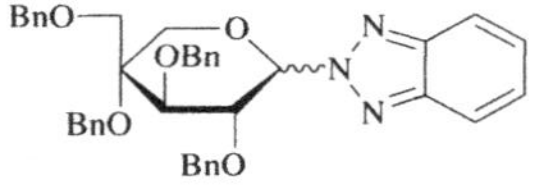 BnO, OBn, BnO, BnO (51) α:β = 1:9 + N^1-β-isomer (24)	1101

TABLE VII. REACTIONS WITH MISCELLANEOUS FRIEDEL-CRAFTS CATALYSTS

Sugar	Base	Conditions	Product(s) and Yield(s) (%)	Refs.
C_6				
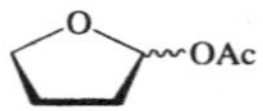 OAc (structure)	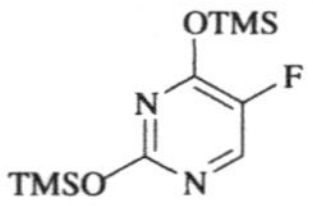 OTMS, F, N, TMSO, N (structure)	$SbCl_5$, CH_2Cl_2, rt, 0.1 h	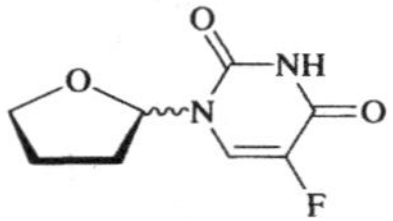 O, NH, O, N, F (structure) (8)	528
C_7				
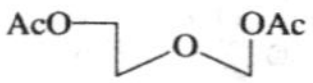 AcO, O, OAc (structure)	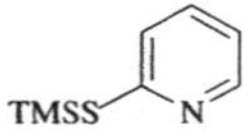 TMSS, N (structure)	ZnI_2, DMF, 100°, 3 h	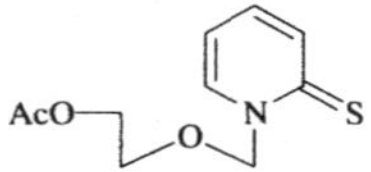 AcO, O, N, S (structure) (67)	113
	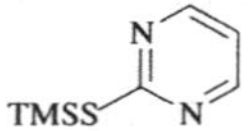 N, TMSS, N (structure)	ZnI_2, DMF, 100°, 3 h	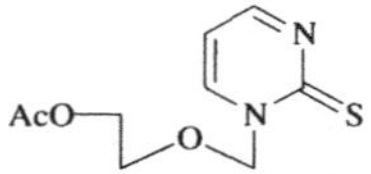 AcO, O, N, N, S (structure) (76)	113
	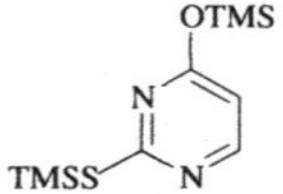 OTMS, N, TMSS, N (structure)	ZnI_2, DMF, 100°, 4 h	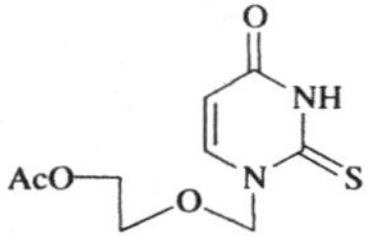 O, NH, AcO, O, N, S (structure) (73)	113
	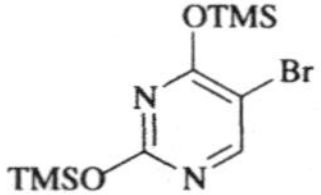 OTMS, Br, N, TMSO, N (structure)	ZnI_2, DMF, 100°, 7 h	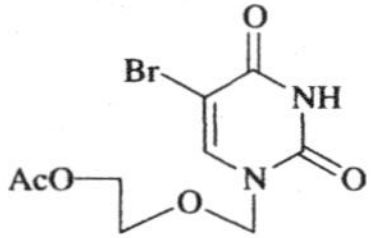 Br, O, NH, AcO, O, N, O (structure) (60)	113
	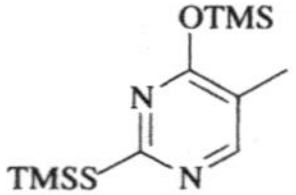 OTMS, N, TMSS, N (structure)	ZnI_2, DMF, 100°, 5 h	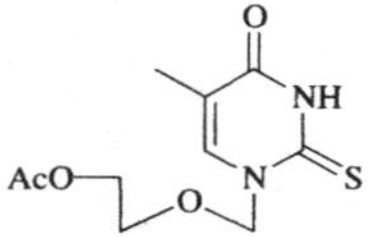 O, NH, AcO, O, N, S (structure) (51)	113

	Conditions	Product(s) and Yield(s) (%)	Refs.
	ZnI_2, DMF, 100°, 7 h	(35)	113
	ZnI_2, DMF, 100°, 7 h	(55) N^7:N^9 = 1:3	113
C_9	$LiClO_4$, $Cl(CH_2)_2Cl$, $Ph_3C^+ClO_4^-$, rt, 10 min	(75)	309
	SiF_4, MeCN, –20°, 2 h	(76) α:β = 40:60	118
	SiF_4, MeCN, –20°, 2 h	(64) α:β = 50:50	118

TABLE VII. REACTIONS WITH MISCELLANEOUS FRIEDEL-CRAFTS CATALYSTS (*Continued*)

Sugar	Base	Conditions	Product(s) and Yield(s) (%)	Refs.
MeO, R^1, F, O, O, O	OTMS, N, TMSO, N	SiF_4, MeCN	O, NH, R^1, MeO, N, O, O, O, O R^1 F 0° (40) R^1 H rt, 10 min (33)	118a
C_{10} AcO, O, D, AcO	Me, O, N, N, Me, N, O, N, TMS	$Ph_3C^+ClO_4^-$, $Cl(CH_2)_2Cl$, rt, 5 min	AcO, O, N, N, D, O, N–Me, N, Me, O (72) α:β = 1:1	123
AcO, O, AcO	Ac, N, TMS, N, TMSO, N	$Ph_3C^+ClO_4^-$, $Cl(CH_2)_2Cl$, MS, rt, 3 h	AcO, O, O, N, N, NHAc (81) α:β = 52:48	123
	OTMS, N, TMSO, N	$Ph_3C^+ClO_4^-$, $Cl(CH_2)_2Cl$, MS, rt, 3 h	AcO, O, O, NH, N, O (63) α:β = 45:55	123
	Cl, N, N, N, N, TMS	$Ph_3C^+ClO_4^-$, $Cl(CH_2)_2Cl$, rt, 5 min	AcO, O, N, N, N, Cl, N (71) α:β = 54:46	123

Reactant	Silylated base	Conditions	Product(s) and Yield(s) (%)	Refs.
	Bz–N(TMS)–, N-TMS	$Ph_3C^+ClO_4^-$, $Cl(CH_2)_2Cl$, rt, 5 min	(61) α:β = 54:46	123
	Me, Me, N-TMS	$Ph_3C^+ClO_4^-$, $Cl(CH_2)_2Cl$, rt, 5 min	(78) α:β = 54:46	123
C_{11} AcO, AcO, OAc	OTMS, TMSO	1. $SbCl_5$, EtOAc, 20°, 20 min 2. NaOMe, MeOH	(17) + α-anomer (16) + N^3-β-isomer (17) + N^3-α-isomer (16)	109
C_{12} BzO, Br	NHTMS, TMSO	TMSBr, CH_2Cl_2, –78° to rt	(44) α:β = 1:1	1102
AcO, OAc, AcO	OTMS, TMSO	$SbCl_5$, EtOAc, rt, 15 min, or 23°, 5 min	(40) + β-anomer (24)	108, 1103

TABLE VII. REACTIONS WITH MISCELLANEOUS FRIEDEL-CRAFTS CATALYSTS (*Continued*)

Sugar	Base	Conditions	Product(s) and Yield(s) (%)	Refs.
	OTMS; N; TMSO; N	$SbCl_5$, EtOAc, rt, 5 min	O; NH; N; O; R (26) + α-anomer (23) R = AcO; O; AcO	107
	Ac; N; TMS; N; TMSO; N	$LiClO_4$, $Cl(CH_2)_2Cl$, $Ph_3C^+ClO_4^-$, rt, 3 h	NHAc; N; N; O; R (56)	123
	OTMS; N; TMSO; N	$SnCl_2$, MeCN	AcO; O; AcO; O; NH; N; O (78)	1104
	OTMS; N; TMSO; N	$LiClO_4$, $Cl(CH_2)_2Cl$, $Ph_3C^+ClO_4^-$, rt, 3 h	AcO; O; AcO; O; NH; N; O (63) α:β = 48:52	123
	Cl; N; N; N; N; TMS	$Ph_3C^+ClO_4^-$, $Cl(CH_2)_2Cl$, rt, 5 min	AcO; O; AcO; N; N; N; N; Cl (76) α:β = 58:42	123
	Bz; N; TMS; N; N; N; N; TMS	$Ph_3C^+ClO_4^-$, $Cl(CH_2)_2Cl$, rt, 5 min	AcO; O; AcO; N; N; N; N; NHBz (61) α:β = 50:50	123

	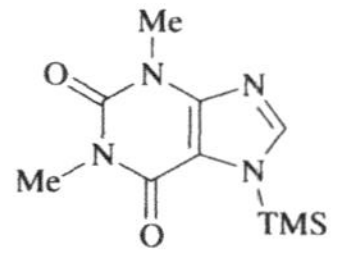	$Ph_3C^+ClO_4^-$, $Cl(CH_2)_2Cl$, rt, 5 min	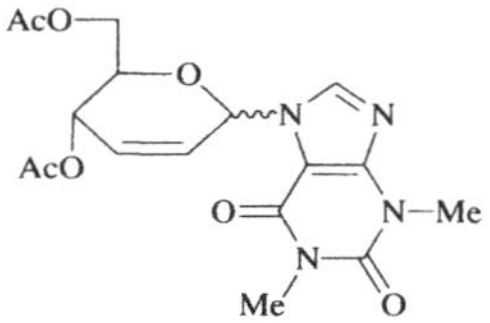 (61) α:β = 49:51	123
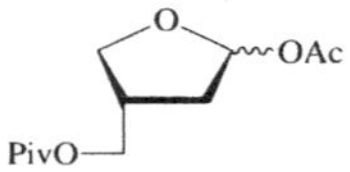	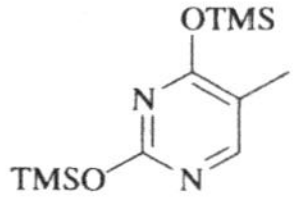	TMSI, MeCN, –5 - 0°, 1 h	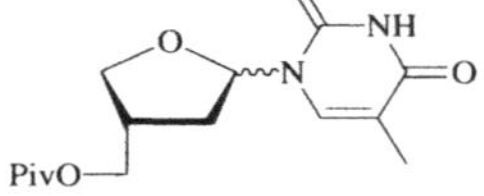 (93) α:β = 5:4	139a
	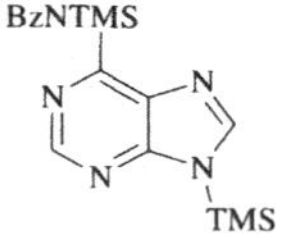	TMSI, MeCN, –5°, 2 h	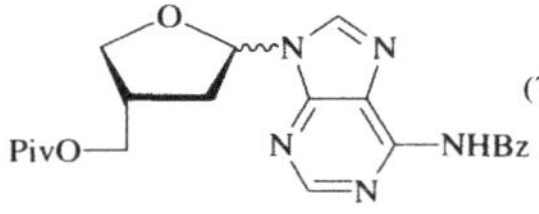 (72) α:β = 2:5	139a
C_{13} 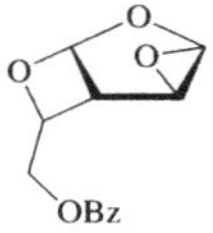	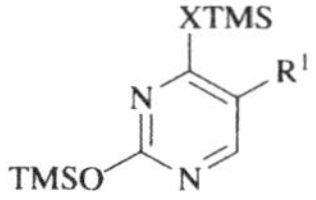	$ZnCl_2$, THF, rt, 18 h	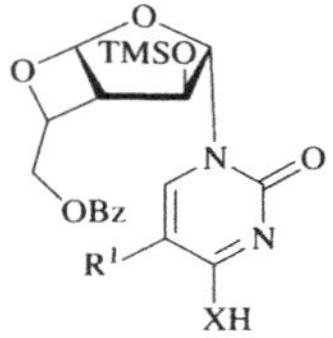 R^1 / X: H / NH (58); Me / O (66)	1105
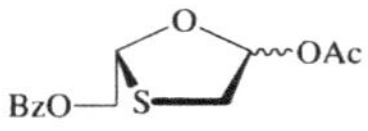	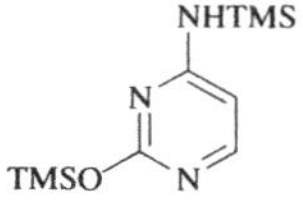	TMSI, $Cl(CH_2)_2Cl$	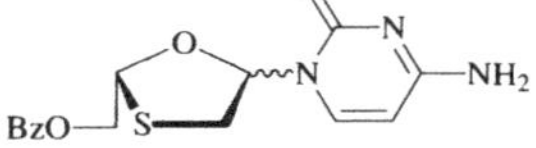 (83) α:β = 1.3:1	1106
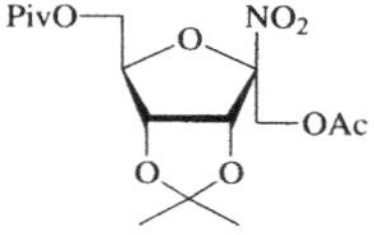	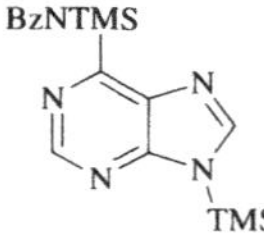	$SnCl_2$, MeCN, 8°, 30 min	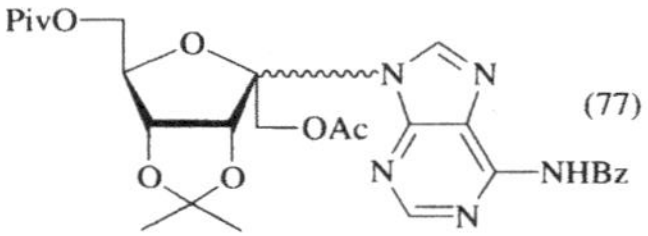 (77) α:β = 1:3.5	106

TABLE VII. REACTIONS WITH MISCELLANEOUS FRIEDEL-CRAFTS CATALYSTS (*Continued*)

Sugar	Base	Conditions	Product(s) and Yield(s) (%)	Refs.
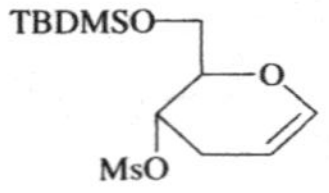 TBDMSO, MsO	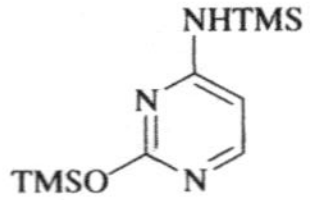 NHTMS, TMSO	$EtAlCl_2$, MeCN, reflux	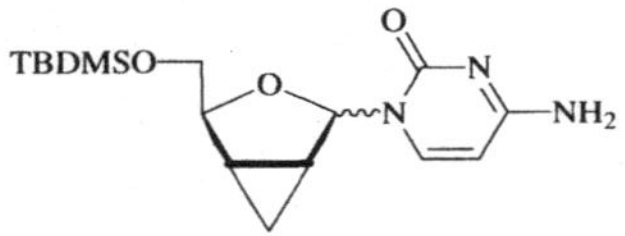 TBDMSO, NH_2 (81) α:β = 1:1	115
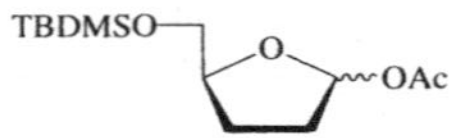 TBDMSO, OAc	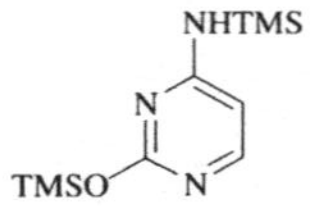 NHTMS, TMSO	$EtAlCl_2$, CH_2Cl_2, 27°, 2.5 h	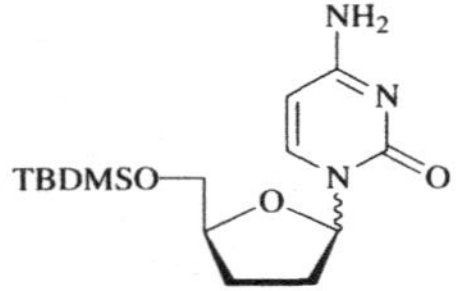 NH_2, TBDMSO (42) + α-anomer (18)	114, 287
C_{14}				
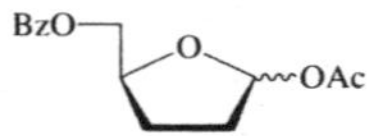 BzO, OAc	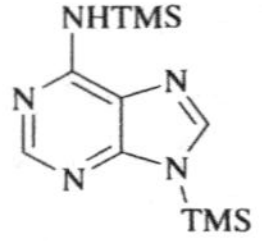 NHTMS, TMS	TMSBr, CH_2Cl_2, 0° to rt	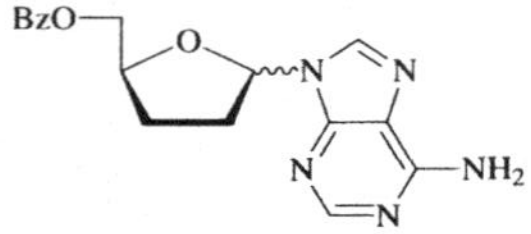 BzO, NH_2 (63)	1102
C_{15}				
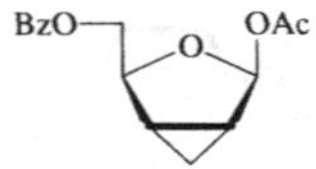 BzO, OAc	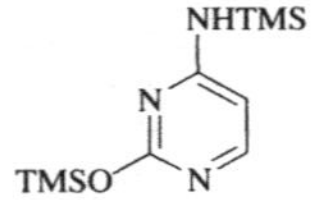 NHTMS, TMSO	$EtAlCl_2$, PhMe, CH_2Cl_2, rt, 40 min	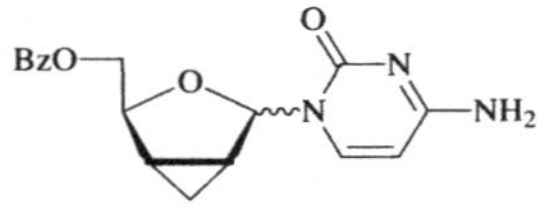 BzO, NH_2 (37) α:β = 1:1	116
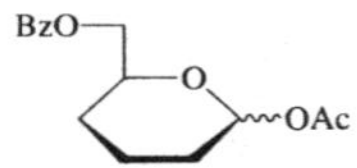 BzO, OAc	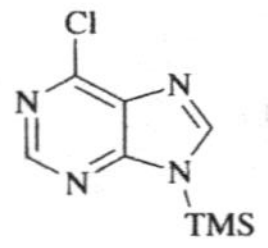 Cl, TMS	TMSBr, $CHCl_3$, rt, 1 h	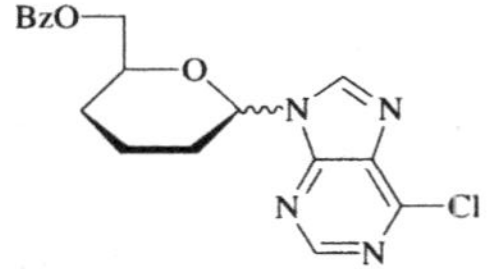 BzO, Cl (63)	1107

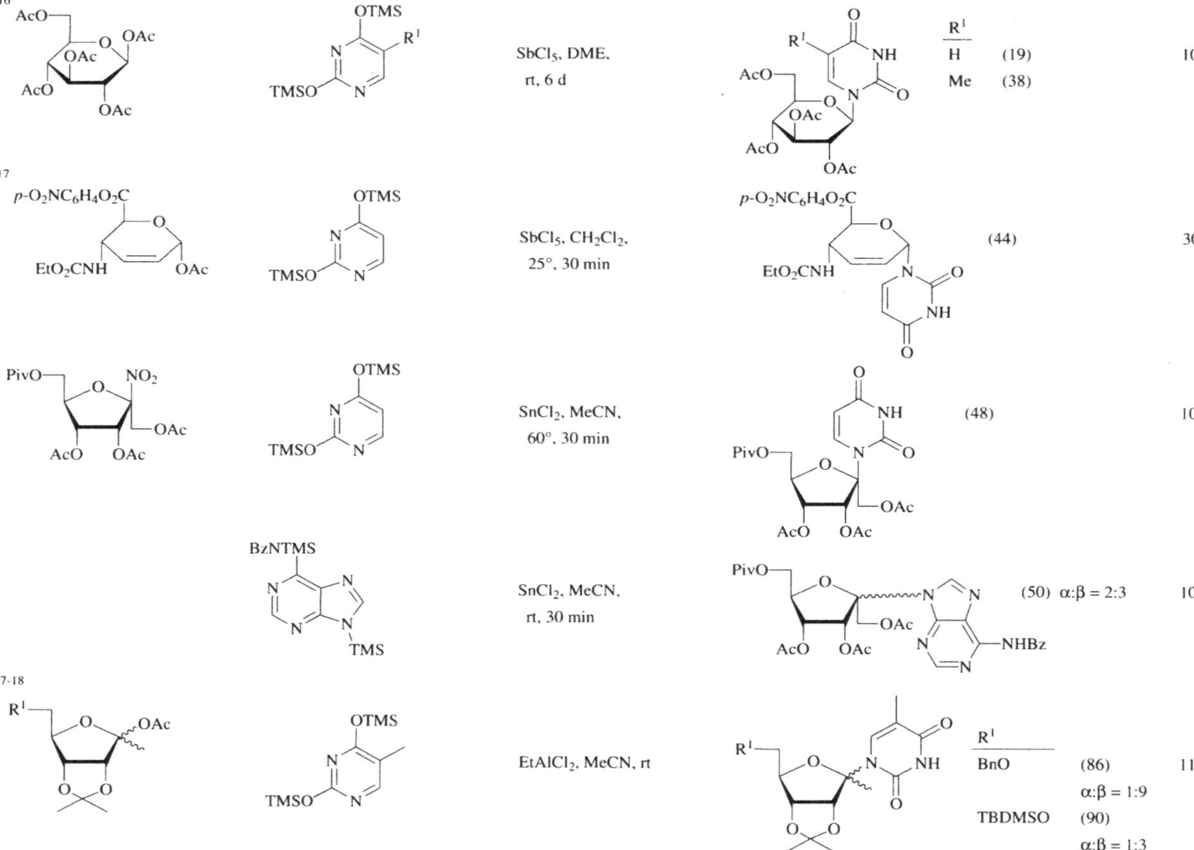

Substrate	Base	Conditions	Product(s) and Yield(s) (%)	Refs.
C_{16}		$SbCl_5$, DME, rt, 6 d	R^1 H (19); Me (38)	102
C_{17}		$SbCl_5$, CH_2Cl_2, 25°, 30 min	(44)	308
		$SnCl_2$, MeCN, 60°, 30 min	(48)	106
		$SnCl_2$, MeCN, rt, 30 min	(50) α:β = 2:3	106
C_{17-18}		$EtAlCl_2$, MeCN, rt	R^1 BnO (86) α:β = 1:9; TBDMSO (90) α:β = 1:3	117

TABLE VII. REACTIONS WITH MISCELLANEOUS FRIEDEL-CRAFTS CATALYSTS (*Continued*)

Sugar	Base	Conditions	Product(s) and Yield(s) (%)	Refs.
C_{19}				
$R^1 = p\text{-}ClC_6H_4CO$		$ZnCl_2$, $CHCl_3$, rt, 12 h	**I** (78) α:β = 25:75	269
		$ZnCl_2$, $CHCl_3$, rt, 15 h sugar:base = 1:4	**I** (58) α:β = 7:93	269
		$ZnCl_2$, $CHCl_3$, rt, 15 h sugar:base = 1:2	**I** (78) α:β = 25:75	269
C_{20}				
		TMSBr, CH_2Cl_2, rt, 12 h	**I** (56) α:β = 54:46	229
		TMSBr, $HgBr_2$, CH_2Cl_2, rt, 12 h	**I** (81) α:β = 68:32	229
		TMSBr, $Sn(OTf)_2$, CH_2Cl_2, rt, 12 h	**I** (84) α:β = 65:35	229
		n-$BuSnCl_3$, CH_2Cl_2, rt, 12 h	**I** (83) α:β = 77:23	229
		ZnI_2, DMF, 100°, 5 h	(71)	113

Reactant	Nucleobase	Conditions	Product(s) and Yield(s) (%)	Refs.
	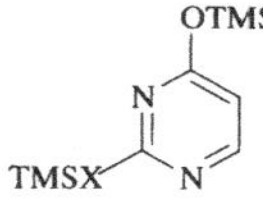 OTMS, TMSX (pyrimidine)	ZnI_2, DMF, 100°, 5 h	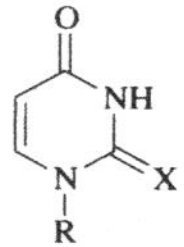 X: O (67); S (88); R 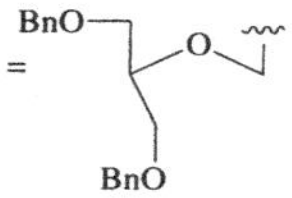 = BnO / O / BnO	113
		ZnI_2, DMF, reflux, 5 h	X = S, (82) N^1:N^3 = 1:1	113
	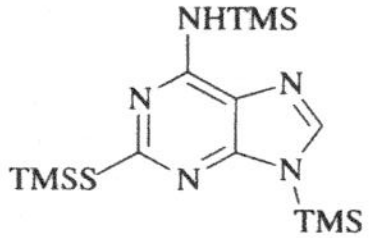 NHTMS, TMSS, TMS (purine)	ZnI_2, DMF, 100°, 7 h	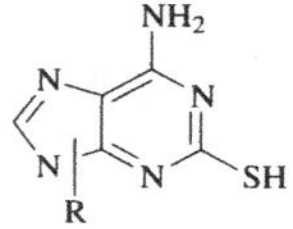 NH_2, SH, R (73) N^9:N^7 = 4:1	113
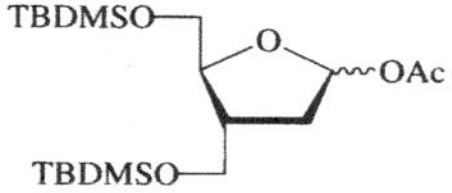 TBDMSO, TBDMSO, OAc	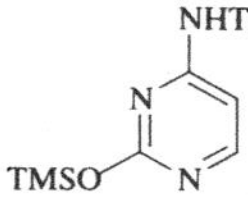 NHTMS, TMSO (pyrimidine)	$EtAlCl_2$, $Cl(CH_2)_2Cl$, 0°, 30 min	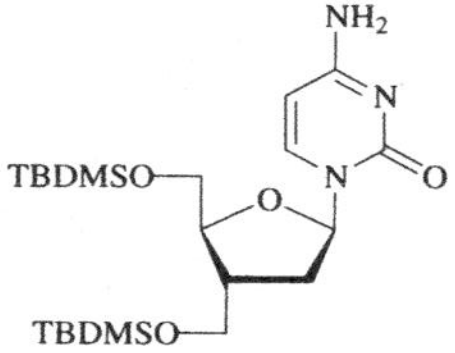 NH_2, TBDMSO, TBDMSO (74)	1108
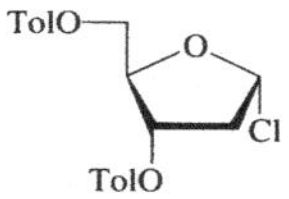 C_{21} TolO, TolO, Cl	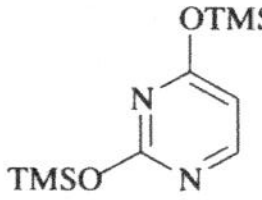 OTMS, TMSO (pyrimidine)	AlI_3, $Cl(CH_2)_2Cl$, 0°	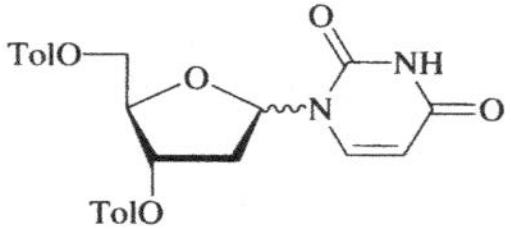 TolO, TolO **I** (—) α:β = 65:35	273
		ZnI_2, EDC, rt	**I** (—) α:β = 49:51	273
		ZnI_2, $CDCl_3$, EDC, 0°	**I** (—) α:β = 54:46	273
	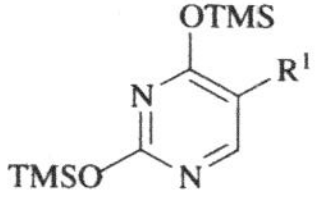 OTMS, R^1, TMSO (pyrimidine)	ZnI_2, $CHCl_3$, rt, 24 h	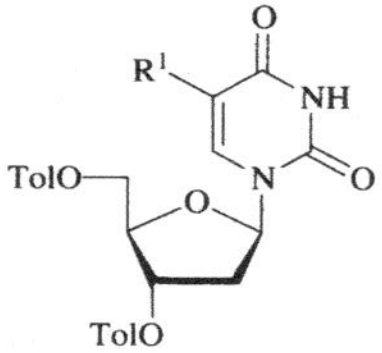 R^1: NO_2 (82); Ac (76); α:β = 1:1.23	261

TABLE VII. REACTIONS WITH MISCELLANEOUS FRIEDEL-CRAFTS CATALYSTS (*Continued*)

Sugar	Base	Conditions	Product(s) and Yield(s) (%)	Refs.
C_{22}				
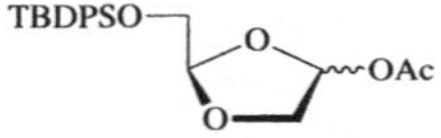 TBDPSO, OAc (structure)	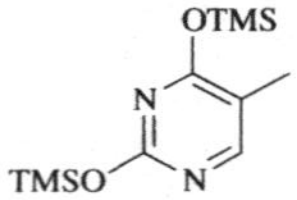 TMSO, N, N, OTMS (structure)	$TiCl_2(OPr\text{-}i)_2$	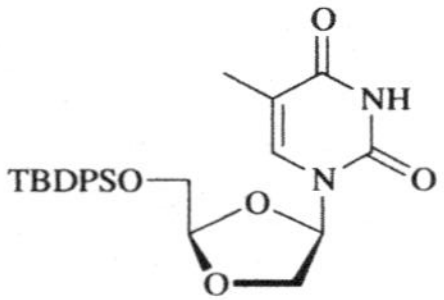 TBDPSO, O, N, NH, O (structure) (40)	235
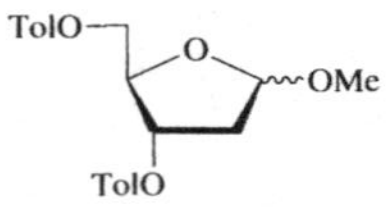 TolO, OMe, TolO (structure)	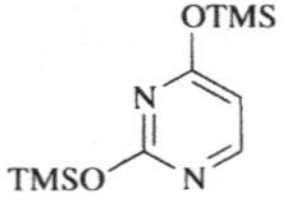 TMSO, N, N, OTMS (structure)	TMSCl, NaI, MeCN	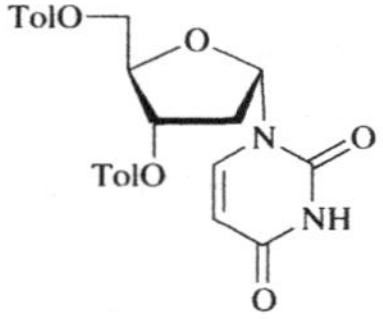 TolO, TolO, N, NH, O (structure) (65)	137
C_{24}				
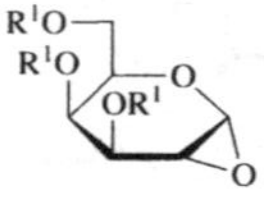 R^1O, R^1O, OR^1 (structure) R^1 = TBDMS	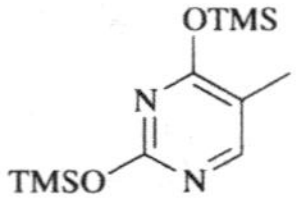 TMSO, N, N, OTMS (structure)	$ZnCl_2$, THF, rt, 12 h	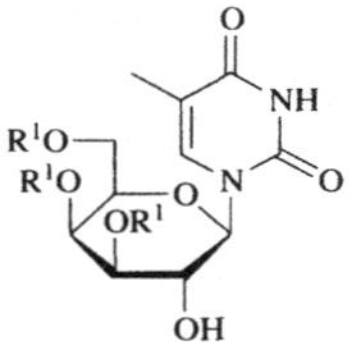 R^1O, R^1O, OR^1, OH, N, NH, O (structure) **I** (44)	112
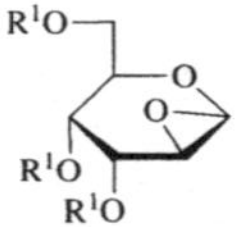 R^1O, R^1O, R^1O (structure) R^1 = TBDMS	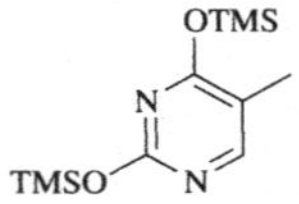 TMSO, N, N, OTMS (structure)	$ZnCl_2$, THF, rt, 17 h	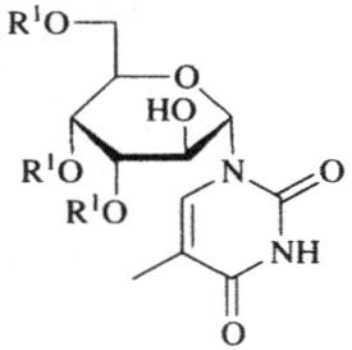 R^1O, HO, R^1O, R^1O, N, NH, O (structure) (79)	112

C_{26}

Substrate	Reagent	Conditions	Product(s) and Yield(s) (%)	Refs.
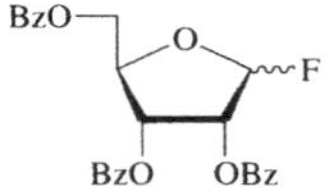 BzO, BzO, OBz, F	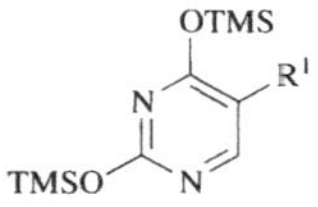 OTMS, R^1, TMSO	SiF_4, MeCN, 0°, 2 h	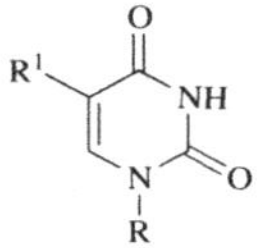 R^1 = H (90); R^1 = F (86); R 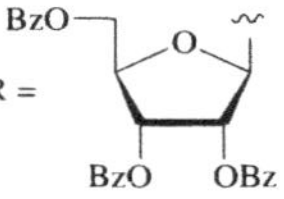 = BzO, BzO, OBz (furanosyl)	118
	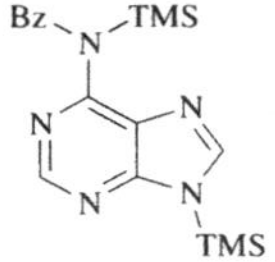 Bz, TMS, TMS	SiF_4, MeCN, 0°, 2 h	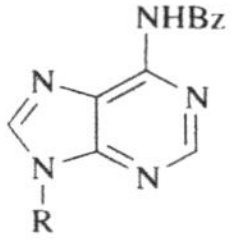 NHBz, R (82)	118
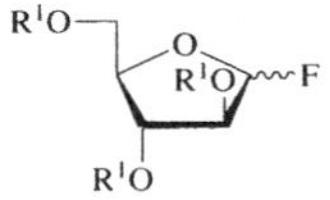 R^1O, R^1O, R^1O, F	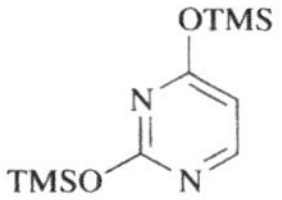 OTMS, TMSO	SiF_4, 0°	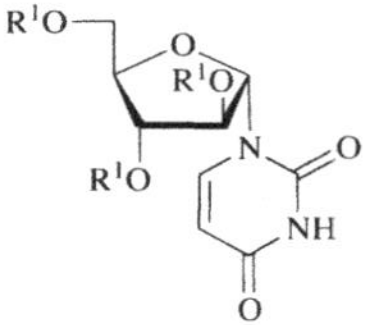 R^1O, R^1O, R^1O, NH, O; R^1 = Bz: MeCN, 1 h (76); Bn: CH_2Cl_2, 2 h (72); Bn: MeCN, 2 h (78)	118
BnO, BnO, OBn, F	OTMS, TMSO	SiF_4, MeCN, rt, 2 h	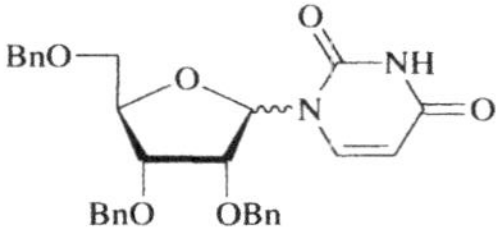 BnO, BnO, OBn, NH, O; **I** (85) α:β = 16:84	1109
		SiF_4, MeCN, 0°, 1 h	**I** (85) α:β = 16:84	118
		SiF_4, CH_2Cl_2, 0°, 0.5 h	**I** (80) α:β = 60:40	118
		SiF_4, Et_2O, 0°, 0.5 h	**I** (82) α:β = 55:45	118
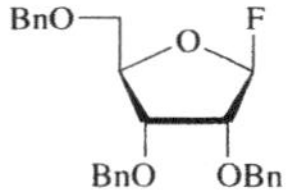 BnO, BnO, OBn, F	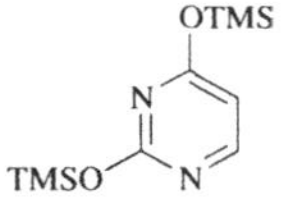 OTMS, TMSO	SiF_4, MeCN, 0°, 1 h	**I** (84) α:β = 18:82	118
		SiF_4, CH_2Cl_2, 0°, 0.5 h	**I** (84) α:β = 68:32	118
		SiF_4, Et_2O, 0°, 0.5 h	**I** (78) α:β = 60:40	118

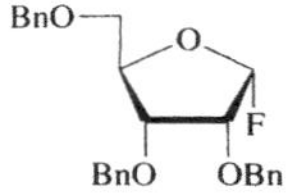

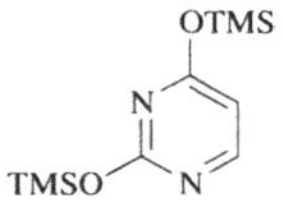

TABLE VII. REACTIONS WITH MISCELLANEOUS FRIEDEL-CRAFTS CATALYSTS (*Continued*)

Sugar	Base	Conditions	Product(s) and Yield(s) (%)	Refs.
C_{27}				
BzO, BzO, BzO, OMe	OTMS, TMSO, N, N	TMSI, MeCN, 70°	BzO, BzO, BzO, N, NH, O, O (65-70)	137
BzO, OMe, BzO, OBz	OTMS, TMSO, N, N	TCS, NaI, MeCN, 70°	O, NH, O, N–R (76) R = BzO, BzO, OBz	137
C_{28}				
BzO, OAc, BzO, OBz	CO_2Me, N, N, N–TMS	TMSCl, MeCN, NaI, 24°, 20 h	CO_2Me, N, N, N–R (82)	1110
	OTMS, TMSO, N, N	TMSI, MeCN, MS, rt	O, NH, O, N–R **I** (89)	136
		TMSCl, MeCN, NaI, rt, 50 min	**I** (64-68)	137, 1111
		TMSCl, MeCN, MS, rt, 100 min	**I** (89)	1111

Silylated base	Conditions	Product(s) and Yield(s) (%)		Refs.
NHTMS / TMSO / N (silylated cytosine)	TMSI, MeCN, MS, rt	NH_2, N, O, N–R (81)		136
OTMS / F / TMSO / N (silylated 5-fluorouracil)	$ZnCl_2$, MeCN, rt, 3 h	O, F, NH, O, N–R **I** (42)	R = BzO, O, BzO, OBz	868
	TMSCl, MeCN, NaI, MS, rt, 3 h	**I** (84)		135
	$AlCl_3$, MeCN, rt, 3 h	**I** (38)		514b
OTMS / TMSO / N (silylated thymine)	TMSCl, MeCN, NaI, MS, rt, 3 h	O, NH, O, N–R **I** (85)		135
	$SbCl_5$, DME, rt, 3.5 h	**I** (36)		102
OTMS / TMSS / N	$AlCl_3$, MeCN, 22°, 4 h	HN, S, N–R, O (5) + *S*-riboside (62)		89a
NHTMS / F / TMSO / N (silylated 5-fluorocytosine)	TMSCl, MeCN, NaI, MS, rt, 3 h	NH_2, F, N, O, N–R (83)		135

TABLE VII. REACTIONS WITH MISCELLANEOUS FRIEDEL-CRAFTS CATALYSTS (*Continued*)

Sugar	Base	Conditions	Product(s) and Yield(s) (%)	Refs.
	TMS–N(COC$_6$H$_4$OMe-*p*), TMSO (pyrimidine)	TMSCl, MeCN, NaI, MS, rt, 3 h	TMS–N(COC$_6$H$_4$OMe-*p*), N–R, O (73)	135
	OTMS, R^1, TMSO (triazine)	TMSI, MeCN, rt, 4-5 h	R^1, O, NH, O, N–R	
	R^1			
	H		(83)	139
	Br		(80)	139
	NMe$_2$		(21)	139
	SMe		(44)	139
	SCHF$_2$		(83)	139
	SCF$_3$	20°, 4-6 h	(73-80)	1111a
	OBn, Me$_2$N, OTMS (pteridine)	ZnCl$_2$, 140-150°, 75 min	OBn, O, N–R, NMe$_2$ (9)	1112
C$_{28}$ BzO, OBz, OAc, OBz (furanose)	OTMS, TMSO (pyrimidine)	TMSI, MeCN, 70°, or TMSI, rt	O, NH, O, BzO, OBz, OBz (58)	137

TABLE VIII. REACTIONS OF SILYLATED BASES WITH PROTECTED SUGARS WITH OR WITHOUT CATALYSTS

Sugar	Base	Conditions	Product(s) and Yield(s) (%)	Refs.
C_2				
 MeO–CH2–Cl	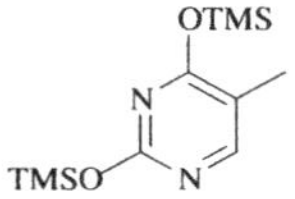 2,4-bis(OTMS)-5-methylpyrimidine (OTMS, TMSO)	Bu_4NI, CH_2Cl_2, reflux, 2 h	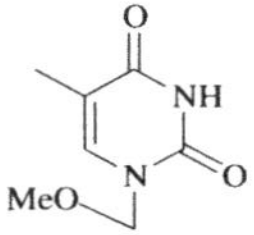 1-(MeOCH2)thymine (54)	399
C_3				
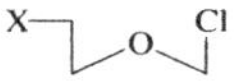 X–CH2CH2–O–CH2–Cl	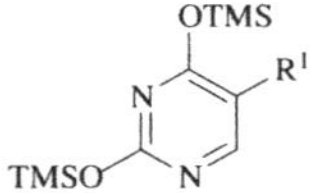 2,4-bis(OTMS)-5-R^1-pyrimidine (OTMS, TMSO)	Bu_4NF, THF, C_6H_6, reflux, 3 h	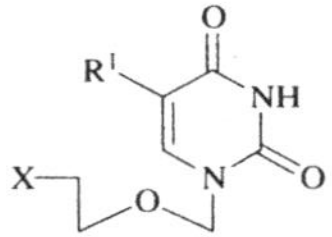 1-(XCH2CH2OCH2)-5-R^1-uracil	1113
X	R^1			
Cl	H		(80-90)	
Cl	Me		(80-90)	
Br	H		(80-90)	
Br	Me		(80-90)	
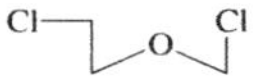 Cl–CH2CH2–O–CH2–Cl	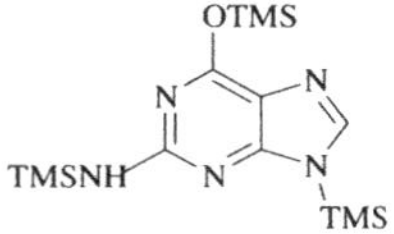 N^2,9-bis(TMS)-6-OTMS-guanine (OTMS, TMSNH, TMS)	Bu_4NF, THF, C_6H_6, reflux, 30 h	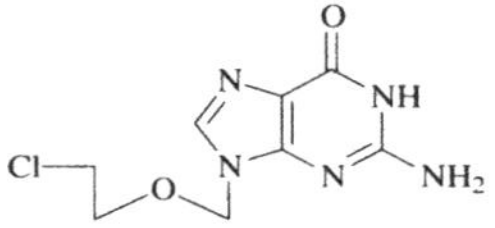 9-(ClCH2CH2OCH2)guanine (90)	1113
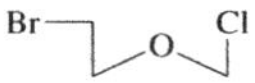 Br–CH2CH2–O–CH2–Cl	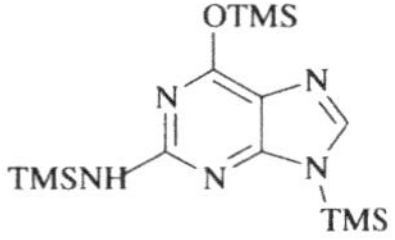 N^2,9-bis(TMS)-6-OTMS-guanine (OTMS, TMSNH, TMS)	Bu_4NF, THF, C_6H_6, reflux, 3 h	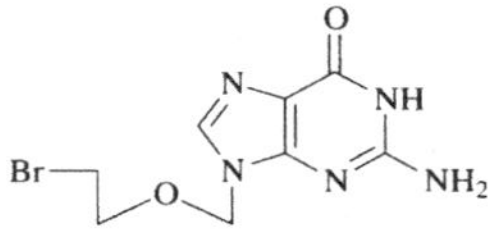 9-(BrCH2CH2OCH2)guanine (95)	1113
	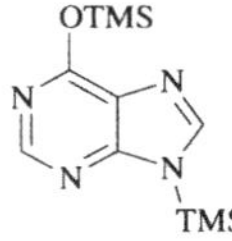 9-TMS-6-OTMS-purine (OTMS, TMS)	Bu_4NF, THF, C_6H_6, reflux, 3 h	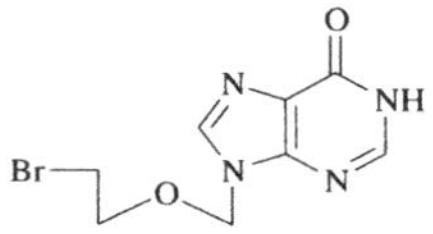 9-(BrCH2CH2OCH2)hypoxanthine (80)	1113

TABLE VIII. REACTIONS OF SILYLATED BASES WITH PROTECTED SUGARS WITH OR WITHOUT CATALYSTS (*Continued*)

Sugar	Base	Conditions	Product(s) and Yield(s) (%)	Refs.
	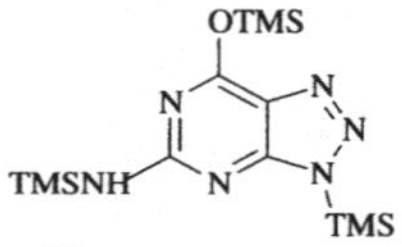 OTMS, TMSNH, N, TMS	Bu_4NF, THF, C_6H_6, reflux, 3 h	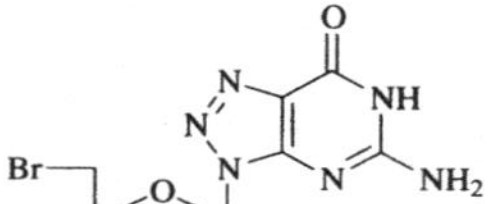 Br, O, N, NH, NH_2 (98)	1113
	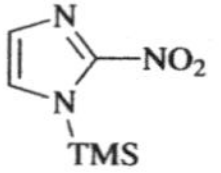 N, NO_2, TMS	Bu_4NF, $Cl(CH_2)_2Cl$ rt, 16 h	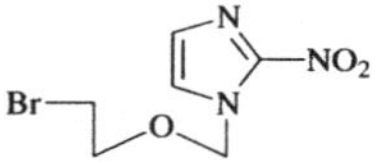 Br, O, N, NO_2 (73)	1114
 EtO Cl	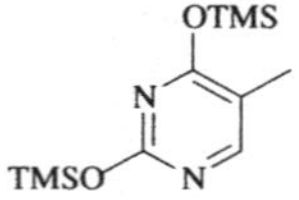 OTMS, TMSO, N	Bu_4NI, CH_2Cl_2, reflux, 2 h	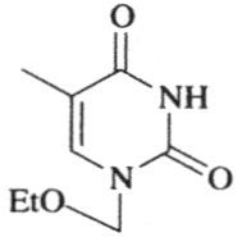 O, NH, N, EtO (62)	399
	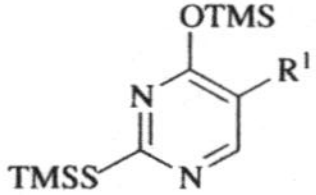 OTMS, R^1, TMSS, N	CsI, MeCN, reflux, 2 h	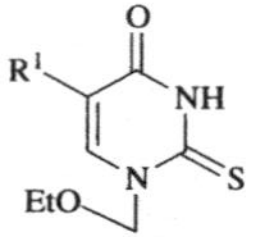 R^1, O, NH, S, N, EtO; R^1: Et (37), *i*-Pr (31), *c*-C_3H_5 (43)	399
C_4, 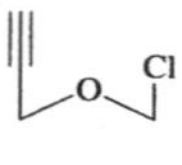 O, Cl	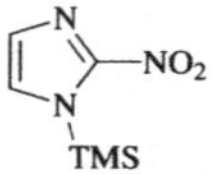 N, NO_2, TMS	Bu_4NF, $Cl(CH_2)_2Cl$ rt, 16 h	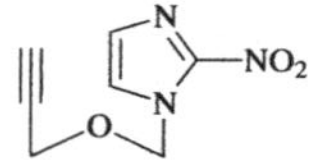 O, N, NO_2 (76)	1114
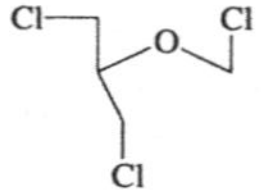 Cl, O, Cl, Cl	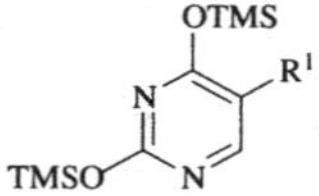 OTMS, R^1, TMSO, N	Bu_4NI, CH_2Cl_2, reflux, 45 min	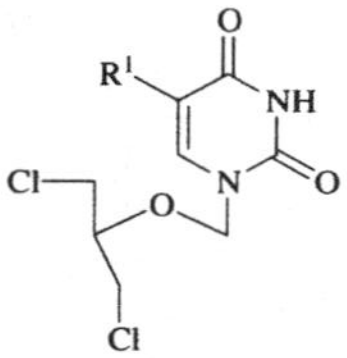 R^1, O, NH, N, Cl, Cl; R^1: H (80), F (88), Br (81), I (67)	786

Reagent	Conditions	Product(s) and Yield(s) (%)	Refs.
OTMS / TMSO (pyrimidine)	1. Bu_4NI, CH_2Cl_2, 24°, 12 h 2. KOAc, DMF	(42)	1115
OTMS / TMSO / F (pyrimidine)	1. H^+ 2. NaI, MeCN, reflux 4 h	(90)	639
OTMS / TMSO (pyrimidine)	CH_2Cl_2, 24°, 3 h	(65)	1116
OTMS / TMSO / F / ^{14}C (pyrimidine)	1. CH_2Cl_2, MS, rt, 1 h 2. NH_3, MeOH, rt, 1 h	(80)	1117
OTMS / TMSO / R^1 (pyrimidine)	C_6H_6, –20°, 4 h	R^1: Br (51); Me (66)	1118
OTMS / TMSO / R^1 (triazine)	C_6H_6, –20°, 4 h	R^1: H (—); Br (—); Me (—)	1118

TABLE VIII. REACTIONS OF SILYLATED BASES WITH PROTECTED SUGARS WITH OR WITHOUT CATALYSTS (*Continued*)

Sugar	Base	Conditions	Product(s) and Yield(s) (%)	Refs.
n-PrO–CH₂Cl	OTMS; N; TMSO; N	Bu_4NI, CH_2Cl_2, reflux, 2 h	O; NH; N; O; n-PrO (63)	399
i-PrO–CH₂Cl	OTMS; Et; N; TMSS; N	CsI, MeCN, reflux, 2 h	O; Et; NH; N; S; i-PrO (25)	399
C_5				
AcO–(CH₂)₂–O–CH₂Br	OTMS; N; R^2; N; TMSO; N; N; R^1	Bu_4NI, MeCN, rt, 4-5 h	O; HN; N; R^2; AcO; O; N; N; R^1; O	1119
	R^1 = H, R^2 = H		(53)	
	R^1 = Ph, R^2 = H		(66)	
	R^1 = p-ClC_6H_4, R^2 = H		(52)	
	R^1 = Me, R^2 = Me		(62)	
	R^1 = Ph, R^2 = Ph		(60)	
	R^1 = 2-pyridyl, R^2 = 2-pyridyl		(55)	
	OTMS; R^1; N; TMSO; N	MeCN, rt, 2 d	O; R^1; NH; AcO; N; O; O — R^1 = Bn (44)	655
			R^1 = OTf (73)	1120
	OTMS; N; TMSO; N; OTMS	MeCN, reflux, 6 h	O; NH; AcO; O; N; O; O (73)	1121

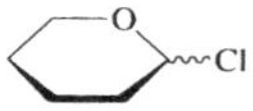

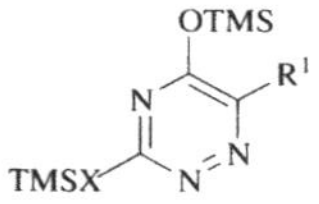

Halide	Silylated base	Conditions	Product(s) and Yield(s) (%)	Refs.
	TMSX-triazine (OTMS, R^1)	MeCN, rt	AcO–CH$_2$CH$_2$O–CH$_2$-triazinone	1122
	X / R^1: O / Me		(62)	
	S / Me		(58)	
	O / Ph		(87)	
	S / Ph		(77)	
	O / Bn		(75)	
	S / Bn		(61)	
	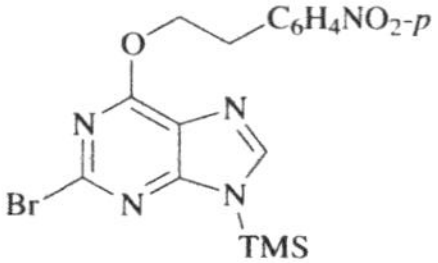 2-Br-6-O(CH$_2$)$_2$C$_6$H$_4$NO$_2$-*p*-9-TMS-purine	MeCN, reflux, 1 h	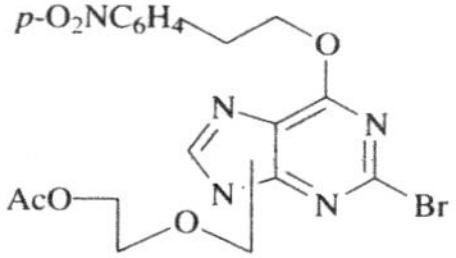 (76) N^7:N^9 = 1:99	143
2-chlorotetrahydropyran (Cl)	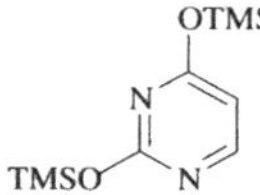 2,4-bis(TMSO)pyrimidine	CH_2Cl_2, rt, 3.5 h	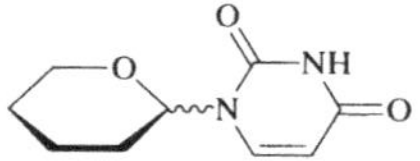 (65)	1116
	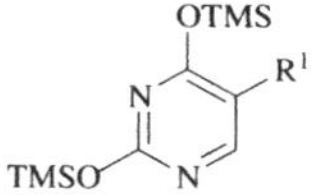 2,4-bis(TMSO)-5-R^1-pyrimidine	C_6H_6, –20°, 4 h	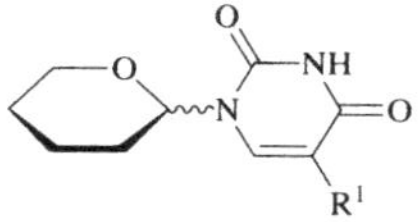 R^1: H (—), F (—), Cl (—), I (—), Me (—)	1118
	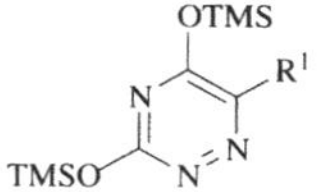 3,5-bis(TMSO)-6-R^1-1,2,4-triazine	C_6H_6, –20°, 4 h	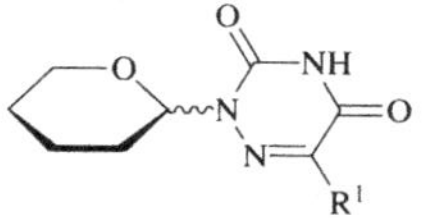 R^1: H (—), Br (—), Me (—)	1118

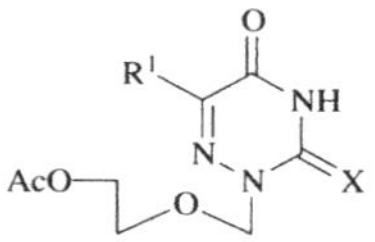

TABLE VIII. REACTIONS OF SILYLATED BASES WITH PROTECTED SUGARS WITH OR WITHOUT CATALYSTS (*Continued*)

Sugar	Base	Conditions	Product(s) and Yield(s) (%)	Refs.
OMe	OTMS, F, N, TMSO, N	NaI, MeCN	O, NH, N, O, F; 2.5 h, 160° (77)	1123
			5 h, 60-80° (84)	1124
 BuO, Cl	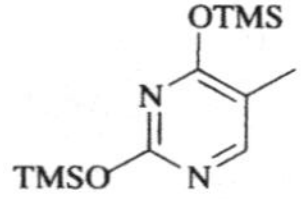 OTMS, N, TMSO, N	Bu_4NI, CH_2Cl_2, reflux, 2 h	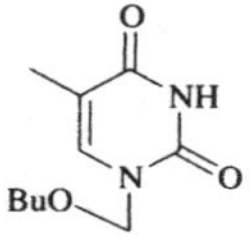 O, NH, N, O, BuO (49)	399
C_6				
 OAc	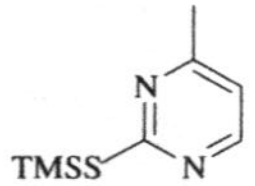 N, TMSS, N	CsCl, MeCN, 25°, 1.5 h	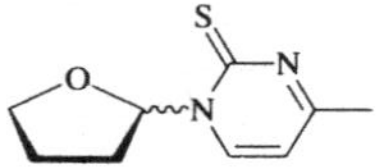 S, N, N (91)	401
	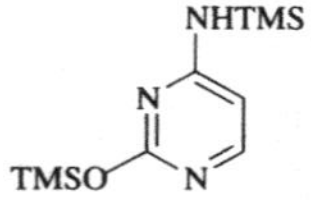 NHTMS, N, TMSO, N	CsCl, MeCN, 25°, 25 h	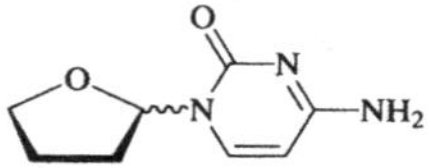 O, N, N, NH_2 (83)	401
	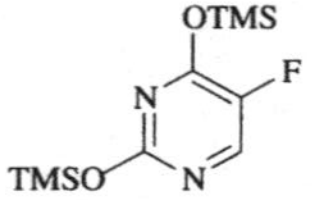 OTMS, F, N, TMSO, N	NaI (1 eq), MeCN, rt, 9 h	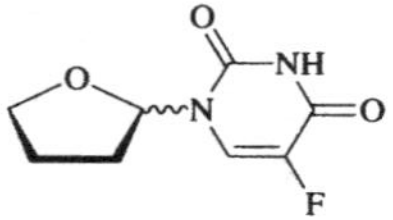 O, NH, N, O, F **I** (81)	528
		NaI, MeCN, 50-60°, 2 h	**I** (95)	394

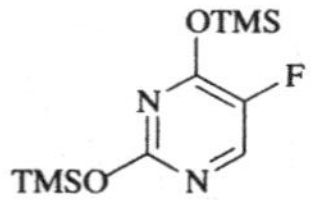

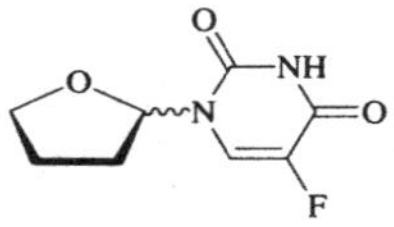

Reactant	Conditions	Product(s) and Yield(s) (%)	Refs.
OTMS, F, N, TMSO, N (2,4-bis(trimethylsilyloxy)-5-fluoropyrimidine)	NaI (1 eq), MeCN, 60°, 0.5 h	O, N, O, N, O, F (N¹,N³-bis(tetrahydrofuranyl)-5-fluorouracil) (78)	528
OTMS, R¹, N, TMSX, N	CsCl, MeCN, 25°, 3 h	X, NH, O, N, O, R¹	401
X = O, R¹ = H		(74) + N^1,N^3-bisproduct	
X = O, R¹ = F		(87)	
X = O, R¹ = Br		(88)	
X = O, R¹ = NH_2		(50)	
X = O, R¹ = Me	50-55°	(73)	
X = S, R¹ = Me	50-55°	(96)	
R¹, N, N, N, N, TMS	CsCl, MeCN	O, N, N, N, N, R²	401
R¹ = Cl	25°, 3 h	R² = Cl (96)	
R¹ = SMe	25°, 1.5 h	R² = SMe (98)	
R¹ = NHTMS	25°, 5 h	R² = NH_2 (92)	
OTMS, N, N, R¹, N, N, TMS	CsCl, MeCN, 25°, 4 h	O, N, N, N, O, NH, R²	401
R¹ = H		R² = H (97)	
R¹ = TMSNH		R² = NH_2 (50)	

TABLE VIII. REACTIONS OF SILYLATED BASES WITH PROTECTED SUGARS WITH OR WITHOUT CATALYSTS (*Continued*)

Sugar	Base	Conditions	Product(s) and Yield(s) (%)	Refs.
	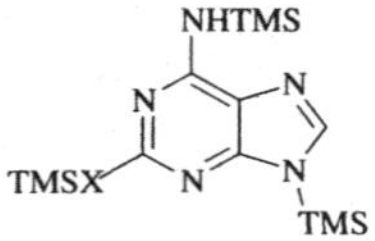 NHTMS, TMSX, N, TMS	CsCl, MeCN, 25°, 4 h	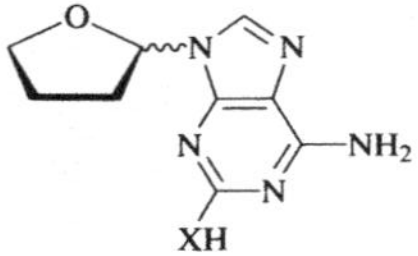 O, N, NH$_2$, XH; X: S (70), NH (98)	401
 TMS, O, Cl	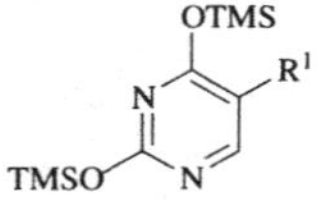 OTMS, R^1, TMSO, N; R^1	Bu_4NI	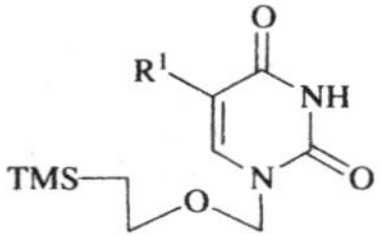 O, R^1, NH, TMS, O, N, O	
	F	MeCN, 20°, 10 min	(43)	1125
	Me	CH_2Cl_2, reflux, 2 h	(87)	399
C_7 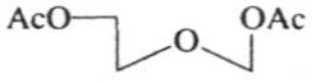 AcO, O, OAc	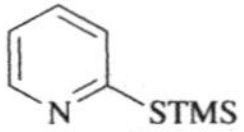 N, STMS	CsI, MeCN, reflux, 2 h	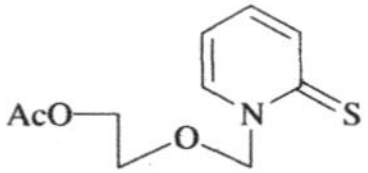 AcO, O, N, S (76)	400
	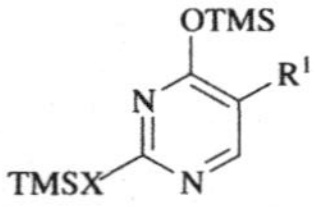 OTMS, R^1, TMSX, N	CsI, MeCN, reflux	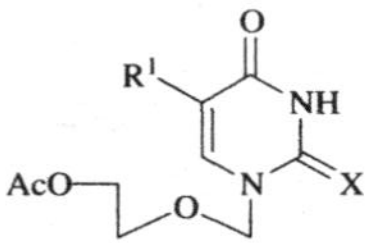 O, R^1, NH, AcO, O, N, X; X / R^1: O H 3 h (75); S H 1 h (96); O F 4 h (51); S Me 12 h (80)	400
	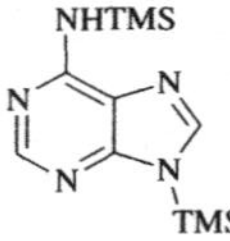 NHTMS, N, TMS	CsI, MeCN, reflux, 2 h	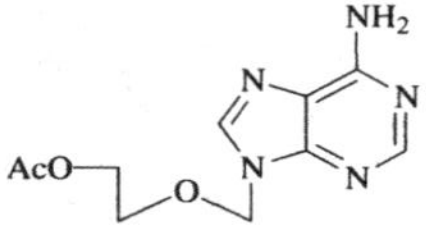 NH_2, N, AcO, O (70)	400

Reactant	Silylated base	Conditions	Product(s) and Yield(s) (%)	Refs.
	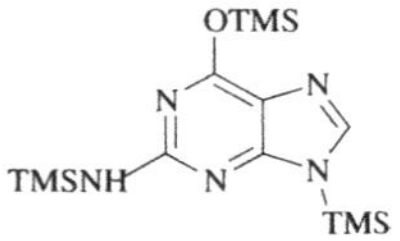	CsI, MeCN, reflux, 3 h	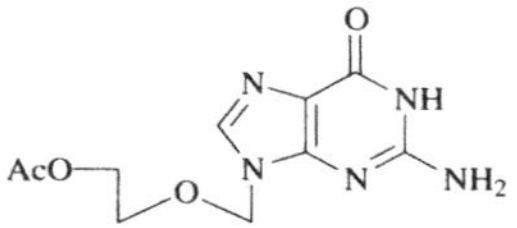 (72)	400
		CsI, MeCN, reflux, 12 h	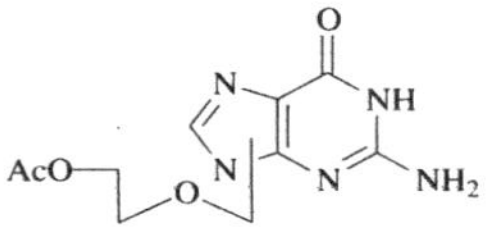 (47) N^9:N^7 = 10:1	400
	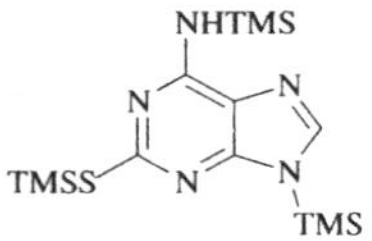	CsI, MeCN, reflux, 2 h	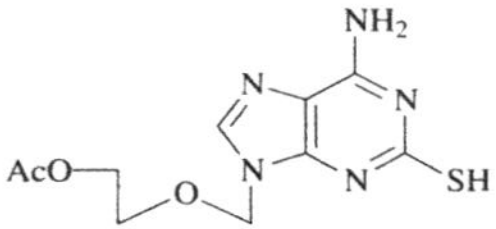 (92)	400
		CsI, MeCN, reflux, 12 h	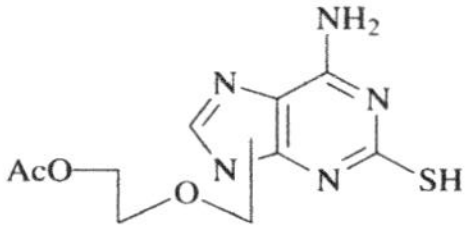 (42) N^9:N^7 = 7:1	400
	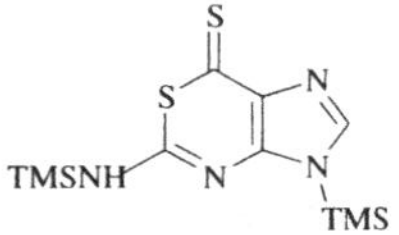	CsI, MeCN, reflux, 4 h	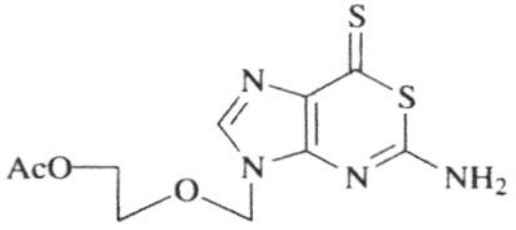 (18) + N^7-isomer (15)	1126
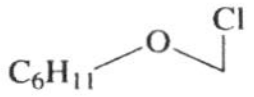	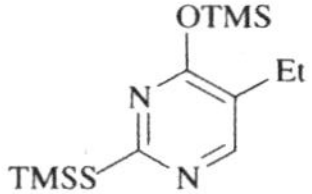	CsI, MeCN, reflux, 2 h	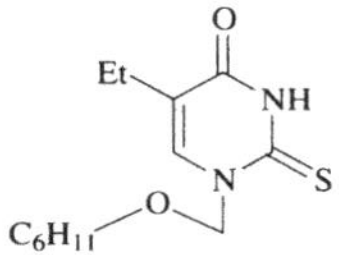 (34)	399
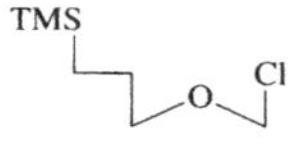	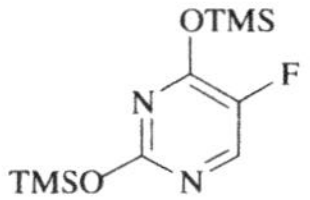	NaI, MeCN, 20°, 10 min	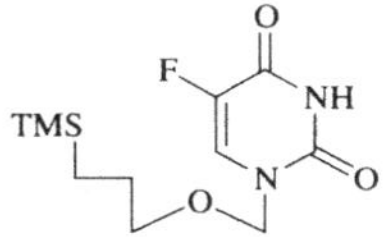 (62)	1125

TABLE VIII. REACTIONS OF SILYLATED BASES WITH PROTECTED SUGARS WITH OR WITHOUT CATALYSTS (*Continued*)

Sugar	Base	Conditions	Product(s) and Yield(s) (%)	Refs.
C_8				
Bn, O, Cl	OTMS, R[1], N, N, TMSX	Reflux, 2 h	O, R[1], NH, N, X, O, Bn	399
	X / R[1]: O / Me	Bu_4NI, CH_2Cl_2	(86)	
	S / Et	CsI, MeCN	(36)	
	S / *i*-Pr	CsI, MeCN	(31)	
Br, O, AcO, OAc	Cl, N, N, N, N, TMSNH, TMS	PhMe, reflux, 18 h	Cl, N, N, N, N, NH_2, O, AcO, OAc (52)	1127
HO, O, $NHNH_2$, O, O	NC, CN, TMS, S, S, TMS	EtOH, reflux, 10 h	NC, S, TMS, N, N, NH_2, HO, O, O, O (34)	447
R[1], O, Cl	OTMS, Et, N, N, TMSS	CsI, MeCN, reflux, 2 h	O, Et, NH, N, S, R[1], O; R[1]: *p*-ClC_6H_4 (16), Ph (32), C_6H_{11} (36)	399

C_9

		$Hg(CN)_2$, C_6H_6, reflux, 30 min	(63)	650
		1. CH_2Cl_2, rt, 48 h 2. HCl, MeOH	(33)	1128
		CH_2Cl_2, rt, 7 d	R^1: Br (61); I (—); Me (—)	1128
		CH_2Cl_2, rt	(65) $\alpha:\beta = 1:6$	219
		MeCN, rt	(22) + α-anomer (20)	219

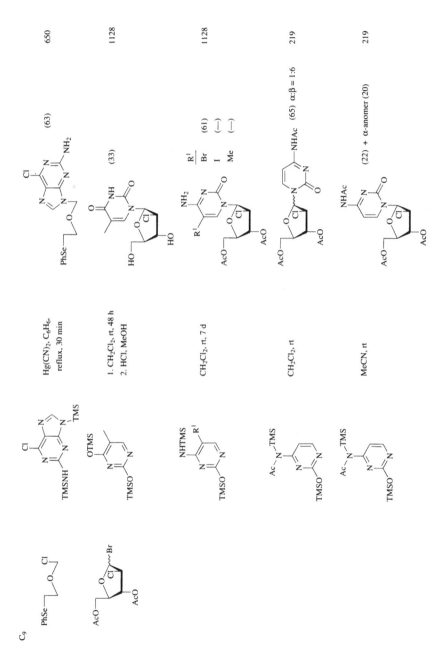

TABLE VIII. REACTIONS OF SILYLATED BASES WITH PROTECTED SUGARS WITH OR WITHOUT CATALYSTS (*Continued*)

Sugar	Base	Conditions	Product(s) and Yield(s) (%)	Refs.
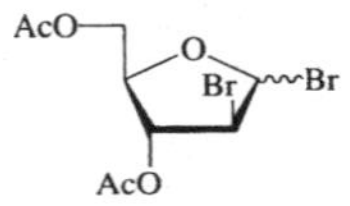 AcO, AcO, O, Br, Br	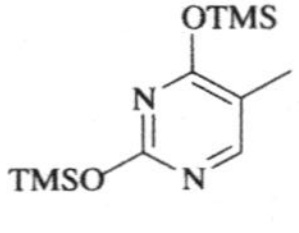 OTMS, N, TMSO, N	1. CH_2Cl_2, rt, 48 h 2. HCl, MeOH	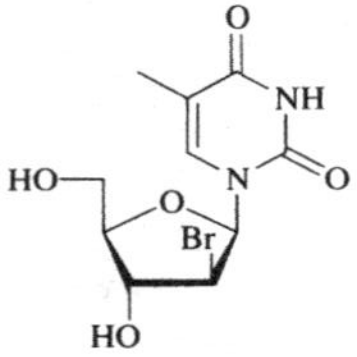 O, NH, HO, O, N, O, Br, HO (—)	1128
	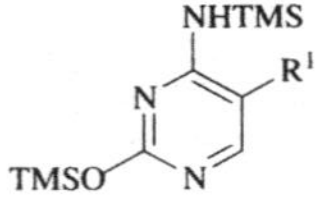 NHTMS, R^1, N, TMSO, N	CH_2Cl_2, rt, 48 h	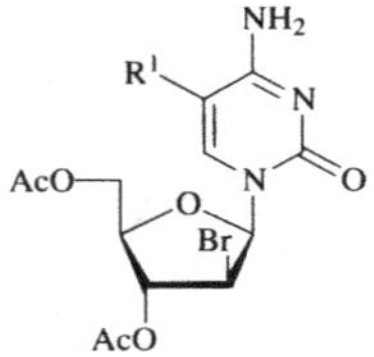 NH_2, R^1, N, AcO, N, O, O, Br, AcO R^1: H (—); Me (—); Br (28); I (—)	1128
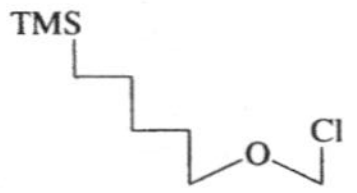 TMS, O, Cl	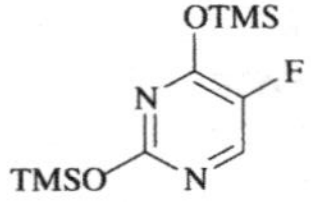 OTMS, F, N, TMSO, N	NaI, MeCN, 20°, 10 min	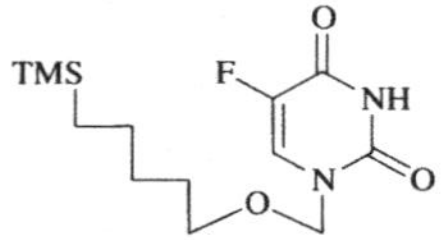 O, TMS, F, NH, N, O, O (73)	1125
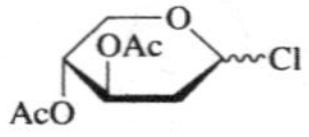 O, OAc, Cl, AcO	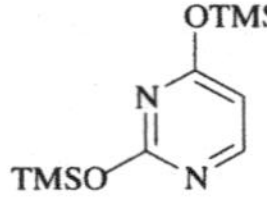 OTMS, N, TMSO, N	HgO, $HgBr_2$, C_6H_6, 80-100°, 5 h	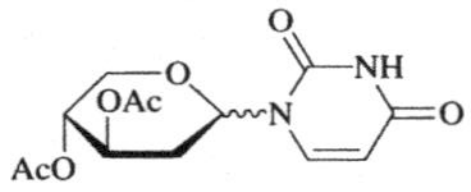 O, NH, O, OAc, N, O, AcO (88)	927
C_{10}				
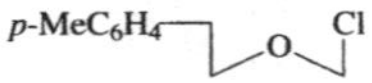 p-MeC_6H_4, O, Cl	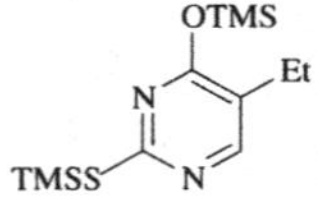 OTMS, Et, N, TMSS, N	CsI, MeCN, reflux, 2 h	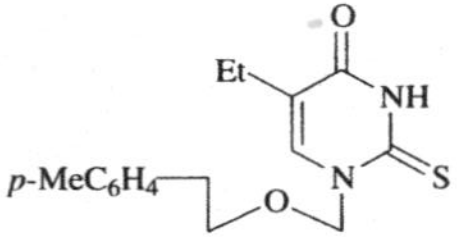 O, Et, NH, N, S, p-MeC_6H_4, O (23)	399

Reactant	Silylated base	Conditions	Product(s) and Yield(s) (%)	Refs.
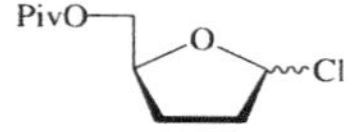	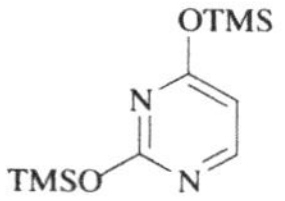	Py, $CHCl_3$, rt, 12 h	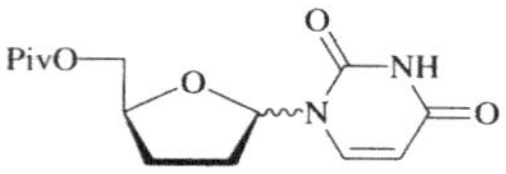 **I** (72) $\alpha:\beta = 34:66$	276
		$CHCl_3$, rt, 12 h	**I** (33) $\alpha:\beta = 38:62$	276
	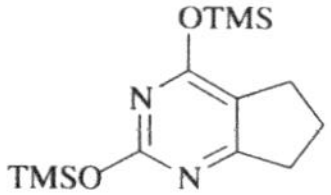	CsI, MeCN, reflux, 2 h	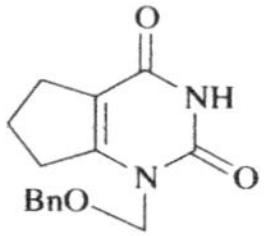 (11)	656
	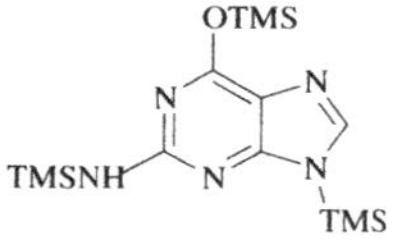	Bu_4NF, THF, C_6H_6, reflux, 3 h	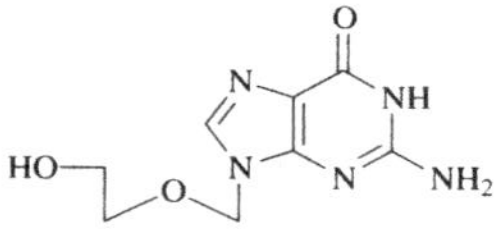 (92)	1113
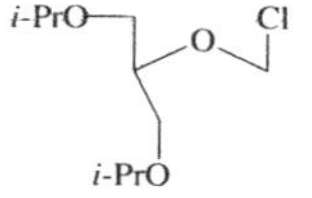	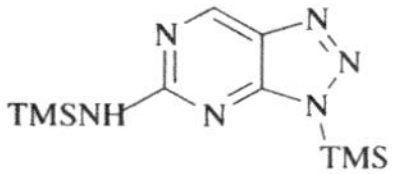	Et_3N, PhMe, 90°, 18 h	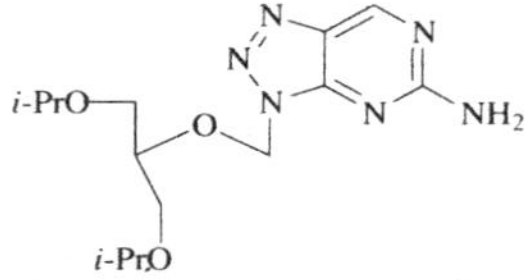 (40)	1129
C_{11}		MeCN, rt, 48 h	see below	1130
		$Hg(CN)_2$, C_6H_6, reflux 1 h	(50)	1131

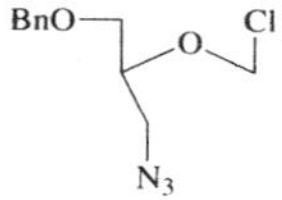

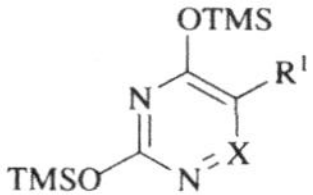

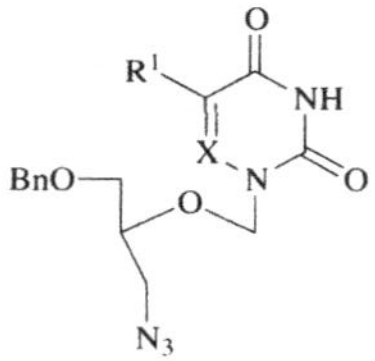

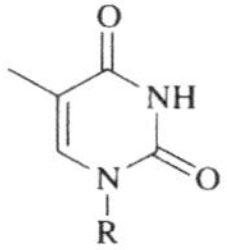

Yields for the MeCN, rt, 48 h entry (ref. 1130):

R^1	X	Yield (%)
H	CH	(55)
Me	CH	(58)
H	N	(51)
Me	N	(52)
Ph	N	(66)
Bn	N	(67)

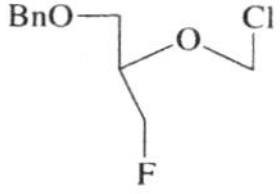

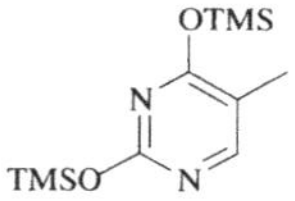

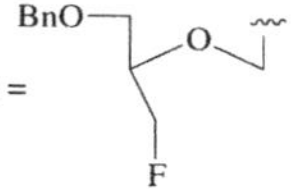

TABLE VIII. REACTIONS OF SILYLATED BASES WITH PROTECTED SUGARS WITH OR WITHOUT CATALYSTS (*Continued*)

Sugar	Base	Conditions	Product(s) and Yield(s) (%)	Refs.
	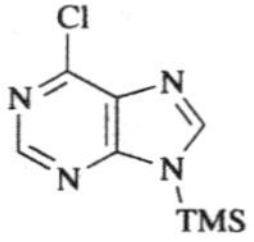 Cl, N, N, N, N, TMS	$Hg(CN)_2$, C_6H_6, reflux, 1 h	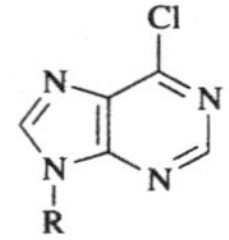 Cl, N, N, N, N, R (20) + N^7-isomer (12)	1131
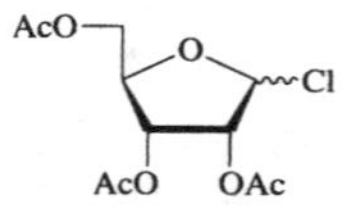 AcO, O, Cl, AcO, OAc	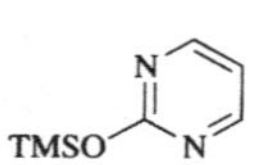 TMSO, N, N	MeCN, reflux, 4 h	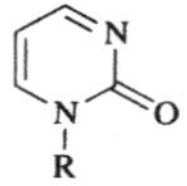 N, O, N, R **I** (51) R 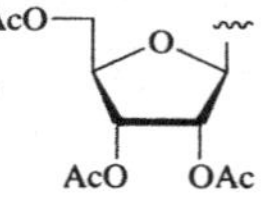 = AcO, O, AcO, OAc	691
		MeCN, reflux, 8 h	**I** (79)	691
	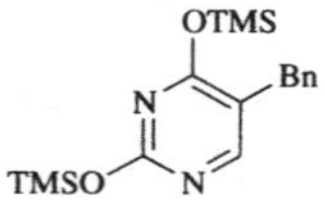 OTMS, Bn, N, N, TMSO	1. $Hg(CN)_2$, PhMe, 60-70°, 6 h 2. rt, 12 h	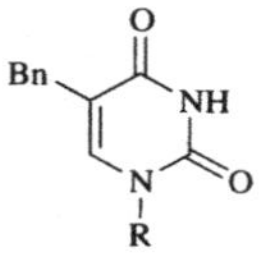 O, Bn, NH, N, O, R (43)	1132
	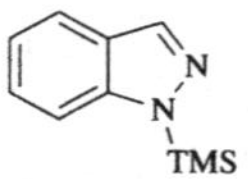 N, N, TMS	MeCN, rt	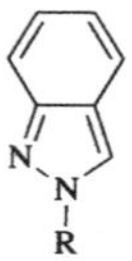 N, N, R **I** (82)	728
		MeCN, reflux	**I** (46)	728
	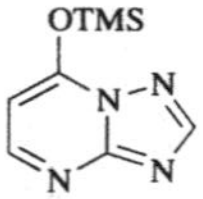 OTMS, N, N, N, N	1. MeCN, rt, 15 d 2. NH_3, MeOH, rt, 1 h	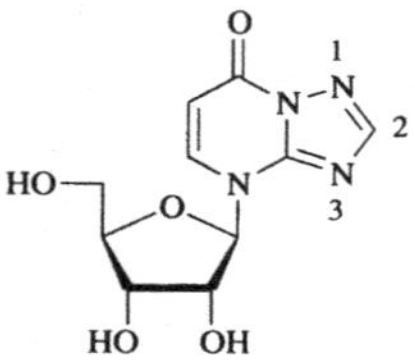 O, N, N, N, N, 1, 2, 3, HO, O, HO, OH (23) + N^3-isomer (15)	257

Conditions	Product(s) and Yield(s) (%)	Refs.
1. $Hg(CN)_2$, PhMe, 60-70°, 6 h 2. rt, 12 h	(37)	1132
MeCN, rt, 3 d	(48) + N^3-isomer (6)	707
MeCN, 25°, 3 d	(51) + N^2-isomer (46)	259
1. $Cl(CH_2)_2Cl$, reflux, 6 h; rt, 12 h 2. NH_3, MeOH, 0°, 4 d	(72)	714
MeCN, rt, 3 d	(16)	89
MeCN, rt, 16 h	(50)	1133
1. MeCN, rt, 3 d 2. NH_3, MeOH, rt, 12 h	(17)	89

TABLE VIII. REACTIONS OF SILYLATED BASES WITH PROTECTED SUGARS WITH OR WITHOUT CATALYSTS (*Continued*)

Sugar	Base	Conditions	Product(s) and Yield(s) (%)	Refs.
	OTMS; H_2N; N; TMS	1. MeCN, rt, 3 d 2. NH_3, MeOH	O; NH; NH_2; N; R (55) N^7:N^9 = 1:8	705
	OTMS; R^1; N; S	1. $HgBr_2$, HgO, C_6H_6, reflux, 20 h 2. NH_3, MeOH, rt, 20 h	S; R^1; N; O; R R^1: H (30); Me (40)	888
AcO; O; OAc; Br; OAc	Cl; N; AcN; TMS; TMS	$Hg(CN)_2$, C_6H_6, reflux, 17 h	Cl; N; NHAc; AcO; O; OAc; OAc (62)	1134
AcO; O; OAc; Br; OAc	OTMS; N; TMSO	$Hg(OAc)_2$, HgO, or $AgClO_4$	AcO; O; OAc; OAc; N; NH; O (90)	77
	Ac; N; TMS; N; TMSO	$Hg(OAc)_2$, HgO, or $AgClO_4$	AcO; O; OAc; OAc; N; O; NHAc (53)	77
	OEt; N; TMSO	$Hg(OAc)_2$, HgO, or $AgClO_4$	AcO; O; OAc; OAc; N; O; OEt (53)	77

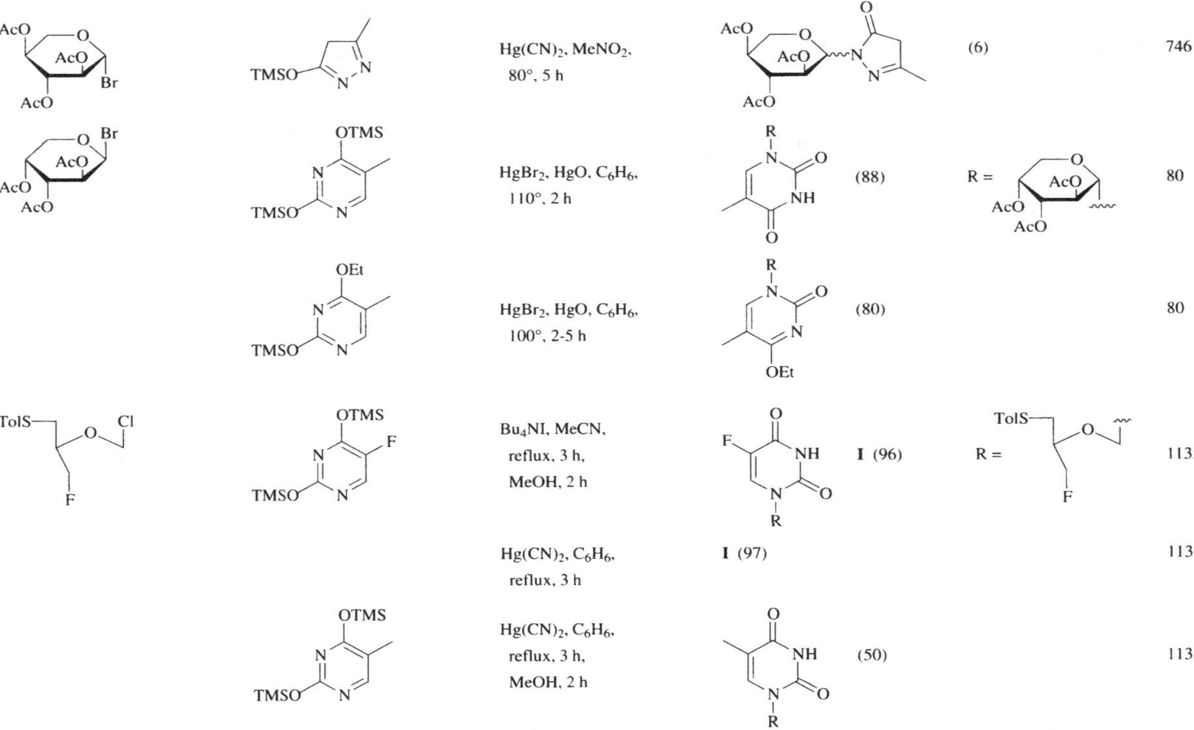

Glycosyl donor	Silylated base	Conditions	Product(s) and Yield(s) (%)		Refs.
		$Hg(CN)_2$, $MeNO_2$, 80°, 5 h	(6)		746
		$HgBr_2$, HgO, C_6H_6, 110°, 2 h	(88)	R =	80
		$HgBr_2$, HgO, C_6H_6, 100°, 2-5 h	(80)		80
		Bu_4NI, MeCN, reflux, 3 h, MeOH, 2 h	**I** (96)	R =	1135
		$Hg(CN)_2$, C_6H_6, reflux, 3 h	**I** (97)		1135
		$Hg(CN)_2$, C_6H_6, reflux, 3 h, MeOH, 2 h	(50)		1135

TABLE VIII. REACTIONS OF SILYLATED BASES WITH PROTECTED SUGARS WITH OR WITHOUT CATALYSTS (*Continued*)

Sugar	Base	Conditions	Product(s) and Yield(s) (%)	Refs.
	6-Cl-9-TMS-purine	Bu_4NI, C_6H_6, reflux, 3 h	6-Cl-9-R-purine (60)	1135
	6-NHTMS-9-TMS-purine	$Hg(CN)_2$, MeCN, reflux, 3 h, MeOH, 2 h	6-NH_2-9-R-purine **I** (30)	1135
		Bu_4NI, C_6H_6, reflux, 3 h	**I** (20)	1131
C_{12} BzO-(2-F-furanosyl bromide)	2-TMSO-pyrimidine (R^1, R^2)	$Cl(CH_2)_2Cl$, reflux	BzO-(2-F-furanosyl)-pyrimidin-2-one (R^2, R^3)	
	R^1 = H, R^2 = H	12 h	R^3 = H (46)	389
	R^1 = NHTMS, R^2 = H	48 h	R^3 = NH_2 (62) + α-anomer (6)	390
	R^1 = NHTMS, R^2 = F	4 h	R^3 = NH_2 (66)	389
	2-TMSO-4-OTMS-pyrimidine (R^1)	$Cl(CH_2)_2Cl$, reflux, 48 h	BzO-(2-F-furanosyl)-uracil (R^1): H (49), F (76), Cl (56), Me (50)	389

Glycosyl donor	Base	Conditions	Product(s) and Yield(s) (%)	Refs.
	6-chloro-9-TMS-purine	MeCN, MS, rt, 14 h	(41) + N^9-α-isomer (12); R = BzO– (2-fluoro sugar)	390
	2,6-dichloro-9-TMS-purine	$Hg(CN)_2$, MeCN, MS, reflux, 20 h	(32) + α-anomer (23)	1136
	TMSNH (2-amino-6-chloro-9-TMS-purine)	$Hg(CN)_2$, MeCN, reflux, 3 h	(31) + N^7–isomer (7) + N^9-α-isomer (7)	1136
R^1 (2,3-dideoxy sugar chloride)	2,4-bis(OTMS)pyrimidine (TMSO, OTMS)	Py, $CHCl_3$, rt, 12 h	R^1 (uracil nucleoside)	
R^1 = p-$ClC_6H_4CO_2$			(85) α:β = 34:66	276
R^1 = p-$O_2NC_6H_4CO_2$			(78) α:β = 39:61	276
BzO (2,3-dideoxy sugar chloride)	TMSO, NHTMS (silylated cytosine)	Py, $CHCl_3$, rt, 12 h	(51) + α-anomer (25); R = BzO– (2,3-dideoxy sugar)	276
	OTMS (silylated triazolopyrimidinone)	Py, $CHCl_3$, rt, 16 h	BzO– nucleoside (51) α:β = 44:56	957

TABLE VIII. REACTIONS OF SILYLATED BASES WITH PROTECTED SUGARS WITH OR WITHOUT CATALYSTS (*Continued*)

Sugar	Base	Conditions	Product(s) and Yield(s) (%)	Refs.
	OTMS, R[1], TMSO, N, N	Py, $CHCl_3$, rt, 12 h	O, R^1, NH, O, N, R; R^1 / α-anomer: H (53) (24)	276, 1137
			Me (52) (25)	276
	OTMS, TMSO, N, N	$CHCl_3$, rt, 12 h	BzO, O, N, O, NH, O **I**	276
		Py	**I** (77) α:β = 43:57	
		—	**I** (62) α:β = 42:58	
		$PhNMe_2$	**I** (73) α:β = 38:62	
		2-Picoline	**I** (70) α:β = 31:69	
		2,6-Lutidine	**I** (65) α:β = 32:68	
		Et_3N	**I** (58) α:β = 30:70	
		2,4,6-Collidine	**I** (49) α:β = 35:65	
		DMAP	**I** (37) α:β = 31:69	
		DBU	**I** (36) α:β = 33:67	
S, OAc, O, B, O, Ph	Cl, N, N, R^1, N, N, TMS	TsOH, $MeNO_2$, rt, 12 h	Cl, N, N, N, N, R, R^1; R^1: H (51), Cl (54); R = S, O, B, O, Ph	687
	O, TMS, Me, N, N, N, O, N, Me	TsOH, $MeNO_2$, rt, 12 h	Me, N, N, O, N, N, Me, R, O (31)	687

Glycosyl donor	Silylated base	Conditions	Product(s) and Yield(s) (%)	Refs.
BnO, O, Cl	OTMS, TMSO, N	1. MeCN, rt 2. H_2, cat.	HO, O, N, NH, O (82)	1122
	NHTMS, TMSO, N	Bu_4NI, $Cl(CH_2)_2Cl$, reflux, 4.5 h	NH_2, N, O, R (48) R = BnO, O	1138
	Cl, N, TMSNH, TMS	$Hg(CN)_2$, C_6H_6, reflux, 3 h	Cl, N, NH_2, R (43)	1138
PivO, O, OAc	NHTMS, TMSO, N	TMSI, MeCN, 0°, 3 h	PivO, O, N, NH_2 (76) α:β = 1:1	139a
C_{13} *p*-$MeOC_6H_4CO_2$, O, Cl	OTMS, TMSO, N	Py, $CHCl_3$, rt, 12 h	*p*-$MeOC_6H_4CO_2$, O, N, NH, O **I** (77) α:β = 31:69	276
		$CHCl_3$, rt, 12 h	**I** (76) α:β = 40:60	276
MeO_2C, O, OAc, AcO, Br, OAc	OTMS, R^1, TMSO, N	$Hg(CN)_2$, $HgBr_2$, C_6H_6, reflux, 3.5 h	R^1, O, NH, MeO_2C, N, OAc, AcO, OAc	767

R^1		bis(riboside)
H	(23)	(4)
F	(38)	(11)
Cl	(57)	(—)
Br	(67)	(11)
I	(72)	(—)

TABLE VIII. REACTIONS OF SILYLATED BASES WITH PROTECTED SUGARS WITH OR WITHOUT CATALYSTS (*Continued*)

Sugar	Base	Conditions	Product(s) and Yield(s) (%)	Refs.
C_{14} 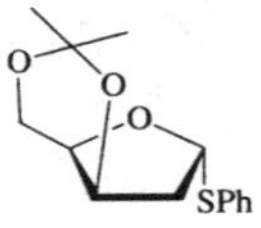	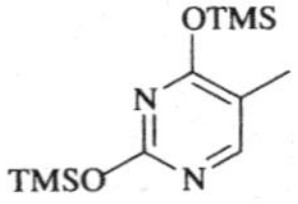	NBS, CH_2Cl_2, MS, rt, 20 min	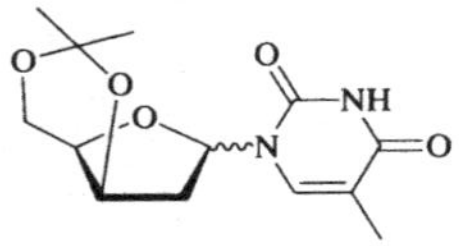 (85) α:β = 1:30	405
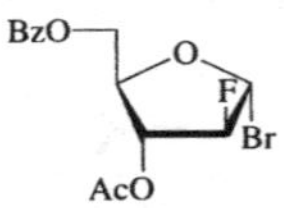	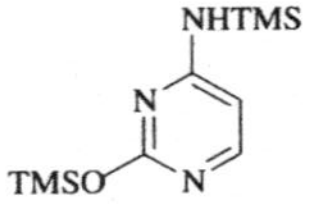	$CHCl_3$, reflux, 20 h	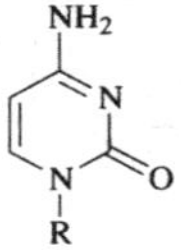 (61) + α-anomer (21); R 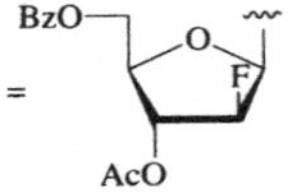 =	386
	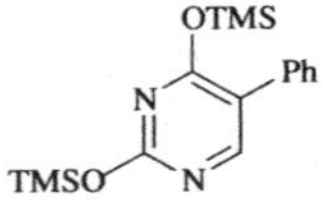	CH_2Cl_2, rt, 5 d	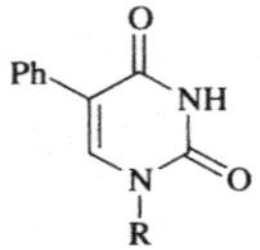 (40)	1139
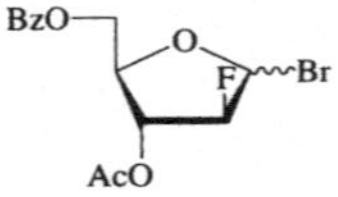	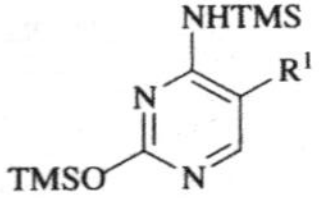	CH_2Cl_2, rt, 7 d	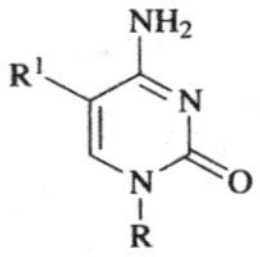 R^1: Me (57); I (32) + α-anomer (2)	1128
	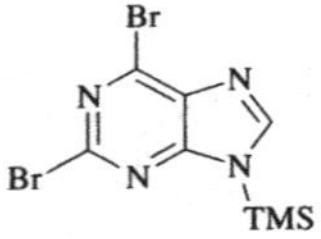	$Cl(CH_2)_2Cl$, MS, 100°, 32 h	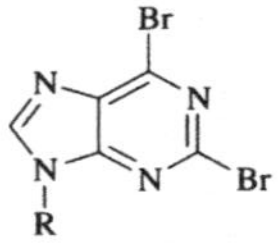 (32)	1140

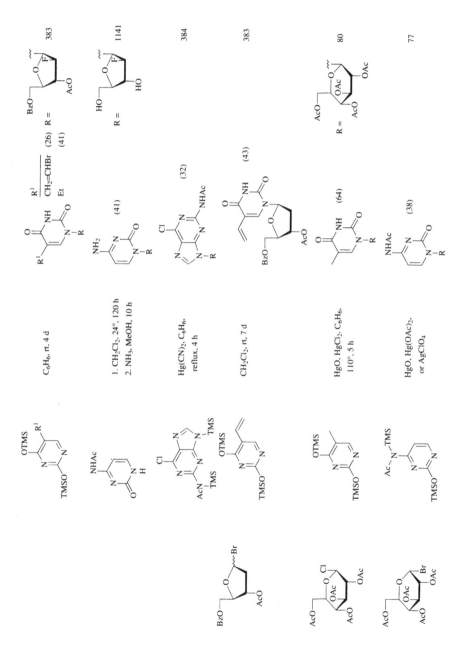

	C_6H_6, rt, 4 d	R^1: CH_2=CHBr (26); Et (41)	383
	1. CH_2Cl_2, 24°, 120 h 2. NH_3, MeOH, 10 h	(41)	1141
	$Hg(CN)_2$, C_6H_6, reflux, 4 h	(32)	384
	CH_2Cl_2, rt, 7 d	(43)	383
	HgO, $HgCl_2$, C_6H_6, 110°, 5 h	(64)	80
	HgO, $Hg(OAc)_2$, or $AgClO_4$	(38)	77

TABLE VIII. REACTIONS OF SILYLATED BASES WITH PROTECTED SUGARS WITH OR WITHOUT CATALYSTS (*Continued*)

Sugar	Base	Conditions	Product(s) and Yield(s) (%)	Refs.
	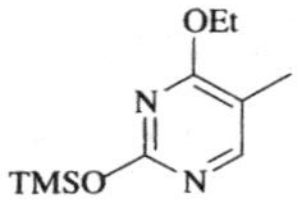 TMSO, N, OEt, N	HgO, $HgBr_2$, C_6H_6, 100°, 2-5 h	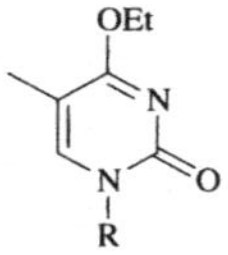 OEt, N, O, N–R (81) R = AcO, OAc, O, AcO, OAc 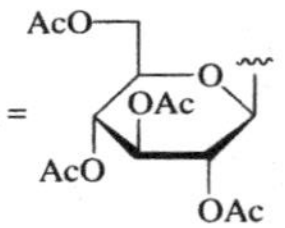	80
	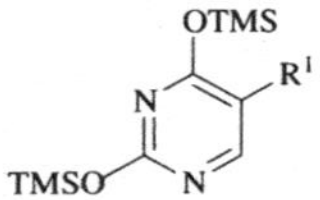 TMSO, N, OTMS, R[1], N		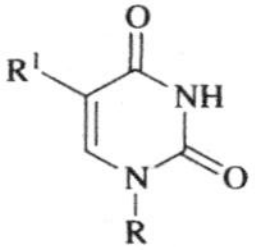 R[1], O, NH, O, N–R	
	R[1]			
	H	$HgBr_2$, C_6H_6, rt	(93)	1142
	H	HgO, $Hg(OAc)_2$, or $AgClO_4$	(44)	77
	Me	HgO, $HgCl_2$, C_6H_6, 110°, 5 h	(86)	80
	Me	HgO, $HgCl_2$, C_6H_6, 50°, 4 h	(72)	80
	Me	HgO, $Hg(OAc)_2$, or $AgClO_4$	(48)	77
	Me	C_6H_6, 100°, 20 h	(27)	80
	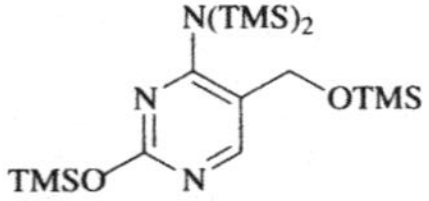 TMSO, N, $N(TMS)_2$, OTMS, N	HgO, $Hg(OAc)_2$, or $AgClO_4$	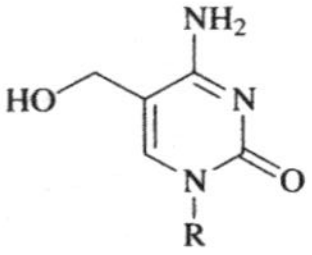 HO, NH_2, N, O, N–R (79)	77
	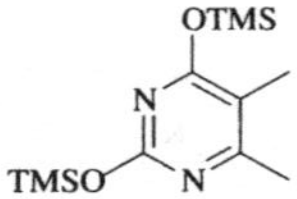 TMSO, N, OTMS, N	HgO, $HgBr_2$, reflux	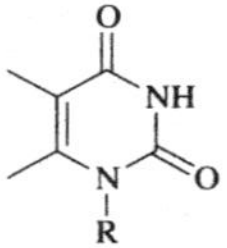 O, NH, O, N–R (90)	1143

	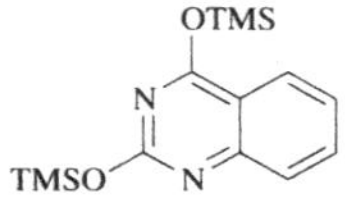 OTMS, N, TMSO, N	HgO, $HgBr_2$, reflux	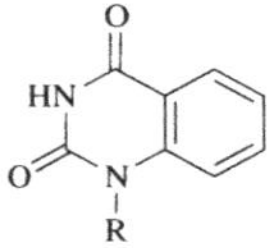 O, HN, O, N–R (65)	1143
	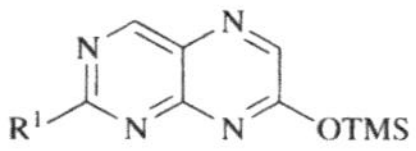 N, N, R^1, N, N, OTMS	HgO, $HgBr_2$, C_6H_6, reflux	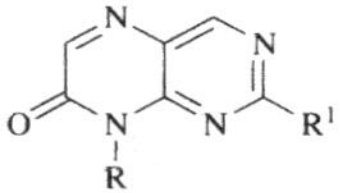 N, N, O, N–R, N, R^1; R^1: H 5 h (8); NMe_2 4 h (66)	1144
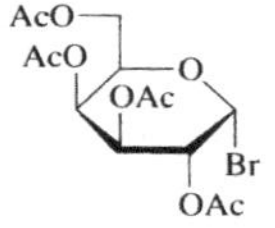 AcO, AcO, O, OAc, Br, OAc	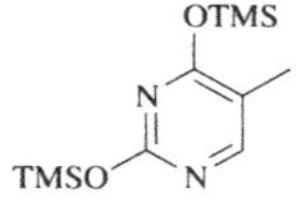 OTMS, N, TMSO, N	HgO, $HgBr_2$, C_6H_6, 100°, 2 h	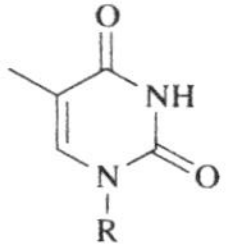 O, NH, O, N–R (45); R = AcO, AcO, O, OAc, OAc 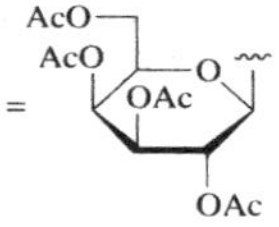	80
	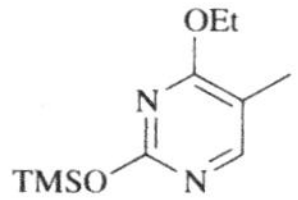 OEt, N, TMSO, N	HgO, $HgBr_2$, C_6H_6, 100°, 2-5 h	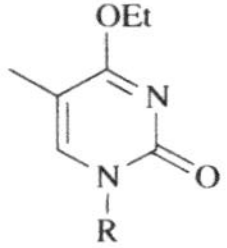 OEt, N, O, N–R (85)	80
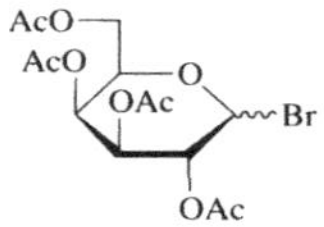 AcO, AcO, O, OAc, Br, OAc	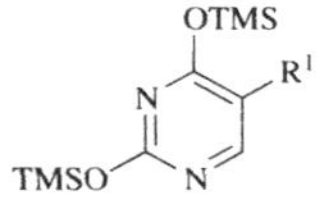 OTMS, R^1, N, TMSO, N	HgO, $Hg(OAc)_2$, or $AgClO_4$	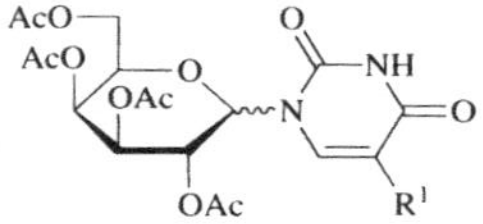 AcO, AcO, O, OAc, OAc, N, O, NH, O, R^1; R^1: H (86); Me (50)	77
	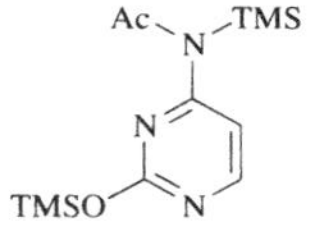 Ac, N, TMS, N, TMSO, N	HgO, $Hg(OAc)_2$, or $AgClO_4$	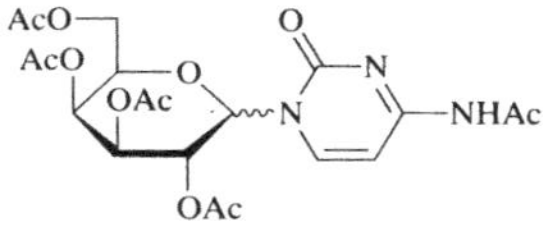 AcO, AcO, O, OAc, OAc, N, O, N, NHAc (70)	77
	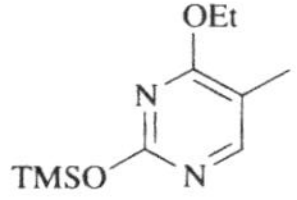 OEt, N, TMSO, N	HgO, $Hg(OAc)_2$, or $AgClO_4$	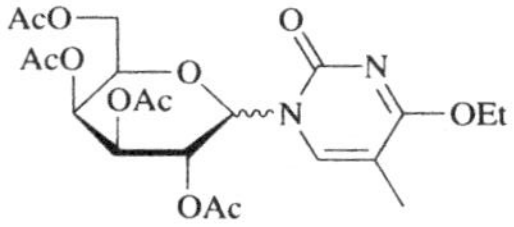 AcO, AcO, O, OAc, OAc, N, O, N, OEt (91)	77

TABLE VIII. REACTIONS OF SILYLATED BASES WITH PROTECTED SUGARS WITH OR WITHOUT CATALYSTS (*Continued*)

Sugar	Base	Conditions	Product(s) and Yield(s) (%)	Refs.
TBDMSO–, F, OAc	OTMS, TMSO, N, N	KI, dibenzo-18-crown-6, PhMe, MeCN, reflux, 4 h	TBDMSO–, F, N, NH, O, O; α-anomer (29), β-anomer (8)	392
C_{15} BnO–, O, Cl, O	OTMS, TMSO, N, N	Bu_4NI, CH_2Cl_2, reflux, 12 h	O, NH, O, N–R (88); R = BnO–, O, O	1145
BnO–, O, Cl, O	OTMS, TMSO, N, N	Bu_4NI, CH_2Cl_2, reflux, 12 h	O, NH, O, N–R (81); R = BnO–, O, O	1145
AcO–, O, OAc, SPh	OTMS, TMSO, N, N	NBS, CH_2Cl_2, MS, rt, 20 min	AcO–, O, OAc, N, NH, O, O (93) α:β = 1:1.5	405
BzO–, O, OH, O, O	N, N–TMS	1. FMPT 2. $EtN(Pr\text{-}i)_2$, CH_2Cl_2 –30°, 3 h, 0°, 1 d; rt, 1 d	BzO–, O, N, N, O, O (82) α:β = 86:14	428
	N, N–TMS	1. FMPT 2. 1-ethylpiperidine, CH_2Cl_2, –30°, 3 h; 0°, 1 d; rt, 1 d	BzO–, O, N, N, O, O (99) α:β = 84:16	428

Reactant	Heterocycle	Conditions	Product(s) and Yield(s) (%)	Refs.
	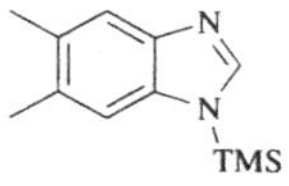 5,6-dimethyl-1-TMS-benzimidazole	1. FMPT 2. 1-ethylpiperidine, CH_2Cl_2, −30°, 3 h; 0°, 1 d; rt, 1 d	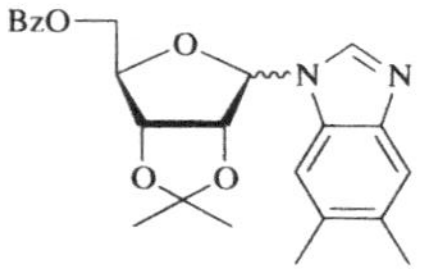 (80) α:β = 90:10	428
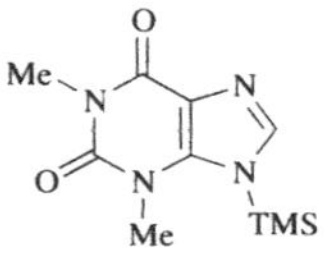	1,3-dimethyl-9-TMS-xanthine (Me, O, N, TMS)	1. FMPT 2. 1-ethylpiperidine, CH_2Cl_2, −30°, 3 h; 0°, 1 d; rt, 1 d	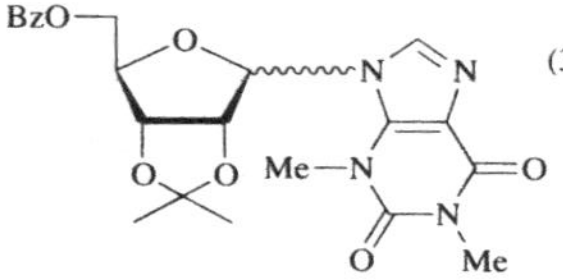 (32) α:β = 53:47	428
	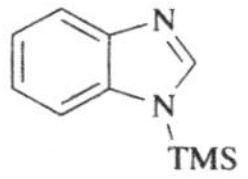 1-TMS-benzimidazole	1. FMPT 2. $EtN(Pr\text{-}i)_2$, CH_2Cl_2 −30°, 3 h; 0°, 1 d; rt, 1 d	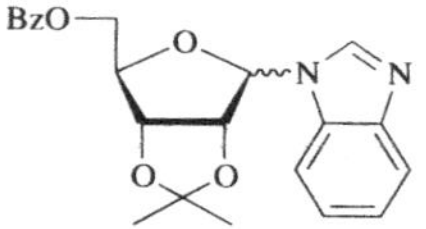 (82) α:β = 89:11	428
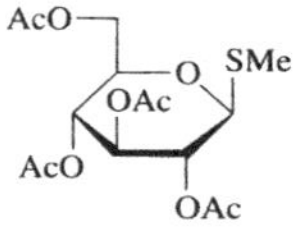 AcO, OAc, AcO, OAc, SMe (peracetylated methyl thioglycoside)	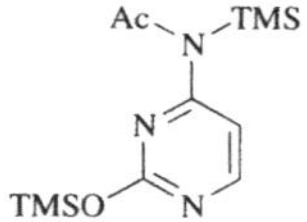 Ac(TMS)N-, TMSO-pyrimidine	NIS, CF_3SO_3H, $Cl(CH_2)_2Cl$, rt, 1 h	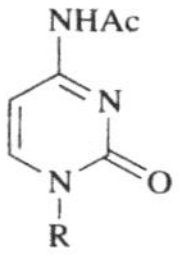 NHAc, R (83) R = (AcO, OAc, AcO, OAc glycosyl)	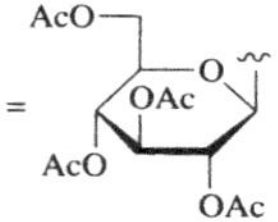 408
	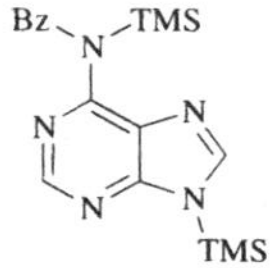 Bz(TMS)N-, 9-TMS-purine	NIS, CF_3SO_3H, $Cl(CH_2)_2Cl$, rt, 2 h	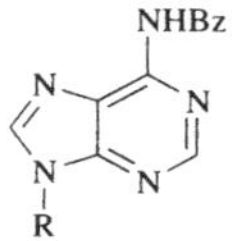 NHBz, R (79)	408
C_{16} 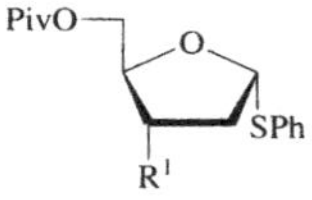 PivO, R[1], SPh	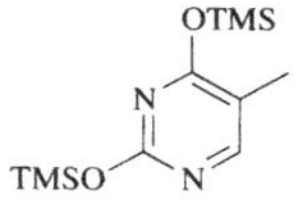 OTMS, TMSO-5-methylpyrimidine	NBS, CH_2Cl_2, MS, rt, 20-30 min	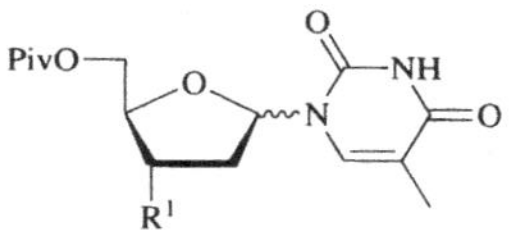 PivO, R[1] thymine nucleoside R[1]: F (96) α:β = 6:1; N_3 (100) α:β = 4:1	404

TABLE VIII. REACTIONS OF SILYLATED BASES WITH PROTECTED SUGARS WITH OR WITHOUT CATALYSTS (*Continued*)

Sugar	Base	Conditions	Product(s) and Yield(s) (%)	Refs.
 R^1, Cl	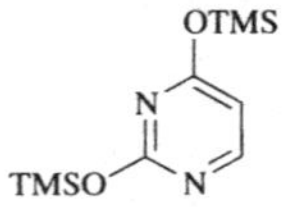 OTMS, TMSO		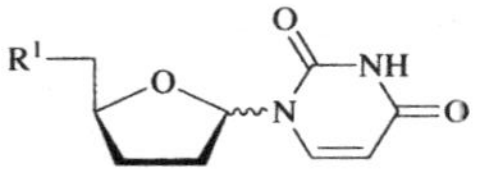 R^1, O, NH, N, O	276
R^1				
2-$C_{10}H_7CO_2$		Py, $CHCl_3$, rt, 12 h	(89) $\alpha:\beta = 33:67$	
		$CHCl_3$, rt, 12 h	(94) $\alpha:\beta = 42:58$	
1-$C_{10}H_7CO_2$		Py, $CHCl_3$, rt, 12 h	(51) $\alpha:\beta = 30:70$	
		$CHCl_3$, rt, 12 h	(57) $\alpha:\beta = 40:60$	
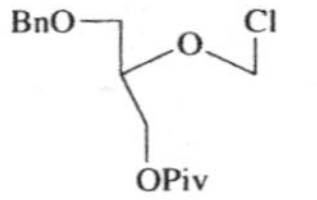 BnO, O, Cl, OPiv	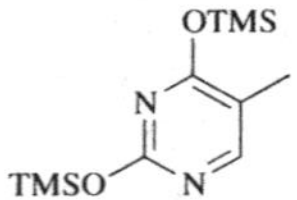 OTMS, TMSO	Bu_4NI, $Cl(CH_2)_2Cl$, 50°, 1 h	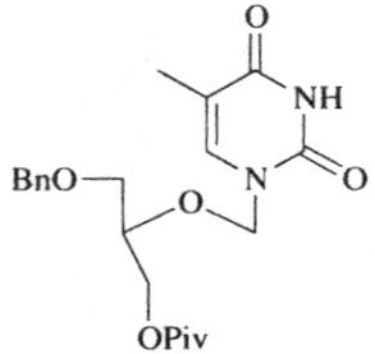 O, NH, O, BnO, OPiv (40)	1146
C_{17} 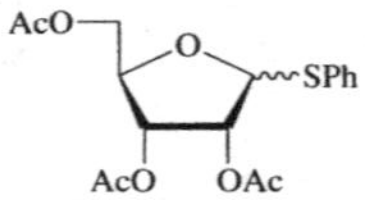 AcO, SPh, AcO, OAc	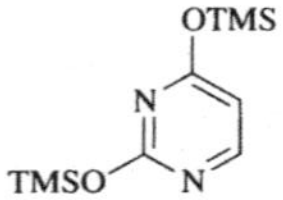 OTMS, TMSO	NIS, CF_3SO_3H, CH_2Cl_2, rt, 10 min	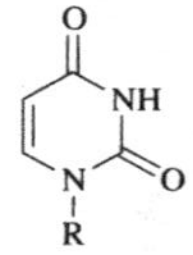 O, NH, O, R (56); R = AcO, AcO, OAc 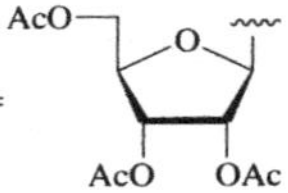	408
	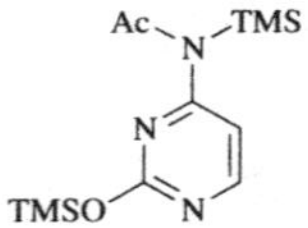 Ac, N, TMS, TMSO	NIS, CF_3SO_3H, CH_2Cl_2, rt, 45 min	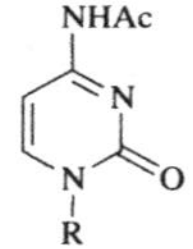 NHAc, N, O, R (81)	408

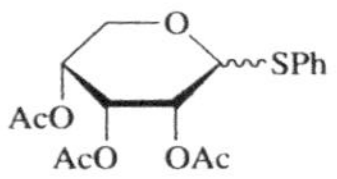 AcO, AcO, OAc, O, SPh	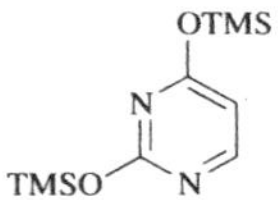 OTMS, N, N, TMSO	NIS, CF_3SO_3H, CH_2Cl_2, rt	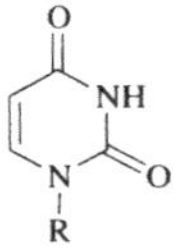 O, NH, O, N–R (80); R = AcO, AcO, OAc, O 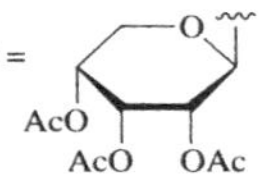	408
	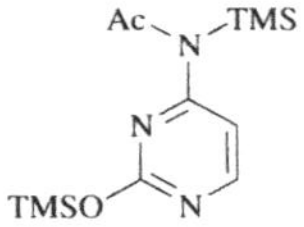 Ac–N–TMS, N, N, TMSO	NIS, CF_3SO_3H, CH_2Cl_2, rt, 30 min	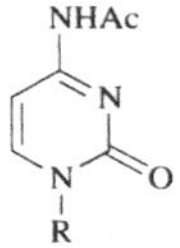 NHAc, N, O, N–R **I** (87)	408
		DMTST, MeCN, reflux, 30 min	**I** (83)	408
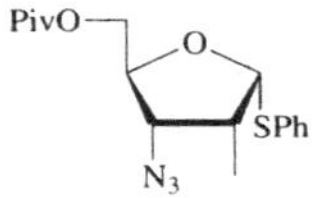 PivO, O, SPh, N_3	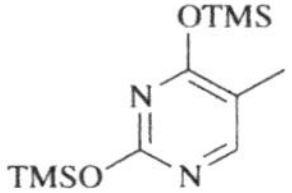 OTMS, N, N, TMSO	NBS, CH_2Cl_2, MS, rt, 20-30 min	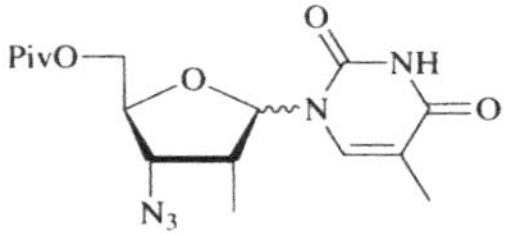 PivO, O, N_3, N, NH, O, O (98) α:β = 9:1	404
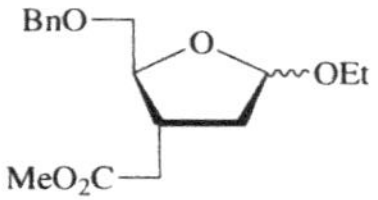 BnO, O, OEt, MeO_2C	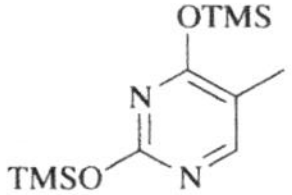 OTMS, N, N, TMSO	1. Me_2BBr, $CHCl_3$, –78 to 0° 2. $Cl(CH_2)_2Cl$, 0°	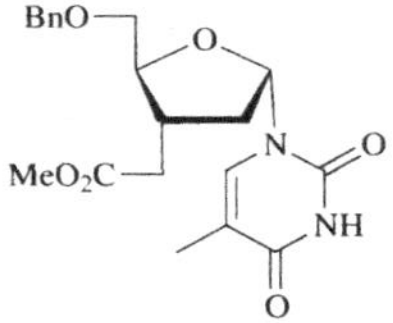 BnO, O, MeO_2C, N, O, NH, O (70) α:β = 1:3	278
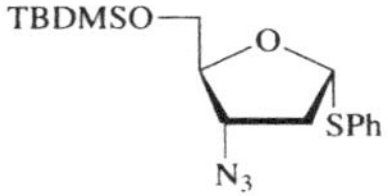 TBDMSO, O, SPh, N_3	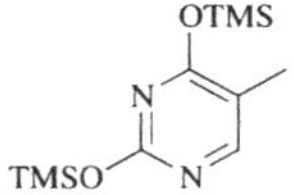 OTMS, N, N, TMSO	NBS, CH_2Cl_2, MS, rt, 20-30 min	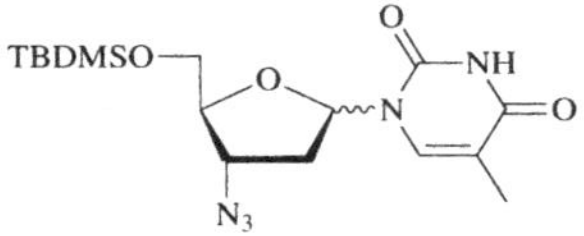 TBDMSO, O, N_3, N, NH, O, O (100) α:β = 2:1	404

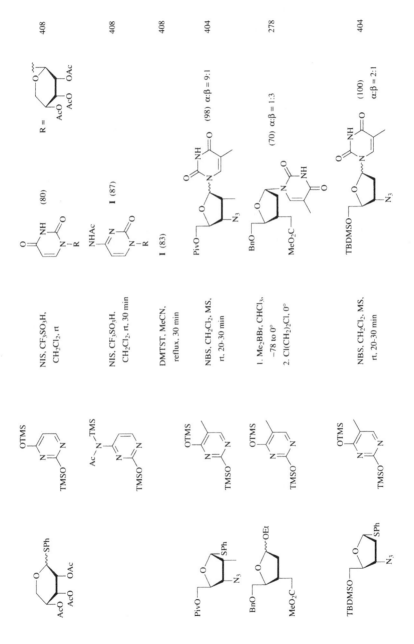

TABLE VIII. REACTIONS OF SILYLATED BASES WITH PROTECTED SUGARS WITH OR WITHOUT CATALYSTS (*Continued*)

Sugar	Base	Conditions	Product(s) and Yield(s) (%)	Refs.
C_{18} 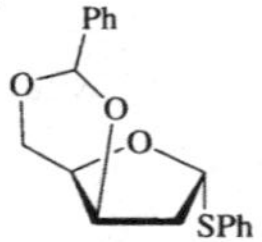 Ph, O, O, O, SPh	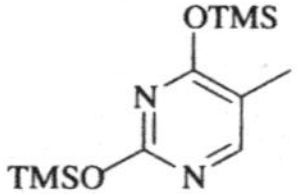 OTMS, N, N, TMSO	NBS, CH_2Cl_2, MS, rt, 20 min	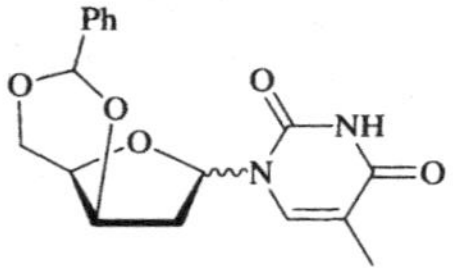 Ph, O, O, O, N, O, NH, O (88) α:β = 1:12	405
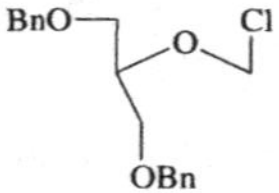 BnO, O, Cl, OBn	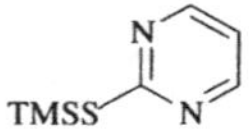 N, N, TMSS	CsI, MeCN, reflux, 0.5 h	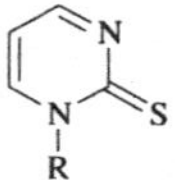 N, N, S, R (80) R = BnO, O, OBn 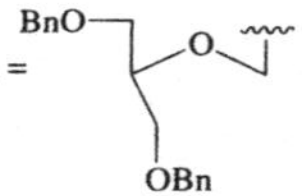	400
	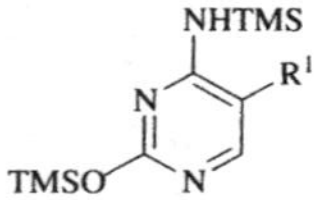 NHTMS, R^1, N, N, TMSO	Bu_4NI, $Cl(CH_2)_2Cl$, reflux, 45 min	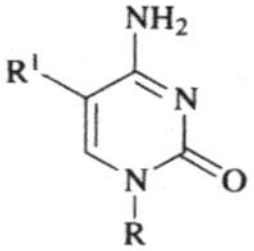 NH_2, R^1, N, N, O, R; R^1 = H (89), F (84)	786
	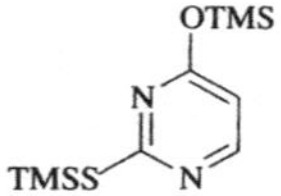 OTMS, N, N, TMSS	CsI, MeCN, reflux, 0.5 h	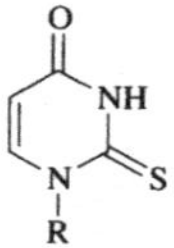 O, NH, N, S, R **I** (95)	400, 1147
		$Hg(CN)_2$, C_6H_6, reflux, 3 h	**I** (23)	400, 1147
	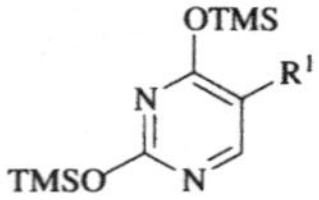 OTMS, R^1, N, N, TMSO	CsI, MeCN, reflux, 1-2 h	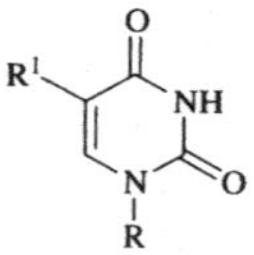 O, R^1, NH, N, O, R; R^1 = H (82), F (80), NH_2 (83)	400

Reactant	Conditions	Product(s) and Yield(s) (%)	Refs.
OTMS, Bn, N, TMSO, N	PhMe, reflux, 24 h	O, Bn, NH, N, O, R (47)	1148
OTMS, R[1], N, TMSX, N=N	MeCN	O, R[1], NH, N, N, X, R	
X: O, R^1: Me	rt	(54) + N^1,N^3-bis(product) (10)	1122
X: S, R^1: Me	rt	(53)	1122
X: O, R^1: Ph	rt	(83)	1122
X: O, R^1: Bn	rt	(77)	1122
Cl, N, N, HN, TMS, N, N, TMS	Bu_4NI, MeCN, reflux, 2 h	Cl, N, N, N, N, NH_2, R **I** (38)	788
	$Hg(CN)_2$, C_6H_6, reflux, 3 h	**I** (80)	788
OTMS, N, N, AcN, TMS, N, N, TMS	Bu_4NI, MeCN, reflux, 12 h	N, N, NHAc, NH, N, R, O **I** (28) + N^9-isomer (28)	788
	$Hg(CN)_2$, C_6H_6, reflux, 1.25 h	**I** (22) + N^9-isomer (14)	788
	$Cl(CH_2)_2Cl$, MS, reflux, 4 h	N, N, NHAc, NH, N, R, O (—) N^7:N^9 = 1:1	788

TABLE VIII. REACTIONS OF SILYLATED BASES WITH PROTECTED SUGARS WITH OR WITHOUT CATALYSTS (*Continued*)

Sugar	Base	Conditions	Product(s) and Yield(s) (%)	Refs.
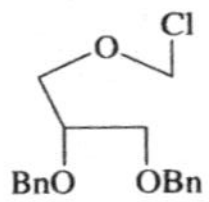 BnO, OBn, O, Cl	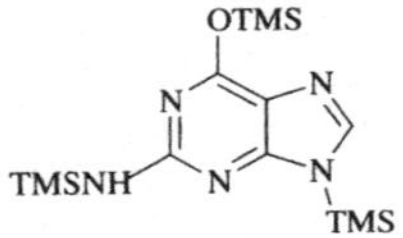 OTMS, TMSNH, N, TMS	1. Xylene, 125°, 12 h 2. *n*-PrOH, AcOH, reflux, 1 h	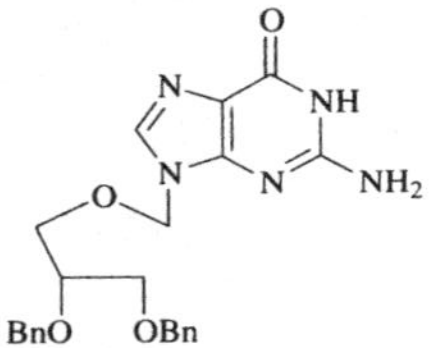 O, NH, NH_2, BnO, OBn (50)	1149
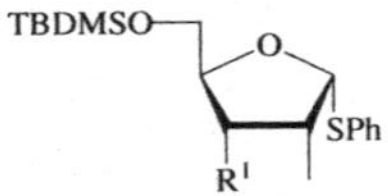 TBDMSO, O, SPh, R^1	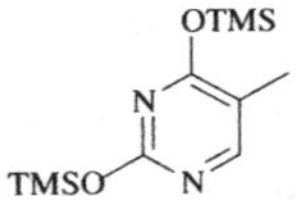 OTMS, TMSO, N	NBS, CH_2Cl_2, MS, rt, 20-30 min	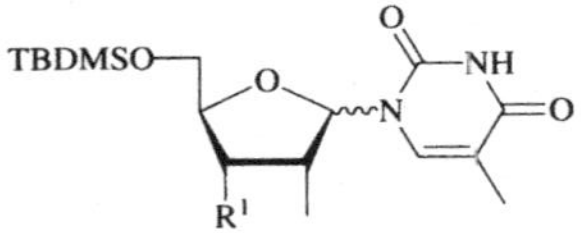 TBDMSO, O, NH, R^1 R^1 — — αβ F (84) 3:1 N_3 (96) 4:1	404
C_{19} 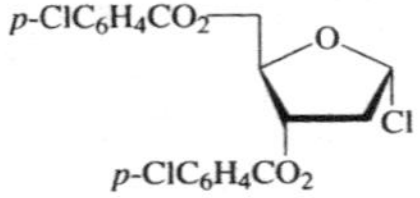 *p*-$ClC_6H_4CO_2$, O, Cl, *p*-$ClC_6H_4CO_2$	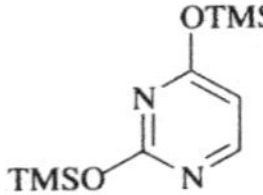 OTMS, TMSO, N	*p*-$O_2NC_6H_4OH$, Py, $CHCl_3$, 30°, 12 h	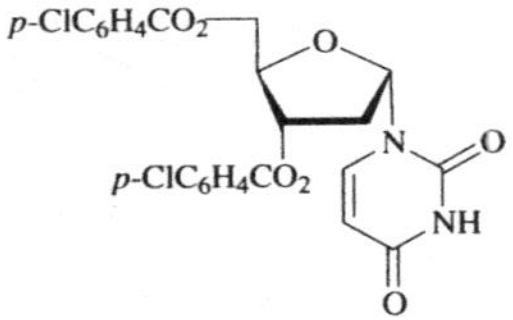 *p*-$ClC_6H_4CO_2$, *p*-$ClC_6H_4CO_2$, NH (75) + β-anomer (21)	267
	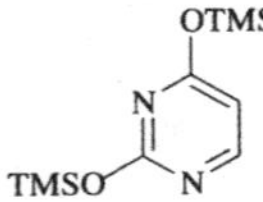 OTMS, TMSO, N	*p*-$O_2NC_6H_4OH$, $CHCl_3$, 30°, 12 h	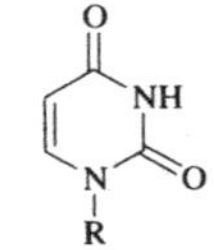 O, NH, N, R (97) + α-anomer (2) R 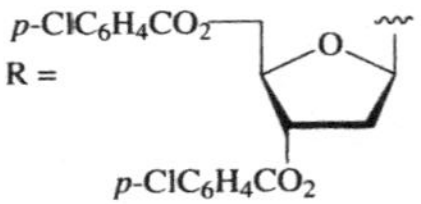 = *p*-$ClC_6H_4CO_2$, O, *p*-$ClC_6H_4CO_2$	267
	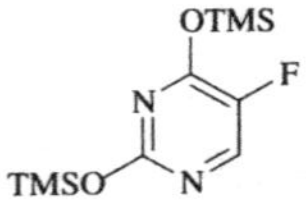 OTMS, F, TMSO, N	$CHCl_3$, rt, 24 h	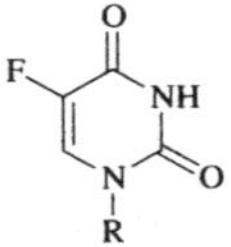 F, O, NH, N, R **I** (80)	1150

Reactant	Conditions	Product(s) and Yield(s) (%)	Refs.
	p-$O_2NC_6H_4OH$, $CHCl_3$, 30°, 12 h	**I** (92) + α-anomer (5)	267
	KI, CH_2Cl_2, rt, 18 h	**I** (70)	1150
	CsF, CH_2Cl_2, rt, 20 h	**I** (—)	1151
	p-$O_2NC_6H_4OH$, $CHCl_3$, 25°, 12 h	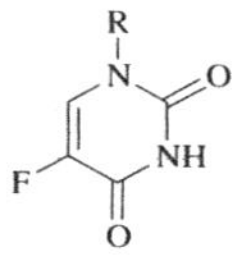 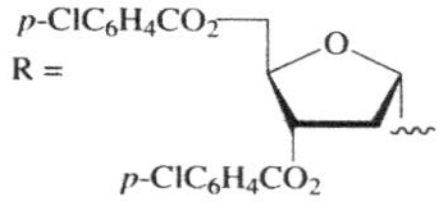 5-fluoro-1-R-uracil **I**; R = 3,5-di-O-(p-$ClC_6H_4CO_2$)-2-deoxyribofuranosyl	267
	β-picoline	**I** (78)	
	α-picoline	**I** (77)	
	2,6-lutidine	**I** (62)	
	DMF	**I** (40)	
	Et_3N	**I** (73)	
	$AcNH_2$	**I** (51)	
	Py	**I** (79) + β-anomer (15)	
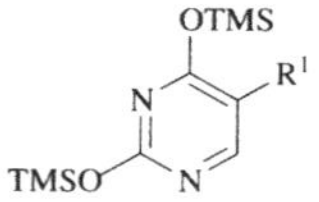 2,4-bis(TMSO)-5-R^1-pyrimidine	p-$O_2NC_6H_4OH$, Py, $CHCl_3$, 30°, 12 h	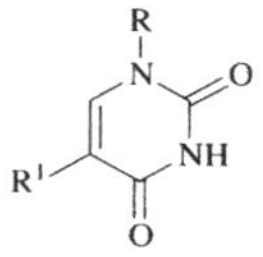 5-R^1-1-R-uracil	267
		R^1 Cl (77), β-anomer (10)	
		R^1 Br (77), β-anomer (21)	
		R^1 I (82), β-anomer (15)	
		R^1 Me (67), β-anomer (33)	
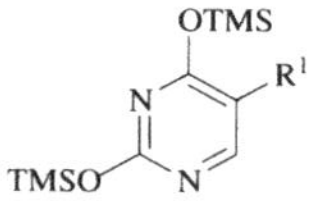 2,4-bis(TMSO)-5-R^1-pyrimidine	p-$O_2NC_6H_4OH$, $CHCl_3$, 30°, 12 h	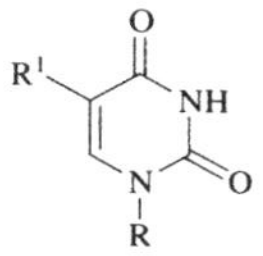 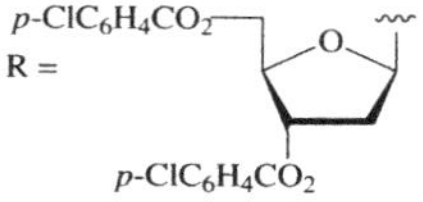 5-R^1-1-R-uracil; R = 3,5-di-O-(p-$ClC_6H_4CO_2$)-2-deoxyribofuranosyl	267
R^1 = Cl		(92), α-anomer (5)	
R^1 = Br		(85), α-anomer (10)	
R^1 = I		(87), α-anomer (9)	
R^1 = Me		(96), α-anomer (—)	

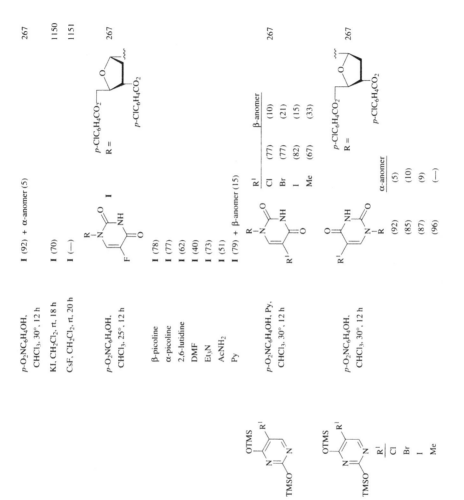

TABLE VIII. REACTIONS OF SILYLATED BASES WITH PROTECTED SUGARS WITH OR WITHOUT CATALYSTS (*Continued*)

Sugar	Base	Conditions	Product(s) and Yield(s) (%)	Refs.
	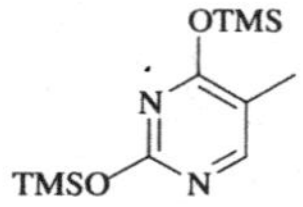 OTMS, TMSO, N, N	C_6H_6, 37°, 90 h	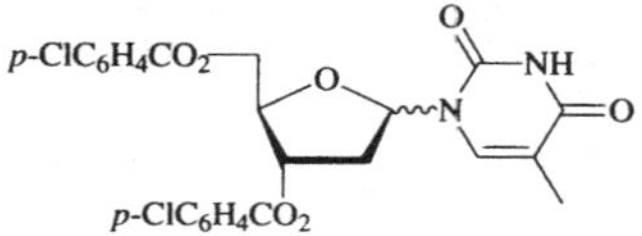 p-$ClC_6H_4CO_2$, p-$ClC_6H_4CO_2$, O, N, NH, O, O (—) α:β = 1:1	266
	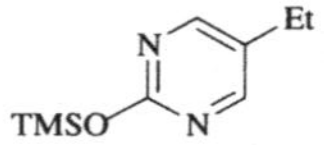 Et, TMSO, N, N	HgO, $HgBr_2$, PhMe, 0°, 35 min; rt, 2.5 h	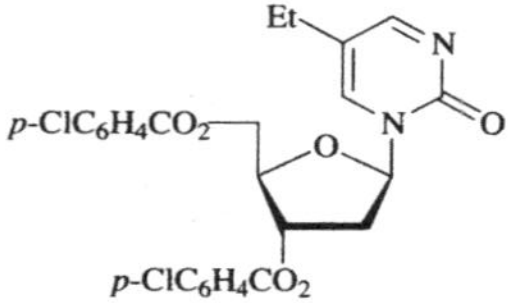 Et, N, N, O, O, p-$ClC_6H_4CO_2$, p-$ClC_6H_4CO_2$ (16) + α-anomer (14)	1152
	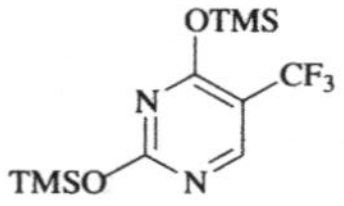 OTMS, CF_3, TMSO, N, N	$CHCl_3$, rt, 12-15 h	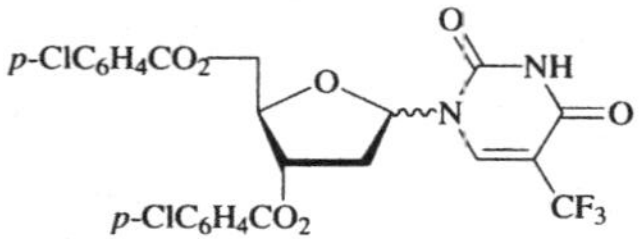 p-$ClC_6H_4CO_2$, p-$ClC_6H_4CO_2$, O, N, NH, O, O, CF_3	269
		p-$O_2NC_6H_4OH$	(77) α:β = 47:53	
		—	(82) α:β = 26:74	
		—, sugar:base = 1:8	(82) α:β = 26:74	
		—, sugar:base = 1:4	(78) α:β = 35:65	
		—, sugar:base = 1:2	(66) α:β = 44:56	
	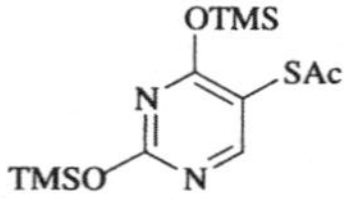 OTMS, SAc, TMSO, N, N	C_6H_6, 37°, 90 h	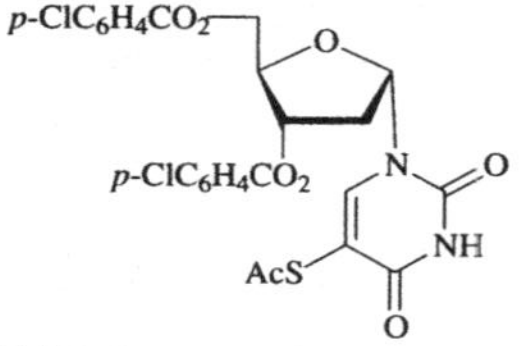 p-$ClC_6H_4CO_2$, p-$ClC_6H_4CO_2$, O, N, O, NH, O, AcS **I** (—)	266
		C_6H_6, reflux, 1.5 h	**I** (54) β-anomer only	266

Reactant	Reactant	Conditions	Product(s) and Yield(s) (%)	Refs.
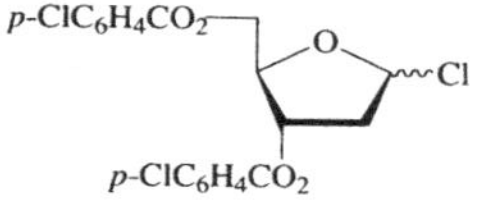 p-$ClC_6H_4CO_2$, p-$ClC_6H_4CO_2$, Cl	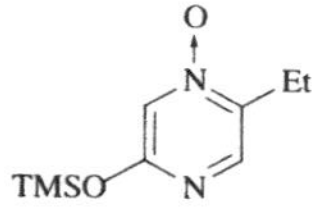 TMSO, Et, O←N	HgO, $HgBr_2$, $Cl(CH_2)_2Cl$, 0 - 5°, 30 min; rt, 15 min	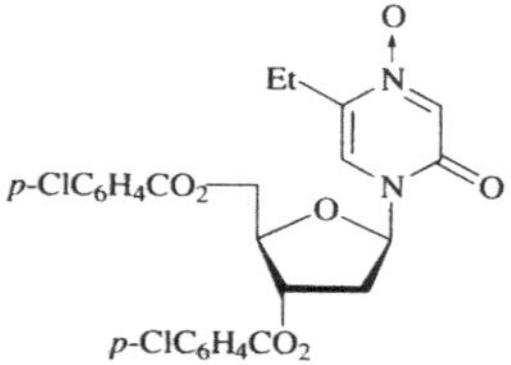 (44) + α-anomer (31)	621
	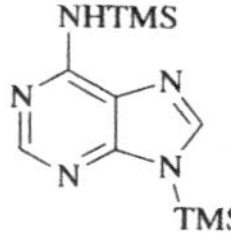 NHTMS, TMS	HMDS, 1,2,4-$Cl_3C_6H_3$, rt, 18 h	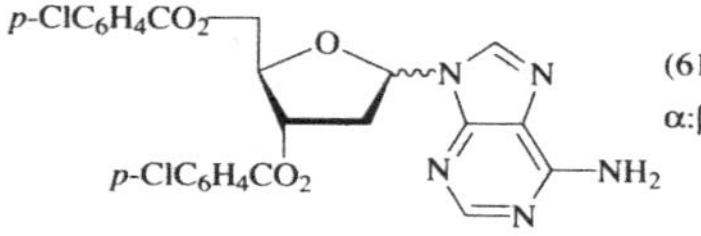 NH_2 (61) $\alpha:\beta = 43:57$	790
		$Cl_2CHCHCl_2$, rt, 18 h	(58) $\alpha:\beta = 71:29$	
		$CHCl_3$, rt, 18 h	(57) $\alpha:\beta = 55:45$	
	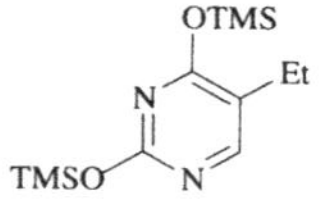 OTMS, Et, TMSO	1. $CHCl_3$, 10-15°, 2.5 h 2. MeOH	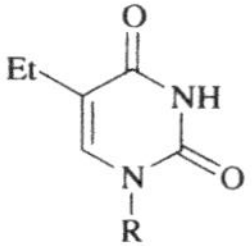 Et, NH, R (95) R = p-$ClC_6H_4CO_2$, p-$ClC_6H_4CO_2$ 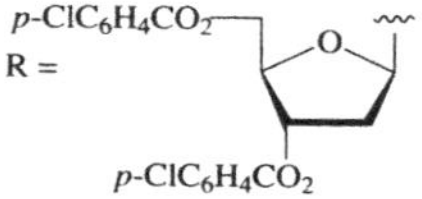	1153
	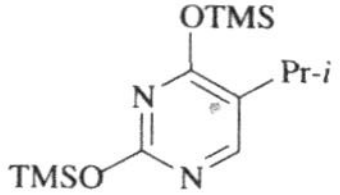 OTMS, Pr-i, TMSO	$CHCl_3$, rt, 30 min	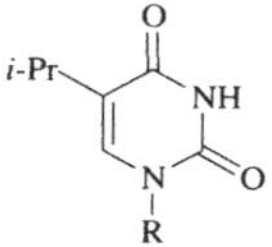 i-Pr, NH, R (90) + α-anomer (8)	1153
	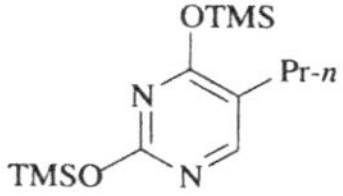 OTMS, Pr-n, TMSO	$Hg(OAc)_2$, PhMe, rt, 24 h	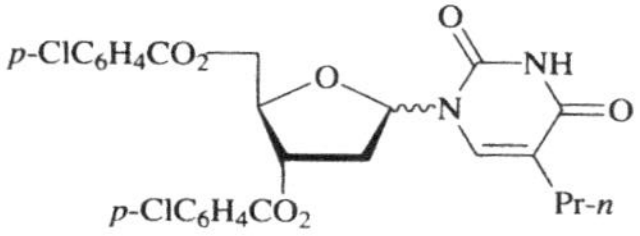 p-$ClC_6H_4CO_2$, p-$ClC_6H_4CO_2$, NH, Pr-n (82)	1154

TABLE VIII. REACTIONS OF SILYLATED BASES WITH PROTECTED SUGARS WITH OR WITHOUT CATALYSTS (*Continued*)

Sugar	Base	Conditions	Product(s) and Yield(s) (%)	Refs.
	Me$_2$N, N, N, N, N, OTMS	HgO, HgBr$_2$, C_6H_6, reflux, 5 h	p-ClC$_6$H$_4$CO$_2$, O, p-ClC$_6$H$_4$CO$_2$, O, N, N, NMe$_2$, N, N (37)	1144
p-O$_2$NC$_6$H$_4$CO$_2$, O, Cl, p-O$_2$NC$_6$H$_4$CO$_2$	OTMS, CF$_3$, N, X, TMSO, N	Hg(OAc)$_2$, C_6H_6	O, CF$_3$, NH, X, N, O, p-O$_2$NC$_6$H$_4$CO$_2$, O, p-O$_2$NC$_6$H$_4$CO$_2$	
	X			
	CH	25°, 18 h	(33) + α-anomer (23)	1155, 529
	N	rt, 3 d	(36)	1156
O, p-O$_2$NC$_6$H$_4$CO$_2$, p-O$_2$NC$_6$H$_4$CO$_2$, Cl	OTMS, R^1, N, TMSO, N	HgBr$_2$, HgO	O, NH, O, N, O, p-O$_2$NC$_6$H$_4$CO$_2$, p-O$_2$NC$_6$H$_4$CO$_2$, R^1 — R^1: H (80); Me (80)	997; 998
BzO, O, F, Br, BzO	N, TMSO, N	Cl(CH$_2$)$_2$Cl, reflux, 18 h	N, N, O, R (65); R = BzO, O, F, BzO	388

Reactant	Conditions	Product(s) and Yield(s) (%)	Refs.
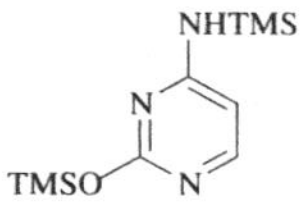 2,4-bis(trimethylsilyl)cytosine (NHTMS, TMSO)	$CHCl_3$, reflux, 20 h	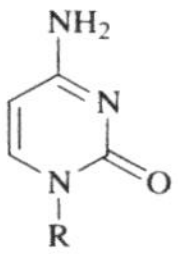 1-R-cytosine (61)	386
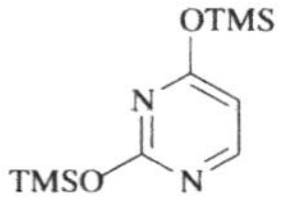 2,4-bis(trimethylsilyloxy)pyrimidine (OTMS, TMSO)	MeCN, reflux, 5 h	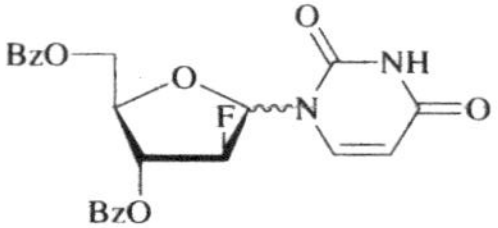 I (—) α:β = 1:4	385
	CH_2Cl_2, reflux, 8.5 h	I (—) α:β = 1:8.5	385
	$CHCl_3$, reflux, 36 h	I (—) α:β = 1:20	385
	CCl_4, reflux, 60 h	I (—) α:β = 1:39	385
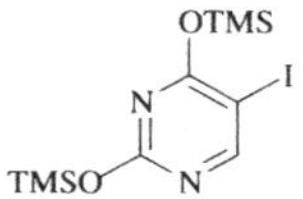 5-iodo-2,4-bis(trimethylsilyloxy)pyrimidine (OTMS, I, TMSO)	NaI, CH_2Cl_2, MeCN, rt, 5 d	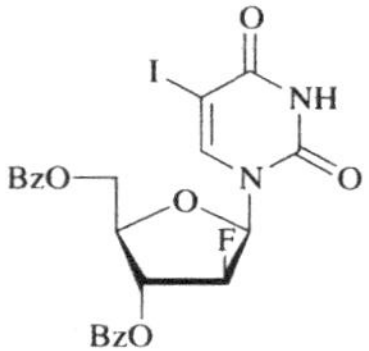 5-iodouracil nucleoside (41)	1157
	NaI, MeCN, rt, 7 d	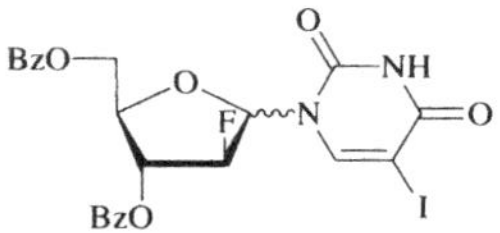 I (96) α:β = 1:3	387
	MeCN, rt, 7 d	I (95) α:β = 1:7	387
	MeCN, reflux, 5 h	I (—) α:β = 1:2	385
	MeCN, reflux, 22 h	I (58) α:β = 1:9	387
	CH_2Cl_2, reflux, 44 h	I (—) α:β = 1:14	385
	$CHCl_3$, reflux, 36 h	I (—) α:β = 1:29	385
	CCl_4, reflux, 60 h	I (—) α:β = 1:34	385

TABLE VIII. REACTIONS OF SILYLATED BASES WITH PROTECTED SUGARS WITH OR WITHOUT CATALYSTS (*Continued*)

Sugar	Base	Conditions	Product(s) and Yield(s) (%)	Refs.
	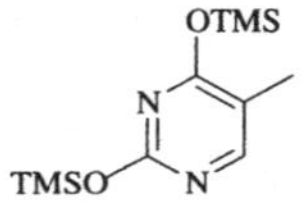 OTMS, TMSO, N, N	MeCN, reflux, 1.5 h	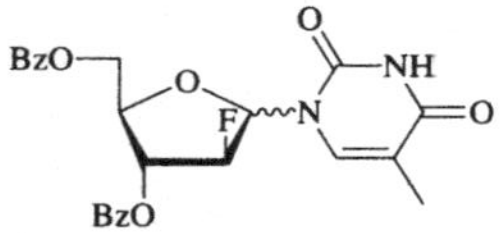 BzO, O, F, N, O, NH, O, BzO **I** (99) α:β = 1:4	387
		MeCN, reflux, 5 h	**I** (—) α:β = 1:3	385
		CH_2Cl_2, reflux, 44 h	**I** (—) α:β = 1:8.5	385
		$CHCl_3$, reflux, 36 h	**I** (—) α:β = 1:19	385
		CCl_4, reflux, 60 h	**I** (—) α:β = 1:36	385
	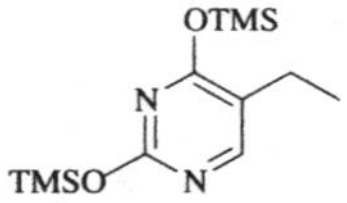 OTMS, TMSO, N, N	MeCN, reflux, 2 h	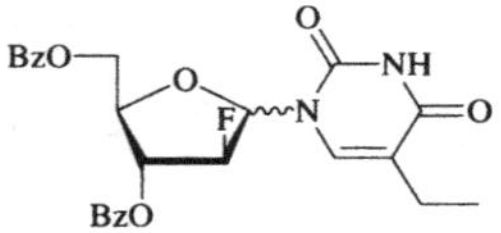 BzO, O, F, N, O, NH, O, BzO **I** (63) α:β = 2:3	1158
		MeCN, reflux, 5 h	**I** (—) α:β = 1:2	385
		CH_2Cl_2, reflux, 44 h	**I** (—) α:β = 1:8.5	385
		$CHCl_3$, reflux, 36 h	**I** (—) α:β = 1:16	385
		CCl_4, reflux, 60 h	**I** (—) α:β = 1:41	385
		$CHCl_3$, reflux, 20 h	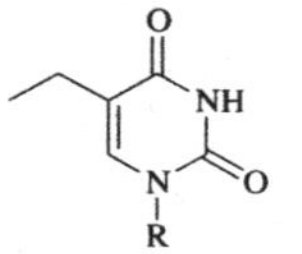 O, NH, N, O, R (76) R = BzO, O, F, BzO 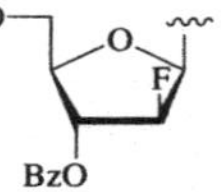	1158
	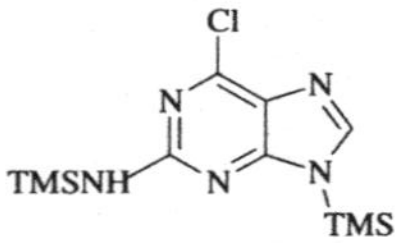 Cl, N, N, N, N, TMSNH, TMS	$Hg(CN)_2$, C_6H_6, reflux, 3 h	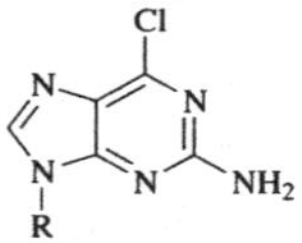 Cl, N, N, N, N, R, NH_2 (28) + α-anomer (6)	1136

Conditions	Product(s) and Yield(s) (%)	Refs.
MeCN, reflux, 18 h	(65)	1159
$Hg(CN)_2$, rt, 18 h	(80) R =	383a
$(TfO)_2O$, Py, CH_2Cl_2, reflux, 2 d	(50)	1160
PhMe, reflux, 24 h	(58) R =	1161, 1148
MeCN, rt	R^1: Ph (74); Bn (63)	1122
CCl_4, 90-100°, 1 h	(6) + α-anomer (3) R =	1162

TABLE VIII. REACTIONS OF SILYLATED BASES WITH PROTECTED SUGARS WITH OR WITHOUT CATALYSTS (*Continued*)

Sugar	Base	Conditions	Product(s) and Yield(s) (%)	Refs.
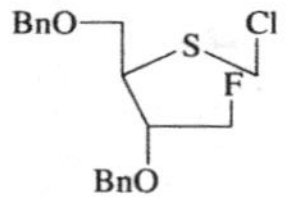 BnO, BnO, S, Cl, F	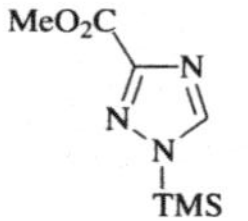 MeO_2C, N, N, N, TMS	Bu_4NI, MeCN, reflux, 4.5 h; rt, 3 d	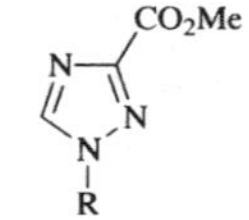 CO_2Me, N, N, N, R (47) R = BnO, S, F, BnO 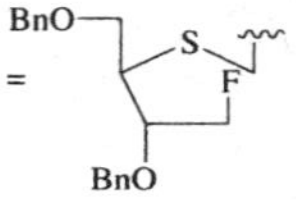	1163
	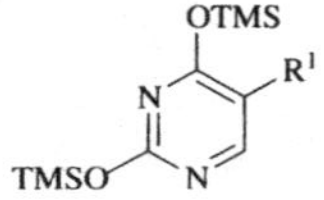 OTMS, R^1, N, N, TMSO	Bu_4NI, CH_2Cl_2, reflux	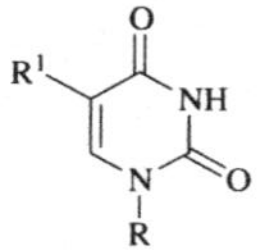 O, R^1, NH, N, O, R; R^1: H 4.5 h (99); Me 24 h (76)	1163
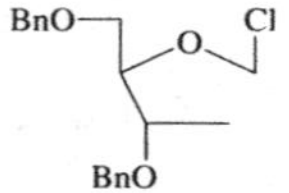 BnO, O, Cl, BnO	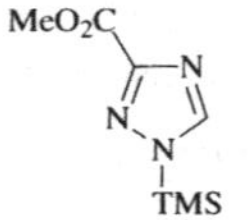 MeO_2C, N, N, N, TMS	Bu_4NI, MeCN, reflux, 4.5 h; rt, 3 d	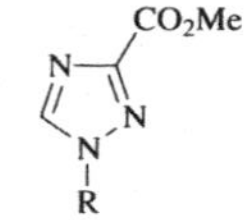 CO_2Me, N, N, N, R (64) R = BnO, O, BnO 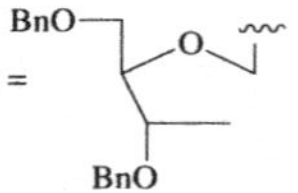	1164, 751
	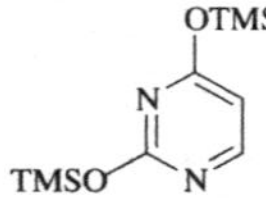 OTMS, N, N, TMSO	Et_4NI, CH_2Cl_2, reflux, 4.5 h	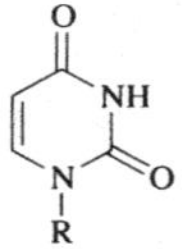 O, NH, N, O, R (59) + N^3-anomer (7)	1165
	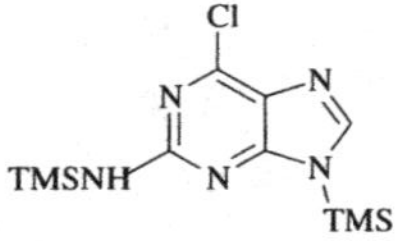 Cl, N, N, N, N, TMSNH, TMS	$Hg(CN)_2$, C_6H_6, reflux, 3 h	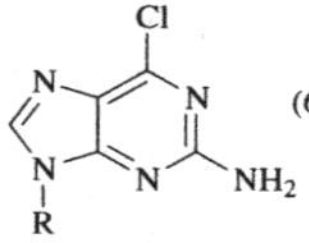 Cl, N, N, N, N, R, NH_2 (65) + N^7-anomer (5)	1165
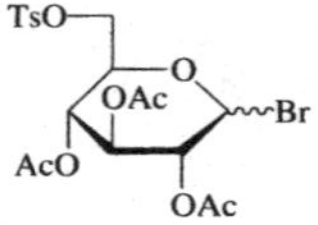 TsO, O, OAc, Br, AcO, OAc	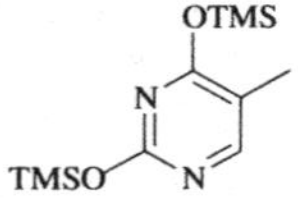 OTMS, N, N, TMSO	$HgCl_2$, HgO, C_6H_6, 60°, 4 h	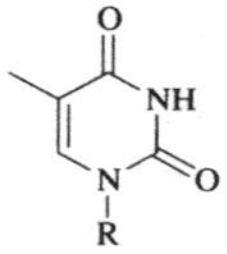 O, NH, N, O, R (63) R = TsO, O, OAc, AcO, OAc 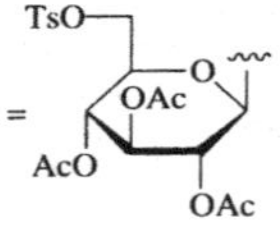	753

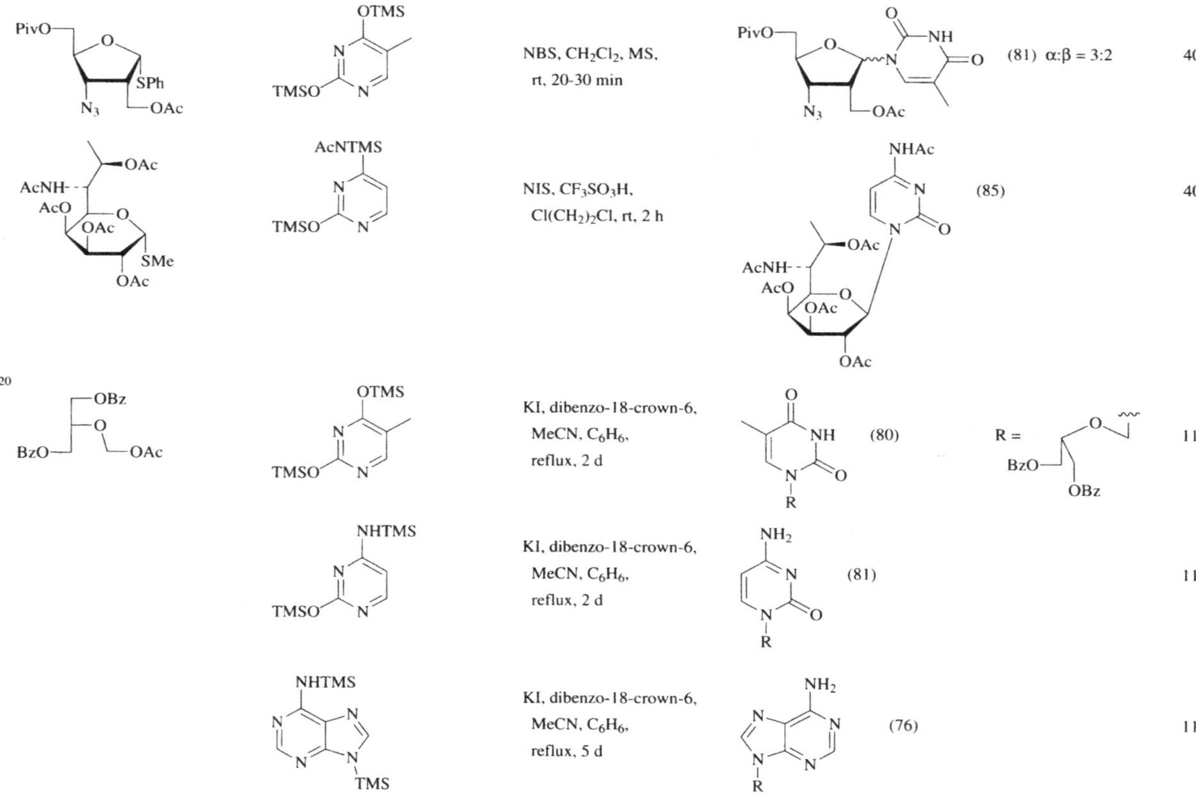

	Conditions	Product (yield)	Refs.
	NBS, CH_2Cl_2, MS, rt, 20-30 min	(81) α:β = 3:2	404
	NIS, CF_3SO_3H, $Cl(CH_2)_2Cl$, rt, 2 h	(85)	408
C_{20}	KI, dibenzo-18-crown-6, MeCN, C_6H_6, reflux, 2 d	(80) R =	1166
	KI, dibenzo-18-crown-6, MeCN, C_6H_6, reflux, 2 d	(81)	1166
	KI, dibenzo-18-crown-6, MeCN, C_6H_6, reflux, 5 d	(76)	1166

TABLE VIII. REACTIONS OF SILYLATED BASES WITH PROTECTED SUGARS WITH OR WITHOUT CATALYSTS (*Continued*)

Sugar	Base	Conditions	Product(s) and Yield(s) (%)	Refs.
	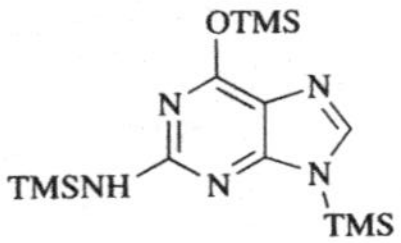 OTMS, TMSNH, N, N, N, N, TMS	KI, dibenzo-18-crown-6, MeCN, C_6H_6, reflux, 1 d	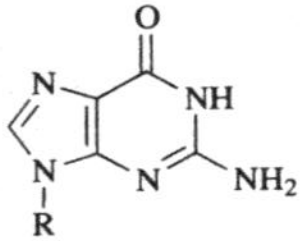 O, N, NH, NH_2, N, N, R (80)	1166
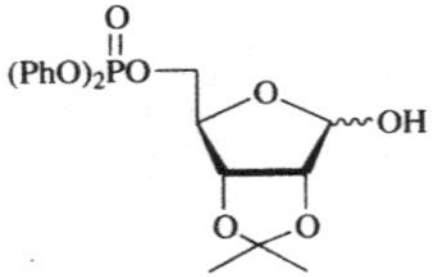 $(PhO)_2\overset{O}{\overset{\parallel}{P}}O$, O, OH, O, O	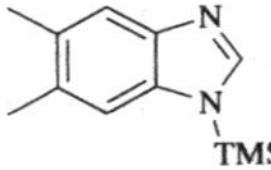 N, N, TMS	1. FMPT 2. $EtN(Pr\text{-}i)_2$	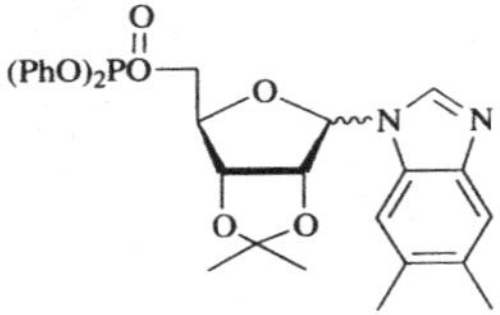 $(PhO)_2\overset{O}{\overset{\parallel}{P}}O$, O, N, N, O, O (72) α:β = 89:11	433
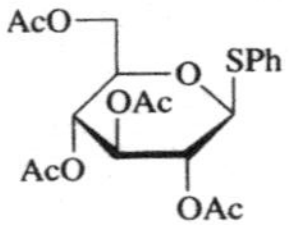 AcO, O, SPh, OAc, AcO, OAc	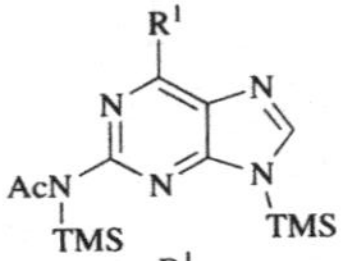 R^1, N, N, N, AcN, N, N, TMS, TMS R^1: H, Cl	NIS, $Cl(CH_2)_2Cl$, TfOH, rt, 3 h	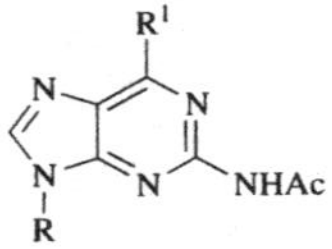 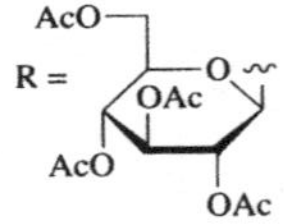 R^1, N, N, N, N, NHAc, R **I**; R = AcO, O, OAc, AcO, OAc **I** (59) N^7:N^9 = 2:98 **I** (63) N^7:N^9 = 2:98	409
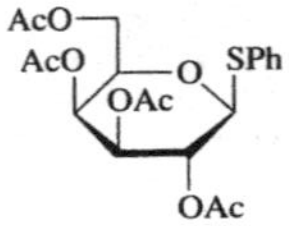 AcO, AcO, O, SPh, OAc, OAc	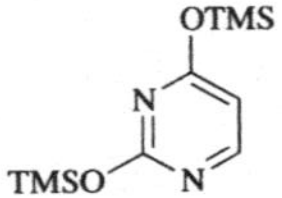 OTMS, N, TMSO, N	NIS, CH_2Cl_2, TfOH, rt, 45 min	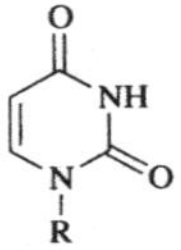 O, NH, O, N, R (83); R = AcO, AcO, O, OAc, OAc 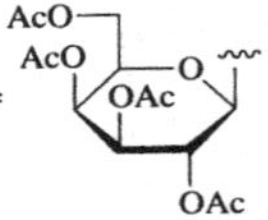	408
	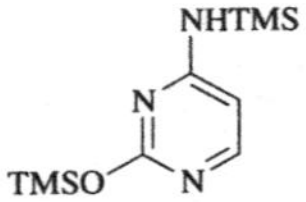 NHTMS, N, TMSO, N	NIS, CH_2Cl_2, TfOH, rt, 30 min	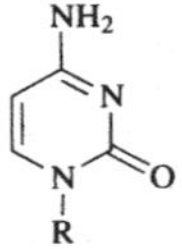 NH_2, N, O, N, R (90)	408

Sugar	Base	Conditions	Product (yield, %)	Ref.
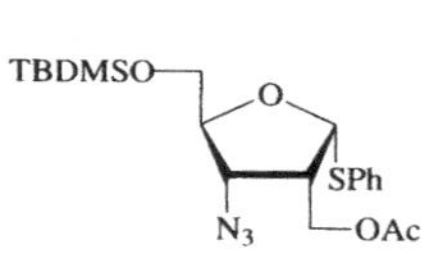	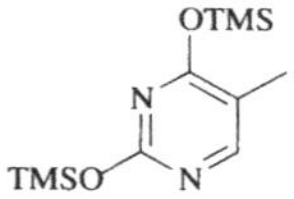	NBS, CH_2Cl_2, MS, rt, 20-30 min	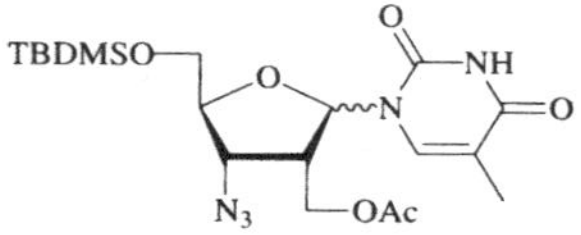 (72) $\alpha:\beta = 4:3$	404
C_{21} 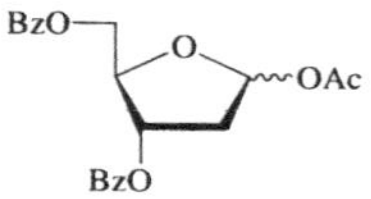	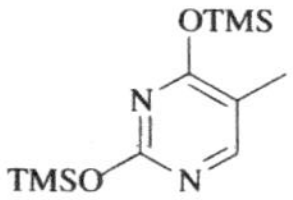	KI, dibenzo-18-crown-6, MeCN:PhMe (1:1), reflux, 2-4 h	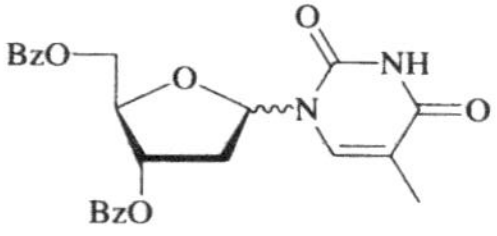 (70)	396
	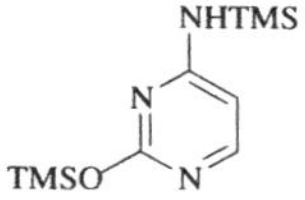	KI, dibenzo-18-crown-6, MeCN:PhMe (1:1), reflux, 2-4 h	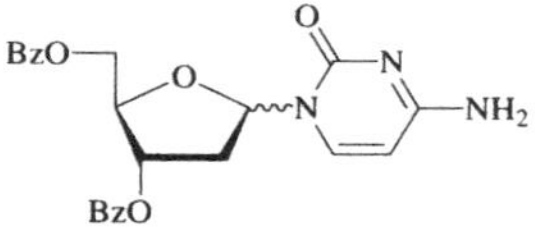 (77) $\alpha:\beta = 54:46$	396
	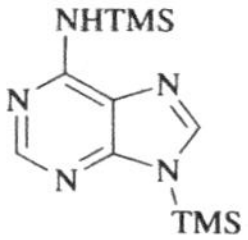	KI, dibenzo-18-crown-6, MeCN:PhMe (1:1), reflux, 2-4 h	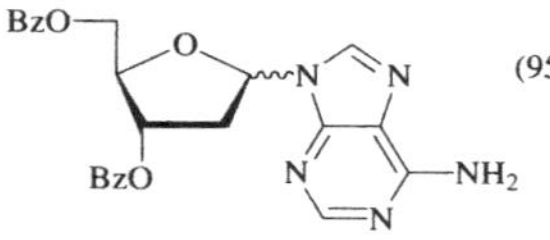 (95) $\alpha:\beta = 62:38$	396
	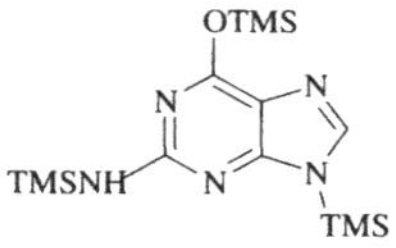	KI, dibenzo-18-crown-6, MeCN:PhMe (1:1), reflux, 2-4 h	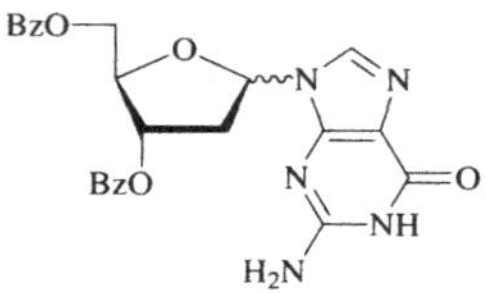 (90) $\alpha:\beta = 68:32$	396
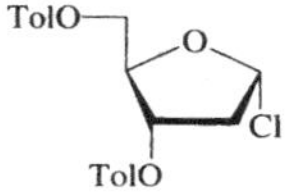	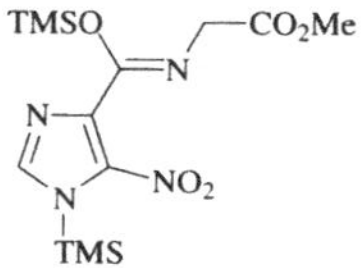	Et_3N, $CHCl_3$, rt, 2 h	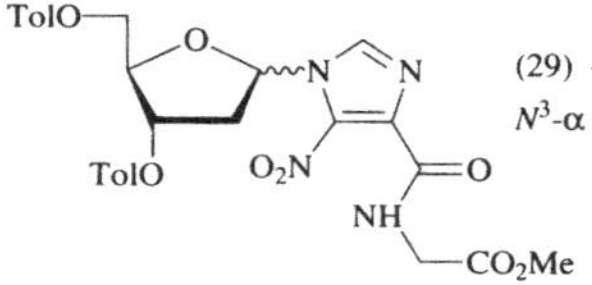 (29) + N^3-α + β-isomers (18)	1167

TABLE VIII. REACTIONS OF SILYLATED BASES WITH PROTECTED SUGARS WITH OR WITHOUT CATALYSTS (*Continued*)

Sugar	Base	Conditions	Product(s) and Yield(s) (%)	Refs.
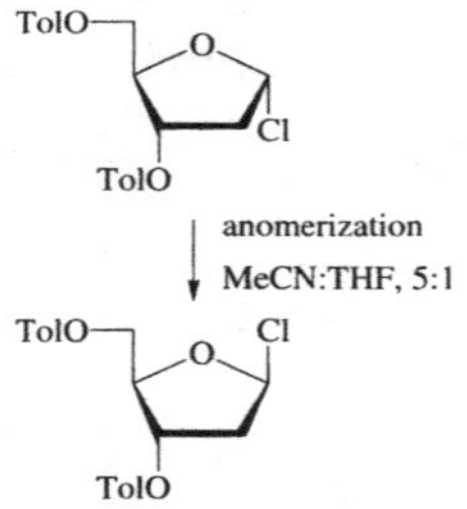 TolO, TolO, Cl; anomerization MeCN:THF, 5:1; TolO, TolO, Cl	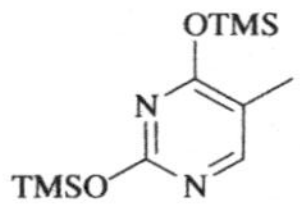 OTMS, TMSO, N, N	MeCN, THF, 4°, 15 h	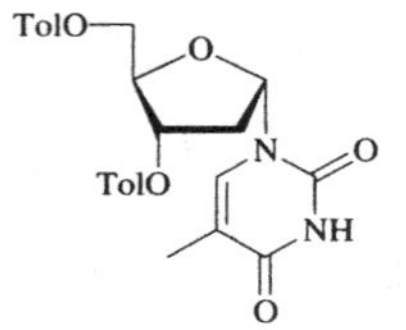 TolO, TolO, N, O, NH, O (61) + β-anomer (19)	1168
		MeCN:THF (5:1), 4°, 18 h	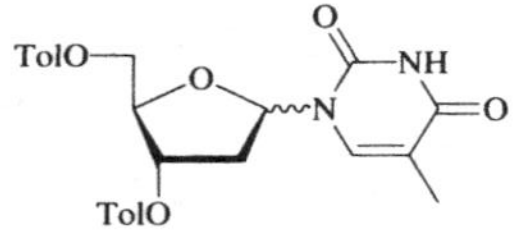 TolO, TolO, N, O, NH, O (—) α:β = 3:1	1169
	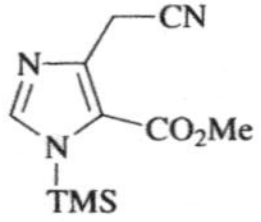 CN, N, N, TMS, CO_2Me	$Cl(CH_2)_2Cl$, rt, 72 h	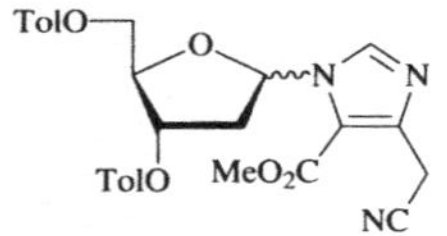 TolO, TolO, N, N, MeO_2C, NC (36)	803
	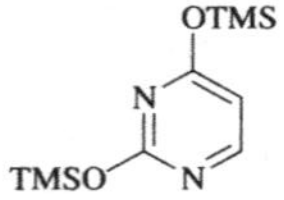 OTMS, TMSO, N, N	CuI, $CHCl_3$, rt, 2 h	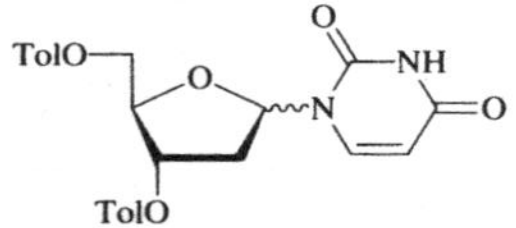 TolO, TolO, N, O, NH, O **I** (93) α:β = 8:92	273
		HgO, $HgBr_2$, C_6H_6, 20°, 10 h	**I** (38) + α-anomer (17)	80
		$CHCl_3$, rt, 24 h	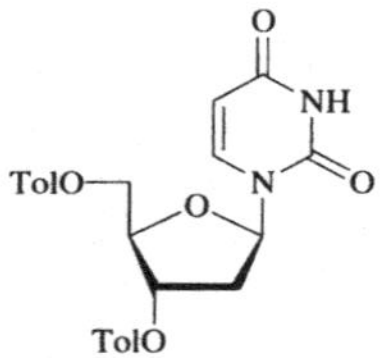 TolO, TolO, N, O, NH, O (92)	261

Reactant	Conditions	Product(s) and Yield(s) (%)	Refs.
TMSO, OTMS, F pyrimidine	CuI, $CHCl_3$, rt, 2 h	(90) α:β = 27:73	273, 273a
TMSO, OTMS, I pyrimidine	CuI, $CDCl_3$, rt, 20 h	**I** (97) α:β = 12:88	273
	$CDCl_3$, rt, 20 h	**I** (97) α:β = 35:65	273
	CuI, $CDCl_3$, rt, 5 h	**I** (87) α:β = 10:90	273
	$CDCl_3$, rt, 5 h	**I** (77) α:β = 28:72	273
	CuI, $CHCl_3$	**I** (—) α:β = 12:88	273a
TMSO, OTMS, methyl pyrimidine	CuI, $CHCl_3$, rt, 2 h	**I** (92) α:β = 7:93	273, 273a
	MeCN, rt, 24 h[c]	**I** (60) α:β = 1:0.2	261
	MeCN, THF, 10°, 10 h	**I** (61) + β-anomer (20)	803
	$Hg(OAc)_2$, DMF, 20°, 48 h	**I** (27) + β-anomer (21)	80

TABLE VIII. REACTIONS OF SILYLATED BASES WITH PROTECTED SUGARS WITH OR WITHOUT CATALYSTS (*Continued*)

Sugar	Base	Conditions	Product(s) and Yield(s) (%)	Refs.
	OTMS, N, TMSO, N	$CHCl_3$, rt, 12-24 h	O, NH, TolO, O, N, O, TolO (78-83)	261, 1170
	OTMS, N, 13 C, TMSO, N	$CHCl_3$, rt, 12 h	O, NH, 13 C, TolO, N, O, O, TolO (73)	1150, 1171
	OTMS, Et, N, TMSO, N	CuI, $CHCl_3$, rt, 2 h	TolO, O, NH, O, N, O, TolO, Et **I** (92) α:β = 7:93	273, 273a
	OTMS, Pr-*i*, N, TMSO, N	$Cl(CH_2)_2Cl$, MS, rt, 12 h	TolO, O, N, O, TolO, NH, *i*-Pr, O (—)	289
	OTMS, R[1], N, TMSO, N		O, R[1], NH, N, O, R; R = TolO, O, TolO	

Reactant	Conditions	Product(s) and Yield(s) (%)	Refs.
R^1			
Pr-*i*	$CHCl_3$, rt, 20-24 h	(95)	1172
Bu-*t*	$CHCl_3$, rt, 20-24 h	(98)	1172
C≡CH	$HgBr_2$, $Cl(CH_2)_2Cl$, 23°, 23 h	(67) $\alpha:\beta = 1:1.45$	872
c-C_3H_5	$CHCl_3$, rt, 20-24 h	(92)	1172
C_5H_9	$CHCl_3$, rt, 20-24 h	(96)	1172
C_6H_{11}	$CHCl_3$, rt, 20-24 h	(99)	1172
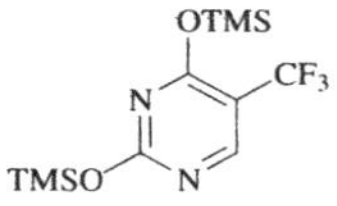	$Hg(OAc)_2$, C_6H_6, rt, 18 h	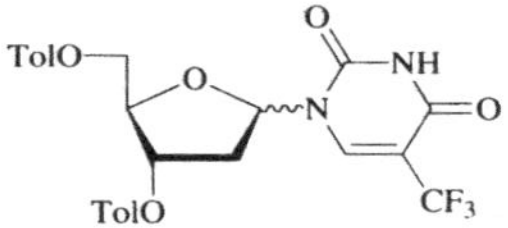 **I** (86) $\alpha:\beta = 1:2$	268
	$Hg(OAc)_2$, C_6H_6, 25°, 3 d	**I** (91) $\alpha:\beta = 1:2$	529
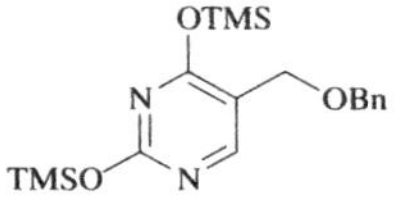	MeCN, MS, 25°, 2 d	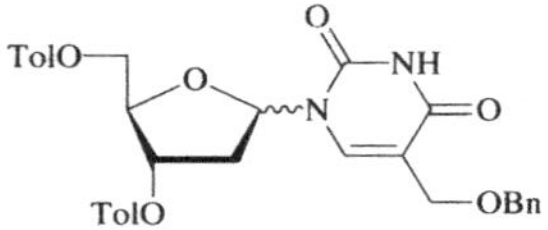 **I** (71) $\alpha:\beta = 19:10$	1174
	C_6H_6, MS, 25°, 2 d	**I** (66) $\alpha:\beta = 10:13$	1174
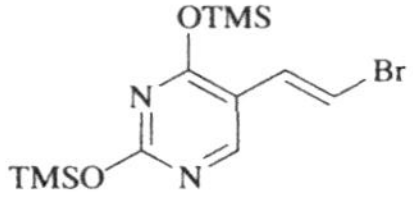	$CHCl_3$, rt, 24 h	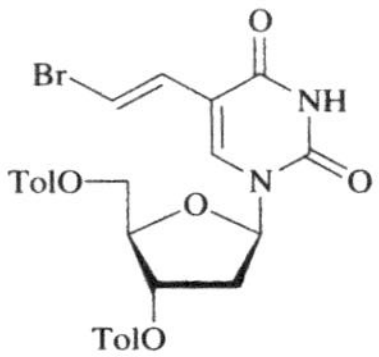 (72)	261
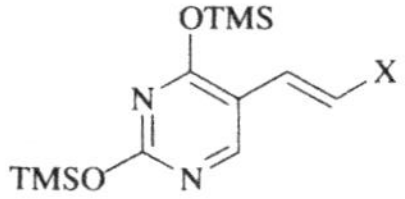	$Cl(CH_2)_2Cl$, 22°, 8 h	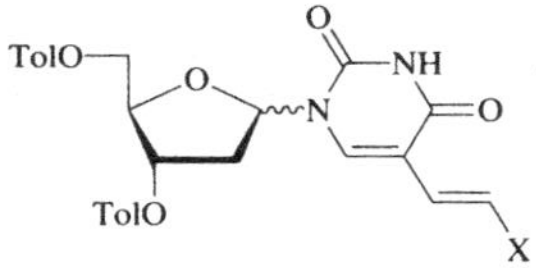 X: Br (88) $\alpha:\beta$ 1.3:1; I (—) —	1175

TABLE VIII. REACTIONS OF SILYLATED BASES WITH PROTECTED SUGARS WITH OR WITHOUT CATALYSTS (*Continued*)

Sugar	Base	Conditions	Product(s) and Yield(s) (%)	Refs.
	OTMS; OTMS; TMSO; N; N	$HgCl_2$, C_6H_6, rt, 30 h	HO; O; NH; O; N; TolO; O; TolO (64) + α-anomer (15)	1176
	OTMS; OTMS; TMSO; N; N	$CHCl_3$, rt, 12 h	HO; O; NH; O; N; TolO; O; TolO (68)	1006
	OTMS; CHO; TMSO; N; N	MeCN, MS, 25°, 2 d	TolO; O; O; NH; N; O; CHO; TolO (31)	1174
	OTMS; OMe; TMSO; N; N	1. MeCN, rt, 64 h 2. NaOMe, MeOH, 6 h	MeO; O; NH; O; N; HO; O; HO (31)	1178
	OTMS; R[1]; TMSO; N; N		TolO; O; O; NH; N; O; R[1]; TolO	

R[1]	Conditions	Product(s) and Yield(s) (%)	Refs.
NO_2	CuI, $CHCl_3$, rt, 5 h	(90) mainly β	273
2-furyl	MeCN, rt, 12 h	(—) α:β = 2:1	750
2-thienyl	$Cl(CH_2)_2Cl$, MS, rt, 12 h	(—) α:β = 1:1.68	750
3-methyl-2-thienyl	$Cl(CH_2)_2Cl$, rt, 12 h	(—)	275
3-methyl-2-thienyl	CuI, $Cl(CH_2)_2Cl$, rt, 12 h	(73) α:β = 4.4:1	275
3-*n*-hexyl-2-thienyl	$Cl(CH_2)_2Cl$, rt, 12 h	(—)	275
3-*n*-hexyl-2-thienyl	CuI, $Cl(CH_2)_2Cl$, rt, 12 h	(87) α:β = 1.6:1	275
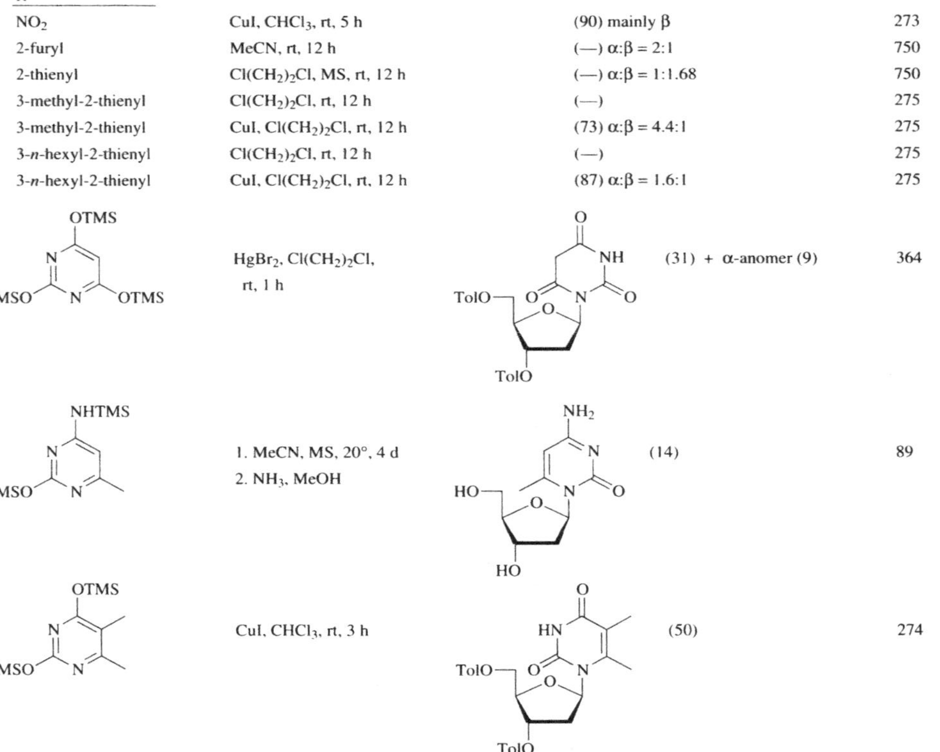	$HgBr_2$, $Cl(CH_2)_2Cl$, rt, 1 h	(31) + α-anomer (9)	364
	1. MeCN, MS, 20°, 4 d 2. NH_3, MeOH	(14)	89
	CuI, $CHCl_3$, rt, 3 h	(50)	274

TABLE VIII. REACTIONS OF SILYLATED BASES WITH PROTECTED SUGARS WITH OR WITHOUT CATALYSTS (*Continued*)

Sugar	Base	Conditions	Product(s) and Yield(s) (%)	Refs.
	NHTMS, TMSO, N, N, R^1		TolO, O, N, N, O, NH_2, TolO, R^1	
	R^1			
	c-C_3H_5	$Cl(CH_2)_2Cl$, rt, 12 h	(—)	275
	c-C_3H_5	CuI, $Cl(CH_2)_2Cl$, rt, 12 h	(77) $\alpha:\beta = 3.6:1$	275
	2-furyl	$Cl(CH_2)_2Cl$, rt, 12 h	(—)	275
	2-furyl	CuI, $Cl(CH_2)_2Cl$, rt, 12 h	(75) $\alpha:\beta = 3.2:1$	275
	3-furyl	$Cl(CH_2)_2Cl$, rt, 12 h	(—)	275
	3-furyl	CuI, $Cl(CH_2)_2Cl$, rt, 12 h	(76) $\alpha:\beta = 1:1.6$	275
	2-thiophenyl	$Cl(CH_2)_2Cl$, rt, 12 h	(—)	275
	2-thiophenyl	CuI, $Cl(CH_2)_2Cl$, rt, 12 h	(79) $\alpha:\beta = 1:1.14$	275
	2-selenophenyl	$Cl(CH_2)_2Cl$, rt, 12 h	(—)	275
	2-selenophenyl	CuI, $Cl(CH_2)_2Cl$, rt, 12 h	(68) $\alpha:\beta = 1.2:1$	275
	2-thiazolyl	$Cl(CH_2)_2Cl$, rt, 12 h	(—)	275
	2-thiazolyl	CuI, $Cl(CH_2)_2Cl$, rt, 12 h	(69) $\alpha:\beta = 1:1.3$	275
	2-*N*-methylpyrryl	$Cl(CH_2)_2Cl$, rt, 12 h	(—)	275
	2-*N*-methylpyrryl	CuI, $Cl(CH_2)_2Cl$, rt, 12 h	(82) $\alpha:\beta = 1.4:1$	275
	2-(5-phenyl)-thiophenyl	$Cl(CH_2)_2Cl$, rt, 12 h	(—)	275
	2-(5-phenyl)-thiophenyl	CuI, $Cl(CH_2)_2Cl$, rt, 12 h	(59) $\alpha:\beta = 1:1.14$	275
	Ph	$Cl(CH_2)_2Cl$, rt, 12 h	(—)	275
	Ph	CuI, $Cl(CH_2)_2Cl$, rt, 12 h	(49) $\alpha:\beta = 1:1.1$	275
	Ph	$HgBr_2$, MS, MeCN, rt, 10 d	(44) $\alpha:\beta = 4:1$	272
	2-pyridyl	$Cl(CH_2)_2Cl$, rt, 12 h	(—)	275
	2-pyridyl	CuI, $Cl(CH_2)_2Cl$, rt, 12 h	(78) $\alpha:\beta = 1:1.88$	275
	3-pyridyl	$Cl(CH_2)_2Cl$, rt, 12 h	(—)	275
	3-pyridyl	CuI, $Cl(CH_2)_2Cl$, rt, 12 h	(58) $\alpha:\beta = 2.4:1$	275
	4-pyridyl	$Cl(CH_2)_2Cl$, rt, 12 h	(—)	275

Reactant	Conditions	Product(s) and Yield(s) (%)	Refs.
OTMS, R[1], TMSO (triazine)	CuI, $CHCl_3$, rt, 2 h	TolO, TolO; R[1] H (92) α:β 8:92; Me (92) α:β 3:97	273, 273a
OTMS, R[1], TMSO (triazine)	CuI, $CHCl_3$, rt, 3 h	R[1] H (70); Me (76); R = TolO, TolO (sugar)	274
	$Hg(CN)_2$, C_6H_6, rt, 3 d	R[1] = H, (25)	1179
OTMS, R[1], TMSO (triazine)	$CHCl_3$, rt, 12 h	R[1] c-C_3H_5 (81); 2-thienyl (94)	1180
Cl, AcN(TMS), TMS (purine)	$Hg(CN)_2$, C_6H_6, reflux, 1.5 h	TolO, TolO, Cl, AcNH; (84) α:β = 1:1	1134
OTMS, TMSO, TMS, TMS	1. Et_3N, $CHCl_3$, 0°, 1 h, H_2O 2. NaOMe, rt, 2-3 h	HO, HO; (41) + α-anomer (26)	1167

TABLE VIII. REACTIONS OF SILYLATED BASES WITH PROTECTED SUGARS WITH OR WITHOUT CATALYSTS (*Continued*)

Sugar	Base	Conditions	Product(s) and Yield(s) (%)	Refs.
	OTMS, TMSO, N, N, N, N, R^1, R^1		O, HN, N, R^1, O, N, N, R^1, R — **I** — R = TolO, O, TolO	
	R^1			
	H	CH_2Cl_2, 20°, 6 d	**I** (37) + α-anomer (37)	104
	H	CaI_2, $CHCl_3$, rt, 15 min	**I** (72) + α-anomer (7)	1029
	Me	C_6H_6, 20°, 7 d	**I** (43) + α-anomer (25)	104
	Ph	C_6H_6, 20°, 4 d	**I** (58) + α-anomer (22)	104
			R, O, N, N, R^1, HN, N, R^1, O — **I** — R = TolO, O, TolO	
	R^1			
	H	HgO, $HgBr_2$, C_6H_6, reflux, 2 h	**I** (17) + β-anomer (5) + bis(riboside) (3)	905
	Me	"	**I** (23) + β-anomer (8) + N^1-α,β-isomers (2)	905
	Ph	"	**I** (29) + β-anomer (12)	905
	Ph	CH_2Cl_2, 20 °, 2 h	**I** (48) + β-anomer (37)	104
	NHTMS, N, N, N, N, Ph, OTMS	HgO, $HgBr_2$, MS, C_6H_6, rt, 48 h	R, Ph, N, N, O, N, N, NH_2 (43)	1098
TolO, S, Br, TolO	OTMS, N, Br, TMSO, N	CCl_4, 90-100°, 1 h	TolO, S, O, NH, N, O, TolO, Br (50) α:β = 1.8:1	1162

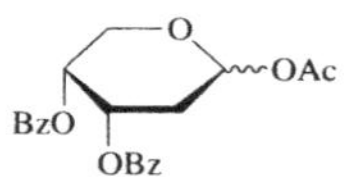	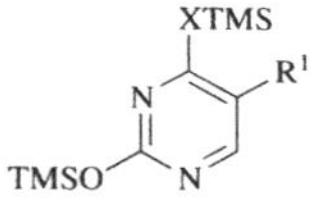	KI, 18-crown-6, MeCN:PhMe (1:1), reflux, 2-4 h	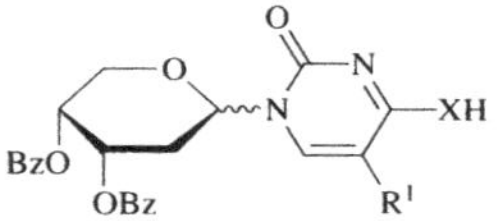	R^1 X α:β H NH (71) 58:42 Me O (72) 60:40	396
	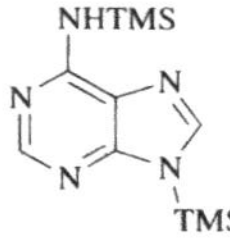	KI, 18-crown-6, MeCN:PhMe (1:1), reflux, 2-4 h	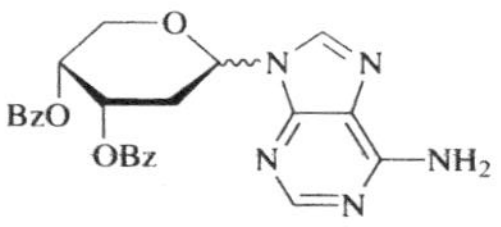	(71) α:β = 44:56	396
	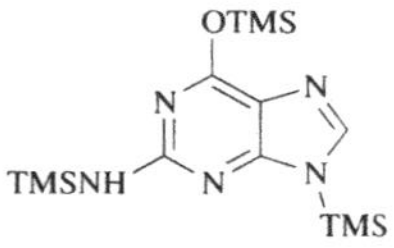	KI, 18-crown-6, MeCN:PhMe (1:1), reflux, 2-4 h	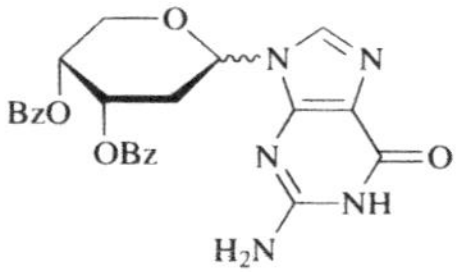	(85) α:β = 58:42	396
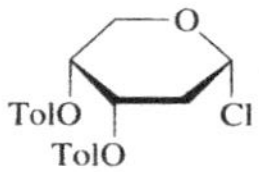	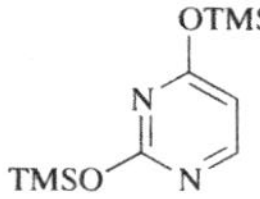	$HgBr_2$, HgO	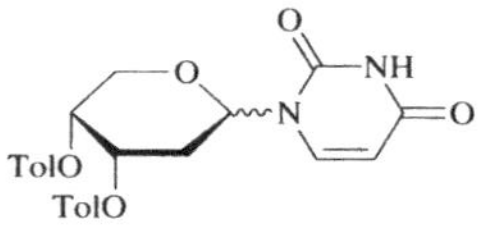	(78)	997
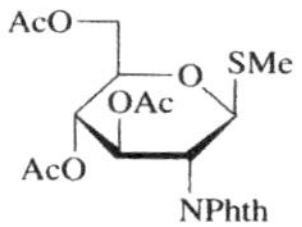	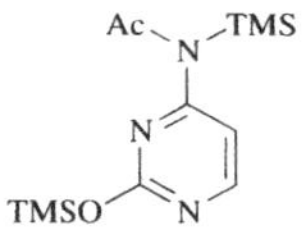	NIS, $Cl(CH_2)_2Cl$, TfOH, rt, 1 h	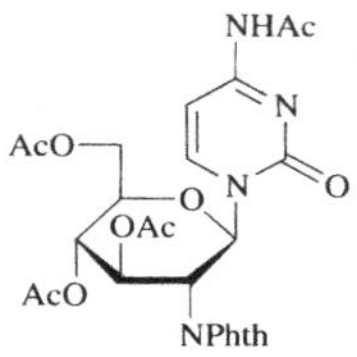	(70)	408
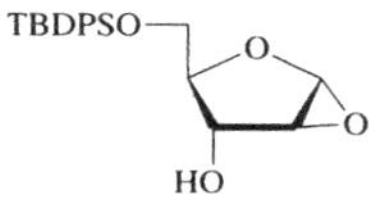	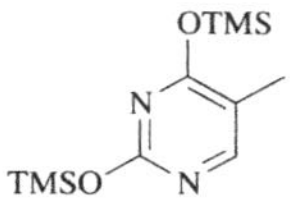	1. MeCN, rt, 19.5 h 2. TBAF 3. Ac_2O, DMAP, rt, 12 h	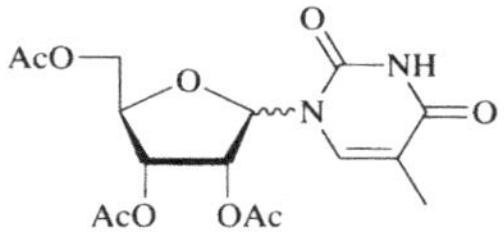	(36) α:β = 1:4	112

TABLE VIII. REACTIONS OF SILYLATED BASES WITH PROTECTED SUGARS WITH OR WITHOUT CATALYSTS (*Continued*)

Sugar	Base	Conditions	Product(s) and Yield(s) (%)	Refs.
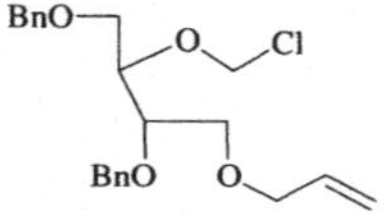 BnO, O, Cl, BnO, O	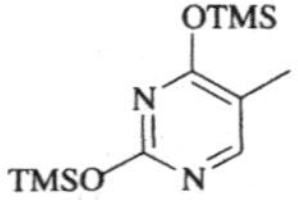 OTMS, N, TMSO, N	Bu_4NI, CH_2Cl_2, reflux, 12 h	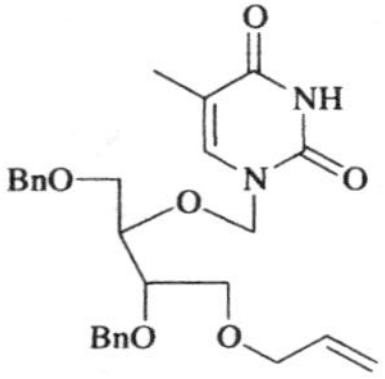 O, NH, N, O, BnO, O, BnO, O (99)	1145
C_{23}				
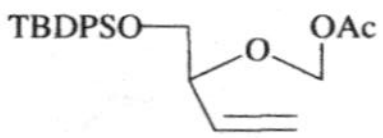 TBDPSO, O, OAc	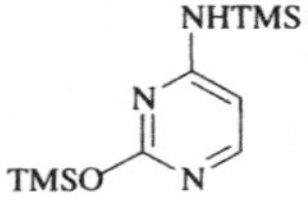 NHTMS, N, TMSO, N	KI, dibenzo-18-crown-6, MeCN, PhMe, 80°, 2-8 h	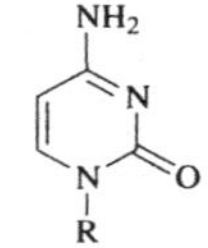 NH_2, N, N, O, R (73); R = TBDPSO, O 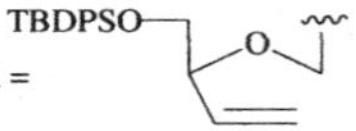	1181
	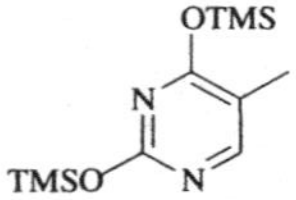 OTMS, N, TMSO, N	KI, dibenzo-18-crown-6, MeCN, PhMe, 80°, 2-8 h	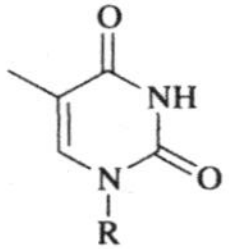 O, NH, N, O, R (90)	1181
	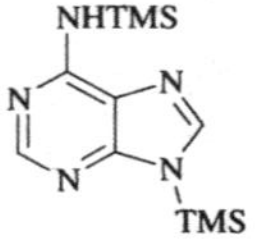 NHTMS, N, N, N, N, TMS	KI, dibenzo-18-crown-6, MeCN, PhMe, 80°, 2-8 h	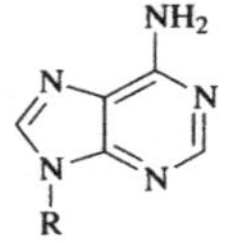 NH_2, N, N, N, N, R (65)	1181
	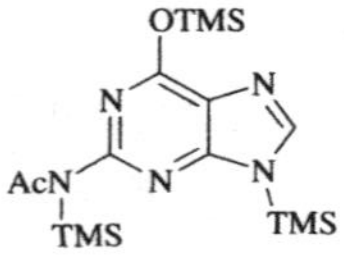 OTMS, N, N, AcN, TMS, N, N, TMS	KI, dibenzo-18-crown-6, MeCN, PhMe, 80°, 2-8 h	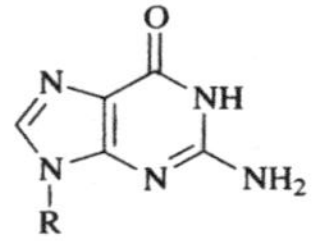 O, N, NH, N, R, N, NH_2 (98)	1181

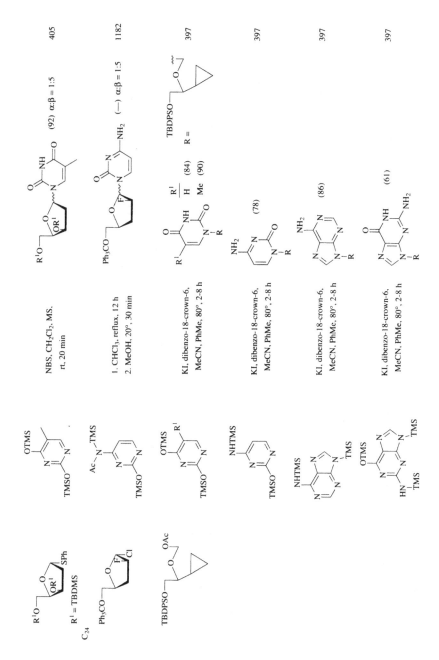

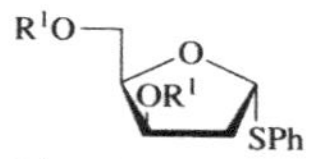 $R^1 = TBDMS$	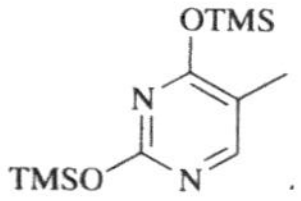 OTMS, TMSO, N	NBS, CH_2Cl_2, MS, rt, 20 min	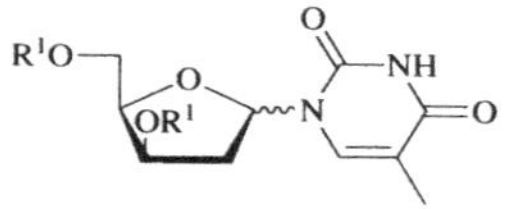 R^1O, OR^1, O, NH, N (92) α:β = 1:5	405
C_{24}				
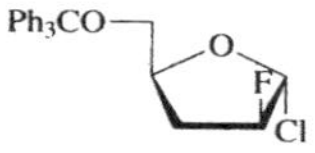 Ph_3CO, F, Cl	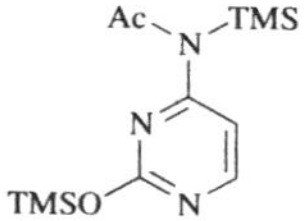 Ac, N, TMS, TMSO	1. $CHCl_3$, reflux, 12 h 2. MeOH, 20°, 30 min	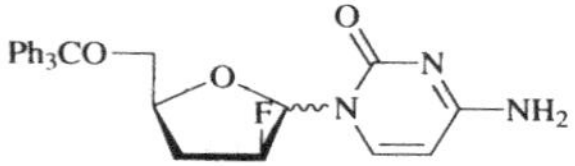 Ph_3CO, F, NH_2 (—) α:β = 1:5	1182
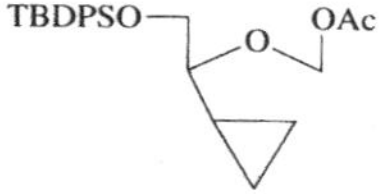 TBDPSO, OAc	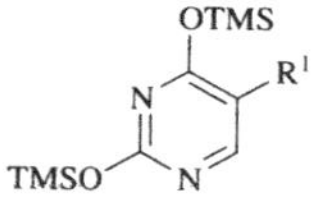 OTMS, R^1, TMSO	KI, dibenzo-18-crown-6, MeCN, PhMe, 80°, 2-8 h	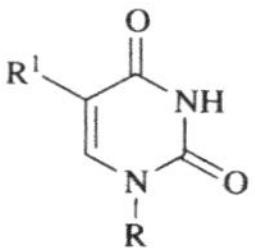 R^1: H (84); Me (90) R 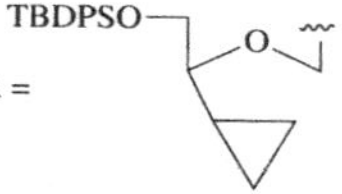 = TBDPSO	397
	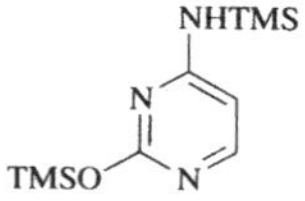 NHTMS, TMSO	KI, dibenzo-18-crown-6, MeCN, PhMe, 80°, 2-8 h	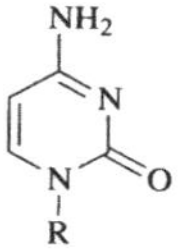 NH_2, R (78)	397
	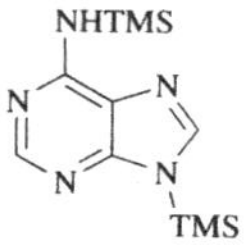 NHTMS, TMS	KI, dibenzo-18-crown-6, MeCN, PhMe, 80°, 2-8 h	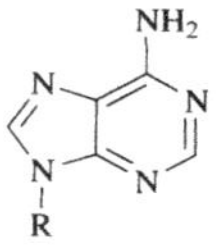 NH_2, R (86)	397
	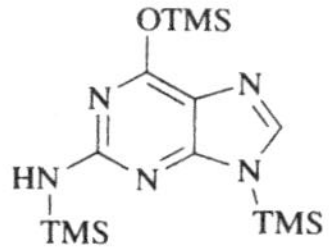 OTMS, HN, TMS, TMS	KI, dibenzo-18-crown-6, MeCN, PhMe, 80°, 2-8 h	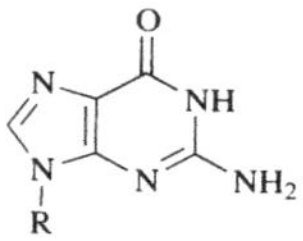 O, NH, NH_2, R (61)	397

TABLE VIII. REACTIONS OF SILYLATED BASES WITH PROTECTED SUGARS WITH OR WITHOUT CATALYSTS (*Continued*)

Sugar	Base	Conditions	Product(s) and Yield(s) (%)	Refs.
TBDPSO, TMSO (anhydro sugar)	OTMS, TMSO (silylated thymine)	MeCN, rt, 18 h	TBDPSO, R[1]O, TMSO, N, O, NH, O — R[1]: TMS (52); H (30)	112
C_{25} R[1]O, OR[1], SPh	OTMS, TMSO (silylated thymine)	NBS, CH_2Cl_2, MS, rt, 20 min	R[1]O, OR[1], O, NH, O — R[1] / α:β: Bz (86) 1:2; Bn (92) 1:4	405
C_{26} BzO, Cl, BzO, OBz	TMSNH, CO_2Et, R[1], N, N, TMS	1. MeCN, rt, 5 d 2. NaOMe, MeOH	H_2N, CO_2Et, R^2, N, N, R — **I** (22); R = HO, HO, OH (ribofuranosyl)	1183
	R[1]: H	Base:Sugar 4:5	R^2: H (22)	
	R[1]: H	Base:Sugar 23.5:20	R^2: H (64)	
	R[1]: NHTMS	Base:Sugar 4:5	R^2: NH_2 (60)	
	R[1]: NHTMS	Base:Sugar 1:1	R^2: NH_2 (53)	
	MeO_2C, MeO_2C, N, N, TMS	$CHCl_3$, rt, 18 h	CO_2Me, CO_2Me, N, N, BzO, BzO, OBz — (—)	853

Reactant	Conditions	Product(s) and Yield(s) (%)	Refs.
	$Hg(OAc)_2$, C_6H_6, 20°, 60 h	(65)	80
	HgO, $HgBr_2$, C_6H_6, 110°, 5 h	**I** (80) R =	80
	$Hg(OAc)_2$, C_6H_6, 20°, 60 h	**I** (65)	80
	HgO, $HgBr_2$, C_6H_6, 100°, 2-5 h	(86)	80
	$Hg(OAc)_2$, MeCN, MS, reflux, 2 h	(11) + bis(riboside) (33) + N^1-riboside (4)	60
	$Hg(OAc)_2$, MeCN, MS, reflux	R^1: H, 2 h (6); Me, 3 h (8)	60
	$Hg(OAc)_2$, MeCN, MS, reflux, 4 h	(4)	60

TABLE VIII. REACTIONS OF SILYLATED BASES WITH PROTECTED SUGARS WITH OR WITHOUT CATALYSTS (*Continued*)

Sugar	Base	Conditions	Product(s) and Yield(s) (%)	Refs.
	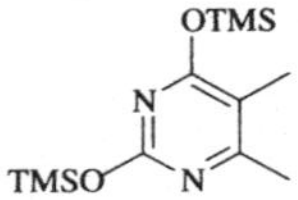 TMSO, OTMS	HgO, $HgBr_2$, C_6H_6, 100°, 7 h	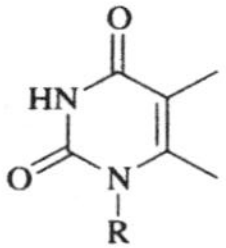 HN, O, N–R (42)	1143
	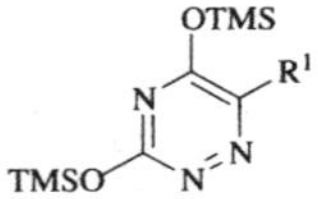 TMSO, OTMS, R^1	$Hg(OAc)_2$, PhMe, rt	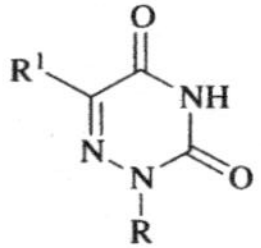 R^1, NH, O, N–R; R^1: Br (48), *N*-morpholinyl (50)	885
	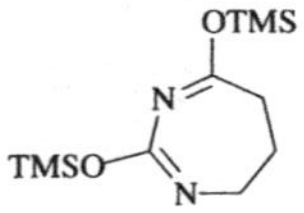 TMSO, OTMS	HgO, $HgBr_2$, C_6H_6, reflux, 18 h	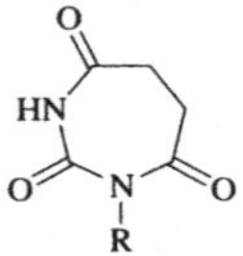 HN, O, N–R (33) 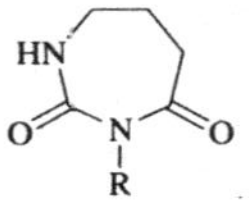 + HN, O, N–R (21)	1184
	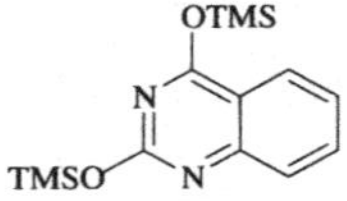 TMSO, OTMS	HgO, $HgBr_2$, C_6H_6, 100°, 7 h	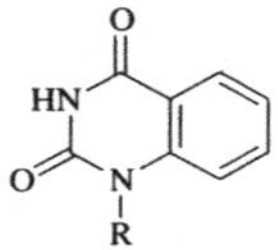 HN, O, N–R (84) R = BzO, O, BzO, OBz 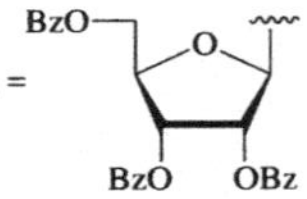	1143
	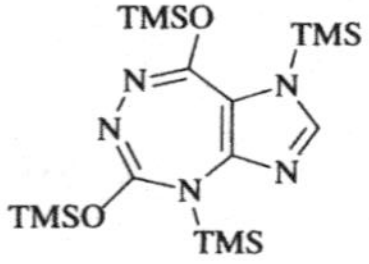 TMSO, TMSO, TMS, TMS	$Hg(CN)_2$, PhMe, reflux, 2-3 h	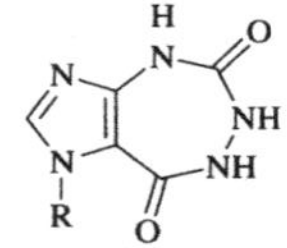 H, N, O, NH, NH, N–R, O (39) + N^3-isomer (4)	507

Glycosyl donor	Silylated base	Conditions	Product(s) and Yield(s) (%)		Refs.
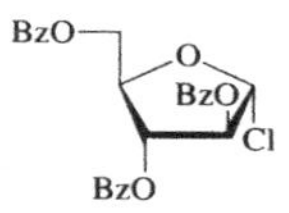 BzO—, BzO, BzO, Cl (furanose)	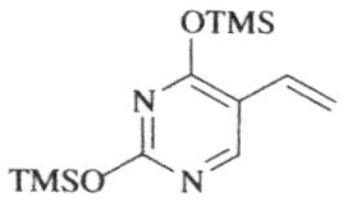 OTMS, TMSO (pyrimidine, vinyl)	$Cl(CH_2)_2Cl$ or MeCN	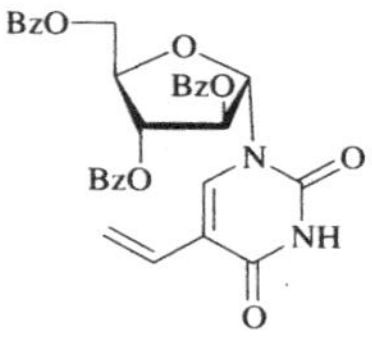 BzO—, BzO, BzO (nucleoside, NH, O, O)	(—)	363
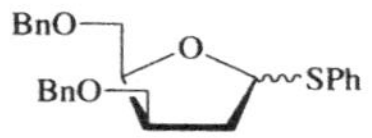 BnO—, BnO, SPh (furanose)	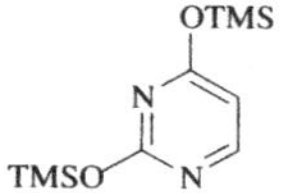 OTMS, TMSO (pyrimidine)	NBS, CH_2Cl_2, rt, 20 min	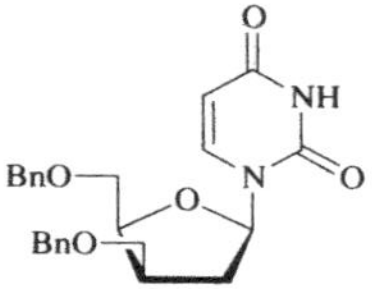 BnO—, BnO (nucleoside, NH, O, O)	(72) α:β = 1:4	1185
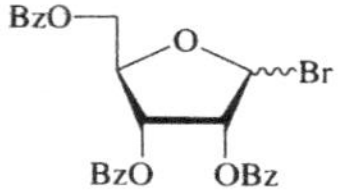 BzO—, Br, BzO, OBz (furanose)	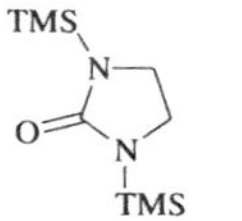 TMS, O, TMS (imidazolidinone)	HgO, $HgBr_2$, C_6H_6, reflux, 18 h	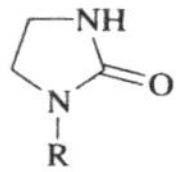 NH, O, N—R	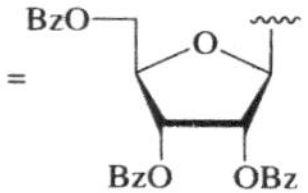 (49) R = BzO—, BzO, OBz (furanosyl)	511
	MeS, N, N—TMS (imidazole)	MeCN, rt, 4 d	R—N, S, N—R	(50)	701
	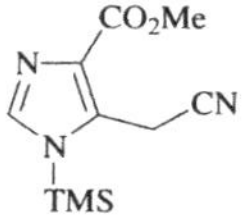 CO_2Me, CN, N—TMS (imidazole)	MeCN, rt, 3 d	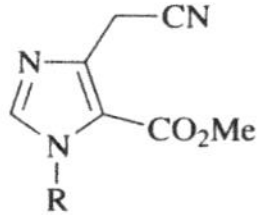 CN, CO_2Me, N—R	(48) + N^3-isomer (6)	704, 707
	N, N, N—TMS (triazole)	MeCN, rt, 4 d	N, N, N—R	(54)	258

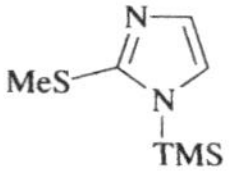

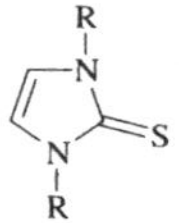

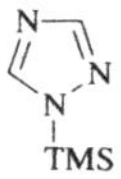

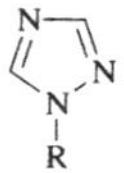

TABLE VIII. REACTIONS OF SILYLATED BASES WITH PROTECTED SUGARS WITH OR WITHOUT CATALYSTS (*Continued*)

Sugar	Base	Conditions	Product(s) and Yield(s) (%)	Refs.
	TMS, TMS	HgO, $HgBr_2$, C_6H_6, reflux, 18 h	NH, O, N, R (76)	511
	H, N, TMS, MeO, N	1. MeCN, rt, 2.5 d 2. NH_3, MeOH	BzNH, HO, O, N, O, HO, OH (32)	1186
	X, R^1, TMSO, N	MeCN	R^1, X, N, O, R; R = BzO, O, BzO, OBz	
	X = CH, R^1 = NO_2	rt, 2 d	(46)	1186
	X = CH, R^1 = CO_2Me	rt, 3 d	(76)	1186
	X = N, R^1 = H	reflux, 5 h	(27)	1187
	OTMS, N, TMSO, N	$HgBr_2$, rt	O, NH, N, O, R (90)	1142
	NHTMS, N, TMSO, N	MeCN, rt, 3 d	NH_2, N, O, N, R (50)	89

Conditions	Yield (%)	Refs.
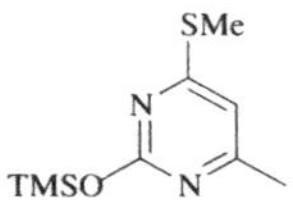 1. MeCN, rt, 3 d 2. NH_3, MeOH	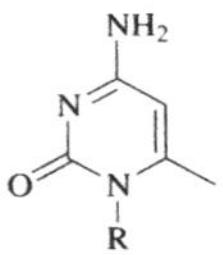 (8)	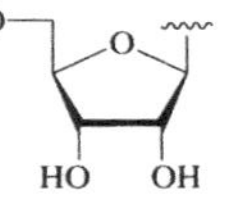 89
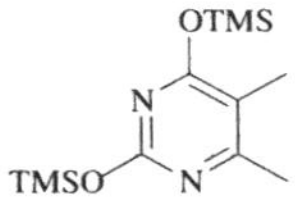 1. MeCN, rt, 3 d 2. NH_3, MeOH, rt, 12 h	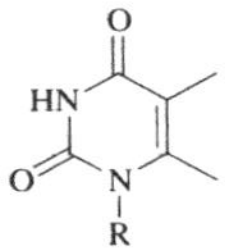 (11)	89
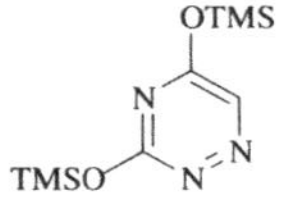 $HgBr_2$	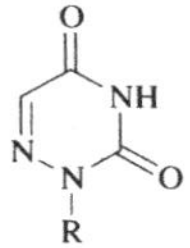 (92)	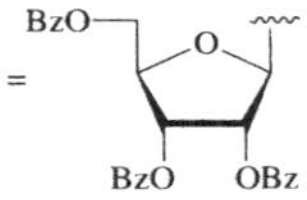 1142
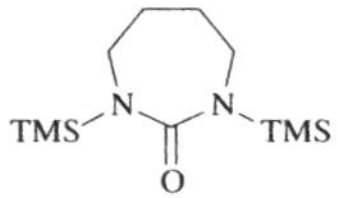 HgO, $HgBr_2$, C_6H_6, reflux, 18 h	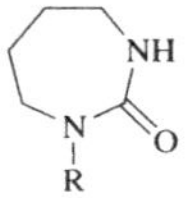 (46)	511
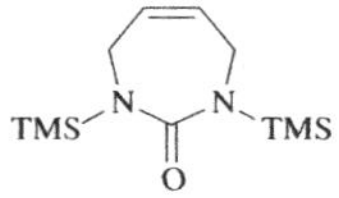 HgO, $HgBr_2$, C_6H_6, reflux, 18 h	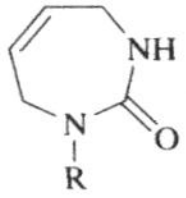 (60)	511
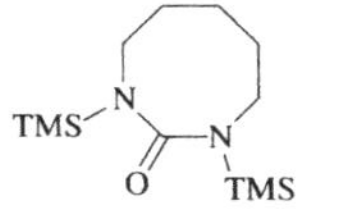 HgO, $HgBr_2$, C_6H_6, reflux, 18 h	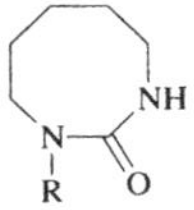 (42)	511
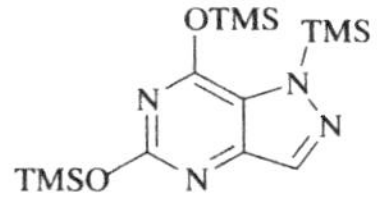 MeCN, rt, 18 h	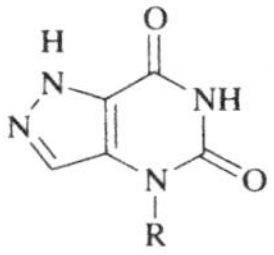 (60)	362

TABLE VIII. REACTIONS OF SILYLATED BASES WITH PROTECTED SUGARS WITH OR WITHOUT CATALYSTS (*Continued*)

Sugar	Base	Conditions	Product(s) and Yield(s) (%)	Refs.
	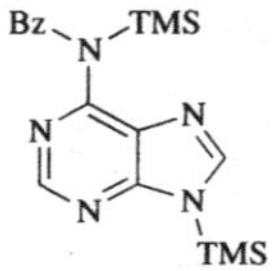 Bz(TMS)N-, N-TMS	$HgBr_2$, C_6H_6, rt, 3 h	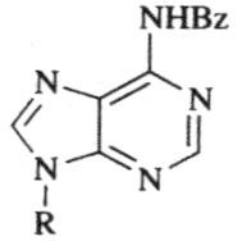 NHBz, N-R (40) + N^7-isomer (30)	1142
	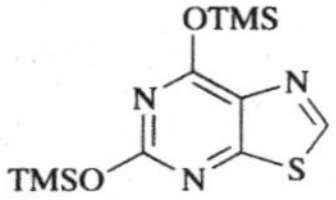 OTMS, TMSO	DMF, rt, 3 d	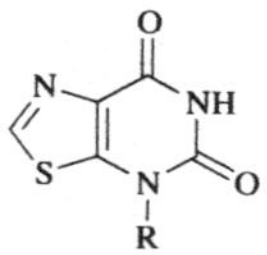 O, NH, O, N-R (46) + α-anomer (19)	901
	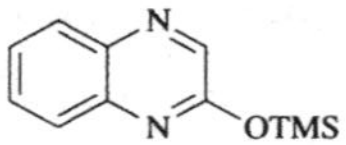 OTMS	HgO, $HgBr_2$, C_6H_6, reflux, 8 h	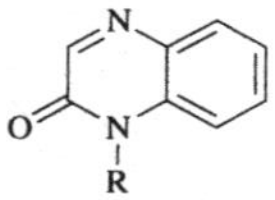 O, N-R (8)	903
	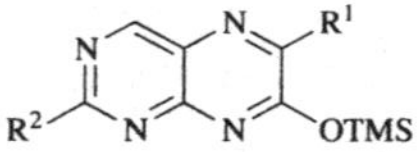 R^1, R^2, OTMS	HgO, $HgBr_2$, C_6H_6, reflux	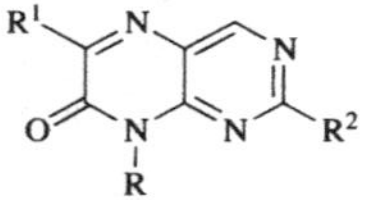 R^1, O, N-R, R^2	1144
	R^1 H, R^2 H	5 h	(41)	
	R^1 H, R^2 Me_2N	5 h	(49)	
	R^1 Me, R^2 Me_2N	3 h	(77)	
	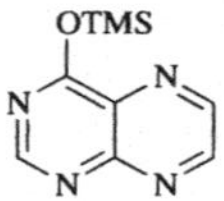 OTMS	HgO, $HgBr_2$, C_6H_6, reflux, 4 h	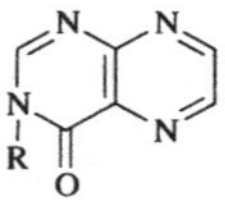 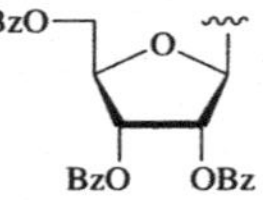 N-R, O (69) R = BzO, O, BzO, OBz	1188

Reactant	Conditions	Product(s) and Yield(s) (%)	Refs.
2,4-bis(trimethylsilyloxy)pteridine (OTMS, TMSO)	HgO, $HgBr_2$, C_6H_6, reflux, 4 h	(50) + bis(riboside) (3)	905
R^1, R^2, OTMS pteridine	HgO, $HgBr_2$, C_6H_6,	R^1, R^3 pteridinone, R = 2,3,5-tri-O-benzoyl-ribofuranosyl (BzO, BzO, OBz)	1144
R^1 = NMe_2, R^2 = NHTMS	reflux, 10 h	R^3 = NH_2 (36)	
R^1 = H, R^2 = NMe_2	80°, 5 h	R^3 = NMe_2 (49)	
TMSO, OTMS, R^1, R^2 pteridine		HN, O, N–R, R^1, R^2 lumazine	104

R^1	R^2	Conditions	Product(s) and Yield(s) (%)	Refs.
H	Ph	HgO, $HgBr_2$, C_6H_6, 80°, 4 h	(6) + N^1,N^3-bis(riboside) (22)	104
H	Ph	CH_2Cl_2, 20°, 3 d	(19) + α-anomer (17)	104
Ph	H	HgO, $HgBr_2$, C_6H_6, 80°, 4 h	(47) + N^1,N^3-bis(riboside) (12)	104
Ph	H	CH_2Cl_2, 20°, 3 d	(52) + α-anomer (9)	104
H	p-ClC_6H_4	CH_2Cl_2, 20°, 5 d	(21)	104
p-ClC_6H_4	H	CH_2Cl_2, 20°, 4 d	(46)	104
Me	Me	HgO, $HgBr_2$, C_6H_6, 80°, 4 h	(44) + bis(riboside) (3)	905
Ph	Ph	HgO, $HgBr_2$, C_6H_6, 80°, 4 h	(74) + N^3-β-isomer (5) + N^1,N^3-bis(riboside) (5)	104, 905
p-ClC_6H_4	p-ClC_6H_4	CH_2Cl_2, 20°, 4 d	(51) + α-anomer (7)	104

TABLE VIII. REACTIONS OF SILYLATED BASES WITH PROTECTED SUGARS WITH OR WITHOUT CATALYSTS (*Continued*)

Sugar	Base	Conditions	Product(s) and Yield(s) (%)	Refs.
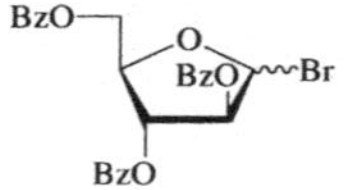	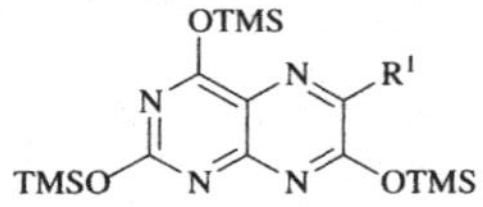	HgO, $HgBr_2$, C_6H_6, reflux, 4 h	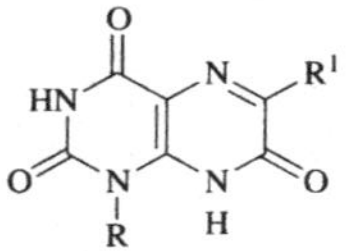 R^1: H (46); Me (34); Ph (18)	904
	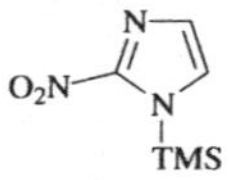	$Hg(CN)_2$, MeCN, rt, 2 d	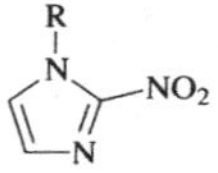 (33) + β-anomer (7) R = 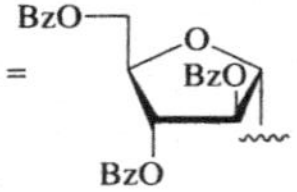	760
	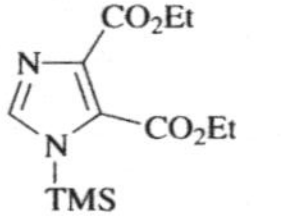	PhMe, reflux, 5 h	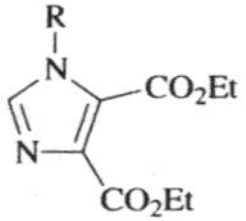 (68) + β-anomer (7)	852
	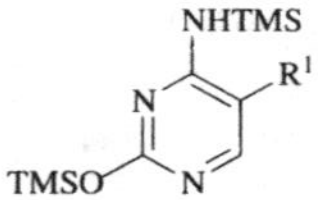	PhMe, reflux, 2 h	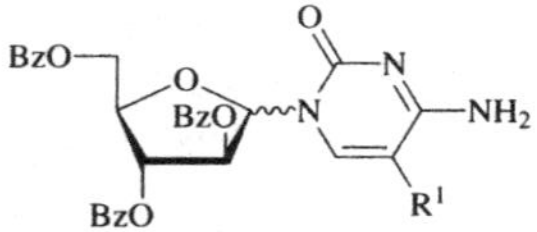 R^1: H (—); F (—)	499
	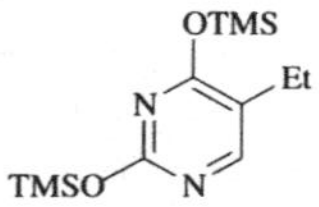	$Cl(CH_2)_2Cl$, MS, rt, 24 h	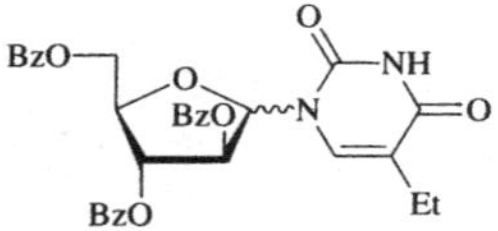 (71) α:β = 5:2	358
	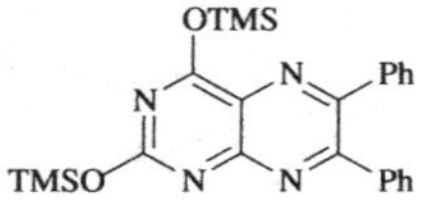	HgO, $HgBr_2$, C_6H_6, reflux, 4.5 h	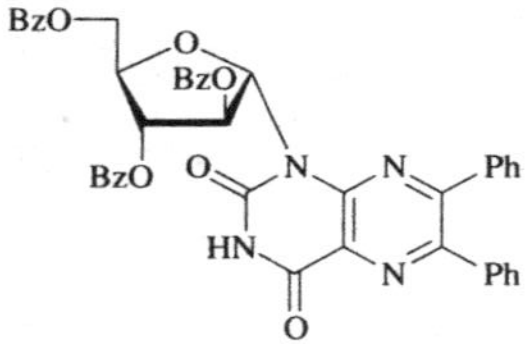 (45)	917

Conditions	Yield (%)	Refs.
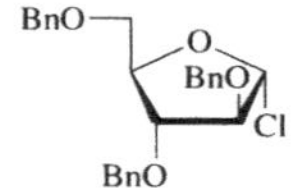 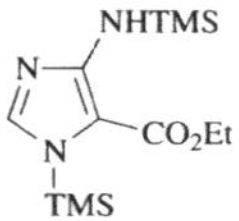 MeCN, reflux, 1 h	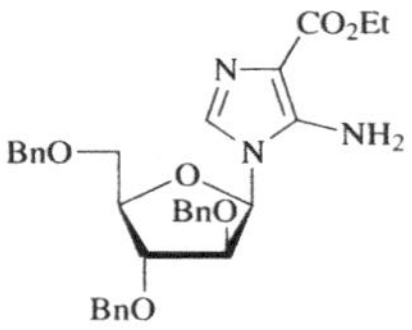 (15)	352
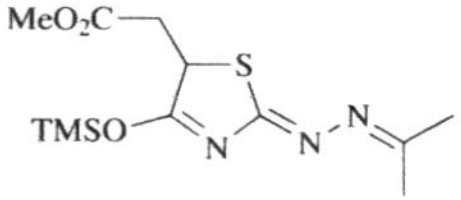 CH_2Cl_2, MS, rt, 12 h	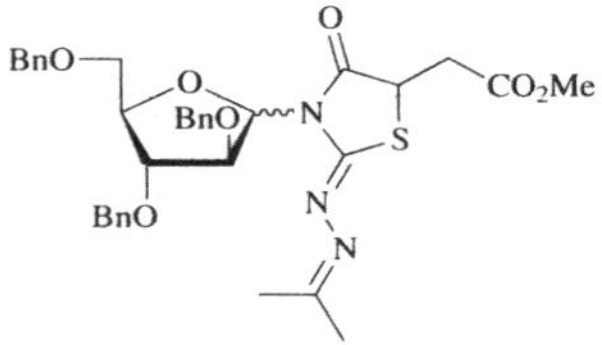 (—) α:β = 2:1	839
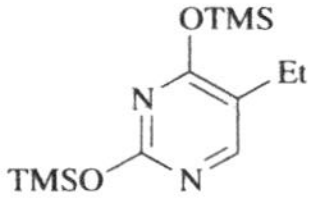 $Cl(CH_2)_2Cl$, MS, rt, 24 h	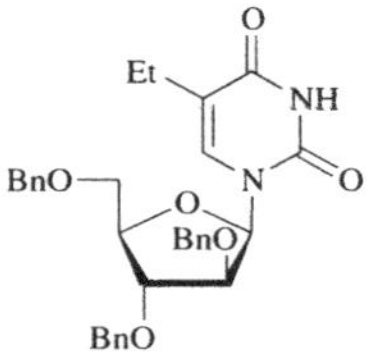 (75)	358
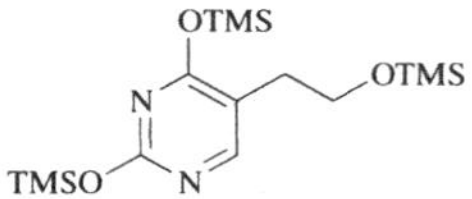 1. CH_2Cl_2, xylene, MS, rt, 2 d 2. H_2, Pd/C, EtOH	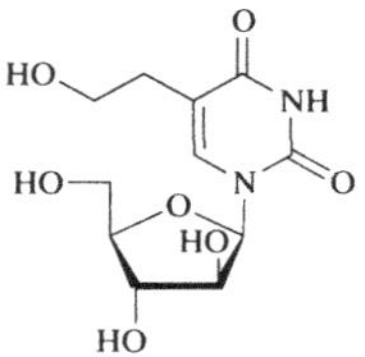 (58)	360
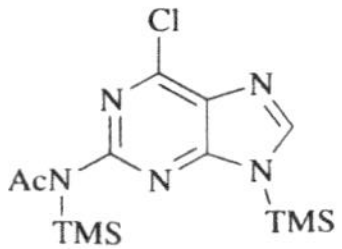 $Hg(CN)_2$, C_6H_6, reflux, 16 h	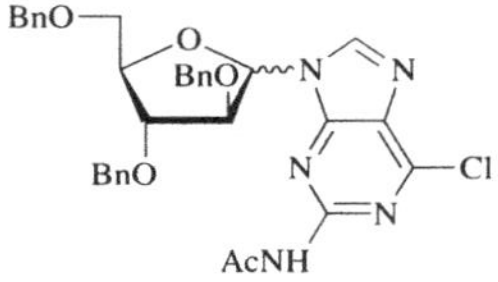 (91)	1134

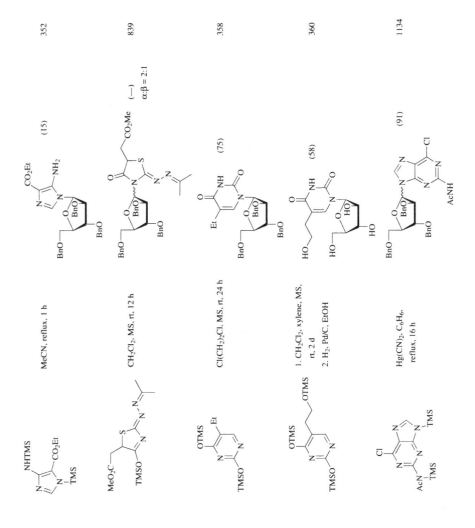

TABLE VIII. REACTIONS OF SILYLATED BASES WITH PROTECTED SUGARS WITH OR WITHOUT CATALYSTS (*Continued*)

Sugar	Base	Conditions	Product(s) and Yield(s) (%)	Refs.
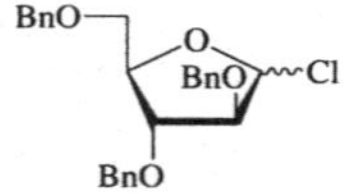 BnO, BnO, BnO, Cl	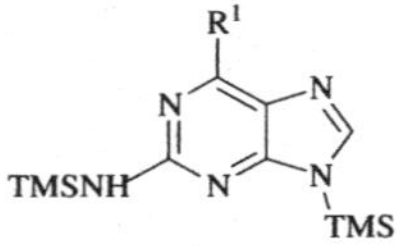 TMSNH, R^1, TMS		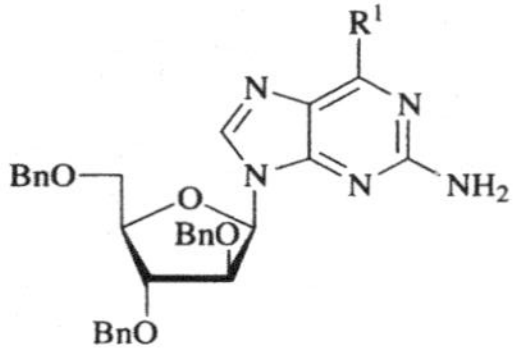 BnO, BnO, BnO, R^1, NH_2	
	R^1 = Cl	$Cl(CH_2)_2Cl$, MS, rt, 5 d	(33)	1189
	R^1 = NHTMS	1. $Cl(CH_2)_2Cl$, reflux, 18 h 2. MeOH, reflux, 15-30 min	(65)	346
	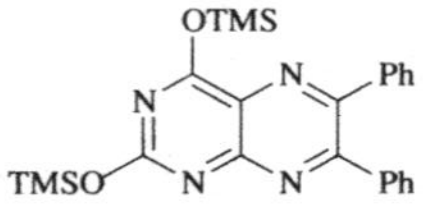 OTMS, TMSO, Ph, Ph	HgO, $HgBr_2$, C_6H_6, reflux, 4.5 h	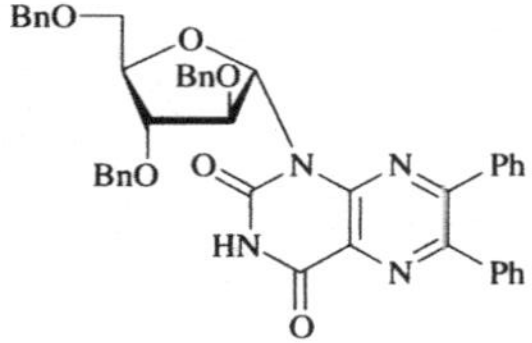 BnO, BnO, BnO, HN, Ph, Ph (29) + β-N^1-isomer (5) + N^1,N^3-α,α-bis(riboside) (34)	917
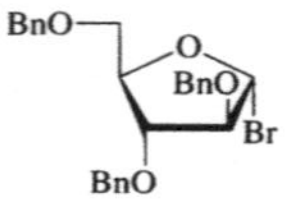 BnO, BnO, BnO, Br	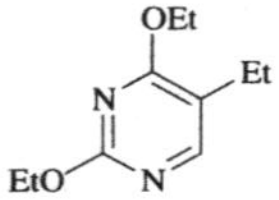 OEt, Et, EtO	MeCN, MS, rt, 5 d	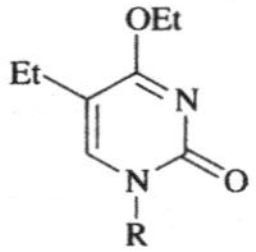 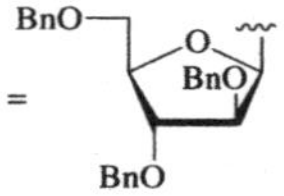 Et, OEt, N–R (49); R = BnO, BnO, BnO	358
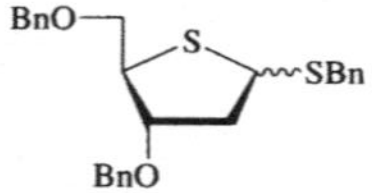 BnO, BnO, S, SBn	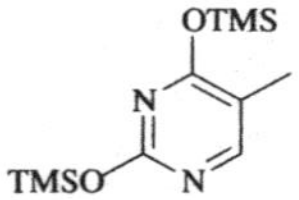 OTMS, TMSO	$HgBr_2$, $CdCO_3$, PhMe, reflux, 24 h	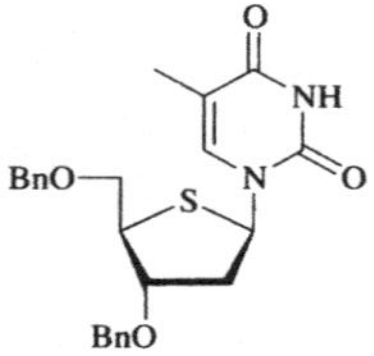 BnO, BnO, S, NH (8) + α-anomer (28)	1162

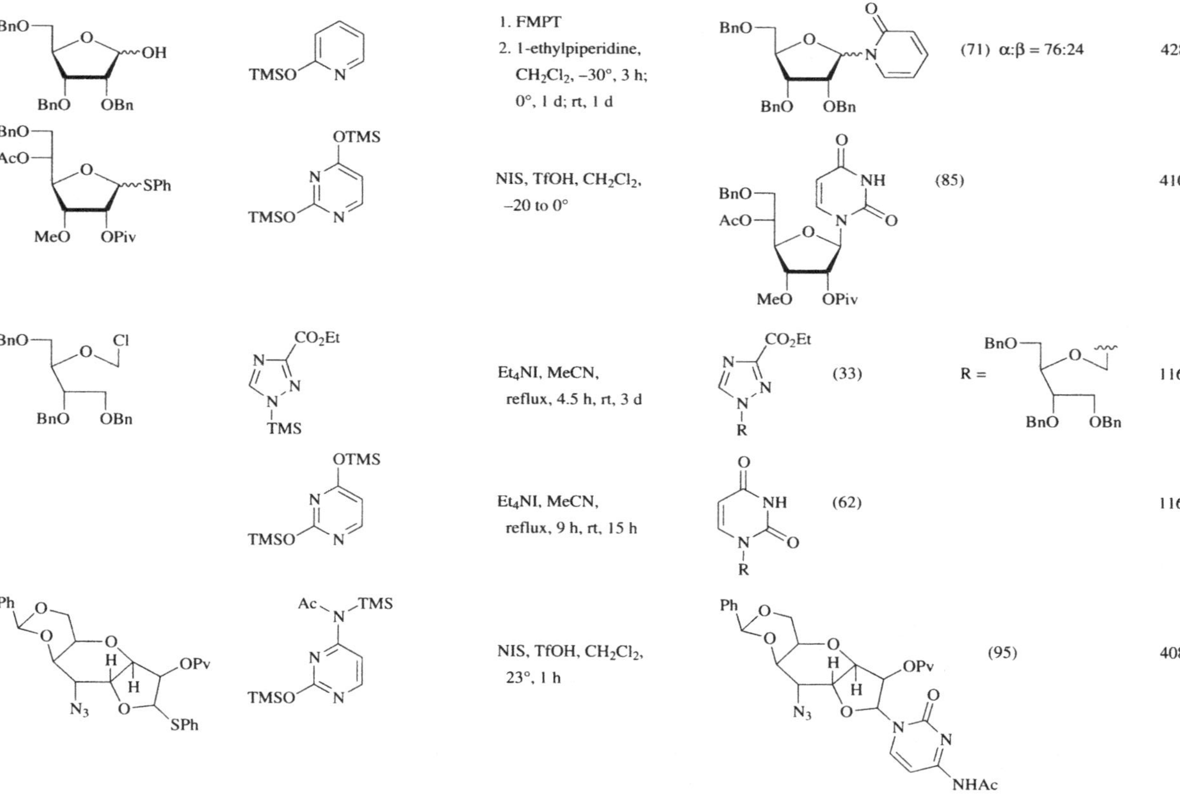
1. FMPT
2. 1-ethylpiperidine, CH_2Cl_2, –30°, 3 h; 0°, 1 d; rt, 1 d
(71) α:β = 76:24
428
NIS, TfOH, CH_2Cl_2, –20 to 0°
(85)
410
Et_4NI, MeCN, reflux, 4.5 h, rt, 3 d
(33)
R =
1164
Et_4NI, MeCN, reflux, 9 h, rt, 15 h
(62)
1165
NIS, TfOH, CH_2Cl_2, 23°, 1 h
(95)
408a

TABLE VIII. REACTIONS OF SILYLATED BASES WITH PROTECTED SUGARS WITH OR WITHOUT CATALYSTS (*Continued*)

Sugar	Base	Conditions	Product(s) and Yield(s) (%)	Refs.
C_{27} 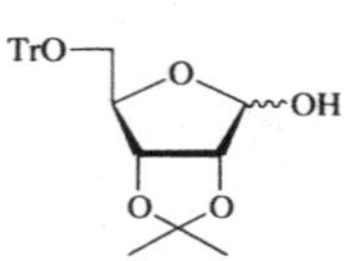 TrO– ... –OH (2,3-O-isopropylidene)	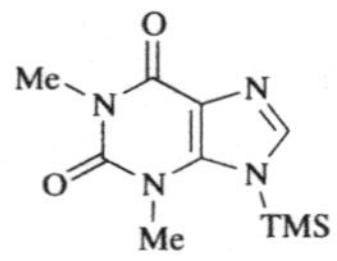 Me, N, O, N, TMS, Me	1. FMPT 2. 1-ethylpiperidine, CH_2Cl_2, –30°, 3 h; 0°, 1 d; rt, 1 d	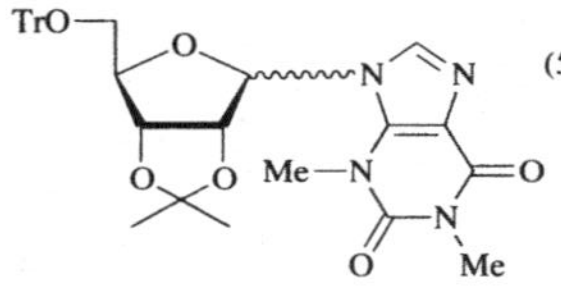 TrO– ... Me–N, O, N, Me (53) α:β = 32:68	428
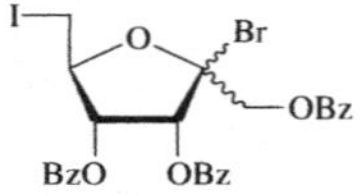 I– ... Br, OBz, BzO, OBz	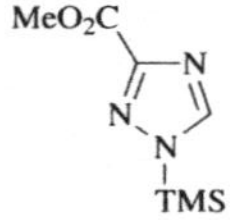 MeO_2C, N, N, N, TMS	$Hg(CN)_2$, C_6H_6, 60°, 2.5 h	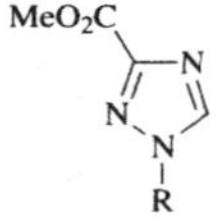 MeO_2C, N, N, N, R (49) R = I– ... OBz, BzO, OBz 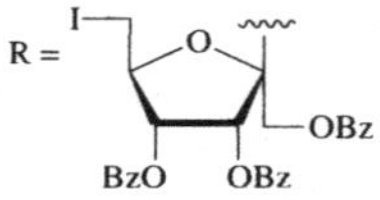	632
	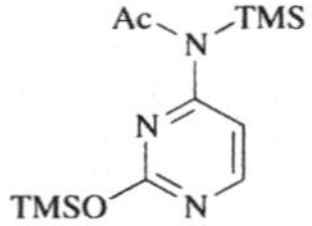 Ac, N, TMS, N, TMSO, N	$Hg(CN)_2$, C_6H_6, 60°, 19 h	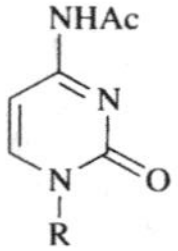 NHAc, N, N, O, R (20)	632
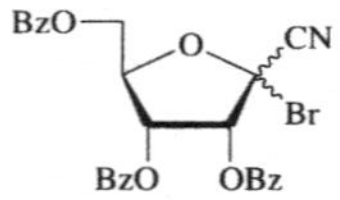 BzO– ... CN, Br, BzO, OBz	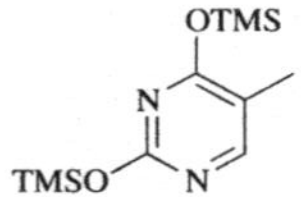 OTMS, N, TMSO, N	$Hg(CN)_2$, $MeNO_2$, rt, 3 d	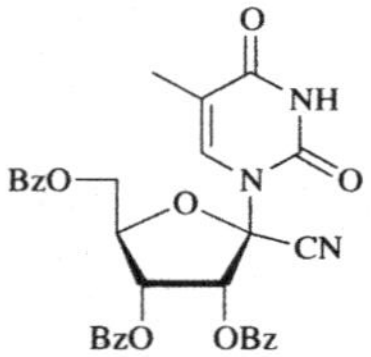 O, NH, N, O, BzO– ... CN, BzO, OBz (56)	1190
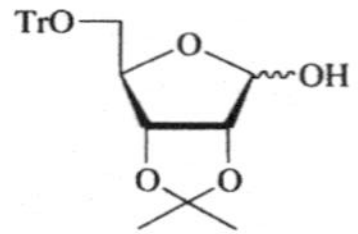 TrO– ... –OH (2,3-O-isopropylidene)	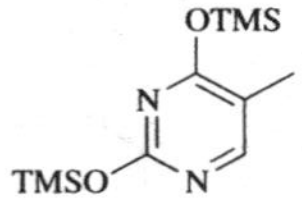 OTMS, N, TMSO, N	1. FMPT 2. 1-ethylpiperidine, CH_2Cl_2, –30°, 3 h; 0°, 1 d; rt, 1 d	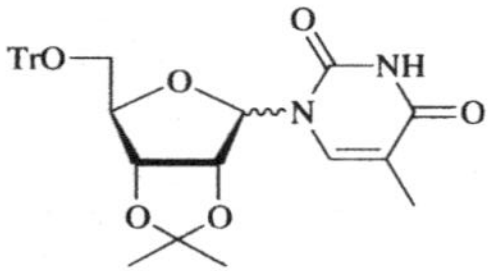 TrO– ... O, NH, N, O (71) α:β = 76:24	428

Glycosyl donor	Base	Conditions	Product(s) and Yield(s) (%)	Refs.
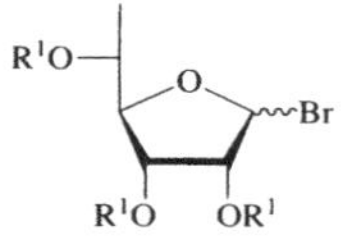 $R^1 = p\text{-}O_2NC_6H_4CO$	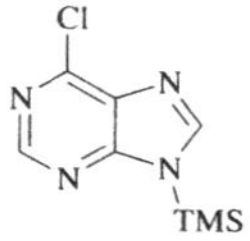	$HgBr_2$, $Hg(CN)_2$, $Cl(CH_2)_2Cl$, MS, rt, 12 h	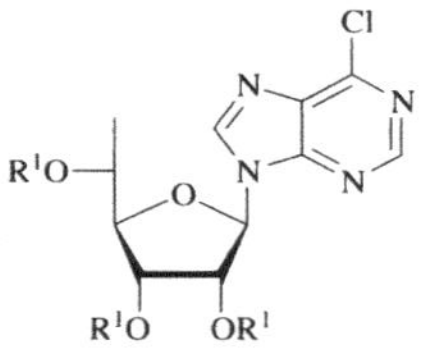 (91)	844
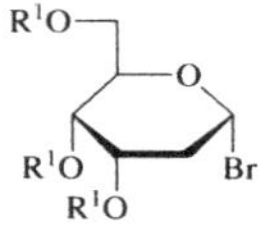 $R^1 = p\text{-}O_2NC_6H_4CO$	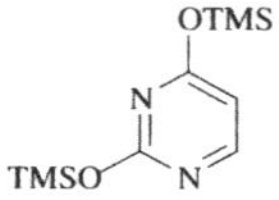	MeCN, $Cl(CH_2)_2Cl$, rt, 2 h	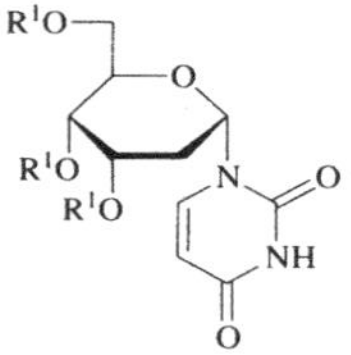 (57)	1065
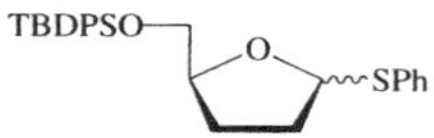	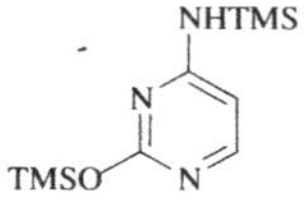	NBS, CH_2Cl_2, MS, –78°, 135 min	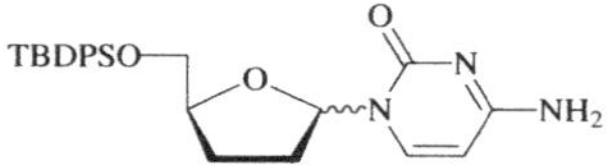 (79) α:β = 1:3	406
	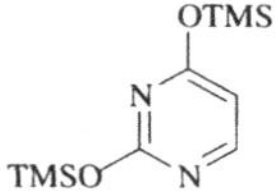	NBS, CH_2Cl_2, MS, –78°, 135 min	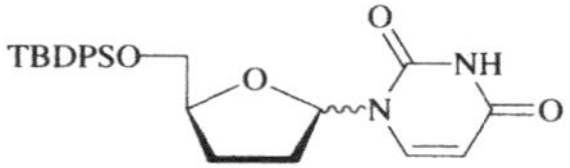 (87) α:β = 1:5.3	406
	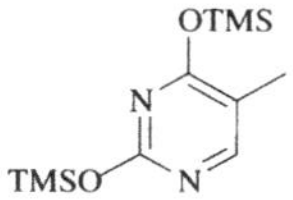	NBS, CH_2Cl_2, MS[d]	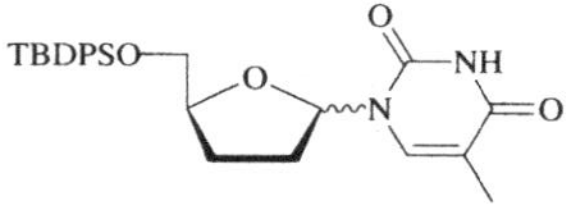	
		–78° to rt, 2 h	(83) α:β = 1:3.6	406
		–30°, 135 min	(93) α:β = 1:3	406
		–0°, 135 min	(93) α:β = 1:2.6	406
		rt, 5 min	(94) α:β = 1:1.6	406

TABLE VIII. REACTIONS OF SILYLATED BASES WITH PROTECTED SUGARS WITH OR WITHOUT CATALYSTS (*Continued*)

Sugar	Base	Conditions	Product(s) and Yield(s) (%)	Refs.
C_{28} 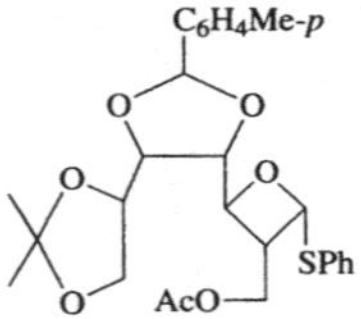 C_6H_4Me-*p*, AcO, SPh	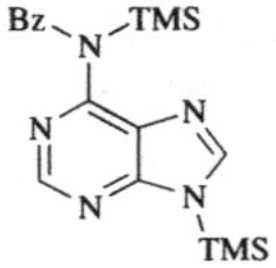 Bz, N, TMS, N, N, N, N, TMS	NBS, CH_2Cl_2, MS, rt, 30 min	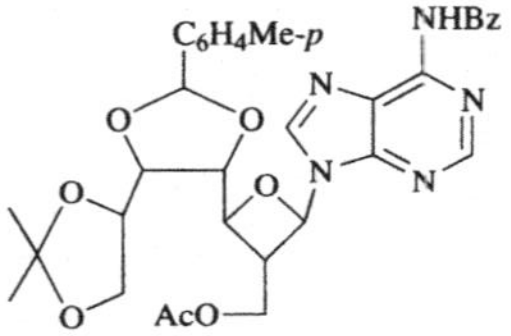 C_6H_4Me-*p*, NHBz, N, N, N, N, AcO (58)	403
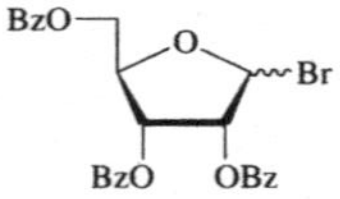 BzO, O, Br, BzO, OBz	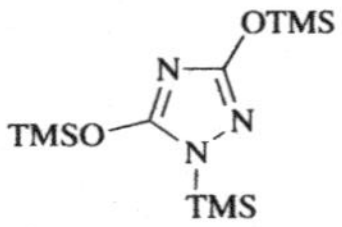 OTMS, N, N, TMSO, N, TMS	MeCN, rt, 3 d	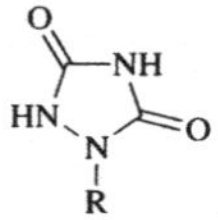 O, NH, HN, N, O, R (84) R = BzO, O, BzO, OBz 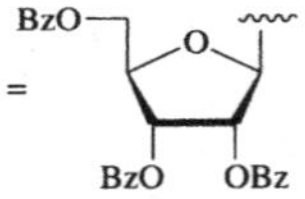	258
	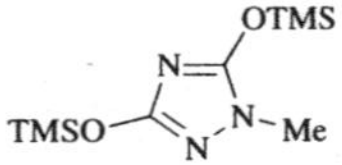 OTMS, N, TMSO, N, N, Me	MeCN, 110°, 15 min	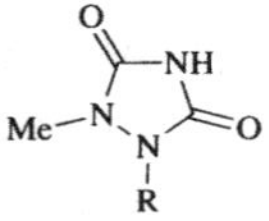 O, NH, Me, N, N, O, R (77)	258
	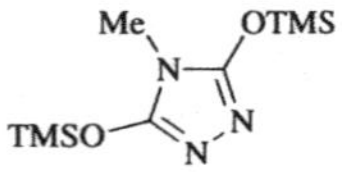 Me, OTMS, N, N, TMSO, N	MeCN, rt, 3 d	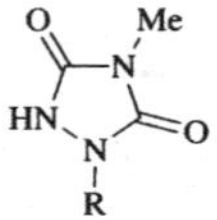 O, Me, N, HN, N, O, R (46)	258
	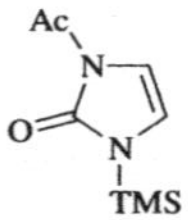 Ac, N, O, N, TMS	$SnCl_4$, MeCN, 20°, 6 h	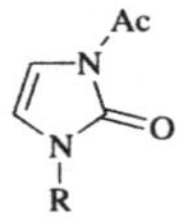 Ac, N, O, N, R (51)	777

		KI, 18-crown-6, MeCN:PhMe (1:1), reflux, 2-4 h		(61) α:β = 24:76	396
		MeCN, rt, 48 h		(75)	1178
		KI, 18-crown-6, MeCN:PhMe (1:1), reflux, 2-4 h		(70) α:β = 32:68	396
		KI, dibenzo-18-crown-6, MeCN, C_6H_6, reflux, 3 d		(87) R =	1166
		KI, dibenzo-18-crown-6, MeCN, C_6H_6, reflux, 5 d		(87)	1166

TABLE VIII. REACTIONS OF SILYLATED BASES WITH PROTECTED SUGARS WITH OR WITHOUT CATALYSTS (*Continued*)

Sugar	Base	Conditions	Product(s) and Yield(s) (%)	Refs.
	NHTMS, N, N, N, N, TMS	KI, dibenzo-18-crown-6, MeCN, C_6H_6, reflux, 4 d	NH_2, N, N, N, N, R (78)	1166
	OTMS, N, N, TMSNH, N, N, TMS	KI, dibenzo-18-crown-6, MeCN, C_6H_6, reflux, 4 d	O, N, NH, N, N, NH_2, R (88)	1166
BnO, O, BnO, SEt, BnO	OTMS, N, TMSO, N	NBS, CH_2Cl_2, MS, rt, 1 h	BnO, O, BnO, N, O, NH, O, BnO (89) α:β = 13:87	407
C_{31} O, Ph_2P, O, Br, BzO, OBz	Ac, N, TMS, N, TMSO, N	$HgBr_2$	NHAc, N, O, N, R (87) R = O, Ph_2P, O, BzO, OBz	1142
	OTMS, N, TMSO, N	$HgCl_2$	O, NH, O, N, R (92)	1142
	Bz, N, TMS, N, N, N, N, TMS	$HgBr_2$, C_6H_6, rt, 3 h	NHBz, N, N, N, N, R (39) + N^7-β-isomer (28)	1142

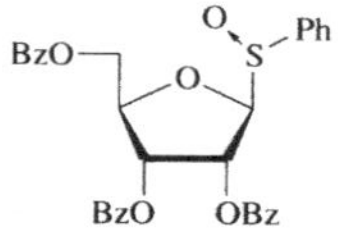 C_{32} BzO, O, S(O)Ph, BzO, OBz	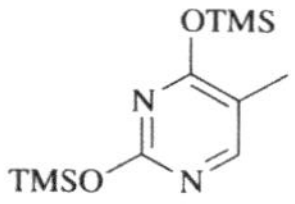 OTMS, N, TMSO, N	$(TfO)_2O$, $Cl(CH_2)_2Cl$	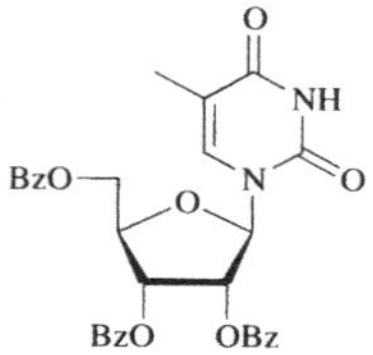 O, NH, O, BzO, O, N, BzO, OBz (30)	412
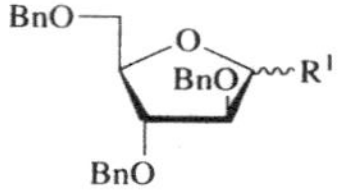 BnO, O, BnO, R^1, BnO	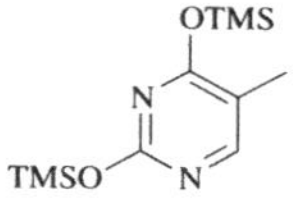 OTMS, N, TMSO, N	NBS, CH_2Cl_2, MS, rt	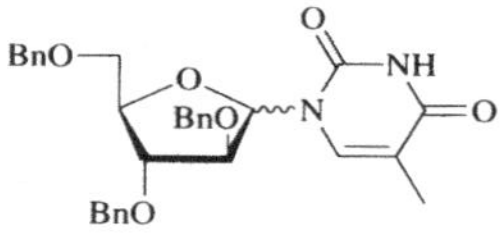 BnO, O, BnO, N, O, NH, O, BnO	407
R^1				
α-SC_6H_4Cl-*p*		6 h	(90) α:β = 25:75	
SPh		7 h	(92) α:β = 92:8	
β-SPh		0.5 h	(87) α:β = 8:92	
α-SPh		5 h	(92) α:β = 26:74	
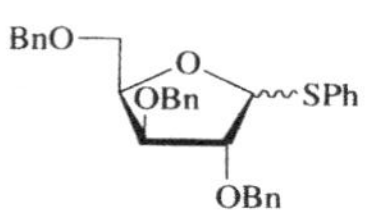 BnO, O, OBn, SPh, OBn	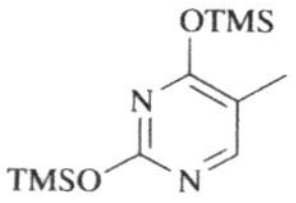 OTMS, N, TMSO, N	NBS, CH_2Cl_2, MS, rt, 6 h	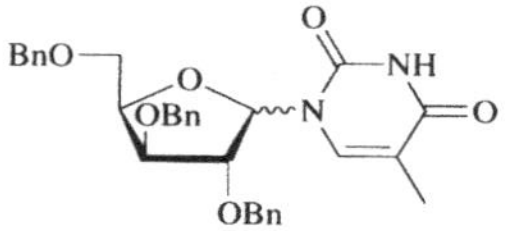 BnO, O, OBn, N, O, NH, O, OBn (88) α:β = 84:16	407
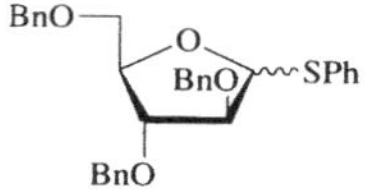 BnO, O, BnO, SPh, BnO	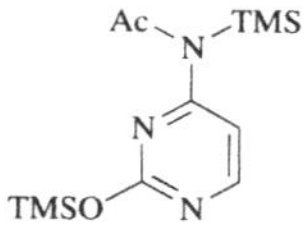 Ac, N, TMS, N, TMSO, N	NBS, CH_2Cl_2, MS, rt, 2.5 h	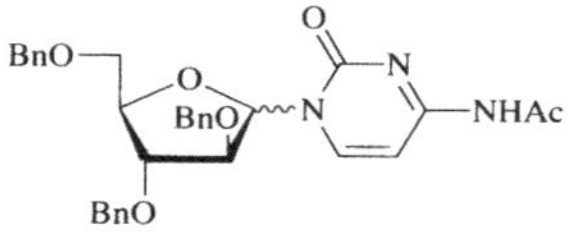 BnO, O, BnO, N, O, N, NHAc, BnO (81) α:β = 10:90	407
	OTMS, N, TMSO, N	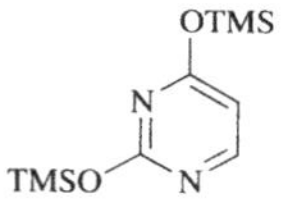 NBS, CH_2Cl_2, MS, rt, 2.5 h	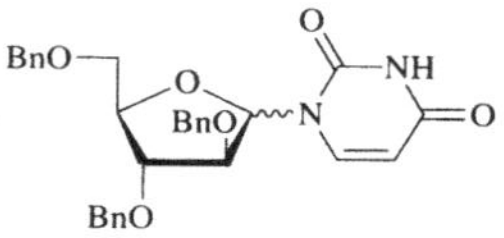 BnO, O, BnO, N, O, NH, O, BnO (85) α:β = 11:89	407

TABLE VIII. REACTIONS OF SILYLATED BASES WITH PROTECTED SUGARS WITH OR WITHOUT CATALYSTS (*Continued*)

Sugar	Base	Conditions	Product(s) and Yield(s) (%)	Refs.
	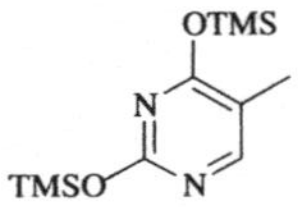 OTMS, TMSO, N, N	NBS, MS, rt	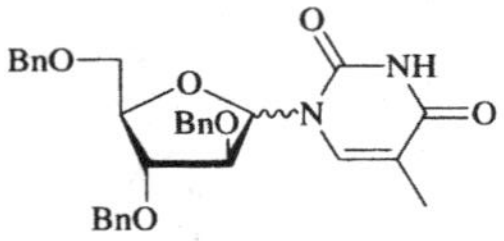 BnO, O, BnO, BnO, O, NH, O, N	407
		CH_2Cl_2, 4.5 h	(89) α:β = 13:87	
		Et_2O, 2.5 h	(85) α:β = 21:79	
		THF, 2.5 h	(91) α:β = 47:53	
		MeCN, 5 h	(92) α:β = 26:74	
		C_6H_6, 5 h	(93) α:β = 16:84	
		$CHCl_3$, 3.5 h	(91) α:β = 9:91	
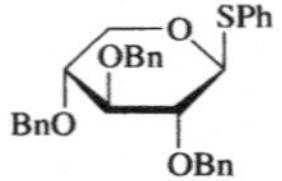 SPh, O, OBn, BnO, OBn	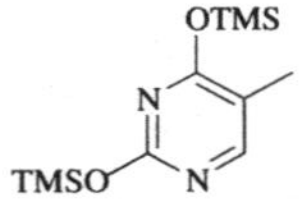 OTMS, TMSO, N, N	NBS, CH_2Cl_2, MS, rt, 6.5 h	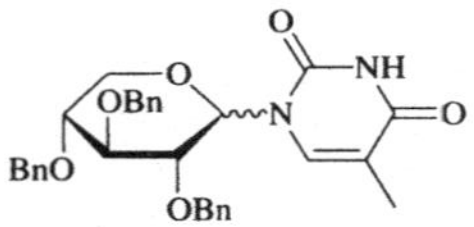 O, OBn, BnO, OBn, O, NH, O, N (88) α:β = 93:7	407
C_{33}				
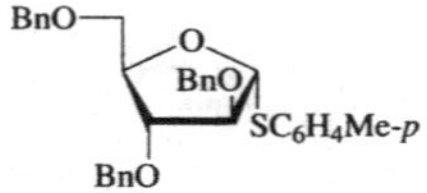 BnO, O, BnO, SC_6H_4Me-*p*, BnO	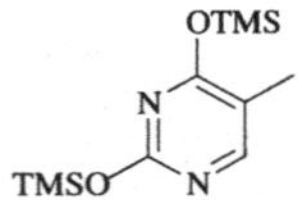 OTMS, TMSO, N, N	NBS, CH_2Cl_2, MS, rt, 2.5 h	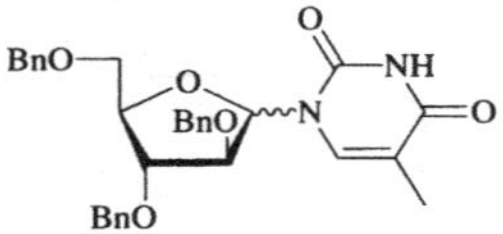 BnO, O, BnO, BnO, O, NH, O, N (95) α:β = 26:74	407
C_{34}				
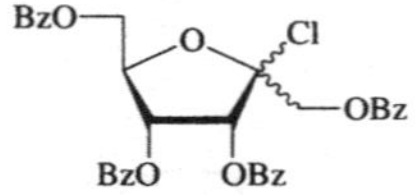 BzO, O, Cl, OBz, BzO, OBz	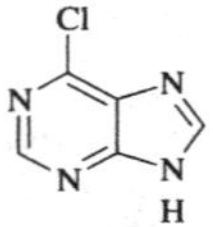 Cl, N, N, N, N, H	$Hg(CN)_2$, $MeNO_2$, MS, 110°, 5 h	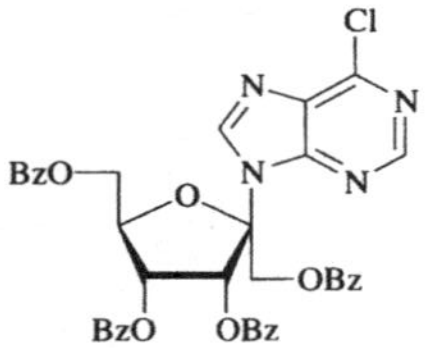 Cl, N, N, N, N, BzO, O, OBz, BzO, OBz (61)	920, 918

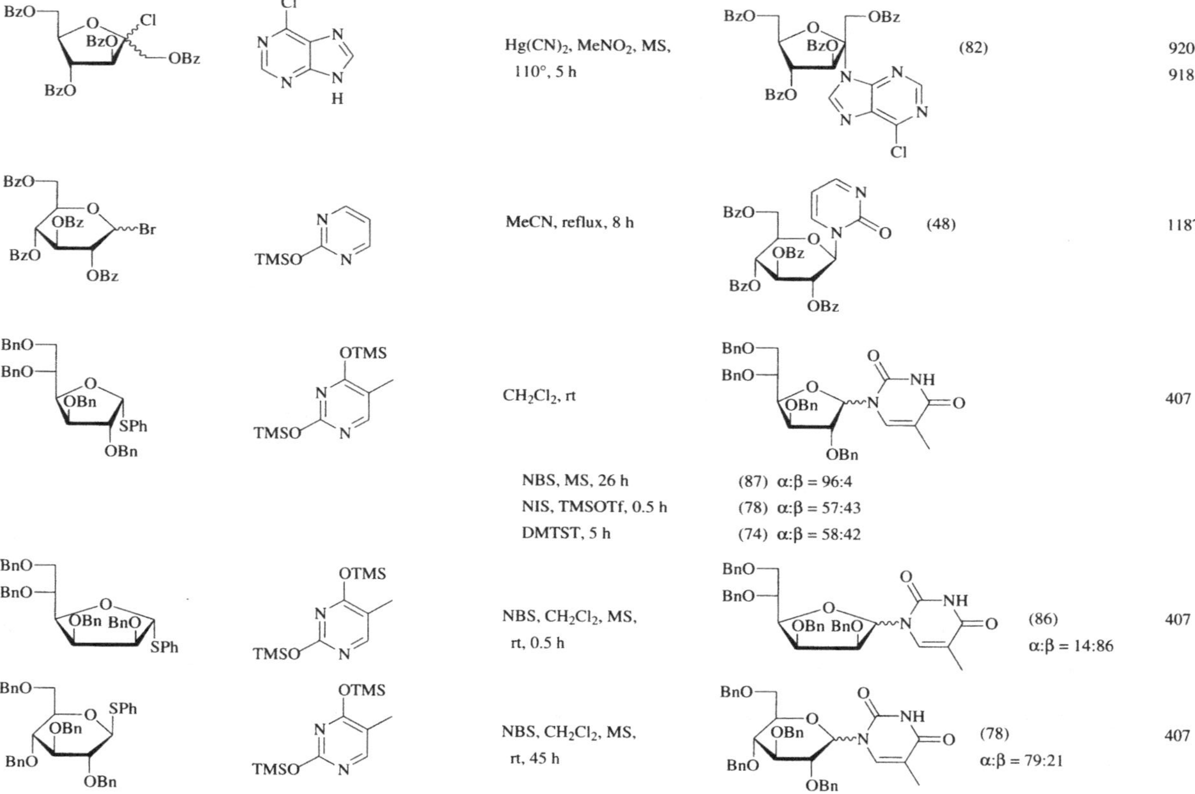

	Conditions	Yield (%)	Refs.
	$Hg(CN)_2$, $MeNO_2$, MS, 110°, 5 h	(82)	920, 918
	MeCN, reflux, 8 h	(48)	1187
C_{40}	CH_2Cl_2, rt		407
	NBS, MS, 26 h	(87) α:β = 96:4	
	NIS, TMSOTf, 0.5 h	(78) α:β = 57:43	
	DMTST, 5 h	(74) α:β = 58:42	
	NBS, CH_2Cl_2, MS, rt, 0.5 h	(86) α:β = 14:86	407
	NBS, CH_2Cl_2, MS, rt, 45 h	(78) α:β = 79:21	407

TABLE VIII. REACTIONS OF SILYLATED BASES WITH PROTECTED SUGARS WITH OR WITHOUT CATALYSTS (*Continued*)

Sugar	Base	Conditions	Product(s) and Yield(s) (%)	Refs.
C_{41} TBDPSO, OAc, O, OTBDPS	NHTMS, N, TMSO, N	KI, dibenzo-18-crown-6, MeCN, PhMe, 80°, 1 h	NH_2, N, N, O, R (95) R= TBDPSO, O, OTBDPS	398
	OTMS, N, TMSO, N	KI, dibenzo-18-crown-6, MeCN, PhMe, 80°, 1 h	O, NH, N, O, R (64)	398
	NHTMS, N, N, N, N, TMS	KI, dibenzo-18-crown-6, MeCN, PhMe, 80°, 2 h	NH_2, N, N, N, N, R (55)	398
	OTMS, N, N, N, N, HN, TMS, TMS	KI, dibenzo-18-crown-6, MeCN, PhMe, 80°, 2 h	O, N, NH, N, N, NH_2, R (90)	398

TABLE IX. FUSION REACTIONS

Sugar	Base	Conditions	Product(s) and Yield(s) (%)	Refs.
C_5				
[structure]	[structure]	100-110°, 2 h	[structure] (76)	552
C_6				
[structure]	[structure]	95-105°, 12 h	[structure] (63)	528
C_7				
[structure]	[structure]	140°, 2 h	[structure] (19) + N^3-isomer (27)	1191
	[structure]		[structure] R^1	
		150°, 20 min	Me (94)	1126
		160°, 20 min	NHBn (56)	
		150°, 20 min	Ph (55) + N^7-isomer (38)	
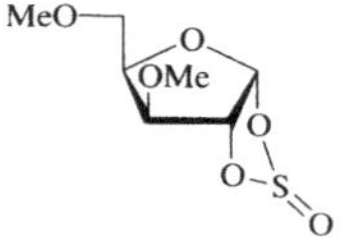 [structure]	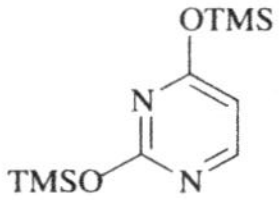 [structure]	90-120°, 18 h	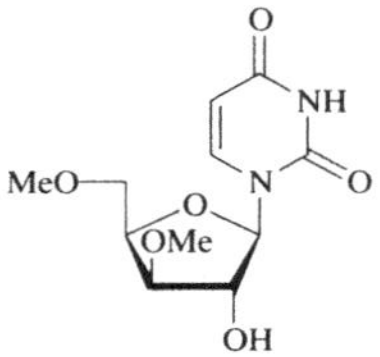 [structure] (80-89)	1192

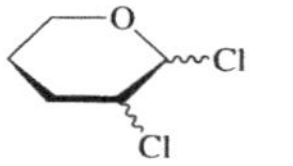

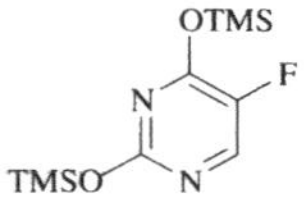

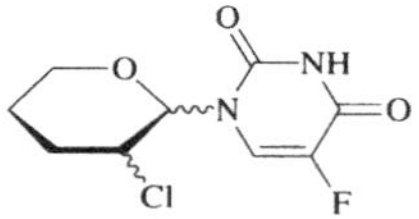

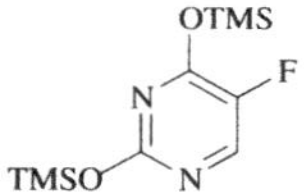

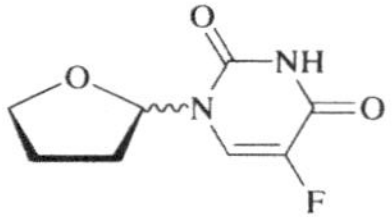

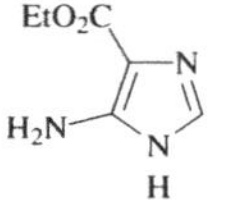

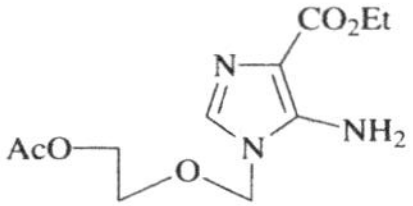

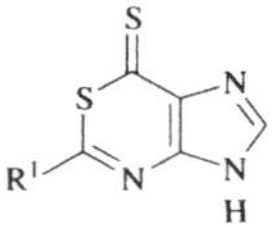

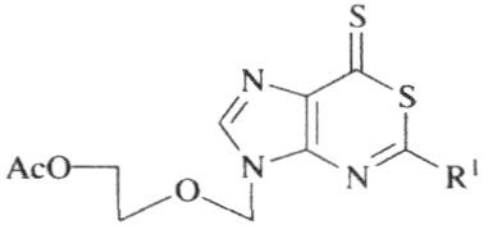

TABLE IX. FUSION REACTIONS (*Continued*)

Sugar	Base	Conditions	Product(s) and Yield(s) (%)	Refs.
C_8 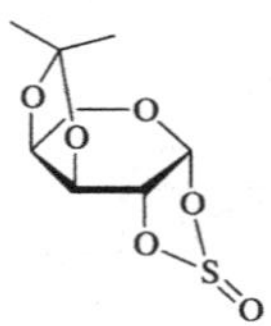	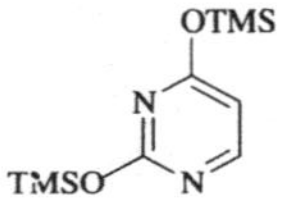	120°, 18 h	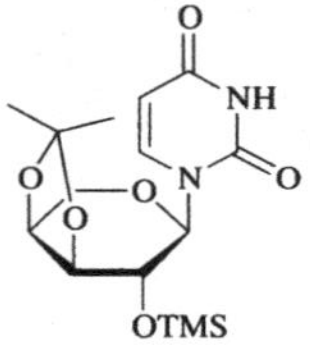 (83)	1192
C_9 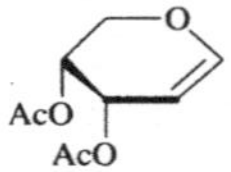	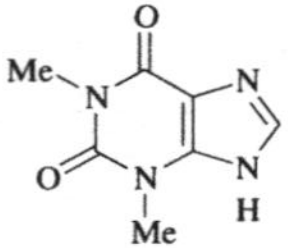	TsOH, 120°	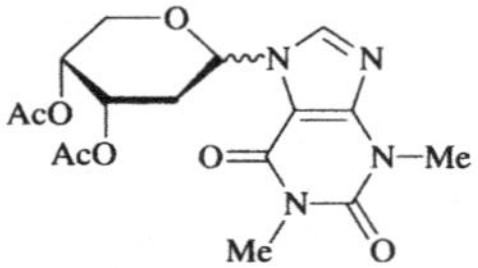 (—)	307
	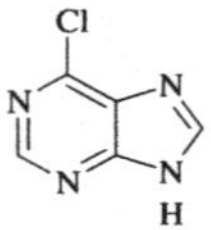	TsOH, 120°	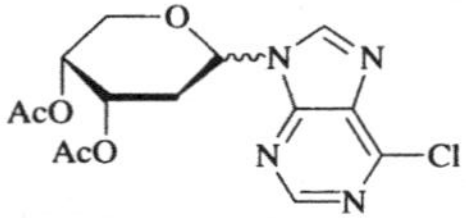 (—)	307
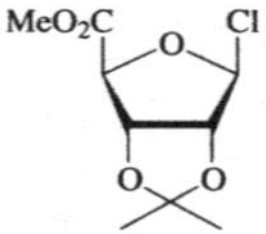	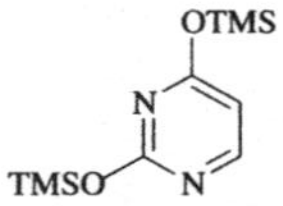	140°, 3 d	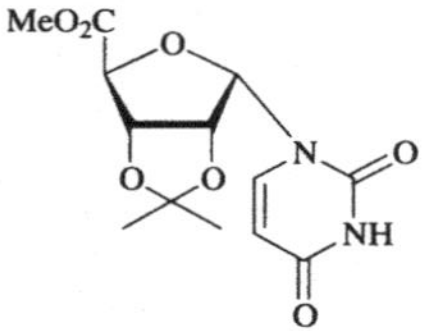 (58)	224
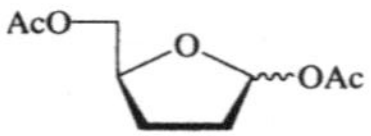	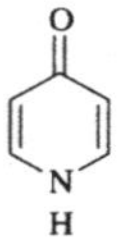	115°, 1 h	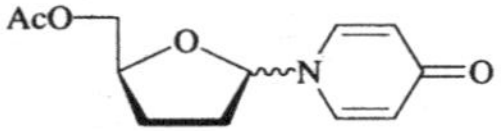 (34) α:β = 1:1	55

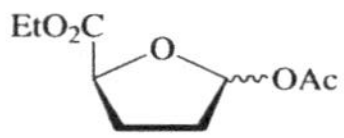

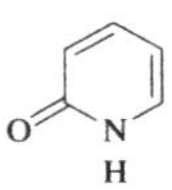

Substrate	Base	Conditions	Product(s) and Yield(s) (%)	Refs.
		110°, 1 h	AcO (38) $\alpha:\beta = 1:1$	55
		115°, 20 min	AcO (35) $\alpha:\beta = 1:1$	55
	Cl	115°	AcO, R^2, Cl, R^1	55
	R^1 / R^2: H / H	25 min	(84) $\alpha:\beta = 1:1$	
	Cl / H	15 min	(70) $\alpha:\beta = 1:1$	
	Cl / Cl	15 min	(82) $\alpha:\beta = 2:1$	
EtO_2C … OAc		120°, 70 min	EtO_2C (30) $\alpha:\beta = 1:1$	55
	Cl		EtO_2C, R^2, Cl, R^1	55
	R^1 / R^2: H / H	110-115°, 25 min	(59) $\alpha:\beta = 1:1$	
	Cl / H	110°, 20 min	(42) $\alpha:\beta = 1:1$	
	Cl / Cl	110°, 20 min	(51) $\alpha:\beta = 1:1$	

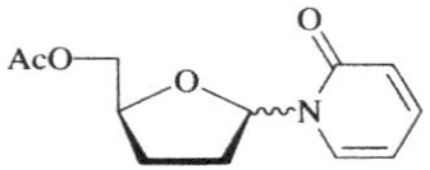

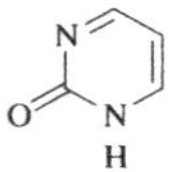

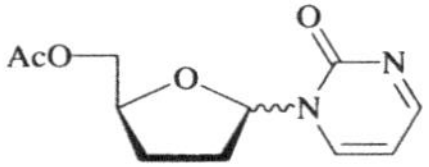

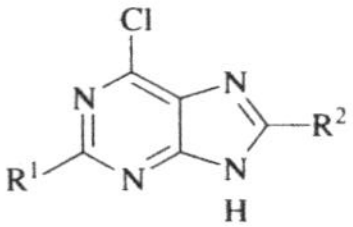

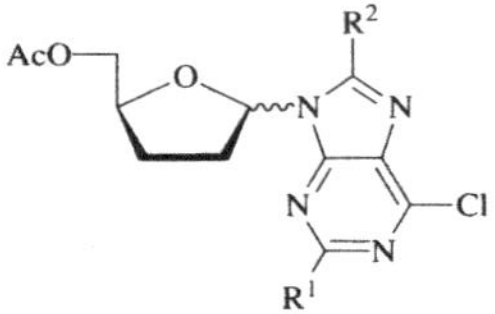

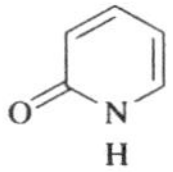

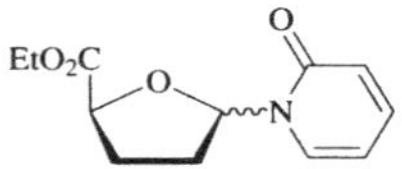

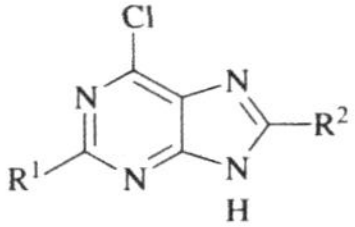

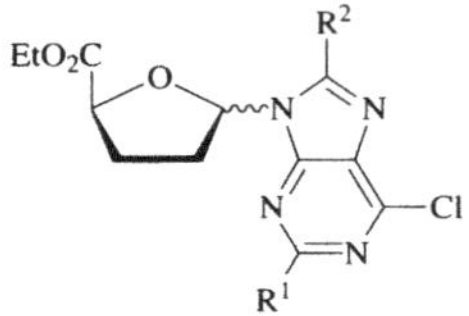

TABLE IX. FUSION REACTIONS (*Continued*)

Sugar	Base	Conditions	Product(s) and Yield(s) (%)	Refs.
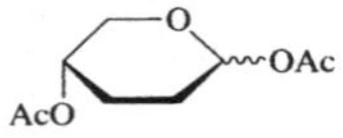 AcO, OAc	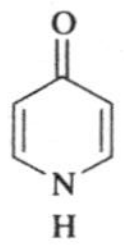 O, N, H	125°, 90 min	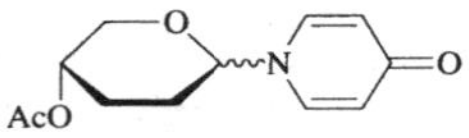 AcO, O, N, O (20) $\alpha:\beta = 1:1$	55
	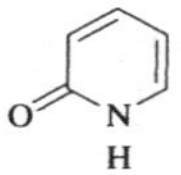 O, N, H	120°, 45 min	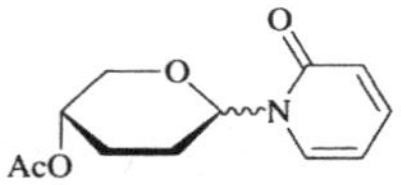 O, AcO, O, N (24) $\alpha:\beta = 1:1$	55
	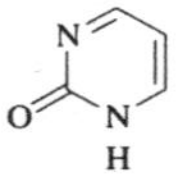 N, O, N, H	120°, 90 min	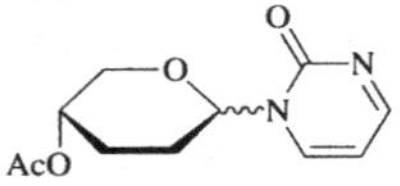 O, N, AcO, O, N (42) $\alpha:\beta = 1:1$	55
	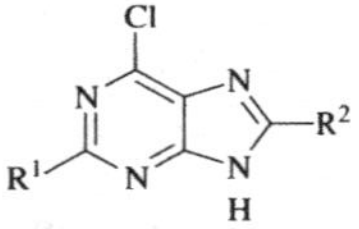 Cl, N, N, R^1, N, N, H, R^2		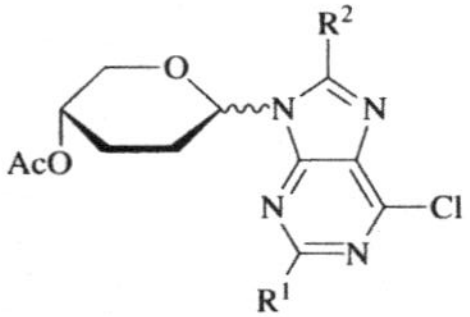 R^2, AcO, O, N, N, N, Cl, N, R^1	55
	R^1 R^2			
	H H	118°, 30 min	(83) $\alpha:\beta = 1:1$	
	Cl H	120°, 20 min	(64) $\alpha:\beta = 1:1$	
	Cl Cl	120°, 20 min	(72)	
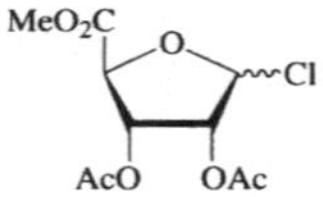 MeO_2C, O, Cl, AcO, OAc	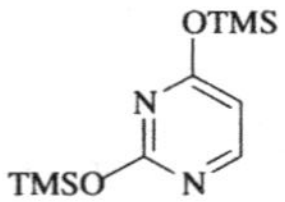 OTMS, N, TMSO, N	110°, 18 h	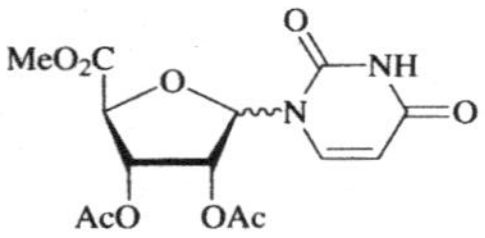 MeO_2C, O, N, O, NH, O, AcO, OAc (34) $\alpha:\beta = 1:3$	224

Reactant	Silylated base	Conditions	Product(s) and Yield(s) (%)		Refs.
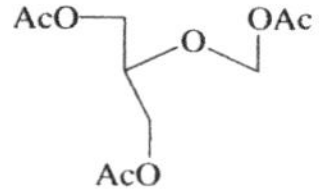 AcO, OAc, AcO (acyclic sugar)	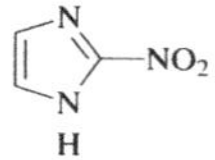 2-nitroimidazole (H)	TsOH, 130-140°, 15 min	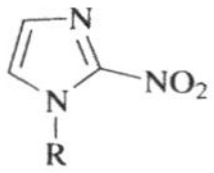 N-R 2-nitroimidazole (89)	R = AcO, O, AcO (acyclic) 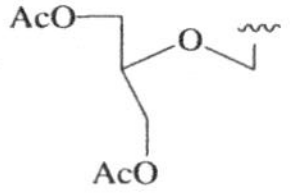	1193
	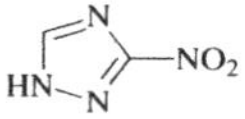 3-nitro-1,2,4-triazole (HN)	TsOH, 130-140°, 15 min	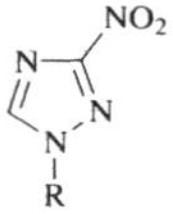 N-R 3-nitro-1,2,4-triazole (83)		1193
C_{11} 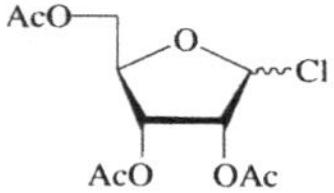 AcO, Cl, AcO, OAc (furanosyl chloride)	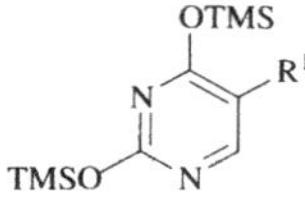 TMSO, OTMS, R[1] pyrimidine	1. 180-190° 2. NaOMe, MeOH	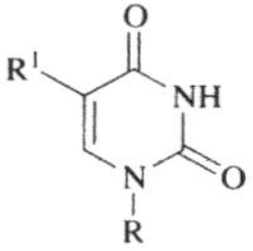 R[1] uracil N-R; R[1] = H (35), Me (46)	R 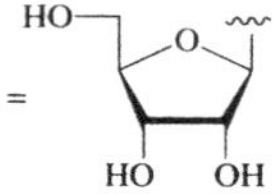 = HO, O, HO, OH (furanosyl)	75
	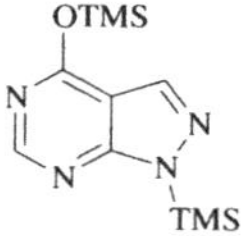 OTMS pyrazolopyrimidine, N-TMS	180-190°	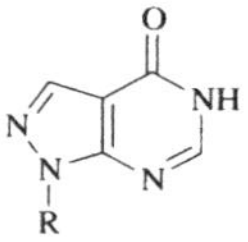 pyrazolopyrimidinone NH, N-R (12)	R = AcO, O, AcO, OAc (furanosyl) 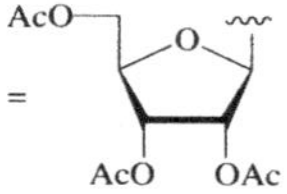	75
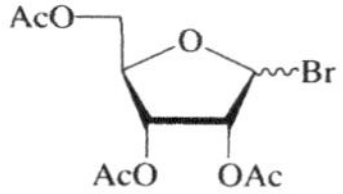 AcO, Br, AcO, OAc (furanosyl bromide)	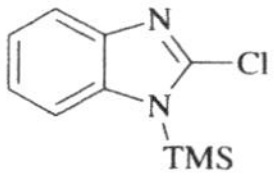 2-Cl benzimidazole, N-TMS	110°, 30 min	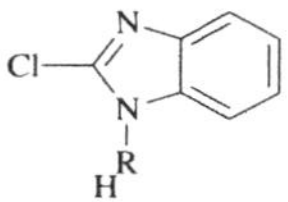 Cl benzimidazole N-R, H (69)		1194
	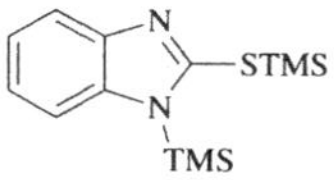 2-STMS benzimidazole, N-TMS	110°, 45 min	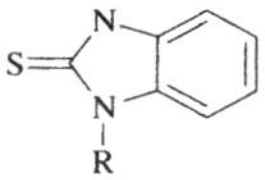 S= benzimidazole N-R (34)		1194

TABLE IX. FUSION REACTIONS (*Continued*)

Sugar	Base	Conditions	Product(s) and Yield(s) (%)	Refs.
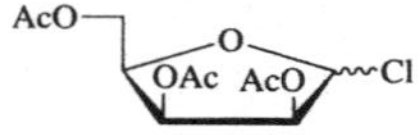 AcO, OAc, AcO, Cl	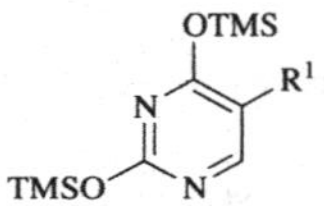 OTMS, R^1, TMSO; R^1: H, Me	1. 190°, 40 min 2. NaOMe, MeOH	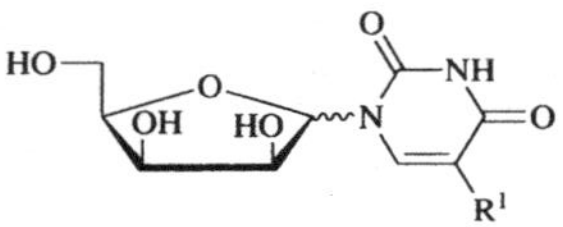 HO, OH, HO, O, NH, O, R^1 (—) α:β = 78:22 (—) α:β = 2:1	1195
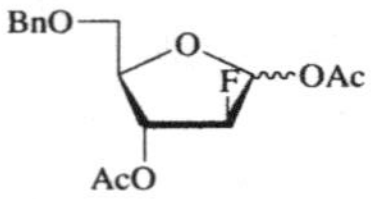 BnO, F, OAc, AcO	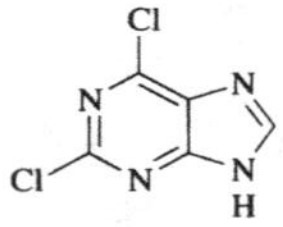 Cl, Cl, H	TsOH, 160°, 20 min	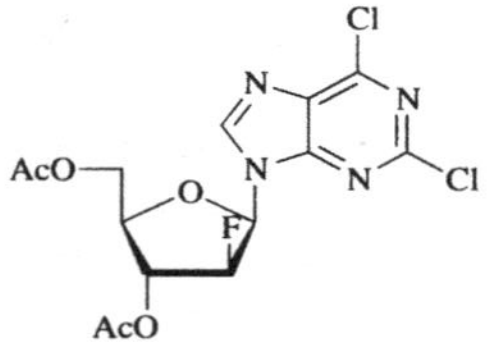 Cl, Cl, AcO, F, AcO (30) + α-anomer (29)	1196
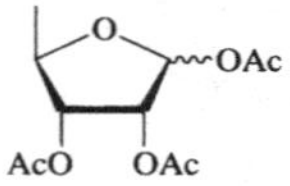 OAc, AcO, OAc	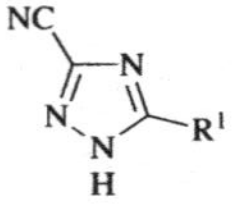 NC, R^1, H	$(p\text{-}O_2NC_6H_4O)_2P(O)OH$, 120°	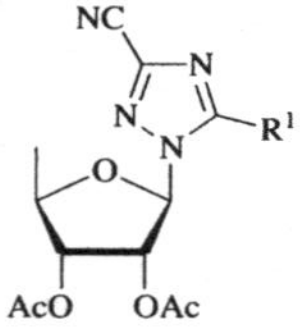 NC, R^1, AcO, OAc R^1: H (84); Cl (34) + N^2-isomer (25)	1197
	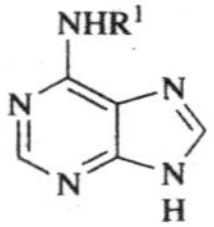 NHR^1, H	1. H_2SO_4, 180-195°, 4 h 2. NaOMe, MeOH, rt, 4 h	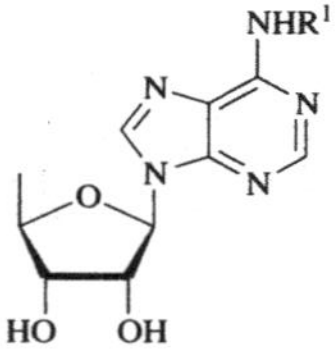 NHR^1, HO, OH	1198

Reactant	R	Conditions	Product(s) and Yield(s) (%)	Refs.
	R^1			
	(*S*)-$PhCH(Me)CH_2$		(19)	
	C_6H_{11}		(18)	
	Ph_2CHCH_2		(8)	
	2-indanyl		(19)	
3,5-di-O-acetyl-2-deoxyribofuranosyl acetate	purine	1. $ClCH_2CO_2H$, 140-145°, 18 min 2. NH_3, MeOH	9-(3,5-di-O-acetyl-2-deoxyribofuranosyl)purine (18) + α-anomer (15)	1199
	2-R^2-6-R^1-purine, R^1 / R^2		9-R-2-R^2-6-R^1-purine, R = 3,5-di-O-acetyl-2-deoxyribofuranosyl	
	Cl / H	$ClCH_2CO_2H$, 125-127°, 4 min	(21) + β-anomer (6)	1199
	Me / H	Cl_2CHCO_2H, 150°, 15-20 min	(25)	1199
	NHBz / H	Cl_2CHCO_2H, 160-165°, 25 min	(20) $R^1 = NH_2$	1199
	Cl / Cl	130-140°, 15 min	(31) + β-anomer (26)	343
	2,6,8-trichloropurine	$ClCH_2CO_2H$, 105°, 5 min	2,6,8-trichloro-9-(3,5-di-O-acetyl-2-deoxyribofuranosyl)purine (30)	1199

TABLE IX. FUSION REACTIONS (*Continued*)

Sugar	Base	Conditions	Product(s) and Yield(s) (%)	Refs.
C_{12} 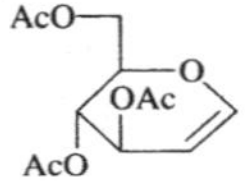 AcO, OAc, AcO (structure)	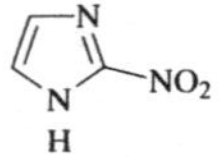 NO_2, N, H (structure)	$ZnCl_2$, 140-160°, 45 min, water aspirator	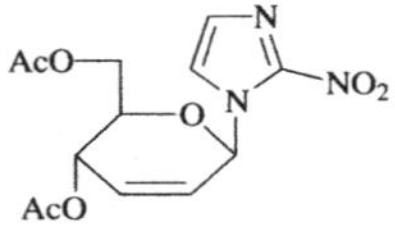 AcO, AcO, NO_2 (structure) (14) + α-anomer (10)	760
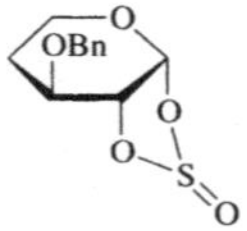 OBn, O, S=O (structure)	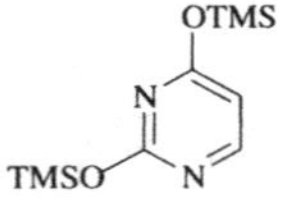 OTMS, TMSO (structure)	90°, 4 h	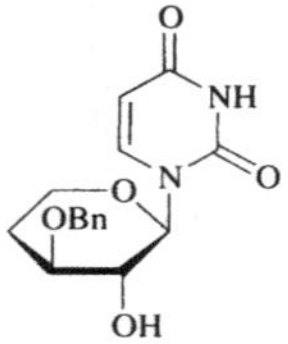 O, NH, O, OBn, OH (structure) (94)	1192
C_{13} 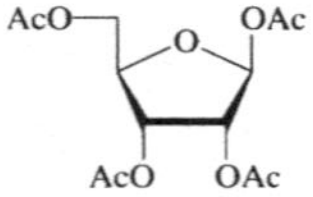 AcO, OAc, AcO, OAc (structure)	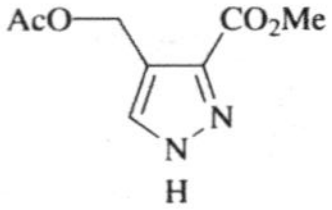 AcO, CO_2Me, N-N-H (structure)	Br_2CHCO_2H, 155-160°, 50 min	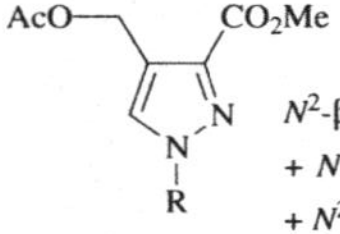 AcO, CO_2Me, N-R (structure) (41) + N^2-β-isomer (16) + N^1-α-isomer (3) + N^2-α-isomer (2); R 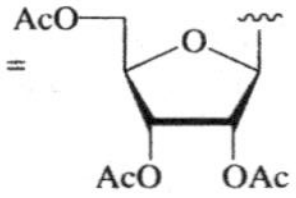 = AcO, AcO, OAc (structure)	1200
	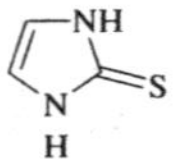 NH, S, N-H (structure)	I_2 (0.07 eq), sugar:base = 2:1, 183°, 20 min	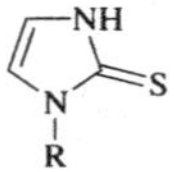 NH, S, N-R (structure) **I** (68) + bis(riboside) (5)	701
		I_2 (0.052 eq), sugar:base = 1:1, 183°, 20 min	**I** (75) + bis(riboside) (11)	701
	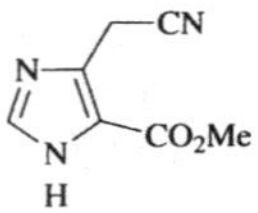 CN, CO_2Me, N-H (structure)	1. 170°, 7-8 min 2. $ClCH_2CO_2H$, 190°, 5 min	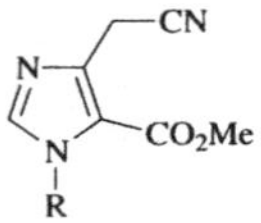 CN, CO_2Me, N-R (structure) **I** (37) + N^3-isomer (<10)	707

Substrate	Conditions	Product(s) and Yield(s) (%)	Refs.
CO_2Me, CN, N, N, H	1. 170°, 7-8 min 2. $ClCH_2CO_2H$, 190°, 25 min	**I** (—) + N^3-β-isomer (10)	705, 707
	$(p\text{-}O_2NC_6H_4O)_2P(O)OH$, 170-175°, 25 min	**I** (47) + α-anomer (5) + N^3-β-isomer (19) + N^3-α-isomer (12)	705, 704
HO, CO_2Et, EtO_2C, N, N, H	I_2, 165°	HO, CO_2Et, EtO_2C, N, N, R (80)	1201
CO_2Me, N, N, N, H	$(p\text{-}O_2NC_6H_4O)_2P(O)OH$, 160-165°, 15-20 min	CO_2Me, N, N, N, R (78) + N^4-isomer (7)	259
HN, O, O, O, N, H	$(p\text{-}O_2NC_6H_4O)_2P(O)OH$, 120-125°, 1 h	HN, O, O, O, N, R (75)	699
N, N, H	I_2, 160°, 10 min	N, N, R **I** (48) + α-anomer (12) + N^2-β-isomer (23) + N^2-α-isomer (2)	1202
	$(p\text{-}O_2NC_6H_4O)_2P(O)OH$, 160°, 10 min	**I** (45) + α-anomer (10) + N^2-β-isomer (27) + N^2-α-isomer (4)	1202
N, SH, N, H	I_2, 200°, 12 mm vac., 12 min	HS, N, N, R (44)	727

TABLE IX. FUSION REACTIONS (*Continued*)

Sugar	Base	Conditions	Product(s) and Yield(s) (%)	Refs.
	(benzazole-2-thione, X, NH)	I_2, 160°, 12 mm vac., 20 min	(N-R benzazole-2-thione) X: O (47), S (75); R = 2,3,5-tri-O-acetylribofuranosyl (AcO, AcO, OAc)	727
	$NHC_6H_4R^1$ (2-methylpurine, 9-H)	1. 250°, 5 min 2. TsOH, 190° (10 mm Hg), 45 min 3. NH_3, MeOH	$NHC_6H_4R^1$ (2-methylpurine, 9-R)	1203
	R^1			
	H		(49)	
	p-F		(49)	
	p-Cl		(55)	
	m-CF_3		(28)	
	p-Me		(41)	
	p-Et		(47)	
	$NHC_6H_3Cl_2$-3,4 (2-methylpurine, 9-H)	1. 250°, 5 min 2. TsOH, 190° (10 mm Hg), 45 min 3. NH_3, MeOH	$NHC_6H_3Cl_2$-3,4 (2-methylpurine, 9-R) (33); R = ribofuranosyl (HO, HO, OH)	1203
	Cl (6-chloro-2-R^1-purine, 9-H)		Cl (6-chloro-2-R^1-purine, 9-R); R = 2,3,5-tri-O-acetylribofuranosyl (AcO, AcO, OAc)	
	R^1			
	Cl	I_2, 130-135°, 15 min	(90)	1204
	SO_2F	$(p\text{-}O_2NC_6H_4O)_2P(O)OH$, 145°, 1 h	(71)	950

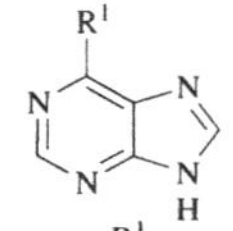

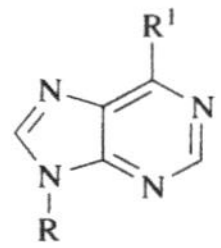

Reactant	Conditions	Product(s) and Yield(s) (%)	Refs.
R^1			
SO_2F	120°, 5 min	(22)	951
SO_2NH_2	$(p\text{-}O_2NC_6H_4O)_2P(O)OH$, 150°, 1 h	(22)	951
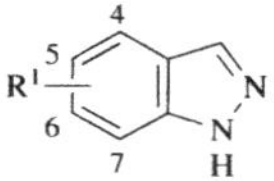	1. TsOH, 160°, 20 min 2. NaOMe, MeOH, rt, 12 h	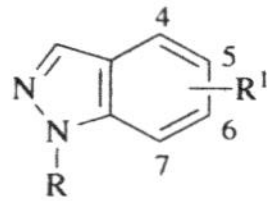 R 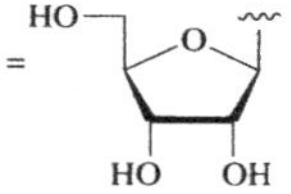 =	1205
R^1			
4-Cl		(32)	
5-Cl		(45)	
5-Br		(47)	
6-Cl		(47)	
5,6-Cl_2		(21)	
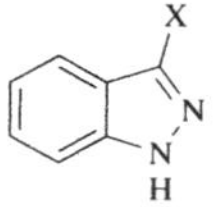	1. TsOH, 20 min 2. NaOMe, MeOH, rt, 12 h	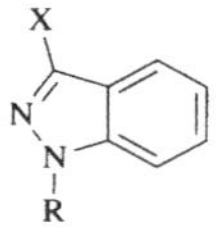	1205
X			
Cl	1. 160°	(50)	
Cl	1. 150°	(31)	
Br	1. 160°	(35)	
I	1. 160°	(47)	
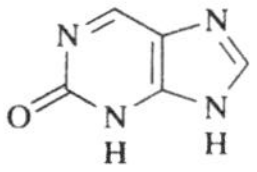	1. H^+, 160°, 15 min 2. NH_3, MeOH	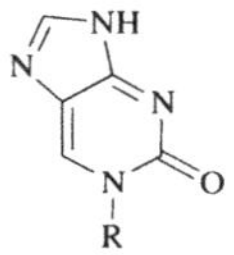 (9)	1206

TABLE IX. FUSION REACTIONS (*Continued*)

Sugar	Base	Conditions	Product(s) and Yield(s) (%)	Refs.
	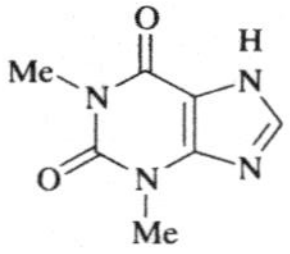	1. 130-150° 2. NH_3, MeOH	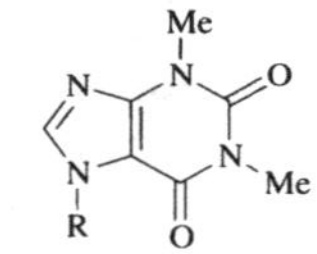 (39)	1207
	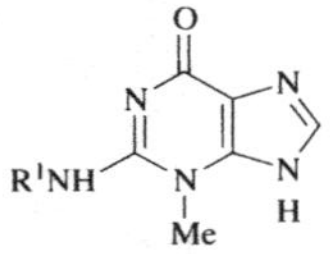	TsOH	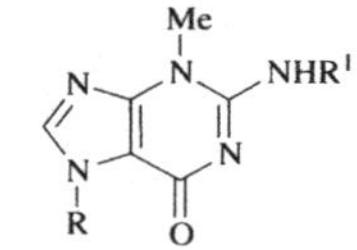 R^1: Ac, 220°, 15 min (39); *i*-Bu, 190°, 45 min (32)	1208
	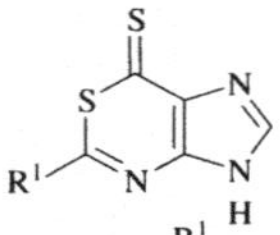 R^1		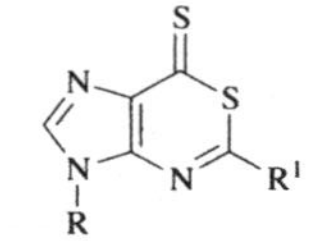 R = 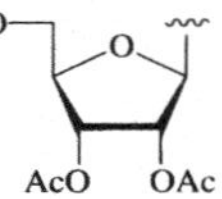	1209
	NH_2	150°, 3 h	(46) + N^7-isomer (10)	
	NHBn	160°, 3 h	(47) + N^7-α-anomer (14) + N^7-β-isomer (8)	
	Me	150°, 6 h	(49) + α-anomer (32) + N^7-α-isomer (9)	
	Ph	150°, 6 h	(44) + α-anomer (41) + N^7-isomer (5)	
	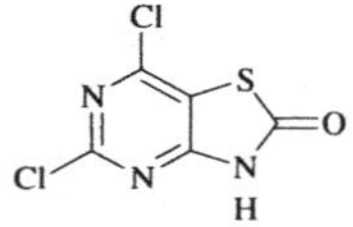	$(p\text{-}O_2NC_6H_4O)_2P(O)OH$, 170°, 10 min	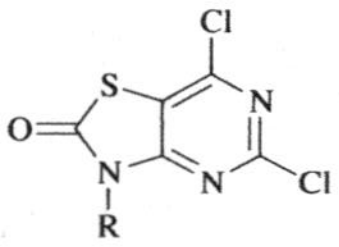 (80)	1058

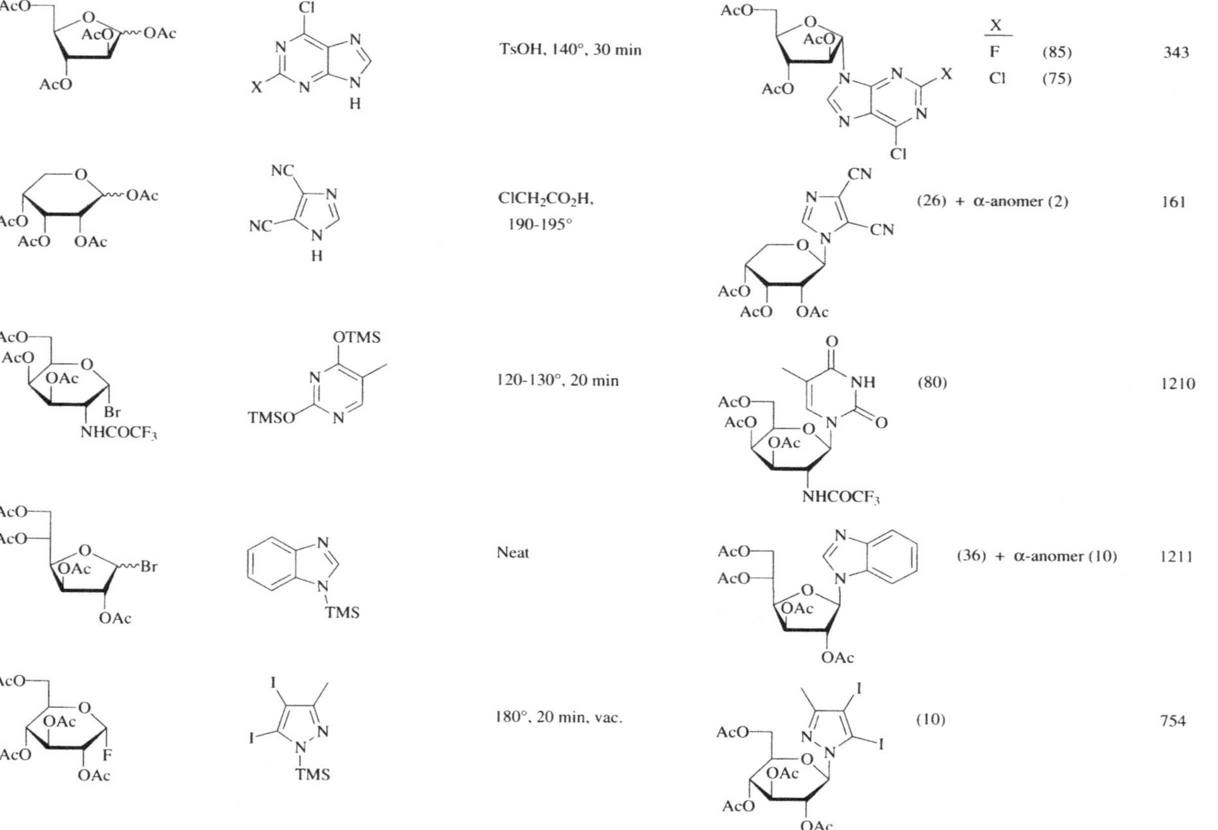

Class	Conditions	Product(s) and Yield(s) (%)	Refs.
	TsOH, 140°, 30 min	X: F (85); Cl (75)	343
C_{13}	$ClCH_2CO_2H$, 190-195°	(26) + α-anomer (2)	161
C_{14}	120-130°, 20 min	(80)	1210
	Neat	(36) + α-anomer (10)	1211
	180°, 20 min, vac.	(10)	754

TABLE IX. FUSION REACTIONS (*Continued*)

Sugar	Base	Conditions	Product(s) and Yield(s) (%)	Refs.
AcO, OAc, AcO, Br, OAc	OTMS, TMSO, N, N	1. 180-190° 2. NaOMe	O, NH, HO, O, N, O, OH, HO, OH (28)	75
	OEt, EtO, N, N	65°, 4 d, reduced pressure	OEt, N, N, O, R **I** (50) + α-anomer (19); R = AcO, O, OAc, AcO, OAc	1212
		65°, 12 h	**I** (—)	71
	R^1, TMSO, N, N		R^2, N, N, O, R	
	R^1 = NHTMS	180-190°	R^2 = NH_2 (15)	75
	R^1 = AcNTMS	185-190°	R^2 = NHAc (15)	76
	OTMS, TMSO, N, N	185-190°	O, NH, N, O, R **I** (52)	76
		100°, 3 h	**I** (41)	80

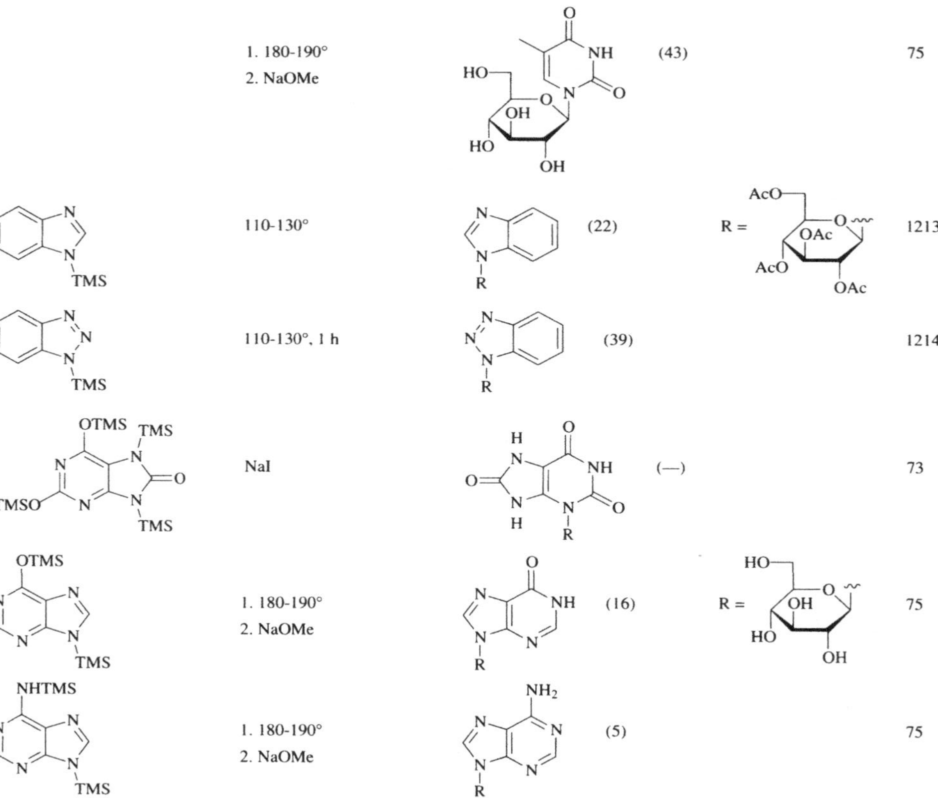

Conditions	Yield (%)	R	Refs.
1. 180-190° 2. NaOMe	(43)		75
110-130°	(22)	R = (AcO, OAc, AcO, OAc sugar)	1213
110-130°, 1 h	(39)		1214
NaI	(—)		73
1. 180-190° 2. NaOMe	(16)	R = (HO, OH, HO, OH sugar)	75
1. 180-190° 2. NaOMe	(5)		75

TABLE IX. FUSION REACTIONS (*Continued*)

Sugar	Base	Conditions	Product(s) and Yield(s) (%)	Refs.
		110-130°	(33) + α-anomer (20)	1215
		130-140°	(40)	1210
C_{16}				
		H_2NSO_3H, 200-210°, 10 min	(30)	1210a
		TsOH, 160°, 4 h	(32) R =	1071
				1207

Substrate	Reagent	Conditions	Product(s) and Yield(s) (%)		Refs.
		p-$O_2NC_6H_4OH$, TsOH, 130-135°, 5 h	(90)		
		$TsNH_2$, 160°, 3 h	(71)		
		o-$O_2NC_6H_4CO_2H$, 150-155°, 3 h	(70)		
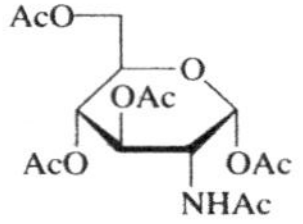 AcO, OAc, AcO, OAc, NHAc	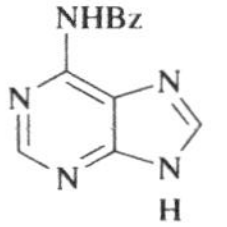 NHBz, N, N, N, N, H	TsOH, 155-160°, 4 h	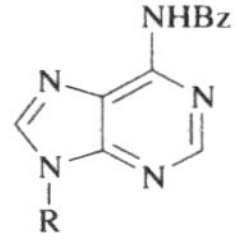 NHBz, N, N, N, N, R (8-14)	R 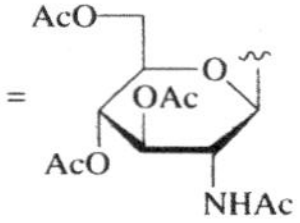 = AcO, O, OAc, AcO, NHAc	1071
	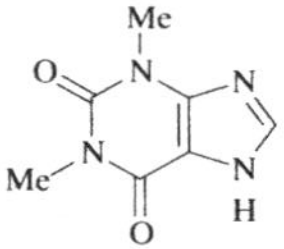 Me, O, N, N, N, Me, N, O, H	p-$O_2NC_6H_4OH$, TsOH, 130°, 4 h	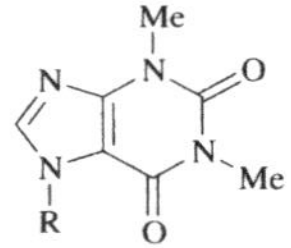 Me, N, N, O, N, Me, N, R, O (45)		1071
C_{17} 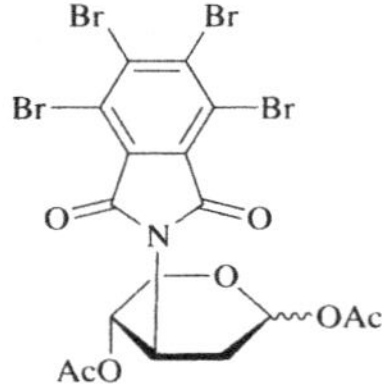 Br, Br, Br, Br, O, N, O, O, OAc, AcO	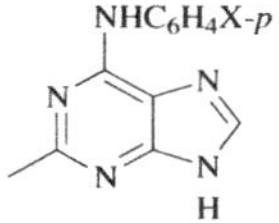 NHC_6H_4X-p, N, N, N, N, H	TsOH, 190°, 45 min	Br, Br, Br, Br, O, N, O, O, AcO, NHC_6H_4X-p, N, N, N, N	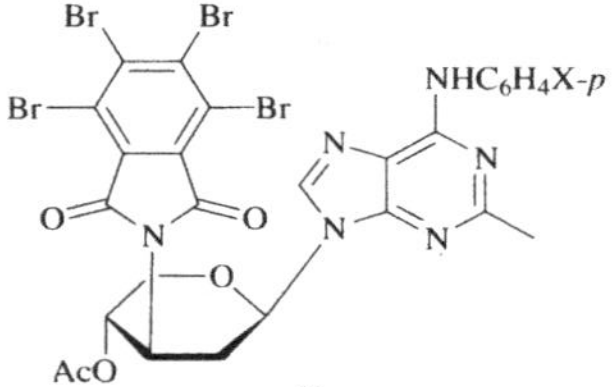 X: Cl (49); H (37)	1216
C_{19} 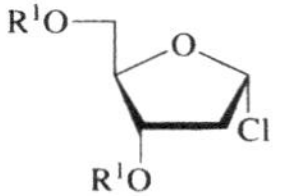 R^1O, O, Cl, R^1O; $R^1 = p$-ClC_6H_4CO	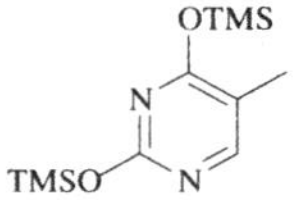 OTMS, N, TMSO, N	100-110°, 15-20 min	R^1O, O, N, O, NH, O, R^1O	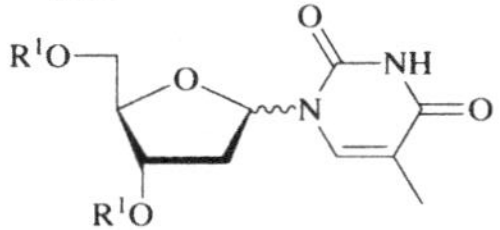 (34) α:β = 1:3.3	266
	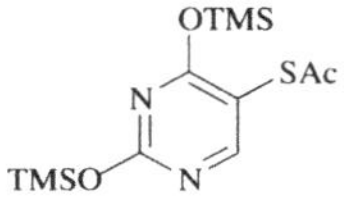 OTMS, SAc, N, TMSO, N	100-110°, 15-20 min	O, AcS, NH, N, O, R (87)	R = R^1O, O, R^1O	266

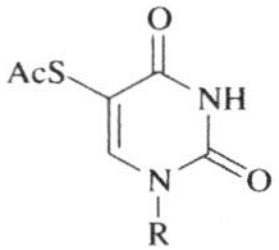

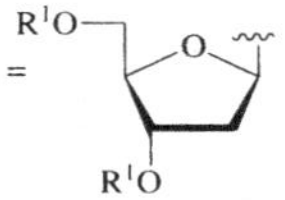

TABLE IX. FUSION REACTIONS (*Continued*)

Sugar	Base	Conditions	Product(s) and Yield(s) (%)	Refs.
	TMSO, N, N, SBn	95-110°, 0.05 mm Hg, 20 min	R^1O, O, N, N, SBn, R^1O (70) α:β = 4:1	1217
	O, N, N, N, TMSO	100-110°, 20 min 2.6-3.3 kPa	R^1O, O, N, N, N→O, R^1O (27)	1218
R^1O, O, Cl, R^1O $R^1 = p\text{-}O_2NC_6H_4CO$	OTMS, R^2, N, N, N, TMSO		R^2, O, NH, N, N, O, R^1O, O, R^1O	
	R^2			
	Br	Heat	(86)	1219
	Br	130°, 0.5 h	(—)	1220
	CF_3	Heat	(51)	1219
	CF_3	150°, 15 min	(33)	1220
	Me	150°, 30 min, 25 mm Hg	(86)	1221
BzO, O, F, Br, BzO	N, N, N, N, TMS	100°, 45 min	N, N, N, N, R (48) + N^7-isomer (16) R = BzO, O, F, BzO	1222

Substrate	Base	Conditions	Product(s) and Yield(s) (%)	Refs.
	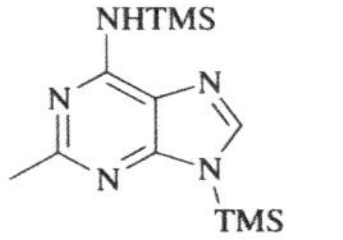 NHTMS, TMS	100°, 45 min	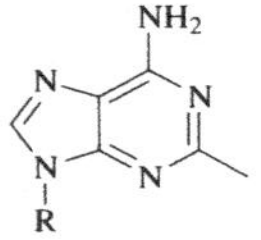 NH_2, R (55) + N^7-isomer (16)	1222
	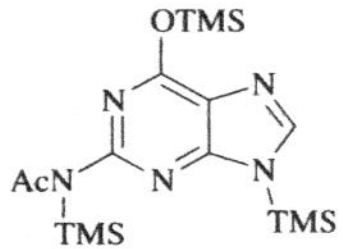 OTMS, AcN, TMS, TMS	—	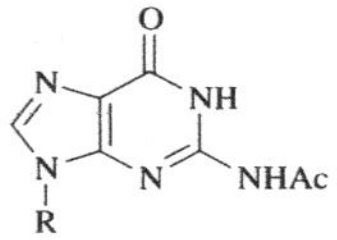 NHAc, R (—) + α-anomer (—) + N^7-α & β-isomers (—)	1223
R^1	TMSO, OTMS		R^1, OH	1192
BzO		125°, 18 h	(92)	
BzO		90°, 20 h	(68) + α-anomer (20)	
BnO		115°, 18 h	(92)	
BnO		90°, 18 h	(70) + α-anomer (10)	
C_{20} 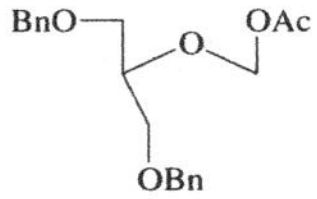 BnO, OAc, OBn	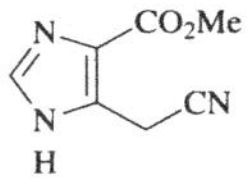 CO_2Me, CN, H	$(p\text{-}O_2NC_6H_4O)_2P(O)OH$, 110°, 4 h	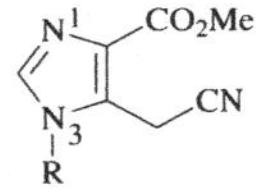 CO_2Me, CN, R (19) + N^1-isomer (31); R 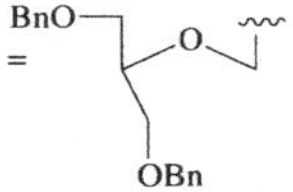 = BnO, OBn	1224
	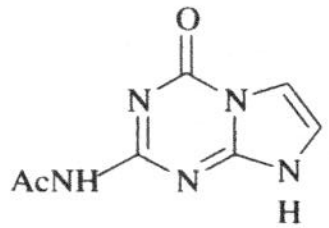 AcNH, H	H^+, 95°, 12 h; 120°, 16 h	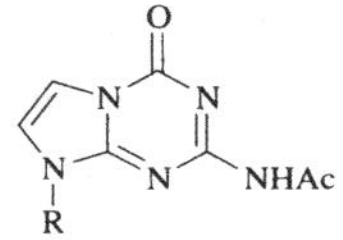 NHAc, R (40)	838

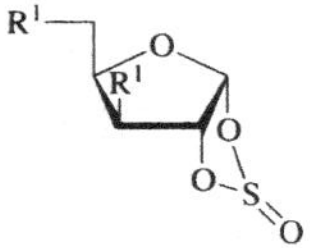

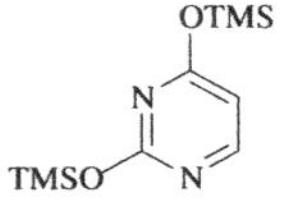

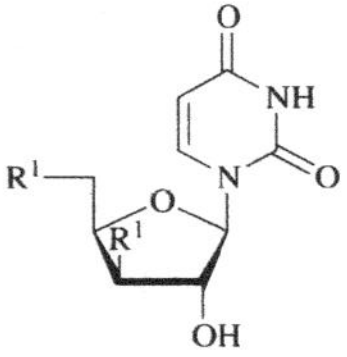

TABLE IX. FUSION REACTIONS (*Continued*)

Sugar	Base	Conditions	Product(s) and Yield(s) (%)	Refs.
C_{21} 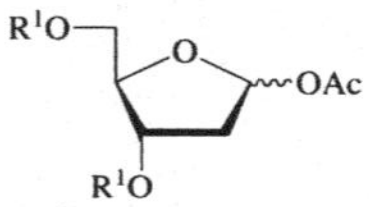 $R^1 = p\text{-}O_2NC_6H_4CO$	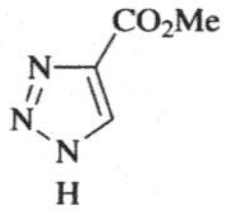	$(p\text{-}O_2NC_6H_4O)_2P(O)OH$, 143°, 20 min, reduced pressure	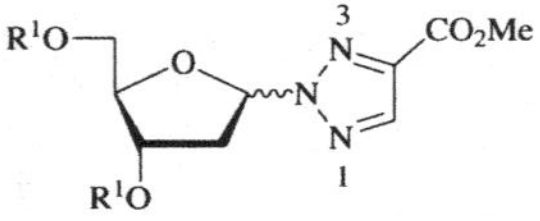 (52) + N^1-β-isomer (12)	842
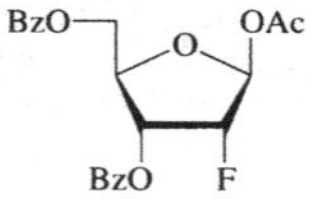	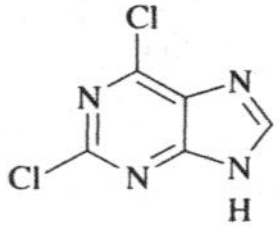	TsOH, 160°, 50 min	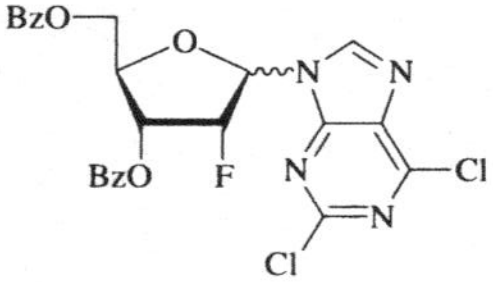 (45) α:β = 1:1	1225
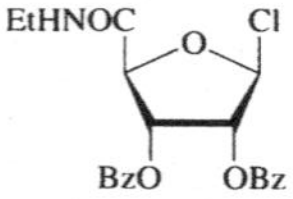	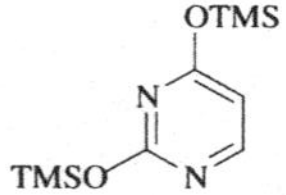	110°, 3 h	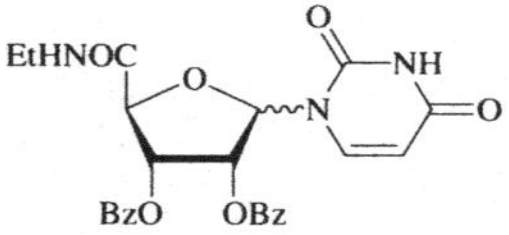 (71) α:β = 3:7	224
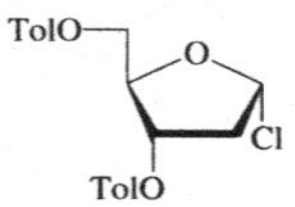	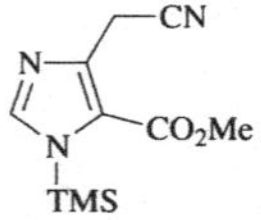	179-180°	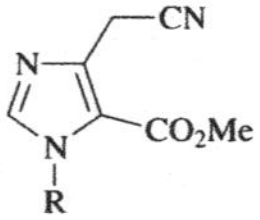 (25) R = 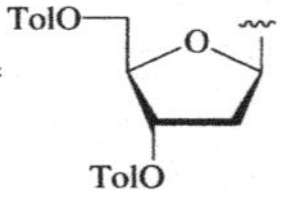	802a
	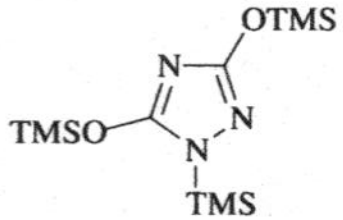	110°, 30 min	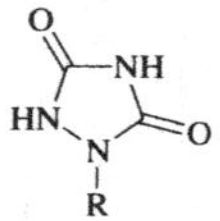 (67)	258
	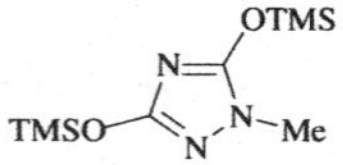	110°, 15 min	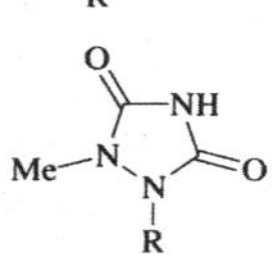 (41)	258

NHTMS, F, TMSO, N	120°, 10 mm, 25 min	TolO, O, TolO, N, O, F, NH_2 (15) + β-anomer (13)	1226
OTMS, N, OR^1, TMSO, N	185-195°, 30 min	R^1O, O, NH, N, O, R; R = TolO, O, TolO	1227
R^1			
Me		(70) + α-anomer (—)	
Et		(73) + α-anomer (—)	
Pr-*n*		(86)	
Bu-*n*		(88)	
Bn		(61)	
OTMS, N, OTMS, TMSO, N	185-195°, 30 min	HO, O, NH, N, O, R (58)	1177
OTMS, N, N, TMSO, N	95-100°, 30 min	O, NH, N, N, O, R (25)	1179
N, N, TMS	1. 115°, 17 min 2. NaOMe, MeOH	HO, O, HO, N, N (24) + β-anomer (9)	816

TABLE IX. FUSION REACTIONS (*Continued*)

Sugar	Base	Conditions	Product(s) and Yield(s) (%)	Refs.
C_{26} 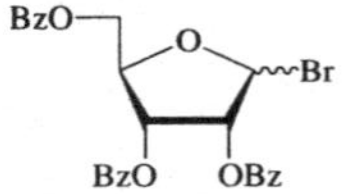 BzO, BzO, OBz, Br	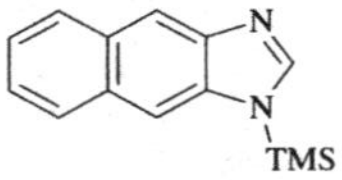 N, N, TMS	KI, 130°, 1 h	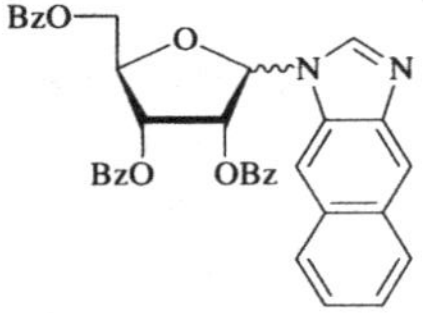 BzO, BzO, OBz, N, N (90)	1228
	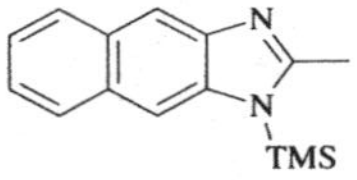 N, N, TMS	KI, 130°, 1 h	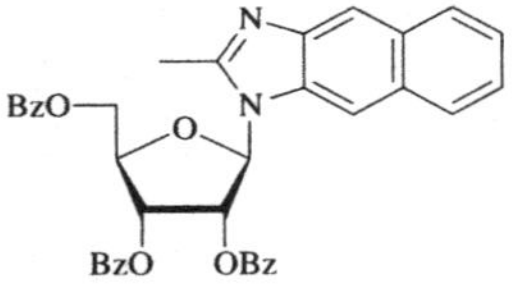 N, N, BzO, BzO, OBz (41) + α-anomer (16)	1228
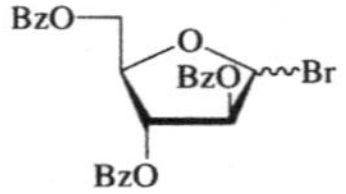 BzO, BzO, BzO, Br	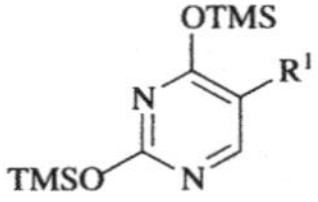 OTMS, R^1, N, TMSO, N	190°, 40 min	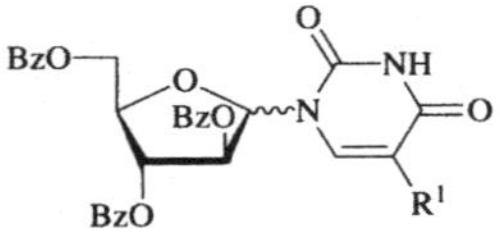 BzO, BzO, BzO, O, NH, N, O, R^1; R^1: H (—) α:β ~ 3:1; Me (—) α:β = 7:2	1195
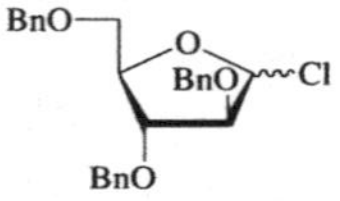 BnO, BnO, BnO, Cl	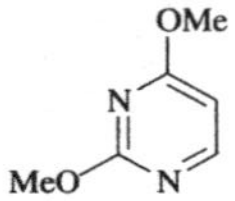 OMe, N, MeO, N	100°, 20 h	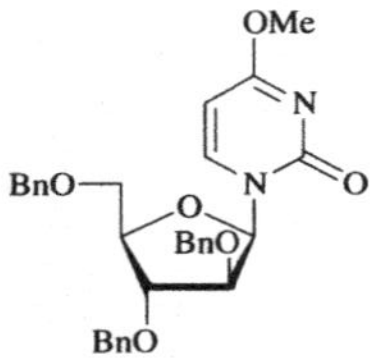 OMe, N, N, O, BnO, BnO, BnO (—)	356

Glycosyl donor	Heterocycle	Conditions	Product(s) and Yield(s) (%)	Refs.
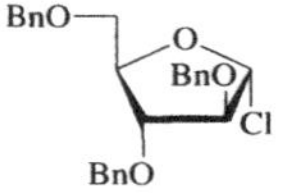 BnO, BnO, BnO, Cl	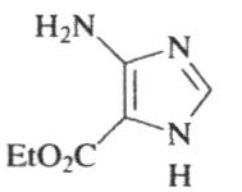 H_2N, EtO_2C, N, N, H	110°, 1 min	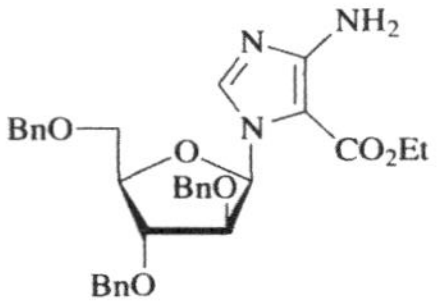 NH_2, CO_2Et, BnO, BnO, BnO (2)	352
C_{27}				
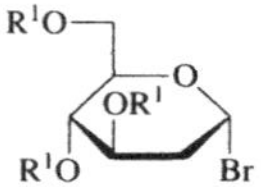 R^1O, OR^1, R^1O, Br; $R^1 = p\text{-}O_2NC_6H_4CO$	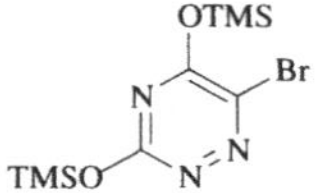 OTMS, Br, TMSO, N	105°, 2 d	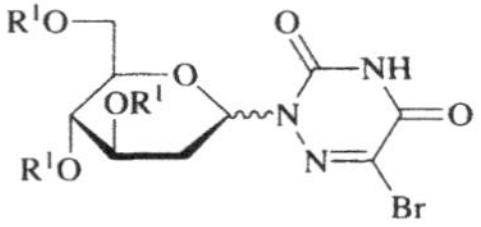 R^1O, OR^1, R^1O, NH, Br (43)	1229
	OTMS, R^1, TMSO, N		R^1, NH, R^1O, OR^1, R^1O	
			R^1 H: 100°, 25 h (32)	1230
			R^1 Me: 90°, 12 h (44)	1230
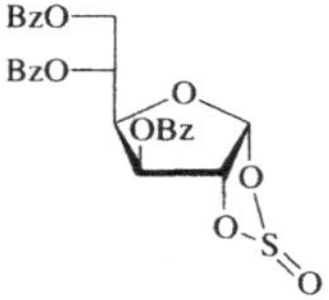 BzO, BzO, OBz, O–S=O	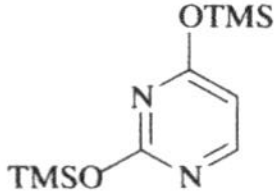 OTMS, TMSO, N	120°, 6 h	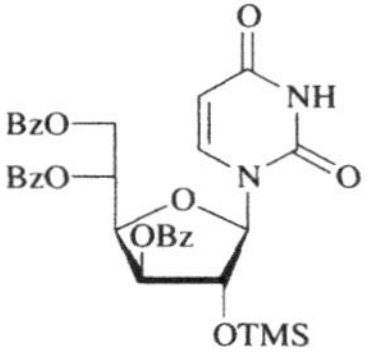 BzO, BzO, OBz, NH, OTMS (93)	1192
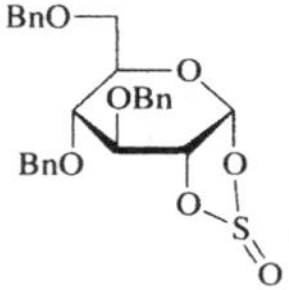 BnO, OBn, BnO, O–S=O	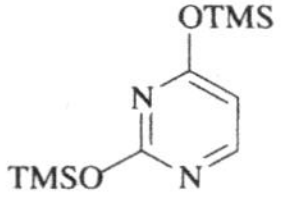 OTMS, TMSO, N	125°, 18 h	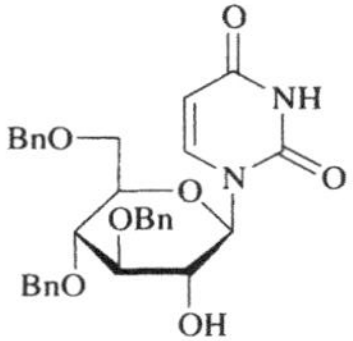 BnO, OBn, BnO, NH, OH **I** (91)	1192

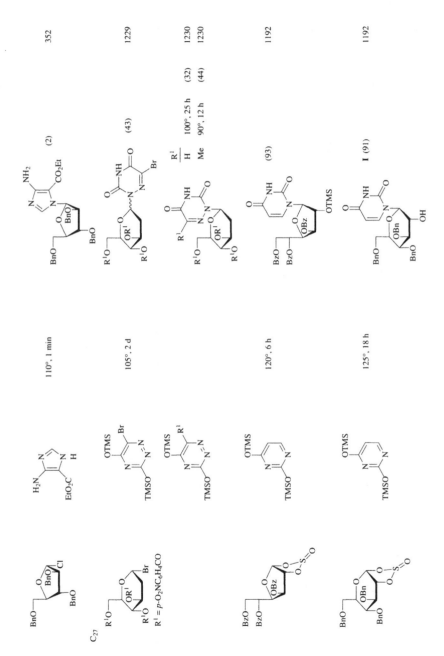

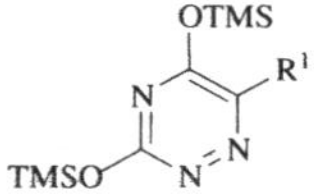

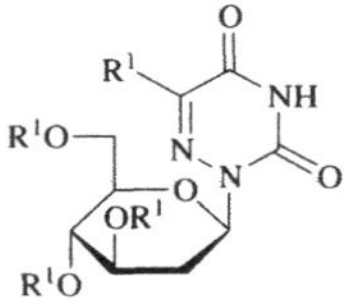

TABLE IX. FUSION REACTIONS (*Continued*)

Sugar	Base	Conditions	Product(s) and Yield(s) (%)	Refs.
BzO, OAc, BzO, OBz	NC, MeO$_2$C, N, N, H	1. 179°, 7-8 min 2. ClCH$_2$CO$_2$H, 190°, 25 min	3 N, CN, 1 N, CO$_2$Me, R; $N^1 + N^3$ isomers (47), R = BzO, BzO, OBz	705
	MeO$_2$C, MeO$_2$C, N, N, H	(p-$O_2NC_6H_4O)_2P(O)OH$	N, CO$_2$Me, N, CO$_2$Me, R; (42) + α-anomer (6)	852
	CO$_2$Me, N, CO$_2$Me, N, TMS	200°, 30 min	CO$_2$Me, N, CO$_2$Me, N, R; (—)	853
	MeO$_2$C, N, N–N, H	(p-$O_2NC_6H_4O)_2P(O)OH$, 160-165°, 15-20 min	N, CO$_2$Me, N–N, R; (~78) + N^2-isomer (~8)	259
	O$_2$N, N, N–N, H	190°, 30 min	N, NO$_2$, N–N, R; (88)	258
	O$_2$N, N, Br, N–N, H	1. 150°, 30 min 2. H$_2$, Pd/C	Br, N, NH$_2$, N–N, R; (60)	258
	Cl, Cl, N, N, H	TsOH, 160-165°, 40 min	N, Cl, N, Cl, R; (25)	849

Reactant	Conditions	Product (Yield %)	Refs.
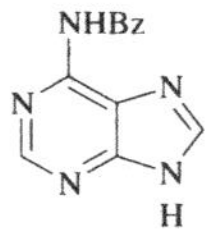 NHBz purine (N-H)	1. I_2, 140-145°, 20 min 2. NaOMe, $CHCl_3$, reflux, 30 min	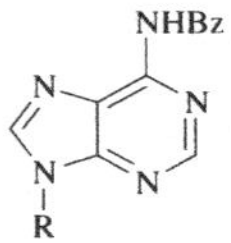 NHBz purine, N-R (45) R = HO–CH₂ furanose, HO, OH	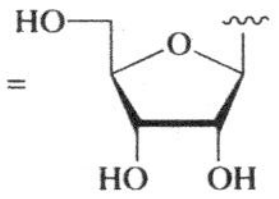 1204
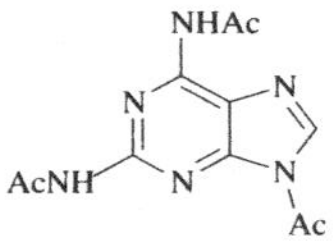 NHAc, AcNH purine, N-Ac	1. I_2, 170-175° 2. NaOMe, $CHCl_3$, NH_3 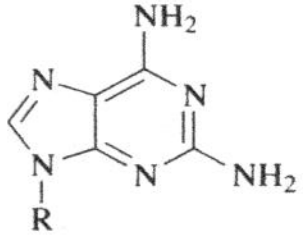	NH_2, NH_2 purine, N-R (59)	1204
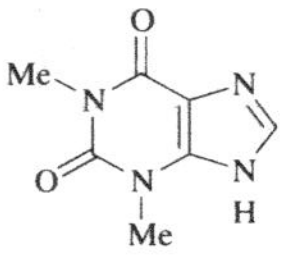 Me, O, O, Me, N-H purine	1. I_2, 160-165° 2. NaOMe, $CHCl_3$ 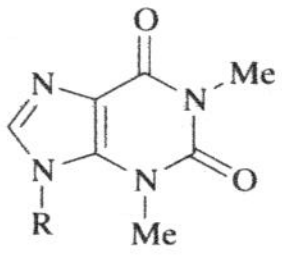	O, Me, O, Me purine, N-R (81)	1204
AcO, AcO, OAc, OAc, AcO, OAc, OAc, Cl disaccharide + OR^1, R^1O pyrimidine		AcO, AcO, OAc, OAc, AcO, OAc, OAc, N, O, OR^1	
R^1 = Me	125-130°, 60 h	(42)	1231
R^1 = Et	120-125°, 6 d	(46)	1231

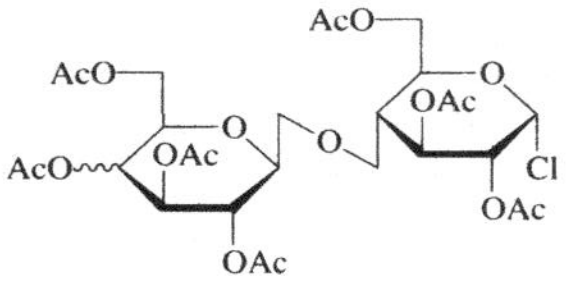

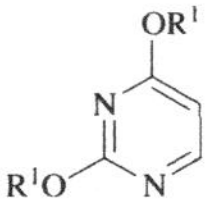

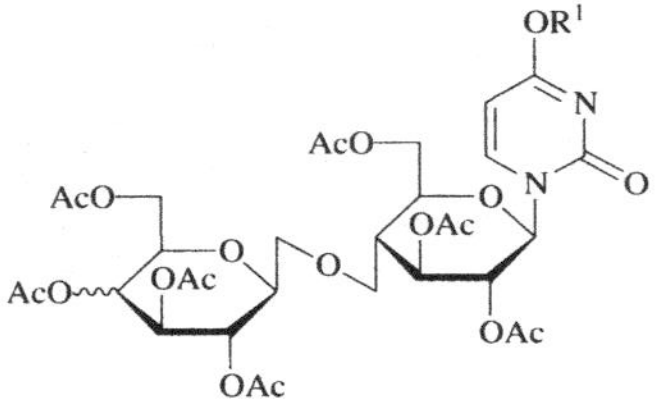

TABLE X. MISCELLANEOUS REACTIONS OF HETEROCYCLIC BASES WITH PROTECTED SUGARS

Sugar	Base	Conditions	Product(s) and Yield(s) (%)	Refs.
C_4 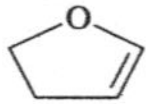	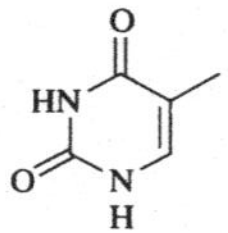	Me_2SiCl_2, Et_3N, MeCN, rt, 1 h	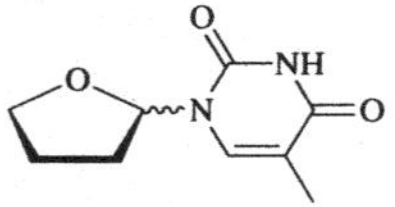 (84)	1232
	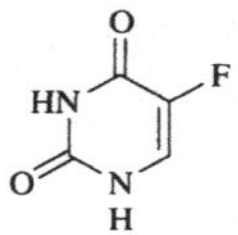		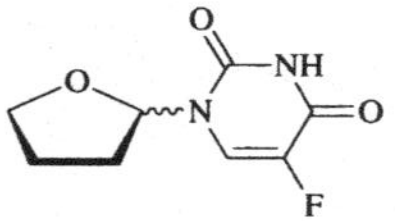	
		Py•HCl, Py, 120°, 20 h	(90)[a]	1067
		Al_2O_3, Py, 130°, 24 h	(89)	1067
		Py, 150°, 8 h, autoclave	(88)	1067
		DMA, 180°, 5 h, autoclave	(86)	1067
		Me_2SiCl_2, Et_3N, MeCN, 30-35°, 2 h	(86)	1232
	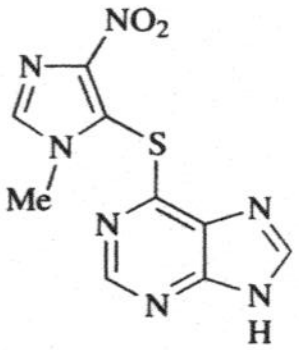	TsOH, EtOAc, 50°, 24 h	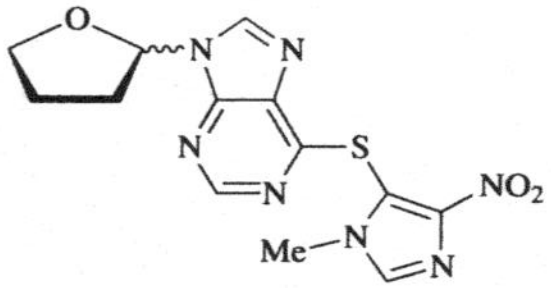 (86)	1092
	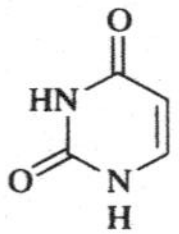	PCl_5, HMPA, rt, 5 h	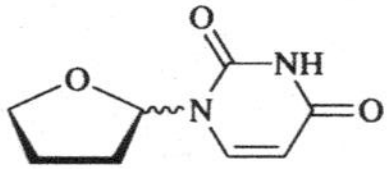 (92)	1233

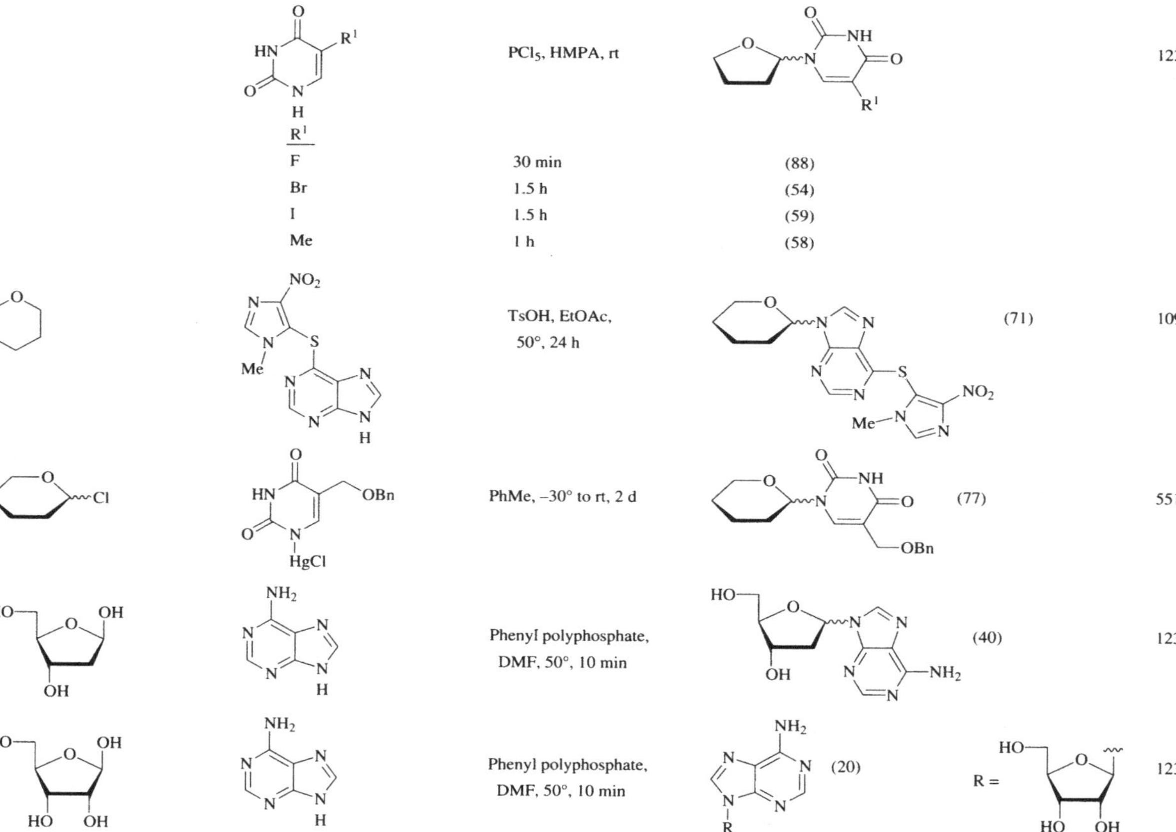

TABLE X. MISCELLANEOUS REACTIONS OF HETEROCYCLIC BASES WITH PROTECTED SUGARS (*Continued*)

Sugar	Base	Conditions	Product(s) and Yield(s) (%)	Refs.
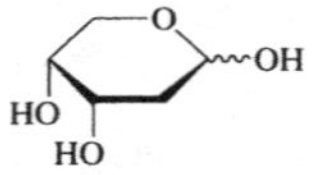	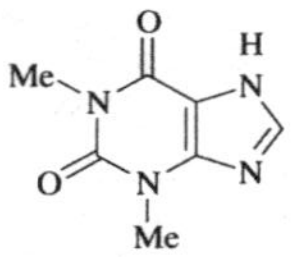	P_4O_{10}, Bu_3N, $CHCl_3$, 40°, 3 d	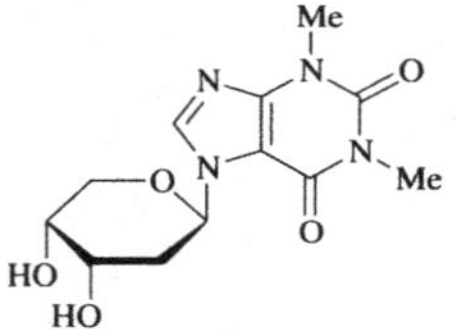 (25) + α-anomer (20)	1235
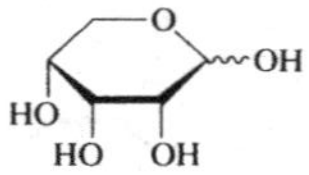	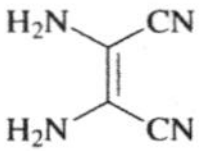	AcOH, MeOH, rt, 24 h	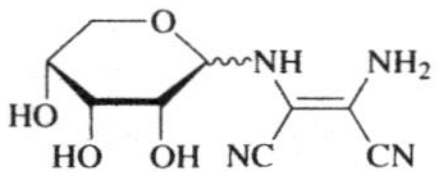 (51)	161
C_6				
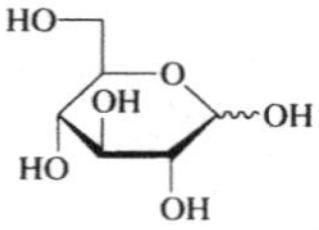	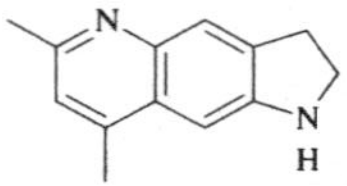	EtOH, reflux, 6 h	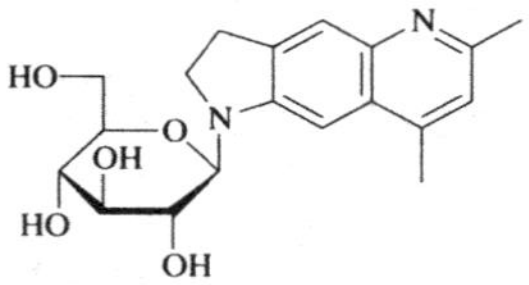 (9)	1236
C_7				
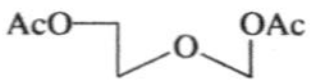	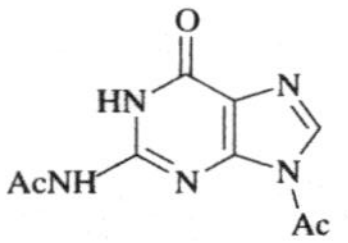	DMSO, 100°	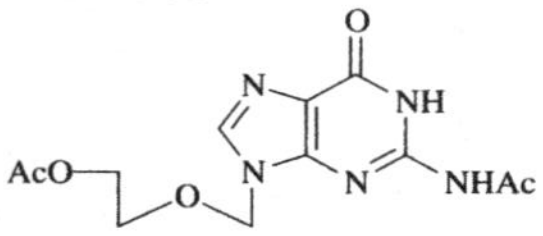	1237
		p-$H_2NC_6H_4SO_3H$[b]	(66) + N^7-isomer (28)	
		TsOH	(66) + N^7-isomer (26)	
		p-$O_2NC_6H_4SO_3H$	(59) + N^7-isomer (24)	
	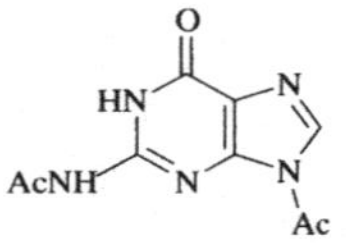	TsOH, DMSO, 80°	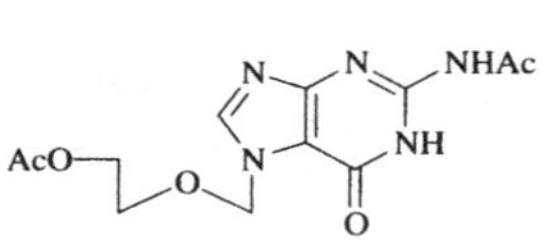 (53) + N^9-isomer (31)	1237

Substrate	Reagent	Conditions	Product and Yield(s) (%)	Refs.
	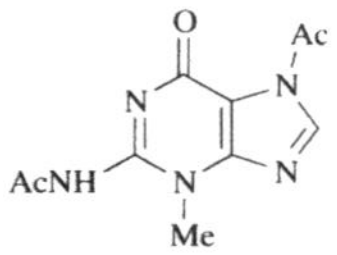	TsOH, PhMe, reflux, 20 h	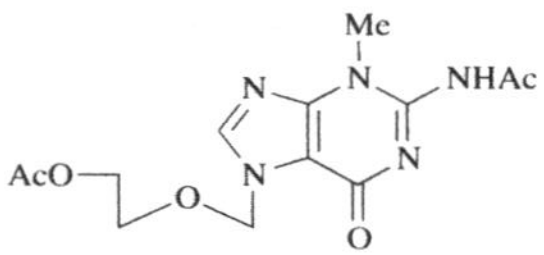 (69)	1238
C_9 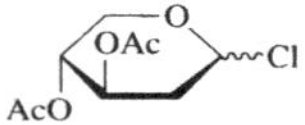	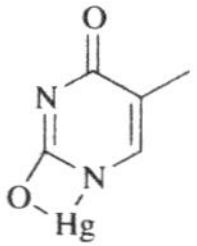	1. DMF, PhMe, rt, 2 h 2. NH_3, MeOH	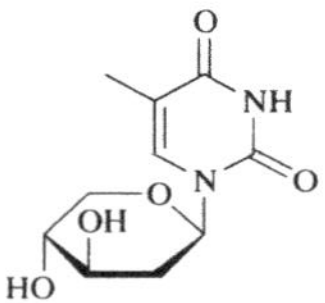 (8) + α-anomer (7)	927
C_{10} 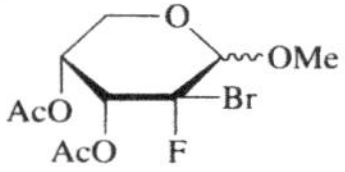	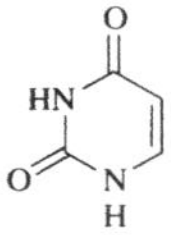	HgO, $HgBr_2$, CH_2Cl_2	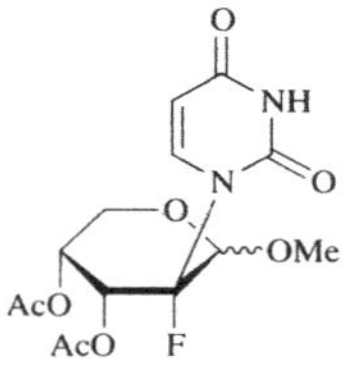 (23)	1239
C_{11} 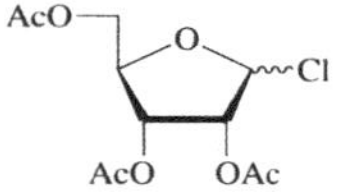	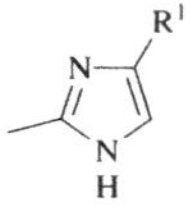	$Hg(CN)_2$, $MeNO_2$, reflux, 3-4 h	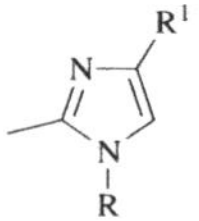 R^1: CO_2Et (35); $CONH_2$ (45) 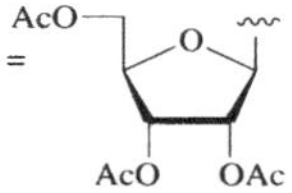	1240
	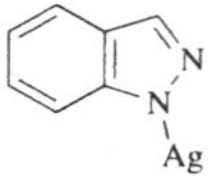	Xylene, reflux	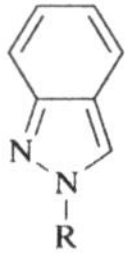 (42)	728
	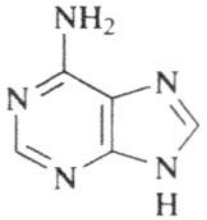	DMF, rt, 4.5 d	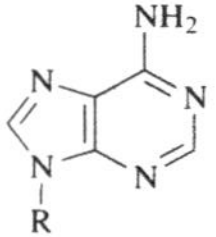 (18)	1241

TABLE X. MISCELLANEOUS REACTIONS OF HETEROCYCLIC BASES WITH PROTECTED SUGARS (*Continued*)

Sugar	Base	Conditions	Product(s) and Yield(s) (%)	Refs.
	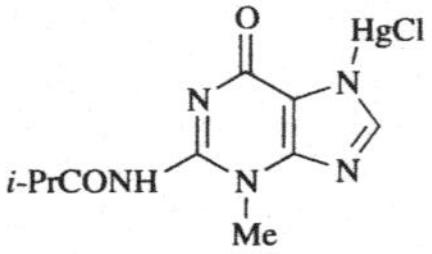 O, HgCl, N, N, N, *i*-PrCONH, N, Me	Xylene, reflux, 1 h	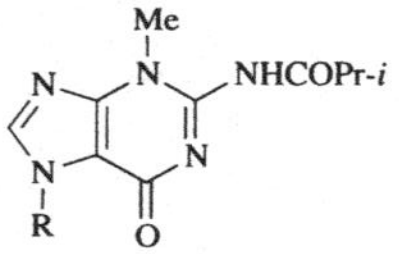 Me, N, N, NHCOPr-*i*, N, N, R, O (49) R = AcO, O, AcO, OAc 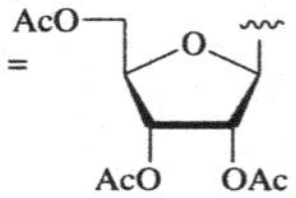	1208
	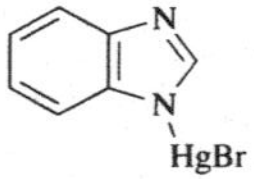 N, N, HgBr	1. Xylene, reflux, 2 h 2. NH_3, MeOH, 18 h	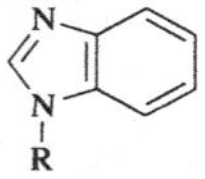 N, N, R (53) R = HO, O, HO, OH 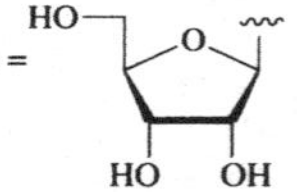	1242
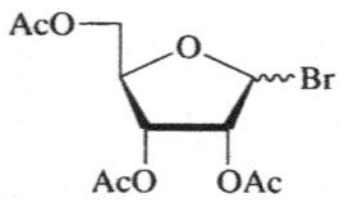 AcO, O, Br, AcO, OAc	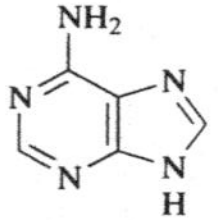 NH_2, N, N, N, N, H	MeCN, 50°, 36 h	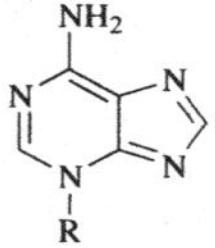 NH_2, N, N, N, N, R **I** (25) + N^9-isomer (18)	1243
		DMF, 90°, 1 h	**I** (20)	1241
	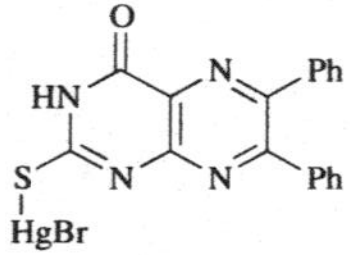 O, HN, N, Ph, S, N, N, Ph, HgBr	PhMe. reflux, 6 h	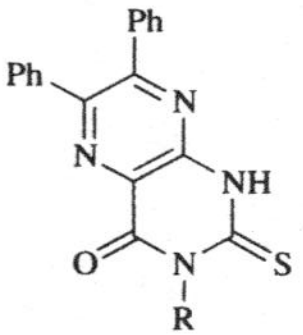 Ph, Ph, N, N, NH, O, N, S, R (17)	1244
C_{12} 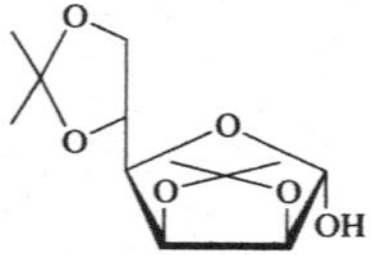 O, O, O, O, O, OH	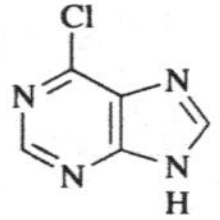 Cl, N, N, N, N, H	Ph_3PMe, DEAD, THF, rt, 12 h	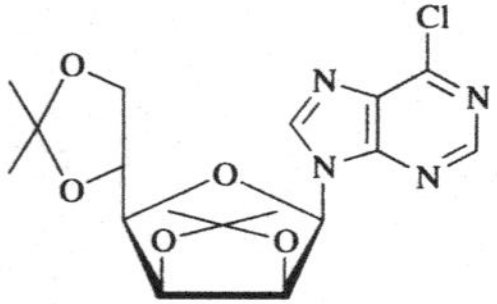 O, O, Cl, N, N, N, N, O, O, O (63) + α-anomer (16)	430

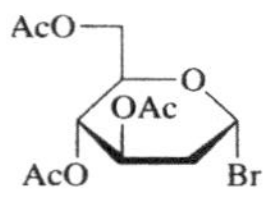	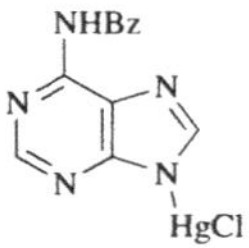 NHBz, HgCl	DMF, PhMe, rt, 8 h	NHBz	(—)	63
C_{13}	NBz, Bn, Hg/2	1. Xylene, reflux, 1 h 2. NaOMe, MeOH 3. H_2, Pd/C, 80°, 16 h	NH_2	(14) + β-anomer (8)	1245
	F, Cbz	DMF		(—)	770
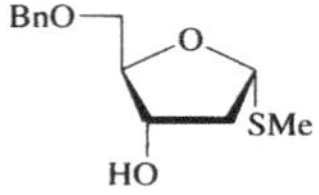	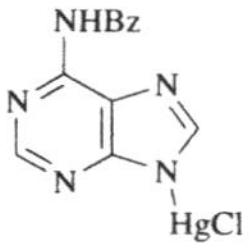 NHBz, HgCl	$Cl(CH_2)_2Cl$, reflux, 5 h	NHBz	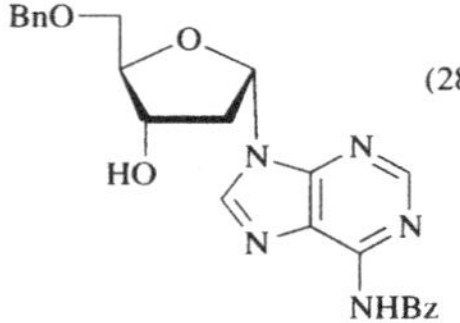 (28) + β-anomer (20)	413
C_{14}		CH_2Cl_2, MS, reflux		R =	
	R^1				
	Cl	15 h	(8)		391
	NHBz	3 d	(34)		384, 1246

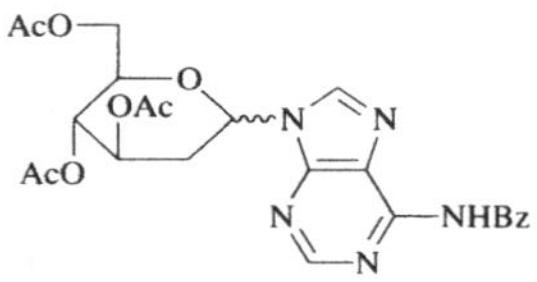

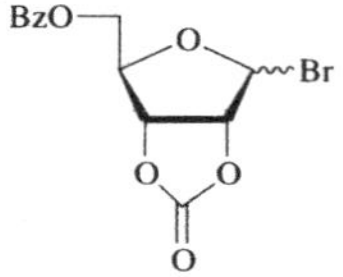

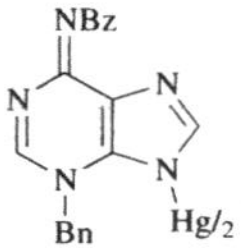

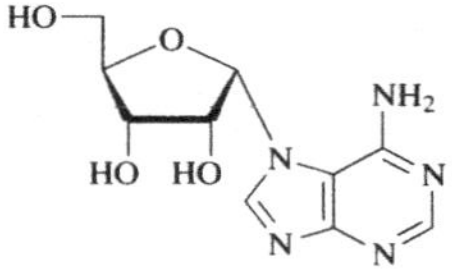

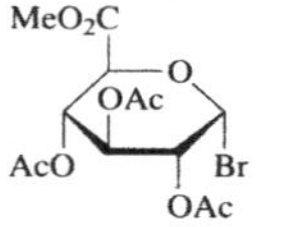

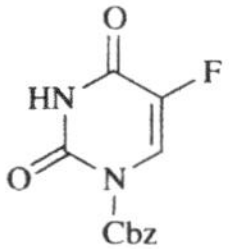

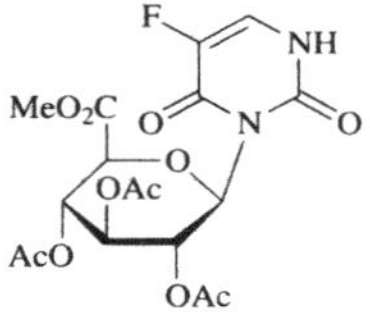

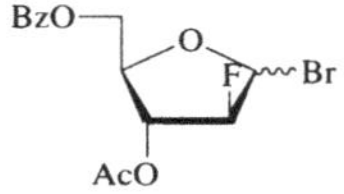

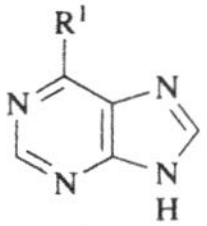

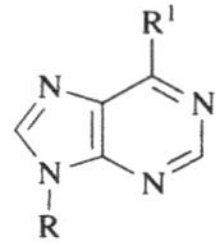

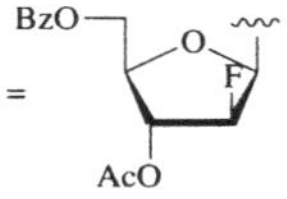

TABLE X. MISCELLANEOUS REACTIONS OF HETEROCYCLIC BASES WITH PROTECTED SUGARS (*Continued*)

Sugar	Base	Conditions	Product(s) and Yield(s) (%)	Refs.
	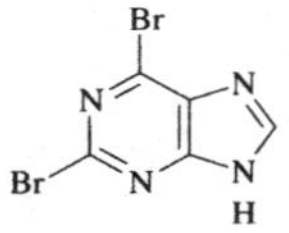	$Cl(CH_2)_2Cl$, MS, 100°, 32 h	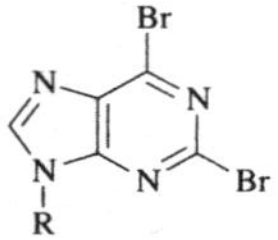 (32)	539
	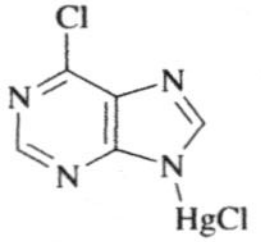 HgCl	1. Xylene, reflux, 12 h 2. thiouracil	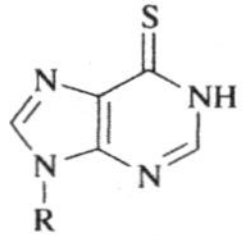 (10)	384
	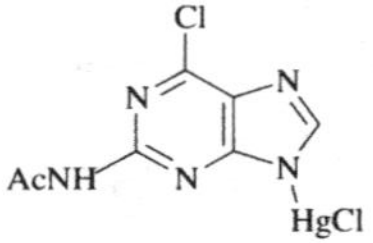 AcNH, HgCl	Xylene, reflux, 15 h	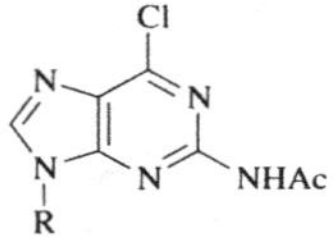 NHAc (17)	384, 1246
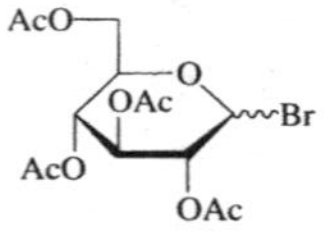 AcO, OAc, Br, AcO, OAc	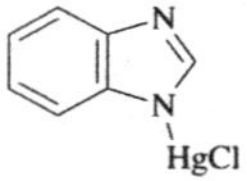 HgCl	1. Xylene, reflux, 1.5-2 h 2. NH_3, MeOH	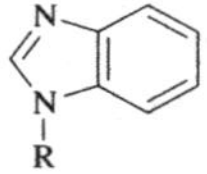 (34) R 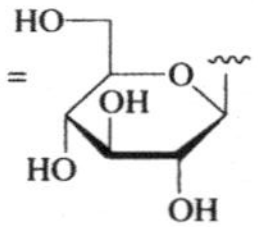 = HO, OH, HO, OH	1247
	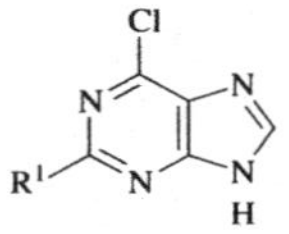 R^1	$Hg(CN)_2$, $MeNO_2$, reflux, 2.5 h	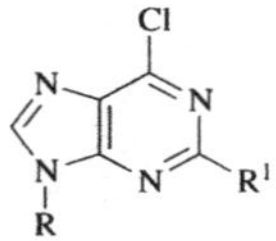 R^1: H (55); Cl (79) R 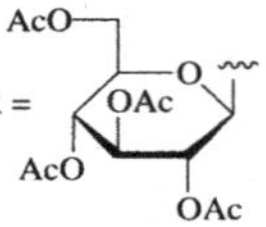 = AcO, OAc, AcO, OAc	1247
	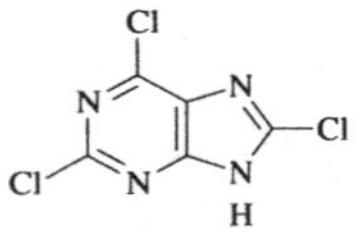 Cl	$Hg(CN)_2$, $MeNO_2$, reflux, 2 h	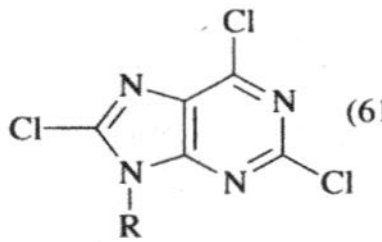 (61)	1247

Glycosyl halide	Aglycone	Conditions	Product(s) and Yield(s) (%)	Refs.
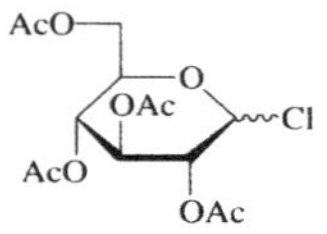 AcO, OAc, AcO, OAc, Cl	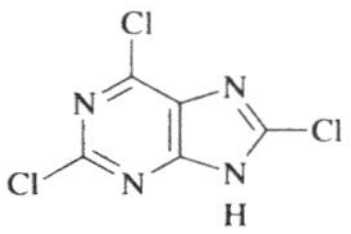 Cl, N, N, N, N, Cl, Cl, H	$Hg(CN)_2$, $MeNO_2$, reflux, 3 h	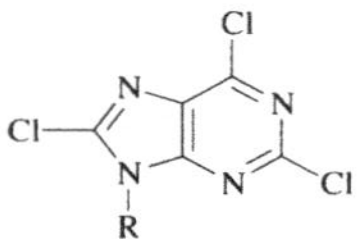 Cl, N, N, N, N, Cl, Cl, R (68)	1247
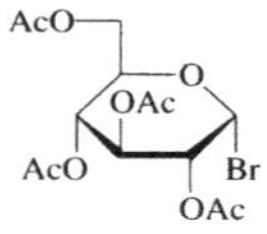 AcO, OAc, AcO, OAc, Br	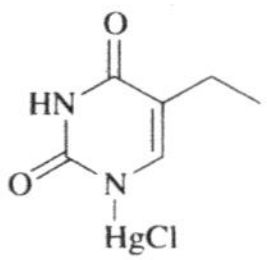 O, HN, O, N, HgCl	PhMe, reflux, 2 h	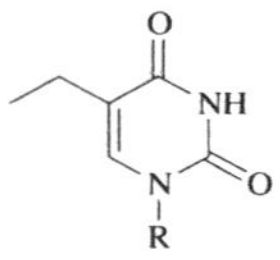 O, NH, O, N, R (—)	1248
	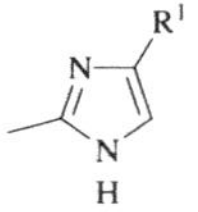 R^1, N, N, H	$Hg(CN)_2$, $MeNO_2$, MS, reflux, 3-4 h	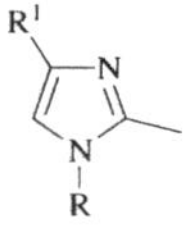 R^1, N, N, R; R^1 = $CONH_2$ (46); CO_2Et (36) + N^3-isomer (12)	1240
	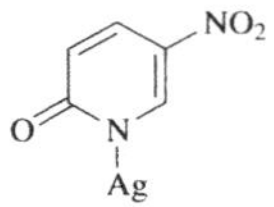 NO_2, O, N, Ag	PhMe, reflux, 20 min	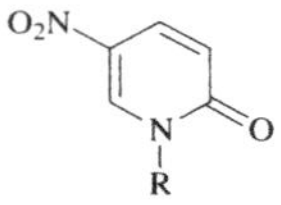 O_2N, O, N, R (~5)	58
	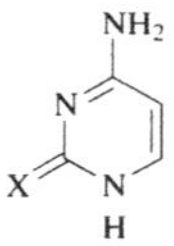 NH_2, N, X, N, H	$Hg(CN)_2$, $MeNO_2$, MS, reflux, 1 h	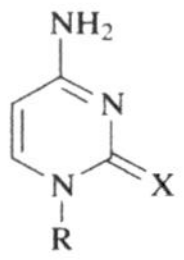 NH_2, N, X, N, R; X = O (50); S (40)	1249
	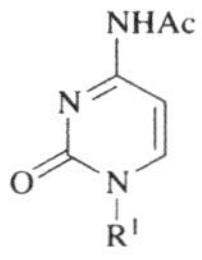 NHAc, N, O, N, R^1	$Hg(CN)_2$, $HgBr_2$, $MeNO_2$, PhMe, reflux, 20 min	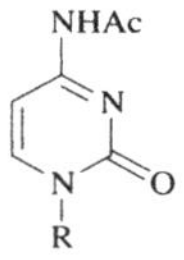 NHAc, N, O, N, R; R^1 = H, (55)	1249

TABLE X. MISCELLANEOUS REACTIONS OF HETEROCYCLIC BASES WITH PROTECTED SUGARS (*Continued*)

Sugar	Base	Conditions	Product(s) and Yield(s) (%)	Refs.
		$HgBr_2$, PhMe, reflux	R^1 = Ag, (55)	57
	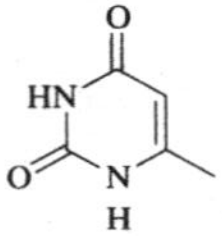	$Hg(CN)_2$, $MeNO_2$, reflux, 4 h	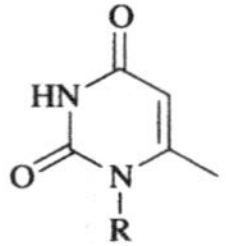 (94)	1249
	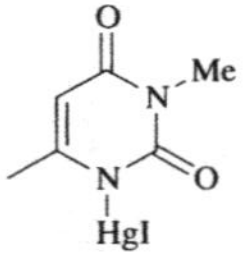 HgI	1. PhMe, reflux, 4 h 2. KI	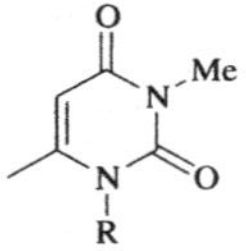 (7)	60
	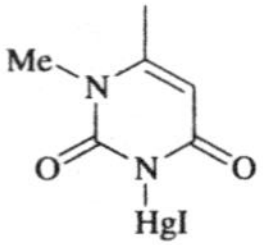 HgI	PhMe, MeCN, reflux, 4 h	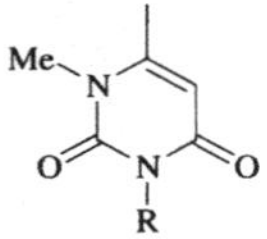 (43) R 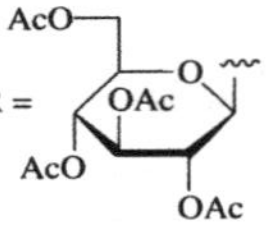 =	60
	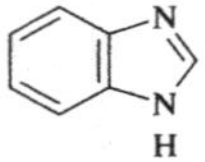	$Hg(CN)_2$, $MeNO_2$, reflux, 4 h	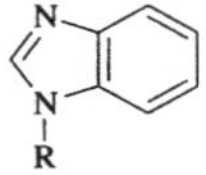 (63)	1247
	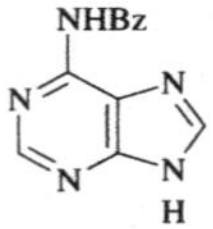 NHBz	$Hg(CN)_2$, $MeNO_2$, reflux, 3 h	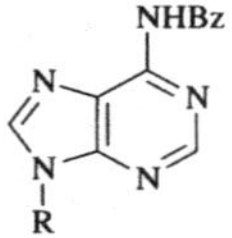 NHBz (75)	1247

Reactant	Conditions	Product(s) and Yield(s) (%)	Refs.
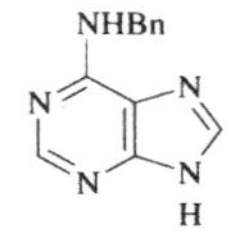	DMF, 100°, 20 h	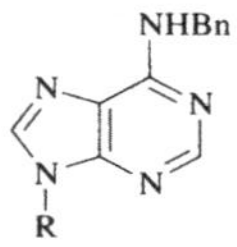 (30) + N^3-isomer (—)	1250
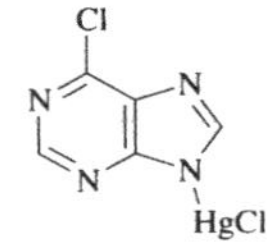	Xylene, reflux, 2 h	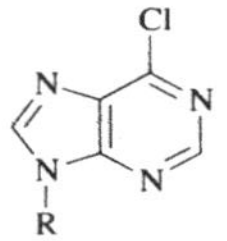 (86)	1251
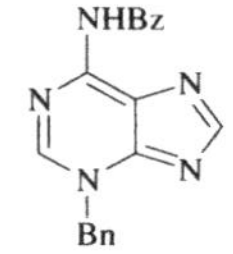	Xylene, reflux, 1.5 h	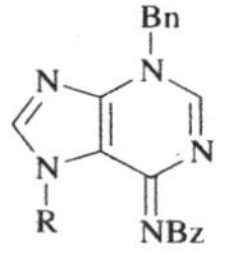 (28)	1251
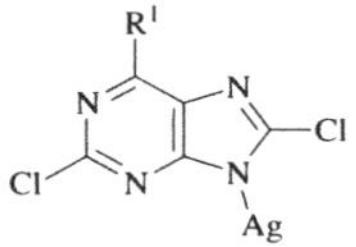	Xylene, reflux	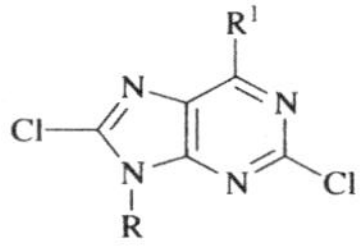 R^1 / Cl — (—) / NH_2 6 h (29)	56
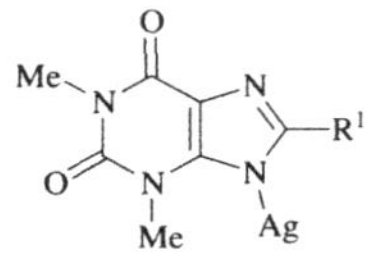	Xylene, reflux	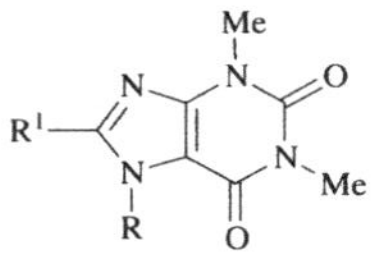 R^1 / H 1 min (75) / Cl 10 min (40)	56
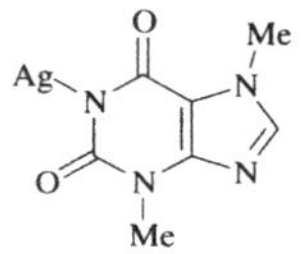	PhMe, reflux, 30 min	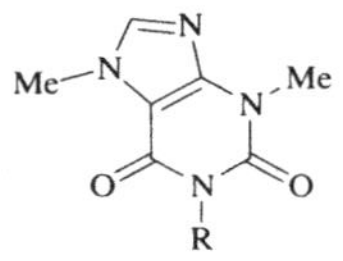 (23)	56

TABLE X. MISCELLANEOUS REACTIONS OF HETEROCYCLIC BASES WITH PROTECTED SUGARS (*Continued*)

Sugar	Base	Conditions	Product(s) and Yield(s) (%)	Refs.
	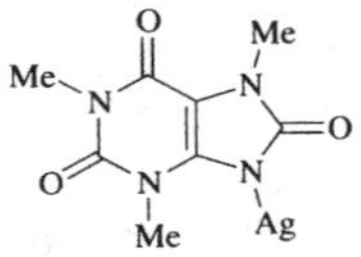	Xylene, reflux, 5 min	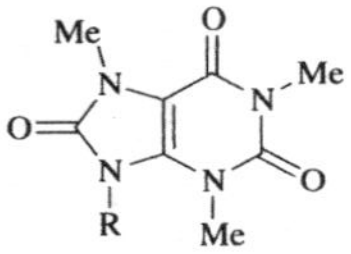 (—)	56
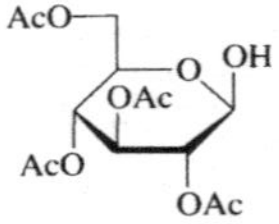	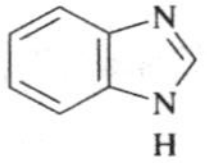	$Et_2O{\bullet}BF_3$, $Cl(CH_2)_2Cl$, 60°, 10 h	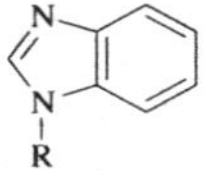 **I** (73)	1252
		1. CETF, $Cl(CH_2)_2Cl$, $Et_3O^+BF_4^-$ 2. $Cl(CH_2)_2Cl$, DME, 60°, 10 h	**I** (73)	429
	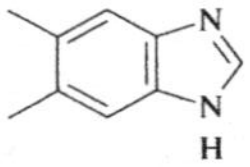	1. $Cl(CH_2)_2Cl$, $Et_3O^+BF_4^-$ 2. $Cl(CH_2)_2Cl$, DME, 60°, 10 h	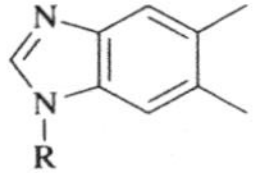 **I** (86) R = 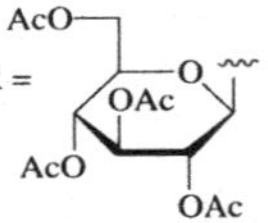	429
		$Et_2O{\bullet}BF_3$	**I** (86)	1252
	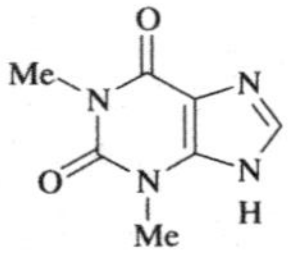	1. CETF, $Cl(CH_2)_2Cl$, $Et_3O^+BF_4^-$ 2. $Cl(CH_2)_2Cl$, DME, 60°, 10 h	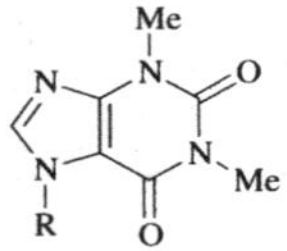 (93)	429
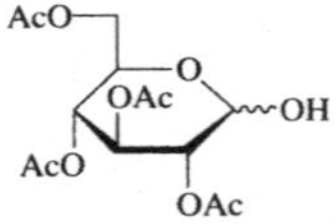	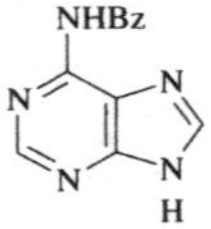	1. P_4O_{10}, DMF, 50-60°, 75 h 2. NH_3, MeOH	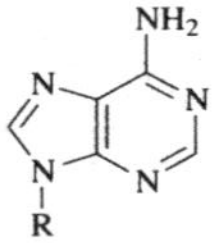 (10) R = 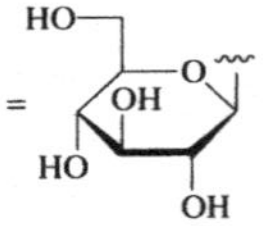	437

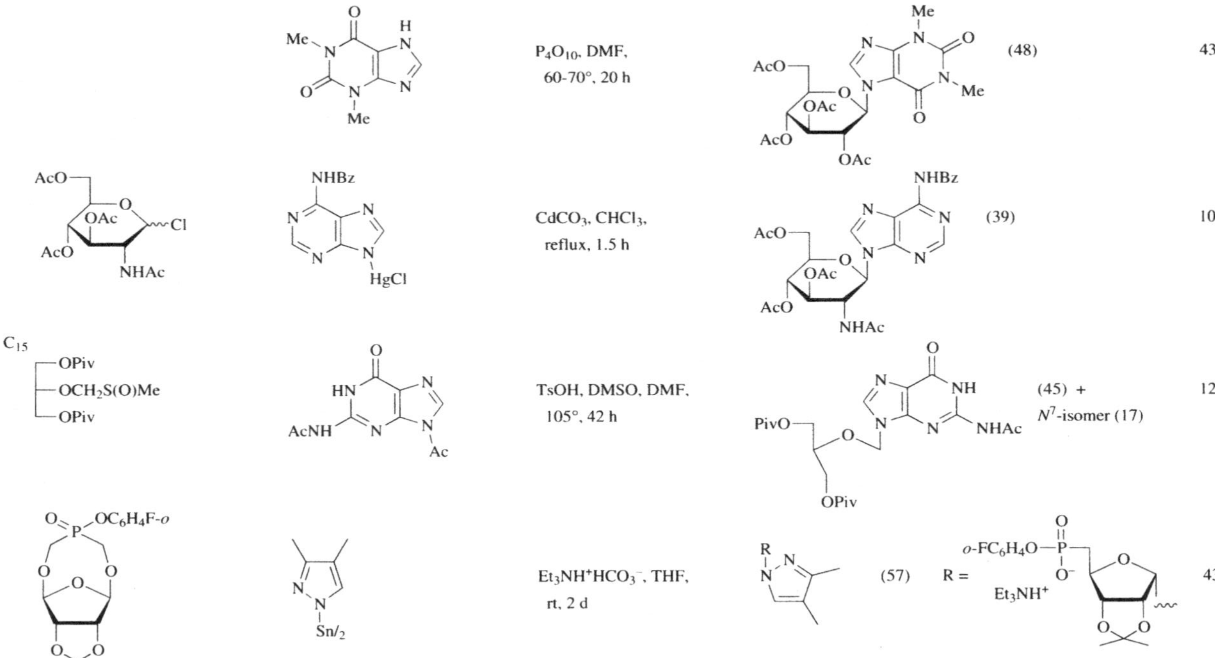

Me
O
H
N
Me
O
N
N
N
Me
P_4O_{10}, DMF,
60-70°, 20 h
AcO
O
OAc
AcO
OAc
(48)
437
AcO
O
OAc
Cl
AcO
NHAc
NHBz
HgCl
$CdCO_3$, $CHCl_3$,
reflux, 1.5 h
NHBz
NHAc
(39)
1071
C_{15}
OPiv
$OCH_2S(O)Me$
OPiv
HN
AcNH
Ac
TsOH, DMSO, DMF,
105°, 42 h
NH
PivO
OPiv
NHAc
(45) +
N^7-isomer (17)
1253
OC_6H_4F-o
P
$Sn/_2$
$Et_3NH^+HCO_3^-$, THF,
rt, 2 d
R
(57)
R =
o-FC_6H_4O
O^-
Et_3NH^+
433

TABLE X. MISCELLANEOUS REACTIONS OF HETEROCYCLIC BASES WITH PROTECTED SUGARS (*Continued*)

Sugar	Base	Conditions	Product(s) and Yield(s) (%)	Refs.
	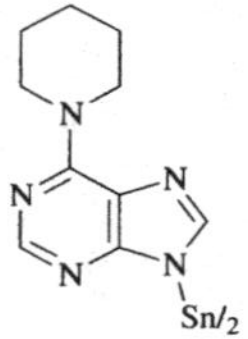 Sn/2	$Et_3NH^+HCO_3^-$, THF, rt, 4 d	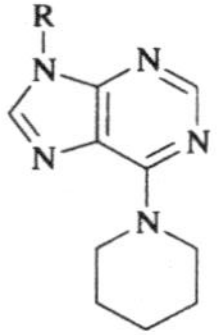 R (30)	433
	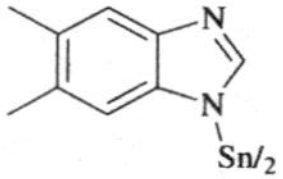 Sn/2	$Et_3NH^+HCO_3^-$, THF, rt, 12 h	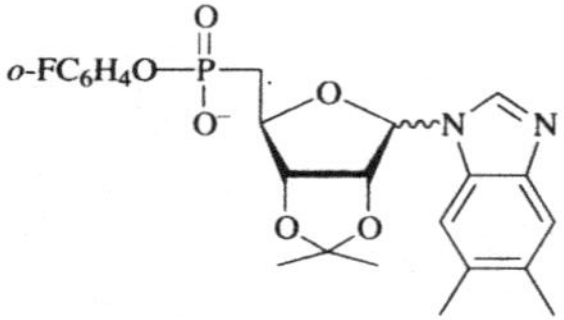 o-FC_6H_4O (81) α:β = 95:5	433
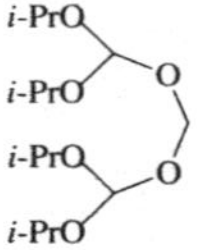 *i*-PrO, *i*-PrO, *i*-PrO, *i*-PrO	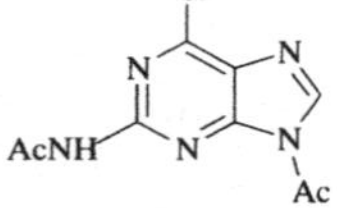 Cl, AcNH, Ac	$Al_2(SO_4)_3$, *N*-methyl-pyrrolidin-2-one, 100°, 6 h	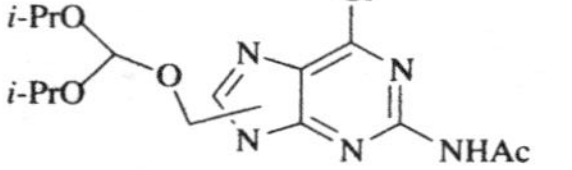 *i*-PrO, *i*-PrO, Cl, NHAc (61) N^9:N^7 = 74:8	521
C_{16} 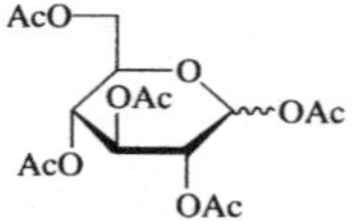 AcO, OAc, OAc, AcO, OAc	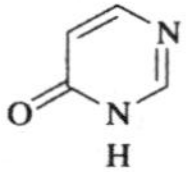 H	$Hg(CN)_2$, $MeNO_2$, reflux, 6 h	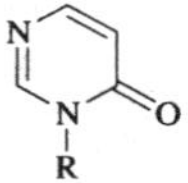 R (52) R = AcO, OAc, AcO, OAc 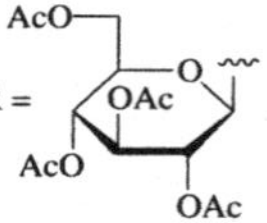	516
	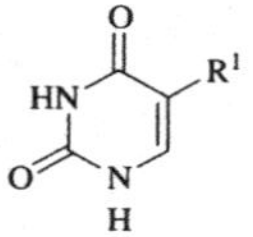 R^1, HN, H	$Hg(CN)_2$, $MeNO_2$, reflux, 6 h	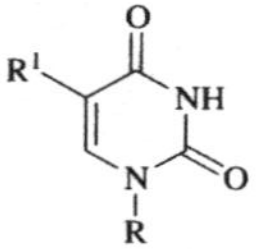 R^1, NH, R; R^1: Me (87), F (65), CO_2Et (85)	516

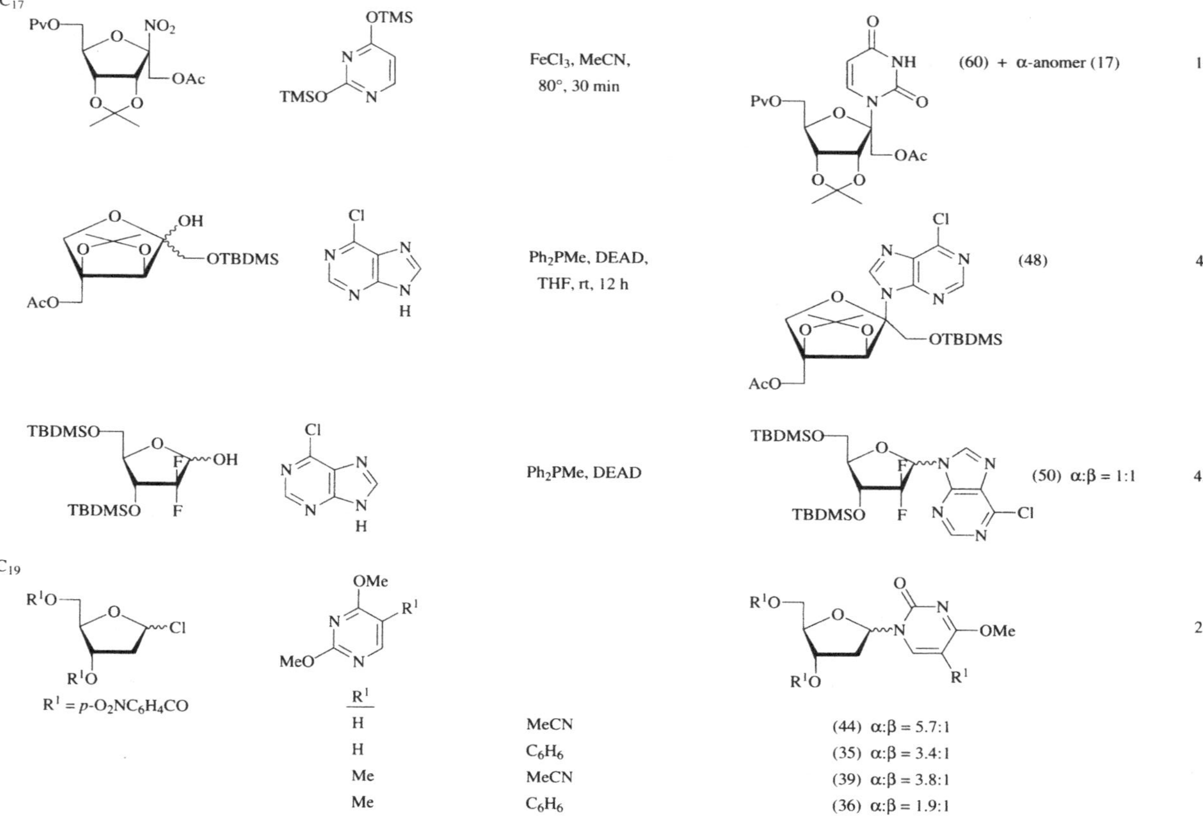

Sugar	Base	Conditions	Product(s) and Yield(s) (%)	Refs.
C_{17} PvO, NO_2, OAc	OTMS, TMSO	$FeCl_3$, MeCN, 80°, 30 min	PvO, NH, O, OAc (60) + α-anomer (17)	106
OH, OTBDMS, AcO	Cl, N, H	Ph_2PMe, DEAD, THF, rt, 12 h	Cl, OTBDMS, AcO (48)	430
TBDMSO, F, OH, TBDMSO, F	Cl, N, H	Ph_2PMe, DEAD	TBDMSO, F, TBDMSO, F, Cl (50) α:β = 1:1	431
C_{19} R^1O, Cl, R^1O; $R^1 = p\text{-}O_2NC_6H_4CO$	OMe, R^1, MeO; R^1		R^1O, O, OMe, R^1O, R^1	2
	H	MeCN	(44) α:β = 5.7:1	
	H	C_6H_6	(35) α:β = 3.4:1	
	Me	MeCN	(39) α:β = 3.8:1	
	Me	C_6H_6	(36) α:β = 1.9:1	

TABLE X. MISCELLANEOUS REACTIONS OF HETEROCYCLIC BASES WITH PROTECTED SUGARS (*Continued*)

Sugar	Base	Conditions	Product(s) and Yield(s) (%)	Refs.
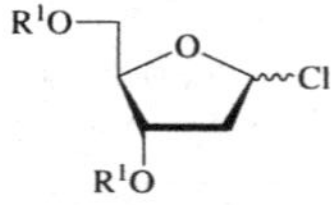 R^1O, Cl; R^1O $R^1 = p\text{-}ClC_6H_4CO$	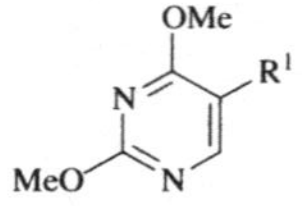 OMe, R^1, N, MeO, N		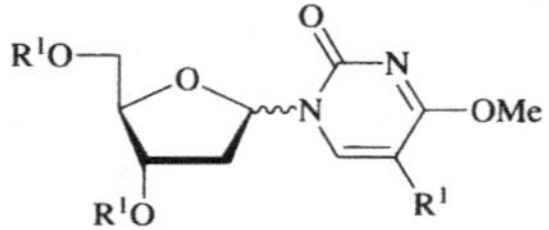 R^1O, O, N, OMe, R^1O, R^1	2
	R^1			
	H	MeCN	(50) α:β = 4.5:1	
	H	C_6H_6	(31) α:β = 2.4:1	
	Me	MeCN	(46) α:β = 3.6:1	
	Me	C_6H_6	(39) α:β = 1.9:1	
	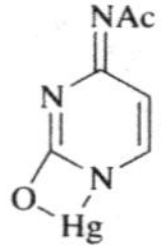 NAc, N, N, O, Hg	1. Xylene, reflux 2. 0°	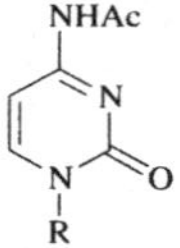 NHAc, N, N, O, R (32) + α-anomer (22) R 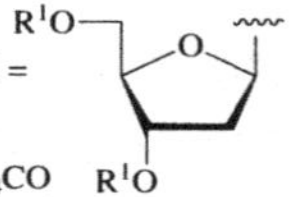 = R^1O, O, R^1O $R^1 = p\text{-}ClC_6H_4CO$	264
	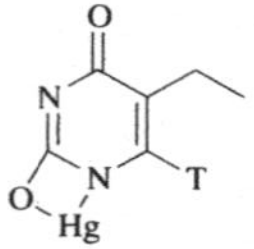 O, N, N, T, O, Hg	1. PhMe, rt, 3 h 2. 60°, 1 h	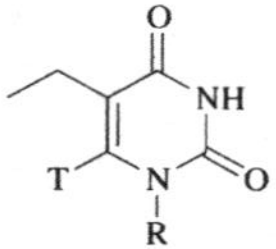 O, NH, T, N, O, R (80)	1254
	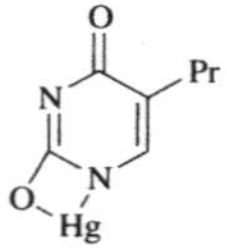 O, Pr, N, N, O, Hg	1. PhMe, reflux 2. NaOMe, MeOH, reflux, 2 h	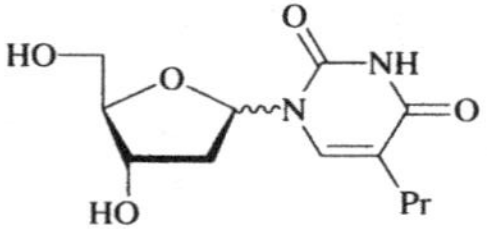 HO, O, N, O, NH, O, HO, Pr (89-90)	1255
	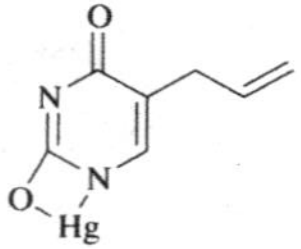 O, N, N, O, Hg	PhMe, reflux, 8 min $R^1 = p\text{-}ClC_6H_4CO$	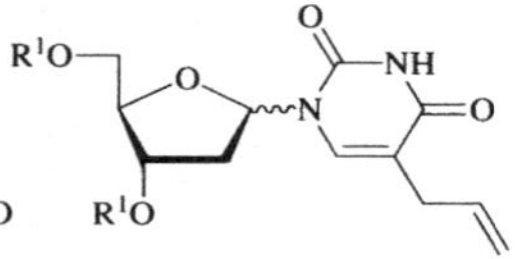 R^1O, O, N, O, NH, O, R^1O (89-90)	265

Glycosyl donor	Base	Conditions	Product(s) and Yield(s) (%)	Refs.
	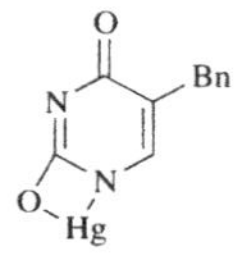 O, N, Bn, O, N, Hg	PhMe, reflux, 1 h	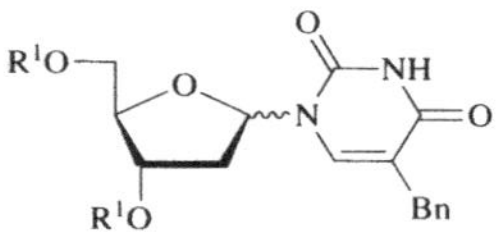 R^1O, O, N, O, NH, O, Bn, R^1O (80)	1256
	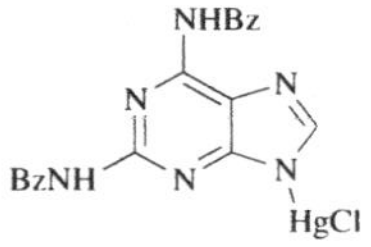 NHBz, N, N, BzNH, N, N, HgCl	C_6H_6, reflux	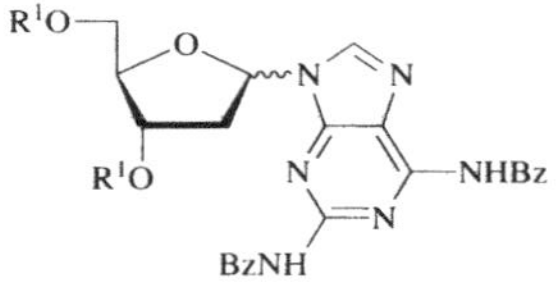 R^1O, O, N, N, R^1O, N, NHBz, N, BzNH (38)	343
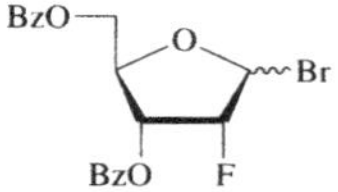 BzO, O, Br, BzO, F	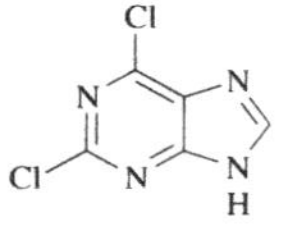 Cl, N, N, Cl, N, N, H	$Hg(CN)_2$, $Cl(CH_2)_2Cl$, rt, 20 min	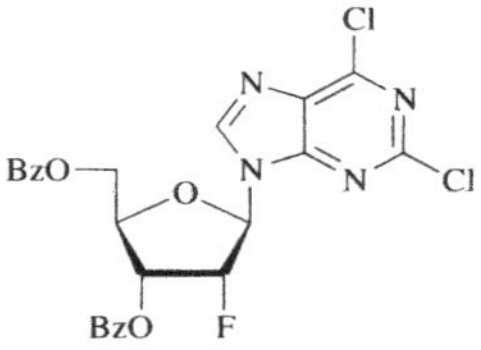 Cl, N, N, BzO, O, N, N, Cl, BzO, F (59) + α-anomer (2)	1225
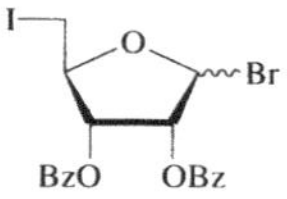 I, O, Br, BzO, OBz	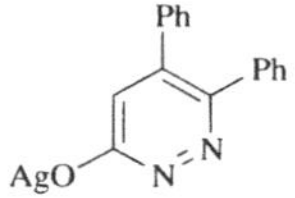 Ph, Ph, AgO, N, N	1. PhMe, reflux, 5 min 2. $HgBr_2$, PhMe, reflux, 5 h	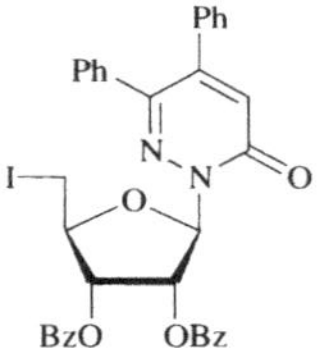 Ph, Ph, N, N, O, I, O, BzO, OBz (36-42)	1257
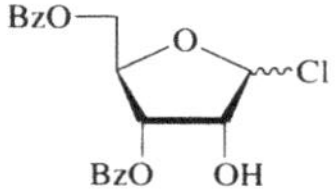 BzO, O, Cl, BzO, OH	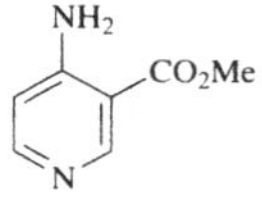 NH_2, CO_2Me, N	CH_2Cl_2, rt, 12 h	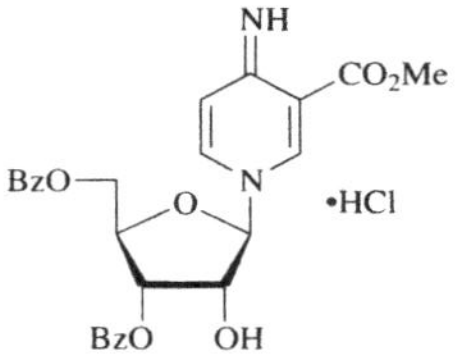 NH, CO_2Me, BzO, N, •HCl, O, BzO, OH (83)	1258

TABLE X. MISCELLANEOUS REACTIONS OF HETEROCYCLIC BASES WITH PROTECTED SUGARS (*Continued*)

Sugar	Base	Conditions	Product(s) and Yield(s) (%)	Refs.
	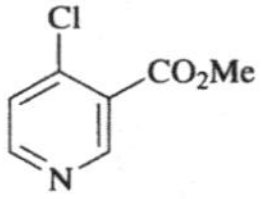 Cl, CO_2Me, N	1. CH_2Cl_2, 4°, 12 h 2. NH_3, MeOH, rt, 12 h	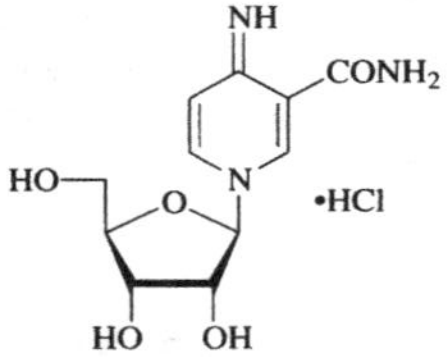 NH, $CONH_2$, N, HO, O, HO, OH, •HCl (26)	1258
C_{21} 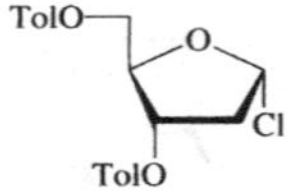 TolO, O, Cl, TolO	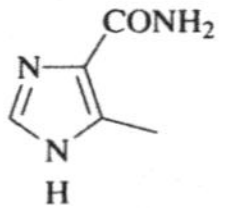 $CONH_2$, N, N, H	$Hg(CN)_2$, $MeNO_2$, MS, reflux, 6 h	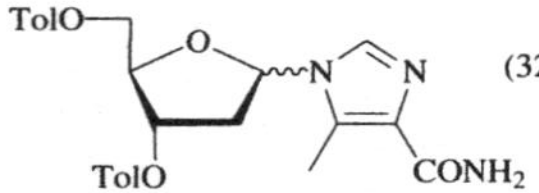 TolO, O, N, N, TolO, $CONH_2$ (32) + N^3-isomer (6)	708
	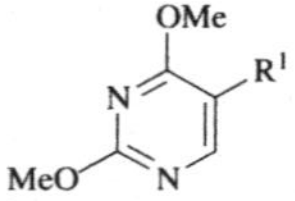 OMe, R^1, N, MeO, N R^1: H, Me	MeCN	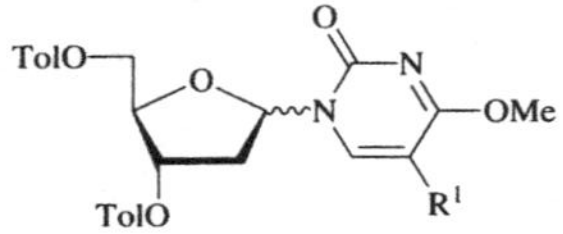 TolO, O, O, N, N, OMe, TolO, R^1 (53) α:β = 7.7:1 (72) α:β = 5.7:1	72
	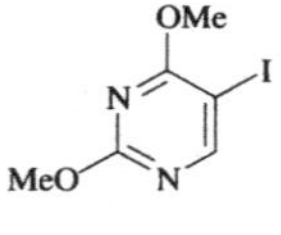 OMe, I, N, MeO, N	$MeNO_2$, rt, 70 h	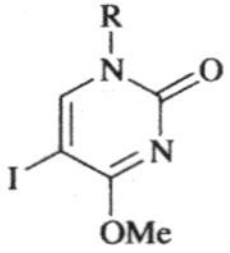 R, N, O, N, I, OMe **I** (18) + β-anomer (3) R 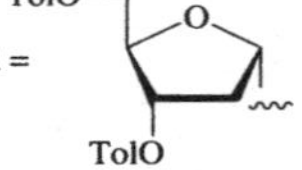 = TolO, O, TolO	263
		MeCN, rt, 5 d	**I** (20) + β-anomer (4)	263
	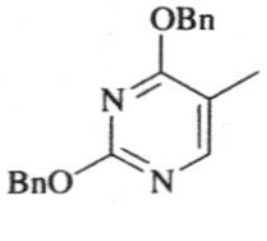 OBn, N, BnO, N	CH_2Cl_2, rt, 24 h	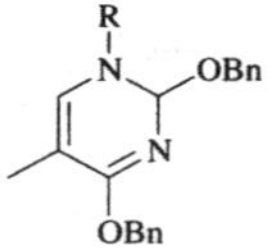 R, N, OBn, N, OBn (55)	261

Reactant	Conditions	Product(s) and Yield(s) (%)	Refs.
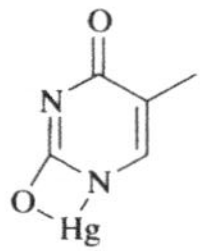 O–Hg (thymine derivative)	DMF, PhMe, rt, 1 h	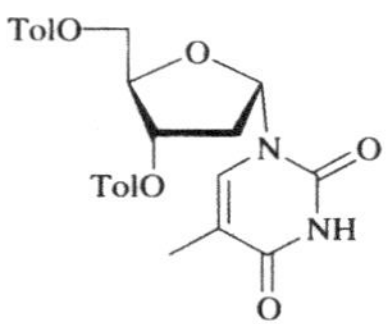 TolO, TolO (68) + β-anomer (23)	51
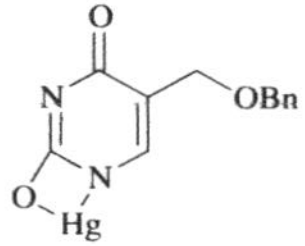 OBn, O–Hg	PhMe, rt, 16 h	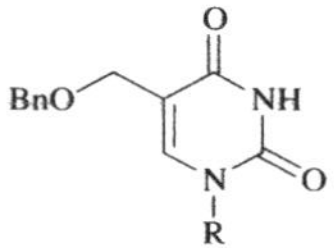 BnO, NH, N–R (49) + α-anomer (35); R 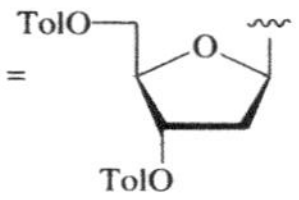 = TolO, TolO	1259
R^1, O–Hg		TolO, TolO, NH, R^1	
R^1			
I	PhMe, DMF, rt, 18 h	(—)	263
CN	MeCN, MS, 95°, 1 h	(—)	1260
NO_2	PhMe, DMF, rt, 4 d	(—)	88
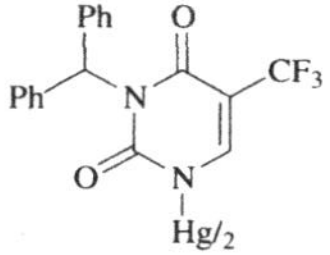 Ph, Ph, CF_3, $Hg/_2$	PhMe, rt, 12 h 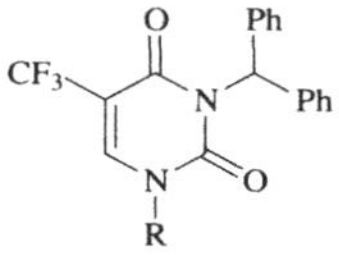	CF_3, Ph, Ph, N–R (24)	1262
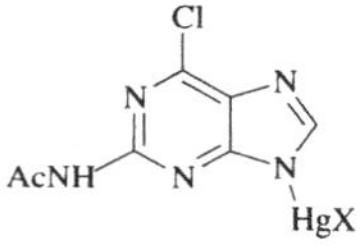 Cl, AcNH, HgX	C_6H_6, reflux, 1 h	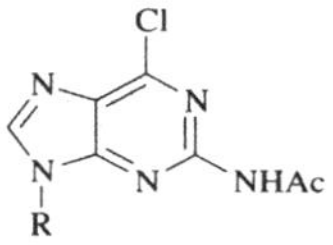 Cl, NHAc, N–R (25) + α-anomer (16)	1263

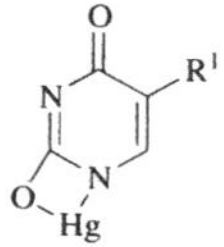

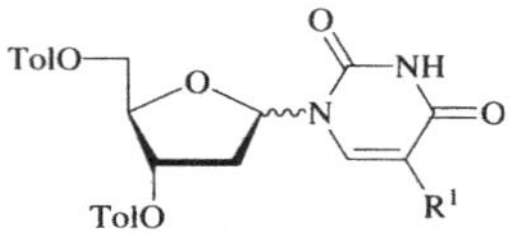

TABLE X. MISCELLANEOUS REACTIONS OF HETEROCYCLIC BASES WITH PROTECTED SUGARS (*Continued*)

Sugar	Base	Conditions	Product(s) and Yield(s) (%)	Refs.
	HN, O, N, N, Br, N, H		TolO, O, N, N, TolO, N, O, NH, Br	1264
		DMF, rt, 24 h	(75) α:β = 1:37.5	
		CH_2Cl_2, rt, 24 h	(51) α:β = 1:2.6	
		DMSO, rt, 24 h	(70) α:β = 1:14	
C_{22}				
i-Pr, Si, *i*-Pr, O, O, O, O, Si, *i*-Pr, Pr-*i*	Cl, N, N, N, N, H	IDCP, MeCN, MS, rt, 2 h	Cl, N, N, N, N, *i*-Pr, Si, *i*-Pr, O, O, O, Si, O, *i*-Pr, Pr-*i* (20) + α-anomer (20)	421
C_{26}				
BzO, O, Cl, BzO, OBz	O, N, PhHgS, N, H	1. $Hg(CN)_2$, $MeNO_2$, reflux, 4 h 2. NaOMe, MeOH	O, NH, N, S, HO, O, HO, OH (45)	1265
	NH_2, N, MeS, N	1. MeCN, MS, 30° 2. H_2S, Py	NH_2, N, N, S, R (40) R = BzO, O, BzO, OBz	1266

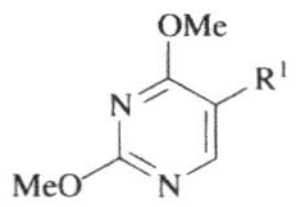

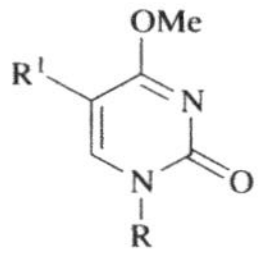

Reactant	Conditions	Product(s) and Yield(s) (%)	Refs.
2,4-dimethoxy-5-R^1-pyrimidine (MeO, OMe, R^1)		4-OMe-5-R^1-1-R-pyrimidin-2-one; R = 2,3,5-tri-O-benzoyl-ribofuranosyl (BzO, BzO, OBz)	
R^1			
F	$HgBr_2$, PhMe, MS, 70-72°, 50 h; rt, 15 h	(39)	1267
F	C_6H_6, MeCN, 70-72°, 50 h	(35)	1267
Cl	C_6H_6, MS, 70-72°, 50 h	(43)	1267
Cl	C_6H_6, MeCN, 70°, 48 h	(37)	1267
Br	$HgBr_2$, C_6H_6, MS, 70-72°, 50 h	(45)	1267
Br	MeCN, C_6H_6, 70°, 48 h	(43)	1267
I	$HgBr_2$, PhMe, C_6H_6, 70°, 48 h	(61)	1267
NHAc	MeCN, MS, 50°, 4 d	(83)	1260
CN	MeCN, MS, 70°, 95 h	(16)	1260
CH_2OBn	MeCN, 70°, 2 d	(37) + α-anomer (8)	1268
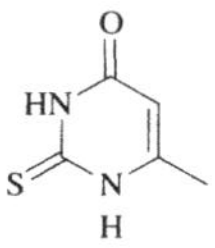 6-methyl-2-thiouracil (HN, S, N, H, O)	1. $Hg(CN)_2$, $MeNO_2$, reflux, 4 h 2. H_2S 3. NaOMe, MeOH 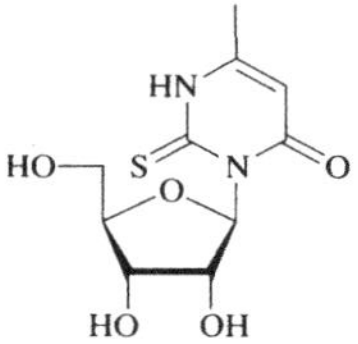	6-methyl-2-thio-uridine analogue (HO, HO, OH, HN, S, N, O) (34)	1265

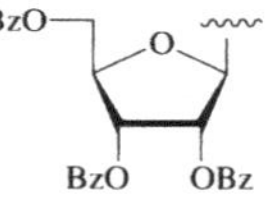

TABLE X. MISCELLANEOUS REACTIONS OF HETEROCYCLIC BASES WITH PROTECTED SUGARS (*Continued*)

Sugar	Base	Conditions	Product(s) and Yield(s) (%)	Refs.
	OMe, N, O, N, HgCl	MeCN, reflux, 25 min	OMe, N, O, N, R (26) + bis(riboside) (13) + N^3-isomer (19)	60
	OMe, N, O, N, CN	MeCN, reflux, 8 h	CN, N, O, N, O, R (59) R = BzO, O, BzO, OBz	254
	O, HN, S, N, HgCl	1. $HgBr_2$, C_6H_6, reflux, 2 h 2. 25°, 16 h	HN, S, N, O, R (13) + *S* isomer (18)	1269
	NHAc, N, S, N, HgCl	1. $HgBr_2$, C_6H_6, reflux, 2 h 2. 25°, 16 h	NHAc, N, N, S, R (—)	1269

Reactant	Conditions	Product(s) and Yield(s) (%)		Refs.
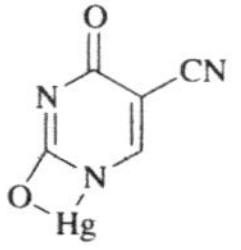 O, N, CN, N, O, Hg	MeCN, MS, 100°, 4 h	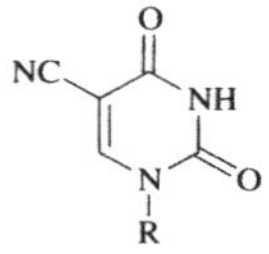 O, NC, NH, N, O, R	(41)	1260
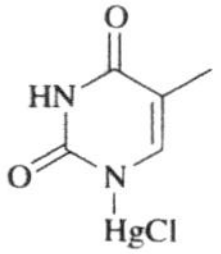 O, HN, O, N, HgCl	PhMe, reflux, 12 h	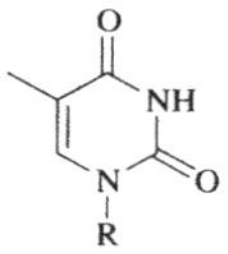 O, NH, N, O, R	(—)	1248
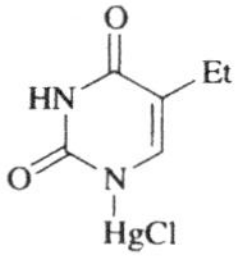 O, HN, Et, O, N, HgCl	C_6H_6, reflux, 1 h	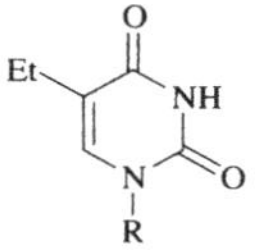 O, Et, NH, N, O, R	(—)	1248
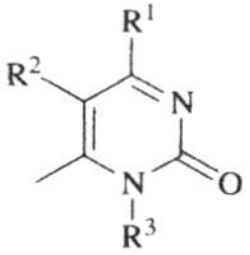 R^1, R^2, N, N, O, R^3	MeCN, reflux, time	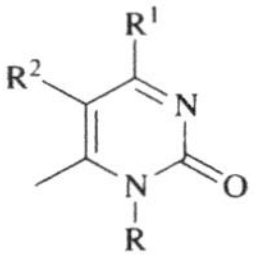 R^1, R^2, N, N, O, R	see below	60

R^1	R^2	R^3	time (min)	Yield (%)
NHAc	H	Hg/$_2$	80	(20)
OMe	Me	Hg/$_2$	7	(43)
OMe	H	HgCl	20	(26) + bis(riboside) (13)

TABLE X. MISCELLANEOUS REACTIONS OF HETEROCYCLIC BASES WITH PROTECTED SUGARS (*Continued*)

Sugar	Base	Conditions	Product(s) and Yield(s) (%)	Refs.
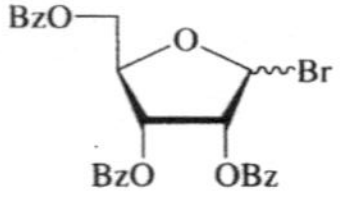	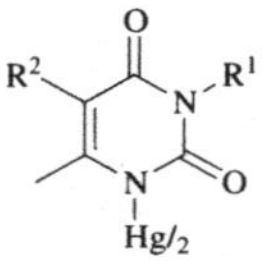	MeCN, reflux	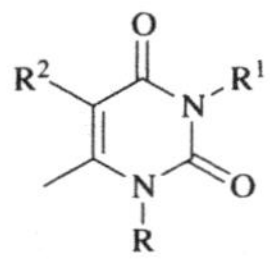 R^1 / R^2: H / Me, 2 h (13); Me / H, 2 h (69); Bn / H, 20 min (69)	60
	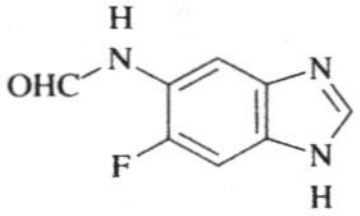	$Hg(CN)_2$, $MeNO_2$, MS, reflux, 4-8 h	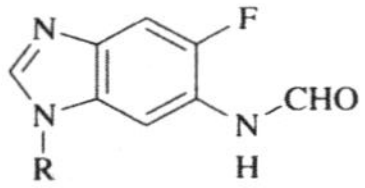 (38) + N^1-isomer (36)	1270
	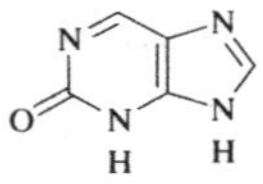	$Hg(OAc)_2$, $HgBr_2$, PhMe, reflux, 22 h	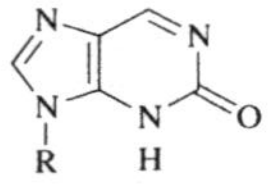 (36) R = 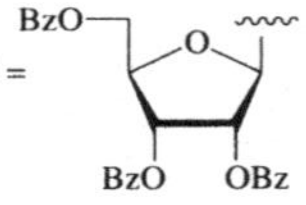	1206
	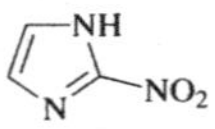	$Hg(CN)_2$, MeCN, 60°, 2.5 h	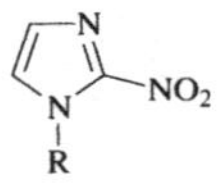 (68)	213
	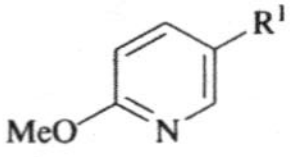	1. MeCN, rt, 3 d 2. NH_3, MeOH	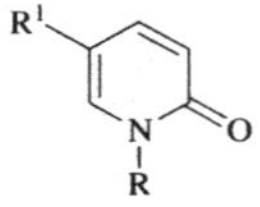 R^1: NHAc (61); NHCbz (52) R 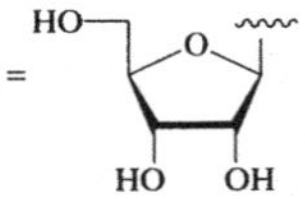 =	1186
	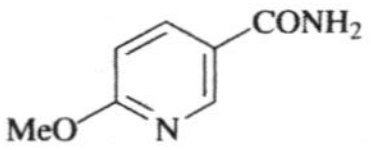	DMF, rt, 3 d	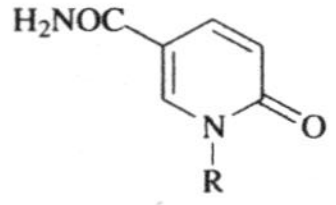 (63) R = 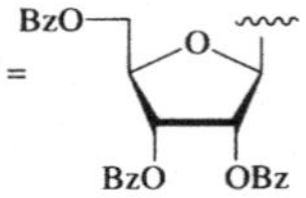	1186

Substrate	Conditions	Product(s) and Yield(s) (%)	Refs.
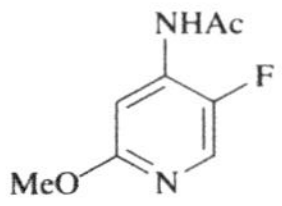 NHAc, F, MeO, N	MeCN, MS, rt, 4 d 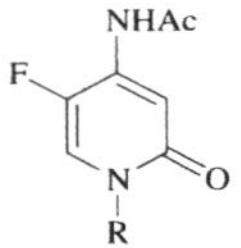	NHAc, F, N, O, R (38)	1226
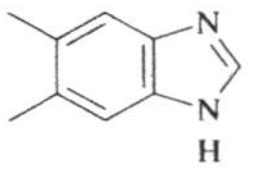 N, N, H	Dioxane, reflux, 2.5 h 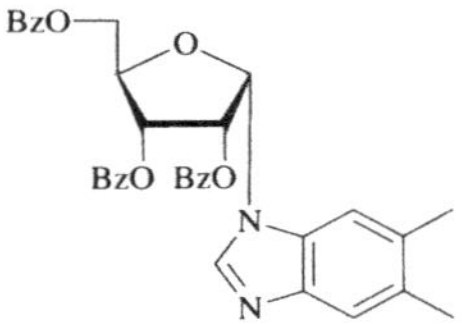	BzO, O, BzO, BzO, N, N (28)	50
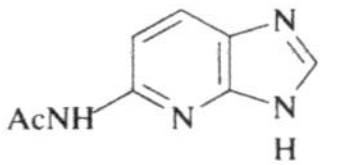 AcNH, N, N, N, H	$Hg(CN)_2$, MS, reflux, 3 h	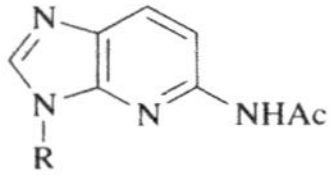 N, N, R, N, NHAc (72) + α-anomer (18); R 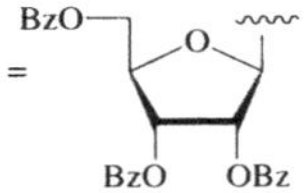 = BzO, O, BzO, OBz	889
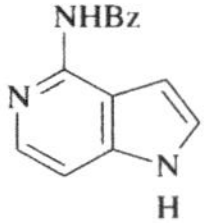 NHBz, N, N, H	$Hg(CN)_2$, MeCN, MS, 50°, 7.5 h 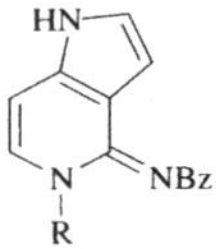	HN, N, R, NBz (50)	1271
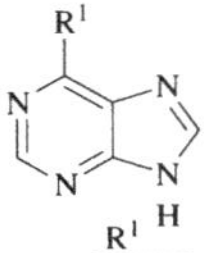 R^1, N, N, N, N, H	MeCN	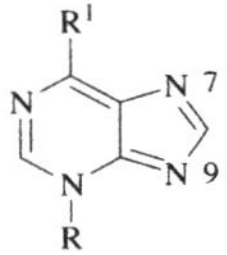 R^1, N, N 7, N 9, N, R	
R^1			
NH_2	50°, 36 h	(25)	1243
	50°	(21)	1272
NMe_2	60-65°, 18 h	(46) + N^9-anomer (11)	61

TABLE X. MISCELLANEOUS REACTIONS OF HETEROCYCLIC BASES WITH PROTECTED SUGARS (*Continued*)

Sugar	Base	Conditions	Product(s) and Yield(s) (%)	Refs.
	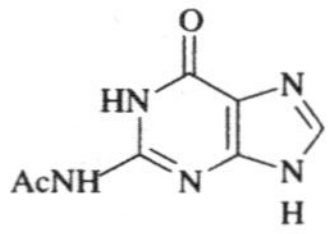 AcNH, HN, O, N, N, H	DMA, 60-65°, 40 h	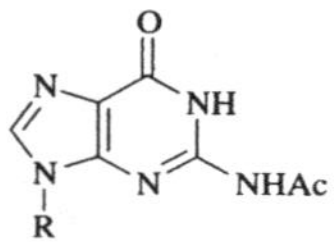 O, NH, NHAc, N, N-R (19) + N^7-isomer (21); R 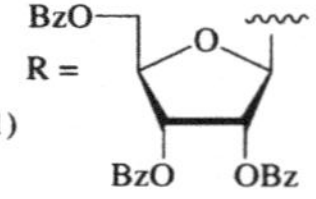 = BzO, O, BzO, OBz	61
	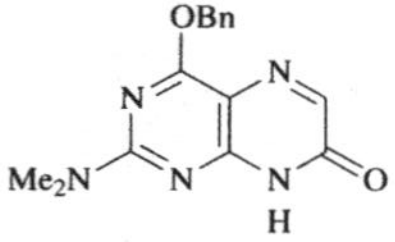 OBn, Me_2N, N, N, N, N, H, O	$Hg(CN)_2$, $MeNO_2$, MS, 101°, 30 min	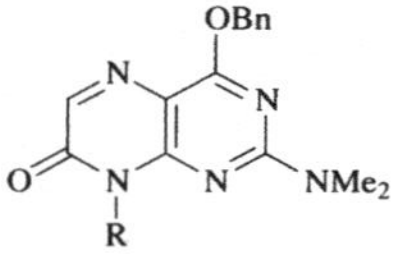 OBn, N, N, NMe_2, O, N-R **I** (76)	1112
		$EtNO_2$, MS, 114°, 30 min	**I** (75)	1112
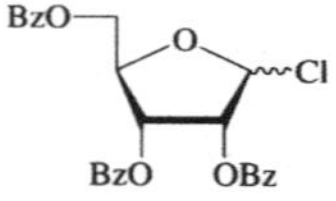 BzO, O, Cl, BzO, OBz	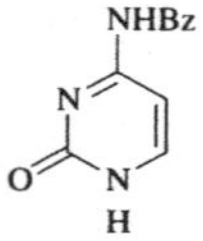 NHBz, N, O, N, H	$Hg(CN)_2$, $MeNO_2$, reflux, 6 h	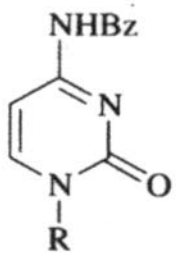 NHBz, N, O, N-R (77); R = BzO, O, BzO, OBz 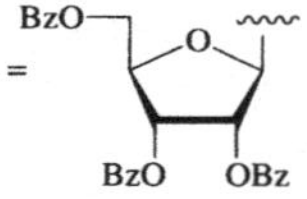	516
	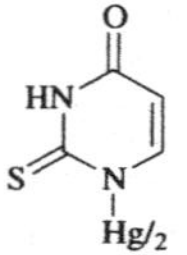 O, HN, S, N-$Hg/_2$	Xylene, reflux, 1 h	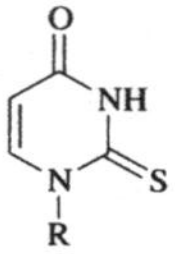 O, NH, S, N-R (—) + N^3-isomer (—) + N^1,N^3-bis(isomer) (—)	1273

Reactant	Conditions	Product(s) and Yield(s) (%)	Refs.
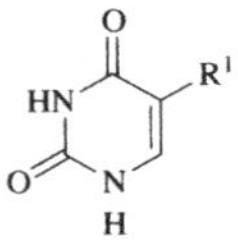 5-R^1-uracil (HN, O, N–H)	$Hg(CN)_2$, $MeNO_2$, reflux, 6 h	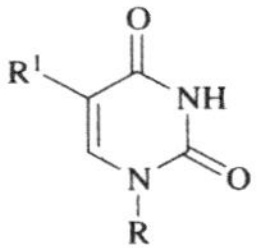 1-R-5-R^1-uracil	516
R^1 = H		(77)	
F		(88)	
CN		(70)	
NO_2		(80)	
Me		(85)	
CO_2Et		(30)	
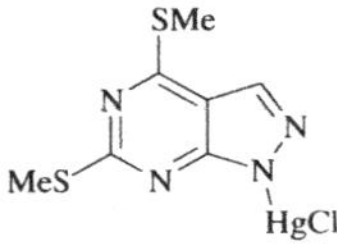 4,6-bis(SMe)-pyrazolo[3,4-d]pyrimidine, N-HgCl	PhMe, 140°, 4 h	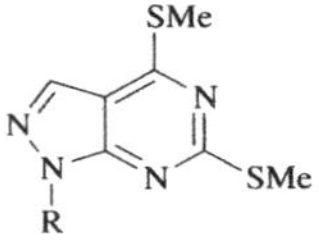 N-R, SMe, SMe (60) + α-anomer (5)	1094
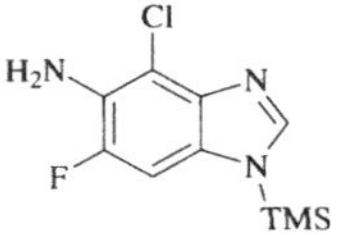 5-amino-4-chloro-6-fluoro-1-TMS-benzimidazole (H_2N, Cl, F, N-TMS)	$Hg(CN)_2$, $MeNO_2$, MS, reflux, 6-8 h	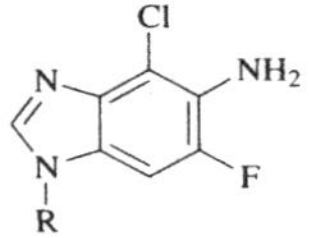 Cl, NH_2, F, N-R (26) + N^3-isomer (19)	1270
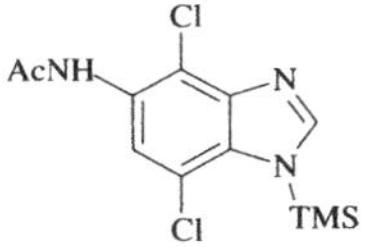 AcNH, Cl, Cl, N-TMS	$Hg(CN)_2$, $MeNO_2$, MS, reflux, 4-8 h	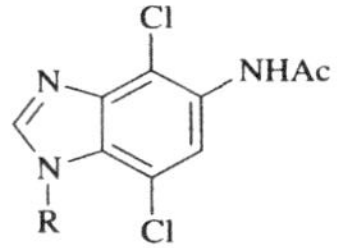 Cl, NHAc, Cl, N-R (22) + N^3-isomer (18)	1270
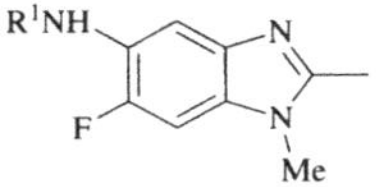 R^1NH, F, N-Me (2-methylbenzimidazole)	$Hg(CN)_2$, $MeNO_2$, MS, reflux, 4-8 h	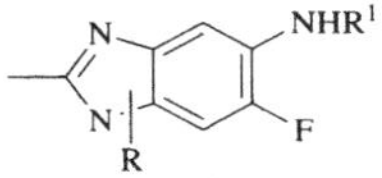 NHR^1, F, N-R; R^1 = H (3) N^1:N^3 = 74:26; CHO (5) N^1:N^3 = 64:36	1270

TABLE X. MISCELLANEOUS REACTIONS OF HETEROCYCLIC BASES WITH PROTECTED SUGARS (*Continued*)

Sugar	Base	Conditions	Product(s) and Yield(s) (%)	Refs.
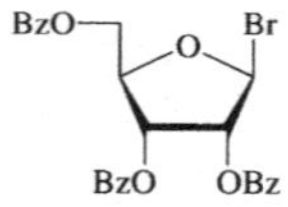	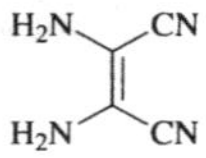	Dioxane, rt, 24 h	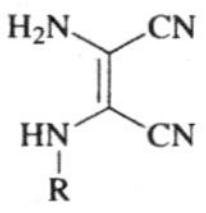 (44)	161
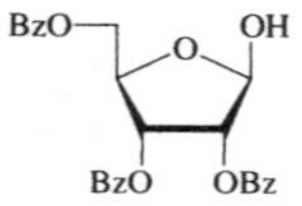	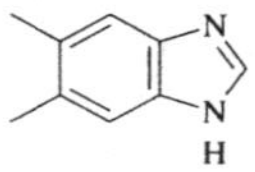	1. CETF, $Cl(CH_2)_2Cl$, Et_3N, $EtO_3^+ BF_4^-$ 2. DME, 60°, 10 h	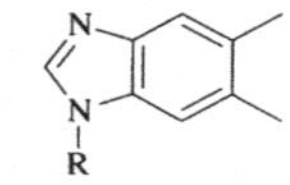 (66)	429
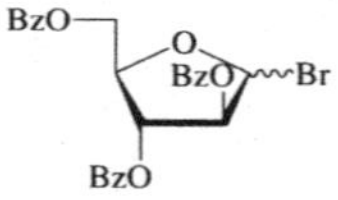	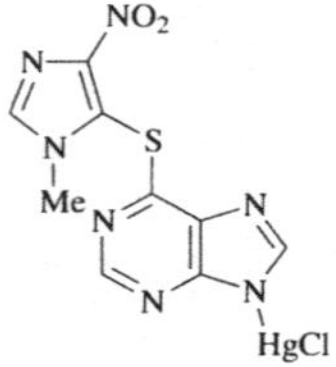	Xylene, reflux, 4 h	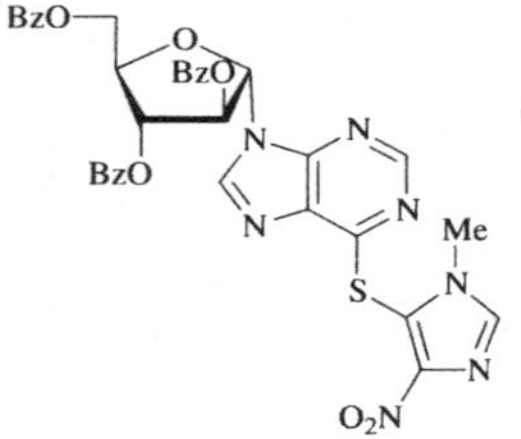 (—)	1193
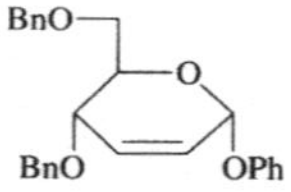	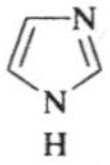	$Pd(dba)_3$, $[Ph_2PCH_2CH_2]_2$, THF, 60-70°	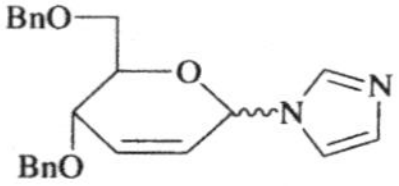 (72) α:β = 90:10	440
	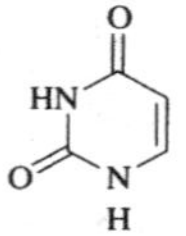	$Pd(dba)_3$, $[Ph_2PCH_2CH_2]_2$, THF, 60-70°	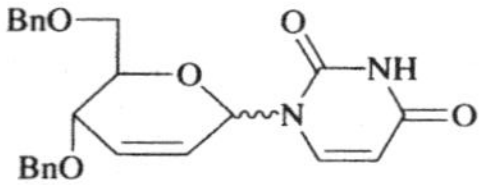 (36) α:β = 90:10	440
	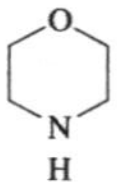	$Pd(dba)_3$, $[Ph_2PCH_2CH_2]_2$, THF, 60-70°	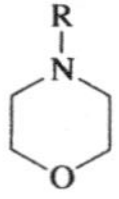 (85) R 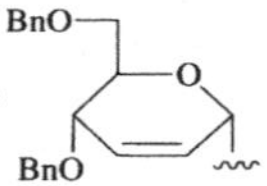 =	440

Sugar	Base	Conditions	Product(s) and Yield(s) (%)	Refs.
BzO, BzO, BzO, Br	OEt, N, O, N–HgCl	$HgBr_2$, xylene	BzO, BzO, BzO, N, O, N, OEt (40)	1195
	Cl, N, N, N, N–H	Pd(dba)$_3$, $[Ph_2PCH_2CH_2]_2$, THF, 60-70°	R–N, N, N, N, Cl (41)	440
	NH_2, N, N, N, N–H	Pd(dba)$_3$, $[Ph_2PCH_2CH_2]_2$, THF, 60-70°	BnO, BnO, O, N, N, N, N, NH_2 (80) α:β = 77:23	440
BnO, BnO, OBn, Cl	N, N–H	Dioxane, 100°, 1.5 h	BnO, BnO, BnO, N, N (66)	50
BnO, BnO, BnO, Cl	OMe, R^1, N, MeO, N	CH_2Cl_2, rt, 3 d	OMe, R^1, N, O, N–R; R^1: H (—), CF_3 (100); R = BnO, BnO, BnO (furanosyl)	356

TABLE X. MISCELLANEOUS REACTIONS OF HETEROCYCLIC BASES WITH PROTECTED SUGARS (*Continued*)

Sugar	Base	Conditions	Product(s) and Yield(s) (%)	Refs.
	OMe, F, N, MeO, N	1. CH_2Cl_2, MS, rt, 3 d 2. H_2, $PdCl_2$, MeOH, 15 min	OMe, F, N, HO, O, HO, N, O, HO (25)	1274
	NHAc, N, O, N, HgCl	PhMe, reflux, 30 min	NHAc, N, O, N, R (—) R = BnO, O, BnO, BnO	356
	NHBz, N, N, N, N, H	CH_2Cl_2, MS, rt, 1 week	NHBz, N, N, N, N, R (46)	341
	NHAc, N, N, AcNH, N, N, H	1. $Cl(CH_2)_2Cl$, MS, reflux 2. $MeNH_2$, EtOH, 55-60°, 18-42 h	NH_2, N, N, N, N, NH_2, R (41)	345
BnO, O, BnO, Cl, BnO	Cl, N, N, Cl, N, N, H	1. CH_2Cl_2, MS, rt, 1 week 2. NH_3, MeOH. 0°	Cl, N, N, N, N, Cl, R **I** (28)	342
		$Hg(CN)_2$, $MeNO_2$, reflux, 3 h	**I** (11) + α-anomer (25)	343

Glycosyl donor	Base	Conditions	Product	Yield (%)	Refs.
	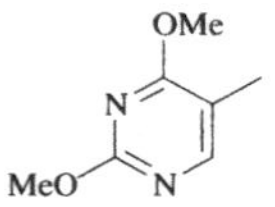 MeO, N, OMe	1. CH_2Cl_2, rt, 5 d 2. H_2, $PdCl_2$	OMe, N, O, R; R = (HO, O, HO, HO)	(49)	357
	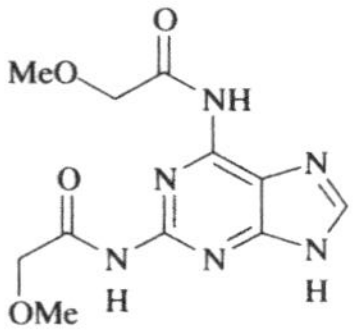 MeO, O, NH, N, O, N, H, OMe, H	$Cl(CH_2)_2Cl$, reflux, 10 h	O, HN, OMe, N, N, O, OMe, R, H	(100)	347
	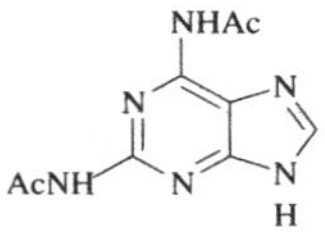 NHAc, N, AcNH, N, H	1. $Cl(CH_2)_2Cl$, reflux, 5 d 2. NaOMe, MeOH	NH_2, N, HO, O, HO, NH_2, HO	(40)	344
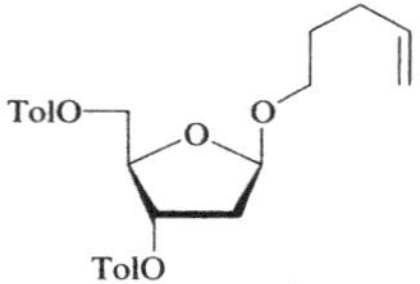 TolO, O, TolO	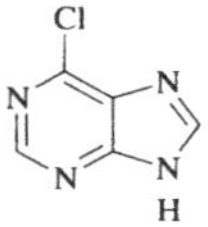 Cl, N, N, H	IDCP, MeCN, MS, rt, 2 h	Cl, N, TolO, O, TolO	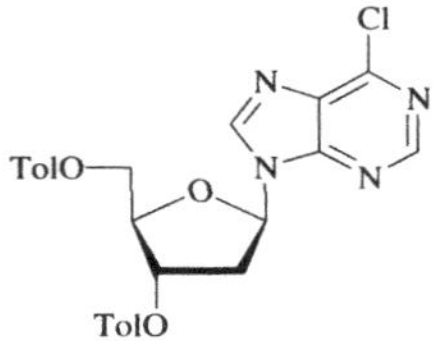 (35) + α-anomer (17) + N^7-(α,β)-isomer (39)	421
C_{27} 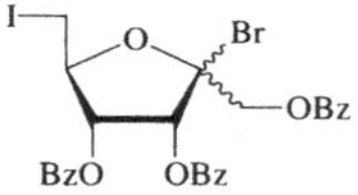 I, O, Br, OBz, BzO, OBz	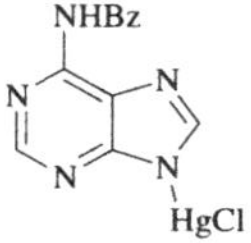 NHBz, N, HgCl	$MeNO_2$, rt, 48 h	NHBz, N, I, O, OBz, BzO, OBz	(49)	632

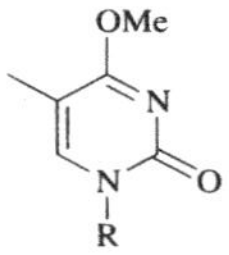

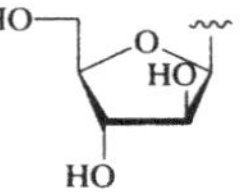

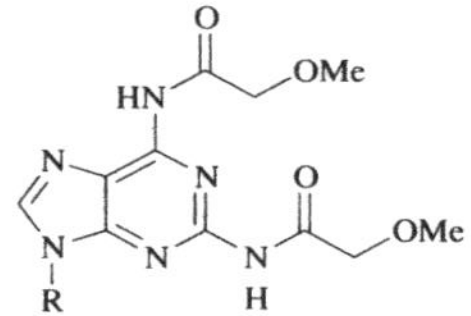

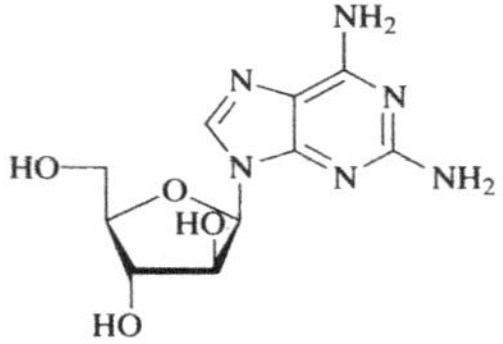

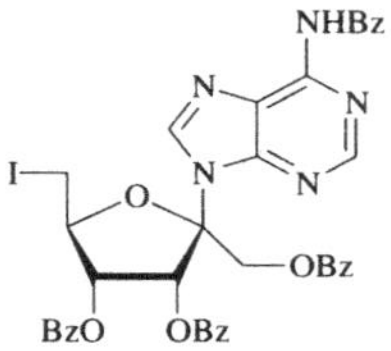

TABLE X. MISCELLANEOUS REACTIONS OF HETEROCYCLIC BASES WITH PROTECTED SUGARS (*Continued*)

Sugar	Base	Conditions	Product(s) and Yield(s) (%)	Refs.
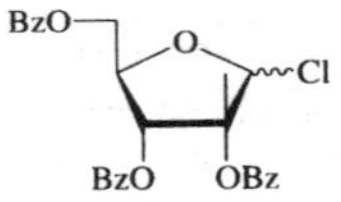 BzO, O, Cl, BzO, OBz	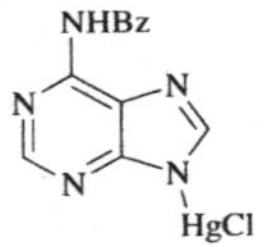 NHBz, N, N, N, N, HgCl	Xylene, 145°, 1.3 h	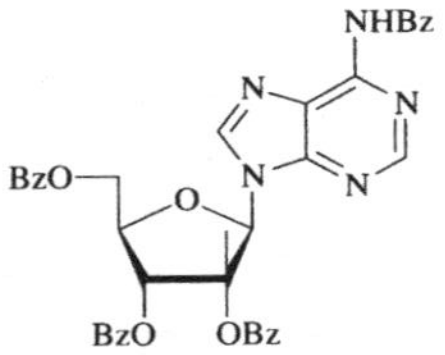 NHBz, N, N, N, N, BzO, O, BzO, OBz (—)	1275
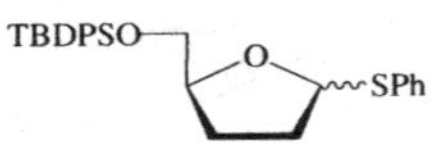 TBDPSO, O, SPh	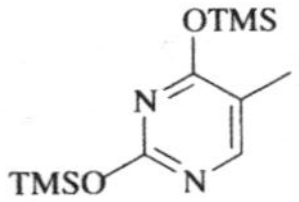 OTMS, N, TMSO, N	NBS, CH_2Cl_2, MS, rt, 5 min	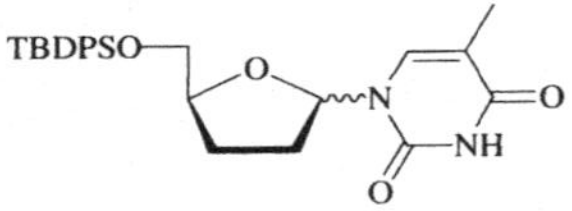 TBDPSO, O, N, O, NH, O (94) α:β = 6:1	406
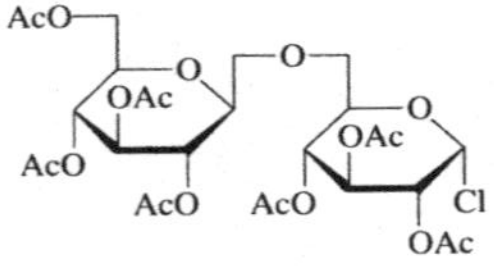 AcO, O, OAc, AcO, AcO, O, O, OAc, AcO, Cl, OAc	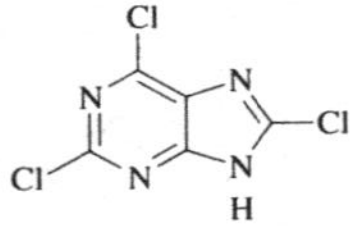 Cl, N, N, Cl, Cl, N, N, H	$Hg(CN)_2$, $MeNO_2$, reflux, 3 h	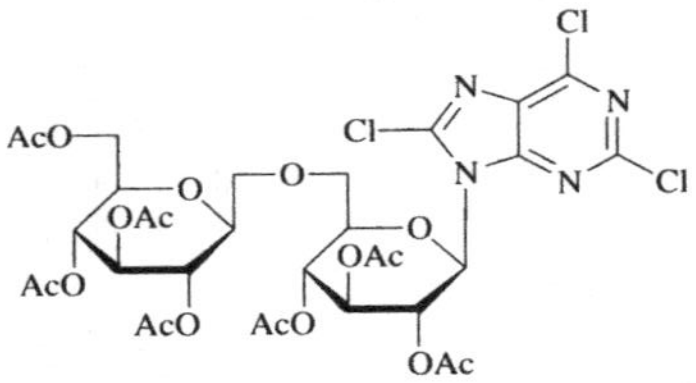 AcO, O, OAc, AcO, AcO, O, O, OAc, AcO, OAc, Cl, N, N, Cl, N, N, Cl (45)	1247
C_{28} BzO, O, OAc, BzO, OBz	CO_2Et, N, H	HOAc (100%), EtOH, reflux, 5 h	EtO_2C, N, R (92) R = BzO, O, BzO, OBz	1276
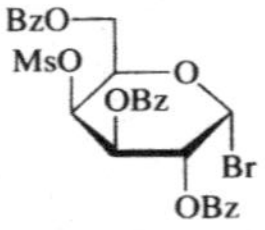 BzO, MsO, O, OBz, Br, OBz	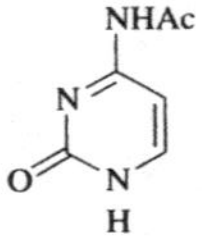 NHAc, N, O, N, H	$Hg(CN)_2$, $MeNO_2$, CH_2Cl_2, reflux, 3.5 h	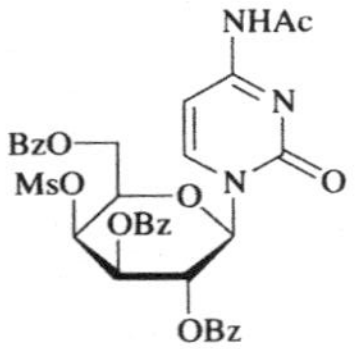 NHAc, N, O, N, BzO, MsO, O, OBz, OBz (96)	517

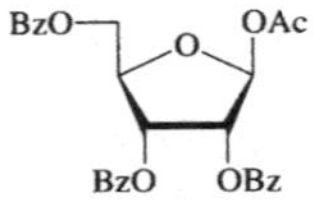

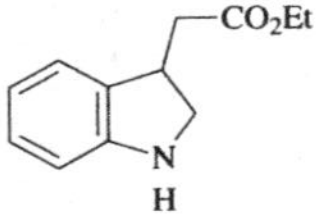

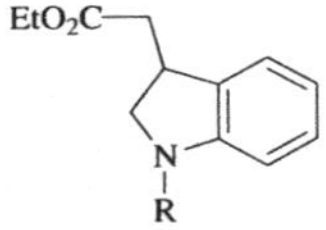

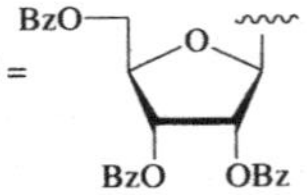

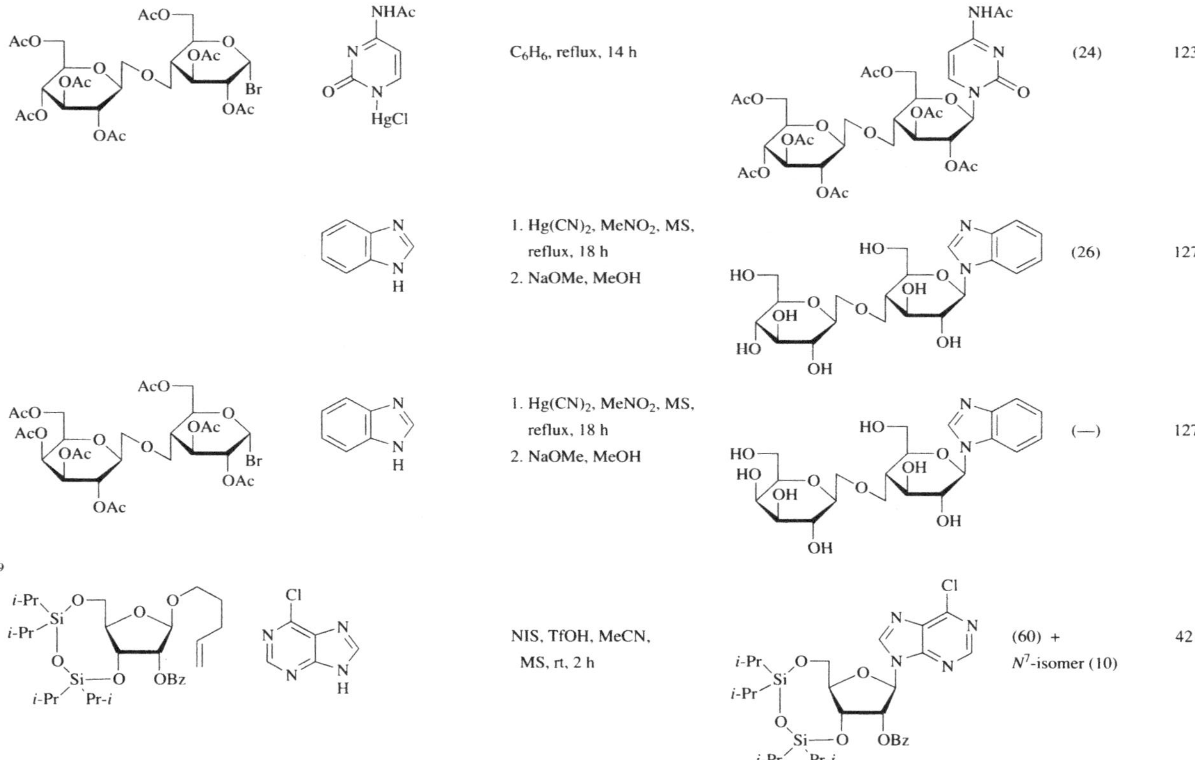

	Conditions	Yield (%)	Refs.
	C_6H_6, reflux, 14 h	(24)	1231
	1. $Hg(CN)_2$, $MeNO_2$, MS, reflux, 18 h 2. NaOMe, MeOH	(26)	1277
	1. $Hg(CN)_2$, $MeNO_2$, MS, reflux, 18 h 2. NaOMe, MeOH	(—)	1277
C_{29}	NIS, TfOH, MeCN, MS, rt, 2 h	(60) + N^7-isomer (10)	421

TABLE X. MISCELLANEOUS REACTIONS OF HETEROCYCLIC BASES WITH PROTECTED SUGARS (*Continued*)

Sugar	Base	Conditions	Product(s) and Yield(s) (%)	Refs.
C_{31} BzO, BzO, O–P(O)(OPh)$_2$, Br	pyrazole (N–H)	Bu_3SnH, *hν*	BzO, BzO, N-pyrazolyl (82) α:β = 1.4:1	434
	imidazole, R[1], R[1] (N–H)	Bu_3SnH, *hν*	BzO, BzO, N-imidazolyl, R[1], R[1]	434
	R[1] = H		(60) α:β = 1:1	
	R[1] = Cl		(66) α:β = 2.4:1	
	R[1] = Ph		(54) α:β = 1.8:1	
	benzimidazole (N–H)	Bu_3SnH, *hν*	BzO, BzO, N-benzimidazolyl (47) α:β = 1.5:1	434
	NHBz purine (N–H)	Bu_3SnH, *hν*	BzO, BzO, N-purinyl, NHBz (42) α:β = 1:1	434
BzO, BzO, OBz, O-pentenyl	R[1] purine (N–H)	NIS, TfOH, MeCN, MS, rt, 40 min	R[1] purine, N–R; R = BzO, BzO, OBz furanosyl	421

	R^1: Cl; NHMe; NHC_6H_{13}-*n*; NHBz		(53); (65); (70); (50)	
	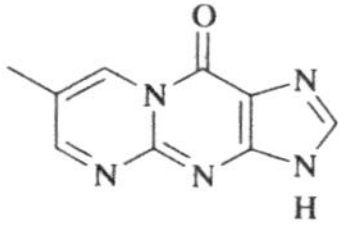	NIS, TfOH, MeCN, MS, rt, 40 min	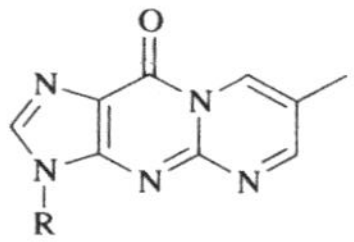 (60)	421
C_{34} 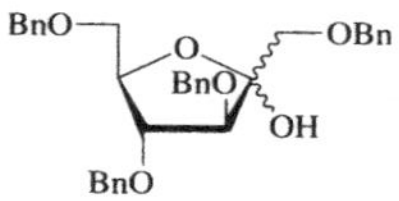	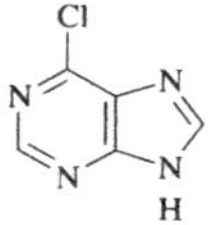	Ph_3P, DEAD, THF, rt, 12 h	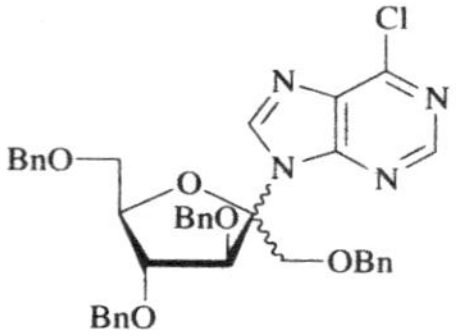 (37) + α-anomer (12)	432
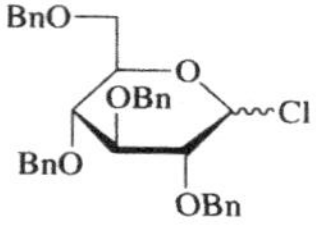	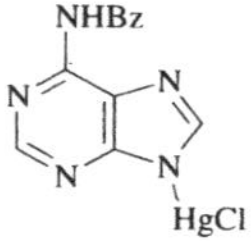	$CdCO_3$, xylene	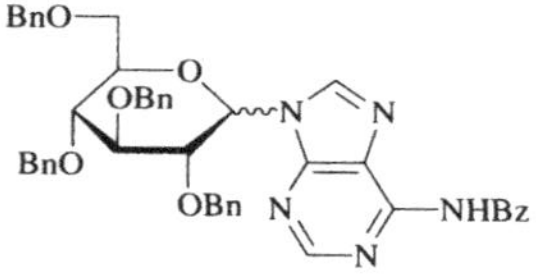 (59) α:β = 1:1	1071
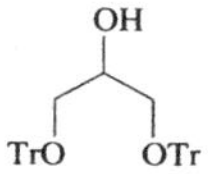	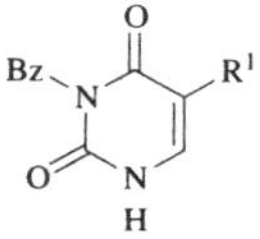	1. PPh_3, DEAD, dioxane, rt, 20 h 2. $MeNH_2$, EtOH	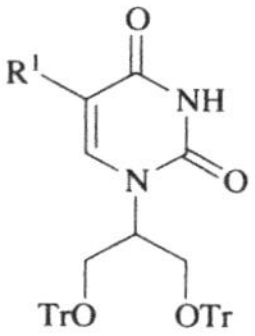 R^1: H (88); Me (85)	1261

TABLE XI. REACTIONS OF ACIDIC HETEROCYCLES WITH 1-HALOSUGARS IN THE PRESENCE OF BASES

Sugar	Base	Conditions	Product(s) and Yield(s) (%)	Refs.
C_2				
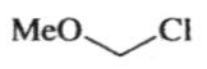 MeO Cl	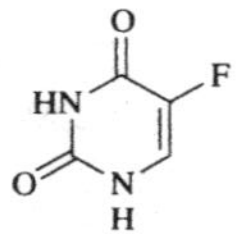 O, F, HN, O, N, H	K_2CO_3, CH_3CONMe_2, 30°, 5 h	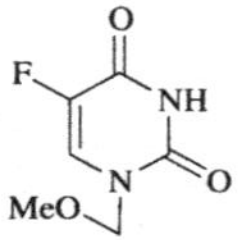 O, F, NH, N, O, MeO (11) + N^3-isomer (9) + N^1,N^3-bis(isomer) (19)	638
C_3				
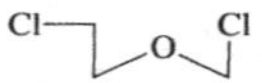 Cl O Cl	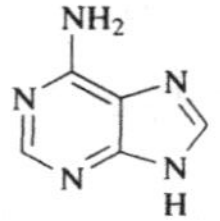 NH_2, N, N, N, N, H	LiH, DMF	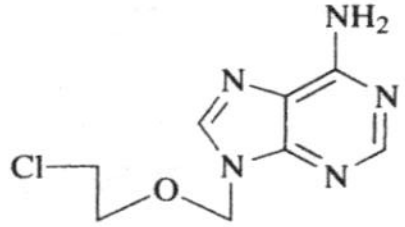 NH_2, N, N, Cl, O, N, N (—)	1278
C_4				
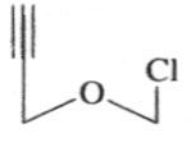 O Cl	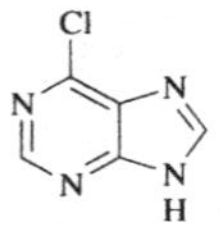 Cl, N, N, N, N, H	NaH, MeCN, 65°, 12 h	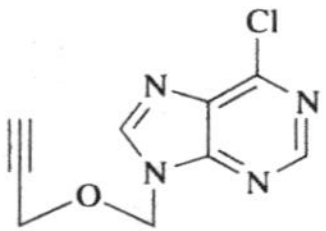 Cl, N, N, O, N, N (59) + N^7-isomer (6)	1279
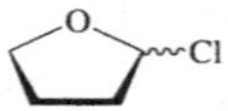 O Cl	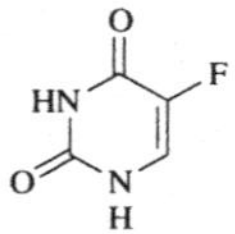 O, F, HN, O, N, H	K_2CO_3, DMA, 30°, 5 h	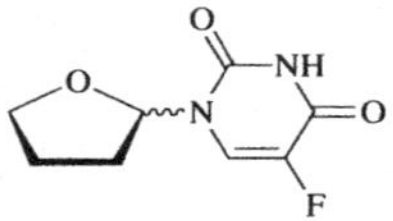 O, O, NH, N, O, F (75)	638
C_5				
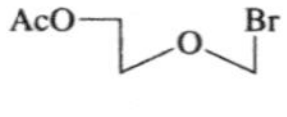 AcO O Br	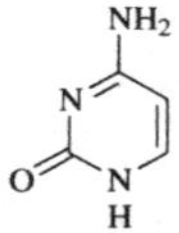 NH_2, N, O, N, H	18-crown-6 or tetraglyme, THF, KOBu-*t*, 45 min	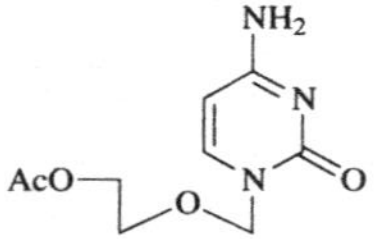 NH_2, N, AcO, O, N, O 0° (50) 25° (40) + N^3-isomer (20)	1280
	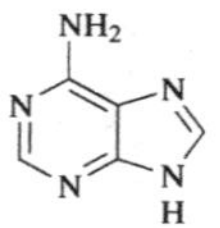 NH_2, N, N, N, N, H	18-crown-6 or tetraglyme, THF, KOBu-*t*, 45 min	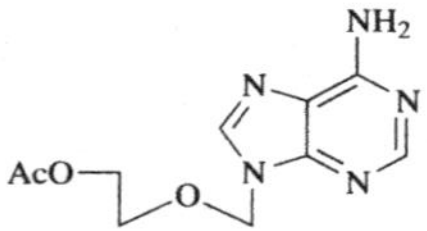 NH_2, N, N, AcO, O, N, N 0° (50) 25° (31) + N^9-isomer (19)	1280

Reactant	Substrate	Conditions	Product(s) and Yield(s) (%)	Refs.
AcO–CH_2CH_2–O–CH_2Br	8-bromoadenine (NH$_2$, Br, N–H)	18-crown-6 or tetraglyme, THF, KOBu-*t*, 45 min	9-[(2-acetoxyethoxy)methyl]-8-bromoadenine: 0° (50); 25° (41) + N^9-isomer (27)	1280
	uracil, R^1 at C-5	18-crown-6 or tetraglyme, THF, KOBu-*t*, 45 min	1-[(2-acetoxyethoxy)methyl]uracil, R^1 at C-5: R^1 = H: 0° (50), 25° (34), N^3-isomer (17); R^1 = Me: 0° (60), 25° (33), N^3-isomer (16)	1280
	pyrrolo[2,3-*d*]pyrimidine, 4-Cl, R^1 at C-2, R^2 at C-5, N–H		7-[(2-acetoxyethoxy)methyl] product, 4-Cl, R^1 at C-2, R^2 at C-5	
	R^1 = H, R^2 = H	NaH, DMF, rt, 3 h	(40)	1281
	R^1 = H, R^2 = Cl	NaH, DMF, rt, 30 min	(51)	1281
	R^1 = Cl, R^2 = H	NaH, DMF, rt, 2 h	(50)	1282
	R^1 = H, R^2 = Br	NaH, DMF, rt, 45 min	(45)	1281
	R^1 = NH$_2$, R^2 = H	NaOH (50%), Bu_4NHSO_4, C_6H_6, rt, 1 min	(50)	1282
	R^1 = AcNH, R^2 = H	NaH, DMF, rt, 5 h	(42)	1282
	R^1 = SMe, R^2 = H	NaH, DMF, rt, 1 h	(68)	1282
	4-amino-6-bromo-5-cyanopyrrolo[2,3-*d*]pyrimidine (NH$_2$, CN, Br, N–H)	NaH, DMF, 100°, 6 h	7-[(2-acetoxyethoxy)methyl] product (NH$_2$, NC, Br): (65) + N^7 & N^1-isomers (—)	1283

TABLE XI. REACTIONS OF ACIDIC HETEROCYCLES WITH 1-HALOSUGARS IN THE PRESENCE OF BASES (*Continued*)

Sugar	Base	Conditions	Product(s) and Yield(s) (%)	Refs.
	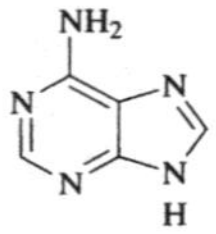 [structure: adenine]	NaH, MeCN, rt, 2 h	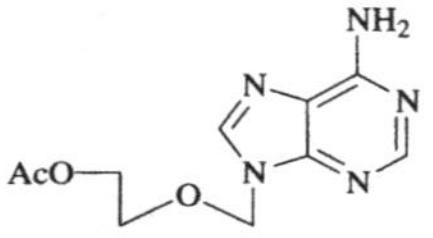 [structure: AcO–CH2CH2–O–CH2–adenine] (75)	1284
	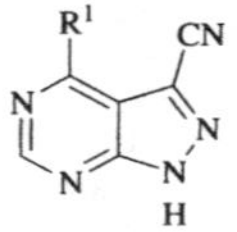 [structure: 4-R^1-3-cyanopyrazolo[3,4-d]pyrimidine]	NaH, DMF, 78°	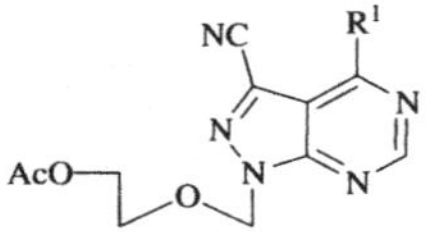 [structure] R^1: NH_2 4 h (41); NHAc 2 h (—)	1285
C_6  TMS–CH2CH2–O–CH2–Cl	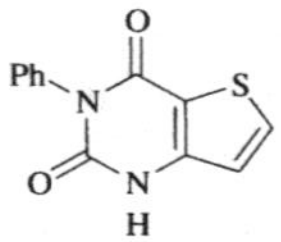 [structure: 3-phenylthieno[3,2-d]pyrimidine-2,4-dione]	NaH, DMF, rt, 12 h	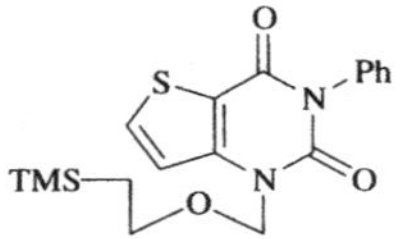 [structure] (88)	894
	[structure: thieno[2,3-d]pyrimidine-2,4-dione] R^1, R^2, R^3: Me, H, H; Me, Et, H; Ph, Ph, H; Me, Me, Me	NaH, DMF, rt, 12 h	[structure] (17); (19); (57); (43)	894
	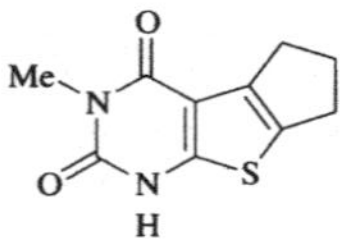 [structure: 3-methyl cyclopenta-thieno[2,3-d]pyrimidine-2,4-dione]	NaH, DMF, rt, 12 h	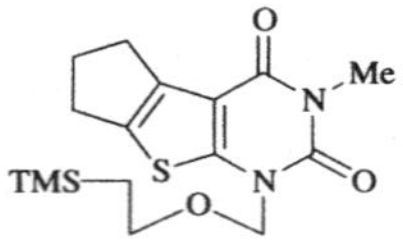 [structure] (50)	894

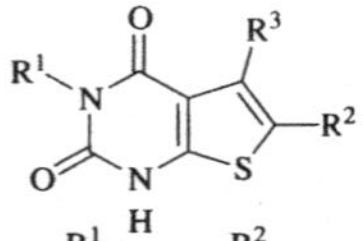

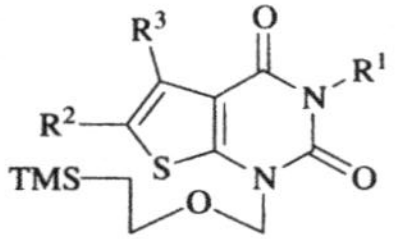

Reactant	Conditions	Product(s) and Yield(s) (%)	Refs.
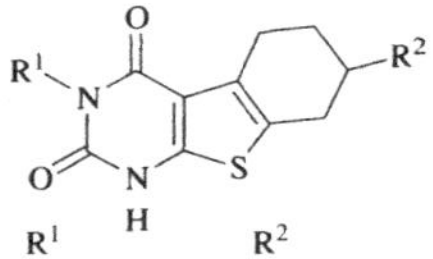 R[1] = Me, R[2] = H	NaH, DMF, rt, 12 h	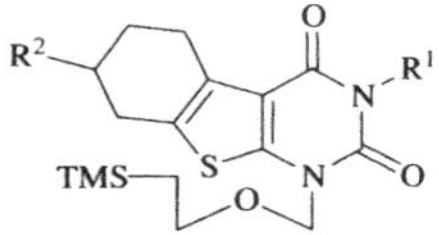 (50)	894
R[1] = Ph, R[2] = Me		(63)	
R[1] = $p\text{-ClC}_6\text{H}_4$, R[2] = H		(54)	
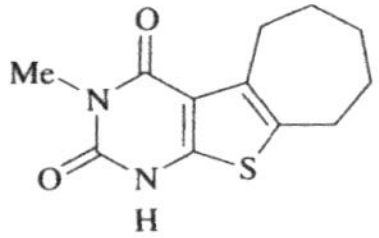	NaH, DMF, rt, 12 h	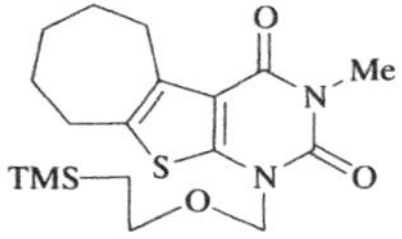 (66)	894
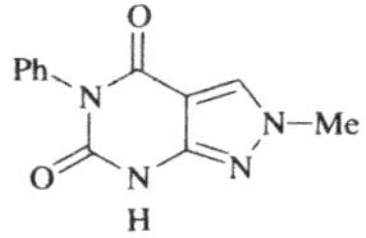	NaH, DMF, rt, 12 h	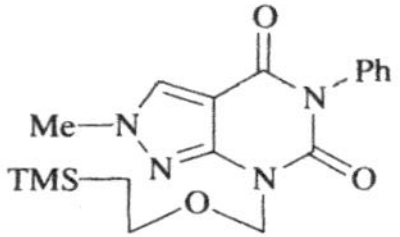 (52)	894
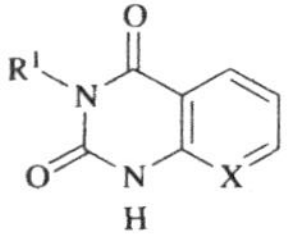 R[1] = Me, X = CH	NaH, DMF, rt, 12 h	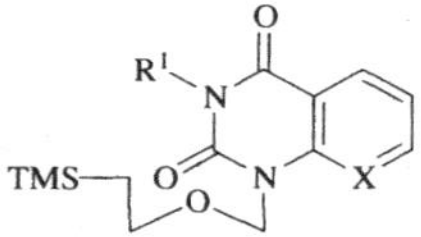 (36)	894
R[1] = Ph, X = N		(36)	
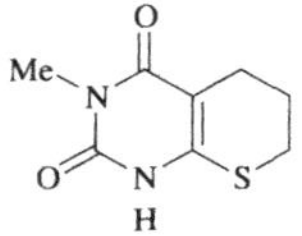	NaH, DMF, rt, 12 h	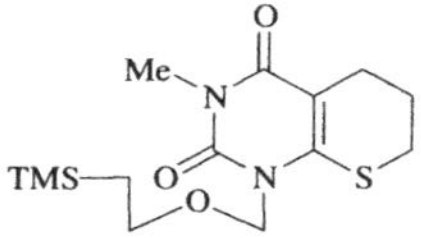 (25)	894
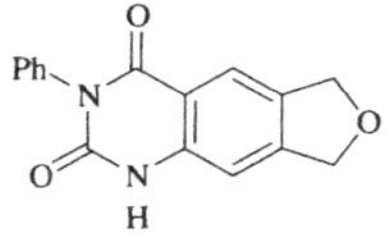	NaH, DMF, rt, 12 h	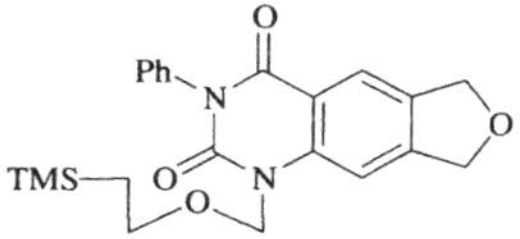 (42)	894

TABLE XI. REACTIONS OF ACIDIC HETEROCYCLES WITH 1-HALOSUGARS IN THE PRESENCE OF BASES (*Continued*)

Sugar	Base	Conditions	Product(s) and Yield(s) (%)	Refs.
C_7				
$Et_2P(O)CH_2OCH_2OCH_2Cl$	cytosine	NaH, DMF, rt, 15 h	(22)	650
	purine (R^1, R^2, R^3)	NaH, DMF, 25°, 15 h		
	R^1 = NH_2, R^2 = H, R^3 = H		(50)	650
	R^1 = Cl, R^2 = NH_2, R^3 = H		(23)	650
	R^1 = NH_2, R^2 = H, R^3 = Na		(—)	977
	R^1 = Cl, R^2 = NH_2, R^3 = Na		(—)	977
C_{10}				
$BzOCH_2CH_2OCH_2Cl$	6-(1-methyl-4-nitroimidazol-5-ylthio)purine	Et_3N, DMF, rt, 4 h	**I** (86)	1092
		1. Et_3N, DMF, 10-15° 2. rt, 4 h	**I** (86)	1092
	triazolopyrimidinone	Et_3N, DMF, rt, 18 h	(60)	743

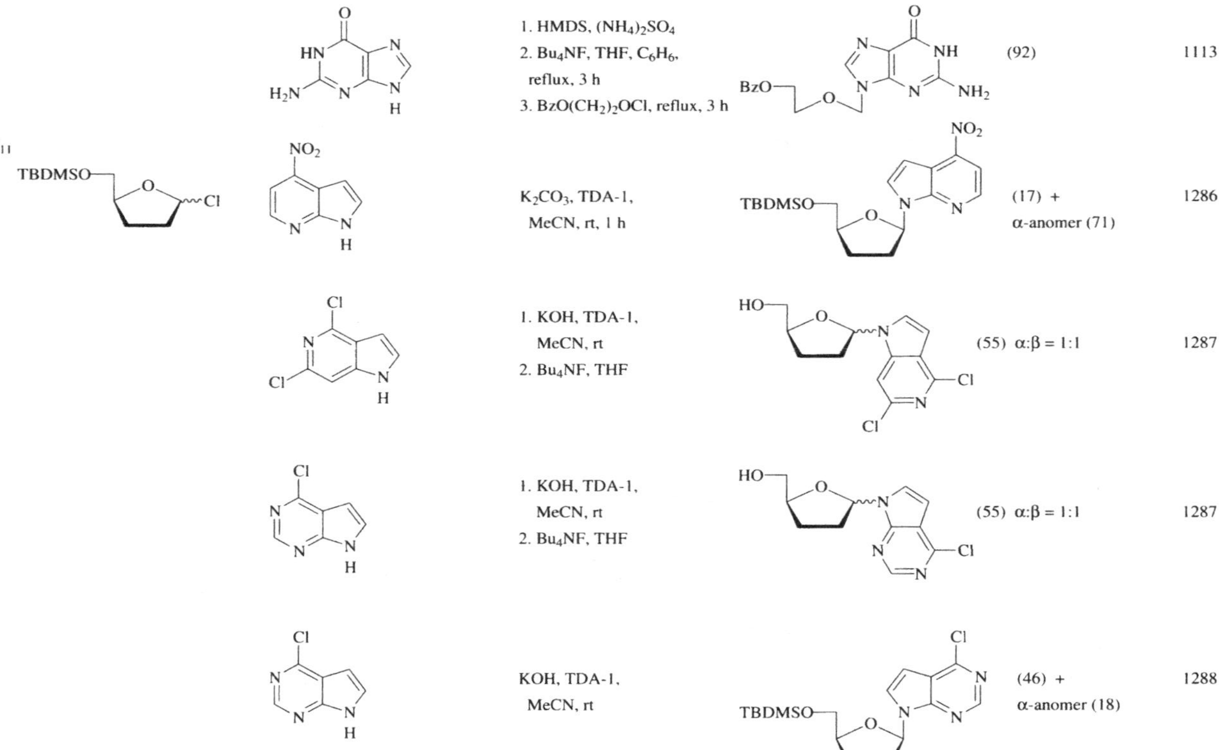
C_{11}
TBDMSO
1. HMDS, $(NH_4)_2SO_4$
2. Bu_4NF, THF, C_6H_6, reflux, 3 h
3. $BzO(CH_2)_2OCl$, reflux, 3 h
(92)
1113
K_2CO_3, TDA-1, MeCN, rt, 1 h
(17) + α-anomer (71)
1286
1. KOH, TDA-1, MeCN, rt
2. Bu_4NF, THF
(55) α:β = 1:1
1287
1. KOH, TDA-1, MeCN, rt
2. Bu_4NF, THF
(55) α:β = 1:1
1287
KOH, TDA-1, MeCN, rt
(46) + α-anomer (18)
1288

TABLE XI. REACTIONS OF ACIDIC HETEROCYCLES WITH 1-HALOSUGARS IN THE PRESENCE OF BASES (*Continued*)

Sugar	Base	Conditions	Product(s) and Yield(s) (%)	Refs.
	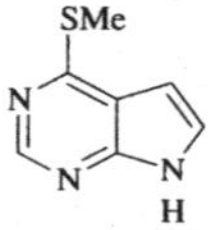	1. KOH, TDA-1, MeCN, rt 2. Bu_4NF, THF	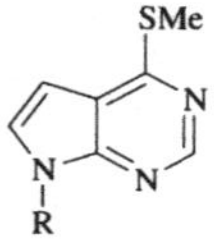 (57) R = 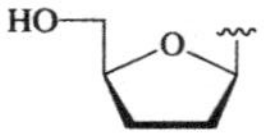	1287
	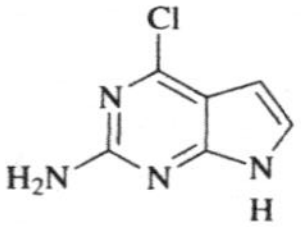	KOH, TDA-1, MeCN, rt, 10 min	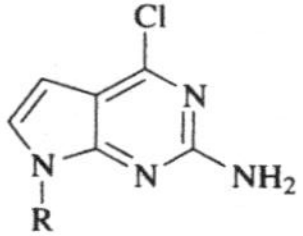 (22) + α-anomer (20)	1288
	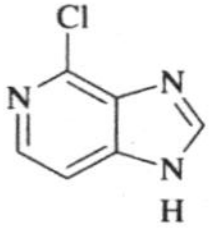	KOH, TDA-1, MeCN, rt, 10 min	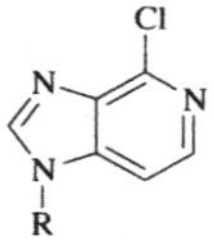 (22) + α-anomer (26) + N^3-β-isomer (17) + N^3-α-isomer (13)	1289
	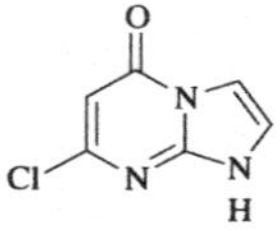	K_2CO_3, TDA-1, MeCN	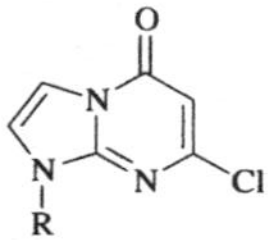 (46) + α-anomer (35)	1290, 1291
	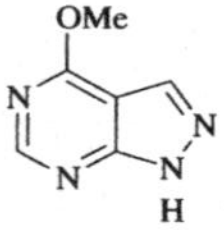	KOH, TDA-1, MeCN	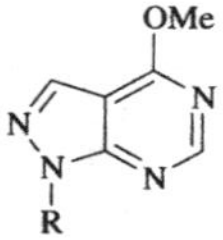 (22) + α-anomer (18) + N^8-β-isomer (9) + N^8-α-isomer (9)	1292
	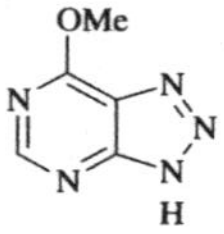	KOH, TDA-1, MeCN	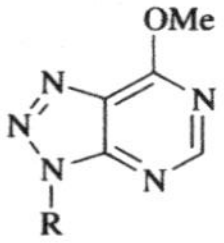 (—)	1291

Glycosyl donor	Base	Conditions	Product(s) and Yield(s) (%)	Refs.
	[structure]	KOH, TDA-1, MeCN	[structure] (14) + N^3-isomer (12) + N^2-isomer (21) + N^1-isomer (5)	1293
C_{12} [structure]	[structure]	K_2CO_3, TDA-1, MeCN, rt, 50 min	[structure] (45) + α-anomer (43)	1290
C_{13} [structure]	[structure]	1. NaH, DMF, 80°, 2 h 2. $Me_2S(SMe)BF_4$, CH_2Cl_2, MS, rt or –20°, 4 h 3. OH^-	[structure] (—)	97
C_{14} [structure]	[structure]	$CdCO_3$, xylene, reflux, 24 h	[structure] (47) R = [structure]	1071
[structure]	[structure]	$(i\text{-}Pr)_2NEt$, DMF, rt, 12 h	[structure] (75)	1294
	[structure]	KOH, Me_2CO, H_2O, rt, 12 h	[structure] R^1: Bn (64); $CH_2C_6H_4OMe$-*p* (73); $CH_2C_6H_4Cl$-*p* (10)	1295

TABLE XI. REACTIONS OF ACIDIC HETEROCYCLES WITH 1-HALOSUGARS IN THE PRESENCE OF BASES (*Continued*)

Sugar	Base	Conditions	Product(s) and Yield(s) (%)	Refs.
	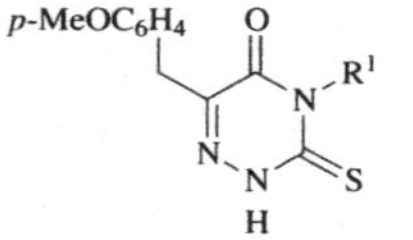 R^1 Ph p-$MeOC_6H_4$ p-MeC_6H_4	KOH, Me_2CO, H_2O, rt, 12 h	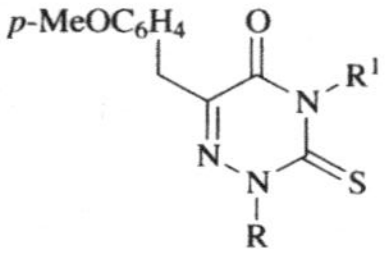 R = 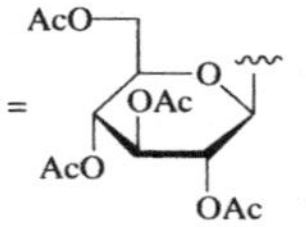 (82) (57) (78)	1295
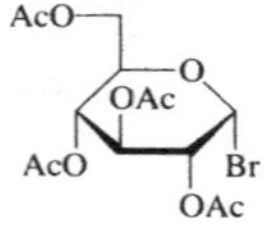	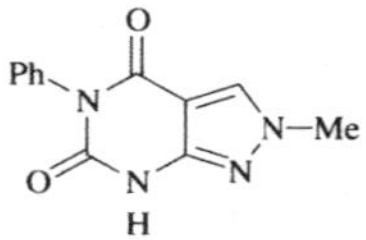	NaH, DMF, rt, 12 h	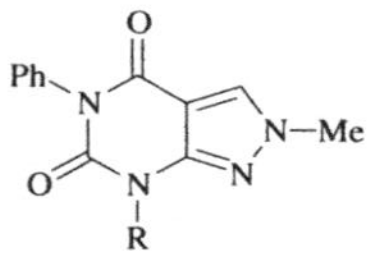 (17)	894
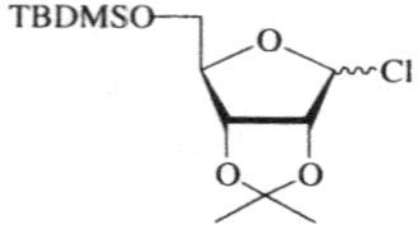	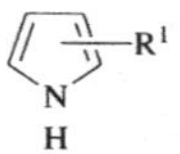	NaH, MeCN, rt	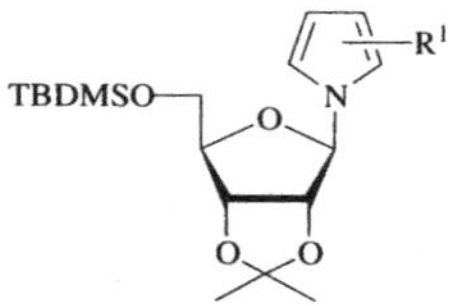 R^1 2-CN 12 h (61) 3-CN 15 h (66)	68 1296, 1297
	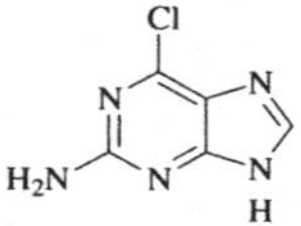	K_2CO_3, NMP, 80°, 3 h	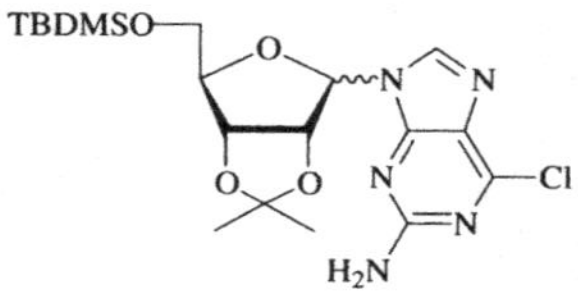 **I** (23) $\alpha:\beta = 2.5:1$	1298
		Cs_2CO_3, NMP, 55°, 3 h	**I** (25-38) $\alpha:\beta = 7.7:1$	1298

Sugar	Base	Conditions	Product(s) and Yield(s) (%)	Refs.
TBDMSO, Cl, O, O (sugar)	Cl, CN, N, H_2N, N, N, H	NaH, MeCN, rt, 12 h	NC, Cl, N, N, NH_2, N, R (58); R = TBDMSO, O, O, O (sugar)	1299
	Cl, N, N, N, H	KOH, TDA-1, MeCN, rt, 10 min	Cl, N, N, N, R **I** (65)	1300
		NaH, MeCN, rt, 30 min	**I** (67)	1297
	Cl, N, R^1, N, N, H	KOH, TDA-1, MeCN, rt, 20 h	Cl, N, N, R^1, N, R	
	R^1		Base:Sugar	
	NH_2		1:1 (34)	1301
	NH_2		2:1 (65)	1301
	SMe		1:1 (53)	1301
	SMe		2:1 (78)	1301
	OMe, N, R^1, N, N, H	KOH, TDA-1, MeCN, rt, 20 h	OMe, N, N, R^1, N, R	
	R^1		Base:Sugar	
	NH_2		1:1 (21)	1301
	NH_2		2:1 (31)	1301
	SMe		1:1 (59)	1301
	SMe		2:1 (82)	1301

TABLE XI. REACTIONS OF ACIDIC HETEROCYCLES WITH 1-HALOSUGARS IN THE PRESENCE OF BASES (*Continued*)

Sugar	Base	Conditions	Product(s) and Yield(s) (%)	Refs.
	SMe; N; N; N; H	KOH, TDA-1, MeCN, rt, 10 min	SMe; N; N; N; R (65) R = TBDMSO; O; O; O	1300
	Cl; CN; N; H_2N; N; N; H	NaH, MeCN, rt, 12 h	NC; Cl; N; N; N; NH_2; R (58)	1297
	Cl; H; N; N; Cl; N	NaH, MeCN, rt, 24 h	N; Cl; N; N; R; Cl (63) + α-anomer (13)	1302
	N_3; N; N; N; H	NaH, MeCN, rt, 12 h	N_3; N; N; N; R (50)	1303
	NHTr; CN; N; N; N; H	NaH, MeCN, MS, 50°, 17 h	NC; NHTr; N; N; N; R (>62)	580
	R^1; N; N; N; H	1. NaH, MeCN, rt, 12 h 2. TFA, H_2O	R^1; N; N; N; R R = HO; O; OH; OH	1303

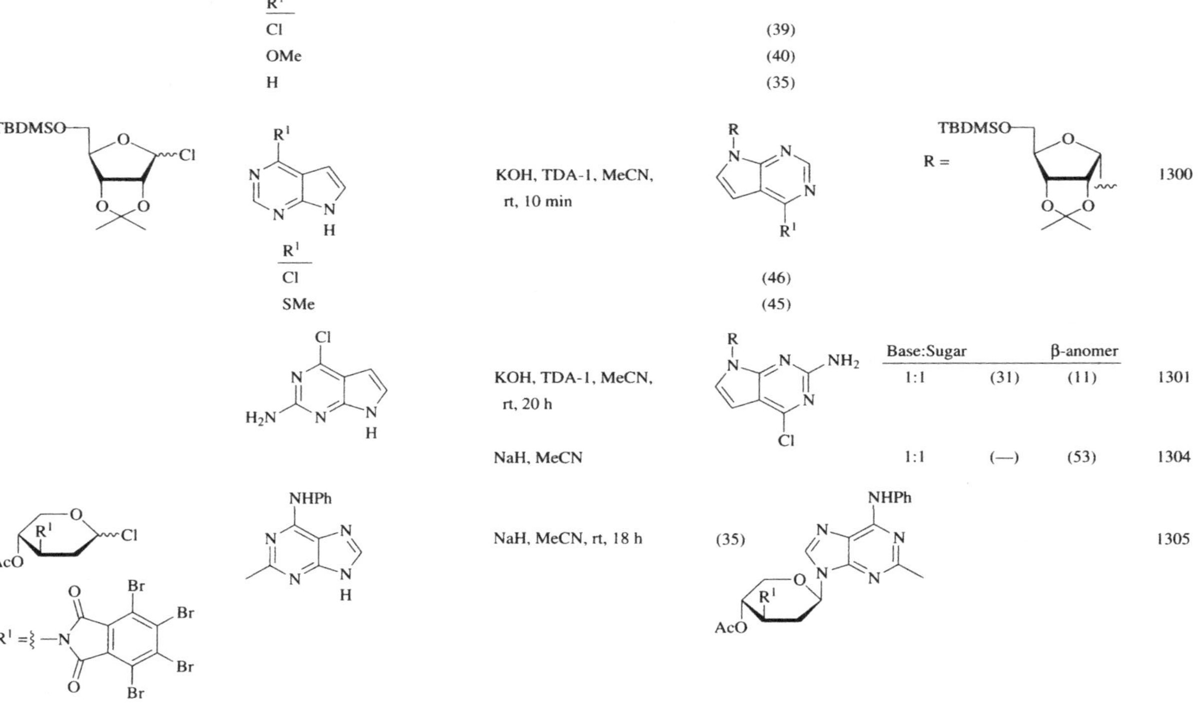

Substrate	Base	Conditions	Product(s) and Yield(s) (%)	Refs.
C_{15} TBDMSO-protected 2,3-O-isopropylidene furanosyl chloride	4-R^1-pyrrolo[2,3-d]pyrimidine; R^1 = Cl, OMe, H	KOH, TDA-1, MeCN, rt, 10 min	R = 5-O-TBDMS-2,3-O-isopropylidene furanosyl; R^1 = Cl (39), OMe (40), H (35)	1300
	R^1 = Cl, SMe		R^1 = Cl (46), SMe (45)	
	2-amino-4-chloropyrrolo[2,3-d]pyrimidine	KOH, TDA-1, MeCN, rt, 20 h	Base:Sugar 1:1 (31), β-anomer (11)	1301
		NaH, MeCN	Base:Sugar 1:1 (—), β-anomer (53)	1304
AcO, R^1 pyranosyl chloride; R^1 = tetrabromophthalimido	6-NHPh-2-methylpurine	NaH, MeCN, rt, 18 h	(35)	1305

TABLE XI. REACTIONS OF ACIDIC HETEROCYCLES WITH 1-HALOSUGARS IN THE PRESENCE OF BASES (*Continued*)

Sugar	Base	Conditions	Product(s) and Yield(s) (%)	Refs.
C_{16} 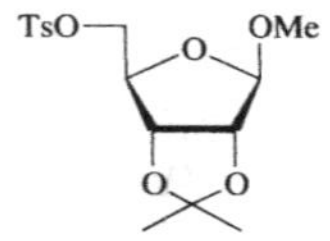 TsO, OMe	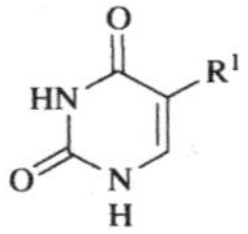 R[1]	NaH, DMF	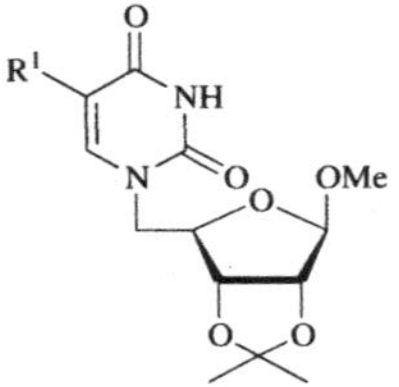 R[1]: H (37); I (58)	1033
C_{17} CO_2Me, Cl; R = OPiv	OMe, CN	NaH, MeCN, 0°	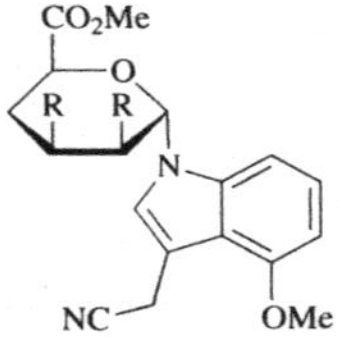 CO_2Me, NC, OMe (44)	1306, 1307
C_{18} 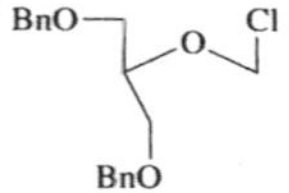 BnO, Cl, BnO	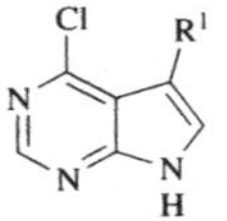 Cl, R[1]	1. NaH, DMF, rt, 40 min 2. BCl_3, CH_2Cl_2, –78°, 15 min	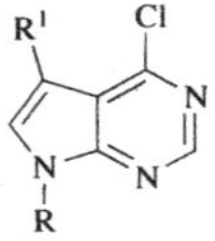 R[1], Cl, R; R = HO, HO 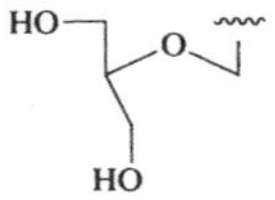	
	R[1] = H		(39)	1308
	Cl		(34)	1308
	Br		(41)	1308
	I		(38)	1308
	Me		(44)	1308
	CN		(30)	1309
	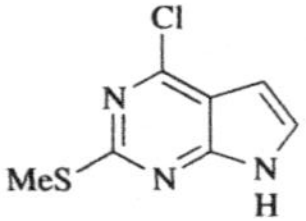 Cl, MeS	NaH, DMF, rt, 20 h	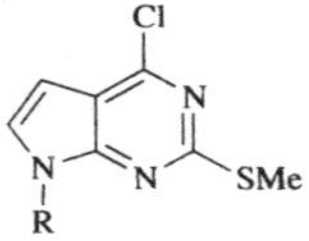 Cl, SMe, R (75)	1309

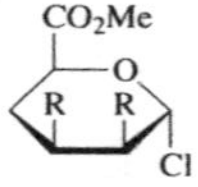

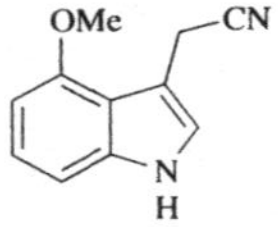

	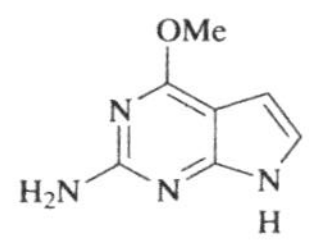	NaOH (50%), CH_2Cl_2, rt, 25 min	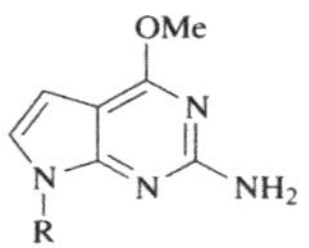 (27)	838
	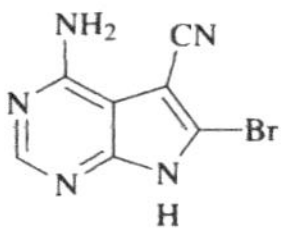	NaH, DMF, 80°, 5 h	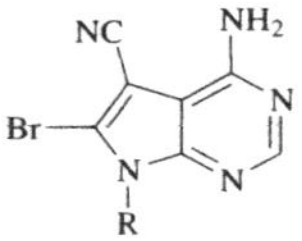 (55)	1309
	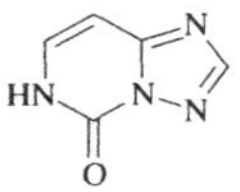	Et_3N, DMF, rt, 18 h	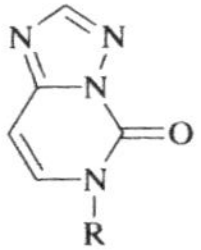 (56)	743
C_{19} 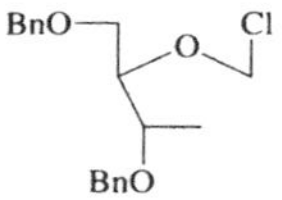	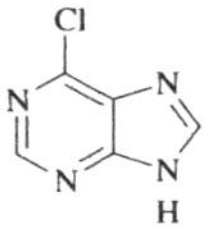	NaH, MeCN	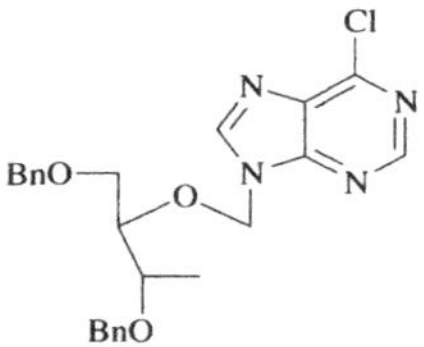 (—) + N^7-isomer (—)	1310
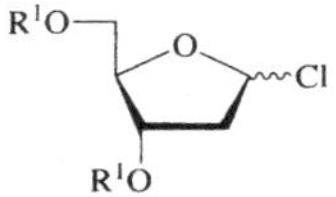 $R^1 = p\text{-}ClC_6H_4CO$	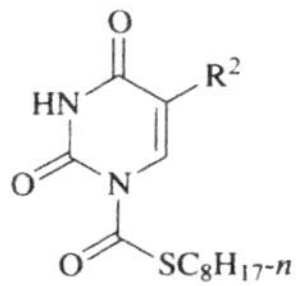	$(i\text{-}Pr)_2NEt$, MeCN, rt, 12 h	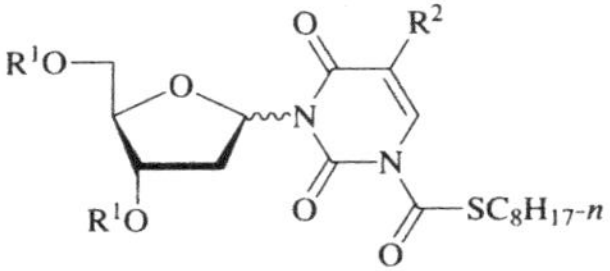	256a
	R^2: H		(82) $\alpha:\beta = 48:52$	
	F		(100) $\alpha:\beta = 60:40$	
	Me		(87) $\alpha:\beta = 45:55$	

TABLE XI. REACTIONS OF ACIDIC HETEROCYCLES WITH 1-HALOSUGARS IN THE PRESENCE OF BASES (*Continued*)

Sugar	Base	Conditions	Product(s) and Yield(s) (%)	Refs.
	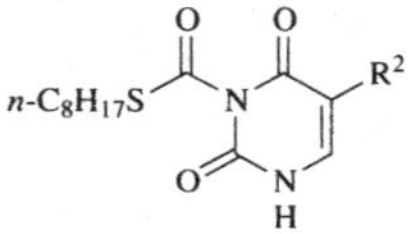 n-$C_8H_{17}S$, O, O, N, R^2, O, N, H	$(i\text{-}Pr)_2NEt$, MeCN, rt, 12 h	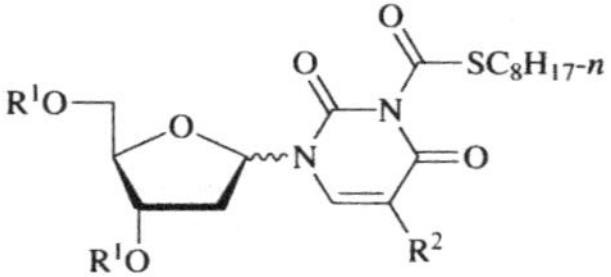 R^1O, O, N, O, SC_8H_{17}-n, O, R^1O, R^2; R^2: H (75), F (86), Me (76)	256a
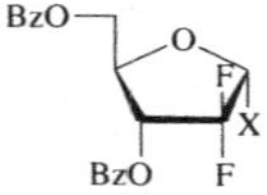 BzO, O, F, X, BzO, F	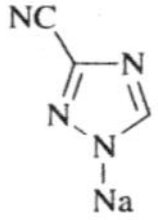 NC, N, N, N, Na	MeCN, 82°	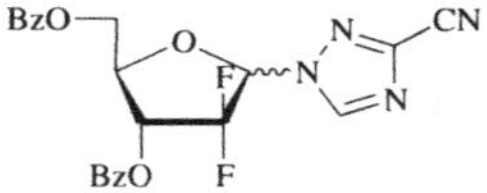 BzO, O, F, N, N, CN, N, BzO, F	1311
	X: Br	78 h	(6) $\alpha:\beta = 1:1.2$	
	X: I	48 h	(48) $\alpha:\beta = 1:3.5$	
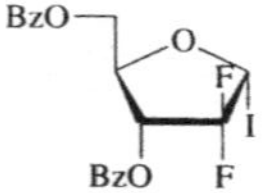 BzO, O, F, I, BzO, F	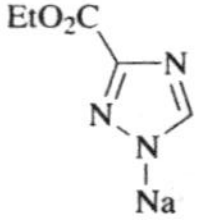 EtO_2C, N, N, N, Na	DMA, 23°, 24 h	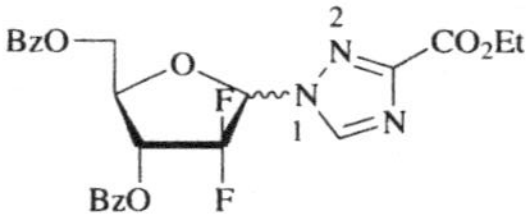 BzO, O, F, N (1), N (2), CO_2Et, N, BzO, F; (67) $\alpha:\beta = 1:3$, $N^1:N^2 = 13:41.5$	1311
	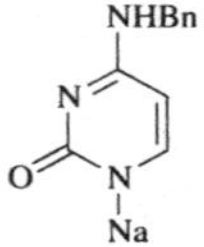 NHBn, N, O, N, Na	DMA, 23°, 5 h	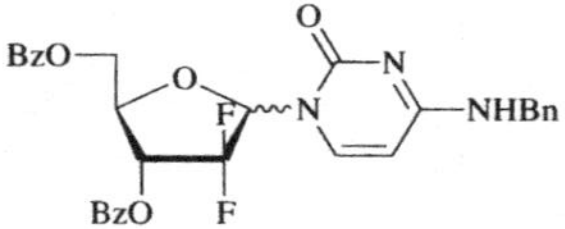 BzO, O, F, N, O, N, NHBn, BzO, F; (7) $\alpha:\beta = 1:1.9$	1311
	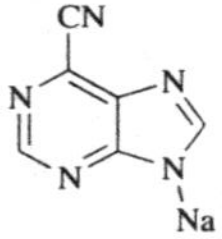 CN, N, N, N, N, Na	DMA, 70°, 5 h	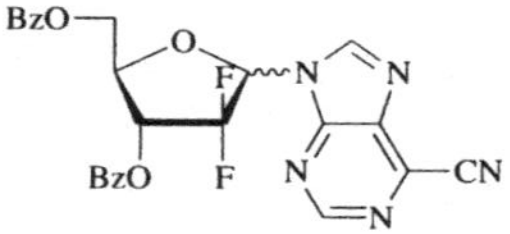 BzO, O, F, N, N, N, N, CN, BzO, F; (7) $\alpha:\beta = 1:1.2$	1311

	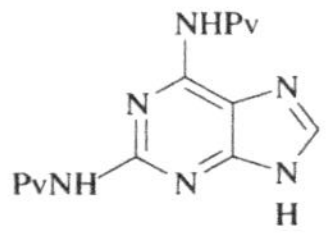	KOBu-*t*, MeCN, 60°, 16 h	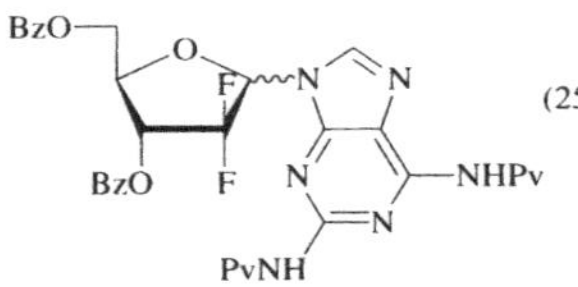	(25) α:β = 1:2.2	1311
	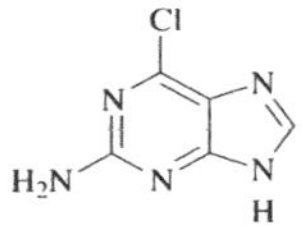	KOH, DMA, 70°, 5 h	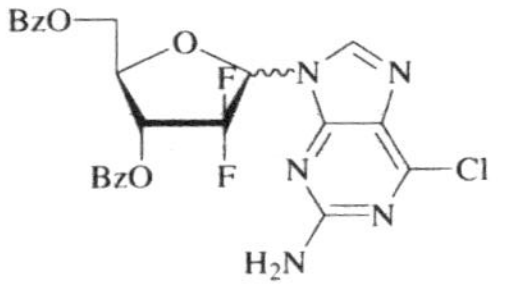	(—) α:β = 1:3	1311
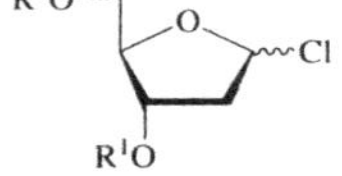 R^1 = *p*-$O_2NC_6H_4CO$	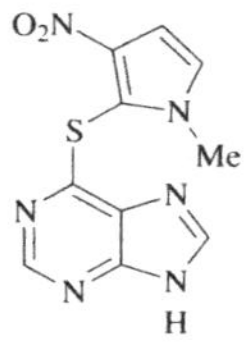	NaH, MeCN, rt, 15 h	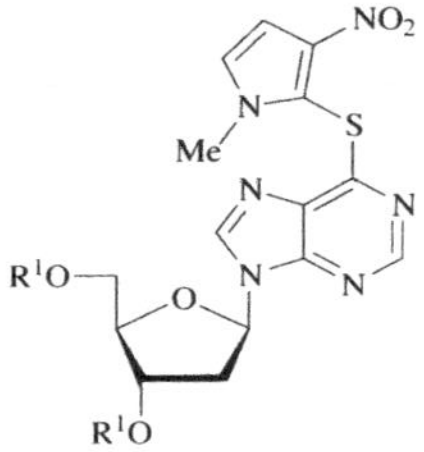	(55)	1092
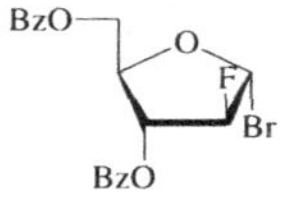	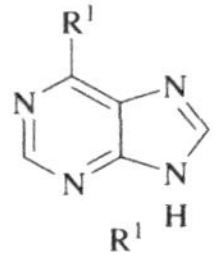	NaH, mineral oil	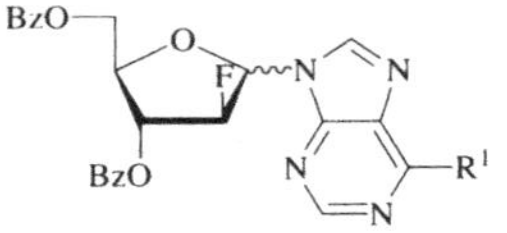		393
	R^1				
	Cl	55°, 3 h	(73) α:β = 1:10		
	NHBz	reflux, 3 h	(—) α:β ~ 1:15		
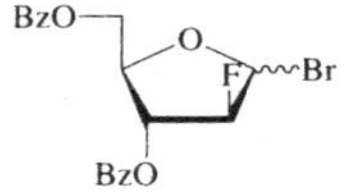	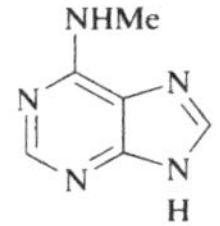	NaH, DMF, rt, 12 h	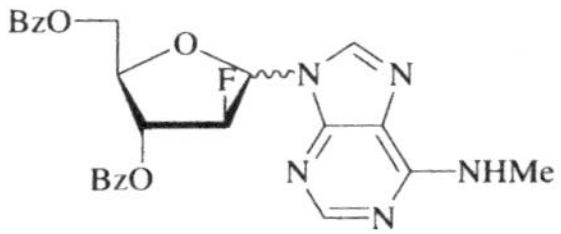	(62)	298

TABLE XI. REACTIONS OF ACIDIC HETEROCYCLES WITH 1-HALOSUGARS IN THE PRESENCE OF BASES (*Continued*)

Sugar	Base	Conditions	Product(s) and Yield(s) (%)	Refs.
C_{21}				
TolO, O, D, D, Cl, TolO, D	Cl, N, N, N, N, H, R^1	NaH, MeCN, rt	Cl, N, N, N, N, R^1, TolO, O, D, D, TolO, D R^1 / Cl 4 h (60) / NH_2 20 h (61)	1312
TolO, O, D, Cl, TolO, D	Cl, N, N, N, N, H, R^1	NaH, MeCN, rt, 20 h	Cl, N, N, N, N, R^1, TolO, O, D, TolO, D R^1 / H (48) / NH_2 (51)	1313
TolO, O, D, Cl, TolO	Cl, N, N, N, N, H	NaH, MeCN, rt, 4 h	Cl, N, N, N, N, R (53) R = TolO, O, D, TolO	1312
	Cl, N, N, N, N, H, $(n\text{-Bu})_2N$, N	1. NaH, MeCN, rt. 10 min 2. Pyridine-2-carbaldoxime, rt, 2 h	Cl, N, N, N, N, R, N, $N(Bu\text{-}n)_2$ (26) + N^7-isomer (—) + N^3-isomer (—) + N^3,N^7-bis(isomer) (—)	1312
TolO, O, Cl, TolO, D	Cl, N, N, N, N, H, R^1	NaH, MeCN, rt, 20 h	Cl, N, N, N, N, R^1, TolO, O, TolO, D R^1 / H (56) / NH_2 (57)	1314

Substrate	Reactant	Conditions	Product(s) and Yield(s) (%)	Refs.
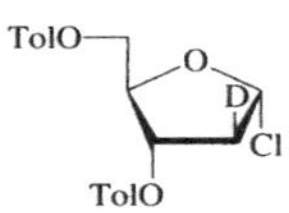 TolO, TolO, D, Cl (furanose)	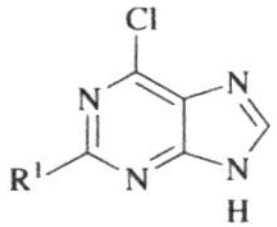 Cl, R^1, N, H (purine)	NaH, MeCN, rt, 24 h	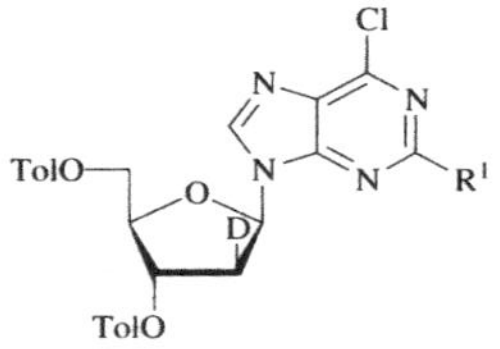 TolO, TolO, D, Cl, N, R^1 — R^1: H (54); NH_2 (52)	1314
TolO, TolO, Cl (furanose)	R^3, R^2, R^1, N, H (pyrrole)	NaH, MeCN, rt	R^3, R^2, R^1, N–R; R = TolO, TolO, O (furanosyl)	
	R^1 = CN, R^2 = H, R^3 = H		(67)	1315
	R^1 = H, R^2 = CN, R^3 = H		(65)	1297
	R^1 = CN, R^2 = H, R^3 = CN		(68)	1315
	R^1 = H, R^2 = $CONH_2$, R^3 = CO_2Me		(—)	1315
	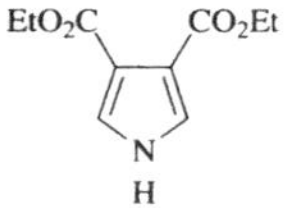 EtO_2C, CO_2Et, N, H	NaH, MeCN, rt, 0.5 h; 50°, 1 h	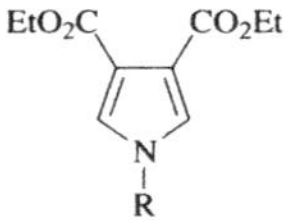 EtO_2C, CO_2Et, N–R (70)	1316
	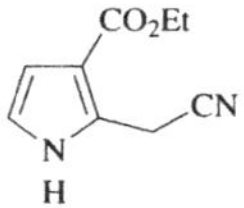 CO_2Et, CN, N, H	NaH, MeCN, rt	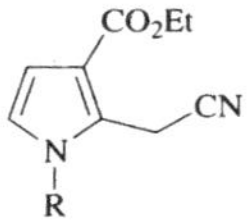 CO_2Et, CN, N–R (95)	1317, 1318
	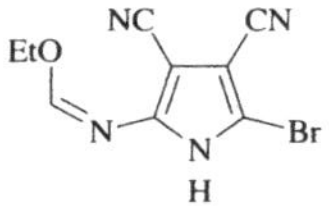 EtO, N, NC, CN, Br, N, H	NaH, MeCN, rt, 0.5 h; 50°, 1 h	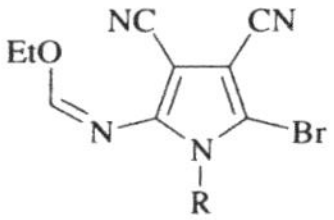 EtO, N, NC, CN, Br, N–R (75)	1316

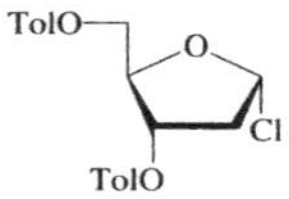

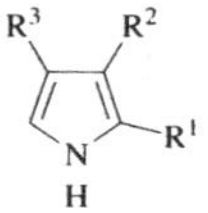

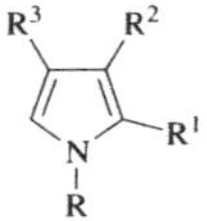

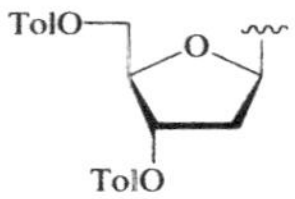

TABLE XI. REACTIONS OF ACIDIC HETEROCYCLES WITH 1-HALOSUGARS IN THE PRESENCE OF BASES (*Continued*)

Sugar	Base R^1	Base R^2	Base R^3	Conditions	Product(s) and Yield(s) (%)	Refs.
	H	Ph	H	NaH, MeCN, rt, 3 h	(53)	1319
	H	m-$O_2NC_6H_4$	H	NaH, MeCN, rt, 3 h	(65)	1319
	NH_2	CN	H	NaH, MeCN, rt, 15 min	(45) + N^3-isomer (21)	1320
	NH_2	CO_2Et	H	NaH, MeCN, rt, 3 h	(10) + N^3-α-isomer (1) + N^3-β-isomer (28)	1321
	CH_2CN	CO_2Me	H	NaH, MeCN, rt, 30 min	(49) + N^3-isomer (45)	1297
	CH_2CN	CO_2Me	H	NaH, MeCN, rt, 3 h	(48) + N^3-β-isomer (45)	1322
	$CH(OEt)_2$	CO_2Me	H	NaH, MeCN, rt, 2 h	(70)	1323
	Ac	H	ONa	DME, rt, 40 h	(43) α:β = 1:1.7	929
	CO_2Me	CH_2CN	Cl	NaH, MeCN, rt, 3 h	(65)	1324
	CO_2Me	CH_2CN	SMe	NaH, MeCN, rt, 3 h	(68)	1324
	CH_2CN	CO_2Me	H	NaH, MeCN, rt, 4 h	(38) + N^2-isomer (18)	1325
	SMe 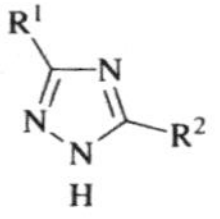	CN	NH_2	NaH, MeCN, rt, 20 min	(47) + N^2-isomer (34) 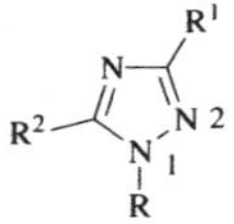	1326

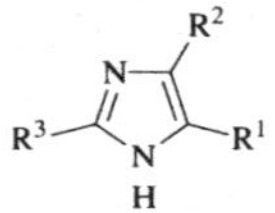

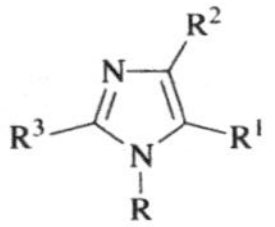

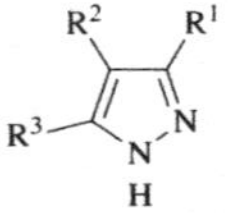

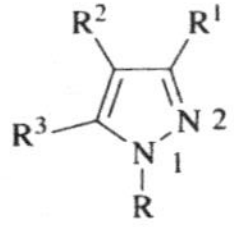

Reactant	Conditions	Product(s) and Yield(s) (%)	Refs.
R^1 / R^2			
CN / H	NaH, MeCN, rt, 30 min	(35)	1327
CO_2Me / H	NaH, MeCN, rt, 30 min	(28)	1327
SO_2NH_2 / H	NaH, MeCN, rt, 20 h	(40) + N^2-isomer (13)	1328
NO_2 / Br	NaH, MeCN, rt, 12 h	(63)	1329
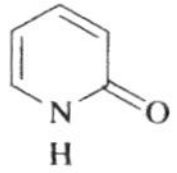 2-pyridone (N–H)	1. KOH, MeCN 2. TsOH, rt, 48 h	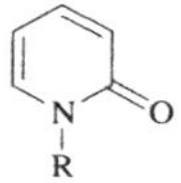 N-R-2-pyridone (33) + α-anomer (22); R 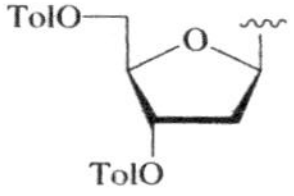 = 3,5-di-O-toluoyl-2-deoxyribofuranosyl (TolO)	1330
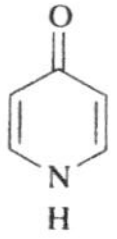 4-pyridone (N–H)	1. KOH, MeCN 2. H^+, 60°, 10 min	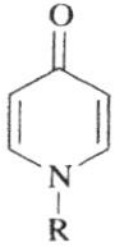 N-R-4-pyridone (41) + α-anomer (42)	1330
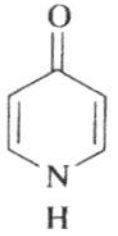 4-pyridone (N–H)	1. KOH, MeCN 2. H^+, rt, 10 min	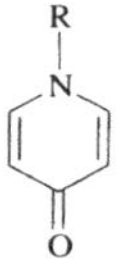 N-R-4-pyridone (—); R = 3,5-di-O-toluoyl-2-deoxyribofuranosyl (TolO)	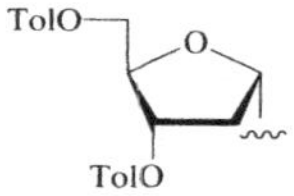 1330
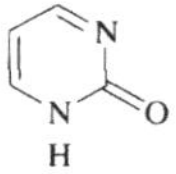 2-pyrimidinone (N–H)	LiOH, MeCN, 60°, 30 min 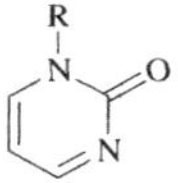	N-R-2-pyrimidinone (64)	1330
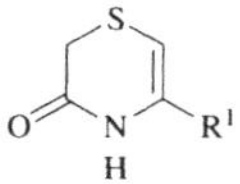 1,4-thiazin-3-one, 5-R^1 (N–H)	NaH, MeCN, rt	N-R-thiazinone; R = 3,5-di-O-toluoyl-2-deoxyribofuranosyl (TolO)	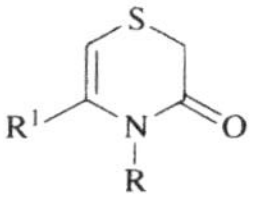 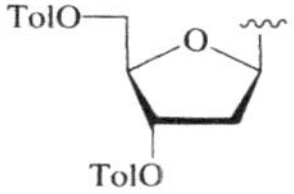 1331
R^1			
H	3 h	(79)	
Me	30 min	(38)	
CO_2Et	3 h	(35)	

TABLE XI. REACTIONS OF ACIDIC HETEROCYCLES WITH 1-HALOSUGARS IN THE PRESENCE OF BASES (*Continued*)

Sugar	Base	Conditions	Product(s) and Yield(s) (%)	Refs.
	HN, O, N–$OCSC_8H_{17}$-*n*, F	$(i\text{-}Pr)_2NEt$, THF or DMF, rt, 12 h	TolO, O, TolO, O, F, N, N, O, $OCSC_8H_{17}$-*n* (100)	256a
	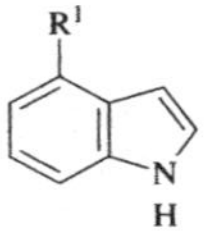 R^1, N, H	NaH, MeCN	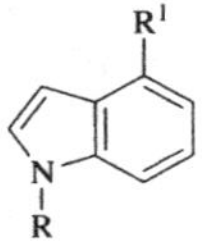 R^1, N–R; R = TolO, O, TolO 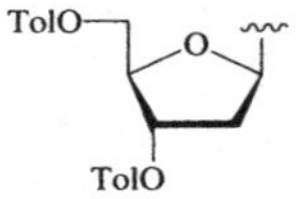	
	R^1 = H	rt, 12 h	(75)	1332
	NO_2	rt, 12 h	(82)	1332
	CN	rt, 12 h	(79)	1332
	CO_2Me	0-25°	(86)	1333
	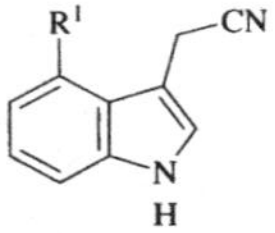 R^1, CN, N, H	NaH, MeCN, rt	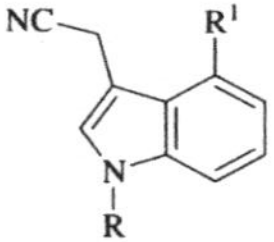 NC, R^1, N–R; R^1 = H, 12 h (70); OMe, 30 min (90)	1332; 1297, 1332
	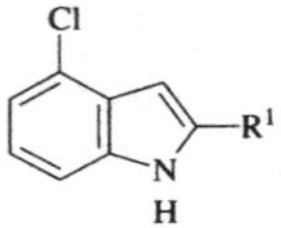 Cl, R^1, N, H	NaH, MeCN, rt, 12 h	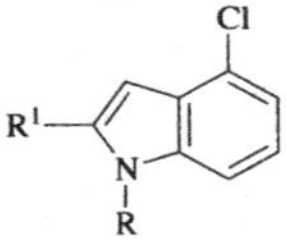 Cl, R^1, N–R; R^1 = CN (57); $CONH_2$ (60)	1332
	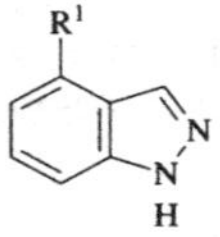 R^1, N, N, H	KOH, TDA-1, MeCN, rt, 20 min	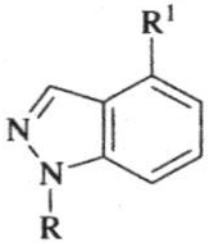 R^1, N, N–R; R^1 = H (34) + N^2-isomer (21); NO_2 (34) + N^2-isomer (34)	1334

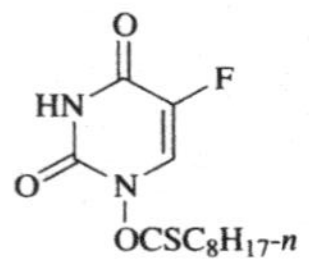

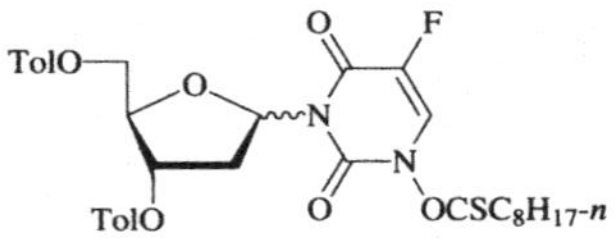

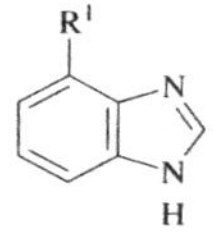

Reactant	Conditions	Product(s) and Yield(s) (%)	Refs.
4-R^1-benzimidazole (N–H)	KOH, MeCN	1-R-4-R^1-benzimidazole; R = 3,5-di-O-TolO-2-deoxyribofuranosyl (TolO—, O, TolO, ～)	
R^1 = H	rt, 15 min	(86)	1335
R^1 = NO_2	TDA-1, rt, 30 min	(45) + N^3-β-isomer (30) + N^3-α-isomer (6)	1336
R^1 = NO_2	18-crown-6, rt, 30 min	(49) + N^3-β-isomer (32) + N^3-α-isomer (8)	1336
R^1 = NO_2	K_2CO_3, TDA-1, MeCN, rt, 2 h	(12) + α-anomer (26)	1336
R^1 = NO_2	K_2CO_3, 18-crown-6, MeCN, rt, 2 h	(14) + α-anomer (27)	1336
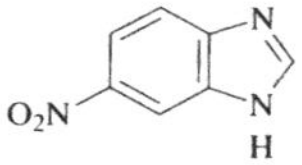 5-O_2N-benzimidazole (N–H)	KOH, TDA-1, MeCN, rt, 2 h	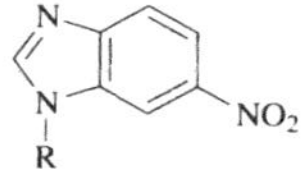 1-R-6-NO_2-benzimidazole (39) + N^3-isomer (46)	1336
5,6-R^1-2-R^2-benzimidazole (N–H)	NaH, MeCN, rt, 15 min	1-R-2-R^2-5,6-R^1-benzimidazole	1335
R^1 = F, R^2 = H		(91)	
R^1 = Cl, R^2 = H		(84)	
R^1 = Cl, R^2 = CF_3		(57)	
R^1 = Cl, R^2 = Me		(72)	
R^1 = Cl, R^2 = Pr-*i*		(49)	
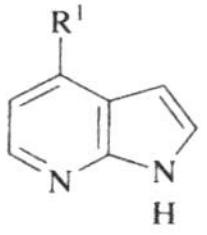 4-R^1-7-azaindole (N–H)	KOH, TDA-1, MeCN, rt, 20 min	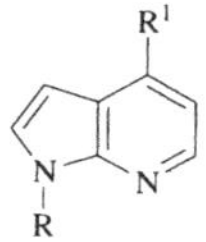 1-R-4-R^1-7-azaindole: R^1 = H (55); R^1 = NO_2 (78)	67, 1337

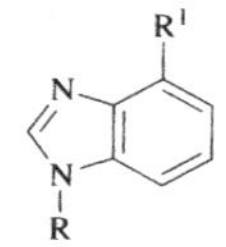

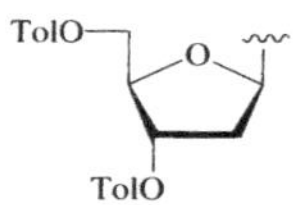

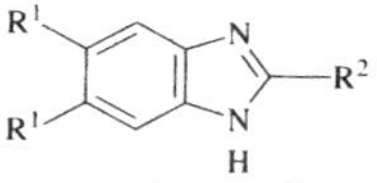

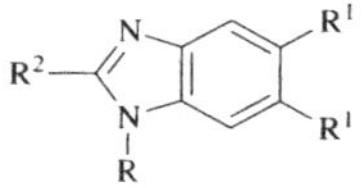

TABLE XI. REACTIONS OF ACIDIC HETEROCYCLES WITH 1-HALOSUGARS IN THE PRESENCE OF BASES (*Continued*)

Sugar	Base	Conditions	Product(s) and Yield(s) (%)	Refs.
	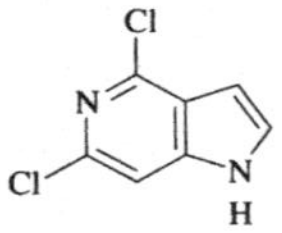 4,6-dichloro-1H-pyrrolo[3,2-c]pyridine (Cl, N, Cl, N–H)	KOH, TDA-1, MeCN, rt, 15 min	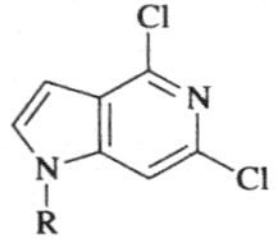 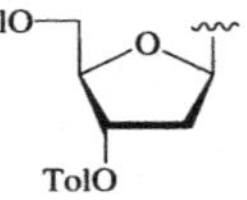 N–R product (Cl, N, Cl) (90); R = TolO–CH₂ / TolO 2-deoxyribofuranosyl	1338
	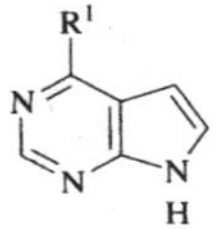 4-R[1]-pyrrolo[2,3-d]pyrimidine (N–H)	KOH, TDA-1, MeCN, rt	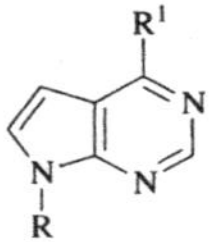 N–R product (R[1])	
	R[1] = Cl	15 min	(81)	1339
	R[1] = NHMe	10 min	(66)	1340
	R[1] = SMe	10 min	(80)	1340
	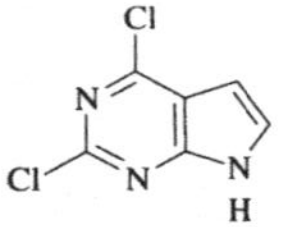 2,4-dichloropyrrolo[2,3-d]pyrimidine (N–H)	KOH, TDA-1, MeCN, rt, 15 min	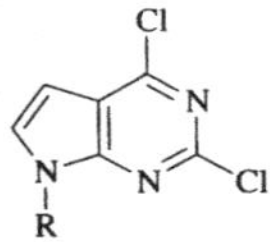 N–R product (Cl, Cl) (61)	1341
	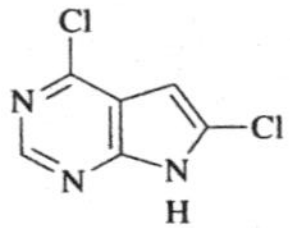 4,6-dichloropyrrolo[2,3-d]pyrimidine (N–H)	NaH, MeCN, rt, 12 h	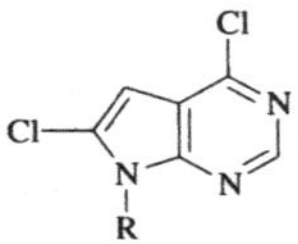 N–R product (Cl, Cl) (63) + N^1-isomer (19)	1342
	4-chloro-2-R[1]-pyrrolo[2,3-d]pyrimidine (N–H)	NaH, MeCN, rt, 2 h	N–R product, R[1] = NH_2 (78)	1343
			R[1] = SMe (66)	64
		KOH, Bu_4NHSO_4, MeCN	R[1] = NH_2 (46)	1344

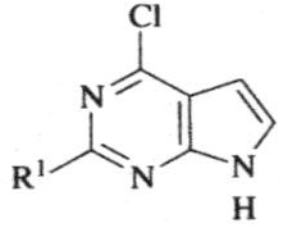

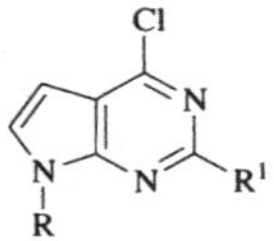

Substrate	Conditions	Product (yield %)		Ref.
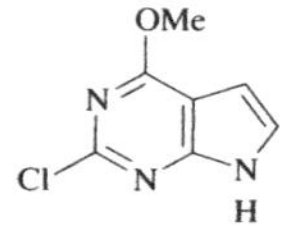	NaOH (50%), TEBA, CH_2Cl_2, $CHCl_3$, rt, 5 min	(67) + α-anomer (7)		1345
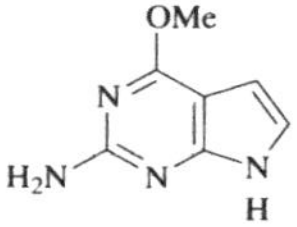	KOH, TDA-1, MeCN, rt, 5 min	(61)		1346, 262
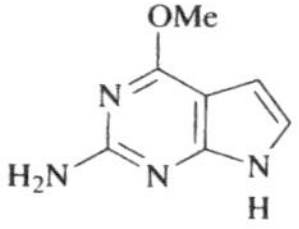	1. KOH, Bu_4NHSO_4, CH_2Cl_2, rt, 3 min 2. NaOMe, MeOH, rt, 3 h	(63)		1339, 1186
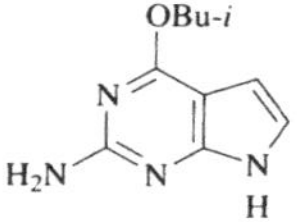	NaOH, CH_2Cl_2, rt, 3 min	(54)	R 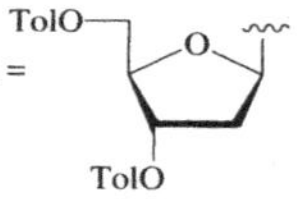 =	262
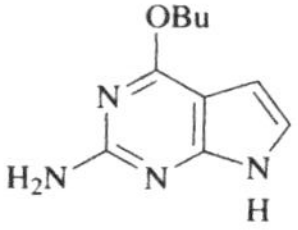	KOH, TDA-1, MeCN, rt, 5 min	(73)		262
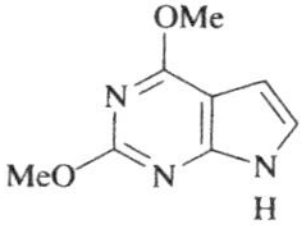	1. NaOH (50%), TEBA, CH_2Cl_2, $CHCl_3$, rt, 30 min 2. NaOMe, MeOH	(67)	R 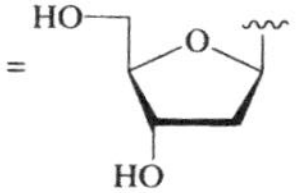 =	1345

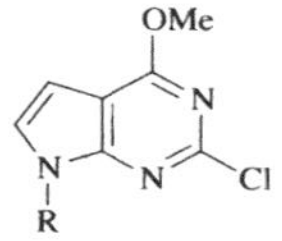

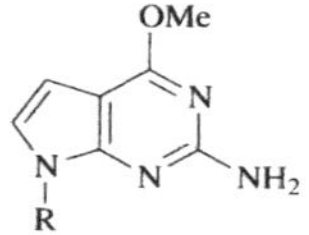

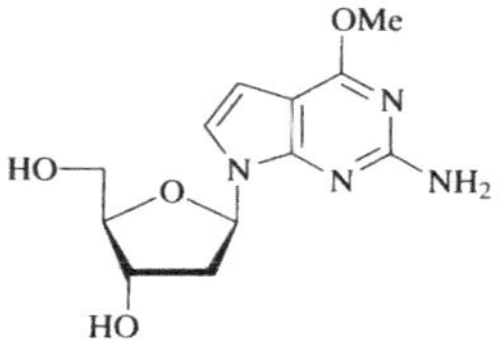

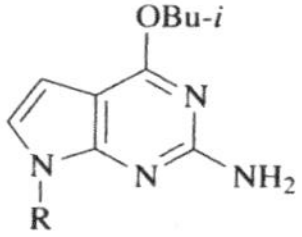

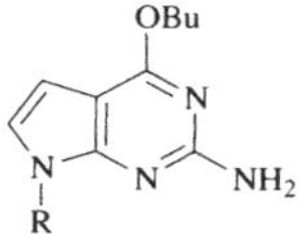

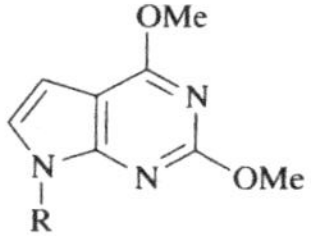

TABLE XI. REACTIONS OF ACIDIC HETEROCYCLES WITH 1-HALOSUGARS IN THE PRESENCE OF BASES (*Continued*)

Sugar	Base	Conditions	Product(s) and Yield(s) (%)	Refs.
	OMe, MeS, N, N, N, H	1. NaH, MeCN, rt, 45 min 2. Rexyn™ 201, MeOH, rt, 1.5 h	OMe, N, N, N, SMe, R (58)	1347
	Cl, R^3, N, R^1, N, N, H, R^2	NaH, MeCN, rt	R^3, Cl, R^2, N, N, N, R, R^1; R = TolO, O, TolO	
	R^1 / R^2 / R^3			
	Me / Cl / H	12 h	(59)	1342
	SMe / Cl / H	30 min	(87)	1342
	H / Br / CN	2 h	(53)	1348
	NH_2 / H / CN	—	(75)	1349
	NH_2 / H / CO_2Me	30 min	(87)	1297
	SMe / Me / H	30 min	(80)	1342
	OMe, N, H_2N, N, N, H	NaOH (50%), TEBA, CH_2Cl_2, rt, 5 min	OMe, N, N, N, NH_2, R (47) + α-anomer (40)	1350
	SMe, CN, N, MeS, N, N, H	NaH, MeCN, 50°, 2 h	NC, SMe, N, N, N, SMe, R (72)	1351

Reactant	Conditions	Product(s) and Yield(s) (%)	Refs.
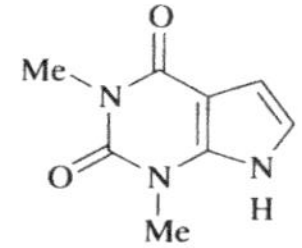 Me, N, O, N, N, H, Me	NaOH (50%), TEBA, CH_2Cl_2, rt	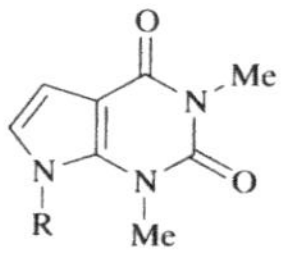 O, N, Me, N, R, N, O, Me (50) + α-anomer (21)	1352
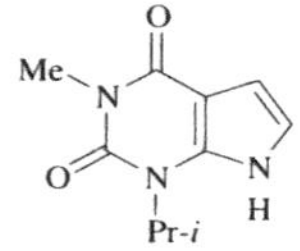 Me, N, O, N, N, H, Pr-*i*	KOH, TDA-1, MeCN, rt, 15 min	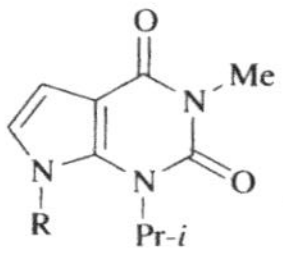 O, N, Me, N, R, N, O, Pr-*i* (42) + α-anomer (14)	1352
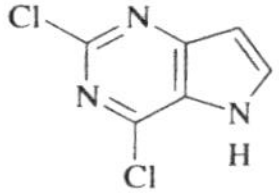 Cl, N, N, N, H, Cl	NaH, MeCN, rt, 30 min	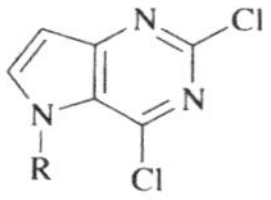 N, Cl, N, R, N, Cl (84)	1297, 1353
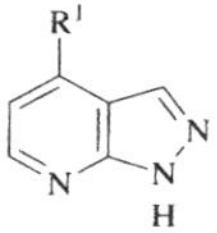 R[1], N, N, N, H	NaH, MeCN, rt 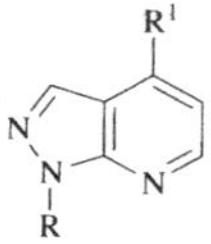	R[1], N, N, R, N	
R[1]			
H	2 h	(95)	1303
Cl	30 min	(63)	1297
N_3	2 h	(61)	1303
OMe	2 h	(64)	1303
OBn	2 h	(78)	1303
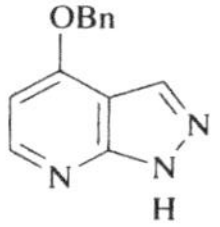 OBn, N, N, N, H	1. NaH, MeCN 2. NH_3, MeOH 3. H_2, Pd/C 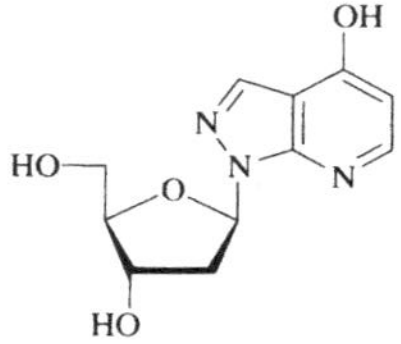	OH, N, N, N, HO, O, HO (60)	1354

TABLE XI. REACTIONS OF ACIDIC HETEROCYCLES WITH 1-HALOSUGARS IN THE PRESENCE OF BASES (*Continued*)

Sugar	Base	Conditions	Product(s) and Yield(s) (%)	Refs.
	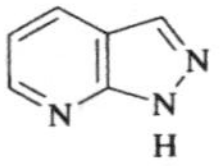 N, N, H	1. NaH, MeCN 2. NH_3, MeOH	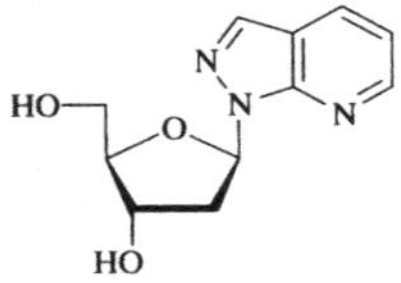 HO, N, N, N, O, HO (—)	1354
	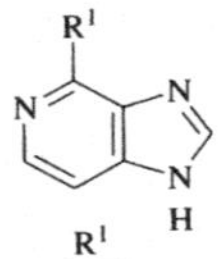 R[1], N, N, N, H	KOH, TDA-1, MeCN, rt	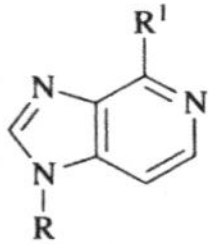 R[1], N, N, N, R; R = TolO, O, TolO 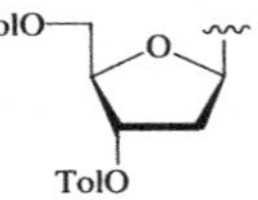	
	R[1]: Cl	1 h	(61) + N^3-isomer (26)	1355
	R[1]: SMe	15 min	(68) + N^3-isomer (29)	1356
	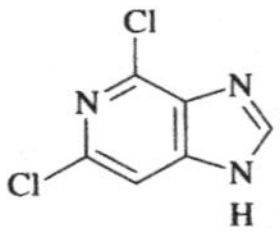 Cl, N, N, Cl, N, H	NaH, MeCN, rt, 30 min	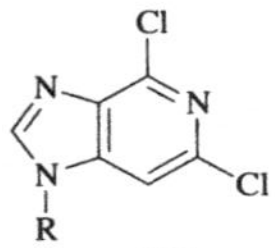 Cl, N, N, N, Cl, R (66)	1297, 64
	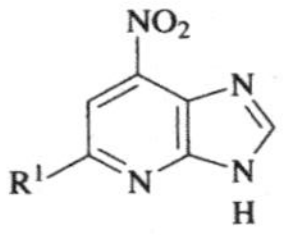 NO_2, N, R[1], N, N, H	NaH, MeCN, 0°, 3 h	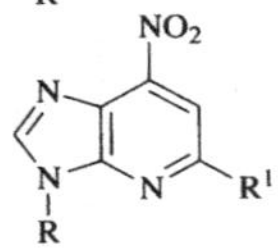 NO_2, N, N, N, R[1], R; R[1]: H (60), Cl (50)	1357

Substrate	Conditions	Product (yield, %)	Ref.
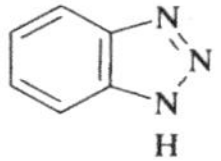 1H-benzotriazole	KOH, TDA-1, MeCN, rt, 15 min	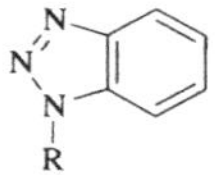 1-R-benzotriazole (57) + N^2-isomer (30)	271
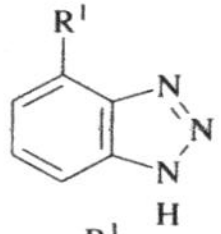 4-R^1-1H-benzotriazole	NaH, MeCN, rt, 20 min	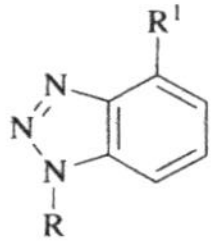 4-R^1-1-R-benzotriazole	
R^1			
NH_2		(19) + N^2-isomer (20) + N^3-isomer (8)	271
NO_2		(8) + N^2- + N^3-isomer (56)	
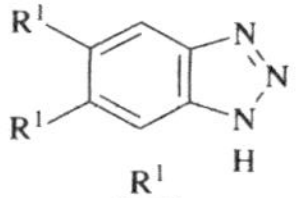 5,6-di-R^1-1H-benzotriazole	NaH, MeCN, rt	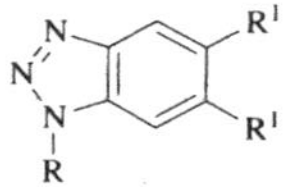 5,6-di-R^1-1-R-benzotriazole	271
R^1			
Cl	20 min	(36) + N^2-isomer (24)	
Me	10 min	(45) + N^2-isomer (25)	
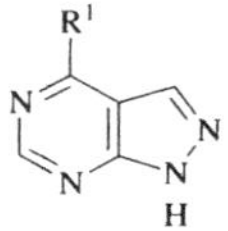 4-R^1-1H-pyrazolo[3,4-d]pyrimidine		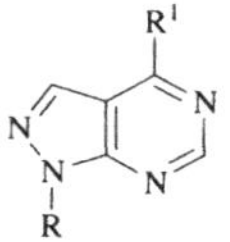 4-R^1-1-R-pyrazolo[3,4-d]pyrimidine	
R^1			
Cl	NaOH (50%), THF, rt, 3 min	(42)	1358
OMe	KOH, 18-crown-6, glyme, 20°, 10 min	(62) + N^2-β-isomer (31)	1292
SMe	KOH, TDA-1, MeCN, rt, 15 min	(65) + N^1-α-isomer 95) + N^2-isomer (9)	1326

TABLE XI. REACTIONS OF ACIDIC HETEROCYCLES WITH 1-HALOSUGARS IN THE PRESENCE OF BASES (*Continued*)

Sugar	Base	Conditions	Product(s) and Yield(s) (%)	Refs.
	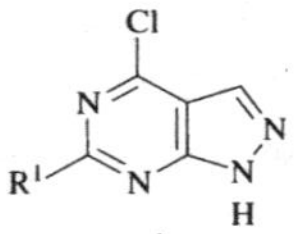 4-Cl-6-R^1-pyrazolo[3,4-d]pyrimidine (1H)		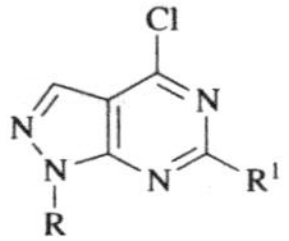 4-Cl-6-R^1-1-R-pyrazolo[3,4-d]pyrimidine	
	R^1			
	Cl	NaOH (50%), CH_2Cl_2, Bu_4NHSO_4	(36)	1359
	NH_2	KOH, (30%), THF, CH_2Cl_2	(39) + N^2-isomer (39)	1360
	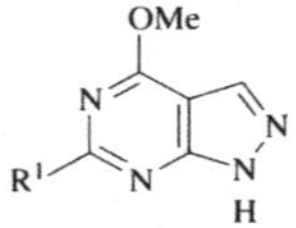 4-OMe-6-R^1-pyrazolo[3,4-d]pyrimidine (1H)		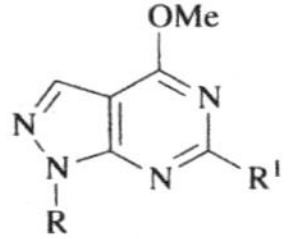 4-OMe-6-R^1-1-R-pyrazolo[3,4-d]pyrimidine	
	R^1			
	Cl	NaOH (50%), CH_2Cl_2, Bu_4NHSO_4	(48)	1359
	NH_2	KOH (30%), THF, CH_2Cl_2, rt, 2 min	(47) + N^2-isomer (13)	1360
	NH_2	KOH (50%), Bu_4NHSO_4, CH_2Cl_2, rt, 2 min	(37) + α-isomer (6) + N^2-β-isomer (15)	1360
	OMe	NaOH (50%), CH_2Cl_2, Bu_4NHSO_4	(40)	1359
	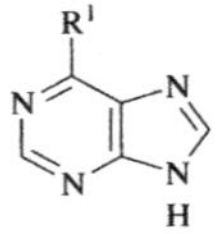 6-R^1-purine (9H)		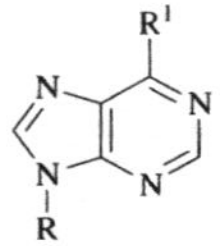 6-R^1-9-R-purine; R^1	
		NaH, MeCN, rt,	Cl 30 min (61) + N^7-isomer (13)	1297
		Me_2CO	Br 1 h (45) + N^7-β-isomer (13)	1361

Reactant	Conditions	Product(s) and Yield(s) (%)	Refs.
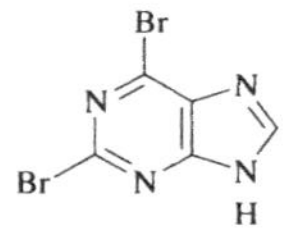 6,2-dibromopurine (9-H)	NaH, MeCN, rt, 15 min	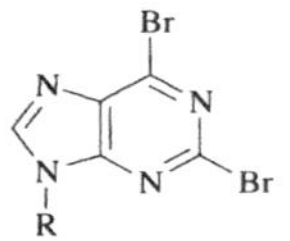 (50) + α-anomer (11)	1361, 1362
6-chloro-2-R^1-purine (9-H)		R = 3,5-di-O-TolO-2-deoxyribofuranosyl	
R^1			
Cl	NaH, MeCN, rt, 30 min	(59) + N^7-isomer (11)	1297
NH_2	NaH, MeCN, rt, 30 min	(55) + N^7-isomer (9)	1297
NHPh	NaH, MeCN, rt, 1 h	(43) + N^7-isomer (10)	1363
NHC_6H_4Bu-*p*	NaH, MeCN, rt, 1 h	(64) + N^7-isomer (14)	1364
6-SMe-2-R^1-purine (9-H)	NaH, MeCN, rt, 1 h	R^1 / N^7-β / N^9-α / N^7-α	
		SMe (62) (19) (0.3) (0.2)	1361, 1365
		Br (69) (15) (2) (0.8)	1361
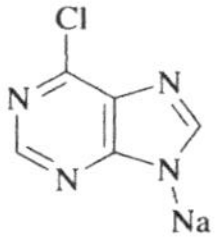 6-chloropurine, 9-Na	MeCN/THF (5:1), 10°, 10 h	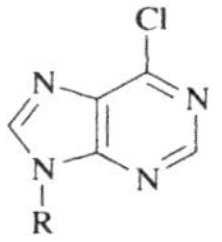 (18) + α-anomer (52)	1168
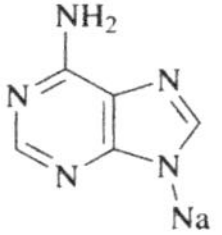 adenine, 9-Na (NH_2)	Me_2CO, rt, 19 h	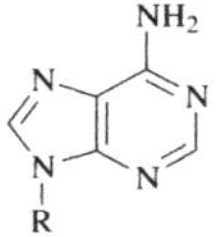 (43)	65, 1366

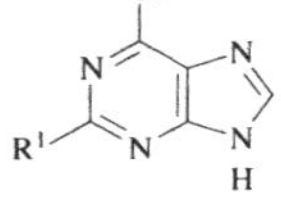

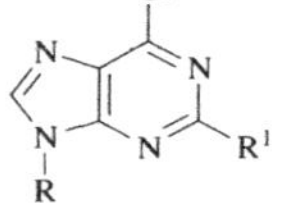

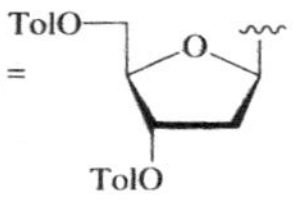

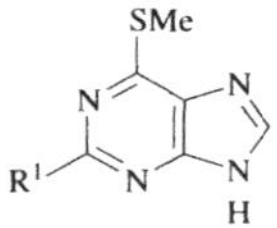

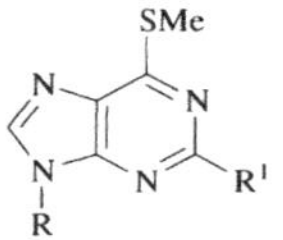

TABLE XI. REACTIONS OF ACIDIC HETEROCYCLES WITH 1-HALOSUGARS IN THE PRESENCE OF BASES (*Continued*)

Sugar	Base	Conditions	Product(s) and Yield(s) (%)	Refs.
	NHBz, N, N, N, N, Na	DME, rt, 24 h	TolO, O, TolO, N, N, N, N, NHBz (46) α:β = 5:95	65
		MeCN. THF, 10°, 10 h	TolO, O, TolO, N, N, N, N, NHBz (52) + β-anomer (18)	65
	NH$_2$, N, N, N, N, M	MeCN, rt, 24 h	TolO, O, TolO, N, N, N, N, NH$_2$	65
	M			
	K		(18) α:β = 57:43	
	Li		(17) α:β = 71:29	
	Me, O, N, N, Me, N, N, O, H	NaH, MeCN, rt, 10 min	Me, N, O, N, N, N, Me, R, O (74) R = TolO, O, TolO	1367
	O, N, N, H$_2$N, N, N, H	K_2CO_3 (10%), CH_2Cl_2, Bu_4NHSO_4, rt	O, N, N, N, N, NH$_2$, R (42) + α-anomer (21)	1368, 1369

Sugar	Base	Conditions	Product(s) and Yield(s) (%)								Refs.
	i-PrCONH (imidazotriazinone, N-H)	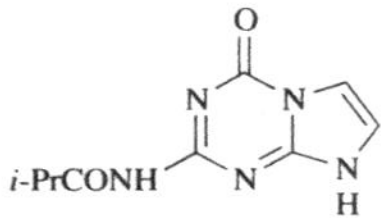 K_2CO_3, TDA-1, MeCN, rt, 1 h	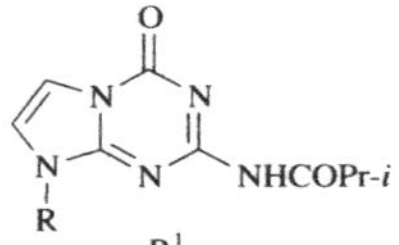 (29) + α-anomer (59)								1369
	(triazolopyridine/triazolopyrimidine; positions 1, 2, 3, 4, 7; R^1, R^2, X)			N^3-α	N^2-β	N^2-α	N^1-β	N^4-α	N^4-β		
	X = CH, R^1 = Cl, R^2 = H	NaH, MeCN, rt, 30 min	(29)	(11)	(21)	(5)	(13)	(—)	(—)		1370
	X = N, R^1 = NH_2, R^2 = H	K_2CO_3, TDA-1, DMF, rt, 20 min	(14)	(13)	(11)	(12)	(—)	(4)	(3)		270
	X = N, R^1 = OMe, R^2 = H	NaH, MeCN, rt, 30 min	(29)	(—)	(32)	(—)	(23)	(—)	(—)		270
	X = N, R^1 = OMe, R^2 = NH_2	KOH, TDA-1, MeCN, rt, 20 min	(20)	(6)	(19)	(4)	(13)	(—)	(—)		1371
	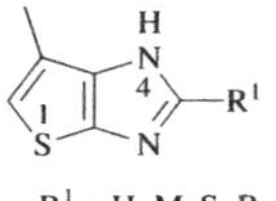 (thienoimidazole, R^1) R^1 = H, MeS, BnS	NaH, MeCN, rt, 1-2 h	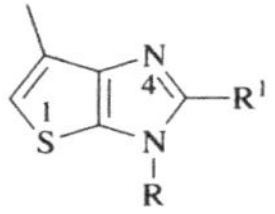 (63-88) + N^4-isomer (—)								350
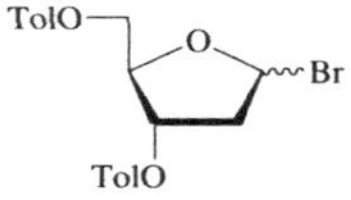 C_{23} TolO (Br sugar)	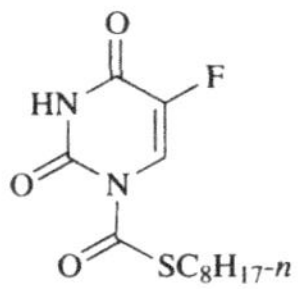 5-fluorouracil N-COSC$_8$H$_{17}$-n	$(i\text{-}Pr)_2NEt$, THF, rt, 12 h	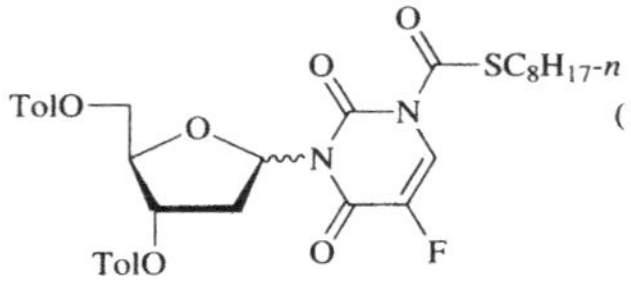 (15)								256a
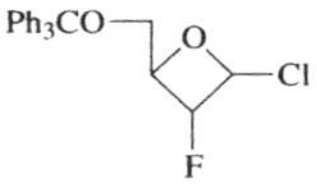 Ph_3CO (Cl, F oxetane)	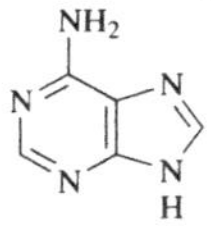 adenine (NH_2)	K_2CO_3, 18-crown-6, DMF, 100°	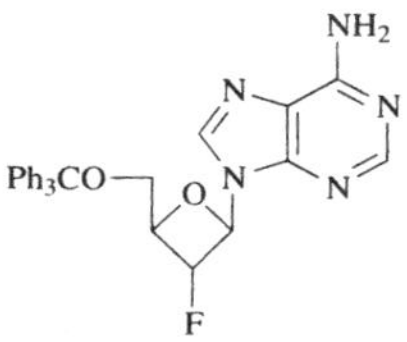 (40) + α-anomer (—)								1372

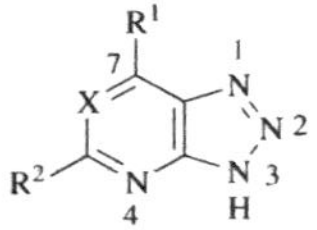

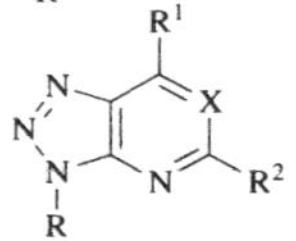

TABLE XI. REACTIONS OF ACIDIC HETEROCYCLES WITH 1-HALOSUGARS IN THE PRESENCE OF BASES (*Continued*)

Sugar	Base	Conditions	Product(s) and Yield(s) (%)	Refs.
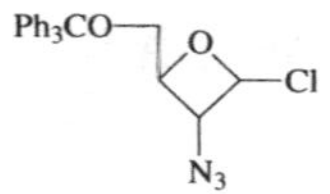 Ph$_3$CO, N$_3$, Cl (oxetane)	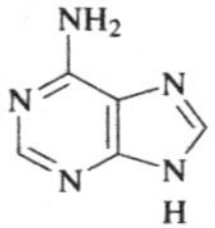 NH$_2$, N, H (adenine)	K_2CO_3, 18-crown-6, DMF, 100°	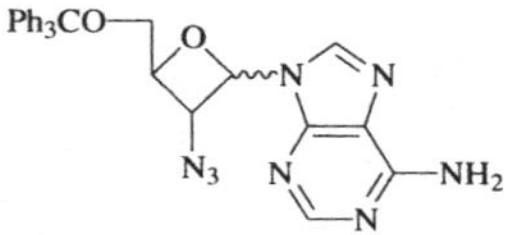 Ph$_3$CO, N$_3$, NH$_2$ (55) α:β = 1:2	1372
C$_{26}$				
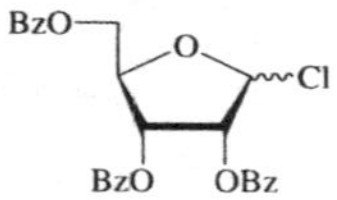 BzO, Cl, BzO, OBz	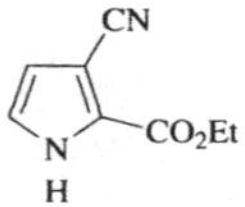 CN, CO$_2$Et, N, H	NaH, MeCN, 70°, 17 h	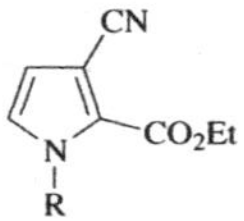 CN, CO$_2$Et, N–R (96) R = BzO, BzO, OBz 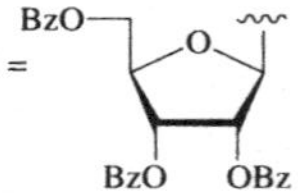	1373, 1374
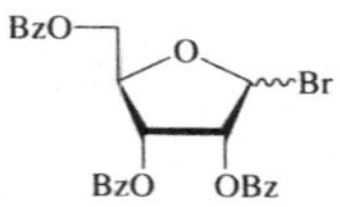 BzO, Br, BzO, OBz	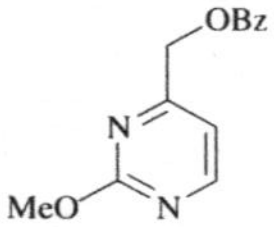 OBz, MeO, N	Na_2CO_3, CH_2Cl_2, MeCN, rt, 24 h	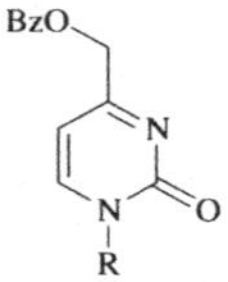 BzO, N, O, N–R (63)	867
	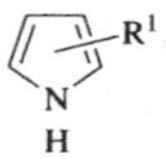 R^1, N, H	NaH, dioxane, rt, 30 min, 50°, 1 h	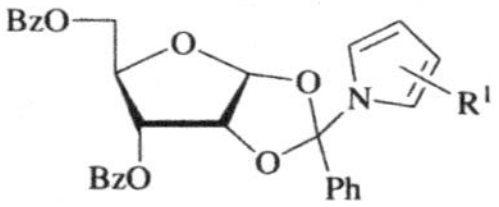 BzO, O, N, R^1, BzO, O, Ph; R^1: 2-CN (84); 3-CN (85)	68; 1297
	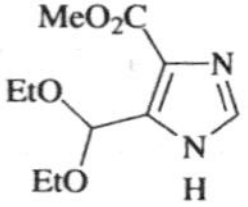 MeO$_2$C, EtO, EtO, N, H	NaH, MeCN, 50°, 4 h	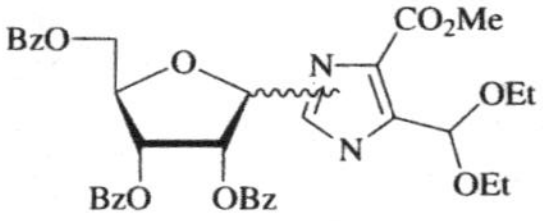 BzO, CO$_2$Me, OEt, OEt, BzO, OBz (70) mixture of N^1 and N^3 isomers	1323
	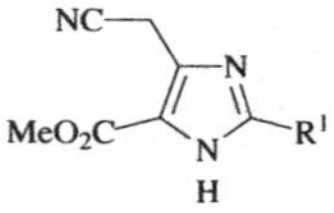 NC, MeO$_2$C, R^1, N, H	NaH, MeCN, rt, 4 h	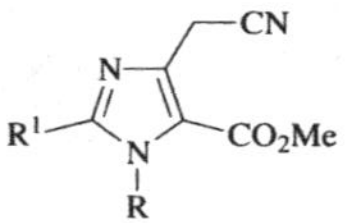 CN, R^1, CO$_2$Me, N–R; R^1: Cl (69), SMe (72); R 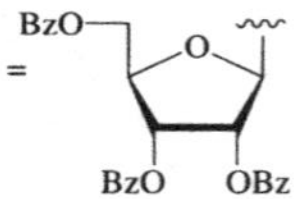 = BzO, BzO, OBz	1324

Substrate	Conditions	Product(s) and Yield(s) (%)		Refs.
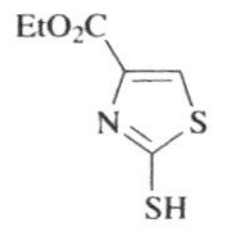 EtO_2C, N, S, SH	NaH, MeCN, rt, 3 h	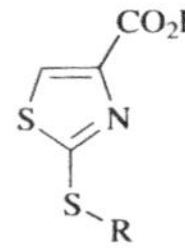 CO_2Et, S, N, S–R (38)	R = BzO, O, BzO, OBz 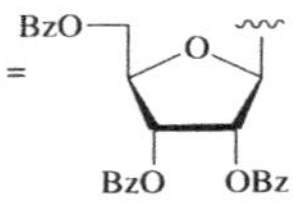	1375
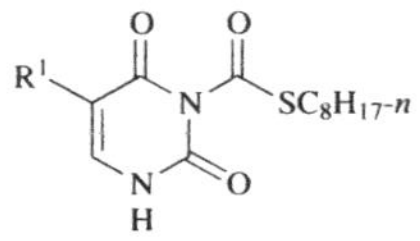 R^1, O, O, N, SC_8H_{17}-*n*, N, H, O	(*i*-Pr)$_2$NEt, DMF, rt, 12 h	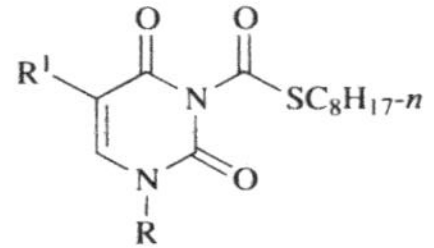 R^1, O, O, N, SC_8H_{17}-*n*, N, R, O	R^1: H (74); F (74)	1294
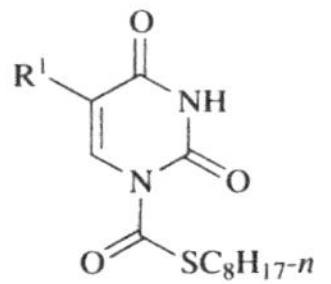 R^1, O, NH, N, O, O, SC_8H_{17}-*n*	(*i*-Pr)$_2$NEt, DMF, rt, 12 h	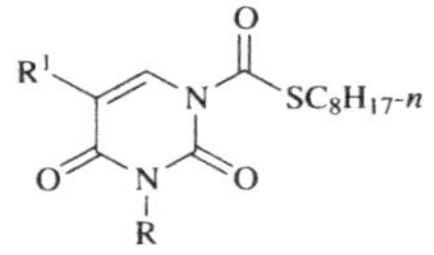 R^1, O, N, SC_8H_{17}-*n*, O, N, R, O	R^1: H (78); F (79); Me (73)	1294
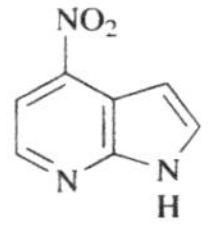 NO_2, N, N, H	NaH, MeCN, rt, 3 h	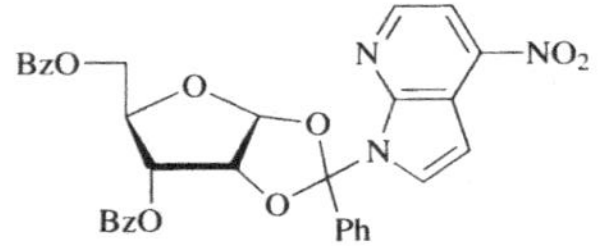 BzO, O, O, N, NO_2, N, BzO, O, Ph (31)		1376
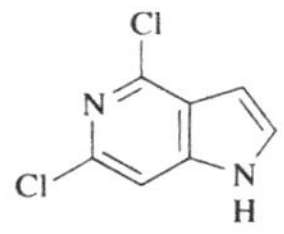 Cl, N, Cl, N, H	NaH, MeCN, rt, 3 h	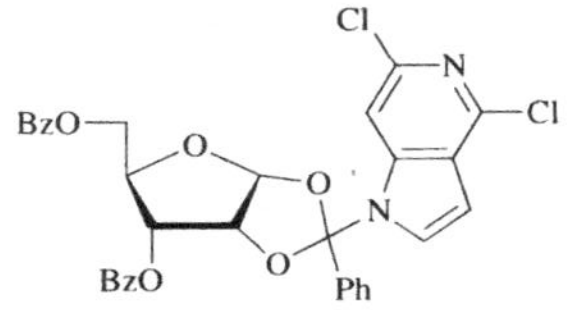 Cl, N, Cl, BzO, O, O, N, BzO, O, Ph (61)		1377
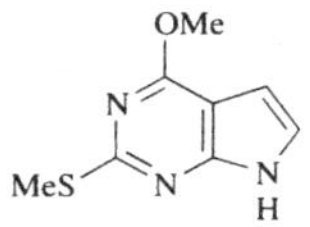 OMe, N, MeS, N, N, H	NaOH (X%), CH_2Cl_2, Bu_4NHSO_4	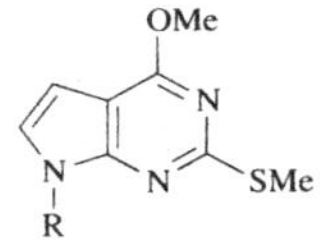 OMe, N, N, N, SMe, R	X 10: (25), α-anomer (10)	1378
			X 50: (25), α-anomer (66)	1378

TABLE XI. REACTIONS OF ACIDIC HETEROCYCLES WITH 1-HALOSUGARS IN THE PRESENCE OF BASES (*Continued*)

Sugar	Base	Conditions	Product(s) and Yield(s) (%)	Refs.
	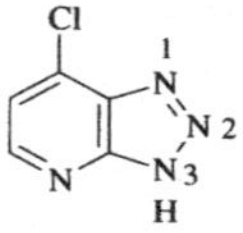	NaH, MeCN, rt, 24 h	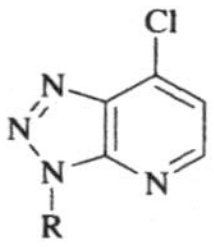 (30) + N^2-β-isomer (26) + N^1-β-isomer (21)	1370
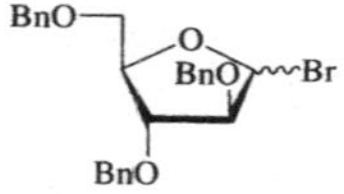	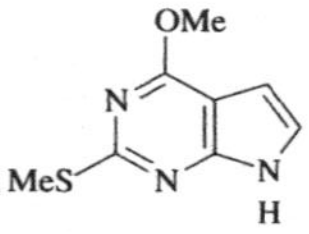	NaOH (50%), TEBA, CH_2Cl_2, rt, 15 min	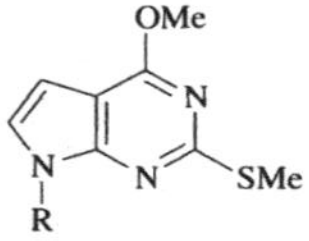 (63) + α-anomer (21) R = 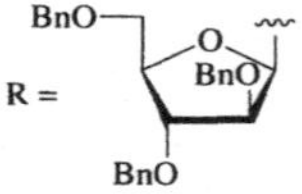	66
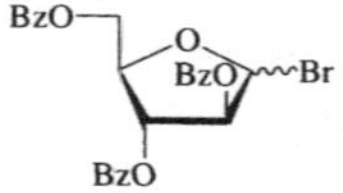	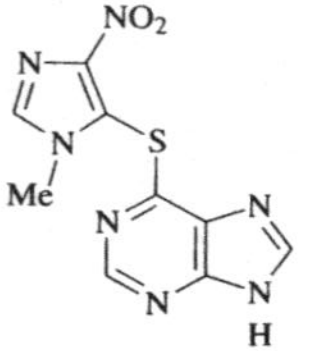	NaH, MeCN, rt, 15 h	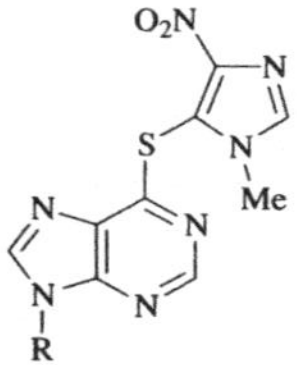 (55) R = 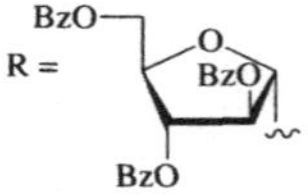	1092
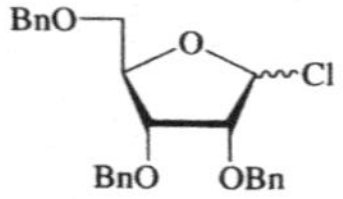	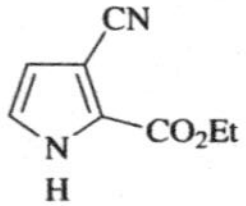	NaH, DMF, rt, 21 h	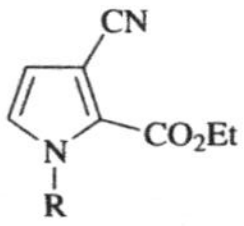 (39) R = 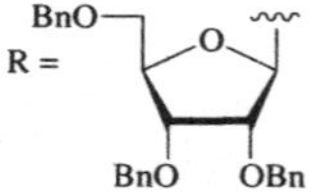	1373
	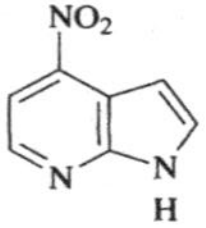	NaH, MeCN, rt, 6 h	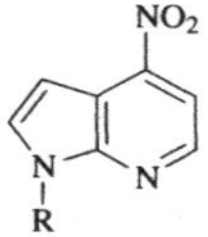 (42) + α-anomer (44)	1376

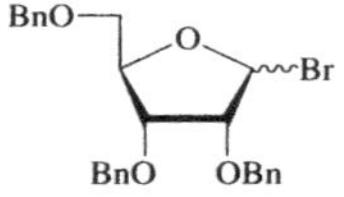	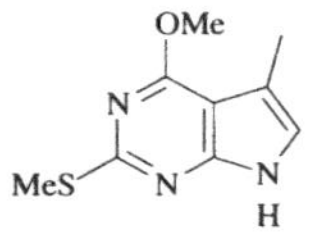	NaOH (50%), TEBA, CH_2Cl_2, rt, 3 h	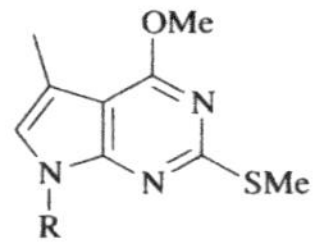 (25) + α-anomer (45)	1379
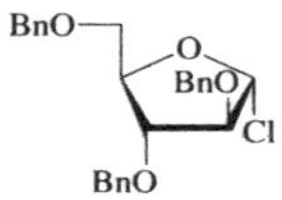	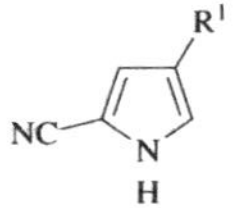	NaH, MeCN, rt, 12 h	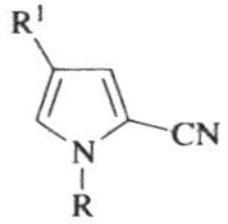 R[1]: H (74); CN (87) — R = 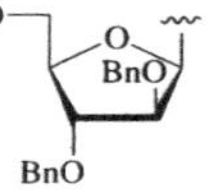	68, 1315
	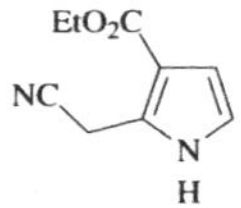	NaH, MeCN, rt, 2 h	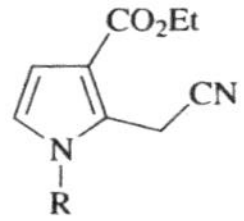 (82)	1317
	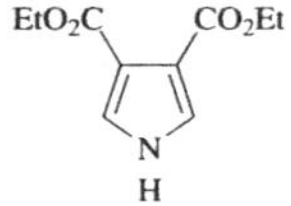	NaH, MeCN, rt	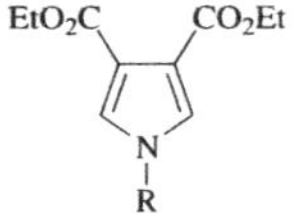 (69)	1316
	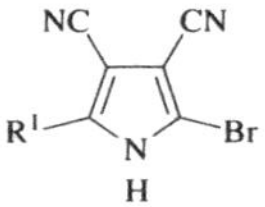	1. $(CH_3)_2NCH(OEt)_2$ 2. NaH, MeCN, rt	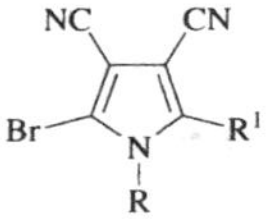 R[1]: NH_2, 30 min (—); N=CHOEt, 12 h (79)	1297 1316, 1380
	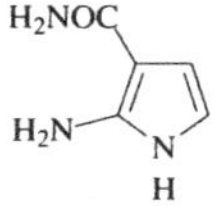	Et_3N, MeCN, 100°, 1.25 h	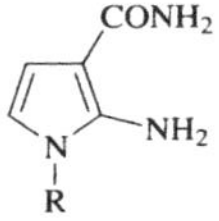 (40)	352
	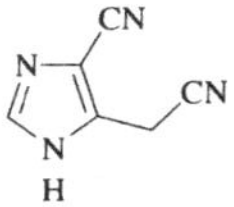	NaH, DMF, rt, 2 h	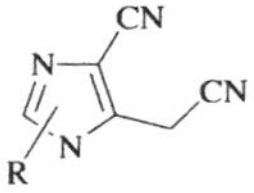 (—)	354

TABLE XI. REACTIONS OF ACIDIC HETEROCYCLES WITH 1-HALOSUGARS IN THE PRESENCE OF BASES (*Continued*)

Sugar	Base	Conditions	Product(s) and Yield(s) (%)	Refs.
	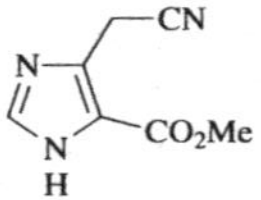 [structure]	NaH, DMF, rt, 2 h	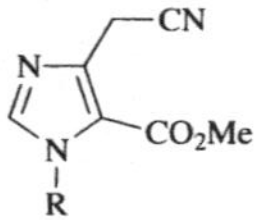 [structure] (—) + N^3-isomer; N^1:N^3 = 5:1.2	354
	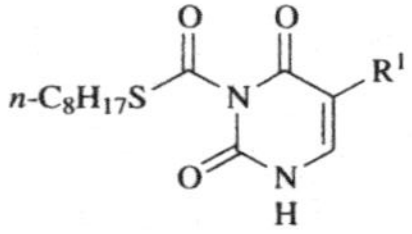 [structure] R^1: F; Me	*i*-Pr_2NEt, MeCN, rt, 12 h	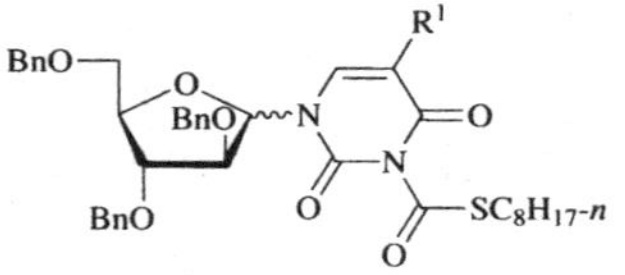 [structure] (99) α:β = 51:48; (84) α:β = 34:54	256a
	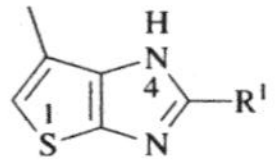 [structure] R^1 = H, SMe, SBn	NaH, MeCN, rt, 48 h	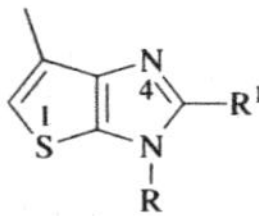 [structure] (50-60) + N^4-isomer (—)	350
	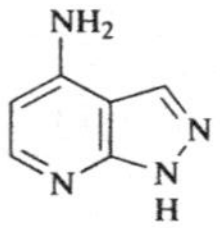 [structure]	NaH, MeCN	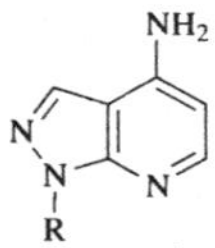 [structure] (—) R = [structure] 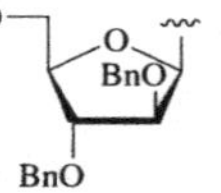	1354
	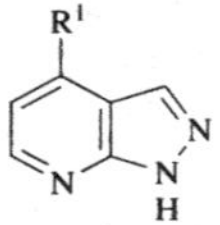 [structure] R^1: H	1. NaH, MeCN, rt, 12 h; 2. $Pd(OH)_2$, cyclohexene, EtOH, reflux, 48 h	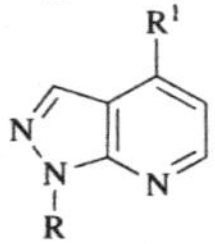 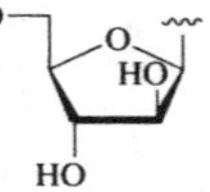 [structure] R = [structure] (60)	1303, 1354

Reactant	Substrate	Conditions	Product(s) and Yield(s) (%)	Refs.
	Cl	2. BCl_3, MeOH, rt, 30 min	(63)	1303, 1354
	OMe	2. $Pd(OH)_2$, cyclohexene, EtOH, reflux, 48 h	(44)	1303, 1354
	N_3	2. $Pd(OH)_2$, cyclohexene, EtOH, reflux, 48 h	(41)	1303, 1354
	OBn	2. $Pd(OH)_2$, cyclohexene, EtOH, reflux, 48 h	(64)	
	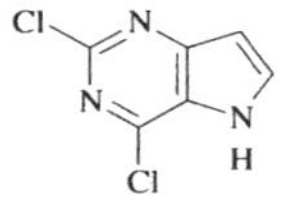	NaH, MeCN, rt, 6 h	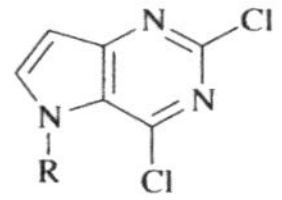 (81)	1353
	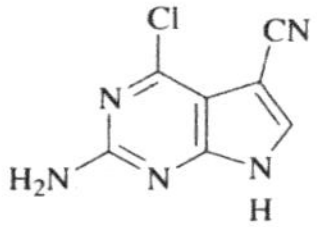	NaH, MeCN, rt, 12 h	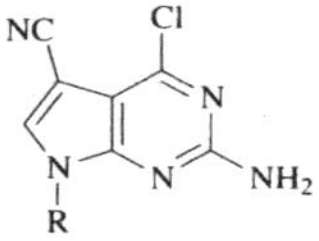 (58)	1299
	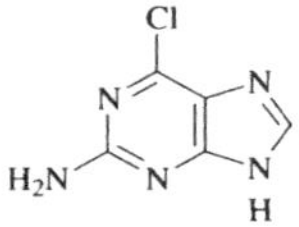	NaH, MeCN, rt, 15 h	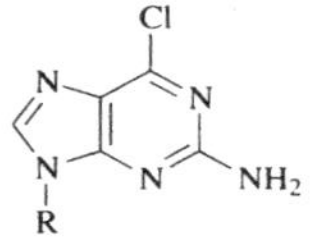 (68) + N^7-isomer (11)	1381
	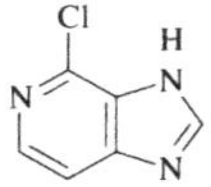	NaH, DMF, rt, 2 h	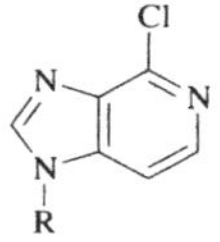 (—)	354
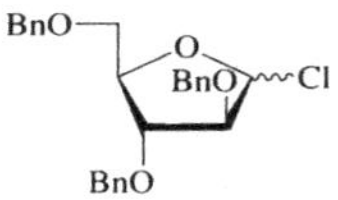	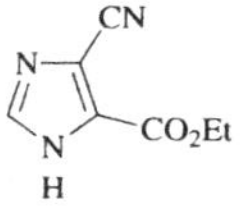	NaH, DMF, rt, 27 h	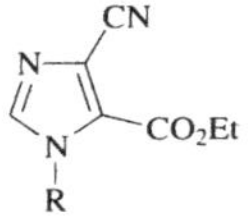 (59)	355

TABLE XI. REACTIONS OF ACIDIC HETEROCYCLES WITH 1-HALOSUGARS IN THE PRESENCE OF BASES (*Continued*)

Sugar	Base	Conditions	Product(s) and Yield(s) (%)	Refs.
BnO, BnO, BnO, Cl	SMe, N, O, N, H	NaH, DMF, rt, 18 h	SMe, N, N, O, R (—)	356
	Cl, N, Cl, N, H	KOH, TDA-1, MeCN, 24°	Cl, N, N, Cl, R (78) + α-anomer (6)	349
	OMe, N, N, N, N, H	KOH, TDA-1, MeCN, rt, 50 min	OMe, N, N, N, N, R (43) + α-anomer (6) + N^7-isomer (25)	340
	Cl, N, Cl, N, H	KOH, TDA-1, MeCN	Cl, N, N, Cl, R (69)	1382
	OMe, N, MeS, N, N, H	NaOH (50%), Bu_4NHSO_4, C_6H_6, DME, 15 min	OMe, N, N, N, SMe, R (35) + α-anomer (57); R = BnO, BnO, BnO	348
	NH_2, N, MeS, N, N, H	NaOH (50%), TEBA, CH_2Cl_2, rt, 15 min	NH_2, N, N, N, SMe, R (69) + α-anomer (14)	1383

C_{27}

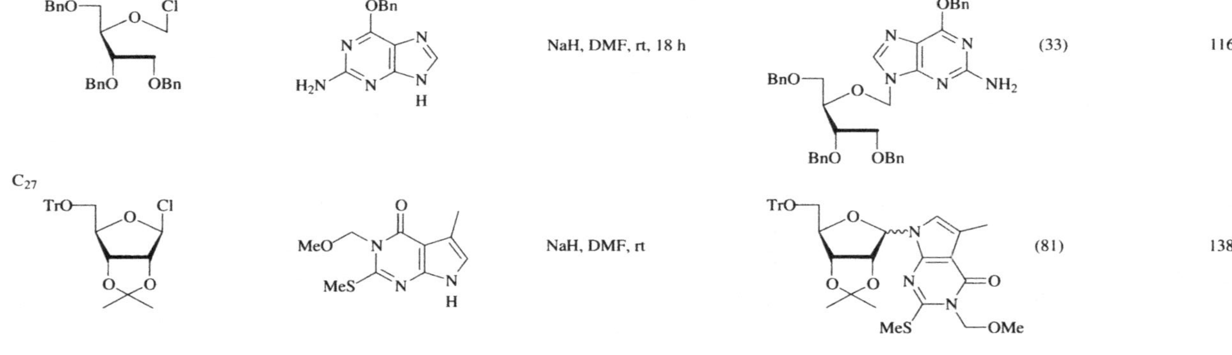

Conditions	Yield (%)	Ref.
NaH, DMF, rt, 18 h	(33)	1165
NaH, DMF, rt	(81)	1384

REFERENCES

[1] Fox, J. J.; Wempen, I. *Adv. Carbohydr. Chem. Biochem.* **1959**, *14*, 283.
[2] Plimi, J.; Prystaš, M. *Adv. Heterocycl. Chem.* **1967**, *8*, 115.
[3] Shimizu, B. *Ann. Sankyo Res. Lab.* **1967**, *19*, 1 [*Chem. Abstr.* **1968**, *68*, 96054y].
[4] Zorbach, W. W. *Synthesis* **1970**, 329.
[5] Goodman, L. In *Basic Principles in Nucleic Acid Chemistry*; Ts'o, P. O. P., Ed.; Academic: New York; 1974, p. 93.
[6] Vorbrüggen, H. In *Nucleoside Analogues*; Walker, R. T., De Clercq, E., Eckstein, F., Eds.; Plenum: New York; 1979, p. 35.
[7] Lukevics, E.; Tablecka, A. In *Nucleoside Synthesis Organosilicon Methods*; Horwood, E., Ed., New York, 1991.
[7a] Wilson, L. J.; Hager, M. W.; El-Kattan, Y. A.; Liotta, D. C. *Synthesis* **1995**, 1465.
[8] Fuller, W. D.; Sanchez, R. A.; Orgel, L. E. *J. Mol. Evol.* **1972**, *1*, 249.
[9] Fuller, W. D.; Sanchez, R. A.; Orgel, L. E. *J. Mol. Biol.* **1972**, *67*, 25.
[10] Müller, D.; Pitsch, S.; Kittaka, A.; Wagner, E.; Wintner, C. E.; Eschenmoser, A. *Helv. Chim. Acta* **1990**, *73*, 1410.
[11] Orgel, L. E. *J. Theor. Biol.* **1986**, *123*, 127.
[12] (a) Shapiro, R. *Orig. Life Evol. Biosph.* **1988**, *18*, 71. (b) Shapiro, R. *Proc. Natl. Acad. Sci. USA* **1999**, *96*, 4396.
[13] Joyce, G. F. *Nature* **1989**, *338*, 217.
[14] Waldrop, M. M. *Science* **1989**, *246*, 1248.
[15] Strazewski, P.; Tamm, C. *Angew. Chem., Int. Ed. Engl.* **1990**, *29*, 36.
[16] Cech, T. R. *Angew. Chem., Int. Ed. Engl.* **1990**, *29*, 759.
[17] Pace, N. R. *Science* **1992**, *256*, 1402.
[18] Piccirilli, J. A.; McConnell, T. S.; Zaug, A. J.; Noller, H. F.; Cech, T. R. *Science* **1992**, *256*, 1420.
[19] Eschenmoser, A.; Loewenthal, E. *Chem. Soc. Rev.* **1992**, *21*, 1.
[20] Waldrop, M. M. *Science* **1992**, *256*, 1396.
[20a] Gesterland, R. F.; Atkins, J. F. *The RNA World*, Cold Spring Harbor Laboratory Press, 1993.
[20b] Kolb, V. M.; Dworkin, J. P.; Miller, S. L. *J. Mol. Evol.* **1994**, *38*, 549.
[20c] Eschenmoser, A.; Kisakürek, M. V. *Helv. Chim. Acta* **1996**, *79*, 1249.
[21] Stubbe, J. *J. Biol. Chem.* **1990**, *265*, 5329.
[22] Nordlund, P.; Sjöberg, B.-M.; Eklund, H. *Nature* **1990**, *345*, 593.
[23] Reichard, P. *Science* **1993**, *260*, 1773.
[24] Suhadolnik, R. J. *Nucleoside Antibiotics*; Wiley-Interscience: New York, 1970.
[25] Suck, D.; Saenger, W.; Vorbrüggen, H. *Nature* **1972**, *235*, 333.
[26] Suck, D.; Saenger, W. *J. Am. Chem. Soc.* **1972**, *94*, 6520.
[27] Egert, E.; Lindner, H. J.; Hillen, W.; Gassen, H. G.; Vorbrüggen, H. *Acta Cryst.* **1979**, *35*, 122.
[28] Kasai, H.; Nishimura, S.; Vorbrüggen, H.; Iitaka, Y. *FEBS Lett.* **1979**, *103*, 270.
[29] Jeffrey, G. A.; Sundaralingam, M. *Adv. Carbohydr. Chem. Biochem.* **1985**, *43*, 203.
[30] Granzin, J.; Saenger, W.; Stolarski, R.; Shugar, D. *Z. Naturforsch.* **1990**, *45c*, 915.
[31] Camerman, N.; Mastropaolo, D.; Camerman, A. *Proc. Natl. Acad. Sci. USA* **1990**, *87*, 3534.
[32] Harte, Jr., W. E.; Starrett, Jr., J. E.; Martin, J. C.; Mansuri, M. M. *Biochem. Biophys. Res. Commun.* **1991**, *175*, 298.
[33] Schweizer, M. P.; Banta, E. B.; Witkowski, J. T.; Robins, R. K. *J. Am. Chem. Soc.* **1973**, *95*, 3770.
[33a] Kreishman, G. P.; Witkowski, J. T.; Robins, R. K.; Schweizer, M. P. *J. Am. Chem. Soc.* **1972**, *94*, 4894.
[34] Akhrem, A. A.; Mikhailopulo, I. A.; Abramov, A. F. *Org. Magn. Reson.* **1979**, *12*, 247.
[35] Serianni, A. S.; Kline, P. C.; Snyder, J. R. *J. Am. Chem. Soc.* **1990**, *112*, 5886.
[36] Kline, P. C.; Serianni, A. S. *J. Am. Chem. Soc.* **1990**, *112*, 7373.
[37] Rosemeyer, H.; Tóth, G.; Golankiewicz, B.; Kazimierczuk, Z.; Bourgeois, W.; Kretschmer, U.; Muth, H.-P.; Seela, F. *J. Org. Chem.* **1990**, *55*, 5784.

[38] Cao, X; Pfleiderer, W.; Rosemeyer, H.; Seela, F.; Bannwarth, W.; Schönholzer, P. *Helv. Chim. Acta* **1992**, *75*, 1267.
[39] Hanessian, S.; Pernet, A. G. *Adv. Carbohydr. Chem. Biochem.* **1976**, *33*, 111.
[40] James, S. R. *J. Carbohydr., Nucleosides, Nucleotides* **1979**, *6*, 417.
[41] Borthwick, A. D.; Biggadike, K. *Tetrahedron* **1992**, *48*, 571.
[42] Postema, M. H. D. *Tetrahedron* **1992**, *48*, 8545.
[43] Agrofoglio, L.; Suhas, E.; Farese, A.; Condom, R.; Challand, S. R.; Earl, R. A.; Guedj, R. *Tetrahedron* **1994**, *36*, 10611.
[44] Secrist, III, J. A. *Org. React.*, in preparation.
[45] Altmann, K.-H. *Tetrahedron Lett.* **1993**, *34*, 7721.
[45a] Huang, B.; Chen, B.; Hui, Y. *Synthesis* **1993**, 769.
[45b] Altmann, K.-H.; Freier, S. M.; Pieles, U.; Winkler, T. *Angew. Chem., Int. Ed. Engl.* **1994**, *33*, 1654.
[45c] Pickering, L.; Malhi, B. S.; Coe, P. L.; Walker, R. T. *Tetrahedron* **1995**, *51*, 2719.
[46] Zinner, H. *Chem. Ber.* **1950**, *83*, 153.
[47] Guthrie, R. D.; Smith, S. C. *Chem. Ind.* **1968**, 547.
[48] Recondo, E. F.; Rinderknecht, H. *Helv. Chim. Acta* **1959**, *42*, 1171.
[49] Yung, N.; Fox, J. J. In *Methods in Carbohydrate Chemistry*; Academic: New York; 1963, Vol. 2, p. 108.
[50] Stevens, J. D.; Ness, R. K.; Fletcher, Jr., H. G. *J. Org. Chem.* **1968**, *33*, 1806.
[51] Hoffer, M. *Chem. Ber.* **1960**, *93*, 2777.
[52] Barker, R.; Fletcher, Jr., H. G. *J. Org. Chem.* **1961**, *26*, 4605.
[53] Glaudemans, C. P. J.; Fletcher, Jr., H. G. *J. Org. Chem.* **1971**, *36*, 3598.
[54] Sato, T. In *Synthetic Procedures in Nucleic Acid Chemistry*; Zorbach, W. W., Tipson, R. S., Eds.; Wiley-Interscience: New York; 1968; p. 264.
[55] Diekmann, E.; Friedrich, K.; Fritz, H.-G. *J. Prakt. Chem.* **1993**, *335*, 415.
[56] Fischer, E.; Helferich, B. *Chem. Ber.* **1914**, *47*, 210.
[57] Ulbricht, T. L. V. *Proc. Chem. Soc.* **1962**, 298.
[58] Thacker, D.; Ulbricht, T. L. V. *J. Chem. Soc. C* **1968**, 333.
[59] Ukita, T.; Hayatsu, H.; Tomita, Y. *Chem. Pharm. Bull.* **1963**, *11*, 1068.
[60] Prystaš, M.; Šorm, F. *Collect. Czech. Chem. Commun.* **1969**, *34*, 2316.
[61] Miyaki, M.; Shimizu, B. *Chem. Pharm. Bull.* **1970**, *18*, 1446.
[61a] Watanabe, K. A.; Hollenberg, D. H.; Fox, J. J. *J. Carbohydr., Nucleosides, Nucleotides* **1974**, *1*, 1.
[62] Škoda, J.; Bartošek, I.; Šorm, F. *Collect. Czech. Chem. Commun.* **1962**, *27*, 906.
[63] Novák, J. J. K.; Šorm, F. *Collect. Czech. Chem. Commun.* **1962**, *27*, 902.
[64] Kazimierczuk, Z.; Cottam, H. B.; Revankar, G. R.; Robins, R. K. *J. Am. Chem. Soc.* **1984**, *106*, 6379.
[65] Kawakami, H.; Matsushita, H.; Naoi, Y.; Itoh, K.; Yoshikoshi, H. *Chem. Lett.* **1989**, 235.
[66] Seela, F.; Winkeler, H.-D. *J. Org. Chem.* **1982**, *47*, 226.
[67] Seela, F.; Gumbiowski, R. *Heterocycles* **1989**, *29*, 795.
[68] Ramasamy, K.; Robins, R. K.; Revankar, G. R. *J. Heterocycl. Chem.* **1987**, *24*, 863.
[69] Hilbert, G. E.; Johnson, T. B. *J. Am. Chem. Soc.* **1930**, *52*, 2001.
[70] Hilbert, G. E.; Johnson, T. B. *J. Am. Chem. Soc.* **1930**, *52*, 4489.
[71] Hilbert, G. E.; Jansen, E. F. *J. Am. Chem. Soc.* **1936**, *58*, 60.
[72] Prystaš, M.; Farkaš, J.; Šorm, F. *Collect. Czech. Chem. Commun.* **1965**, *30*, 3123.
[73] Birkofer, L.; Ritter, A.; Kühlthau, H. P. *Angew. Chem.* **1963**, *4*, 209.
[74] Birkofer, L.; Ritter, A.; Kühlthau, H. P. *Chem. Ber.* **1964**, *97*, 934.
[75] Nishimura, T.; Shimizu, B.; Iwai, I. *Chem. Pharm. Bull.* **1963**, *11*, 1470.
[76] Nishimura, T.; Iwai, I. *Chem. Pharm. Bull.* **1964**, *12*, 357.
[77] Wittenburg, E. *Z. Chem.* **1964**, *4*, 303.
[78] Colvin, E. N. *Silicon in Organic Synthesis*; Butterworths: London, 1981; p. 4.
[78a] Klebe, J. F. *Acc. Chem. Res.* **1970**, *3*, 299.
[79] Vorbrüggen, H.; Krolikiewicz, K.; Niedballa, U. *Liebigs Ann. Chem.* **1975**, 988.

[80] Wittenburg, E. *Chem. Ber.* **1968**, *101*, 1095.
[81] Vorbrüggen, H.; Strehlke, P. *Chem. Ber.* **1973**, *106*, 3039.
[82] Wanagat, U.; Liehr, W. *Angew. Chem.* **1957**, *24*, 783.
[83] Niedballa, U.; Vorbrüggen, H. *Angew. Chem., Int. Ed. Engl.* **1970**, *9*, 461.
[84] Niedballa, U.; Vorbrüggen, H. *J. Org. Chem.* **1974**, *39*, 3654.
[85] Baker, B. R.; Schaub, R. E.; Kissman, H. M. *J. Am. Chem. Soc.* **1955**, *77*, 5911.
[86] Furukawa, Y.; Honjo, M. *Chem. Pharm. Bull.* **1968**, *16*, 1076.
[86a] Tiwari, K. N.; Secrist III, J. A.; Montgomery, J. A. *Nucleosides & Nucleotides* **1994**, *13*, 1819.
[86b] Peterli-Roth, P.; Maguire, M. P.; León, E.; Rapoport, H. *J. Org. Chem.* **1994**, *59*, 4186.
[87] Mukaiyama, T. *Angew. Chem., Int. Ed. Engl.* **1977**, *16*, 817.
[88] Prystaš, M.; Šorm, F. *Collect. Czech. Chem. Commun.* **1965**, *30*, 1900.
[89] Winkley, M. W.; Robins, R. K. *J. Org. Chem.* **1968**, *33*, 2822.
[89a] Niedballa, U.; Vorbrüggen, H. *J. Org. Chem.* **1974**, *39*, 3660.
[90] Niedballa, U.; Vorbrüggen, H. *J. Org. Chem.* **1976**, *41*, 2084.
[91] Yoshimura, Y.; Sano, T.; Matsuda, A.; Ueda, T. *Chem. Pharm. Bull.* **1988**, *36*, 162.
[92] Yoshimura, Y.; Matsuda, A.; Ueda, T. *Nucleosides & Nucleotides* **1988**, *7*, 409.
[93] Yoshimura, Y.; Matsuda, A.; Ueda, T. *Chem. Pharm. Bull.* **1989**, *37*, 660.
[94] El Subbagh, H. I.; Ping, L.-J.; Abushanab, E. *Nucleosides & Nucleotides* **1992**, *11*, 603.
[95] Hsu, L.-Y.; Wise, D. S.; Kucera, L. S.; Drach, J. C.; Townsend, L. B. *J. Org. Chem.* **1992**, *57*, 3354.
[96] Jung, M. E.; Castro, C. *J. Org. Chem.* **1993**, *58*, 807.
[97] Sujino, K.; Sugimura, H. *Chem. Lett.* **1993**, 1187.
[97a] Bévierre, M.-O.; De Mesmaeker, A.; Wolf, R. M.; Freier, S. M. *Bioorg. Med. Chem. Lett.* **1994**, *4*, 237.
[97b] Chemla, P. *Tetrahedron Lett.* **1993**, *34*, 7391.
[97c] Lipshutz, B. H.; Hayakawa, H.; Kato, K.; Lowe, R. F.; Stevens, K. L. *Synthesis* **1994**, *12*, 1476.
[97d] Sujino, K.; Sugimura, H. *J. Chem. Soc., Chem. Commun.* **1994**, 2541.
[97e] Sujino, K.; Sugimura, H. *Tetrahedron Lett.* **1994**, *35*, 1883.
[97f] Danel, K.; Pedersen, E. B.; Nielsen, C. *J. Med. Chem.* **1998**, *41*, 191.
[98] Niedballa, U.; Vorbrüggen, H. *J. Org. Chem.* **1974**, *39*, 3664.
[99] Niedballa, U.; Vorbrüggen, H. *J. Org. Chem.* **1974**, *39*, 3668.
[100] Lichtenthaler, F. W.; Voss, P.; Heerd, A. *Tetrahedron Lett.* **1974**, 2141.
[101] Akhrem, A. A.; Adarich, E. K.; Kulinkovich, L. N.; Mikhailopulo, I. A.; Posshchast'eva, E. B.; Timoshchuk, V. A. *Dokl. Akad. Nauk SSSR* **1974**, *219*, 99 [*Chem. Abstr.* **1975**, *82*, 086532].
[102] Ogawa, T.; Yasui, M.; Matsui, M. *Agric. Biol. Chem.* **1972**, *36*, 913.
[103] Ienaga, K.; Pfleiderer, W. *Chem. Ber.* **1977**, *110*, 3449.
[104] Ritzmann, G.; Ienaga, K.; Pfleiderer, W. *Liebigs Ann. Chem.* **1977**, 1217.
[105] Suzuki, T.; Sakaguchi, S.; Myata, Y.; Mori, T. JP 01139596, 1989 [*Chem. Abstr.* **1990**, *112*, 36386a].
[106] Mahmood, K.; Vasella, A.; Bernet, B. *Helv. Chim. Acta* **1991**, *74*, 1555.
[106a] Mukaiyama, T.; Matsutani, T.; Shimomura, N. *Chem. Lett.* **1994**, 2089.
[106b] Mukaiyama, T.; Nagai, M.; Matsutani, T.; Shimomura, N. *Nucleosides & Nucleotides* **1996**, *15*, 17.
[106c] Nagai, M.; Matsutani, T.; Mukaiyama, T. *Heterocycles* **1996**, *42*, 57.
[107] Kondo, T.; Nakai, H.; Goto, T. *Agric. Biol. Chem.* **1971**, *35*, 1990.
[108] Kondo, T.; Nakai, H.; Goto, T. *Tetrahedron* **1973**, *29*, 1801.
[109] Cadet, J. *J. Carbohydr., Nucleosides, Nucleotides* **1975**, *2*, 459.
[110] Ott, M.; Pfleiderer, W. *Chem. Ber.* **1974**, *107*, 339.
[111] Autenrieth, D.; Schmid, H.; Ott, M.; Pfleiderer, W. *Liebigs Ann. Chem.* **1977**, 1194.
[112] Chow, K.; Danishefsky, S. *J. Org. Chem.* **1990**, *55*, 4211.
[113] Kim, Y. H.; Kim, J. Y. *Heterocycles* **1988**, *27*, 71.
[114] Okabe, M.; Sun, R.-C.; Tam, S. Y.-K.; Todaro, L. J.; Coffen, D. L. *J. Org. Chem.* **1988**, *53*, 4780.

[115] Okabe, M.; Sun, R.-C. *Tetrahedron Lett.* **1989**, *30*, 2203.
[116] Beard, A. B.; Butler, P. I.; Mann, J.; Partlett, N. K. *Carbohydr. Res.* **1990**, *205*, 87.
[117] Faivre-Buet, V.; Grouiller, A.; Descotes, G. *Nucleosides & Nucleotides* **1992**, *11*, 1651.
[118] Noyori, R.; Hayashi, M. *Chem. Lett.* **1987**, 57.
[118a] Lloyd, A. E.; Coe, P. L.; Walker, R. T.; Howarth, O. W. *J. Fluor. Chem.* **1993**, *60*, 239.
[119] Mukaiyama, T.; Shimpuku, T.; Takashima, T.; Kobayashi, S. *Chem. Lett.* **1989**, 145.
[120] Mukaiyama, T.; Uchiro, H.; Shiina, I.; Kobayashi, S. *Chem. Lett.* **1990**, 1019.
[121] Matsumoto, T.; Katsuki, M.; Suzuki, K. *Tetrahedron Lett.* **1989**, *30*, 833.
[122] Mukaiyama, T.; Kobayashi, S.; Shoda, S. *Chem. Lett.* **1984**, 907.
[123] Herscovici, J.; Montserret, R.; Antonakis, K. *Carbohydr. Res.* **1988**, *176*, 219.
[124] Mukaiyama, T.; Matsutani, T.; Shimomura, N. *Chem. Lett.* **1993**, 1627.
[125] Vorbrüggen, H.; Krolikiewicz, K. *Angew. Chem., Int. Ed. Engl.* **1975**, *14*, 255.
[126] Vorbrüggen, H.; Krolikiewicz, K. *Liebigs Ann. Chem.* **1980**, 1438.
[127] Carbon, J.; David, H.; Studier, M. H. *Science* **1968**, *161*, 1146.
[128] Vorbrüggen, H.; Krolikiewicz, K. *Angew. Chem., Int. Ed. Engl.* **1975**, *14*, 818.
[129] Hamada, Y.; Kato, S.; Shioiri, T. *Tetrahedron Lett.* **1985**, *26*, 3223.
[130] Hamada, Y.; Shioiri, T. *J. Org. Chem.* **1986**, *51*, 5489.
[131] Marsmann, H. C.; Horn, H. G. *Z. Naturforsch.* **1972**, *27b*, 1448.
[132] Vorbrüggen, H.; Krolikiewicz, K. *Angew. Chem., Int. Ed. Engl.* **1975**, *14*, 421.
[133] Vorbrüggen, H.; Krolikiewicz, K.; Bennua, B. *Chem. Ber.* **1981**, *114*, 1234.
[133a] Mahling, J.-A.; Schmidt, R. R. *Synthesis* **1993**, 325.
[133b] Mahling, J.-A.; Jung, K.-H.; Schmidt, R. R. *Liebigs Ann. Chem.* **1995**, 461.
[133c] Mahling, J.-A.; Schmidt, R. R. *Liebigs Ann. Chem.* **1995**, 467.
[134] Olah, G. A.; Narang, S. C.; Balaram Gupta, B. G.; Malhotra, R. *J. Org. Chem.* **1979**, *44*, 1247.
[135] Matsuda, A.; Kurasawa, Y.; Watanabe, K. A. *Synthesis* **1981**, 748.
[136] Tocik, Z.; Earl, R. A.; Beránek, J. *Nucleic Acids Res.* **1980**, *8*, 4755.
[137] Tocik, Z.; Beránek, J. *Nucleic Acids Res., Symp. Ser. 9* **1981**, 37.
[138] Huang, Z.; Schneider, K. C.; Benner, S. A. *J. Org. Chem.* **1991**, *56*, 3869.
[139] Kobylinskaya, V. I.; Dashevskaya, T. A.; Shalamai, A. S.; Levitskaya, Z. V. *Zh. Obshch. Khim.* **1992**, *62*, 1115 [*Chem. Abstr.* **1993**, *118*, 39326r].
[139a] Ohsawa, K.; Shiozawa, T.; Achiwa, K.; Terao, Y. *Chem. Pharm. Bull.* **1993**, *41*, 1906.
[139b] Murata, S.; Noyori, R. *Tetrahedron Lett.* **1980**, *21*, 767.
[139c] Dwyer, O. *Synlett* **1995**, 1163.
[139d] Tse, A.; Mansour, T. S. *Tetrahedron Lett.* **1995**, *36*, 7807.
[140] Zou, R.; Robins, M. J. *Can. J. Chem.* **1987**, *65*, 1436.
[141] Robins, M. J.; Zou, R.; Hansske, F.; Madej, D.; Tyrrell, D. L. J. *Nucleosides & Nucleotides* **1989**, *8*, 725.
[141a] Robins, M. J.; Zou, R.; Guo, Z.; Wnuk, S. F. *J. Org. Chem.* **1996**, *61*, 9207.
[142] Garner, P.; Ramakanth, S. *J. Org. Chem.* **1988**, *53*, 1294.
[143] Raju, N.; Robins, R. K.; Vaghefi, M. M. *J. Chem. Soc., Chem. Commun.* **1989**, 1769.
[143a] El-Kattan, Y.; Gosselin, G.; Imbach, J.-L. *Bull. Soc. Chim. Fr.* **1994**, *131*, 118.
[143b] El-Kattan, Y.; Gosselin, G.; Imbach, J.-L. *J. Chem. Soc., Perkin Trans. 1* **1994**, 1289.
[144] Clausen, F. P.; Juhl-Christensen, J. *Org. Prep. Proced. Int.* **1993**, *25*, 375.
[145] Ferrier, R. J. *Top. Curr. Chem.* **1970**, *14*, 389.
[146] Vorbrüggen, H.; Bennua, B. unpublished results.
[147] Altermatt, R.; Tamm, C. *Helv. Chim. Acta* **1985**, *68*, 475.
[148] Walczak, K.; Pedersen, E. B. *Acta Chem. Scand.* **1991**, *45*, 930.
[149] Efange, S. M. N.; Dutta, A. K. *Nucleosides & Nucleotides* **1991**, *10*, 1451.
[150] Prystaš, M.; Šorm, F. *Collect. Czech. Chem. Commun.* **1964**, *29*, 131.
[151] Yamaguchi, T.; Saneyoshi, M. *Chem. Pharm. Bull.* **1984**, *32*, 1441.
[151a] Koga, M.; Wilk, A.; Moore, M. F.; Scremin, C. L.; Zhou, L.; Beaucage, S. L. *J. Org. Chem.* **1995**, *60*, 1520.
[152] Ward, D. I.; Jeffs, S. M.; Coe, P. L.; Walker, R. T. *Tetrahedron Lett.* **1993**, *34*, 6779.
[152a] Ward, D. I.; Coe, P. L.; Walker, R. T. *Collect. Czech. Chem. Commun., Special Issue* **1993**, *58*, 1.

[152b] Larsen, E.; Jørgensen, P. T.; Sofan, M. A.; Pedersen, E. B. *Synthesis* **1994**, 1037.
[152c] Cameron, M. A.; Cush, S. B.; Hammer, R. P. *J. Org. Chem.* **1997**, *62*, 9065.
[153] Dyatkina, N. B.; Arzumanov, A. A. In *Biophosphates and Their Analogues-Synthesis, Structure, Metabolism, and Activity*; Bruzik, K. S., Stec, W. J., Eds.; Elsevier: Amsterdam; 1987; p. 395.
[154] Danishefsky, S.; Hungate, R. *J. Am. Chem. Soc.* **1986**, *108*, 2486.
[155] Danishefsky, S.; Hungate, R.; Schulte, G. *J. Am. Chem. Soc.* **1988**, *1110*, 7434.
[156] Kozaki, S.; Sakanaka, O.; Yasuda, T.; Shimizu, T.; Ogawa, S.; Suami, T. *J. Org. Chem.* **1988**, *53*, 281.
[157] Sakanaka, O.; Ohmori, T.; Kozaki, S.; Suami, T. *Bull. Chem. Soc. Jpn.* **1987**, *60*, 1057.
[158] Maier, S.; Preuss, R.; Schmidt, R. R. *Liebigs Ann. Chem.* **1990**, 483.
[159] Ikemoto, N.; Schreiber, S. L. *J. Am. Chem. Soc.* **1990**, *112*, 9657.
[160] Ikemoto, N.; Schreiber, S. L. *J. Am. Chem. Soc.* **1992**, *114*, 2524.
[160a] Knapp, S. *Chem. Rev.* **1995**, *95*, 1859.
[161] Ferris, J. P.; Devadas, B.; Huang, C.-H.; Ren, W.-Y. *J. Org. Chem.* **1985**, *50*, 747.
[162] Marquez, V. E. *Nucleosides & Nucleotides* **1983**, *2*, 81.
[162a] Ochoa, C.; Provencio, R.; Jimeno, M. L.; Balzarini, J.; De Clercq, E. *Nucleosides & Nucleotides* **1998**, *17*, 901.
[163] Miyaki, M.; Saito, A.; Shimizu, B. *Chem. Pharm. Bull.* **1970**, *18*, 2459.
[164] Lichtenthaler, F. W.; Kitahara, K. *Angew. Chem., Int. Ed. Engl.* **1975**, *14*, 815.
[165] Azuma, T.; Isono, K.; Crain, P. F.; McCloskey, J. A. *Tetrahedron Lett.* **1976**, 1687.
[166] Azuma, T.; Isono, K. *Chem. Pharm. Bull.* **1977**, *25*, 3347.
[167] Azuma, T.; Isono, K.; Crain, P. F.; McCloskey, J. A. *J. Chem. Soc., Chem. Commun.* **1977**, 159.
[168] Mizuno, Y.; Tsuchida, K.; Tampo, H. *Chem. Pharm. Bull.* **1984**, *32*, 2915.
[169] Murofushi, Y.; Kimura, M.; Kuwano, H.; Iijima, Y.; Yamazaki, M.; Kaneko, M. *Nucleic Acids Res., Symp. Ser. 17* **1986**, p. 45.
[170] Imazawa, M.; Eckstein, F. *J. Org. Chem.* **1978**, *43*, 3044.
[171] Robins, M. J.; Wood, S. G.; Dalley, N. K.; Herdewijn, P.; Balzarini, J.; De Clercq, E. *J. Med. Chem.* **1989**, *32*, 1763.
[172] Eckstein, F.; Hunsmann, G.; Hartmann, H. EP 286825, 1988 [*Chem. Abstr.* **1989**, *110*, 18537e].
[173] Eckstein, F.; Hunsmann, G.; Hartmann, H. DE 3,639,780, 1988 [*Chem. Abstr.* **1988**, *109*, 170808f].
[174] Matthes, E.; Lehmann, C.; Scholz, D.; Von, J. L.; Gaertner, K.; Langen, P.; Rosenthal, H. A. EP 254268, 1988 [*Chem. Abstr.* **1988**, *109*, 55173h].
[175] Kowollik, G.; Langen, P.; Kvasyuk, E. I.; Mikhailopulo, I. A. DD 158903, 1983 [*Chem. Abstr.* **1983**, *99*, 158785p].
[176] Dyatkina, N. B.; Alexandrova, L. A.; von Janta-Lipinski, M.; Langen, P. *Z. Chem.* **1985**, *25*, 180.
[177] Atlas, A. EP 322384, 1989; JP 01,151595 [*Chem. Abstr.* **1990**, *112*, 36387b].
[178] Suzuki, T.; Sakaguchi, S.; Myata, Y.; Mori, T. JP 01151595, 1989 [*Chem. Abstr.* **1990**, *112*, 36387b].
[179] Imazawa, M.; Eckstein, F. *J. Org. Chem.* **1979**, *44*, 2039.
[180] Sugiura, Y.; Furuya, S.; Furukawa, Y. *Chem. Pharm. Bull.* **1988**, *36*, 3253.
[181] Boryski, J.; Golankiewicz, B. *Nucleosides & Nucleotides* **1989**, *8*, 529.
[181a] Boryski, J. *Collect. Czech. Chem. Commun., Special Issue* **1993**, *58*, 3.
[182] Shiragami, H.; Koguchi, Y.; Izawa, K. EP 0532878, 1993 [*Chem. Abstr.* **1993**, *119*, 226348g].
[183] Kato, K.; Minami, T.; Takita, T.; Nishiyama, S.; Yamamura, S.; Naganawa, H. *Tetrahedron Lett.* **1989**, *30*, 2269.
[184] Neumann, A.; Cech, D.; Schubert, F. *Z. Chem.* **1989**, *29*, 209.
[185] Spadari, S.; Maga, G.; Focher, F.; Ciarrocchi, G.; Manservigi, R.; Arcamone, F.; Capobianco, M.; Carcuro, A.; Colonna, F.; Iotti, S.; Garbesi, A. *J. Med. Chem.* **1992**, *35*, 4214.
[186] Roig, V.; Kurfurst, R.; Thuong, N. T. *Tetrahedron Lett.* **1993**, *34*, 1601.
[186a] Boryski, J. *Nucleosides & Nucleotides* **1996**, *15*, 771.

[187] Cooper, B. E. *Chem. Ind.* **1978**, 794.
[188] Vorbrüggen, H.; Bennua, B. *Tetrahedron Lett.* **1978**, 1339.
[189] Vorbrüggen, H.; Bennua, B. *Chem. Ber.* **1981**, *114*, 1279.
[190] Dudycz, L. W.; Wright, G. E. *Nucleosides & Nucleotides* **1984**, *3*, 33.
[191] Su, T. L.; Bennua, B.; Vorbrüggen, H.; Lindner, H. J. *Chem. Ber.* **1981**, *114*, 1269.
[192] Böhringer, M.; Roth, H.-J.; Hunziker, J.; Göbel, M.; Krishnan, R.; Giger, A.; Schweizer, B.; Schreiber, J.; Leumann, C.; Eschenmoser, A. *Helv. Chim. Acta* **1992**, *75*, 1416.
[192a] Saneyoshi, M.; Satoh, E. *Chem. Pharm. Bull.* **1979**, *27*, 2518.
[192b] Nakayama, C.; Saneyoshi, M. *Nucleosides & Nucleotides* **1982**, *1*, 139.
[192c] Iyer, V. K.; Butler, W. B.; Horwitz, J. P.; Rozhin, J.; Brooks, S. C.; Corombos, J.; Kessel, D. *J. Med. Chem.* **1983**, *26*, 162.
[192d] Sinhababu, A. K.; Bartel, R. L.; Pochopin, N.; Borchardt, R. T. *J. Am. Chem. Soc.* **1985**, *107*, 7628.
[192e] Gosselin, G.; Bergogne, M.-C.; De Rudder, J.; De Clercq, E.; Imbach, J.-L. *J. Med. Chem.* **1986**, *29*, 203.
[192f] Gosselin, G.; Bergogne, M.-C.; Imbach, J.-L.; De Rudder, J.; De Clercq, E. *Nucleosides & Nucleotides* **1987**, *6*, 65.
[192g] Maguire, M. P.; Feldman, P. L.; Rapoport, H. *J. Org. Chem.* **1990**, *55*, 948.
[192h] Périgaud, C.; Gosselin, G.; Imbach, J.-L. *J. Chem. Soc., Perkin Trans. 1* **1992**, 1943.
[192i] Gosselin, G.; Bergogne, M.-C.; Imbach, J.-L. *J. Heterocycl. Chem.* **1993**, *30*, 1229.
[192j] Serra, C.; Dewynter, G.; Montero, J.-L.; Imbach, J.-L. *Tetrahedron* **1994**, *50*, 8427.
[192k] Papageorgiou, C.; Tamm, C. *Tetrahedron Lett.* **1986**, 555.
[192l] Ghosh, A. K.; Liu, W. *J. Org. Chem.* **1996**, *61*, 6175.
[192m] Hřebabecký, H.; Dočkal, J.; Holý, A. *Collect. Czech. Chem. Commun.* **1994**, *59*, 1408.
[192n] Hřebabecký, H.; Holý, A. *Collect. Czech. Chem. Commun.* **1994**, *59*, 1654.
[192o] Sheppard, T. L.; Rosenblatt, A. T.; Breslow, R. *J. Org. Chem.* **1994**, *59*, 7243.
[192p] Gosselin, G.; Imbach, J.-L. *J. Heterocycl. Chem.* **1982**, *19*, 597.
[193] Vorbrüggen, H.; Niedballa, U.; Krolikiewicz, K.; Bennua, B.; Höfle, G. In *Chemistry and Biology of Nucleosides and Nucleotides*; Harmon, R. E., Robins, R. K., Townsend, L. B., Eds.; Academic: New York, 1978; p. 251.
[194] Vorbrüggen, H.; Höfle, G. *Chem. Ber.* **1981**, *114*, 1256.
[195] Anderson, C. B.; Friedrich, E. C.; Winstein, S. *Tetrahedron Lett.* **1963**, 2037.
[196] Hart, H.; Tomalia, D. A. *Tetrahedron Lett.* **1966**, 3383.
[197] Paulsen, H.; Meyborg, H. *Chem. Ber.* **1975**, *108*, 3176.
[198] Paulsen, H. *Adv. Carbohydr. Chem. Biochem.* **1971**, *26*, 127.
[198a] Doboszewski, B.; Herdewijn, P. *Tetrahedron* **1996**, *52*, 1651.
[199] Krimen, L. I.; Cota, D. J. *Org. React.* **1969**, *17*, 213.
[200] Klemer, A.; Kohla, M. *J. Carbohydr. Chem.* **1988**, *7*, 785.
[201] Martinez, A. G.; Alvarez, R. M.; Vilar, E. T.; Fraile, A. G.; Hanack, M.; Subramanian, L. R. *Tetrahedron Lett.* **1989**, *30*, 581.
[202] Ratcliffe, A. J.; Fraser-Reid, B. *J. Chem. Soc., Perkin Trans. 1* **1989**, 1805.
[203] Ratcliffe, A. J.; Fraser-Reid, B. *J. Chem. Soc., Perkin Trans. 1* **1990**, 747.
[204] Heinemann, F.; Hiegemann, M.; Welzel, P. *Tetrahedron* **1992**, *48*, 3781.
[205] Baker, B. R.; Joseph, J. P.; Schaub, R. E.; Williams, J. H. *J. Org. Chem.* **1954**, *19*, 1786.
[206] Ogawa, T.; Beppu, K.; Nakabayashi, S. *Carbohydr. Res.* **1981**, *93*, C6.
[207] Toshima, K.; Tatsuta, K. *Chem. Rev.* **1993**, *93*, 1503.
[208] Kimura, Y.; Suzuki, M.; Matsumoto, T.; Abe, R.; Terashima, S. *Chem.* Lett. **1984**, 501.
[208a] Terashima, S. *Synlett* **1992**, 691.
[209] Kolar, C.; Kneissl, G. *Angew. Chem., Int. Ed. Engl.* **1990**, *29*, 809.
[209a] Roush, W. R.; Briner, K.; Sebesta, D. P. *Synlett* **1993**, 264.
[209b] Sliedregt, L. A. J. M.; van der Marel, G. A.; van Boom, J. H. *Tetrahedron Lett.* **1994**, *35*, 4015.
[209c] Teng, K.; Cook, P. D. *J. Org. Chem.* **1994**, *59*, 278.
[209d] Ludin, C.; Weller, T.; Seitz, B.; Meier, W.; Erbeck, S.; Hoenke, C.; Krieger, R.; Keller, M.; Knothe, L.; Pelz, K.; Wittmer, A.; Prinzbach, H. *Liebigs Ann. Chem.* **1995**, 291.

[210] Kochetkov, N. K.; Khorlin, A. J.; Bochkov, A. F. *Tetrahedron* **1967**, *23*, 693.
[211] Kochetkov, N. K.; Bochkov, A. F. *Recent Dev. Chem. Nat. Carbon Compd.* **1971**, *4*, 75 [*Chem. Abstr.* **1972**, *76*, 599144].
[212] Shone, R. L. *Tetrahedron Lett.* **1977**, 993.
[212a] Shone, R. L. *Tetrahedron Lett.* **1977**, 4017.
[213] Prisbe, E. J.; Verheyden, J. P. H.; Moffatt, J. G. *J. Org. Chem.* **1978**, *43*, 4784.
[213a] Frame, A. S.; Wightman, R. H.; Mackenzie, G. *Tetrahedron* **1996**, *52*, 9219.
[214] Riley, T. A.; Larson, S. B.; Avery, T. L.; Finch, R. A.; Robins, R. K. *J. Med. Chem.* **1990**, *33*, 572.
[215] Ototani, N.; Whistler, R. L. *J. Med. Chem.* **1974**, *17*, 535.
[216] Brink, A. J.; Jordaan, A. *Carbohydr. Res.* **1975**, *41*, 355.
[217] Brink, A. J.; de Villiers, O. G.; Jordaan, A. *J. Chem. Soc., Perkin Trans. 1* **1977**, 1608.
[218] Garg, H. G.; Jeanloz, R. W. *Carbohydr. Res.* **1978**, *62*, 185.
[219] Ritzmann, G.; Klein, R. S.; Hollenberg, D. H.; Fox, J. J. *Carbohydr. Res.* **1975**, *39*, 227.
[220] Kiss, J.; D'Souza, R. *J. Carbohydr., Nucleosides, Nucleotides* **1980**, *7*, 141.
[221] Kiss, J.; D'Souza, R.; van Koeveringe, J. A.; Arnold, W. *Helv. Chim. Acta* **1982**, *65*, 1522.
[222] Mikhailov, S. N. *Bioorg. Khim.* **1984**, *10*, 940 [*Chem. Abstr.* **1984**, *102*, 7016t].
[223] Mikhailov, S. N.; Pfleiderer, W. *Synthesis* **1985**, 397.
[223a] Pfleiderer, W. Universitätsstr. 10, D-78464 Konstanz, Germany, personal communication.
[224] Jung, K.-H.; Schmidt, R. R. *Chem. Ber.* **1980**, *113*, 1775.
[225] Haines, A. H. *Tetrahedron* **1973**, *29*, 2807.
[226] Chavis, C.; Dumont, F.; Wightman, R. H.; Ziegler, J. C.; Imbach, J. L. *J. Org. Chem.* **1982**, *47*, 202.
[226a] Pudlo, J. S.; Townsend, L. B. *Nucleosides & Nucleotides* **1992**, *11*, 279.
[226b] Parmentier, G.; Schmitt, G.; Dolle, F.; Luu, B. *Tetrahedron* **1994**, *50*, 5361.
[226c] Chanteloup, L.; Thuong, N. T. *Tetrahedron Lett.* **1994**, *35*, 877.
[227] Wilson, L. J.; Liotta, D. *Tetrahedron Lett.* **1990**, *31*, 1815.
[228] Kawakami, H.; Ebata, T.; Koseki, K.; Matsushita, H.; Naoi, Y.; Itoh, K. *Chem. Lett.* **1990**, 1459.
[229] Kawakami, H.; Ebata, T.; Koseki, K.; Matsumoto, K.; Matsushita, H.; Naoi, Y.; Itoh, K. *Heterocycles* **1991**, *32*, 2451.
[229a] Kawakami, H.; Ebata, T.; Koseki, K.; Okano, K.; Matsumoto, K.; Matsushita, H. *Heterocycles* **1993**, *36*, 2765.
[230] Chu, C. K.; Babu, J. R.; Beach, J. W.; Ahn, S. K.; Huang, H.; Jeong, L. S.; Lee, S. J. *J. Org. Chem.* **1990**, *55*, 1418.
[230a] Beach, J. W.; Kim, H. O.; Jeong, L. S.; Nampalli, S.; Islam, Q.; Ahn, S. K.; Babu, J. R.; Chu, C. K. *J. Org. Chem.* **1992**, *57*, 3887.
[231] Chu, C. K.; Beach, J. W.; Babu, J. R.; Jeong, L. S.; Jeong, H. K.; Ahn, S. K.; Islam, Q.; Lee, S. J.; Chen, Y. *Nucleosides & Nucleotides* **1991**, *10*, 423.
[232] Forestier, M. A.; Ayi, A. I.; Condom, R.; Boyode, B. P.; Colin, J. N.; Selway, J.; Challand, R.; Guedj, R. *Nucleosides & Nucleotides* **1993**, *12*, 915.
[233] Nishiyama, S.; Yamamura, S.; Kato, K.; Takita, T. *Tetrahedron Lett.* **1988**, *29*, 4743.
[234] Perrin, D. D. *Dissociation Constants of Organic Bases in Aqueous Solution*; Butterworth: London, 1965.
[235] Choi, W-B.; Wilson, L. J.; Yeola, S.; Liotta, D. C.; Schinazi, R. F. *J. Am. Chem. Soc.* **1991**, *113*, 9377.
[236] Björsne, M.; Classon, B.; Kvarnström, I.; Samuelsson, B. *Tetrahedron* **1993**, *49*, 8637.
[237] Stothers, J. B. *Carbon-13 NMR Spectroscopy*; Academic: New York, 1972, p. 265.
[238] Wierenga, W.; Skulnick, H. I. *Tetrahedron Lett.* **1979**, 3631.
[239] Skulnick, H. *J. Org. Chem.* **1978**, *43*, 3188.
[240] Bassindale, A. R.; Stout, T. *Tetrahedron Lett.* **1985**, *26*, 3403.
[241] Hayashi, M.; Hirano, T.; Yaso, M.; Mizuno, K.; Ueda, T. *Chem. Pharm. Bull.* **1975**, *23*, 245.
[242] Tarumi, Y.; Moriguchi, K.; Atsumi, T. *J. Heterocycl. Chem.* **1984**, *21*, 529.
[243] Tarumi, Y.; Takebayashi, Y.; Atsumi, T. *J. Heterocycl. Chem.* **1984**, *21*, 849.
[244] Reetz, M. T.; Hüllmann, M.; Seitz, T. *Angew. Chem., Int. Ed. Engl.* **1987**, *26*, 477.

[245] Castellino, S. *J. Org. Chem.* **1990**, *55*, 5197.
[246] Shambayati, S.; Crowe, W. E.; Schreiber, S. L. *Angew. Chem., Int. Ed. Engl.* **1990**, *29*, 256.
[247] Csürös, Z.; Deák, G.; Gyurkovics, I.; Haraszthy-Papp, M.; Zára-Kaczián, E. *Acta Chim. Acad. Sci. Hung.* **1971**, *67*, 93 [*Chem. Abstr.* **1971**, *74*, 112315r].
[248] Csürös, Z.; Deák, G.; Holly, S.; Török-Kalmár, A.; Zára-Kaczián, E. *Acta Chim. Acad. Sci. Hung.* **1969**, *62*, 95 [*Chem. Abstr.* **1970**, *72*, 18051p].
[249] Beach, J. W.; Jeong, L. S.; Alves, A. J.; Pohl, D.; Kim, H. O.; Chang, C.-N.; Doong, S.-L.; Schinazi, R. F.; Cheng, Y.-C.; Chu, C. K. *J. Org. Chem.* **1992**, *57*, 2217.
[249a] Wang, W.; Jin, H.; Mansour, T. S. *Tetrahedron Lett.* **1994**, *35*, 4739.
[249b] Bednarski, K.; Dixit, D. M.; Wang, W.; Evans, C. A.; Jin, H.; Yuen, L.; Mansour, T. S. *Bioorg. Med. Chem. Lett.* **1994**, *4*, 2667.
[249c] Jin. H.; Siddiqui, M. A.; Evans, C. A.; Tse, H. L. A.; Mansour, T. S.; Goodyear, M. D.; Ravenscroft, P; Beels, C. D. *J. Org. Chem.* **1995**, *60*, 2621.
[249d] Liang, C.; Lee, D. W.; Newton, M. G.; Chu, C. K. *J. Org. Chem.* **1995**, *60*, 1546.
[249e] Milton, J.; Brand, S.; Jones, M. F.; Rayner, C. M. *Tetrahedron Lett.* **1995**, *36*, 6961.
[249f] Mansour, T. S.; Jin, H; Wang, W.; Hooker, E. U.; Ashman, C.; Cammack, N.; Salomon, H.; Belmonte, A. R.; Wainberg, M. A. *J. Med. Chem.* **1995**, *38*, 1.
[249g] Huang, J.; Rideout, J. L.; Martin, G. E. *Nucleosides & Nucleotides* **1995**, *14*, 195.
[249h] Breining, T.; Cimpoia, A. R.; Mansour, T. S.; Cammack, N.; Hopewell, P.; Ashman, C. *Heterocycles* **1995**, *41*, 87.
[249i] Mourier, N.; Trabaud, C.; Niddam, V.; Graciet, J. C.; Camplo, M.; Chermann, J.-C.; Kraus, J.-L. *Nucleosides & Nucleotides* **1996**, *15*, 1397.
[249j] Pan, B.-C.; Cheng, Y.-C.; Chu, S.-H. *J. Heterocycl. Chem.* **1997**, *34*, 909.
[249k] Du, J.; Surzhykov, S.; Lin, J. S.; Newton, M. G.; Cheng, Y.-C.; Schinazi, R. F.; Chu, C. K. *J. Med. Chem.* **1997**, *40*, 2991.
[250] Dubois, C. *Bull. Soc. Chim. Fr.* **1978**, I-143.
[251] Beattie, I. R.; Rule, L. *J. Chem. Soc.* **1964**, 3267.
[252] Schönfeld, P.; Döring, I.; Jahnke, D.; Reinheckel, H. *Z. Chem.* **1976**, *16*, 22.
[253] Bambury, R. E.; Feeley, D. T.; Lawton, G. C.; Weaver, J. M.; Wemple, J. *J. Med. Chem.* **1984**, *27*, 1613.
[254] Holý, A. *J. Carbohydr., Nucleosides, Nucleotides* **1978**, *5*, 487.
[255] Itoh, T.; Melik-Ohanjanian, R. G.; Ishikawa, I.; Kawahara, N.; Mizuno, Y.; Honma, Y.; Hozumi, M.; Ogura, H. *Chem. Pharm. Bull.* **1989**, *37*, 3184.
[256] Felczak, K.; Samulewicz, I.; Drabikowska, A. K.; Kulikowski, T.; Shugar, D. *Nucleic Acids Res., Symp. Ser. 18* **1987**, 65.
[256a] Watanabe, Y.; Uemura, A.; Ohrai, K.; Nagase, T.; Ozaki, S. *Nucleic Acids Res., Symp. Ser. 16* **1985**, 53.
[257] Barascut, J. L.; de Marichard, C. O.; Imbach, J. L. *J. Carbohydr., Nucleosides, Nucleotides* **1976**, *3*, 281.
[258] Witkowski, J. T.; Robins, R. K. *J. Org. Chem.* **1970**, *35*, 2635.
[259] Witkowski, J. T.; Robins, R. K.; Sidwell, R. W.; Simon, L. N. *J. Med. Chem.* **1972**, *15*, 1150.
[260] Kotick, M. P.; Szantay, C.; Bardos, T. J. *J. Org. Chem.* **1969**, *34*, 3806.
[261] Hubbard, A. J.; Jones, A. S.; Walker, R. T. *Nucleic Acids Res., Symp. Ser. 12* **1984**, 6827.
[262] Seela, F.; Westermann, B.; Bindig, U. *J. Chem. Soc., Perkin Trans. 1* **1988**, 697.
[263] Prystaš, M.; Šorm, F. *Collect. Czech. Chem. Commun.* **1964**, *29*, 121.
[264] Fox, J. J.; Yung, N. C.; Wempen, I.; Hoffer, M. *J. Am. Chem. Soc.* **1961**, *83*, 4066.
[265] Minnemeyer, H. J.; Clarke, P. B.; Tieckelmann, H.; Holland, J. F. *J. Med. Chem.* **1964**, *7*, 567.
[266] Bardos, T. J.; Kotick, M. P.; Szantay, C. *Tetrahedron Lett.* **1966**, 1759.
[267] Aoyama, H. *Bull. Chem. Soc. Jpn.* **1987**, *60*, 2073.
[268] Ryan, K. J.; Acton, E. M.; Goodman, L. *J. Org. Chem.* **1966**, *31*, 1181.
[269] Kawakami, H.; Ebata, T.; Koseki, K.; Matsushita, H.; Naoi, Y.; Itoh, K.; Mizutani, N. *Heterocycles* **1990**, *31*, 569.
[270] Kazimierczuk, Z.; Binding, U.; Seela, F. *Helv. Chim. Acta* **1989**, *72*, 1527.
[271] Kazimierczuk, Z.; Seela, F. *Helv. Chim. Acta* **1990**, *73*, 316.

[272] Kulikowski, T.; Shugar, D. *J. Med. Chem.* **1974**, *17*, 269.
[272a] Prystas, M. *Collect. Czech. Chem. Commun.* **1975**, *40*, 1786.
[273] Freskos, J. N. *Nucleosides & Nucleotides* **1989**, *8*, 549.
[273a] Freskos, J. N. *Nucleosides & Nucleotides* **1989**, *8*, 1075.
[274] Sanghvi, Y. S.; Hoke, G. D.; Freier, S. M.; Zounes, M. C.; Gonzalez, C.; Cummins, L.; Sasmor, H.; Cook, P. D. *Nucleic Acids Res., Symp. Ser. 21* **1993**, 3197.
[275] Peters, D.; Hörnfeldt, A.-B.; Gronowitz, S.; Johansson, N. G. *Nucleosides & Nucleotides* **1992**, *11*, 1151.
[275a] Wellmar, U.; Hörnfeldt, A.-B.; Gronowitz, S. *Nucleosides & Nucleotides* **1996**, *15*, 1059.
[275b] Michel, J.; Gueguen, G.; Vercauteren, J.; Moreau, S. *Tetrahedron* **1997**, *53*, 8457.
[276] Kawakami, H.; Ebata, T.; Koseki, K.; Matsumoto, K.; Matsushita, H.; Naoi, Y.; Itoh, K. *Heterocycles* **1990**, *31*, 2041.
[276a] Jungmann, O.; Pfleiderer, W. *Tetrahedron Lett.* **1996**, *37*, 8355.
[277] Okauchi, T.; Kubota, H.; Narasaka, K. *Chem. Lett.* **1989**, 801.
[278] Lavallée, J.-F.; Just, G. *Tetrahedron Lett.* **1991**, *32*, 3469.
[278a] Young, R. J.; Shaw-Ponter, S.; Hardy, G. W.; Mills, G. *Tetrahedron Lett.* **1994**, *35*, 8687.
[278b] Mukaiyama, T.; Hirano, N.; Nishida, M.; Uchiro, H. *Chem. Lett.* **1996**, 99.
[278c] Mukaiyama, T.; Ishikawa, T.; Uchiro, H. *Chem. Lett.* **1997**, 389.
[279] Dyatkina, N. B.; Kraevskii, A. A.; Azhaev, A. B. *Bioorg. Khim.* **1986**, *12*, 1048 [*Chem. Abstr.* **1987**, *107*, 78188q].
[280] Chu, C. K.; Beach, J. W.; Ullas, G. V.; Kosugi, Y. *Tetrahedron Lett.* **1988**, *29*, 5349.
[281] Fleet, G. W. J.; Son, J. C.; Derome, A. E. *Tetrahedron* **1988**, *44*, 625.
[282] Chu, C. K.; Raghavachari, R.; Beach, J. W.; Kosugi, Y.; Ullas, G. V. *Nucleosides & Nucleotides* **1989**, *8*, 903.
[283] Bou, V.; Vilarrasa, J. *Tetrahedron Lett.* **1990**, *31*, 567.
[284] Jung, M. E.; Gardiner, J. M. *J. Org. Chem.* **1991**, *56*, 2614.
[285] Wengel, J.; Pedersen, E. B. *Synthesis* **1991**, 451.
[286] El-Torgoman, A. M.; Motawia, M. S.; Kjærsgaard, U.; Pedersen, E. B. *Monatsh. Chem.* **1992**, *123*, 355.
[287] Greengrass, C. W.; Hoople, D. W. T.; Street, S. D. A.; Hamilton, F.; Marriott, M. S.; Bordner, J.; Dalgleish, A. G.; Mitsuya, H.; Broder, S. *J. Med. Chem.* **1989**, *32*, 618.
[288] Etzold, G.; Hintsche, R.; Kowollik, G.; Langen, P. *Tetrahedron* **1971**, *27*, 2463.
[289] Datema, R.; Lindborg, B. G.; Kovacs, Z. M. I.; Ligure, P.; Stening, G. B.; Johansson, K. N. G.; Öberg, B. F. WO 8808001, 1988 [*Chem. Abstr.* **1989**, *110*, 232026f].
[290] Mikhailopulo, I. A.; Pricota, T. I.; Poopeiko, N. E.; Klenitskaya, T. V.; Khripach, N. B. *Synthesis* **1993**, 700.
[291] Motawia, M. S.; Jørgensen, P. T.; Larnkjær, A.; Pedersen, E. B.; Nielsen, C. *Monatsh. Chem.* **1993**, *124*, 55.
[292] Häbich D.; Neumann, R. EP 0507188, 1992 [*Chem. Abstr.* **1993**, *119*, 203761x].
[293] Dueholm, K. L.; Pedersen, E. B. *Synthesis* **1992**, 1.
[294] Huryn, D. M.; Okabe, M. *Chem. Rev.* **1992**, *92*, 1745.
[295] Herdewijn, P.; Balzarini, J.; De Clercq, E. *Adv. Antiviral Drug Design* **1993**, *1*, 233.
[296] Morisawa, Y.; Yasuda, A.; Uchida, K. JP 02270893, 1990 [*Chem. Abstr.* **1991**, *114*, 143939w].
[297] Sahlberg, C. *Tetrahedron Lett.* **1992**, *33*, 679.
[298] Chu, C. K.; Ullas, G. V.; Jeong, L. S.; Ahn, S. K.; Doboszewski, B.; Lin, Z. X.; Beach, J. W.; Schinazi, R. F. *J. Med. Chem.* **1990**, *33*, 1553.
[299] Hager, M. W.; Liotta, D. C. *J. Am. Chem. Soc.* **1991**, *113*, 5117.
[300] Hager, M. W.; Liotta, D. C. *Tetrahedron Lett.* **1992**, *33*, 7083.
[301] *Drugs of the Future* **1990**, 794.
[301a] *Drugs of the Future* **1993**, 764.
[302] Hertel, L. W.; Kroin, J. S.; Misner, J. W.; Tustin, J. M. *J. Org. Chem.* **1988**, *53*, 2406.
[303] Chou, T. S.; Heath, P. C.; Patterson, L. E.; Poteet, L. M.; Lakin, R. E.; Hunt, A. H. *Synthesis* **1992**, 565.
[304] Britton, T. C.; LeTourneau, M. E. EP 0587364, 1993 [*Chem. Abstr.* **1994**, *121*, 180125t].

[305] Kawakami, H.; Matsumoto, K.; Koseki, K.; Ebata, T. WO 9222548, 1991 [*Chem. Abstr.* **1993**, *118*, 213452r].

[305a] Kawakami, H.; Ebata, T.; Koseki, K.; Matsumoto, K.; Okano, K.; Matsushita, H. *Nucleosides & Nucleotides* **1992**, *11*, 1673.

[306] Ueda, T.; Watanabe, S.-I. *Chem. Pharm. Bull.* **1985**, *33*, 3689.

[307] Bowles, W. A.; Robins, R. K. *J. Am. Chem. Soc.* **1964**, *86*, 1252.

[308] Schmidt, R. R.; Wagner, A. *Tetrahedron Lett.* **1983**, *24*, 4661.

[309] Bessodes, M.; Egron, M.-J.; Filippi, J.; Antonakis, K. *J. Chem. Soc., Perkin Trans. 1* **1990**, 3035.

[310] Pedersen, H.; Pedersen, E. B.; Nielsen, C. M. *Heterocycles* **1992**, *34*, 265.

[311] Hansen, H. B.; Pedersen, E. B.; Vestergaard, B. F. *Arch. Pharm. (Weinheim)* **1992**, *325*, 491.

[312] Sharma, A. P.; Ollapally, A. P.; Jones, W.; Lemon, T. *Nucleosides & Nucleotides* **1992**, *11*, 1009.

[312a] Doboszewski, B.; Blaton, N.; Herdewijn, P. *Tetrahedron Lett.* **1995**, *36*, 1321.

[313] Marumoto, R.; Honjo, M. *Chem. Pharm. Bull.* **1974**, *22*, 128.

[314] Starrett, J. E. Jr.; Tortolani, D. R.; Baker, D. C.; Omar, M. T.; Hebbler, A. K.; Wos, J. A.; Martin, J. C.; Mansuri, M. M. *Nucleosides & Nucleotides* **1990**, *9*, 885.

[315] Huang, H.; Chu, C. K. *Synth. Commun.* **1990**, *20*, 1039.

[316] Shiragami, H.; Uchida, Y.; Izawa, K. EP 0519464, 1992 [*Chem. Abstr.* **1993**, *119*, 9106x].

[317] Talekar, R. R.; Coe, P. L.; Walker, R. T. *Synthesis* **1993**, 303.

[318] Honjo, M.; Marumoto, R.; Kobayashi, K., JP 70039705, 1970 [*Chem. Abstr.* **1970**, *75*, 62483].

[319] Ozaki, S.; Katakami, T.; Saneyoshi, M. *Bull. Chem. Soc. Jpn.* **1977**, *50*, 2197.

[320] Adachi, T.; Iwasaki, T.; Inoue, I.; Miyoshi, M. *J. Org. Chem.* **1979**, *44*, 1404.

[321] Johansen, O.; Holan, G.; Marcuccio, S. M.; Mau, A. W.-H. *Aust. J. Chem.* **1991**, *44*, 37.

[321a] Park, M.; Rizzo, C. J. *J. Org. Chem.* **1996**, *61*, 6092.

[322] Markiewicz, W. T.; Wiewiorowski, M. *Nucleic Acids Res., Spec. Publ. 4* **1978**, 185.

[323] Markiewicz, W. T. *J. Chem. Res. (S)* **1979**, *1*, 24.

[324] Robins, M. J.; Wilson, J. S. *J. Am. Chem. Soc.* **1981**, *103*, 932.

[324a] Lessor, R. A.; Leonard, N. J. *J. Org. Chem.* **1981**, *46*, 4300.

[325] Kawasaki, A. M.; Wotring, L. L.; Townsend, L. B. *J. Med. Chem.* **1990**, *33*, 3170.

[326] De Mesmaeker, A.; Lebreton, J.; Hoffmann, P.; Freier, S. M. *Synlett* **1993**, 677.

[327] De Clercq, E.; Inoue, I.; Kondo, K. EP 381335, 1990 [*Chem. Abstr.* **1991**, *114*, 122980n].

[328] Ishido, Y.; Nakazaki, N.; Sakairi, N. *J. Chem. Soc., Perkin Trans. 1* **1979**, 2088.

[329] Ishido, Y.; Sakairi, N.; Okazaki, K.; Nakazaki, N. *J. Chem. Soc., Perkin Trans. 1* **1980**, 563.

[330] Josan, J. S.; Eastwood, F. W. *Aust. J. Chem.* **1968**, *21*, 2013.

[330a] Eastwood, F. W.; Harrington, K. J.; Josan, J. S.; Pura, J. L. *Tetrahedron Lett.* **1970**, 5223.

[331] Hanessian, S.; Bargiotti, A.; LaRue, M. *Tetrahedron Lett.* **1978**, 737.

[332] Barton, D. H. R.; Jang, D. O.; Jaszberenyi, J. Cs. *Tetrahedron Lett.* **1991**, *32*, 2569.

[333] Greenberg, S.; Moffatt, J. G. *J. Am. Chem. Soc.* **1973**, *95*, 4016.

[334] Russell, A. F.; Greenberg, S.; Moffatt, J. G. *J. Am. Chem. Soc.* **1973**, *95*, 4025.

[335] Robins, M. J.; Hansske, F.; Low, N. H.; Park, J. I. *Tetrahedron Lett.* **1984**, *25*, 367.

[335a] Manchand, P. S.; Brelica, P. S.; Holman, M. J.; Huang, T. N.; Maehr, H.; Tam, S. Y. K.; Yang, R. T. *J. Org. Chem.* **1992**, *57*, 3473.

[335b] Chen, B.-C.; Quinlan, S. L.; Reid, J. G.; Spector, R. H. *Tetrahedron Lett.* **1998**, *39*, 729.

[336] *Drugs of the Future* **1990**, *15*, 569.

[337] Ikehara, M.; Maruyama, T.; Miki, H. *J. Carbohydr., Nucleosides, Nucleotides* **1977**, *4*, 409.

[338] Fox, J. J. *Pure Appl. Chem.* **1969**, *18*, 223.

[339] Robins, M. J.; Hawrelak, S. D.; Hernández, A. E.; Wnuk, S. F. *Nucleosides & Nucleotides* **1992**, *11*, 821.

[339a] Binkley, R. W. *J. Org. Chem.* **1991**, *56*, 3892.

[339b] Krolikiewicz, K.; Vorbrüggen, H. *Nucleosides & Nucleotides* **1994**, *13*, 673.

[339c] Bennua-Skalmowski, B.; Vorbrüggen, H. *Tetrahedron Lett.* **1995**, *36*, 2611.

[340] Seela, F.; Winter, H. *Liebigs Ann. Chem.* **1991**, 105.

[341] Glaudemans, C. P. J.; Fletcher, Jr., H. G. *J. Org. Chem.* **1963**, *28*, 3004.

[342] Keller, F.; Botvinick, I. J.; Bunker, J. E. *J. Org. Chem.* **1967**, *32*, 1644.
[343] Montgomery, J. A.; Hewson, K. *J. Med. Chem.* **1969**, *12*, 498.
[344] Montgomery, J. A.; Clayton, S. D.; Shortnacy, A. T. *J. Heterocycl. Chem.* **1979**, *16*, 157.
[345] Montgomery, J. A.; Clayton, S. D.; Shortnacy, A. T. In *Nucleic Acid Chemistry*; Townsend, L. B., Tipson, R. S., Eds.; Wiley-Interscience: New York; 1986, Vol. 3; p. 156.
[346] Markovac, A.; LaMontagne, M. P. EP 461407, 1991 [*Chem. Abstr.* **1992**, *116*, 194798y].
[347] Blumbergs, P.; Khan, M. S.; Kalamas, R. L. WO 9108215, 1991 [*Chem. Abstr.* **1991**, *115*, 115026y].
[348] Seela, F.; Winkeler, H.-D. *Liebigs Ann. Chem.* **1982**, 1634.
[349] Bourgeois, W; Seela, F. *J. Chem. Soc., Perkin Trans. 1* **1991**, 279.
[350] Swayze, E. E.; Drach, J. C.; Wotring, L. L; Townsend, L. B. *J. Med. Chem.* **1997**, *40*, 771.
[351] Mackenzie, G.; Shaw, G. *J. Chem. Soc., Chem. Commun.* **1978**, 882.
[352] Kadir, K.; Mackenzie, G.; Shaw, G. *J. Chem. Soc., Perkin Trans. 1* **1980**, 2304.
[353] Mackenzie, G.; Shaw, G. In *Nucleic Acid Chemistry;* Townsend, L. B., Tipson, R. S., Eds.; Wiley-Interscience: New York; 1986, Vol. 3; p. 117.
[354] Cook, P. D. EP 043722, 1982 [*Chem. Abstr.* **1982**, *96*, 181576t].
[355] Meade, E. A.; Wotring, L. L.; Drach, J. C.; Townsend, L. B. *J. Med. Chem.* **1992**, *35*, 526.
[356] Shen, T. Y.; Lewis, H. M.; Ruyle, W. V. *J. Org. Chem.* **1965**, *30*, 835.
[357] Keller, F.; Tyrrill, A. R. *J. Org. Chem.* **1966**, *31*, 1289.
[358] Kulikowski, T.; Zawadzki, Z.; Shugar, D.; Descamps, J.; De Clercq, E. *J. Med. Chem.* **1979**, *22*, 647.
[359] Kulikowski, T.; Zawadzki, Z.; De Clercq, E.; Shugar, D. *Acta Biochim. Pol.* **1984**, *31*, 341 [*Chem. Abstr.* **1985**, *103*, 6656k].
[360] Griengl, H.; Wanek, E.; Schwarz, W.; Streicher, W.; Rosenwirth, B.; De Clercq, E. *J. Med. Chem.* **1987**, *30*, 1199.
[361] Sudhakar Rao, T.; Revankar, G. R.; Vinayak, R. S.; Robins, R. K. *J. Heterocycl. Chem.* **1991**, *28*, 1779.
[362] Sudhakar Rao, T.; Revankar, G. R.; Vinayak, R. S.; Robins, R. K. *J. Heterocycl. Chem.* **1992**, *29*, 343.
[363] Kulikowski, T.; Zawadski, Z.; Shugar, D. *Nucleic Acids Res., Symp. Ser. 9* **1981**, p. 103.
[364] Harnden, M. R.; Jarvest, R. L. *Nucleosides & Nucleotides* **1985**, *4*, 465.
[365] Adams, A. D.; Petrie, C. R.; Meyer, Jr., R. B. *Nucleic Acids Res.* **1991**, *19*, 3647.
[366] Jørgensen, P. T.; El-Barbary, A. A.; Pedersen, E. B. *Liebigs Ann. Chem.* **1993**, 1.
[367] Sharma, A. P.; Blair, M.; Ollapally, A. P. *Nucleosides & Nucleotides* **1990**, *9*, 713.
[368] Ozola, V.; Ramzaeva, N.; Maurinsh, Y.; Lidaks, M. *Nucleosides & Nucleotides* **1993**, *12*, 479.
[369] Kumar, P.; Wiebe, L. I. *Nucleosides & Nucleotides* **1990**, *9*, 861.
[370] Jørgensen, P. T.; Pedersen, E. B.; Nielsen, C. *Synthesis* **1992**, 1299.
[370a] Sofan, M. A.; Abdel-Megied, A. E.-S.; Pedersen, M. B.; Pedersen, E. B.; Nielsen, C. *Synthesis* **1994**, 516.
[370b] Janardhanam, S.; Nambiar, K. P. *Tetrahedron Lett.* **1994**, *35*, 3657.
[370c] Gallo, M. A.; Espinosa, A.; Campos, J.; Entrena, A., Domínguez, J. F.; Camacho, E., Pineda, M. J.; Gómez, J. A. *Synlett* **1993**, 389.
[370d] El-Barbary, A. A.; El-Brollosy, N.; Abdel-Bary, H.; Pedersen, E. B.; Stein, S.; Nielsen, C. *Liebigs Ann. Chem.* **1995**, 1371.
[370e] El-Barbary, A. A.; El-Brollosy, N. R.; Pedersen, E. B.; Nielsen, C. *Monatsh. Chem.* **1995**, *126*, 593.
[370f] Cristalli, G.; Volpini, R.; Vittori, S.; Camaioni, E.; Rafaiani, G.; Potenza, S.; Vita, A. *Nucleosides & Nucleotides* **1996**, *15*, 1567.
[371] Chattopadhyaya, J. B.; Reese, C. B. *Nucleic Acids Res., Spec. Publ. 4* **1978**, 67.
[372] Sanchez, R. A.; Orgel, L. E. *J. Mol. Biol.* **1970**, *47*, 531.
[373] Gish, D. T.; Neil, G. L.; Wechter, W. J. *J. Med. Chem.* **1971**, *14*, 882.
[374] Tolman, R. L.; Robins, R. K. *J. Med. Chem.* **1971**, *14*, 1112.
[375] Holý, A. *Tetrahedron Lett.* **1971**, 189.

[376] Holý, A. *Collect. Czech. Chem. Commun.* **1973**, *38*, 428.
[377] Holý, A. *Collect. Czech. Chem. Commun.* **1972**, *37*, 4072.
[378] Holý, A. *Tetrahedron Lett.* **1973**, 1147.
[379] Holý, A. *Collect. Czech. Chem. Commun.* **1973**, *38*, 3912.
[380] Holý, A. *Nucleic Acids Res.* **1974**, *1*, 289.
[381] Yoshimura, Y.; Ueda, T.; Matsuda, A. *Tetrahedron Lett.* **1991**, *32*, 4549.
[382] Averett, D. R.; Koszalka, G. W.; Fyfe, J. A.; Roberts, G. B.; Purifoy, D. J. M.; Krenitsky, T. A. *Antimicrob. Agents Chemother.* **1991**, *35*, 851.
[382a] Koszalka, G. W.; Averett, D. R.; Fyfe, J. A.; Roberts, G. B.; Spector, T.; Biron, K.; Krenitsky, T. A. *Antimicrob. Agents Chemother.* **1991**, *35*, 1437.
[383] Watanabe, K. A.; Su, T.-L.; Reichman, U.; Greenberg, N.; Lopez, C.; Fox, J. J. *J. Med. Chem.* **1984**, *27*, 91.
[383a] Smee, D. F.; Chernow, M.; Kraft, M.; Okamoto, P. M.; Prisbe, E. J. *Nucleosides & Nucleotides* **1988**, *7*, 155.
[384] Chu, C. K.; Matulic-Adamic, J.; Huang, J.-T.; Chou, T.-C.; Burchenal, J. H.; Fox, J. J.; Watanabe, K. A. *Chem. Pharm. Bull.* **1989**, *37*, 336.
[385] Howell, H. G.; Brodfuehrer, P. R.; Brundidge, S. P.; Benigni, D. A.; Sapino, Jr., C. *J. Org. Chem.* **1988**, *53*, 85.
[386] Martin, J. A.; Bushnell, D. J.; Duncan, I. B.; Dunsdon, S. J.; Hall, M. J.; Machin, P. J.; Merrett, J. H.; Parkes, K. E. B.; Roberts, N. A.; Thomas, G. J.; Galpin, S. A.; Kinchington, D. *J. Med. Chem.* **1990**, *33*, 2137.
[387] Tann, C. H.; Brodfuehrer, P. R.; Brundidge, S. P.; Sapino, Jr., C.; Howell, H. G. *J. Org. Chem.* **1985**, *50*, 3644.
[388] Driscoll, J. S.; Marquez, V. E.; Plowman, J.; Liu, P. S.; Kelley, J. A.; Barchi, Jr., J. J. *J. Med. Chem.* **1991**, *34*, 3280.
[389] Siddiqui, M. A.; Driscoll, J. S.; Marquez, V. E.; Roth, J. S.; Shirasaka, T.; Mitsuya, H.; Barchi, Jr., J. J.; Kelley, J. A. *J. Med. Chem.* **1992**, *35*, 2195.
[390] Wysocki, Jr., R. J.; Siddiqui, M. A.; Barchi, Jr., J. J.; Driscoll, J. S.; Marquez, V. E. *Synthesis* **1991**, 1005.
[391] Marquez, V. E.; Tseng, C. K.-H.; Mitsuya, H.; Aoki, S.; Kelley, J. A.; Ford, Jr., H.; Roth, J. S.; Broder, S.; Johns, D. G.; Driscoll, J. S. *J. Med. Chem.* **1990**, *33*, 978.
[392] Araki, K.; Yun, W. Q.; O'Toole, J.; Toscano, P. J.; Welch, J. T. *Carbohydr. Res.* **1993**, *249*, 139.
[393] Vemishetti, P.; Howell, H. G.; Walker, D. G.; Brodfuehrer, P. R.; Shih, K.M. EP 428109, 1991 [*Chem. Abstr.* **1991**, *115*, 92840t].
[394] Yasumoto, M.; Marunaka, T.; Hashimoto, S.; Harima, K.; Suzue, T. JP 76146482, 1976 [*Chem. Abstr.* **1977**, *87*, 53355x].
[395] Vorbrüggen, H.; Krolikiewicz, K. unpublished results.
[396] Baud, M. V.; Chavis, C.; Lucas, M.; Imbach, J.-L. *Tetrahedron Lett.* **1990**, *31*, 4437.
[397] Nechab, M.; Chavis, C.; Lucas, M.; Imbach, J.-L. *Synth. Commun.* **1992**, *22*, 1115.
[398] Azymah, M.; Chavis, C.; Fruchier, A.; Lucas, M.; Imbach, J.-L. *Nucleosides & Nucleotides* **1992**, *11*, 1607.
[399] Tanaka, H.; Takashima, H.; Ubasawa, M.; Sekiya, K.; Nitta, I.; Baba, M.; Shigeta, S.; Walker, R. T.; De Clercq, Ė.; Miyasaka, T. *J. Med. Chem.* **1992**, *35*, 4713.
[400] Kim, Y. H.; Kim, J. Y.; Lee, C. H. *Chem. Lett.* **1988**, 1045.
[401] Lee, C. H.; Kim, J. Y.; Kim, W. J.; Kim, Y. H. *Heterocycles* **1990**, *31*, 211.
[402] Zuurmond, H. M.; van der Klein, P. A. M.; Veeneman, G. H.; van Boom, J. H. *Recl. Trav. Chim. Pays-Bas* **1990**, *109*, 437.
[403] Sugimura, H. JP 2237986, 1990 [*Chem. Abstr.* **1991**, *114*, 122986u].
[404] Sugimura, H.; Osumi, K.; Yamazaki, T.; Yamaya, T. *Tetrahedron Lett.* **1991**, *32*, 1813.
[404a] Sugimura, H.; Osumi, K.; Kodaka, Y.; Sujino, K. *J. Org. Chem.* **1994**, *59*, 7653.
[405] Sugimura, H.; Sujino, K.; Osumi, K. *Tetrahedron Lett.* **1992**, *33*, 2515.
[406] Sujino, K.; Sugimura, H. *Synlett* **1992**, 553.
[407] Sugimura, H.; Muramoto, I.; Nakamura, T.; Osumi, K. *Chem. Lett.* **1993**, 169.
[408] Knapp, S.; Shieh, W.-C. *Tetrahedron Lett.* **1991**, *32*, 3627.

[408a] Knapp, S.; Shieh, W.-C.; Jaramillo, C.; Trilles, R. V.; Nandan, S. R. *J. Org. Chem.* **1994**, *59*, 946.
[409] Garner, P.; Yoo, J. U.; Sarabu, R. *Tetrahedron* **1992**, *48*, 4259.
[410] Knapp, S.; Nandan, S. R. *J. Org. Chem.* **1994**, *59*, 281.
[410a] Sujino, K.; Yoshida, T.; Sugimura, H. *Tetrahedron Lett.* **1996**, *37*, 6133.
[411] Kahne, D.; Walker, S.; Cheng, Y.; Van Engen, D. *J. Am. Chem. Soc.* **1989**, *111*, 6881.
[412] Chanteloup, L.; Beau, J.-M. *Tetrahedron Lett.* **1992**, *33*, 5347.
[413] Trost, B. M.; Nübling, C. *Carbohydr. Res.* **1990**, *202*, 1.
[414] O'Neil, I. A.; Hamilton, K. M. *Synlett* **1992**, 791.
[414a] Nishizono, N.; Koike, N.; Yamagata, Y.; Fujii, S.; Matsuda, A. *Tetrahedron Lett.* **1996**, *37*, 7569.
[414b] Yoshimura, Y.; Watanabe, M.; Satoh, H.; Ashida, N.; Ijichi, K.; Sakata, S.; Machida, H.; Matsuda, A. *J. Med. Chem.* **1997**, *40*, 2177.
[414c] Yoshimura, Y.; Kitano, K.; Yamada, K.; Satoh, H.; Watanabe, M.; Miura, S.; Sakata, S.; Sasaki, T.; Matsuda, A. *J. Org. Chem.* **1997**, *62*, 3140.
[414d] Leydier, C.; Bellon, L.; Barascut, J.-L.; Deydier, J.; Maury, G.; Pelicano, H.; El Alaoui, M. A.; Imbach, J. L. *Nucleosides & Nucleotides* **1994**, *13*, 2035.
[414e] Jeong, L. S.; Nicklaus, M. C.; George, C.; Marquez, V. E. *Tetrahedron Lett.* **1994**, *35*, 7569.
[414f] Mann, J.; Tench, A. J.; Weymouth-Wilson, A. C.; Shaw-Ponter, S.; Young, R. J. *J. Chem. Soc., Perkin Trans. 1* **1995**, 677.
[414g] Secrist, J. A.; Parker, W. B.; Tiwari, K. N.; Messini, L.; Shaddix, S. C. R.; Rose, L. M.; Bennett, Jr., L. L.; Montgomery, J. A. *Nucleosides & Nucleotides* **1995**, *14*, 675.
[414h] Leydier, C.; Bellon, L.; Barascut, J.-L.; Imbach, J.-L. *Nucleosides & Nucleotides* **1995**, *14*, 1027.
[414i] Jandu, K. S.; Selwood, D. L. *J. Org. Chem.* **1995**, *60*, 5170.
[414j] Young, R. J.; Shaw-Ponter, S.; Thomson, J. B.; Miller, J. A.; Cumming, J. G.; Pugh, A. W.; Rider, P. *Bioorg. Med. Chem. Lett.* **1995**, *5*, 2599.
[414k] Hancox, E. L.; Walker, R. T. *Nucleosides & Nucleotides* **1996**, *15*, 135.
[414l] Shaw-Ponter, S.; Mills, G.; Robertson, M.; Bostwick, R. D.; Hardy, G. W.; Young, R. J. *Tetrahedron Lett.* **1996**, *37*, 1867.
[414m] Wnuk, S. F. *Tetrahedron* **1993**, *49*, 9877.
[414n] Walker, R. T. *Royal Soc. Chem., Special Publication* **1997**, *198*, 203.
[415] Ratcliffe, A. J.; Mootoo, D. R.; Andrews, C. W.; Fraser-Reid, B. *J. Am. Chem. Soc.* **1989**, *111*, 7661.
[416] Konradsson, P.; Udodong, U. E.; Fraser-Reid, B. *Tetrahedron Lett.* **1990**, *31*, 4313.
[417] Llera, J. M.; Lopez, J. C.; Fraser-Reid, B. *J. Org. Chem.* **1990**, *55*, 2997.
[418] Fraser-Reid, B.; Wu, Z.; Udodong, U. E.; Ottosson, H. *J. Org. Chem.* **1990**, *55*, 6068.
[419] Lopez, J. C.; Fraser-Reid, B. *J. Chem. Soc., Chem. Commun.* **1991**, 159.
[420] Ratcliffe, A. J.; Konradsson, P.; Fraser-Reid, B. *Carbohydr. Res.* **1991**, *216*, 323.
[421] Chapeau, M.-C.; Marnett, L. J. *J. Org. Chem.* **1993**, *58*, 7258.
[421a] Wang, J.; Wurster, J. A.; Wilson, L. J.; Liotta, D. *Tetrahedron Lett.* **1993**, *34*, 4881.
[421b] Kawakami, H.; Ebata, T.; Koseki, K.; Okano, K.; Matsumoto, K.; Matsushita, H. *Heterocycles* **1993**, *36*, 665.
[421c] El-Laghdach, A.; Diaz, Y.; Castillón, S. *Tetrahedron Lett.* **1993**, *34*, 2821.
[422] Umemoto, T.; Fukami, S.; Tomizawa, G.; Harasawa, K.; Kawada, K.; Tomita, K. *J. Am. Chem. Soc.* **1990**, *112*, 8563.
[423] Kim, C. U.; Misco, P. F. *Tetrahedron Lett.* **1992**, *33*, 5733.
[423a] Diaz, Y.; El-Laghdach, A.; Matheu, M. I.; Castillon, S. *J. Org. Chem.* **1997**, *62*, 1501.
[423b] Diaz, Y.; El-Laghdach, A.; Castillon, S. *Tetrahedron* **1997**, *53*, 10921.
[424] Kawada, M.; Matsumoto, K.; Tsurushima, M. EP 193903, 1985 [*Chem. Abstr.* **1986**, *105*, 209351w].
[424a] Furukawa, Y. EP 136693, 1984 [*Chem. Abstr.* **1985**, *103*, 71623m].
[425] Nováček, A.; Hesoun, D.; Gut, J. *Collect. Czech. Chem. Commun.* **1965**, *30*, 1890.
[425a] Cruickshank, K. A.; Jiricny, J.; Reese, C. B. *Tetrahedron* **1994**, *50*, 3603.
[425b] Botta, M.; Saladino, R.; Gentile, G.; Summa, V.; Nicoletti, R.; Verri, A.; Focher, F.; Spadari, S. *Tetrahedron* **1994**, *50*, 3603.

[425c] Podzigun, G. I.; Petrova, V. V.; Reznik, V. S.; Shagidullin, R. R.; Awakumova, L. V. *Izv. Akad. Nauk, Ser. Khim.* **1997**, *2*, 397 [*Chem. Abstr.* **1997**, *127*, 161781m].
[426] Ogawa, T.; Matsui, M. *J. Organometal. Chem.* **1978**, *145*, C37.
[427] Keilen, G.; Benneche, T.; Undheim, K. *Acta Chem. Scand., Ser. B* **1987**, *41*, 577.
[428] Mukaiyama, T.; Hashimoto, Y.; Hayashi, Y.; Shoda, S. *Chem. Lett.* **1984**, 557.
[429] Mukaiyama, T.; Shoda, S.; Nakatsuka, T.; Narasaka, K. *Chem. Lett.* **1978**, 605.
[430] Szarek, W. A.; Depew, C.; Jarrell, H. C.; Jones, J. K. N. *J. Chem. Soc., Chem. Commun.* **1975**, 648.
[431] Hertel, L. W.; Grossman, C. S.; Kroin, J. S.; Mineishi, S.; Chubb, S.; Nowak, B.; Plunkett, W. *Nucleosides & Nucleotides* **1989**, *8*, 951.
[432] Bouali, A., Ewing, D. F.; Mackenzie, G. *Nucleosides & Nucleotides* **1994**, *13*, 491.
[432a] Falahatpisheh, N.; Sulikowski, G. A. *Synlett* **1994**, 672.
[432b] Xiang, Y.; Kotra, L. P.; Chu, C. K.; Schinazi, R. F. *Bioorg. Med. Chem. Lett.* **1995**, *5*, 743.
[432c] Andersen, M. W.; Daluge, S. M.; Kerremans, L.; Herdewijn, P. *Tetrahedron Lett.* **1996**, *37*, 8147.
[432d] Kotra, L. P.; Xiang, Y.; Newton, M. G.; Schinazi, R. F.; Cheng, Y. C.; Chu, C. K. *J. Med. Chem.* **1997**, *40*, 3635.
[432e] Zacharie, B.; Gagnon, L.; Attardo, G.; Connolly, T. P.; St-Denis, Y.; Penney, C. L. *J. Med. Chem.* **1997**, *40*, 2883.
[432f] Ohkubo, M.; Nishimura, T.; Jona, H.; Honma, T.; Ito, S.; Morishima, H. *Tetrahedron* **1997**, *53*, 5937.
[432g] Hendrix, C.; Rosemeyer, H.; Verheggen, I.; Seela, F.; Van Aerschot, A.; Herdewijn, P. *Chem. Eur. J.* **1997**, *3*, 110.
[432h] Hossain, N.; Rozenski, J.; De Clercq, E.; Herdewijn, P. *J. Org. Chem.* **1997**, *62*, 2442.
[433] Mukaiyama, T.; Hayashi, Y.; Hashimoto, Y. *Chem. Lett.* **1985**, 1087.
[434] Koch, A.; Lamberth, C.; Wetterich, F.; Giese, B. *J. Org. Chem.* **1993**, *58*, 1083.
[435] Fischer, E.; Delbrück, K. *Chem. Ber.* **1909**, 2776.
[436] Schramm, G.; Grötsch, H.; Pollmann, W. *Angew. Chem., Int. Ed. Engl.* **1962**, *1*, 1.
[437] Onodera, K.; Hirano, S.; Kashimura, N.; Masuda, F.; Yajima, T.; Miyazaki, N. *J. Org. Chem.* **1966**, *31*, 1291.
[438] Peetersen, H.; Pedersen, E. B; Nielsen, C. M. *Chem. Scr.* **1989**, *29*, 375.
[439] Hjuler-Nielsen, H. P.; Pedersen, H.; Hansen, H. B.; Pedersen, E. B.; Nielson, C. *J. Heterocycl. Chem.* **1992**, *29*, 511.
[440] Bolitt, V.; Chaguir, B.; Sinou, D. *Tetrahedron Lett.* **1992**, *33*, 2481.
[440a] Trost, B. M.; Shi, Z. *J. Am. Chem. Soc.* **1996**, *118*, 3037.
[441] Cusack, N. J.; Hildick, B. J.; Robinson, D. H.; Rugg, P. W.; Shaw, G. *J. Chem. Soc., Perkin Trans. 1* **1973**, 1720.
[442] Takeuchi, T.; Umezawa, S.; Tsuchiya, T.; Takahashi, Y. WO 9310137, 1993 [*Chem. Abstr.* **1993**, *119*, 139705c].
[443] Agathocleous, D. C.; Shaw, G. *J. Chem. Soc., Perkin Trans. 1* **1991**, 2317.
[444] Nair, V.; Purdy, D. F. *Heterocycles* **1993**, *36*, 421.
[445] Jung, M. E.; Trifunovich, I. D. *Tetrahedron Lett.* **1992**, *33*, 2921.
[446] Fuentes, J.; Moreda, W.; Ortiz, C.; Robina, I.; Welsh, C. *Tetrahedron* **1992**, *48*, 6413.
[447] Yokoyama, M.; Kumata, K.; Yamada, N.; Noro, H.; Sudo, Y. *J. Chem. Soc., Perkin Trans. 1* **1988**, 2309.
[448] Yokoyama, M.; Ikuma, T.; Sugasawa, S.; Togo, H. *Bull. Chem. Soc. Jpn.* **1991**, *64*, 2306.
[449] Yokoyama, M.; Ikuma, T.; Obara, N.; Togo, H. *J. Chem. Soc., Perkin Trans. 1* **1990**, 3243.
[450] Acevedo, O. L.; Townsend, L. B. In *Nucleic Acid Chemistry;* Townsend, L. B., Tipson, R. S., Eds.; Wiley-Interscience: New York; 1986, Vol. 3; p. 119.
[451] Vasella, A. *Helv. Chim. Acta* **1977**, *60*, 426.
[452] Vasella, A. *Helv. Chim. Acta* **1977**, *60*, 1273.
[453] Nogueras, M.; Melguizo, M.; Quijano, M. L.; Sánchez, A. *J. Heterocycl. Chem.* **1991**, *28*, 1417.
[454] Melguizo, M.; Nogueras, M.; Sánchez, A. *Synthesis* **1992**, 491.
[455] Moss, R. J.; Petrie, C. R.; Meyer, Jr., R. B.; Nord, L. D.; Willis, R. C.; Smith, R. A.; Larson, S. B.; Kini, G. D.; Robins, R. K. *J. Med. Chem.* **1988**, *31*, 786.
[455a] Grouiller, A.; Mackenzie, G.; Najib, B.; Shaw, G.; Ewing, D. *J. Chem. Soc., Chem. Commun.* **1988**, 671.

[456] Geen, G. R.; Kincey, P. M.; Choudary, B. M. *Tetrahedron Lett.* **1992**, *33*, 4609.
[457] Baker, B. R.; Tanna, P. M. *J. Org. Chem.* **1965**, *30*, 2857.
[458] Lira, E. P.; Huffman, C. W. *J. Org. Chem.* **1966**, *31*, 2188.
[459] Yakout, E.-S. M. A.; Pedersen, E. B. *Chem. Scr.* **1989**, *29*, 185.
[460] Prevost, N.; Rouessac, F. *Synth. Commun.* **1997**, *227*, 2325.
[460a] Prevost, N.; Rouessac, F. *Tetrahedron Lett.* **1997**, *38*, 4215.
[461] Hanrahan, J. R.; Hutchinson, D. W. *J. Biotechnol.* **1992**, *23*, 193.
[462] Krenitsky, T. A.; Rideout, J. L.; Chao, E. Y.; Koszalka, G. W.; Gurney, F.; Crouch, R. C.; Cohn, N. K.; Wolberg, G.; Vinegar, R. *J. Med. Chem.* **1986**, *29*, 138.
[463] Morisawa, H.; Utagawa, T.; Miyoshi, T.; Kashima, N.; Nakamatsu, T.; Yamanaka, S.; Yamazaki, A. *Nucleic Acids Res., Symp. Ser. 6* **1979**, s25.
[464] Utagawa, T.; Morisawa, H.; Miyoshi, T.; Yoshinaga, F.; Yamazaki, A.; Mitsugi, K. *FEBS Lett.* **1980**, *109*, 261.
[465] Morisawa, H.; Utagawa, T.; Miyoshi, T.; Yoshinaga, F.; Yamazaki, A.; Mitsugi, K. *Tetrahedron Lett.* **1980**, *21*, 479.
[466] Krenitsky, T. A.; Koszalka, G. W.; Tuttle, J. V.; Rideout, J. L.; Elion, G. B. *Carbohydr. Res.* **1981**, *97*, 139.
[467] Krenitsky, T. A.; Koszalka, G. W.; Jones, L. A.; Averett, D. R.; Moorman, A. R. EP 294114, 1988 [*Chem. Abstr.* **1989**, *111*, 7760s].
[468] Krenitsky, T. A.; Koszalka, G. W.; Tuttle, J. V. *Biochemistry* **1981**, *20*, 3615.
[469] Barbas III, C. F.; Wong, C.-H. *Bioorg. Chem.* **1991**, *19*, 261.
[470] Tisdale, S. M.; Tuttle, J. V.; Slater, M. J.; Daluge, S. M.; Miller, W. H.; Krenitsky, T. A.; Koszalka, G. W. EP 417999, 1991 [*Chem. Abstr.* **1991**, *115*, 230514t].
[471] Tuttle, J. V.; Tisdale, M.; Krenitsky, T. A. *J. Med. Chem.* **1993**, *36*, 119.
[472] Tuttle, J. V.; Krenitsky, T. A. EP 285432, 1988 [*Chem. Abstr.* **1989**, *111*, 55835s].
[473] Burns, Ch. L.; Daluge, S. M.; Koszalka, G. W.; Krenitsky, T. A.; Van Tuttle, J. EP317128, 1989 [*Chem. Abstr.* **1990**, *112*, 151842h].
[474] Morisawa, H.; Utagawa, T.; Nakamatsu, T.; Yamanaka, S.; Yamazaki, A. *Nucleic Acids Res., Symp. Ser. 8* **1980**, S11.
[475] Morisawa, H.; Utagawa, T.; Yamanaka, S.; Yamazaki, A. *Chem. Pharm. Bull.* **1981**, *29*, 3191.
[476] Utagawa, T.; Morisawa, H.; Yamanaka, S.; Yamazaki, A.; Yoshinaga, F.; Hirose, Y. *Agric. Biol. Chem.* **1985**, *49*, 3239.
[477] Utagawa, T.; Morisawa, H.; Yamanaka, S.; Yamazaki, A.; Hirose, Y. *Agric. Biol. Chem.* **1986**, *50*, 121.
[478] Hennen, W. J.; Wong, C.-H. *J. Org. Chem.* **1989**, *54*, 4692.
[479] Stout, M. G.; Hoard, D. E.; Holman, M. J.; Wu, E. S.; Siegel, J. M. *Methods Carbohydr. Chem.* **1976**, *7*, 19.
[480] Huang, M.-C.; Hatfield, K.; Roetker, A. W.; Montgomery, J. A.; Blakley, R. L. *Biochem. Pharmacol.* **1981**, *30*, 2663.
[481] Holguin, J.; Cardinaud, R.; Salemink, C. A. *Eur. J. Biochem.* **1975**, *54*, 515.
[482] Betbeder, D.; Hutchinson, D. W.; Richards, A. O'L. *Nucleic Acids Res.* **1989**, *17*, 4217.
[483] Betbeder, D.; Hutchinson, D. W. *Nucleosides & Nucleotides* **1990**, *9*, 569.
[484] Fischer, W.; Kaun, E.; Genz, U. DE 3840160, 1989 [*Chem. Abstr.* **1990**, *114*, 22457d].
[485] Fischer, W.; Kaun, E.; Genz, U. WO 9006312, 1989 [*Chem. Abstr.* **1990**, *114*, 99958a].
[486] Burns, C. L.; St. Clair, M. H.; Frick, L. W.; Spector, T.; Averett, D. R.; English, M. L.; Holmes, T. J.; Krenitsky, T. A.; Koszalka, G. W. *J. Med. Chem.* **1993**, *36*, 378.
[487] Vasiloiu, R. DE 4020529, 1990 [*Chem. Abstr.* **1991**, *115*, 227097d].
[488] Kojima, E.; Yoshioka, H.; Tomikanehara, H.; Murakami, K. JP 2222695, 1989 [*Chem. Abstr.* **1991**, *114*, 80106z].
[489] Murakami, K.; Shirasaka, T.; Yoshioka, H.; Kojima, E.; Aoki, S.; Ford, Jr., H.; Driscoll, J. S.; Kelley, J. A.; Mitsuya, H. *J. Med. Chem.* **1991**, *34*, 1606.
[489a] Kojima, E.; Yoshioka, H.; Fukinbara, H.; Murakami, K. GB 2228479, 1989 [*Chem. Abstr.* **1991**, *114*, 102707a].
[490] Shirae, H.; Yokozeki, K. *Agric. Biol. Chem.* **1991**, *55*, 609.

[491] Ajinomoto Inc. *Drugs of the Future* **1990**, *15*, 399.
[492] Carson, D. A.; Wasson, D. B. *Biochem. Biophys. Res. Commun.* **1988**, *155*, 829.
[492a] Votruba, I.; Holý, A.; Dvorakova, H.; Günter, J.; Hockova, D.; Hrebabecky, H.; Cihlar, T.; Masojidkova, M. *Collect. Czech. Chem. Commun.* **1994**, *59*, 2303.
[493] Drueckhammer, D. G.; Hennen, W. J.; Pederson, R. L.; Barbas III, C. F.; Gautheron, C. M.; Krach, T.; Wong, C.-H. *Synthesis* **1991**, 499.
[494] Rolland, V.; Kotera, M.; Lhomme, J. *Synth. Commun.* **1997**, *27*, 3505.
[494a] Kuimelis, R. G.; Hope, H.; Nambiar, K. P. *Nucleosides & Nucleotides* **1993**, *12*, 737.
[495] Gold, A.; Sangaiah, R. *Nucleosides & Nucleotides* **1990**, *9*, 907.
[496] Martin, O. R. *Tetrahedron Lett.* **1985**, *26*, 2055.
[496a] Martin, O. R.; Rao, S. P.; El-Shenawy, H. A.; Kurz, K. G.; Cutler, A. B. *J. Org. Chem.* **1988**, *53*, 3287.
[497] Brown, D. M.; Todd, A.; Varadarajan, S. *J. Chem. Soc.* **1956**, 2384.
[498] Kohn, P.; Samaritano, R. H.; Lerner, L. M. In *Synthetic Procedures in Nucleic Acid Chemistry*; Zorbach, W. W., Tipson, R. S., Eds.; Wiley-Interscience: New York, 1968, Vol. 1; p. 117.
[499] Hoffer, M. US 3868373, 1975 [*Chem. Abstr.* **1975**, *82*, 171366w].
[500] Quittmann, W. EP 411467, 1991 [*Chem. Abstr.* **1991**, *114*, 247058z].
[501] Langer, S. H.; Connell, S.; Wender, I. *J. Org. Chem.* **1958**, *23*, 50.
[502] Wyss, P. C.; Schönholzer, P.; Arnold, W. *Helv. Chim. Acta* **1980**, *63*, 1353.
[503] Kristinsson, H.; Nebel, K.; O'Sullivan, A. C.; Struber, F.; Winkler, T.; Yamaguchi, Y. *Tetrahedron* **1994**, *50*, 6825.
[504] Hawkins, L. D.; Hanvey, J. C.; Boyd, Jr., F. L.; Baker, D. C.; Showalter, H. D. H. *Nucleosides & Nucleotides* **1983**, *2*, 479.
[505] Wright, G. E.; Dudycz, L. W. *J. Med. Chem.* **1984**, *27*, 175.
[506] Jarvi, E. T.; Sunkara, P. S.; Bowlin, T. L. *Nucleosides & Nucleotides* **1989**, *8*, 1111.
[507] Bhadti, V. S.; Bhan, A.; Hosmane, R. S.; Hulce, M. *Nucleosides & Nucleotides* **1992**, *11*, 1137.
[507a] Cristalli, G.; Volpini, R.; Vittori, S.; Camaioni, E.; Rafaiani, G.; Potenza, S.; Vita, A. *Nucleosides & Nucleotides* **1996**, *15*, 1567.
[508] Musicki, B.; Widlanski, T. S. *Tetrahedron Lett.* **1991**, *32*, 1267.
[509] O'Hara Dempcy, R.; Skibo, E. B. *J. Org. Chem.* **1991**, *56*, 776.
[509a] Vorbrüggen, H. *Acta Biochem. Pol.* **1996**, *43*, 25 [*Chem. Abstr.* **1996**, *125*, 168452x].
[510] Lichtenthaler, F. W.; Cuny, E. *Chem. Ber.* **1981**, *114*, 1610.
[511] Liu, P. S.; Marquez, V. E.; Driscoll, J. S.; Fuller, R. W. *J. Med. Chem.* **1981**, *24*, 662.
[512] Skulnick, H. I.; Wierenga, W. *J. Carbohydr., Nucleosides, Nucleotides* **1979**, *6*, 262.
[513] Demuth, M.; Mikhail, G. *Synthesis* **1982**, 827.
[514] Corey, E. J.; Cho, H.; Rücker, C.; Hua, D. H. *Tetrahedron Lett.* **1981**, *22*, 3455.
[514a] Boudjouk, P.; So, J-H. *Inorg. Syn.* **1992**, *29*, 108.
[514b] Saneyoshi, M.; Inomata, M.; Fukuoka, F. *Chem. Pharm. Bull.* **1978**, *26*, 2990.
[515] Janardhanam, S.; Nambiar, K. P. *J. Chem. Soc., Chem. Commun.* **1994**, 1009.
[515a] Bednarski, K.; Dixit, D. M.; Mansour, T. S.; Colman, S. G.; Walcott, S. M.; Ashman, C. *Bioorg. Med. Chem. Lett.* **1995**, *5*, 1741.
[515b] Matulic-Adamic, J.; Haeberli, P.; Usman, N. *J. Org. Chem.* **1995**, *60*, 2563.
[515c] Austin, R. E.; Cleary, D. G. *Nucleosides & Nucleotides* **1995**, *14*, 1803.
[515d] Kofoed, T.; Ismail, A. E. A. A.; Pedersen, E. B.; Nielsen, C. *Bull. Soc. Chim. Fr.* **1997**, 59.
[516] Watanabe, K. A.; Fox, J. J. *J. Heterocycl. Chem.* **1969**, *6*, 109.
[517] Watanabe, K. A.; Kotick, M. P.; Fox, J. J. *J. Org. Chem.* **1970**, *35*, 231.
[518] Leonard, N. J.; Sprecker, M. A.; Morrice, A. G. *J. Am. Chem. Soc.* **1976**, *98*, 3987.
[519] Cottam, H. B.; Wasson, D. B.; Shih, H. C.; Raychaudhuri, A.; Di Pasquale, G.; Carson, D. A. *J. Med. Chem.* **1993**, *36*, 3424.
[519a] Mayr, H.; Patz, M. *Angew. Chem., Int. Ed. Engl.* **1994**, *33*, 938.
[520] Wengel, J.; Pedersen, E. B.; Vestergaard, B. F. *Synthesis* **1992**, 319.
[521] Jähne, G. DE 4020481, 1992 [*Chem. Abstr.* **1992**, *116*, 152305e].
[521a] Schmidt, C. L.; Rusho, W. J.; Townsend, L. B. *J. Chem. Soc., Chem. Commun.* **1971**, 1515.
[521b] El Ashry, E. S. H.; El Kilany, Y. *Adv. Heterocycl. Chem.* **1997**, *67*, 391.
[521c] El Ashry, E. S. H.; El Kilany, Y. *Adv. Heterocycl. Chem.* **1997**, *68*, 1.

[521d] El Ashry, E. S. H.; El Kilany, Y. *Adv. Heterocycl. Chem.* **1998**, *69*, 129.
[521e] Yoshimura, Y.; Endo, M.; Sakata, S. *Tetrahedron Lett.* **1999**, *40*, 1937.
[522] Rochow, E. G.; Gingold, K. *J. Am. Chem. Soc.* **1954**, *76*, 4852.
[523] Pitt, M. J. *Chem. Ind. (London)* **1982**, *20*, 804.
[523a] Reichardt, C. *Solvent Effect in Organic Chemistry*; VCH, Weinheim; 1969.
[523b] Branalt, J.; Kvarnström, I.; Classon, B.; Samuelsson, B. *J. Org. Chem.* **1996**, *61*, 3599.
[523c] Branalt, J.; Kvarnström, I.; Classon, B.; Samuelsson, B. *J. Org. Chem.* **1996**, *61*, 3604.
[524] Katz, D. J.; Wise, D. S.; Townsend, L. B. *J. Heterocycl. Chem.* **1983**, *20*, 369.
[525] Patil, V. D.; Wise, D. S.; Townsend, L. B. *J. Chem. Soc., Perkin Trans. 1* **1980**, 1853.
[526] Bennua, B.; Vorbrüggen, H. unpublished results.
[527] Nelson, V.; El Khadem, H. *Carbohydr. Res.* **1983**, *124*, 161.
[528] Yasumoto, M.; Yamawaki, I.; Marunaka, T.; Hashimoto, S. *J. Med. Chem.* **1978**, *21*, 738.
[529] Corey, E. J.; Venkateswarlu, A. *J. Am. Chem. Soc.* **1972**, *94*, 6190.
[529a] Nelson, T. D.; Crouch, R. D. *Synthesis* **1996**, 1031.
[530] Lau, J.; Pedersen, E. B.; Wengel, J. *Nucleosides & Nucleotides* **1989**, *8*, 961.
[531] Walczak, K.; Wengel, J.; Pedersen, E. B. *Monatsh. Chem.* **1992**, *123*, 349.
[532] Hiebl, J.; Zbiral, E. *Nucleosides & Nucleotides* **1996**, *15*, 1649.
[533] Depelley, J.; Granet, R.; Kaouadji, M.; Krausz, P.; Piekarski, S. *Nucleosides & Nucleotides* **1996**, *15*, 995.
[534] Sharma, M.; Alderfer, J. L. *Nucleosides & Nucleotides* **1983**, *2*, 89.
[535] Pathak, V. P. *Synth. Commun.* **1993**, *23*, 83.
[536] Asakura, J. *Nucleosides & Nucleotides* **1993**, *12*, 701.
[536a] Rassu, G.; Zanardi, F.; Cornia, M.; Casiraghi, G. *Nucleosides & Nucleotides* **1996**, *15*, 1113.
[536b] Kierzek, R.; Ito, H.; Bhatt, R.; Itakura, K. *Tetrahedron Lett.* **1981**, *22*, 3761.
[537] Gosselin, G.; Bergogne, M.-C.; De Rudder, J.; De Clercq, E.; Imbach, J.-L. *J. Med. Chem.* **1987**, *30*, 982.
[538] Niedballa, U.; Vorbrüggen, H. unpublished results.
[538a] Ross, B. S.; Springer, R. H.; Vasquez, G.; Andrews, R. S.; Cook, P. D.; Acevedo, O. L. *J. Heterocycl. Chem.* **1994**, *31*, 765.
[539] Montgomery, J. A.; Shortnacy-Fowler, A. T.; Clayton, S. D.; Riordan, J. M.; Secrist III, J. A. *J. Med. Chem.* **1992**, *35*, 397.
[540] Bärwolff, D.; Kowollik, G.; Langen, P. *Collect. Czech. Chem. Commun.* **1974**, *39*, 1494.
[541] Prasad, A. K.; Wengel, J. *Nucleosides & Nucleotides* **1996**, *15*, 1347.
[542] Melander, C.; Horne, D. A. *J. Org. Chem.* **1997**, *62*, 9295.
[543] Cook, P. D.; Dea, P.; Robins, R. K. *J. Heterocycl. Chem.* **1978**, *15*, 1.
[544] Folkers, G.; Junginger, G.; Müller, C. E.; Schloz, U.; Eger, K. *Arch. Pharm. (Weinheim)* **1989**, *322*, 119.
[545] Wilson, J. D. US 4921950, 1990 [*Chem. Abstr.* **1990**, *113*, 132719c].
[546] Mitchell, W. L.; Ravenscroft, P.; Hill, M. L.; Knutsen, L. J. S.; Judkins, B. D.; Newton, R. W.; Scopes, D. I. C. *J. Med. Chem.* **1986**, *29*, 809.
[547] Bobek, M.; Kavai, I.; De Clercq, E. *J. Med. Chem.* **1987**, *30*, 1494.
[548] Frister, H.; Schlimme, E. *Z. Naturforsch.* **1987**, *42c*, 603.
[549] Gelpi, M. E.; Cadenas, R. A. *Adv. Carbohydr. Chem. Biochem.* **1975**, *31*, 81.
[550] Haines, A. H. *Adv. Carbohydr. Chem. Biochem.* **1981**, *39*, 13.
[551] Brossmer, R.; Eschenfelder, V. *Liebigs Ann. Chem.* **1972**, *762*, 160.
[552] Fujii, S.; Unemi, N.; Takeda, S. DE 2814202, 1978 [*Chem. Abstr.* **1979**, *90*, 43803q].
[553] Hronowski; L. J. J.; Szarek, W. A. *J. Med. Chem.* **1982**, *25*, 522.
[554] Rosowsky, A.; Kim, S.-H.; Wick, M. *J. Med. Chem.* **1981**, *24*, 1177.
[555] Smirnov, I. P.; Kochetkova, S. V.; Shchaveleva, I. L.; Tsilevich, T. L.; Gottikh, B. P.; Florent'ev, V. L. *Bioorg. Khim.* **1988**, *14*, 921 [*Chem. Abstr.* **1989**, *110*, 231998n].
[556] Rosowsky, A.; Kim, S.-H.; Trites, D.; Wick, M. *J. Med. Chem.* **1982**, *25*, 1034.
[557] Hollis Showalter, H. D.; Putt, S. R. *Tetrahedron Lett.* **1981**, *22*, 3155.
[558] Kelley, J. L.; Krochmal, M. P.; Schaeffer, H. J. *J. Med. Chem.* **1981**, *24*, 472.
[559] Bessodes, M.; Benamghar, R.; Antonakis, K. *Carbohydr. Res.* **1990**, *200*, 493.

[560] Farr, R. A.; Bey, P.; Sunkara, P. S.; Lippert, B. J. *J. Med. Chem.* **1989**, *32*, 1879.
[561] Sasamori, H.; Hori, C.; Yoshida, M.; Umezawa, H. JP 63115895, 1986 [*Chem. Abstr.* **1988**, *109*, 110864m].
[562] Czernecki, S.; Ezzitouni, A. *Tetrahedron Lett.* **1993**, *34*, 315.
[563] Morr, M. *Liebigs Ann. Chem.* **1982**, 666.
[564] Kulinkovich, L. N.; Timoshchuk, V. A. *Zh. Obshch. Khim.* **1983**, *53*, 1649 [*Chem. Abstr.* **1983**, *99*, 212866f].
[565] Bartolome, N. F. J.; Carreno, G. V.; Castillo, A. I.; Quiroga, E. J. A. ES 2040615, 1993 [*Chem. Abstr.* **1994**, *120*, 317324h].
[566] Esanu, A. DE 3514641, 1985 [*Chem. Abstr.* **1986**, *104*, 110121p].
[567] Hamilton, H. W.; Johnson, S. A. In *Nucleic Acid Chemistry*; Townsend, L. B., Tipson, R. S., Eds.; Wiley: New York; 1991, Vol. 3; p. 286.
[568] Clement, M. A.; Berger, S. H. *Med. Chem. Res.* **1992**, *2*, 154.
[569] Tarköy, M.; Bolli, M.; Schweizer, B.; Leumann, C. *Helv. Chim. Acta* **1993**, *76*, 481.
[570] Schneider, K. C.; Benner, S. A. *Tetrahedron Lett.* **1990**, *31*, 335.
[571] Sterzycki, R. Z.; Mansuri, M. M.; Martin, J. C. EP 391411, 1990 [*Chem. Abstr.* **1990**, *114*, 229305y].
[572] Wheeler, W. J.; Mabry, T. E.; Jones, C. D. *J. Labelled Comp. Radiopharm.* **1991**, *29*, 583.
[573] Gosselin, G.; Puech, F.; Genu-Dellac, C.; Imbach, J.-L. *Carbohydr. Res.* **1993**, *249*, 1.
[574] Lin, T.-S.; Zhu, J.-L.; Dutschman, G. E.; Cheng, Y.-C.; Prusoff, W. H. *J. Med. Chem.* **1993**, *36*, 353.
[575] Beauchamp, L. M.; Serling, B. L.; Kelsey, J. E.; Biron, K. K.; Collins, P.; Selway, J.; Lin, J.-C.; Schaeffer, H. J. *J. Med. Chem.* **1988**, *31*, 144.
[576] Czernecki, S; Ezzitouni, A. *J. Org. Chem.* **1992**, *57*, 7325.
[577] Czernecki, S.; Ezzitouni, A.; Krausz, P. *Synthesis* **1990**, 651.
[578] Almond, M. R.; Lowen, G. T.; Martin, G. E.; Rideout, J. L. *Nucleosides & Nucleotides* **1993**, *12*, 905.
[579] Beer, D.; Meuwly, R.; Vasella, A. *Helv. Chim. Acta* **1982**, *65*, 2570.
[580] Murai, Y.; Shiroto, H.; Ishizaki, T.; Iimori, T.; Kodama, Y.; Ohtsuka, Y.; Oishi, T. *Heterocycles* **1992**, *33*, 391.
[581] Hiebl, J.; Zbiral, E; Balzarini, J.; De Clercq, E. *J. Med. Chem.* **1992**, *35*, 3016.
[582] Hiebl, J.; Zbiral, E. *Tetrahedron Lett.* **1990**, *31*, 4007.
[583] Hiebl, J.; Zbiral, E.; Balzarini, J.; De Clercq, E. *Nucleosides & Nucleotides* **1991**, *10*, 521.
[584] Niedballa, U.; Vorbrüggen, H. DE 1943428, 1971 [*Chem. Abstr.* **1971**, *74*, 100361].
[585] Morr, M.; Ernst, L.; Grotjahn, L. *Z. Naturforsch.* **1983**, *38b*, 1665.
[586] Montgomery, J. A. WO 91 04,033, 1991 [*Chem. Abstr.* **1991**, *115*, 115017w].
[587] Sterzycki, R. Z.; Martin, J. C.; Wittman, M.; Brankovan, V.; Yang, H.; Hitchcock, M. J.; Mansuri, M. M. *Nucleosides & Nucleotides* **1991**, *10*, 291.
[588] Morr, M.; Ernst, L.; Schomburg, D. *Liebigs Ann. Chem.* **1991**, 615.
[589] Pudlo, J. S.; Wadwani, S.; Milligan, J. F.; Matteucci, M. D. *Bioorg. Med. Chem. Lett.* **1994**, *4*, 1025.
[590] Augustyns, K.; Rozenski, J.; Van Aerschot, A.; Janssen, G.; Herdewijn, P. *J. Org. Chem.* **1993**, *58*, 2977.
[591] Ojha, L. M.; Gulati, D.; Seth, N.; Bhakuni, D. S.; Pratap, R.; Agarwal, K. C. *Nucleosides & Nucleotides* **1995**, *14*, 1889.
[592] Halmos, T.; Komiotis, D.; Antonakis, K. *Carbohydr. Res.* **1986**, *156*, 256.
[593] Robins, M. J.; Madej, D.; Hansske, F.; Wilson, J. S.; Gosselin, G.; Bergogne, M. C.; Imbach, J.-L.; Balzarini, J.; De Clercq, E. *Can. J. Chem.* **1988**, *66*, 1258.
[594] Koizumi, F., Oritani, T.; Yamashita, K. *Agric. Biol. Chem.* **1990**, *54*, 3093.
[595] Génu-Dellac, C.; Gosselin, G.; Puech, F.; Henry, J.-C.; Aubertin, A.-M.; Obert, G.; Kirn, A.; Imbach, J.-L. *Nucleosides & Nucleotides* **1991**, *10*, 1345.
[596] Génu-Dellac, C.; Gosselin, G.; Imbach, J.-L. *Tetrahedron Lett.* **1991**, *32*, 79.
[597] Rassu, G.; Spanu, P.; Pinna, L.; Zanardi, F.; Casiraghi, G. *Tetrahedron Lett.* **1995**, *36*, 1941.

[598] Gurjar, M. K.; Lalitha, S. V. S.; Sharma, P. A.; Rama Rao, A. V. *Tetrahedron Lett.* **1992**, *33*, 7945.

[599] Gurjar, M. K.; Kunwar, A. C.; Reddy, D. V.; Islam, A.; Lalitha, S. V. S.; Jagannadh, B.; Rama Rao, A. V. *Tetrahedron* **1993**, *49*, 4373.

[600] Secrist III, J. A.; Riggs, R. M.; Tiwari, K. N.; Montgomery, J. A. *J. Med. Chem.* **1992**, *35*, 533.

[601] Tiwari, K. N.; Montgomery, J. A.; Secrist, III, J. A. *Nucleosides & Nucleotides* **1993**, *12*, 841.

[602] Kulikowski, T.; Zawadzki, Z.; Shugar, D.; De Clercq, E. *Nucleic Acids Res., Symp. Ser. 9* **1981**, 103.

[603] Benson, T. J.; Robinson, B. *J. Chem. Soc., Perkin Trans. 1* **1992**, 211.

[604] Hosmane, R. S.; Bhan, A.; Karpel, R. L. *J. Org. Chem.* **1990**, *55*, 5882.

[605] Mansuri, M. M.; Farina, V.; Starrett Jr., J. E.; Benigni, D. A.; Brankovan, V.; Martin, J. C. *Bioorg. Med. Chem. Lett.* **1991**, *1*, 65.

[606] Lewis, A. F.; Revankar, G. R.; Hogan, M. E. *J. Heterocycl. Chem.* **1993**, *30*, 1309.

[607] Mansuri, M. M.; Starrett, Jr., J. E.; Wos, J. A.; Tortolani, D. R.; Brodfuehrer, P. R.; Howell, H. G.; Martin, J. C. *J. Org. Chem.* **1989**, *54*, 4780.

[608] Vorbrüggen, H.; Bennua, B. In *Nucleic Acid Chemistry*; Townsend, L. B., Tipson, R. S., Eds.; Wiley: New York; 1986, Vol. 3; p. 70.

[609] Hui, C. F.; Hutchinson, D. W. *Biochim. Biophys. Acta* **1981**, *656*, 129.

[610] Dinan, F. J.; Chodkowski, J.; Barren, J. P.; Robinson, D. M.; Reinhardt, D. V. *J. Org. Chem.* **1982**, *47*, 1769.

[611] Kondo, T.; Fukami, T.; Goto, T.; Kawakami, M.; Takemura, S. *Nucleic Acids Res., Symp. Ser. 12* **1983**, 127.

[612] Yavorsky, A. E.; Turov, A. V.; Reshotko, L. N.; Florent'ev, V. L. *Bioorg. Khim.* **1990**, *16*, 963 [*Chem. Abstr.* **1990**, *113*, 224415t].

[613] Mathé, C.; Périgaud, C.; Gosselin, G.; Imbach, J.-L. *Nucleosides & Nucleotides* **1994**, *13*, 437.

[614] Absalon, M. J.; Krishnamoorthy, C. R.; McGall, G.; Kozarich, J. W.; Stubbe, J. *Nucleic Acids Res.* **1992**, *20*, 4179.

[615] Van Galen, P. J. M.; Ijzerman, A. P.; Soudijn, W. *Nucleosides & Nucleotides* **1990**, *9*, 275.

[616] Henry, E. M.; Kini, G. D.; Larson, S. B.; Robins, R. K.; Alaghamandan, H. A.; Smee, D. F. *J. Med. Chem.* **1990**, *33*, 2127.

[617] Hosmane, R. S.; Bhan, A. *Nucleosides & Nucleotides* **1990**, *9*, 913.

[618] Hosmane, R. S.; Vaidya, V. P.; Chung, M. K. *Nucleosides & Nucleotides* **1991**, *10*, 1693.

[619] Czernecki, S.; Le Diguarher, T. *Synthesis* **1991**, 683.

[620] Bellon, L.; Barascut, J.-L.; Imbach, J.-L. *Nucleosides & Nucleotides* **1992**, *11*, 1467.

[621] Trehan, S. B.; MacDiarmid, J. E.; Bardos, T. J.; Cheng, Y. *Med. Chem. Res.* **1991**, *1*, 30.

[622] Wolfe, M. S.; Harry-O'Kuru, R. E. *Tetrahedron Lett.* **1995**, *36*, 7611.

[623] Harry-O'Kuru, R. E.; Smith, J. M.; Wolfe, M. S. *J. Org. Chem.* **1997**, *62*, 1754.

[624] Hobbs, J. B.; Eckstein, F. *J. Org. Chem.* **1977**, *42*, 714.

[625] Itoh, T.; Mizuno, Y. *Heterocycles* **1976**, *5*, 285.

[626] Cristalli, G.; Franchetti, P.; Grifantini, M.; Vittori, S.; Bordoni, T.; Geroni, C. *J. Med. Chem.* **1987**, *30*, 1686.

[627] Cottam, H. B.; Revankar, G. R.; Robins, R. K. *Nucleic Acids Res.* **1983**, *11*, 871.

[628] Salcines, E. M.; Schneller, S. W. *Nucleosides & Nucleotides* **1989**, *8*, 1083.

[629] Youssefyeh, R. D.; Verheyden, J. P. H.; Moffatt, J. G. *J. Org. Chem.* **1979**, *44*, 1301.

[630] Sakaguchi, M.; Webb, M. W.; Agrawal, K. C. *J. Med. Chem.* **1982**, *25*, 1339.

[631] Montgomery, J. A.; Secrist III, J. A. WO 9104033, 1991 [*Chem. Abstr.* **1991**, *115*, 115017].

[632] Prisbe, E. J.; Smejkal, J.; Verheyden, J. P. H.; Moffatt, J. G. *J. Org. Chem.* **1976**, *41*, 1836.

[633] Saneyoshi, M.; Watanabe, S. *Chem. Pharm. Bull.* **1988**, *36*, 2673.

[634] Saneyoshi, M.; Ozaki, S. JP 7652183, 1976 [*Chem. Abstr.* **1977**, *86*, 43973n].

[635] Jois, Y. H. R.; Kwong, C. D.; Riordan, J. M.; Montgomery, J. A.; Secrist III, J. A. *J. Heterocycl. Chem.* **1993**, *30*, 1289.

[636] Itaya, T.; Saito, T.; Harada, T.; Kagatani, S.; Fujii, T. *Heterocycles* **1982**, *19*, 1059.

[636a] Papageorgiou, C.; Tamm, C. *Tetrahedron Lett.* **1986**, *27*, 555.

[637] Ochi, K.; Miyamoto, K.; Mitsui, H.; Tsuruma, Y.; Matsunaga, I.; Matsuno, T.; Takanashi, S.; Shindo, M. EP 46307, 1984 [*Chem. Abstr.* **1982**, *97*, 6068c].

[638] Ozaki, S.; Watanabe, Y.; Hoshiko, T.; Nagase, T.; Ogasawara, T.; Furukawa, H.; Uemura, A.; Ishikawa, K.; Mori, H.; Hoshi, A.; Iigo, M.; Tokuzen, R. *Chem. Pharm. Bull.* **1986**, *34*, 150.
[639] Michael, U.; Sasson, Z.; Schoenberger, E. DE 2744956, 1978 [*Chem. Abstr.* **1978**, *89*, 24357].
[640] Ueda, S.; Takeda, S.; Yamawaki, I.; Yamashita, J.-I.; Yasumoto, M.; Hashimoto, S. *Chem. Pharm. Bull.* **1982**, *30*, 125.
[641] Ochi, K.; Miyamoto, K.; Miura, Y.; Mitsui, H.; Matsunaga, I.; Shindo, M. *Chem. Pharm. Bull.* **1985**, *33*, 1703.
[642] Nishitani, T.; Horikawa, H.; Iwasaki, T.; Matsumoto, K.; Inoue, I.; Miyoshi, M. *J. Org. Chem.* **1982**, *47*, 1706.
[643] Kelley, J. L.; Kelsey, J. E.; Hall, W. R.; Krochmal, M. P.; Schaeffer, H. J. *J. Med. Chem.* **1981**, *24*, 753.
[644] Holshouser, M. H.; Shipp, A. M.; Ferguson, P. W. *J. Med. Chem.* **1985**, *28*, 242.
[645] Banijamali, A. R.; Foye, W. O. *J. Heterocycl. Chem.* **1986**, *23*, 1613.
[646] Takanaka, K.; Muraoka, M; Tsuji, T. *J. Heterocycl. Chem.* **1997**, *34*, 669.
[647] Woodman, R. J. *Drugs of the Future* **1982**, *7*, 107.
[648] Zhuk, R. A.; Berzina, A.; Serins, L.; Kaulina, L.; Hillers, S. *Khim. Geterotsikl. Soedin.* **1977**, *9*, 1258 [*Chem. Abstr.* **1978**, *88*, 22843w].
[649] Yasumoto, M.; Yamashita, J.-I.; Hashimoto, S. *Yakugaku Zasshi* **1978**, *98*, 1551 [*Chem. Abstr.* **1979**, *90*, 121528].
[650] Kim, C. U.; Martin, J. C.; Misco, P. F.; Luh, B. Y. EP 398231, 1990 [*Chem. Abstr.* **1991**, *115*, 208461].
[651] Yamashita, K.; Iumi, K.; Iida, A. JP 0311089, 1991 [*Chem. Abstr.* **1991**, *114*, 207698b].
[652] Yamashita, M.; Ikai, K.; Takahashi, C.; Oshikawa, T. *Phosphorus Sulfur* **1993**, *79*, 293.
[653] Tsilevich, T. L.; Shchaveleva, I. L.; Nosach, L. N.; Zhovnovataya, V. L.; Smirnov, I. P.; Kochetkova, S. V.; Gottikh, B. P.; Florentev, V. L. *Bioorg. Khim.* **1988**, *14*, 689 [*Chem. Abstr.* **1988**, *109*, 66372y].
[654] Keppeler, K.; Kiefer, G.; De Clercq, E. *Arch. Pharm. (Weinheim)* **1986**, *319*, 360.
[655] Lin, T.-S.; Liu, M.-C. *Synth. Commun.* **1988**, *18*, 931.
[656] Renault, J.; Laduree, D.; Robba, M. *Nucleosides & Nucleotides* **1994**, *13*, 891.
[657] Keppeler, K.; de Clercq, E. *Arch. Pharm. (Weinheim)* **1987**, *320*, 271.
[658] Renault, J.; Laduree, D.; Robba, M. *Nucleosides & Nucleotides* **1994**, *13*, 1135.
[659] Jourdan, F.; Ladurée, D.; Robba, M. *J. Heterocycl. Chem.* **1994**, *31*, 305.
[660] Lin, A. J.; Benjamin, R. S.; Rao, P. N.; Loo, T. L. *J. Med. Chem.* **1979**, *22*, 1096.
[661] Mertens, A.; Koch, E.; Bernhard; Harald DE 3905725, 1990 [*Chem. Abstr.* **1991**, *114*, 207694x].
[662] Hangeland, J. J.; De Voss, J. J.; Heath, J. A.; Townsend, C. A.; Ding, W.-D.; Ashcroft, J. S.; Ellestad, G. A. *J. Am. Chem. Soc.* **1992**, *114*, 9200.
[663] Harnden, M. R.; Bailey, S.; Shanks, C. T. EP 49144, 1982 [*Chem. Abstr.* **1982**, *97*, 72384a].
[664] Bailey, S.; Harnden, M. R. *Nucleosides & Nucleotides* **1987**, *6*, 555.
[665] Yavorskii, A. E.; Reshot'ko, L. N.; Kucheryavenko, A. A.; Florent'ev, V. L. *Khim.-Farm. Zh.* **1988**, *7*, 833 [*Chem. Abstr.* **1989**, *110*, 135144].
[666] Yavorskii, A. E.; Kochetkova, S. V.; Smirnov, I. P.; Shchaveleva, I. L.; Tsilevich, T. L.; Gottikh, B. P.; Florent'ev, V. L. *Bioorg. Khim.* **1987**, *13*, 1000 [*Chem. Abstr.* **1988**, *108*, 131399h].
[667] Tanabe Seiyaku Co., Ltd. JP 58216169, 1983 [*Chem. Abstr.* **1984**, *100*, 174851b].
[668] Khripach, N. B.; Mikhailopulo, I. A.; Akhrem, A. A. *Khim. Geterotsikl. Soedin.* **1982**, *1*, 111 [*Chem. Abstr.* **1982**, *96*, 200078a].
[669] Skulnick, H. I. DE 2735378, 1989 [*Chem. Abstr.* **1978**, *89*, 24728].
[670] McCormick, J. E.; McElhinney, R. S. *J. Chem. Res. (S)* **1981**, 12.
[671] Iwasaki, T.; Nishitani, T.; Horikawa, H.; Inoue, I. *Tetrahedron Lett.* **1981**, *22*, 1029.
[672] Parikh, D. K.; Watson, R. R. *J. Med. Chem.* **1978**, *21*, 706.
[673] Yamashita, M.; Kawai, Y.; Uchida, I.; Komori, T.; Kohsaka, M.; Imanaka, H.; Sakane, K.; Setoi, H.; Teraji, T. *J. Antibiotics* **1984**, *37*, 1284.
[674] Szekeres, G. L; Witkowski, J. T.; Robins, R. K. *J. Carbohydr., Nucleosides, Nucleotides* **1977**, *4*, 147.

[675] Varela, O.; De Fina, G.; de Lederkremer, R. M. *J. Chem. Res. (S)* **1990**, 262.
[676] Bailey, S.; Harnden, M. R. *J. Chem. Soc., Perkin Trans. 1* **1988**, 2767.
[677] Lucey, N. M.; McCormick, J. E.; McElhinney, R. S. *J. Chem. Soc., Perkin Trans. 1* **1990**, 795.
[678] Chiu, T. M. K.; Ohrui, H.; Watanabe, K. A.; Fox, J. J. *J. Org. Chem.* **1973**, *38*, 3622.
[679] Kowollik, G.; Demirow, G.; Schütt, M.; Langen, P. *Z. Chem.* **1972**, *12*, 106.
[680] Cook, P. D.; Allen, L. B.; Streeter, D. G.; Huffman, J. H.; Sidwell, R. W.; Robins, R. K. *J. Med. Chem.* **1978**, *21*, 1212.
[681] Efange, S. M. N.; Alessi, E. M.; Shih, H. C.; Cheng, Y.-C.; Bardos, T. J. *J. Med. Chem.* **1985**, *28*, 904.
[682] Shchaveleva, I. L.; Khorlin, A. A.; Gottikh, B. P.; Florentev, V. L. *Bioorg. Khim.* **1988**, *14*, 824 [*Chem. Abstr.* **1988**, *110*, 24210].
[683] Namikawa, J.; Tominaga, M.; Manabe, Y. JP 62258396, 1986 [*Chem. Abstr.* **1988**, *109*, 129604].
[684] Sato, T.; Watamori, N. JP 0269476, 1990 [*Chem. Abstr.* **1990**, *113*, 97977m].
[685] Sato, T.; Watamori, N. JP 02229192, 1990 [*Chem. Abstr.* **1991**, *114*, 102709c].
[686] McCormick, J. E.; McElhinney, R. S. *J. Chem. Res. (S)* **1980**, 126.
[687] McCormick, J. E.; McElhinney, R. S. *J. Chem. Soc., Perkin Trans. 1* **1978**, 500.
[688] Kilevica, M.; Maurins, J.; Paegle, R.; Liepins, E.; Zidermane, A.; Lidak, M. *Khim. Geterotskl. Soedin.* **1981**, *11*, 1532 [*Chem. Abstr.* **1982**, *96*, 85922g].
[689] Schmit, C. *Synlett* **1994**, 241.
[690] Mori, Y.; Morishima, N. *Nucleic Acids Res., Symp. Ser. 29* **1993**, 47.
[691] Holý, A. Czech Patent 185435, 1980 [*Chem. Abstr.* **1981**, *95*, 98244].
[692] Ogura, H.; Iwaki, K.; Furuhata, K. In *Nucleic Acid Chemistry*; Townsend, L. B., Tipson, R. S., Eds.; Wiley: New York; 1991, Vol. 4; p. 109.
[693] Azhayev, A.; Guzaev, A.; Hovinen, J.; Mattinen, J.; Sillanpää, R.; Lönnberg, H. *Synthesis* **1994**, 396.
[694] Maeda, M.; Kajimoto, N.; Yamaizumi, Z.; Okamoto, Y.; Nagahara, K.; Takayanagi, H. *Tetrahedron Lett.* **1997**, *38*, 6841.
[695] Lichtenthaler, F. W.; Morino, T.; Winterfeldt, W. *Nucleic Acids Res., Spec. Publ. 1* **1975**, S33.
[696] Lichtenthaler, F. W.; Morino, T.; Winterfeldt, W.; Sanemitsu, Y. *Tetrahedron Lett.* **1975**, *16*, 3527.
[697] Schlimme, E.; Frister, H.; Raezke, K.-P. *Nucleosides & Nucleotides* **1988**, *7*, 577.
[698] Frister, H.; Schlimme, E. *Liebigs Ann. Chem.* **1985**, 1704.
[699] Srivastava, P. C.; Robins, R. K. *J. Med. Chem.* **1981**, *24*, 1172.
[700] Revankar, G. R.; Robins, R. K. *J. Heterocycl. Chem.* **1976**, *13*, 169.
[701] Gosselin, G.; Imbach, J.-L.; Townsend, L. B.; Panzica, R. P. *J. Heterocycl. Chem.* **1979**, *16*, 1185.
[702] Hasan, A.; Lambert, C. R.; Srivastava, P. C. *J. Heterocycl. Chem.* **1990**, *27*, 1877.
[703] Andrés, J. I.; García-Lopez, M. T. *J. Heterocycl. Chem.* **1986**, *23*, 679.
[704] Cook, P. D.; Rousseau, R. J.; Mian, A. M.; Dea, P.; Meyer, Jr., R. B.; Robins, R. K. *J. Am. Chem. Soc.* **1976**, *98*, 1492.
[705] Robins, R. K.; Rousseau, R. J.; Mian, A. M. DE 2529533, 1977 [*Chem. Abstr.* **1977**, *86*, 171784].
[706] Cook, P. D.; Rousseau, R. J.; Mian, A. M.; Meyer, Jr., R. B.; Dea, P.; Ivanovics, G.; Streeter, D. G.; Witkowski, J. T.; Stout, M. G.; Simon, L. N.; Sidwell, R. W.; Robins, R. K. *J. Am. Chem. Soc.* **1975**, *97*, 2916.
[707] Mian, A. M.; Robins, R. K. US 3919193, 1975 [*Chem. Abstr.* **1976**, *84*, 90520f].
[708] Alonso, R.; Andrés, J. I.; García-López, M. T.; de las Heras, F. G.; Herranz, R.; Alarcón, B.; Carrasco, L. *J. Med. Chem.* **1985**, *28*, 834.
[709] Herranz, R.; Andres Gil, J. I.; García Lopez, M. T.; Gomez de las Heras, F. ES 522927, 1985 [*Chem. Abstr.* **1988**, *108*, 22218c].
[710] Hayashi, M.; Mizuno, K.; Hirano, T.; Yaso, M. JP 75121275, 1975 [*Chem. Abstr.* **1976**, *84*, 122242k].
[711] Fedoseeva, N. M.; Zavadskaya, M. I.; Kvasyuk, E. I.; Drachenova, O. A.; Boreko, E. I. *Khim. Geterosikl. Soedin.* **1983**, *7*, 923 [*Chem. Abstr.* **1983**, *99*, 212863c].

[712] Srivastava, P. C.; Rousseau, R. J.; Robins, R. K. *J. Chem. Soc., Chem. Commun.* **1977**, 151.
[713] Frister, H.; Kemper, K.; Boos, K.-S.; Schlimme, E. *Liebigs Ann. Chem.* **1985**, 510.
[714] Lynch, B. M.; Sharma, S. C. *Can. J. Chem.* **1976**, *54*, 1029.
[715] Schlieper, C. A.; Wemple, J. *Nucleosides & Nucleotides* **1984**, *3*, 369.
[716] Robins, M. J.; Kaneko, C.; Kaneko, M. *Can. J. Chem.* **1981**, *59*, 3356.
[717] Knoblauch, B. H. A.; Geis, U.; Deters, D.; Müller, C. E. *Arch. Pharm. (Weinheim)* **1993**, *326*, 662.
[718] Mel'nik, S. Ya.; Bakhmedova, A. A.; Miniker, T. D.; Preobrazhenskaya, M. N. *Nucleic Acids Res., Symp. Ser. 9* **1981**, 53.
[719] Lönnberg, H. *Tetrahedron* **1982**, *38*, 1517.
[720] Mikhailopulo, I. A.; Kalinichenko, E. N.; Akhrem, A. A. *J. Carbohyd., Nucleosides, Nucleotides* **1981**, *8*, 227.
[721] Bärwolff, D.; Etzold, G.; Langen, P. DD 98930, 1972 [*Chem. Abstr.* **1974**, *80*, 83535d].
[722] Chwang, T. L.; Wood, W. F.; Parkhurst, J. R.; Nesnow, S.; Danenberg, P. V.; Heidelberger, C. *J. Med. Chem.* **1976**, *19*, 643.
[723] Akhrem, A. A.; Pupeiko, N. E.; Kvasyuk, E. I.; Mikhailopulo, I. A.; Reshchikov, V. P.; Fertukova, N. M.; Nikolaeva, N. A.; Grishina, T. P. *Vestsi Akad. Navuk BSSR, Ser. Khim. Navuk* **1979**, *6*, 103 [*Chem. Abstr.* **1980**, *92*, 147110u].
[724] Fu, Y. L.; Parthasarathy, R.; Bobek, M. *J. Carbohydr., Nucleosides, Nucleotides* **1978**, *5*, 79.
[725] Wegner, G.; Jochims, J. C. *Chem. Ber.* **1979**, *112*, 1941.
[726] Berkowitz, P. T.; Robins, R. K.; Dea, P.; Long, R. A. *J. Org. Chem.* **1976**, *41*, 3128.
[727] Gosselin, G.; Loukil, H. F.; Mathieu, A.; Mesli, A. *J. Heterocycl. Chem.* **1978**, *15*, 657.
[728] Barascut, J. L.; Molko, D.; Imbach, J. L. *J. Carbohyd., Nucleosides, Nucleotides* **1980**, *7*, 185.
[729] Pickering, M. V.; Dea, P.; Streeter, D. G.; Witkowski, J. T. *J. Med. Chem.* **1977**, *20*, 818.
[730] Itoh, T.; Ono, K.; Sugawara, T.; Mizuno, Y. *Nucleosides & Nucleotides* **1982**, *1*, 179.
[731] Itoh, T.; Sugawara, T.; Mizuno, Y. *Heterocycles* **1982**, *17*, 305.
[732] Itoh, T.; Sugawara, T.; Mizuno, Y. *J. Heterocycl. Chem.* **1982**, *19*, 513.
[733] Bartholomew, D. G.; Dea, P.; Robins, R. K.; Revankar, G. R. *J. Org. Chem.* **1975**, *40*, 3708.
[734] Lynch, B. M.; Sharma, S. C. *Can. J. Chem.* **1977**, *55*, 831.
[735] Cuny, E.; Lichtenthaler, F. W. *Nucleic Acids Res., Spec. Publ. 1* **1975**, S25.
[736] Steinmaus, H. DE 2224379, 1973 [*Chem. Abstr.* **1974**, *80*, 48338v].
[737] Baraldi, P. G.; Simoni, D.; Periotto, V.; Manfredini, S.; Guarneri, M.; Manservigi, R.; Cassai, E.; Bertolasi, V. *J. Med. Chem.* **1984**, *27*, 986.
[738] Kazmina, E. M.; Bezchinsky, Y. E.; Novikov, N. A.; Rosenberg, S. G.; Bochkarev, A. V.; Zdanov, A. S.; Fedorov, I. I. *Collect. Czech. Chem. Commun.* **1990**, *55*, 41.
[739] Goya, P.; Martínez, A.; Ochoa, C. *Nucleosides & Nucleotides* **1987**, *6*, 631.
[740] de la Cruz, A.; Elguero, J.; Goya, P.; Martinez, A.; De Clercq, E. *J. Chem. Soc., Perkin Trans. 1* **1993**, 845.
[741] Parry, R. J.; Ju, S. *Tetrahedron* **1991**, *47*, 6069.
[742] Rama Rao, A. V.; Gurjar, M. K.; Lalitha, S. V. S. *J. Chem. Soc., Chem. Commun.* **1994**, 1255.
[743] Singh, K.; Hasan, A.; Pratap, R.; Guru, P. Y.; Bhakuni, D. S. *J. Indian Chem. Soc.* **1989**, *66*, 686.
[744] Kelley, J. A.; Driscoll, J. S.; McCormack, J. J.; Roth, J. S.; Marquez, V. E. *J. Med. Chem.* **1986**, *29*, 2351.
[745] Jochims, J. C.; v. Voithenberg, H.; Wegner, G. *Chem. Ber.* **1978**, *111*, 2745.
[746] Mackiewiczowa, D.; Jasinska, J.; Sokolowski, J. *Pol. J. Chem.* **1981**, *55*, 2333 [*Chem. Abstr.* **1983**, *99*, 88504v].
[747] Kochetkova, S. V.; Tsilevich, T. L.; Smirnov, I. P.; Shchaveleva, I. L.; Khorlin, A. A.; Gottikh, B. P.; Florentev, V. L. *Bioorg. Khim.* **1988**, *14*, 820 [*Chem. Abstr.* **1989**, *110*, 8558y].
[748] Suzuki, T.; Sakaguchi, M.; Miyata, Y.; Suzuki, A.; Mori, T. EP 513351, 1992 [*Chem. Abstr.* **1991**, *115*, 256569q]..
[749] Jeong, L. S.; Beach, J. W.; Chu, C. K. *J. Heterocycl. Chem.* **1993**, *30*, 1445.
[750] Johansson, K. N. G.; Malmberg, H. C. G.; Noreen, R.; Sahlberg, S. C.; Sohn, D. D.; Gronowitz, S. EP 357571 1990 [*Chem. Abstr.* **1990**, *112*, 23577e].

751 Kuno, H.; Niihata, S.; Ebata, T.; Matsushita, H. *Heterocycles* **1995**, *41*, 523.

752 Watanabe, K. A.; Matsuda, A.; Itoh, T. *Can. J. Chem.* **1981**, *59*, 468.

753 Etzold, G.; v. Janta-Lipinski, M.; Langen, P. *J. Prakt. Chem.* **1976**, *318*, 79.

754 Randow, H.; Miethchen, R. *J. Prakt. Chem.* **1991**, *333*, 285.

755 Timoshchuk, V. A. *Zh. Obshch. Khim.* **1984**, *54*, 2646 [*Chem. Abstr.* **1985**, *102*, 113861z].

756 Komiotis, D.; Delatre, S.; Holt, L.; Ollapally, A. P.; Balzarini, J.; De Clercq, E.; Iigo, M. *Nucleosides & Nucleotides* **1991**, *10*, 431.

757 Ford, R. A.; Gibson, M.; Rawji, H.; Hughes, L.; Montserret, R.; Ollapally, A. P. *Nucleosides & Nucleotides* **1984**, *3*, 313.

758 Cheng, V.; Hughes, L.; Griffin, V. B.; Montserret, R.; Ollapally, A. P. *Nucleosides & Nucleotides* **1986**, *5*, 223.

759 Alaoui-Jamali, M. A.; Egron, M. J.; Bessodes, M.; Antonakis, K.; Chouroulinkov, I. *Eur. J. Med. Chem.* **1987**, *22*, 305.

760 Sakaguchi, M.; Larroquette, C. A.; Agrawal, K. C. *J. Med. Chem.* **1983**, *26*, 20.

761 Lerner, L. M.; Mennitt, G.; Gaetjens, E.; Sheid, B. *Carbohydr. Res.* **1993**, *244*, 285.

762 Brink, A. J.; De Villiers, O. G.; Jordaan, A. *Carbohydr. Res.* **1977**, *54*, 285.

763 Mikhailopulo, I. A.; Sivets, G. G.; Pricota, T. I.; Poopeiko, N. E.; Balzarini, J.; De Clercq, E. *Nucleosides & Nucleotides* **1991**, *10*, 1743.

764 Murata, M.; Achiwa, K. *Chem. Pharm. Bull.* **1990**, *38*, 836.

765 Brink, A. J.; Coetzer, J.; De Villiers, O. G.; Hall, R. H.; Jordaan, A.; Kruger, G. J. *Tetrahedron* **1976**, *32*, 965.

766 Lichtenthaler, F. W.; Heerd, A.; Strobel, K. *Chem. Lett.* **1974**, 449.

767 Mahrwald, R.; Wagner, G. *Pharmazie* **1980**, *35*, 587.

768 Kulinkovich, L. N.; Timoshchuk, V. A. *Khim. Prir. Soedin.* **1983**, *5*, 617 [*Chem. Abstr.* **1984**, *100*, 156915w].

769 Watanabe, K. A.; Hollenberg, D. H.; Fox, J. J. *J. Antibiotics* **1976**, *29*, 597.

770 Kaneko, M.; Kimura, M.; Tanaka, H.; Shimizu, F.; Arakawa, M.; Shimizu, B. *Nucleic Acids Res., Symp. Ser. 3* **1977**, S35.

771 Chwang, T. L.; Heidelberger, C. *Tetrahedron Lett.* **1974**, *15*, 95.

772 Azhaev, A. V.; Ozols, A. M.; Bushnev, A. S.; Dyatkina, N. B.; Kochetkova, S. V.; Viktorova, L. S.; Kukhanova, M. K.; Kraevskii, A. A.; Gottikh, B. P. *Nucleic Acids Res., Symp. Ser. 6* **1979**, 625.

773 Kumar, A.; Khan, S. I.; Manglani, A.; Khan, Z. K.; Katti, S. B. *Nucleosides & Nucleotides* **1994**, *13*, 1049.

774 Rizzo, C. J.; Dougherty, J. P.; Breslow, R. *Tetrahedron Lett.* **1992**, *33*, 4129.

775 Warnock, D. H.; Watanabe, K. A.; Fox, J. J. *Carbohydr. Res.* **1971**, *18*, 127.

776 Jochims, J. C.; von Voithenberg, H.; Wegner, G. *Chem. Ber.* **1978**, *111*, 1693.

777 Novák, J. J. K. *Collect. Czech. Chem. Commun.* **1978**, *43*, 1511.

778 Sumitomo Chemical Co., Ltd. JP 5931797, 1984 [*Chem. Abstr.* **1984**, *101*, 111351p].

779 Idegami, K.; Ikeda, K.; Achiwa, K. *Chem. Pharm. Bull.* **1990**, *38*, 1766.

780 Lerner, L. M.; Mennitt, G. *Carbohydr. Res.* **1994**, *259*, 191.

781 Anzai, K.; Saita, T. *Nucleic Acids Res., Spec. Publ. 2* **1976**, 87.

782 Murofushi, Y.; Kimura, M.; Kuwano, H.; Iijima, Y.; Yamazaki, M.; Kaneko, M. *Chem. Pharm. Bull.* **1988**, *36*, 3760.

783 Rassu, G.; Pinna, L.; Spanu, P.; Ulgheri, F.; Casiraghi, G. *Tetrahedron Lett.* **1994**, *35*, 4019.

784 Bardos, T. J.; Cheng, Y.-C.; Schroeder, A. C.; Mbua, N. E. S. EP 175004, 1984 [*Chem. Abstr.* **1986**, *105*, 60912v].

785 Paulsen, H.; Brieden, M.; Benz, G. *Liebigs Ann. Chem.* **1987**, 565.

786 Ogilvie, K. K.; Dixit, D. M.; Radatus, B. K.; Smith, K. O.; Galloway, K. S. *Nucleosides & Nucleotides* **1983**, *2*, 147.

787 Ogilvie, K. K. EP 49072, 1982 [*Chem. Abstr.* **1982**, *97*, 56193k].

788 Ogilvie, K. K.; Cheriyan, U. O.; Radatus, B. K.; Smith, K. O.; Galloway, K. S.; Kennell, W. L. *Can. J. Chem.* **1982**, *60*, 3005.

789 Garner, P.; Park, J. M. *J. Org. Chem.* **1990**, *55*, 3772.

[790] Kawakami, H.; Matsushita, H.; Shibagaki, M.; Naoi, Y.; Itoh, K.; Yoshikoshi, H. *Chem. Lett.* **1989**, 1365.
[791] Efange, S. M. N.; Cheng, Y. C.; Bardos, T. J. *Nucleosides & Nucleotides* **1985**, *4*, 545.
[792] Chiacchio, U.; Gumina, G.; Rescifina, A.; Romeo, R.; Uccella, N.; Casuscelli, F.; Piperno, A.; Romeo, G. *Tetrahedron* **1996**, *52*, 8889.
[793] Linn, J. A.; Kelley, J. L. *Nucleosides & Nucleotides* **1993**, *12*, 199.
[794] Wierenga, W.; Skulnick, H. I. *Carbohydr. Res.* **1981**, *90*, 41.
[795] Beigel'mann, L. N.; Karpeiskii, M. Ya.; Mikhailov, S. N.; Rosenthal, A. *Nucleic Acids Res., Symp. Ser. 9* **1981**, 115.
[796] Mikhailov, S. N.; Beigel'mann, L. N.; Gurskaya, G. V.; Padyukova, N. Sh.; Yakovlev, G. I.; Karpeiskii, M. Ya. *Carbohydr. Res.* **1983**, *124*, 75.
[797] Mikhailopulo, I. A.; Zaitseva, G. V.; Mikhailovskaya, N. A. *Khim. Geterotsikl. Soedin.* **1982**, *4*, 542 [*Chem. Abstr.* **1982**, *97*, 24153f].
[798] Iwakawa, M.; Martin, O. R.; Szarek, W. A. *Carbohydr. Res.* **1983**, *121*, 99.
[799] Paegle, R. A.; Lidaka, M. J.; Žhuk, R. A.; Mauriņš, J. A.; Zidermane, A. A.; Kilevica, M. K. DE 2833507, 1979 [*Chem. Abstr.* **1979**, *91*, 39793q].
[800] Saneyoshi, M.; Tohyama, J.; Nakayama, C. *Chem. Pharm. Bull.* **1982**, *30*, 2223.
[801] Bobek, M.; Bloch, A.; Kuhar, S. *Tetrahedron Lett.* **1973**, *14*, 3493.
[802] Crich, D.; Hao, X. *J. Org. Chem.* **1998**, **63**, 3796.
[803] Mian, A. M.; Khwaja, T. A. *J. Med. Chem.* **1983**, *26*, 286.
[804] Hollis Showalter, H. D; Putt, S. R; Borondy, P. E.; Shillis, J. L. *J. Med. Chem.* **1983**, *26*, 1478.
[805] Chan, E.; Putt, E. S.; Hollis Showalter, H. D.; Baker, D. C. *J. Org. Chem.* **1982**, *47*, 3457.
[806] Preobrazhenskaya, M. N.; Mel'nik, S. Ya.; Bakhmedova, A. A.; Mezhevich, Z. I.; Mamaev, V. P.; Zagulyaeva, O. A.; Bektemirov, T. A.; Chekunova, E. V.; Andzhaparidze, O. G. SU-671287, 1983 [*Chem. Abstr.* **1984**, *100*, 45298c].
[807] Sharma, R. A.; Bobek, M. In *Nucleic Acid Chemistry*; Townsend, L. B., Tipson, R. S., Eds.; Wiley: New York; 1986, Vol. 3; p. 53.
[808] Sharma, R. A.; Bobek, M. *J. Org. Chem.* **1975**, *40*, 2377.
[809] Okada, J.; Nakano, K.; Miyake, H. *Chem. Pharm. Bull.* **1985**, *33*, 856.
[810] Draminski, M.; Zgit-Wróblewska, A. *Pol. J. Chem.* **1980**, *54*, 1085 [*Chem. Abstr.* **1981**, *94*, 84418u].
[811] Wise, D. S.; Townsend, L. B. *J. Chem. Soc., Chem. Commun.* **1979**, *6*, 271.
[812] Coe, P. L.; Harnden, M. R.; Jones, A. S.; Noble, S. A.; Walker, R. T. *J. Med. Chem.* **1982**, *25*, 1329.
[813] Perman, J.; Sharma, R. A.; Bobek, M. *Tetrahedron Lett.* **1976**, *17*, 2427.
[814] Pískala, A.; Hanna, N. B.; Budesínsky, M.; Cihák, A.; Vesely, J. *Nucleic Acids Res., Symp. Ser. 9* **1981**, 83.
[815] Mitchell, W. L.; Ravenscroft, P.; Hill, M. L.; Knutsen, L. J. S.; Newton, R. F.; Scopes, D. I. C. *Nucleosides & Nucleotides* **1985**, *4*, 173.
[816] Sweeney, D. L.; Gearien, J. E.; Bauer, L.; Currie, B. L. *J. Carbohydr., Nucleosides, Nucleotides* **1979**, *6*, 387.
[817] Gould, J. H. M.; Mann, J. *J. Chem. Soc., Chem. Commun.* **1997**, 243.
[818] Hollis Showalter, H. D.; Putt, S. R; Baker, D. C. In *Nucleic Acid Chemistry*; Townsend, L. B., Tipson, R. S., Eds.; Wiley: New York; 1991, Vol. 4; p. 308.
[819] Woo, P. W. K.; Lee, H. T. *J. Labelled Comp. Radiopharm.* **1990**, *28*, 445.
[820] David, S.; de Sennyey, G. *Carbohydr. Res.* **1979**, *77*, 79.
[821] Mikhailov, S. N.; Padyukova, N. S.; Karpeiskii, M. Y.; Kolobushkina, L. I.; Beigelman, L. N. *Collect. Czech. Chem. Commun.* **1989**, *54*, 1055.
[822] Holý, A.; Ludziša, A.; Votruba, I.; Šediva, K.; Pischel, H. *Collect. Czech. Chem. Commun.* **1985**, *50*, 393.
[823] Augustyns, K.; Vandendriessche, F.; Van Aerschot, A.; Busson, R.; Urbanke, C.; Herdewijn, P. *Nucleic Acids Res.* **1992**, *20*, 4711.
[824] Funai Pharmaceutical Industries, Ltd. JP 5929699, 1984 [*Chem. Abstr.* **1984**, *101*, 55487y].
[825] Poopeiko, N. E.; Kvasyuk, E. I.; Mikhailopulo, I. A.; Lidaks, M. J. *Synthesis* **1985**, 605.

[826] Tolstikov, G. A.; Mustafin, A. G.; Yghibaeva, G. Kh.; Gataullin, R. R.; Spirikhin, L. V.; Sultanova, V. S.; Abdrakhmanov, I. B. *Mendeleev Commun.* **1993**, 194.
[827] Hřebanecký, H.; Holý, A. EP 301908, 1989 [*Chem. Abstr.* **1989**, *111*, 78550j].
[828] Van Aerschot, A.; Herdewijn, P.; Janssen, G.; Cools, M.; De Clercq, E. *Antiviral Res.* **1989**, *12*, 133.
[829] Schmit, C. *Synlett* **1994**, 238.
[830] Wang, Z.; Rizzo, C. J. *Tetrahedron Lett.* **1997**, *38*, 8177.
[831] Johansson, K. N. G.; Lindborg, B. G.; Noreen, R. EP 352248, 1990 [*Chem. Abstr.* **1990**, *113*, 41231w].
[832] Watanabe, K. A.; Wempen, I. M.; Fox, J. J. *Carbohydr. Res.* **1972**, *21*, 148.
[833] Billich, A.; Stockhove, U.; Witzel, H. *Nucleic Acids Res.* **1983**, *11*, 7611.
[834] David, S.; De Sennyey, G. *CR Hebd. Seances Acad. Sci. Ser. C* **1974**, *279*, 651.
[835] Keller, T. H.; Häner, R. *Helv. Chim. Acta* **1993**, *76*, 884.
[836] Ogura, H.; Furuhata, F.; Iwaki, K.; Takahashi, H. *Nucleic Acids Res., Symp. Ser. 10* **1981**, 23.
[837] Watanabe, K. A.; Matsuda, A.; Halat, M. J.; Hollenberg, D. H.; Nisselbaum, J. S.; Fox, J. J. *J. Med. Chem.* **1981**, *24*, 893.
[838] McGee, D. P. C.; Martin, J. C.; Verheyden, J. P. H. *J. Heterocycl. Chem.* **1985**, *22*, 1137.
[839] Rusjakovski, B.; Kobe, J.; Brdar, B.; Japelj, M.; Valcavi, U. *Il Farmaco, Ed. Sc.* **1982**, *37*, 764 [*Chem. Abstr.* **1983**, *98*, 72650h].
[840] Mikhailov, S. N.; Efimtseva, E. V. *Khim. Geterotsikl. Soedin.* **1988**, *7*, 947 [*Chem. Abstr.* **1989**, *110*, 173660h].
[841] Baud, M.-V.; Chavis, C.; Lucas, M.; Imbach, J.-L. *Tetrahedron* **1991**, *47*, 9993.
[842] Makabe, O.; Suzuki, H.; Umezawa, S. *Bull. Chem. Soc. Jpn.* **1977**, *50*, 2689.
[843] Mazur, A.; Tropp, B. E.; Engel, R. *Tetrahedron* **1984**, *40*, 3949.
[844] Nelson, V.; El Khadem, H. S. *J. Med. Chem.* **1983**, *26*, 1527.
[845] Dunkel, M.; Cook, P. D.; Acevedo, O. L. *J. Heterocycl. Chem.* **1993**, *30*, 1421.
[846] Dumont, F.; Wightman, R. H.; Ziegler, J. C.; Chavis, C.; Imbach, J.-L. *Tetrahedron Lett.* **1979**, *20*, 3291.
[847] Chavis, C.; Dumont, F.; Wightman, R.; Ziegler, J.-C.; Imbach, J.-L. In *Nucleic Acid Chemistry*; Townsend, L. B., Tipson, R. S., Eds.; Wiley: New York; 1986, Vol. 3; p. 76.
[848] Kazimierczuk, Z.; Dudycz, L.; Stolarski, R.; Shugar, D. Z. *Naturforsch.* **1980**, *35c*, 30.
[849] Kazimierczuk, Z.; Stolarski, R.; Dudycz, L.; Shugar, D. *Z. Naturforsch.* **1981**, *36c*, 126.
[850] Haines, D. R.; Leonard, N. J.; Wiemer, D. F. *J. Org. Chem.* **1982**, *47*, 474.
[851] Cook, P. D.; Robins, R. K. *J. Org. Chem.* **1978**, *43*, 289.
[852] Wyss, P. C.; Fischer, U. *Helv. Chim. Acta* **1978**, *61*, 3149.
[853] Fischer, U.; Wyss, P. C. DE 2735458, 1978 [*Chem. Abstr.* **1978**, *89*, 24726e].
[854] Fuertes, M.; Robins, R. K.; Witkowski, J. T. *J. Carbohydr., Nucleosides, Nucleotides* **1976**, *3*, 169.
[855] Meyer, Jr., R. B.; Revankar, G. R.; Cook, P. D.; Ehler, K. W.; Schweizer, M. P.; Robins, R. K. *J. Heterocycl. Chem.* **1980**, *17*, 159.
[856] Ohba, Y.; Nishiwaki, T.; Akagi, H.; Mori, M. *J. Chem. Res. S* **1984**, 276.
[857] Cook, P. D.; Day, R. T.; Robins, R. K. *J. Heterocycl. Chem.* **1977**, *14*, 1295.
[858] Dutta, S. P.; Chheda, G. B. In *Nucleic Acid Chemistry*; Townsend, L. B., Tipson, R. S., Eds.; Wiley: New York; 1991, Vol. 4; p. 152.
[859] Legraverend, M.; Nguyen, C. H.; Zerial, A.; Bisagni, E. *Nucleosides & Nucleotides* **1986**, *5*, 125.
[860] McNamara, D. J.; Cook, P. D. *J. Med. Chem.* **1987**, *30*, 340.
[861] McNamara, D. J.; Cook, P. D. EP 210639, 1987 [*Chem. Abstr.* **1987**, *106*, 176811m].
[862] Katz, D. J.; Wise, D. S.; Townsend, L. B. *J. Heterocycl. Chem.* **1975**, *12*, 609.
[863] Katz, D. J.; Wise, D. S.; Townsend, L. B. *J. Med. Chem.* **1982**, *25*, 813.
[864] Roberts, J. L.; Poulter, C. D. *J. Org. Chem.* **1978**, *43*, 1547.
[865] Sierzputowska-Gracz, H.; Sochacka, E.; Malkiewicz, A.; Kuo, K.; Gehrke, C. W.; Agris, P. F. *J. Am. Chem. Soc.* **1987**, *109*, 7171.
[866] Wise, D. S.; Townsend, L. B. *J. Heterocycl. Chem.* **1972**, *9*, 1461.

[867] Kim, C.-H.; Marquez, V. E.; Mao, D. T.; Haines, D. R.; McCormack, J. J. *J. Med. Chem.* **1986**, *29*, 1374.

[868] Ozaki, S.; Japan. Kokai 75 82 079, 1975 [*Chem. Abstr.* **1975**, *83*, 193647m].

[869] Scott, J. W.; Cook, A. F.; Gregg, J. J.; May, K. B.; Nix, Jr., G.; Parrish, D. R.; Wagner, D. P. *J. Carbohydr., Nucleosides, Nucleotides* **1981**, *8*, 171.

[870] Kvasyuk, E. I.; Kulak, T. I.; Tkachenko, O, V.; Mikhailopulo, I. A.; Zinchenko, A. I.; Barai, V. N.; Bokut, S. B.; Marennikova, S. S.; Chekunova, E. V. *Khim.-Farm. Zh.* **1989**, *23*, 699 [*Chem. Abstr.* **1989**, *111*, 166826j].

[871] Holý, A.; König, J.; Vesely, J.; Cech, D.; Votruba, I.; De Clercq, E. *Collect. Czech. Chem. Commun.* **1987**, *52*, 1589.

[872] Barr, P. J.; Jones, A. S.; Serafinowski, P.; Walker, R. T. *J. Chem. Soc., Perkin Trans. 1* **1978**, 1263.

[873] Bašnák, I.; Farkaš, J. *Collect. Czech. Chem. Commun.* **1976**, *41*, 3635.

[874] Malkiewicz, A. J.; Nawrot, B. *Z. Naturforsch.* **1987**, *42B*, 355.

[875] Malkiewicz, A.; Sochacka, E. *Tetrahedron Lett.* **1983**, *24*, 5387.

[876] Anzai, K.; Miyamoto, T. *Nucleic Acids Res., Spec. Publ. 3* **1977**, S1.

[877] Cook, A. F.; Holman, M. J. *J. Org. Chem.* **1978**, *43*, 4200.

[878] Otvos, L.; Kruppa, G. HU 27646, 1983 [*Chem. Abstr.* **1984**, *100*, 192219f].

[879] Levine, H. L.; Brody, R. S.; Westheimer, F. H. *Biochemistry* **1980**, *19*, 4993.

[880] Tolstikov, G. A.; Baltina, L. A.; Khalilov, L. M.; Spirikhin, L. V.; Sultanmuratova, V. R.; Murinov, Y. I. *Khim. Geterotsikl. Soedin.* **1991**, *6*, 794 [*Chem. Abstr.* **1992**, *116*, 41965a].

[881] Sun, S.; Tang, X. Q.; Merchant, A.; Anjaneyulu, P. S. R.; Piccirilli, J. A. *J. Org. Chem.* **1996**, *61*, 5708.

[882] Lee, T.-C.; Chello, P. L.; Chou, T.-C.; Templeton, M. A.; Parham, J. C. *J. Med. Chem.* **1983**, *26*, 283.

[883] Davis, J.; Benhaddou, R.; Granet, R.; Krausz, P.; Demonte, M.; Aubertin, A. M. *Nucleosides & Nucleotides* **1998**, *17*, 875.

[884] Ognyanik, S. S.; Tarnavskii, S. S.; Kobylinskaya, V. I.; Alekseeva, I. V.; Shalamai, A. S. *Ukr. Khim. Zh. (Russ. Ed.)* **1988**, *54*, 1094 [*Chem. Abstr.* **1989**, *111*, 78513z].

[885] Cristescu, C. *Rev. Roum. Chim.* **1975**, *20*, 1287 [*Chem. Abstr.* **1976**, *84*, 59916r].

[886] Cristescu, C. *Monatsh. Chem.* **1977**, *108*, 1455.

[887] Yoshii, K.; Ohba, Y.; Nishiwaki, T. *J. Heterocycl. Chem.* **1993**, *30*, 141.

[888] Patil, V. D.; Wise, D. S.; Wotring, L. L.; Bloomer, L. C.; Townsend, L. B. *J. Med. Chem.* **1985**, *28*, 423.

[889] Itoh, T.; Inaba, J.; Mizuno, Y. *Heterocycles* **1977**, *8*, 433.

[890] Frankowski, A. *Tetrahedron* **1986**, *42*, 1511.

[891] Okada, J.; Nakano, K.; Miyake, H. *Chem. Pharm. Bull.* **1983**, *31*, 3074.

[892] Fossey, C.; Landelle, H.; Laduree, D.; Robba, M. *Nucleosides & Nucleotides* **1994**, *13*, 925.

[893] Fossey, C.; Landelle, H.; Laduree, D.; Robba, M. *Nucleosides & Nucleotides* **1993**, *12*, 973.

[894] Wamhoff, H.; Wambach, W.; Herrmann, S.; Jansen, M.; Brühne, B. *J. Prakt. Chem./Chem.-Ztg.* **1994**, *336*, 129.

[895] Kawasaki, A. M.; Townsend, L. B. In *Nucleic Acid Chemistry*; Townsend, L. B., Tipson, R. S., Eds.; Wiley: New York; 1991, Vol. 4; p. 298.

[896] Schmidt, C. L.; Townsend, L. B. *J. Chem. Soc., Perkin Trans. 1* **1975**, *13*, 1257.

[897] Voegel, J. J.; Altorfer, M. M.; Benner, S. A. *Helv. Chim. Acta* **1993**, *76*, 2061.

[898] Kim, S.-H.; Bartholomew, D. G.; Allen, L. B.; Robins, R. K.; Revankar, G. R.; Dea, P. *J. Med. Chem.* **1978**, *21*, 883.

[899] Gagnier, R. P.; Halat, M. J.; Otter, B. A. *J. Heterocycl. Chem.* **1982**, *19*, 221.

[900] Gagnier, R. P.; Halat, M. J.; Otter, B. A. *J. Heterocycl. Chem.* **1984**, *21*, 481.

[901] Schmidt, C. L.; Townsend, L. B. *J. Org. Chem.* **1975**, *40*, 2476.

[902] Nishiwaki, T.; Abe, N. *J. Chem. Res. (S)* **1984**, 264.

[903] Pfleiderer, W.; Schranner, M. *Chem. Ber.* **1971**, *104*, 1915.

[904] Ritzmann, G.; Kiriasis, L.; Pfleiderer, W. *Chem. Ber.* **1980**, *113*, 1524.

[905] Ritzmann, G.; Pfleiderer, W. *Chem. Ber.* **1973**, *106*, 1401.

[906] Kobayashi, K.; Pfleiderer, W. *Chem. Ber.* **1976**, *109*, 3194.
[907] Kaz'mina, E. M.; Kenbaeva, R. M.; Stepanenko, B. N. *Zh. Obshch. Khim.* **1978**, *48*, 1892 [*Chem. Abstr.* **1978**, *89*, 180287t].
[908] Kaz'mina, E. M.; Fedorov, I. I.; Bezchinskii, Ya. E.; Kiseleva, N. V.; Novikov, N. A.; Galegov, G. A.; Arzamastsev, A. P. *Khim.-Farm. Zh.* **1989**, *23*, 1217 [*Chem. Abstr.* **1990**, *112*, 235744r].
[909] Volpini, R.; Camaioni, E.; Vittori, S.; Barboni, L.; Lambertucci, C.; Cristalli, G. *Helv. Chim. Acta* **1998**, *81*, 145.
[910] Hřebabecký, H.; Holý, A. CS 274511, 1992 [*Chem. Abstr.* **1994**, *120*, 8932x].
[911] David, S.; de Sennyey, G. *J. Chem. Soc., Perkin Trans. 1* **1982**, 385.
[912] Karpeiskii, M. Ya.; Mikhailov, S. N.; Padyukova, N. Sh.; Smrt, J. *Nucleic Acids Res., Symp. Ser. 9* **1981**, 157.
[913] Ollmann, J. E.; DePasquale, R. J. US 4898936, 1990 [*Chem. Abstr.* **1986**, *105*, 227244t].
[914] Hřebabecký, H.; Holý, A. *Collect. Czech. Chem. Commun.* **1993**, *58*, 409.
[915] Hřebabecký, H.; Holý, A. *Carbohyd. Res.* **1991**, *216*, 179.
[916] Suami, T.; Sasai, H.; Matsuno, K. *Chem. Lett.* **1983**, 819.
[917] Hutzenlaub, W.; Kobayashi, K.; Pfleiderer, W. *Chem. Ber.* **1976**, *109*, 3217.
[918] Grouiller, A.; Chattopadhyaya, J. *Heterocycles* **1984**, *21*, 543.
[919] Ben-Hattar, J.; Jiricny, J. *J. Org. Chem.* **1986**, *51*, 3211.
[920] Grouiller, A.; Chattopadhyaya, J. *Acta Chem. Scand. Ser. B* **1984**, *38*, 367.
[921] Lichtenthaler, F. W.; Ueno, T.; Voss, P. *Bull. Chem. Soc. Jpn.* **1974**, *47*, 2304.
[922] Augustyns, K.; Van Aerschot, A.; Herdewijn, P. *Bioorg. Med. Chem. Lett.* **1992**, *2*, 945.
[923] Tietze, L. F.; Krach, T.; Beller, M.; Arlt, M. *Chem. Ber.* **1991**, *124*, 2019.
[924] Tronchet, J. M. J.; Iznaden, M.; Barbalat-Rey, F.; Dhimane, H.; Ricca, A.; Balzarini, J.; De Clercq, E. *Eur. J. Med. Chem.* **1992**, *27*, 555.
[925] Schmidt, L.; Pedersen, E. B.; Nielsen, C. *Acta Chem. Scand.* **1994**, *48*, 215.
[926] Jung, M. E.; Castro, C.; Gardiner, J. M. *Tetrahedron Lett.* **1991**, *32*, 5717.
[927] Wittenburg, E.; Etzold, G.; Langen, P. *Chem. Ber.* **1968**, *101*, 494.
[928] Kline, P. C.; Serianni, A. S. *J. Org. Chem.* **1992**, *57*, 1772.
[929] Jiang, X.-J.; Kalman, T. I. *Nucleosides & Nucleotides* **1994**, *13*, 379.
[930] D'Sousa, R.; Kiss, J. EP 21231, 1981 [*Chem. Abstr.* **1981**, *95*, 62602t].
[931] Beigelman, L. N.; Mikhailov, S. N. *Carbohydr. Res.* **1990**, *203*, 324.
[932] Kawai, S. H.; Chin, J.; Just, G. *Nucleosides & Nucleotides* **1990**, *9*, 1045.
[933] Dueholm, K. L.; Motawia, M. S.; Pedersen, E. B.; Nielsen, C. *Arch. Pharm. (Weinheim)* **1992**, *325*, 597.
[934] Saischek, G.; Fuchs, F.; Dax, K.; Billiani, G. EP 450585, 1991 [*Chem. Abstr.* **1992**, *116*, 41980b].
[935] Maurins, J. A.; Paegle, R. A.; Lidaks, M. J.; Kvasyuk, E. I.; Mikhailopulo, I. A. *Bioorg. Khim.* **1986**, *12*, 1514 [*Chem. Abstr.* **1987**, *107*, 176394y].
[936] Casara, P. J.; Kenny, M. T.; Jund, K. C. *Tetrahedron Lett.* **1991**, *32*, 3823.
[937] Gautier, C.; Leroy, R.; Monneret, C.; Roger, P. *Tetrahedron Lett.* **1991**, *32*, 3361.
[938] Almond, M. R.; Collins, J. L.; Reitter, B. E.; Rideout, J. L.; Freeman, A.; St. Clair, M. H. *Tetrahedron Lett.* **1991**, *32*, 5745.
[939] Hansen, P.; Lau, J.; Pedersen, E. B.; Nielsen, C. M. *Liebigs Ann. Chem.* **1990**, 1079.
[940] Dionne, G.; WO 9303027, 1992 [*Chem. Abstr.* **1993**, *119*, 226345d].
[941] Same as Ref. 940.
[942] Belleau, B.; Nguyen, B. N. EP 515144, 1992 [*Chem. Abstr.* **1993**, *118*, 169533s].
[943] Bamford, M. J.; Humber, D. C.; Storer, R. *Tetrahedron Lett.* **1991**, *32*, 271.
[944] Evans, C. A.; Dixit, D. M.; Siddiqui, M. A.; Jin, H.; Tse, H. L. A.; Cimpoia, A.; Bednarski, K.; Breining, T.; Mansour, T. S. *Tetrahedron Asymmetry* **1993**, *4*, 2319.
[945] Siddiqui, M. A.; Brown, W. L.; Nguyen-Ba, N.; Dixit, D. M.; Mansour, T. S.; Hooker, E.; Viner, K. C.; Cameron, J. M. *Bioorg. Med. Chem. Lett.* **1993**, *3*, 1543.
[946] Belleau, B.; Belleau, P. Nguyen-Ba, N. EP 382526, 1990 [*Chem. Abstr.* **1991**, *114*, 43492b].
[947] Barrett, A. G. M.; Lebold, S. A. *J. Org. Chem.* **1990**, *55*, 3853.
[948] Hanna, N. B.; Joshi, R. V.; Larson, S. B.; Robins, R. K.; Revankar, G. R. *J. Heterocycl. Chem.* **1989**, *26*, 1835.

[949] Ugarkar, B. G.; Cottam, H. B.; McKernan, P. A.; Robins, R. K.; Revankar, G. R. *J. Med. Chem.* **1984**, *27*, 1026.

[949a] Depellier, J.; Granet, R.; Krausz, P.; Piekarski, S. *Nucleosides & Nucleotides* **1994**, *13*, 1007.

[950] Hanna, N. B.; Larson, S. B.; Robins, R. K.; Revankar, G. R. *J. Heterocycl. Chem.* **1990**, *27*, 1713.

[951] Revankar, G. R.; Hanna, N. B.; Imamura, N.; Lewis, A. F.; Larson, S. B.; Finch, R. A.; Avery, T. L.; Robins, R. K. *J. Med. Chem.* **1990**, *33*, 121.

[952] Maurins, J. A.; Paegle, R. A.; Lídaks, M. J.; Kvasyuk, E. I.; Kuzmichkin, P. V.; Mikhailopulo, I. A. *Nucleosides & Nucleotides* **1986**, *5*, 79.

[953] Ozola, V.; Ramzaeva, N.; Maurinsh, Y.; Lidaks, M. *Nucleosides & Nucleotides* **1993**, *12*, 827.

[954] Kini, G. D.; Anderson, J. D.; Sanghvi, Y. S.; Lewis, A. F.; Smee, D. F.; Revankar, G. R.; Robins, R. K.; Cottam, H. B. *J. Med. Chem.* **1991**, *34*, 3006.

[955] Al-Masoudi, N. A.; Pfleiderer, W. *Tetrahedron* **1993**, *49*, 7579.

[956] Sztaricskai, F.; Gergely, L.; Szegedi, G.; Fachet, J.; Csernus, I.; Jancsó, S.; Marossy, K. WO 9015063, 1990 [*Chem. Abstr.* **1991**, *115*, 9261m].

[957] Nair, V.; Lyons, A. G.; Purdy, D. F. *Tetrahedron* **1991**, *47*, 8949.

[958] Akita, H.; Uchida, K.; Chen, C. Y.; Kato, K. *Chem. Pharm. Bull.* **1998**, *46*, 1034.

[959] Chida, N.; Koizumi, K.; Kitada, Y.; Yokoyama, C.; Ogawa, S. *J. Chem. Soc., Chem. Commun.* **1994**, 111.

[960] Wengel, J.; Lau, J.; Pedersen, E. B.; Nielsen, C. M. *J. Org. Chem.* **1991**, *56*, 3591.

[961] Wengel, J.; Lau, J.; Walczak, K.; Pedersen, E. B. *Nucleosides & Nucleotides* **1991**, *10*, 405.

[962] Classon, B. O.; Samuelsson, B. B.; Kvarnström, I. S. A.; Svansson, L. G.; Svensson, S. C. T. WO 9206102, 1992 [*Chem. Abstr.* **1993**, *118*, 234419a].

[963] Svansson, L.; Kvarnström, I.; Classon, B.; Samuelsson, B. *Nucleosides & Nucleotides* **1992**, *11*, 1353.

[964] Abdel-Megied, A. E.-S.; Pedersen, E. B.; Nielsen, C. M. *Synthesis* **1991**, 313.

[965] El-Barbary, A. A.; Khodair, A. I.; Pedersen, E. B.; Nielsen, C. *Nucleosides & Nucleotides* **1994**, *13*, 707.

[966] Fedorov, I. I.; Bezchinskii, Ya. E.; Novikov, N. A.; Kaz'mina, E. M.; Rosenberg, S. G.; Mikhailov, S. N. *Bioorg. Khim.* **1990**, *16*, 997 [*Chem. Abstr.* **1990**, *113*, 212559r].

[967] Abdel-Megied, A. E.-S.; Pedersen, E. B.; Nielsen, C. M. *Monatsh. Chem.* **1991**, *122*, 59.

[968] Abdel-Megied, A. E.-S.; Pedersen, H.; Pedersen, E. B.; Nielsen, C. *Heterocycles* **1993**, *36*, 681.

[969] Jung, M. E.; Gardiner, J. M. *Tetrahedron Lett.* **1992**, *33*, 3841.

[970] Faury, P.; Camplo, M.; Charvet, A.-S.; Chermann, J.-C.; Kraus, J.-L. *Nucleosides & Nucleotides* **1992**, *11*, 1481.

[971] Vorbrüggen, H.; Strehlke, P.; US 3708469, 1969 [*Chem. Abstr.* **1970**, *73*, 110084v].

[972] Walczak, K.; Pedersen, E. B. *Synthesis* **1991**, 959.

[973] Monneret, C.; Florent, J.-C. *Synlett* **1994**, 305.

[974] Sells, T. B.; Nair, V. *Tetrahedron Lett.* **1993**, *34*, 3527.

[975] Auberson, Y.; Vogel, P. *Tetrahedron* **1990**, *46*, 7019.

[976] Aly, Y. L.; Abdel-Megied, A. E.-S.; Pedersen, E. B.; Nielsen, C. *Liebigs Ann. Chem.* **1992**, 127.

[977] Kim, C. U.; Misco, P. F.; Luh, B. Y.; Martin, J. C. *Heterocycles* **1990**, *31*, 1571.

[978] Jeong, L. S.; Moon, H. R.; Yoo, S. J.; Lee, S. N.; Chun, M. W.; Lim, Y.-H. *Tetrahedron Lett.* **1998**, *39*, 5201.

[979] Visser, G. M.; Tromp, M.; van Westrenen, J.; Schipperus, O.; van Boom, J. H. *Recl. Trav. Chim. Pays-Bas* **1986**, *105*, 85.

[980] Lau, J.; Wengel, J.; Pedersen, E. B.; Vestergaard, B. F. *Synthesis* **1991**, 1183.

[981] Shutalev; A. D.; Mikerin, I. E.; Arshava, B. M.; Nikitenko, A. A.; Raifeld, Y. E.; Vid, G. Y.; Lee, V. J.; Gurskaya, G. V.; Viktorova, L. S.; Jas'ko, M. V.; Semizarov, D. G.; Zavodnik, V. E. *Bioorg. Med. Chem. Lett.* **1994**, *4*, 761.

[982] Yu, C.-F.; Soine, W. H.; Thomas, D. *Med. Chem. Res.* **1992**, *2*, 410.

[983] Motawia, M. S.; Jacobsen, J. P.; Pedersen, E. B. *Chem. Scr.* **1989**, *29*, 51.

[984] Motawia, M. S.; Pedersen, E. B.; Suwinski, J.; Nielsen, C. M. *Arch. Pharm. (Weinheim)* **1990**, *323*, 949.

[985] Motawia, M. S.; Wengel, J.; Abdel-Megid, A. E.-S.; Pedersen, E. B. *Synthesis* **1989**, 384.

[986] Motawia, M. S.; Abdel-Megied, A. E.-S.; Pedersen, E. B.; Nielsen, C. M.; Ebbesen, P. *Acta Chem. Scand.* **1992**, *46*, 77.

[987] Wengel, J.; Lau, J.; Pedersen, E. B. *Synthesis* **1989**, 829.

[988] Dunkel, M.; Pfleiderer, W. *Nucleosides & Nucleotides* **1993**, *12*, 125.

[989] Dunkel, M.; Pfleiderer, W. *Nucleosides & Nucleotides* **1992**, *11*, 787.

[990] Jenny, T. F.; Benner, S. A. *Tetrahedron Lett.* **1992**, *33*, 6619.

[991] Narasaka, K. personal communication, 1988.

[992] Belleau, B.; Dixit, D.; Nguyen-Ba, N. EP 337713, 1989 [*Chem. Abstr.* **1990**, *112*, 198359w].

[993] Visser, G. M.; Schattenkerk, C.; van Boom, J. H. *J. R. Neth. Chem. Soc.* **1984**, *103*, 165.

[994] Visser, G. M.; Keemink, R.; Schattenkerk, C.; Kraal, B.; van Boom, J. H. *Nucleosides & Nucleotides* **1984**, *3*, 277.

[995] Lau, J.; Pedersen, E. B.; Nielsen, C. M. *Acta Chem. Scand.* **1991**, *45*, 616.

[996] Hertel, L. W. EP 0122707, 1987 [*Chem. Abstr.* **1985**, *102*, 113894n].

[997] Wittenburg, E. *Chem. Ber.* **1968**, *101*, 2132.

[998] Wittenburg, E. *Chem. Ber.* **1968**, *101*, 1614.

[999] Pérez-Pérez, M. J.; San-Félix, A.; Balzarini, J.; De Clercq, E.; Camarasa, M. J. *J. Med. Chem.* **1992**, *35*, 2988.

[1000] Tjarks, W.; Anisuzzaman, A. K. M.; Soloway, A. H. *Nucleosides & Nucleotides* **1992**, *11*, 1765.

[1001] Chou, T.-S.; Heath, P. C.; Patterson, L. E. EP 306190, 1989 [*Chem. Abstr.* **1989**, *111*, 134693x].

[1002] Motawia, M. S.; El-Torgoman, A. M.; Pedersen, E. B. *Liebigs Ann. Chem.* **1991**, 879.

[1003] Hjuler-Nielsen, H. P.; Motawia, M. S.; Pedersen, E. B.; Nielsen, C. *Bull. Soc. Chim. Fr.* **1992**, *129*, 523.

[1004] Beigel'man, L. N.; Karpeiskii, M. Ya.; Mikhailov, S. N. *Bioorg. Khim.* **1986**, *12*, 1359 [*Chem. Abstr.* **1987**, *107*, 176391v].

[1005] Griengl, H.; Bodenteich, M.; Hayden, W.; Wanek, E.; Streicher, W.; Stütz, P.; Bachmayer, H.; Ghazzouli, I.; Rosenwirth, B. *J. Med. Chem.* **1985**, *28*, 1679.

[1006] Griengl, H.; Hayden, W.; Schwarz, W.; Bachmayer, H.; Rosenwirth, B. *Eur. J. Med. Chem. Chim. Ther.* **1985**, *20*, 105.

[1007] Haas, A.; Lieb, M.; Steffens, B. *J. Fluorine Chem.* **1992**, *56*, 55.

[1008] Haas, A.; Lieb, M.; Steffens, B. *Chimia* **1989**, *43*, 98.

[1009] Kasnar, B.; Wise, D. S.; Kucera, L. S.; Drach, J. C.; Townsend, L. B. *Nucleosides & Nucleotides* **1994**, *13*, 459.

[1010] Tarussova, N. B.; Dyatkina, N. B.; v. Janta-Lipinski, M.; Langen, P. *Z. Chem.* **1987**, *27*, 366.

[1011] Svansson, L.; Kvarnström, I.; Classon, B.; Samuelsson, B. *J. Org. Chem.* **1991**, *56*, 2993.

[1012] Kvarnström, I.; Svansson, L.; Svensson, C.; Svensson, S. C. T. *Nucleosides & Nucleotides* **1992**, *11*, 1367.

[1013] Chen, A.; Savage, I.; Thomas, E. J.; Wilson, P. D. *Tetrahedron Lett.* **1993**, *34*, 6769.

[1014] Garner, P.; Park, J. M. *Tetrahedron Lett.* **1989**, *30*, 5065.

[1015] Cao, X.; Matteucci, M. D. *Bioorg. Med. Chem. Lett.* **1994**, *4*, 807.

[1016] Brånalt, J.; Kvarnström, I.; Niklasson, G.; Svensson, S. C. T.; Classon, B.; Samuelson, B. *J. Org. Chem.* **1994**, *59*, 1783.

[1017] El-Barbary, A. A.; Khodair, A. I.; Pedersen, E. B. *J. Org. Chem.* **1993**, *58*, 5994.

[1018] El-Barbary, A. A.; Khodair, A. I.; Pedersen, E. B.; Nielsen, C. *Monatsh. Chem.* **1994**, *125*, 593.

[1019] Kikuchi, Y.; Kurata, H.; Nishiyama, S.; Yamamura, S.; Kato, K. *Tetrahedron Lett.* **1997**, *38*, 4795.

[1019a] Follmann, M.; Kunz, H. *Synlett* **1998**, 989.

[1020] Dueholm, K. L.; Aly, Y. L.; Jørgensen, P. T.; El-Barbary, A. A.; Pedersen, E. B.; Nielsen, C. *Monatsh. Chem.* **1993**, *124*, 37.

[1021] Kim, H. O.; Ahn, S. K.; Alves, A. J.; Beach, J. W.; Jeong, L. S.; Choi, B. G.; Van Roey, P.; Schinazi, R. F.; Chu, C. K. *J. Med. Chem.* **1992**, *35*, 1987.

[1022] Chu, C. K.; Ahn, S. K.; Kim, H. O.; Beach, J. W.; Alves, A. J.; Jeong, L. S.; Islam, Q.; Van Roey, P.; Schinazi, R. F. *Tetrahedron Lett.* **1991**, *32*, 3791.

[1023] Jeong, L. S.; Schinazi, R. F.; Beach, J. W.; Kim, H. O.; Nampalli, S.; Shanmuganathan, K.; Alves, A. J.; McMillan, A.; Chu, C. K.; Mathis, R. *J. Med. Chem.* **1993**, *36*, 181.

[1024] Jeong, L. S.; Alves, A. J.; Carrigan, S. W.; Kim, H. O.; Beach, J. W.; Chu, C. K. *Tetrahedron Lett.* **1992**, *33*, 595.

[1025] Chu, C. K.; Beach, J. W.; Jeong, L. S.; Choi, B. G.; Comer, F. I.; Alves, A. J.; Schinazi, R. F. *J. Org. Chem.* **1991**, *56*, 6503.

[1026] Abdel-Megied, A. E.-S.; Hansen, P.; Pedersen, E. B.; Nielsen, C. M. *Acta Chem. Scand.* **1991**, *45*, 1060.

[1027] Kjærsgaard, U.; Pedersen, E. B.; Nielsen, C.; El-Torgoman, A. M. *Acta Chem. Scand.* **1992**, *46*, 1016.

[1028] Abdel-Megied, A. E.-S.; Hansen, P.; Pedersen, E. B.; Nielsen, C.; Nielsen, C. M. *Arch. Pharm. (Weinheim)* **1993**, *326*, 377.

[1029] Cao, X.; Pfleiderer, W. *Nucleosides & Nucleotides* **1994**, *13*, 773.

[1030] Tseng, C. K.-H.; Marquez, V. E.; Milne, G. W. A.; Wysocki, Jr., R. J.; Mitsuya, H.; Shirasaki, T.; Driscoll, J. S. *J. Med. Chem.* **1991**, *34*, 343.

[1031] Benhaddou, R.; Czernecki, S.; Valéry, J. M.; Bellosta, V. *Bull. Soc. Chim. Fr.* **1991**, *127*, 108.

[1032] Agyei-Aye, K.; Baker, D. C. *Carbohydr. Res.* **1988**, *183*, 261.

[1033] Kasnar, B.; Skaric, V.; Klaic, B.; Zinic, M. *Tetrahedron Lett.* **1993**, *34*, 4997.

[1034] Raju, N.; Smee, D. F.; Robins, R. K.; Vaghefi, M. M. *J. Med. Chem.* **1989**, *32*, 1307.

[1035] Lau, J.; Walczak, K.; Pupek, K.; Buch, C.; Nielsen, C. M.; Pedersen, E. B. *Arch. Pharm. (Weinheim)* **1991**, *324*, 953.

[1036] Cheeseman, G. W. H.; Freestone, A. J.; Godwin, R. A.; Hough, T. L. *J. Chem. Soc., Perkin Trans. 1* **1975**, 1888.

[1037] Herdewijn, P.; Van Aerschot, A.; Busson, R.; Claes, P.; De Clercq, E. *Nucleosides & Nucleotides* **1991**, *10*, 1525.

[1038] Herdewijn, P.; Van Aerschot, A.; Balzarini, J.; De Clercq, E. *Nucleosides & Nucleotides* **1991**, *10*, 119.

[1039] Ioannidis, P.; Classon, B.; Samuelsson, B.; Kvarnström, I. *Nucleosides & Nucleotides* **1992**, *11*, 1205.

[1040] Hayakawa, H.; Miyazawa, M.; Tanaka, H.; Miyasaka, T. *Nucleosides & Nucleotides* **1994**, *13*, 297.

[1041] Buchanan, J. G.; McCaig, A. E.; Wightman, R. H. *J. Chem. Soc. Perkin Trans 1* **1990**, 955.

[1042] Wyss, P.-C. EP 004 261, 1982 [*Chem. Abstr.* **1980**, *92*, 147135f].

[1043] Vorbrüggen, H. DE 2508312, 1976 [*Chem. Abstr.* **1977**, *86*, 5772p].

[1044] Sanghvi, Y. S.; Bhattacharya, B. K.; Kini, G. D.; Matsumoto, S. S.; Larson, S. B.; Jolley, W. B.; Robins, R. K.; Revankar, G. R. *J. Med. Chem.* **1990**, *33*, 336.

[1045] Mao, D. T.; Driscoll, J. S.; Marquez, V. E. *J. Med. Chem.* **1984**, *27*, 160.

[1046] McNamara, D. J.; Cook, P. D.; Allen, L. B.; Kehoe, M. J.; Holland, C. S.; Teepe, A. G. *J. Med. Chem.* **1990**, *33*, 2006.

[1047] Franchetti, P.; Cappellacci, L.; Cristalli, G.; Grifantini, M.; Vittori, S. *Nucleosides & Nucleotides* **1991**, *10*, 543.

[1048] Visser, G. M.; van Westrenen, J.; van Boeckel, C. A. A.; van Boom, J. H. *Recl. Trav. Chim. Pays-Bas* **1986**, *105*, 528.

[1049] Wood, S. G.; Dalley, N. K.; George, R. D.; Robins, R. K.; Revankar, G. R. *J. Org. Chem.* **1984**, *49*, 3534.

[1050] Sharma, M.; Bloch, A.; Bobek, M. *Nucleosides & Nucleotides* **1993**, *12*, 643.

[1051] Cuny, E.; Lichtenthaler, F. W.; Jahn, U. *Chem. Ber.* **1981**, *114*, 1624.

[1052] Ikehara, M.; Inaoka, T. *Nucleosides & Nucleotides* **1985**, *4*, 515.

[1053] Abdelal, A. M.; El-Emam, A. A.; Moustafa, M. A. *J. Chin. Chem. Soc. (Taipei)* **1992**, *39*, 257 [*Chem. Abstr.* **1992**, *117*, 131484t].

[1054] Hijazi, A. *Nucleosides & Nucleotides* **1988**, *7*, 537.

1055 Cottam, H. B.; Petrie, C. R.; McKernan, P. A.; Goebel, R. J.; Dalley, N. K.; Davidson, R. B.; Robins, R. K.; Revankar, G. R. *J. Med. Chem.* **1984**, *27*, 1119.

1056 Bontems, R. J.; Anderson, J. D.; Smee, D. F.; Jin, A.; Alaghamandan, H. A.; Sharma, B. S.; Jolley, W. B.; Robins, R. K.; Cottam, H. B. *J. Med. Chem.* **1990**, *33*, 2174.

1057 Kini, G. D.; Hennen, W. J.; Robins, R. K. *Nucleosides & Nucleotides* **1987**, *6*, 581.

1058 Nagahara, K.; Anderson, J. D.; Kini, G. D.; Dalley, N. K.; Larson, S. B.; Smee, D. F.; Jin, A.; Sharma, B. S.; Jolley, W. B.; Robins, R. K.; Cottam, H. B. *J. Med. Chem.* **1990**, *33*, 407.

1058a Kim, H. O.; Ji, X-D.; Melman, N.; Olah, M. E.; Stiles, G. L.; Jacobson, K. A. *J. Med. Chem.* **1994**, *37*, 4020.

1059 Al-Masoudi, N. A.; Pfleiderer, W. *Nucleosides & Nucleotides* **1989**, *8*, 1485.

1060 Al-Masoudi, N. A. L.; Pfleiderer, W. *Pteridines* **1990**, *2*, 9.

1061 Kawai, S. H.; Just, G. *Nucleosides & Nucleotides* **1991**, *10*, 1485.

1062 Földesi, A.; Nilson, F. P. R.; Glemarec, C.; Gioeli, C.; Chattopadhyaya, J. *Tetrahedron* **1992**, *48*, 9033.

1063 Yasumoto, S.; Matsumoto, H.; Tada, Y.; Kobayashi, K.; Noguchi, K. JP 62187483, 1987 [*Chem. Abstr.* **1988**, *108*, 56547z].

1064 Pudlo, J. S.; Townsend, L. B. *Tetrahedron Lett.* **1990**, *31*, 3101.

1065 Nord, L. D.; Dalley, N. K.; McKernan, P. A.; Robins, R. K. *J. Med. Chem.* **1987**, *30*, 1044.

1066 Walczak, K.; Pupek, K.; Pedersen, E. B. *Liebigs Ann. Chem.* **1991**, 1041.

1067 Kojima, A.; Kono, Y.; Ike, Y.; Yokoyama, T.; Odate, M. DE 2656604, 1977 [*Chem. Abstr.* **1977**, *87*, 13539b].

1068 Lerner, L. M. *J. Org. Chem.* **1978**, *43*, 2469.

1069 Bobek, M.; Bloch, A. *J. Med. Chem.* **1972**, *15*, 164.

1070 Camplo, M.; Faury, P.; Charvet, A.-S.; Lederer, F.; Chermann, J.-C.; Kraus, J.-L. *Nucleosides & Nucleotides* **1993**, *12*, 631.

1071 Pravdic, N.; Franjic-Mihalic, I. *Carbohydr. Res.* **1978**, *62*, 301.

1072 Tronchet, J. M. J.; Tronchet, J. *Carbohydr. Res.* **1977**, *59*, 594.

1073 Bobek, M.; Tuntiwachwuttikul, P.; Ismail, M. M.; Bardos, T. J. *Nucleosides & Nucleotides* **1991**, *10*, 1657.

1074 Lerner, L. M. *J. Org. Chem.* **1978**, *43*, 962.

1075 Lerner, L. M. *Carbohydr. Res.* **1977**, *53*, 177.

1076 Lerner, L. M. *J. Org. Chem.* **1979**, *44*, 4359.

1077 Szarek, W. A.; Ritchie, R. G. S.; Vyas, D. M. *Carbohydr. Res.* **1978**, *62*, 89.

1078 Ritchie, R. G. S.; Vyas, D. M.; Szarek, W. A. *Can. J. Chem.* **1978**, *56*, 794.

1079 Lerner, L. M. In *Nucleic Acid Chemistry*; Townsend, L. B., Tipson, R. S., Eds.; Wiley: New York; 1991, Vol. 4; p. 274.

1080 Steinmaus, H. DE 2226673, 1973 [*Chem. Abstr.* **1974**, *80*, 60154q].

1081 Hamilton, H. W.; Bristol, J. A. *J. Med. Chem.* **1983**, *26*, 1601.

1082 de la Cruz, A.; Goya, P.; Martinez, A.; Paez, J. A. *Nucleosides & Nucleotides* **1990**, *9*, 69.

1083 Goya, P.; Martinez, A. *Arch. Pharm. (Weinheim)* **1988**, *321*, 99.

1084 Kiriasis, L.; Pfleiderer, W. *Nucleosides & Nucleotides* **1989**, *8*, 1345.

1085 Goya, P.; Martinez, A.; Jimeno, M. L.; Pfleiderer, W. *Liebigs Ann. Chem.* **1986**, 1872.

1086 Avasthi, K.; Deo, K., Garg, N.; Bhakuni, D. S. *Bioorg. Med. Chem. Lett.* **1991**, *1*, 249.

1087 Misra, S.; Jain, S.; Avasthi, K.; Bhakuni, D. S. *Nucleosides & Nucleotides* **1990**, *9*, 837.

1088 Goya, P.; Martinez, A.; Jimeno, M. L. *J. Chem. Soc., Perkin Trans. 2* **1990**, 783.

1089 Cuberta, A.; Jimeno, M. L.; Ochoa, C. *Nucleosides & Nucleotides* **1987**, *6*, 831.

1090 Azmy, B.; Fernández-Resa, P.; Goya, P.; Nieves, R.; Ochoa, C.; Stud, M.; Jimeno, M. L. *Nucleosides & Nucleotides* **1984**, *3*, 325.

1091 Goya, P.; Martinez, A.; Jimeno, M. L.; Pfleiderer, W. *Liebigs Ann. Chem.* **1987**, 961.

1092 Mishra, A.; Pratap, R.; Bhakuni, D. S. *Indian J. Chem.* **1987**, *26B*, 847.

1093 Anderson, J. D.; Cottam, H. B.; Larson, S. B.; Nord, L. D.; Revankar, G. R.; Robins, R. K. *J. Heterocyl. Chem.* **1990**, *27*, 439.

1094 Tripathi, R. P.; Hasan, A.; Pratap, R.; Bhakuni, D. S. *Indian J. Chem.* **1987**, *26B*, 851.

1095 Anderson, J. D.; Dalley, N. K.; Revankar, G. R.; Robins, R. K. *J. Heterocycl. Chem.* **1986**, *23*, 1869.
1096 Kazimierczuk, Z.; Pfleiderer, W. *Liebigs Ann. Chem.* **1982**, 754.
1097 Lutz, H.; Pfleiderer, W. *Carbohydr. Res.* **1984**, *130*, 179.
1098 Harris, R.; Pfleiderer, W. *Liebigs Ann. Chem.* **1981**, 1457.
1099 Goya, P.; Pfleiderer, W. *Chem. Ber.* **1981**, *114*, 699.
1100 Ienaga, K.; Pfleiderer, W. *Chem. Ber.* **1978**, *111*, 2586.
1101 Schmidt, R. R.; Michel, J. *J. Carbohydr. Chem.* **1985**, *4*, 141.
1102 Farina V.; Benigni, D.A. *Tetrahedron Lett.* **1988**, *29*, 1239.
1103 Kondo, T.; Nakai, H.; Goto, T. *Tetrahedron Lett.* **1972**, *19*, 1881.
1104 Suzuki, T.; Sakaguchi, S.; Myata, Y.; Mori, T. JP 01139596, 1989 [*Chem. Abstr.* **1990**, *112*, 36386a].
1105 Hambalek, R. J.; Just, G. *Nucleosides & Nucleotides* **1992**, *11*, 1539.
1106 Humber, D. C.; Jones, M. F.; Payne, J. J.; Ramsay, M. V. J.; Zacharie, B.; Jin, H.; Siddiqui, A.; Evans, C. A.; Tse, H. L. A.; Mansour, T. S. *Tetrahedron Lett.* **1992**, *33*, 4625.
1107 Kojima, E.; Yoshioka, H.; Murakami, K. JP 03255083, 1991 [*Chem. Abstr.* **1992**, *116*, 106703k].
1108 Mann, J.; Weymouth-Wilson, A. *Synlett* **1992**, 67.
1109 Noyori, R.; Hayashi, M. JP 62267294, 1987 [*Chem. Abstr.* **1988**, *109*, 6905q].
1110 Cristescu, C.; Supuran, C. *Rev. Roum. Chim.* **1987**, *32*, 329 [*Chem. Abstr.* **1988**, *108*, 112880s].
1111 Tocik, Z.; Earl, R. A.; Beranek, J. In *Nucleic Acid Chemistry*; Townsend, L. B., Tipson, R. S., Eds.; Wiley: New York; 1991, Vol. 4, p. 105.
1111a Kobylinskaya, V. I.; Dashevskaya, T. A.; Shalamai, A. S.; Suleimanov, S. P.; Kharchuk, A. N.; Protopopova, G. V. *Khim.-Farm. Zh.* **1987**, *21*, 290 [*Chem. Abstr.* **1988**, *108*, 112876v].
1112 Itoh, T.; Pfleiderer, W. *Chem. Ber.* **1976**, *109*, 3228.
1113 Hakimelahi, G. H.; Khalafi-Nezhad, A. *Helv. Chim. Acta* **1989**, *72*, 1495.
1114 Hasan, A.; Knapp, Jr., F. F.; Kilbourn, M. R.; Buchsbaum, D. J. *J. Heterocycl. Chem.* **1993**, *30*, 1351.
1115 Ogawa, T.; Takaku, H.; Yamamoto, N. *Nucleosides & Nucleotides* **1989**, *8*, 499.
1116 Earl, R. A.; Townsend, L. B. *J. Heterocycl. Chem.* **1972**, *9*, 1141.
1117 Manning, S. J.; Cohen, A. M.; Townsend, L. B. *J. Labelled Comp. Radiopharm.* **1978**, *15*, 723.
1118 Hillers, S.; Lidaks, M.; Zhuk, R. A.; Berzina, A.; Pecs, K.; Getsova, I. N.; Bruk, E. I. *Khim. Geterotsikl. Soedin.* **1969**, *2*, 375 [*Chem. Abstr.* **1969**, *71*, 30436e].
1119 Al-Masoudi, N. A.; Pfleiderer, W.; Al-Masoudi, W. A. *Nucleosides & Nucleotides* **1993**, *12*, 675.
1120 Crisp, G. T.; Flynn, B. L. *Tetrahedron* **1993**, *49*, 5873.
1121 Levesque, D. L.; Wang, E.-C.; Wei, D.-C.; Tzeng, C.-C.; Panzica, R. P.; Naguib, F. N. M.; el Kouni, M. H. *J. Heterocycl. Chem.* **1993**, *30*, 1399.
1122 Han, C. H.; Chen, Y. L.; Tzeng, C. C. *Nucleosides & Nucleotides* **1991**, *10*, 1391.
1123 Kobayashi, K.; Sone, T.; Suzuki, N.; Wakabayashi, M.; Sowa, T. JP 52036673, 1977 [*Chem. Abstr.* **1977**, *87*, 135387y].
1124 Marunaka, T.; Yasumoto, M.; Hashimoto, S.; Harima, K.; Suzue, T. JP 105673, 1975 [*Chem. Abstr.* **1976**, *84*, 59543s].
1125 Kurono, M..; Unno, R.; Kimura, H.; Ozawa, H.; Mitani, T.; Shiromori, T.; Koketsu, M.; Michishita, H.; Sawai, K. EP 280 841, 1988 [*Chem. Abstr.* **1989**, *111*, 96984h].
1126 Kaneko, C.; Hara, S.; Matsumoto, H.; Takeuchi, T.; Mori, T.; Ikeda, K.; Mizuno, Y. *Chem. Pharm. Bull.* **1991**, *39*, 871.
1127 Beauchamp, L. M.; Krenitsky, T. A. EP 203736, 1986 [*Chem. Abstr.* **1987**, *106*, 67010d].
1128 Watanabe, K. A.; Su, T.-L.; Klein, R. S.; Chu, C. K.; Matsuda, A.; Chun, M. W.; Lopez, C.; Fox, J. J. *J. Med. Chem.* **1983**, *26*, 152.
1129 Winkelmann, E.; Winkler, I. EP 298467, 1989 [*Chem. Abstr.* **1989**, *111*, 7146w].
1130 Lee, K.-H.; Chen, Y.-L.; Huang, B.-R.; Tzeng, C.-C.; Zhu, Q.-Y.; Chou, T.-C. *Nucleosides & Nucleotides* **1991**, *10*, 1407.
1131 Bravo, P.; Resnati, G.; Viani, F. *Tetrahedron* **1993**, *49*, 713.
1132 Chu, S. H.; Weng, Z. Y.; Chen, Z. H.; Rowe, E. C.; Chu, E.; Naguib, F. N. M.; el Kouni, M. H.; Cha, S.; Chu, M. Y. *Nucleosides & Nucleotides* **1988**, *7*, 91.

1133 Nedorezova, T. P.; Melnik, S. Ya.; Preobrazhenskaya, M. N. *Bioorg. Khim.* **1976**, *2*, 1205 [*Chem. Abstr.* **1977**, *86*, 55653w]

1134 Lee, W. W.; Martinez, A. P.; Goodman, L.; Henry, D. W. *J. Org. Chem.* **1972**, *37*, 2923.

1135 Bravo, P; Resnati, G.; Fiorenza, C.; Viani, F. *Gazz. Chim. Ital.* **1992**, *122*, 493.

1136 Ford, Jr., H.; Driscoll, J. S.; Siddiqui, M.; Kelley, J. A.; Mitsuya, H.; Shirasaka, T.; Johns, D. G.; Marquez, V. E. *Nucleosides & Nucleotides* **1994**, *13*, 213.

1137 Kawakami, H.; Koseki, Y.; Ehata, T.; Matsushita, H.; Ito, K.; Naoi, Y. JP 3215484, 1991 [*Chem. Abstr.* **1992**, *116*, 84116s].

1138 Abushanab, E.; Sarma, M. S. P. *J. Med. Chem.* **1989**, *32*, 76.

1139 Dziewiszek, K.; Schinazi, R. F.; Chou, T.-C.; Su, T.-L.; Dzik, J. M.; Rode, W.; Watanabe, K. A. *Nucleosides & Nucleotides* **1994**, *13*, 77.

1140 Montgomery, J. A.; Secrist III, J. A. US 5034518, 1991 [*Chem. Abstr.* **1991**, *115*, 9260k].

1141 Reichman, U.; Watanabe, K. A.; Fox, J. J. *Carbohydr. Res.* **1975**, *42*, 233.

1142 Shimizu, B.; Saito, A. *Agr. Biol. Chem.* **1969**, *33*, 119.

1143 Wittenburg, E. *Collect. Czech. Chem. Commun.* **1971**, *36*, 246.

1144 Pfleiderer, W.; Autenrieth, D.; Schranner, M. *Chem. Ber.* **1973**, *106*, 317.

1145 Cichy, A. F.; Saibaba, R.; El Subbagh, H. I.; Panzica, R. P.; Abushanab, E. *J. Org. Chem.* **1991**, *56*, 4653.

1146 Schneider, K. C.; Benner, S. A. *J. Am. Chem. Soc.* **1990**, *112*, 453.

1147 Martin, J. C.; Jeffrey, G. A.; McGee, D. P. C.; Tippie, M. A.; Smee, D. F.; Matthews, T. R.; Verheyden, J. P. H. *J. Med. Chem.* **1985**, *28*, 358.

1148 Lin, T.-S.; Liu, M.-C. *J. Med. Chem.* **1985**, *28*, 971.

1149 Ashton, W. T.; Canning, L. F.; Reynolds, G. F.; Tolman, R. L.; Karkas, J. D.; Liou, R.; Davies, M.-E. M.; DeWitt, C. M.; Perry, H. C.; Field, A. K. *J. Med. Chem.* **1985**, *28*, 926.

1150 Ito, K.; Kaihara, H.; Kawakami, H.; Ebata, T. JP 0559088, 1993 [*Chem. Abstr.* **1993**, *119*, 96075c].

1150a Teijin, Ltd. JP6023395, 1985 [*Chem. Abstr.* **1986**, *104*, 6144n].

1151 Teijin, Ltd. JP6023397, 1985 [*Chem. Abstr.* **1986**, *104*, 6142k].

1152 Trehan, S. B.; MacDiarmid, J. E.; Bardos, T. J. *Nucleosides & Nucleotides* **1994**, *13*, 235.

1153 Otvös, L.; Rakoczi, J.; Szabolcs, A.; Sagi, J.; Dékany, G.; Szemzo, A.; Gruber, L.; Nagy, G.; Nagy, J.; Ivan, P.; Tüdos, H. EP 543404, 1993 [*Chem. Abstr.* **1993**, *119*, 271637m].

1154 Gauri, K. K. GB 1170565, 1969 [*Chem. Abstr.* **1970**, *72*, 79425k].

1155 Ryan, K. J.; Acton E. M.; Goodman, L. US 3531464, 1970 [*Chem. Abstr.* **1971**, *74*, 42597h].

1156 Dipple, A.; Heidelberger, C. *J. Med. Chem.* **1966**, *9*, 715.

1157 Ferguson, A. S. EP 14597, 1980 [*Chem. Abstr.* **1980**, *93*, 241264q].

1158 Mansuri, M. M.; Ghazzouli, I.; Chen, M. S.; Howell, H. G.; Brodfuehrer, P. R.; Benigni, D. A.; Martin, J. C. *J. Med. Chem.* **1987**, *30*, 867.

1159 Marquez, V.; Driscoll, J.; Tseng, C. K. H. US 3288652, 1990 [*Chem. Abstr.* **1990**, *113*, 126590n].

1160 Herdewijn, P.; Van Aerschot, A. *Nucleosides & Nucleotides* **1989**, *8*, 933.

1161 Lin, T.-S. In *Nucleic Acid Chemistry*; Townsend, L. B., Tipson, R. S., Eds.; Wiley: New York; 1991, Vol. 4; p. 73.

1162 Dyson, M. R.; Coe, P. L.; Walker, R. T. *J. Med. Chem.* **1991**, *34*, 2782.

1163 Vemishetti, P.; Saibaba, R.; Panzica, R. P. Abushanab, E. *J. Med. Chem.* **1990**, *33*, 681.

1164 Vemishetti, P.; Leiby, R.W.; Abushanab, E.; Panzica, P. *J. Heterocycl. Chem.* **1988**, *25*, 651.

1165 Vemishetti, P.; Abushanab, E.; Leiby, R. W.; Panzica, R. P. *Nucleosides & Nucleotides* **1989**, *8*, 201.

1166 Azymah, M.; Chavis, C.; Lucas, M.; Imbach, J.-L. *Tetrahedron Lett.* **1989**, *30*, 6165.

1167 Bhan, A.; Hosmane, R. S. *Nucleosides & Nucleotides* **1992**, *11*, 1175.

1168 Shinozuka, K.; Hirota, Y.; Morita, T.; Sawai, H. *Heterocycles* **1992**, *34*, 2117.

1169 Shinozuka, K.; Morita, T.; Hirota, Y.; Sawai, H. *Chem. Lett.* **1991**, 1941.

1170 Urata, H.; Ogura, E.; Shinohara, K.; Ueda, Y.; Akagi, M. *Nucleic Acids Res.* **1992**, *20*, 3325.

1171 Williamson, J. R.; Boxer, S. G. *Nucleic Acids Res.* **1988**, *16*, 1529.

1172 Basnak, I.; Balkan, A.; Coe, P. L.; Walker, R. T. *Nucleosides & Nucleotides* **1994**, *13*, 177.

1173 Rosowsky, A.; Lazarus, H.; Yamashita, A. *J. Med. Chem.* **1976**, *19*, 1265.

[1174] Mertes, M. P.; Shipchandler, M. T. *J. Heterocycl. Chem.* **1971**, *8*, 133.
[1175] Jones, A. S.; Walker, R. T.; Barr, P. J.; De Clercq, E. DE 2915254, 1979 [*Chem. Abstr.* **1980**, *92*, 181586p].
[1176] Lin, P. K. T.; Brown, D. M. *Nucleic Acids Res.* **1989**, *17*, 10373.
[1177] Gupta, V. S.; Bubbar, G. L. *Can. J. Chem.* **1971**, *49*, 719.
[1177a] Griengl, H.; Hayden, W.; Schwarz, W.; Bachmeyer, H.; Rosenwirth, B. *Eur. J. Med. Chem., Chim. Ther.* **1985**, *20*, 105.
[1178] Stout, M. G.; Robins, R. K. *J. Heterocycl. Chem.* **1972**, *9*, 545.
[1179] Tong, G. L.; Lee, W. W.; Goodman, L. *J. Heterocycl. Chem.* **1966**, *3*, 226.
[1180] Basnak, I.; Coe, P. L.; Walker, R. T. *Nucleosides & Nucleotides* **1994**, *13*, 163.
[1181] Azymah, M.; Chavis, C.; Lucas, M.; Imbach, J.-L. *J. Chem. Soc., Perkin Trans. 1* **1991**, 1561.
[1182] Okabe, M.; Sun, R.-C.; Zenchoff, G. B. *J. Org. Chem.* **1991**, *56*, 4392.
[1183] Spassova, M. K.; Holý, A.; Masojídková, M. *Collect. Czech. Chem. Commun.* **1986**, *51*, 1512.
[1184] Kim, C.-H.; Marquez, V. E. *J. Org. Chem.* **1987**, *52*, 1979.
[1185] Kassou, M.; Castillon, S. *J. Org. Chem.* **1997**, *62*, 3696.
[1186] Miyai, K.; Tolman, R. L.; Robins, R. K.; Cheng, C. C. *J. Med. Chem.* **1978**, *21*, 427.
[1187] Holý, A. *Collect. Czech. Chem. Commun.* **1977**, *42*, 902.
[1188] Eistetter, K.; Pfleiderer, W. *Chem. Ber.* **1976**, *109*, 3208.
[1189] Cheriyan, U. O.; Ogilvie, K. K. *Nucleosides & Nucleotides* **1982**, *1*, 233.
[1190] Uteza, V.; Chen, G.-R.; Tuoi, J. L. Q.; Descotes, G.; Fenet, B.; Grouiller, A. *Tetrahedron* **1993**, *49*, 8579.
[1191] Matsumoto, H.; Kaneko, C.; Mori, T.; Mizuno, Y. *Chem. Pharm. Bull.* **1989**, *37*, 229.
[1192] Gagnieu, C. H.; Guiller, A.; Pacheco, H. *Carbohydr. Res.* **1988**, *180*, 233.
[1193] Suzuki, T.; Sakaguchi, M.; Miyata, Y.; Mori, T. EP 312858 [*Chem. Abstr.* **1989**, *111*, 174101j].
[1194] Revankar, G. R.; Townsend, L. B. *J. Heterocycl. Chem.* **1968**, *5*, 477.
[1195] Nishimura, T.; Shimizu, B. *Chem. Pharm. Bull.* **1965**, *13*, 803.
[1196] Wright, J. A.; Taylor, N. F.; Fox, J. J. *J. Org. Chem.* **1969**, *34*, 2632.
[1197] Sanghvi, Y. S.; Hanna, N. B.; Larson, S. B.; Fujitaki, J. M.; Willis, R. C.; Smith, R. A.; Robins, R. K.; Revankar, G. R. *J. Med. Chem.* **1988**, *31*, 330.
[1198] Hamilton, H. W.; Bristol, J. A.; Moos, W.; Trivedi, B. K.; Taylor, M.; Patt, W. C. EP 181129, 1986 [*Chem. Abstr.* **1986**, *105*, 79310u].
[1199] Robins, M. J.; Robins, R. K. *J. Am. Chem. Soc.* **1965**, *87*, 4934.
[1200] Akhrem, A. A.; Garbuz, N. I.; Kvasyuk, E. I.; Mikhailopulo, I. A.; Pupeiko, N. E. *Vestsi Akad. Navuk BSSR, Ser. Khim. Navuk* **1977**, *4*, 97 [*Chem. Abstr.* **1978**, *88*, 7277w].
[1201] Akhrem, A. A.; Kvasyuk, E. I.; Mikhailopulo, I. A.; Pupeiko, N. E. *Vestsi Akad. Navuk BSSR, Ser. Khim. Navuk* **1978**, *6*, 124 [*Chem. Abstr.* **1979**, *90*, 152528z].
[1202] Barascut, J.-L.; Kam, B.; Imbach, J.-L. *Bull. Soc. Chim. Fr.* **1976**, 1983.
[1203] Andersen, J.; Andreassen, E. S.; Pedersen, E. B. *Acta Chem. Scand.* **1987**, *B 41*, 473.
[1204] Honjo, M.; Imai, K. US 3380996, 1968 [*Chem. Abstr.* **1967**, *67*, 108955n].
[1205] Kazimierczuk, Z.; Lönnberg, H.; Vilpo, J.; Pfleiderer, W. *Nucleosides & Nucleotides* **1989**, *8*, 599.
[1206] Holý, A. *Collect. Czech. Chem. Commun.* **1979**, *44*, 2846.
[1207] Sekiya, M.; Yoshino, T.; Tanaka, H.; Ishido, Y. *Bull. Chem. Soc. Jpn.* **1973**, *46*, 556.
[1208] Boryski, J.; Ostrowski, T.; Golankiewicz, B. *Nucleosides & Nucleotides* **1989**, *8*, 1271.
[1209] Hara, S.; Kaneko, C.; Matsumoto, H.; Nishino, T.; Takeuchi, T.; Mori, T.; Mizuno, Y.; Ikeda, K. *Nucleosides & Nucleotides* **1992**, *11*, 571.
[1210] Wolfrom, M. L.; Bhat, H. B.; Conigliaro, P. J. *Carbohydr. Res.* **1971**, *20*, 375.
[1210a] McChesney, J. D.; Buchman, R. *Heterocycles* **1976**, *4*, 1065.
[1211] Bräuniger, H.; Koine, A. *Arch. Pharm. (Weinheim)* **1965**, *298*, 41.
[1212] Stevens, C. L.; Radhakrishnan, R.; Pillai, P. M. *J. Carbohydr., Nucleosides, Nucleotides* **1976**, *3*, 71.
[1213] Bräuniger, H.; Koine, A. *Arch. Pharm. (Weinheim)* **1963**, *296*, 668.
[1214] Bräuniger, H.; Koine, A. *Arch. Pharm. (Weinheim)* **1963**, *296*, 665.
[1215] Bräuniger, H.; Koine, A. *Arch. Pharm. (Weinheim)* **1965**, *298*, 712.
[1216] Motawia, M. S.; Pedersen, E. B. *Chem. Scr.* **1988**, *28*, 339.
[1217] Tuntiwachwuttikul, P.; Bardos, T. J.; Bobek, M. *J. Heterocycl. Chem.* **1991**, *28*, 1131.

1218 Szerkes, G. L.; Robins, R. K. In *Nucleic Acid Chemistry*; Townsend, L. B., Tipson, R. S., Eds.; Wiley: New York; 1986, Vol. 3; p. 131.
1219 Shen, T. Y.; Ruyle, W. V.; Bugianesi, R. L. *J. Heterocycl. Chem.* **1965**, *2*, 495.
1220 van Steen, R. US 3352849, 1968 [*Chem. Abstr.* **1968**, *68*, 3158].
1221 Shiau, G. T.; Prusoff, W. H. *Carbohydr. Res.* **1978**, *62*, 175.
1222 Barchi, Jr., J. J.; Marquez, V. E.; Driscoll, J. S.; Ford, Jr., H.; Mitsuya, H.; Shirasaka, T.; Aoki, S.; Kelley, J. A. *J. Med. Chem.* **1991**, *34*, 1647.
1223 Mansuri, M. M.; Krishnan, B.; Martin, J. C. *Tetrahedron Lett.* **1991**, *32*, 1287.
1224 McGee, D. P. C.; Martin, J. C.; Smee, D. F.; Verheyden, J. P. H. *Nucleosides & Nucleotides* **1990**, *9*, 815.
1225 Thomas, H. J.; Tiwari, K. N.; Clayton, S. J.; Secrist III, J. A.; Montgomery, J. A. *Nucleosides & Nucleotides* **1994**, *13*, 309.
1226 Nesnow, S.; Heidelberger, C. *J. Heterocycl. Chem.* **1975**, *12*, 941.
1227 Bubbar, G. L.; Gupta, V. S. *Can. J. Chem.* **1970**, *48*, 3147.
1228 Hijazi, A.; Pfleiderer, W. *Nucleosides & Nucleotides* **1984**, *3*, 549.
1229 Durr, G. J.; Hammond, S. *J. Heterocycl. Chem.* **1970**, *7*, 743.
1230 Durr, G. J.; Keiser, J. F.; Ierardi III, P. A. *J. Heterocycl. Chem.* **1967**, *4*, 291.
1231 Stevens, C. L.; Blumbergs, P. *J. Org. Chem.* **1965**, *30*, 2723.
1232 Ishibashi, K.; Ishiguro, S.; Komaki, R. DE 2834698, 1979 [*Chem. Abstr.* **1979**, *91*, 20334u].
1233 Sakurai, K.; Aoyagi, S.; Toyofuku, H.; Ohki, M.; Yoshizawa, T.; Kuroda, T. *Chem. Pharm. Bull.* **1978**, *26*, 3565.
1234 Schramm, G.; Lunzmann, G. FR 1569465, 1969 [*Chem. Abstr.* **1970**, *72*, 111786s].
1235 Andersen, L.; Lau, J.; Pedersen, E. B. *Chem. Scr.* **1988**, *28*, 307.
1236 Krasavina, L. S.; Turchina, R. P.; Vigdorchik, M. M.; Turchin, K. F.; Akhvlediani, R. N.; Suvorov, N. N. *Zh. Org. Khim.* **1988**, *24*, 1960 [*Chem. Abstr.* **1989**, *111*, 39722x].
1237 Matsumoto, H.; Kaneko, C.; Yamada, K.; Takeuchi, T.; Mori, T.; Mizuno, Y. *Chem. Pharm. Bull.* **1988**, *36*, 1153.
1238 Golankiewicz, B.; Ostrowski, T.; Boryski, J.; De Clercq, E. *J. Chem. Soc., Perkin Trans. 1* **1991**, 589.
1239 An, S.-H.; Bobek, M. *Tetrahedron Lett.* **1986**, *27*, 3219.
1240 García-López, T.; Herranz, R.; Andrés, J. I. *Eur. J. Med. Chem. Chim. Ther.* **1984**, *19*, 187.
1241 Leonard, N. J.; Laursen, R. A. *Biochemistry* **1965**, *4*, 354.
1242 Davoll, J.; Brown, G. B. *J. Am. Chem. Soc.* **1951**, *73*, 5781.
1243 Leonard, N. J.; Laursen, R. A. *J. Am. Chem. Soc.* **1963**, *85*, 2026.
1244 Southon, I. W.; Pfleiderer, W. *Chem. Ber.* **1978**, *111*, 2571.
1245 Montgomery, J. A.; Thomas, H. J. *J. Am. Chem. Soc.* **1965**, *87*, 5442.
1246 Watanabe, K. A.; Chu, C. K.; Fox, J. J. EP 219829, 1987 [*Chem. Abstr.* **1987**, *107*, 59409w].
1247 Yamaoka, N.; Aso, K.; Matsuda, K. *J. Org. Chem.* **1965**, *30*, 149.
1248 Muraoka, M.; Takada, A.; Ueda, T. *Chem. Pharm. Bull.* **1970**, *18*, 261.
1249 Rogers, G. T.; Ulbricht, T. L. V. *J. Chem. Soc. (C)* **1970**, 1109.
1250 Letham, D. S.; Wilson, M. M.; Parker, C. W.; Jenkins, I. D.; MacLeod, J. K.; Summons, R. E. *Biochim. Biophys. Acta* **1975**, *399*, 61.
1251 Cowley, D. E.; Duke, C. C.; Liepa, A. J.; MacLeod, J. K.; Letham, D. S. *Aust. J. Chem.* **1978**, *31*, 1095.
1252 Mukaiyama, T. *Angew. Chem., Int. Ed. Engl.* **1979**, *18*, 707.
1253 McGee, D. P. C.; Martin, J. C.; Verheyden, J. P. H. *Synth. Commun.* **1988**, *18*, 1651.
1254 Gauri, K. K.; Pflughaupt, K.-W.; Müller, R. *Z. Naturforsch.* **1969**, *24 b*, 833.
1255 Szabolcs, A.; Sagi, J.; Otvos, L. *J. Carbohydr. Nucleosides, Nucleotides* **1975**, *2*, 197.
1256 Gauri, K. K. DE 1935379, 1970 [*Chem. Abstr.* **1970**, *73*, 15179w].
1257 Pischel, H.; Wagner, G. *Arch. Pharm.(Weinheim)* **1969**, *302*, 213.
1258 Konno, K.; Hayano, K.; Shirahama, H.; Saito, H.; Matsumoto, T. *Tetrahedron* **1982**, *38*, 3281.
1259 Brossmer, R.; Röhm, E. *Hoppe-Seyler's Z. Physiol. Chem.* **1967**, *348*, 1431.
1260 Prystas, M.; Sorm, F. *Collect. Czech. Chem. Commun.* **1966**, *31*, 3990.

1261 Capaldi, D. C.; Eleuteri, A.; Chen, Q.; Schinazi, R. F. *Nucleosides & Nucleotides* **1997**, *16*, 403.
1262 Mertes, M. P.; Saheb, S. E.; Miller, D. *J. Med. Chem.* **1966**, *9*, 876.
1263 Keller, F.; Bunker, J. E.; Brown, L. H. *J. Org. Chem.* **1966**, *31*, 3840.
1264 Jhingan, A. K.; Meehan, T. *Synth. Commun.* **1992**, *22*, 3129.
1265 Ueda, T.; Tanaka, H. *Chem. Pharm. Bull.* **1970**, *18*, 1491.
1266 Ueda, T.; Nishino, H. *J. Am. Chem. Soc.* **1968**, *90*, 1678.
1267 Prystas, M.; Sorm, F. *Collect. Czech. Chem. Commun.* **1964**, *29*, 2956.
1268 Prystas, M.; Sorm, F. *Collect. Czech. Chem. Commun.* **1966**, *31*, 1053.
1269 Lee, H.-J.; Wigler, P. W. *Biochemistry* **1968**, *7*, 1427.
1270 Camarasa, M.-J.; Walker, R. T.; Jones, A. S. *Nucleosides & Nucleotides* **1988**, *7*, 181.
1271 Ducrocq, C.; Bisagni, E.; Lhoste, J.-M.; Mispelter, J.; Defaye, J. *Tetrahedron* **1976**, *32*, 773.
1272 Nair, V.; Buenger, G. S.; Leonard, N. J.; Balzarini, J.; De Clercq, E. *J. Chem. Soc., Chem. Commun.* **1991**, 1650.
1273 Pal, B. C. *J. Org. Chem.* **1971**, *36*, 3026.
1274 Keller, F.; Bunker, J. E.; Tyrrill, A. R. *J. Med. Chem.* **1967**, *10*, 979.
1275 Jenkins, S. R.; Arison, B.; Walton, E. *J. Org. Chem.* **1968**, *33*, 2490.
1276 Chu, C. K.; Suh, J.; Cutler, H. G. *J. Heterocycl. Chem.* **1986**, *23*, 1777.
1277 Rao, D. R.; Lerner, L. M. *J. Org. Chem.* **1972**, *37*, 3741.
1278 Damianos, N.; Galons, H.; Miocque, M.; Quero, A. M.; Cotte, J. *Pharm. Weekbl. Sci. Ed.* **1989**, *11*, Suppl. C, C3.
1279 Hasan, A.; Srivastava, P. C. *J. Med. Chem.* **1992,** *35*, 1435.
1280 Lazrek, H. B.; Taourirte, M.; Barascut, J.-L.; Imbach, J.-L. *Nucleosides & Nucleotides* **1991**, *10*, 1285.
1281 Pudlo, J. S.; Saxena, N. K.; Nassiri, M. R.; Turk, S. R.; Drach, J. C.; Townsend, L. B. *J. Med. Chem.* **1988**, *31*, 2086.
1282 Saxena, N. K.; Hagenow, B. M.; Genzlinger, G.; Turk, S. R.; Drach, J. C.; Townsend, L. B. *J. Med. Chem.* **1988**, *31*, 1501.
1283 Gupta, P. K.; Nassiri, M. R.; Coleman, L. A.; Wotring, L. L.; Drach, J. C.; Townsend, L. B. *J. Med. Chem.* **1989**, *32*, 1420.
1284 Robins, M. J.; Hatfield, P. W. *Can. J. Chem.* **1982**, *60*, 547.
1285 Saxena, N. K.; Coleman, L. A.; Drach, J. C.; Townsend, L. B. *J. Med. Chem.* **1990**, *33*, 1980.
1286 Seela, F.; Gumbiowski, R. *Helv. Chim. Acta* **1991**, *74*, 1048.
1287 Seela, F.; Bourgeois, W.; Muth, H.-P.; Rosemeyer, H. *Heterocycles* **1989**, *29*, 2193.
1288 Seela, F.; Muth, H.-P.; Röling, A. *Helv. Chim. Acta* **1991**, *74*, 554.
1289 Seela, F.; Rosemeyer, H.; Fischer, S. *Helv. Chim. Acta* **1990**, *73*, 1602.
1290 Seela, F.; Gumbiowski, R. *Liebigs Ann. Chem.* **1992**, 679.
1291 Seela, F.; Rosemeyer, H.; Gumbiowski, R.; Mersmann, K.; Muth, H.-P.; Röling, A. *Nucleosides & Nucleotides* **1991**, *10*, 409.
1292 Seela, F.; Winter, H.; Möller, M. *Helv. Chim. Acta* **1993**, *76*, 1450.
1293 Seela, F.; Mersmann, K. *Helv. Chim. Acta* **1993**, *76*, 2184.
1294 Ozaki, S.; Watanabe, Y.; Hoshiko, T.; Fujisawa, H.; Uemura, A.; Ohrai, K. *Nucleic Acids Res., Symp. Ser. 15* **1984**, 33.
1295 Eid, M. M.; Addel-Hady, S. A.; Ali, H. A. W. *Arch. Pharm. (Weinheim)* **1990**, *323*, 243.
1296 Ramasamy, K.; Robins, R. K.; Revankar, G. R. *Nucleosides & Nucleotides* **1988**, *7*, 385.
1297 Revankar, G. R.; Robins, R. K. *Nucleosides & Nucleotides* **1989**, *8*, 709.
1298 Reitz, A. B.; Rebarchak, M. C. *Nucleosides & Nucleotides* **1992**, *11*, 1115.
1299 Ramasamy, K.; Joshi, R. V.; Robins, R. K. Revankar, G. R. *J. Chem. Soc., Perkin Trans. 1* **1989**, 2375.
1300 Rosemeyer, H.; Seela, F. *Helv. Chim. Acta* **1988**, *71*, 1573.
1301 Seela, F.; Soulimane, T., Mersmann, K.; Jürgens, T. *Helv. Chim. Acta* **1990**, *73*, 1879.
1302 Girgis, N. S.; Robins, R. K.; Cottam, H. B. *J. Heterocycl. Chem.* **1990**, *27*, 171.
1303 Sanghvi, Y. S.; Larson, S. B.; Willis, R. C.; Robins, R. K.; Revankar, G. R. *J. Med. Chem.* **1989**, *32*, 945.

1304 Ramasamy, K.; Imamura, N.; Robins, R. K.; Revankar, G. R. *Tetrahedron Lett.* **1987**, *28*, 5107.

1305 Motawia, M. S.; Pedersen, E. B. *Synthesis* **1988**, 797.

1306 Buchanan, J. G.; Stoddart, J.; Wightman, R. H. *J. Chem. Soc., Perkin Trans. 1* **1994**, 1417.

1307 Buchanan, J. G.; Stoddart, J.; Wightman, R. H. *J. Chem. Soc., Chem. Commun.* **1989**, 823.

1308 Pudlo, J. S.; Nassiri, M. R.; Kern, E. R.; Wotring, L. L. Drach, J. C.; Townsend, L. B. *J. Med. Chem.* **1990**, *33*, 1984.

1309 Gupta, P. K.; Daunert, S.; Nassiri, M. R.; Wotring, L. L.; Drach, J. C.; Townsend, L. B. *J. Med. Chem.* **1989**, *32*, 402.

1310 Chu, S.-H.; Chen, Z.-H.; Savarese, T. M.; Nakamura, C. E.; Parks, Jr., R. E.; Abushanab, E. *Nucleosides & Nucleotides* **1989**, *8*, 829.

1311 Chou, T.-S.; Grossman, C. S.; Hertel, L. W.; Holmes, R. E.; Jones, C. D.; Mabry, T. E. EP 577304, 1994 [*Chem. Abstr.* **1994**, *121*, 83888m].

1312 Hodge, R. P.; Brush, C. K.; Harris, C. M.; Harris, T. M. *J. Org. Chem.* **1991**, *56*, 1553.

1313 Wu, J.-C.; Bazin, H.; Chattopadhyaya, J. *Tetrahedron* **1987**, *43*, 2355.

1314 Pathak, T.; Bazin, H.; Chattopadhyaya, J. *Tetrahedron* **1986**, *42*, 5427.

1315 Revankar, G. R.; Ramasamy, K.; Robins, R. K. *Nucleosides & Nucleotides* **1987**, *6*, 261.

1316 Ramasamy, K.; Robins, R. K.; Revankar, G. R. *Tetrahedron* **1986**, *42*, 5869.

1317 Girgis, N. S.; Cottam, H. B.; Larson, S. B.; Robins, R. K. *Nucleic Acids Res.* **1987**, *15*, 1217.

1318 Acevedo, O. L.; Andrews, R. S.; Cook, P. D. *Nucleosides & Nucleotides* **1993**, *12*, 403.

1319 Griffin, L. C.; Kiessling, L. L.; Beal, P. A.; Gillespie, P.; Dervan, P. B. *J. Am. Chem. Soc.* **1992**, *114*, 7976.

1320 Kazimierczuk, Z.; Seela, F. *Liebigs Ann. Chem.* **1991**, 695.

1321 Ewing, D. F.; Holý, A.; Votruba, I.; Humble, R. W.; Mackenzie, G.; Hewedi, F.; Shaw, G. *Carbohydr. Res.* **1991**, *216*, 109.

1322 Revankar, G. R.; Gupta, P. K.; Adams, A. D.; Dalley, N. K.; McKernan, P. A.; Cook, P. D.; Canonico, P. G.; Robins, R. K. *J. Med. Chem.* **1984**, *27*, 1389.

1323 Ramesh, K.; Panzica, R. P. *J. Chem. Soc., Perkin Trans. 1* **1989**, 1769.

1324 Gupta, P. K.; Robins, R. K.; Revankar, G. R. *Nucleic Acids Res.* **1985**, *13*, 5341.

1325 Gupta, P. K.; Dalley, N. K.; Robins, R. K.; Revankar, G. R. *J. Heterocycl. Chem.* **1986**, *23*, 59.

1326 Oertel, F.; Winter, H.; Kazimierczuk, Z.; Vilpo, J. A.; Richter, P.; Seela, F. *Liebigs Ann. Chem.* **1992**, 1165.

1327 Sanghvi, Y. S.; Hanna, N. B.; Larson, S. B.; Robins, R. K.; Revankar, G. R. *Nucleosides & Nucleotides* **1987**, *6*, 761.

1328 Hanna, N. B.; Dimitrijevich, S. D.; Larson, S. B.; Robins, R. K.; Revankar, G. R. *J. Heterocycl. Chem.* **1988**, *25*, 1857.

1329 Michael, J.; Larson, S. B.; Vaghefi, M. M.; Robins, R. K. *J. Heterocycl. Chem.* **1990**, *27*, 1063.

1330 Seela, F.; Bindig, U. *Liebigs Ann. Chem.* **1989**, 895.

1331 Marcus, T. E.; Gundy, A.; Levenson, C. H.; Meyer, Jr., R. B. *J. Med. Chem.* **1988**, *31*, 1575.

1332 Girgis, N. S.; Cottam, H. B.; Robins, R. K. *J. Heterocycl. Chem.* **1988**, *25*, 361.

1333 Coleman, R. S.; Dong, Y.; Arthur, J. C. *Tetrahedron Lett.* **1993**, *34*, 6867.

1334 Seela, F.; Bourgeois, W. *Helv. Chim. Acta* **1991**, *74*, 315.

1335 Kazimierczuk, Z.; Stolarski, R.; Shugar, D. *Z. Naturforsch.* **1985**, *40c*, 715.

1336 Seela, F.; Bourgeois, W. *Synthesis* **1989**, 912.

1337 Seela, F.; Bourgeois, W. *Synthesis* **1990**, 94.

1338 Seela, F.; Bourgeois, W. *Synthesis* **1988**, 938.

1339 Seela, F.; Muth, H.-P.; Bindig, U. *Synthesis* **1988**, 670.

1340 Seela, F.; Kaiser, K.; Bindig, U. *Helv. Chim. Acta* **1989**, *72*, 868.

1341 Seela, F.; Steker, H.; Driller, H.; Bindig, U. *Liebigs Ann. Chem.* **1987**, 15.

1342 Cottam, H. B.; Kazimierczuk, Z.; Geary, S.; McKernan, P. A.; Revankar, G. R.; Robins, R. K. *J. Med. Chem.* **1985**, *28*, 1461.

[1343] Ramasamy, K.; Imamura, N.; Robins, R. K.; Revankar, G. R. *J. Heterocycl. Chem.* **1988**, *25*, 1893.
[1344] Seela, F.; Driller, H.; Steker, H. In *Nucleic Acid Chemistry*; Townsend, L. B., Tipson, R. S., Eds.; Wiley: New York; 1991, Vol. 4, p. 302.
[1345] Seela, F.; Driller, H.; Liman, U. *Liebigs Ann. Chem.* **1985**, 312.
[1346] Winkeler, H.-D.; Seela, F. *J. Org. Chem.* **1983**, *48*, 3119.
[1347] Cocuzza, A. J. *Tetrahedron Lett.* **1988**, *29*, 4061.
[1348] Gupta, P. K.; Vittori, S.; Townsend, L. B. *Nucleosides & Nucleotides* **1990**, *9*, 35.
[1349] Ramasamy, K.; Robins, R. K.; Revankar, G. R. *J. Chem. Soc., Chem. Commun.* **1989**, 560.
[1350] Winkeler, H.-D.; Seela, F. *Liebigs Ann. Chem.* **1984**, 708.
[1351] Garaeva, L. D.; Yartseva, I. V.; Melnik, S. Ya. *Nucleosides & Nucleotides* **1991**, *10*, 1295.
[1352] Seela, F.; Kretschmer, U. *J. Heterocycl. Chem.* **1990**, *27*, 479.
[1353] Girgis, N. S.; Cottam, H. B.; Larson, S. B.; Robins, R. K. *J. Heterocycl. Chem.* **1987**, *24*, 821.
[1354] Sanghvi, Y. S.; Larson, S. B.; Robins, R. K.; Revankar, G. R. *Nucleosides & Nucleotides* **1989**, *8*, 887.
[1355] Serafinowski, P. *Synthesis* **1990**, 757.
[1356] Seela, F.; Grein, T.; Samnick, S. *Helv. Chim. Acta* **1992**, *75*, 1639.
[1357] Cristalli, G.; Vittori, S.; Eleuteri, A.; Grifantini, M.; Volpini, R.; Lupidi, G.; Capolongo, L.; Pesenti, E. *J. Med. Chem.* **1991**, *34*, 2226.
[1358] Seela, F.; Steker, H. *J. Chem. Soc., Perkin Trans. 1* **1985**, 2573.
[1359] Seela, F.; Menkhoff, S. *Liebigs Ann. Chem.* **1986**, 1213.
[1360] Seela, F.; Steker, H. *Helv. Chim. Acta* **1986**, *69*, 1602.
[1361] Hildebrand, C.; Wright, G. E. *J. Org. Chem.* **1992**, *57*, 1808.
[1362] Wright, G. E.; Hildebrand, C.; Freese, S.; Dudycz, L. W.; Kazimierczuk, Z. *J. Org. Chem.* **1987**, *52*, 4617.
[1363] Focher, F.; Hildebrand, C.; Freese, S.; Ciarrocchi, G.; Noonan, T.; Sangalli, S.; Brown, N.; Spadari, S.; Wright, G. *J. Med. Chem.* **1988**, *31*, 1496.
[1364] Wright, G. E.; Dudycz, L. W.; Kazimierczuk, Z.; Brown, N. C.; Khan, N. N. *J. Med. Chem.* **1987**, *30*, 109.
[1365] Kazimierczuk, Z.; Vilpo, J.; Hildebrand, C.; Wright, G. *J. Med. Chem.* **1990**, *33*, 1683.
[1366] Kawakami, H.; Matsushita, H.; Yoshikoshi, H.; Itoh, K.; Naoi, Y. US 5262531, 1993; EP 0350292, 1989 [*Chem. Abstr.* **1990**, *113*, 6744u].
[1367] Kondo, K.; Shigemori, H.; Ishibashi, M.; Kobayashi, J. *Tetrahedron* **1992**, *48*, 7145.
[1368] Seela, F.; Bindig, U.; Driller, H.; Herdering, W.; Kaiser, K.; Kehne, A.; Rosemeyer, H.; Steker, H. *Nucleosides & Nucleotides* **1987**, *6*, 11.
[1369] Rosemeyer, H.; Seela, F. *J. Org. Chem.* **1987**, *52*, 5136.
[1370] Franchetti, P.; Cappellacci, L.; Grifantini, M.; Lupidi, G.; Nocentini, G.; Barzi, A. *Nucleosides & Nucleotides* **1992**, *11*, 1059.
[1371] Seela, F.; Lampe, S. *Helv. Chim. Acta* **1993**, *76*, 2388.
[1372] Wang, Y.; Fleet, G. W. J.; Wilson, F. X.; Storer, R.; Myers, P. L.; Wallis, C. J.; Doherty, O.; Watkin, D. J.; Vogt, K.; Witty, D. R.; Peach, J. M. *Tetrahedron Lett.* **1991**, *32*, 1675.
[1373] Meade, E. A.; Wotring, L. L.; Drach, J. C.; Townsend, L. B. *J. Med. Chem.* **1993**, *36*, 3834.
[1374] Meade, E. A.; Townsend, L. B. *Bioorg. Med. Chem. Lett.* **1991**, *1*, 111.
[1375] Sanghvi, Y. S.; Larson, S. B.; Robins, R. K.; Revankar, G. R.; Gupta, P. K.; George, R. D.; Dalley, N. K. *J. Heterocycl. Chem.* **1988**, *25*, 623.
[1376] Cristalli, G.; Vittori, S.; Eleuteri, A.; Volpini, R.; Cola, D.; Camaioni, E.; Gariboldi, P. V.; Lupidi, G. *Nucleosides & Nucleotides* **1993**, *12*, 39.
[1377] Cristalli, G.; Franchetti, P.; Grifantini, M.; Nocentini, G.; Vittori, S. *J. Med. Chem.* **1989**, *32*, 1463.
[1378] Seela, F.; Winkeler, H.-D.; Ott, J.; Tran-Thi, Q.-H.; Hasselmann, D.; Franzen, D.; Bussmann, W. *Nucleosides, Nucleotides, and Their Biological Applications;* Academic: New York; 1983; p. 181.

[1379] Seela, F.; Lüpke, U.; Hasselmann, D. *Chem. Ber.* **1980**, *113*, 2808.
[1380] Ramasamy, K.; Robins, R. K.; Revankar, G. R. *J. Heterocycl. Chem.* **1988**, *25*, 1043.
[1381] Hanna, N. B.; Ramasamy, K., Robins, R. K.; Revankar, G. R. *J. Heterocycl. Chem.* **1988**, *25*, 1899.
[1382] Seela, F.; Bourgeois, W.; Jürgens, T. *Nucleosides & Nucleotides* **1989**, *8*, 1089.
[1383] Seela, F.; Winkeler, H.-D. In *Nucleic Acid Chemistry*; Townsend, L. B., Tipson, R. S., Eds.; Wiley: New York; 1991, Vol. 4, p. 209.
[1384] Okamoto, K; Goto, T. JP 61207400 1986 [*Chem. Abstr.* **1987**, *106*, 156826s].

INDEX